Frontiers in Plant Sciences

Frontiers in Plant Sciences

Editors

K.G. Mukerji
Formerly of
Department of Botany
University of Delhi
Delhi

K.V.B.R. Tilak
Formerly of
Microbiology Division
IARI, PUSA
New Delhi

S.M. Reddy
Dept. of Botany
Kakatiya University, Warangal
Andhra Pradesh

L.V. Gangawane
Dept. of Botany
Dr. B.R. Ambedkar Marathwada Univ.
Aurangabad, Maharashtra

P. Prakash
Joint Secretary
University Grants Commission
Bahadur Shah Zafar Marg
New Delhi

I.K. Kunwar
Botany Department
Osmania University
Hyderabad, A.P.

I.K. International Pvt. Ltd.
New Delhi • Bangalore • Mumbai

Cover photographs: Selected illustrations from "CBS Centenary Studies in Mycology", Centroalburean voor Schoinomeleultures, Uppralaaon, Utrecht, The Netherlands:

Published by

I.K. International Pvt. Ltd.
S-25, Green Park Extension
Uphaar Cinema Market
New Delhi - 110 016 (India)
E-mail : ik_in@vsnl.net

Branch Offices

A-6, Royal Industrial Estate, Naigaum Cross Road
Wadala, Mumbai - 400 031 (India)
E-mail : ik_mumbai@vsnl.net

G-4, "Embassy Centre", 11 Crescent Road, Kumara Park East
Bangalore - 560 001 (India)
E-mail : ik_bang@vsnl.net

ISBN 81-88237-42-6

First reprint 2007

Published by Krishan Makhijani for I.K. International Pvt. Ltd., S-25, Green Park Extension, Uphaar Cinema Market, New Delhi - 110 016 and Printed by Rekha Printers Pvt. Ltd., Okhla Industrial Area, Phase II, New Delhi - 110 020.

Preface

Frontiers in Plant Sciences is a humble dedication of contributions from fellow Botanists, Mycologists—Plant Pathologists and admirers of Professor C. Manoharachary, D.Sc. for his valuable and meaningful contributions in the fields of Mycology, Plant Pathology and Microbial Technology. This volume is being brought out on the occasion of his superannuation. Professor Chary, through his lifetime of devotion to science and teaching, has been a significant factor instrumental in generating a wealth of scientific knowledge that will continue to refresh the minds of scientists for times to come. He has a kind and caring disposition, and his humane nature has touched numerous lives and endeared him to innumerable friends and students.

During recent years great strides have been made in all the fields of plant sciences and to bring the widely different aspects under one cover is indeed a herculean effort. That is precisely what the Editors have done in the present Festschrift volume, which is a collection of review articles as well as original research papers. The book contains wide ranging articles on Taxonomy of Fungi, Microbial Ecology and Applied Microbiology (including aerobiology, mycorrhiza, soil rhizosphere microflora and biological control of plant diseases). Few articles are on algae and one on aquatic angiosperm.

In all there are forty four papers in this book. In developing this volume we decided that the treatment of the subject should be comprehensive but concise enough to enable completion within set time. Further, we wanted to develop a treatise that would offer a detailed overview of the applied aspects of Botany in terms of its theoretical, methodological and empirical contributions. By emphasizing the interdisciplinary aspects of the subject, we hope that the Festschrift will highlight recent prospective linkages among diverse research paradigms. An attempt has been made to arrange the articles in a logical sequence as far as possible. But because of the multidisciplinary nature of the subject, judgement had to be made about the relative importance and significance of the topics. Only selected contributions have been included in the book. However, it may be pointed out that the authors are solely responsible for the views expressed in their articles.

Chapter 1 describes two new genera named after Professor C. Manoharachary, Chapters 1 to 8 are on fungal taxonomy, 9-16 are fungal ecology, 17-20 on microbial physiology, 21-26 on microbial application, 27 and 28 are on biological control, 29 to 35 on mycorrhiza, one chapter i.e. 36 on lichens, 37 and 38 on plant genetics, 39-42 on cyanobacteria and the last Chapter 43 on aquatic angiosperm–Podostemataceae.

In a work of this nature, it is just possible that some mistakes might have crept in the manuscript inadvertently and we owe responsibility for the same. The credit for anything of value goes to the contributors who responded instantaneously to our request for writing comprehensive review articles and for sticking to the time schedule. We thank them all from the core of our heart. It was because of their active cooperation that we could complete this uphill task in a short time.

We take great pleasure in presenting this Festschrift volume to Professor C. Manoharachary as a token of our affection. We wish him many more happy and fruitful years of research and a peaceful and prosperous life. We do hope that all those interested in the pursuit of plant sciences would find this volume useful and stimulating.

Delhi, Hyderabad **Editors**

Contents

SECTION 2: FUNGAL ECOLOGY

SECTION 3: MICROBIAL PHYSIOLOGY

SECTION 4: MICROBIAL APPLICATION

SECTION 5: BIOLOGICAL CONTROL

SECTION 6: MYCORRHIZA

SECTION 7: LICHENS

SECTION 8: GENETICS

SECTION 1

FUNGAL TAXONOMY

Frontiers in Plant Sciences (Pages 3-10)
Eds. K.G. Mukerji, K.V.B.R. Tilak, S.M. Reddy, L.V. Gangwane, P. Prakash, I.K. Kunwar

Some New Conidial Fungi

B. Bhadraiah and Vasant Rao

ABSTRACT

In this communication two new anamorphic fungal genera are being introduced in honour of Prof. C. Manoharachary taking his middle name *Manohara* as *Manoharomyces* with *M. elagans* as type and another new genus viz., *Chari* taking his last name as *Chariomyces* with *C. radiatus* as type species. These fungi are different from other known fungal genera in possessing.

Key words: *Manoharomyces, elagans, Chariomyces, radiatus.*

During the survey of microfungi from Goa and other forest localities the authors have encountered several known and unknown taxa. In this communication two new taxa viz., *Manoharomyces* with *M. elagans* and *Chariomyces* with *C. radiatus* are being proposed.

1. *MANOHAROMYCES* gen. et sp. nov.

Type Species: *Manoharomyces elagans*

Manoharomyces gen. nov.

Deuteromycotina, hyphomycete, anamorph.

Coloniae albae-gresia, conidiomata sporodochia crustati-platformis. Mycelium ex hyphis septatis producentibus conidiophora ramis, laevia incorporentibus in pseudo parenchymatous, stromatica. Conidiphora macronemata, synnemata, rami, hyalina vel sub-hyalina laevia, septata. Cellae conidiogenae sympodici mono vel polyblastici, lageniformae vel

Author's address: Department of Botany, Osmania University, Hyderabad - 500007

cylindricae leavia, rami in uni vel multi directio. Conidia holoblastica, cylindricia usque allantoidea 1-cellae, laevia, hyalina vel subhyalina setulatae terminal vel ad basim.

Species Typica: *Manoharomyces elagans*

Manoharomyces elagans sp. nov. (Fig. 1, Plate 1)

Colonies sporodochial whitish grey, radiating, whitish, pin-head like, globose, stalked or sessile, solitary, rarely in groups of 2-3. Mycelium, partly immersed hyaline to subhyaline, septate, 2-3 µm broad, branched producing pseudoparenchymatous basal fruit body, free ends of the conidiomata becoming conidiophores. No chlamydospores were observed. Conidiophores macronematous, synnematous forming compact sporodochial individual conidiphores, hyaline to subhyaline, smooth, 2-5 septate, simple or branched, 3-4 µm broad and 15-35 µm long. Conidiogenous cells terminal or lateral mono or polyblastic, lageniform or clavate shaped, 1-celled , smooth arising from both sides irregularly 2-3 µm broad at base and up to 1 µm broad at apex producing holoblastic conidia terminally. Conidia holoblastic mono or polyblstic, initially cylindrical or allantoid, 1-celled, smooth, 5-7 µm long and 2-3 µm broad but later on producing setulae on terminal and basal ends. Setulae straight or curved, slightly extensions of the conidium. Conidial secession schizolytic and germination not observed. No teleomorph was found.

Material examined on dead leaves of unidentified plant, Canconum forest, Goa OUMH. Coll. BB No. 201, Jan. 2004.

Manoharomyces elagans sp. nov. (Fig. 1, Plate 1)

Coloniae albae-gresiae. Conidiomata plati-synnematica, capita globosa pedicellata vel sessilis, solataire rarier agregato consisto 2-3. Mycelium partim immersum partim superficiale, hyalina vel subhyalina ad basim pseudoparenchymatica pallide brunnea terminale partim producentibus bonidiophora. Conidiophora macronemata, synnemata formentibus compacta sporodochia, individuale conidifero hyphis 3-4 µm lata, 15-35 mm longa cellae conidiogenae terminal vel lataraliter mono vel polyblastici, lageniformae et clavatae 1-cellae, rami 2-3 µm lati ad basim usque 1 um ad apicem 5-10 µm longa holoblastici. Conidia holoblastici, mono vel plyblastici, cylindricea, allantoida in formis 1-cellae, 5-7 µm longa, 2-3 µm lata laeviae producentibus setulae ad apicem vel ad basim. Conidis secessio schizolytico.

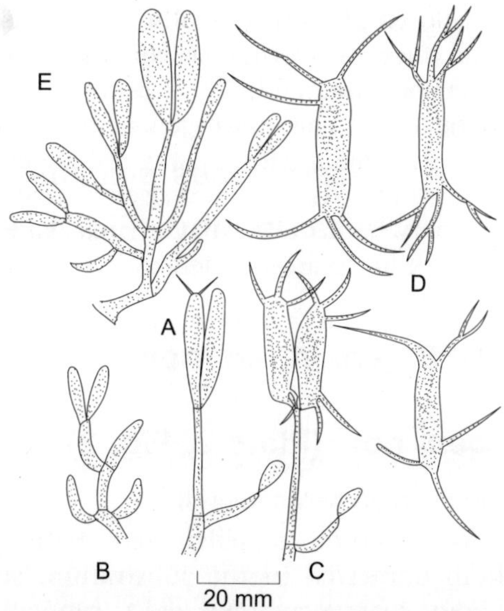

Fig. 1 *Manoharomyces elagans*; A. Conidiophore with conidiogenous cell and B. conidia with setulae. D. Conidia with setulae

Plate 1 *Manoharomyces elagans*. Conidiophores with conidia, 450x

Holotypis in emortio folio, canconum regio, Goa, Indiae, OUMH.leg. BB. Numaero 201. Jan. 2004.

Manoharomyces is very close to *Ephelis* Fries *Bahusutrabeeja* Subramanian and Bhat (1977) and *Thozetellopsis* Agnihothrudu (1958). *Ephelis* is also sporodochial and its conidia produce terminal setulae only. *Bahusutrabeeja*

produces globose phialoconidia with several setulae on its conidial surface. *Thozetellopsis* again produces phialoconidia which are fusiform or crescent shaped bearing setulae on either ends. There is no hyphomycete known to have combined characters of the above genera hence described as new taxon having polyblastic conidia with 2-many setulae on either ends and at times on side. Therefore, it is being proposed as new genus in honour of Prof. C. Manoharachary, Dept. of Botany, Osmania University, Hyderabad, for his meritorious work in Botanical sciences.

2. *CHARIOMYCES gen.* et. sp. nov.

Chariomyces gen. nov. (Plate 2, Fig. 2)

Deuteromycotina, hyphomycete, anamorph.

Colonies sporodochial, erumpent, pulvinate, radiating disc-like, greyish to brown. Mycelium immersed in the substratum, septate, branched hyphae, Conidiophores macronematous, erect, synnematous, splashing outwards, simple. Conidiogenous cells polyblastic, geniculate, with distinct scars, sympodic, blunt, denticulate. Conidia holoblastic, acropleurogenous, 1-celled, elongated, cylindrical, smooth, rostrate. Conidial secession schizolytic, germination through the basal scars. Teleomorph could not be observed.

Type Species: *Chariomyces radiatus.*

Plate 2 Chariomyees radiatus. gen et sp. nov. a. Whole colony b. Conidophores with conidia.

Chariomyces gen. nov.

Deuteromycotina, hyphomycete, anamorphae.

Coloniae sporodochiae, erumpenta, pulvinata, radiata, platibus, gresis vel brueni. Mycelium immersum, septatis, ramosis ex hyphis. Conidiophora macronemata, recta, radiata, simplicea, ramosa, ascenda. Cellae conidiogenae, polyblastici, geniculata, sympodica et denticulata. Conidia holoblastica, acrogenae, 1-cellae, aseptata, elongata, cylindrica laevial, rostrata ad basim. Conidio secessio schizolytica. Germinato ad basim Telomorpha nulla.

Species Typica: *Chariomyces radiatus*

Chariomyces radiatus sp. nov. (Plate 2 Fig. 2)

Colonies sporodochial, conidiomata disc like with radiating conidiferous hyphae, erumpent, pulvinate, greyish to brown, 50-100 µm broad at base, 300-500 µm at apex and 250-400 µm tall. Mycelium immersed in the substratum, visible hyphae septate, smooth, 3-5 µm broad. Conidiophores macronematous, synnematous, erect, branched splashing outwards arising from the base of the conidiomata, 150-450 µm long, 3-5 µm broad, septate, pale to light brown, smooth. Conidiogenous, cells polyblastic geniculate, with distinct scars sympodic, blunt, denticulate 150-300 µm long and 3-5 µm broad.

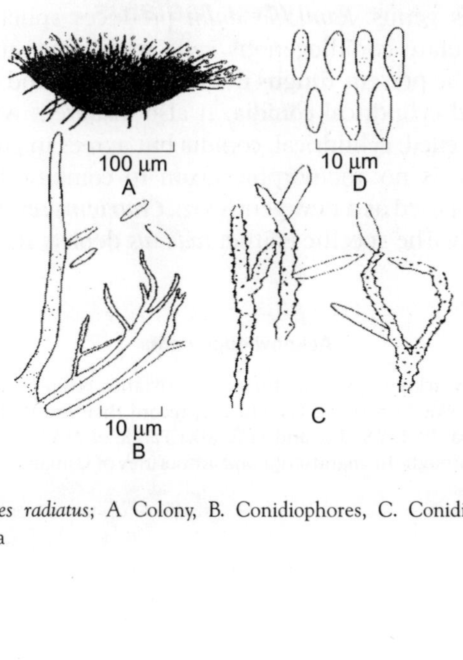

Fig. 2 *Chariomyces radiatus*; A Colony, B. Conidiophores, C. Conidiophores with conidia D. Conidia

Conidia holoblastic, acropleurogenous, smooth cylindrical 5-15 μm long, 3-5 μm broad, 1-celled with rostrate basal scar. Conidial sucession schizolytic. Germination through base.

Material examined on unidentified leaves from Canconum forest Goa, OUMH. Coll. BB. No. 227, Jan. 2004.

Chariomyces radiatus sp. nov. (Plate 2, Fig. 2).

Coloniae erumpenti pulvinata sporodochia. Conidiomata discosa radiata, gresia. Mycelium superficiale 50-100 μm crassi ad basim, 300-500 μm lati ad apicem, 250-400 μm longa. Myceilium ex hyphis septati, laevia 3-5 μm lata.

Conidiophora aggregata synnemata, macronemata, erecta, simplicea vel ramosa, radiata, 150-450 μm longa, 3-5 μm lata, septata laevia. Cellae conidiogenae polyblasticae, geniculatae, sympodicio, denticulatae 150-300 μm longa et 3-5 μm lata. Conidia holoblastica, acropleurogenae 1-cellae laevia cylindricia 5-15 μm longa, 3-5 μm lata, rostrata ad basim scaro (demancato promino). Secessio conidio schizolytico.

Typus emortio folio Canconum, Goa regio, Indiae, OUMH. leg BB. Num, 227, Jan. 2004.

Chariomyces is being proposed as a new genus in Hyphomycetes for its characteristic fruit body, macronematous conidiophores, polyblastic conidiogenous cells and 1-celled elongated conidia. The known genera *Ramichloridium* Stahl ex de Hoog, *Spadicoides* Hughes, *Pseudospiropes* Ellis do not resemble this genus. *Ramichloridium* produces spheroid conidia and polyblastic geniculate conidiogenous cells with synanamorphic state of *Exophiala* which the present fungus does not produce and differs in having 1-celled elongated cylindrical conidia. It also disagrees with *Pseudospiropes* Ellis in having 1-celled, cylindrical, conidia but agrees in possessing distinct scars. Thus, there is no anamorphic taxon to combine these characters, therefore, it is proposed as a new genus viz., ***Chariomyces*** in honour of Prof. C. Manoharachary. The specific epithet *radiatus* depicts its radiating colony character.

Acknowledgements

The authors gratefully acknowledge the financial assistance given by UGC, New Delhi and UGC, O.U., DST, FIST SAP, New Delhi. They further record their gratitude to Prof. D.J. Bhat, University of Goa, Goa, Prof, Y.S. Rao and Prof. Alka Pande of MACS, Pune for constructive criticism and going through the manuscript and authorities of Osmania University Hyderabd, AP.

References

Carmichal, J.W., Kerdrick, W.B. Conners, I.L. and Sigler, Lysine 1980. *Genera of Hyphomycetes* University of Alberta, Alberta Pr. Education.

Ellis, M.B. 1971. *Dematiacceous Hyphomycetes* IMI, CAB, Survey, UK.

Ellis, M.B. 1976. *More Dematiaceous Hyphomycetes*, IMI, CAB, Survey, UK.

Mahsushima, T. 1975. *Icones Fungorum Matsushima lectorium*, Kobe, Japan.

Subramanium, C.V. 1971. *Hyphomycetes*, ICAR, New Delhi.

Frontiers in Plant Sciences (Pages 11-20)
Eds. K.G. Mukerji, K.V.B.R. Tilak, S.M. Reddy, L.V. Gangwane, P. Prakash, I.K. Kunwar

Fungi of Delhi, XXXVII. Studies on Indian Corticolous Ascomycetes: Six Unrecorded Species

Surinder Kaur, Geeta Malhotra and K.G. Mukerji

ABSTRACT

The present paper describes six corticolous ascomycetes isolated from tree barks. These are *Podospora ellisiana, Sordaria tetraspora, S. macrospora, Coniochaeta velutina, Gelasinospora seminuda, Gnomonia fructicola*. These are reported here for the first time from India.

Key words: Ascomycetes, Bark, Corticolous, India, New records

INTRODUCTION

During our studies on the taxonomy of the corticolous fungi occuring on bark of different trees *viz. Eucalyptus globulus* Labil, *Mangifera indica* Linn., *Azadirachta indica* A Juss., and *Ficus religiosa* Linn. many new rarely occurring fungi were isolated and reported earlier (Rikhy & Mukerji, 1973, Malhotra & Mukerji, 1978, 1979, Saxena & Muker ji, 1973). In total, 182 species belonging to 149 genera were collected from tree barks of which six ascomycetes were found to be interesting and are reported here. These fungi were found to be occurring as saprophytes when bark pieces were incubated in moist chambers at 25 ± 1°C. These species were cultured on

Author's address: Department of Botany, University of Delhi, Delhi-110007, India.

different media (Czapek's Dox Yeast Extract agar, Potato dextrose agar, potato carrot agar, Com meal agar, Malt agar, Abram's agar, Abram's cellulose agar) for sporulation.

Living cultures and type specimens of all these forms have been deposited in Centraalbureau Voor Schimmelcultures, Baarn, Netherlands (CBS) and in Indian type culture collection, New Delhi (ITCC). The fungi described here tally with the type descriptions.

RESULTS

Podospora ellisiana (Griff.) Mirza & Cain, Can. J. Bot. 47: 2020, 1969 Fig. 1 (14-25), Fig. 4 (1-4)

Perithecia scattered, single to aggregated in groups of 5–6 on bark surface, semi-immersed, obpyriform, 580-650 × 480-510 µm, sparsely covered with flexuous septate olivaceous brown hair, hairs 3-3.5 µm wide, with obtuse tip, neck dark, straight, cylindrical, thick walled, 168-175 × 132-140 µm. Peridium embranous, 7-layered, semi-transparent, 10-15 µm wide, olivaceous brown except in neck region, where it is black, thick, with angular hyaline to yellowish brown outer cells, ostiole 20-30 µm wide, periphysis yellowish brown, septate, 10-12.5 × 5-6.5 µm. Asci octosporous, broadly clavate to cylindrical, 160-216 × 15-60 µm, narrowing towards the stipe, tapering above in a somewhat truncate apex, apical ring hyaline. 3-3.5 µm in diameter, rounded in cross section, calotte slightly thickened, light refractive, 5-7.5 µm in diameter, hyaline, broader in the middle portion. Stipe narrow, short, hyaline, 20-30 × 10-15 µm. Ascospores uniseriate to biseriate, initially one celled, hyaline, narrowing, clavate, 25-40 × 10-15 µm, then swelling above become transversely uniseptate, a little inequilarial with a truncate base and an apical germ pore, pedicel cylindrical, subhyaline to light brown, 15-27.5 × 5-6 µm, constriction is distinct below the septum, tail collapsing, cauda present on both sides, very small, hyaline, broader at the upper end and long cylindrical towards tail region.

This was isolated from *Eucalyptus* bark when incubated in moist chambers. Perithecia appeared after 30-32 days of incubation. Part of this specimen has been deposited at CBS under reference number DU/KG 374.

This isolate shows marked similarities with the type description except perithecia and asci are smaller in size, and absence that an agglutinated hair at the neck region.

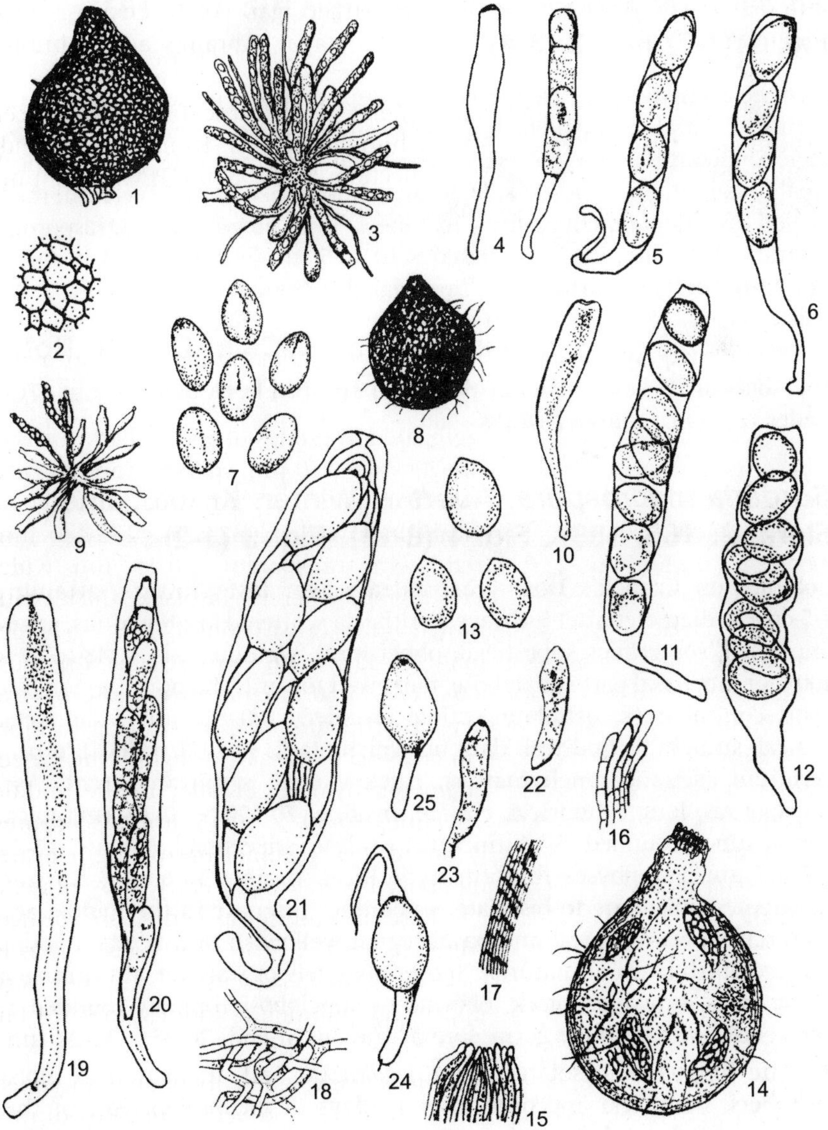

Fig. 1 1-7 *Sordaria tetraspora* Wint, 1. Perithecium × 100, 2. Peridium × 750, 3. Asci and Paraphysis × 150, 4-6 Young and mature asci, 7 Ascospores × 600 , 8-13 *Sordaria macrospora* (Auersw) Winter, 8 Perithecium × 30, 9 Group of asci × 80, 10-12 Young and mature asci × 440, 13 Ascospores × 400, 14-25 *Podospora ellisiana* (Speg.) Lundq., 14 Perithecium × 66, 15-16 Periphysis, 17 Peridium × 330, 18 Hair × 280, 19-21 Young and mature asci × 465, 22-25 Young and mature ascospores × 500

Sordaria tetraspora **Wint. Hedwigia 10: 161, 1871 Fig. 1 (1-7) Fig. 3 (3-4)**

Perithecia scatterted to gregarious in groups of 5-6 or more, sub globose to pyriform, dark brownish black, 280-330 × 220-250 µm, conical in the upper region, globose to subglobose at the basal part, ostiolate, neck conical, small, straight, black, 40-60 × 70-80 µm, ostiole 45.5-50 µm. Peridium thick walled, pseudoparenchymatous, hyaline to light brown. Asci tetrasporous, cylindrical, clavate, hyaline, 70-100 × 10-10.5 µm in diameter. Ascospores uniseriate, obliquely arranged, ellipsoidal, dark brown, germ pore one at the apical region, 15-17.5 × 7.5-10 µm.

This fungus was isolated from bark of *Azadirachta* incubated in moist chambers at 25 + 1°C. Part of this specimen has been deposited at ITCC under reference number DU/KG 385.

Sordaria macrospora **(Auersw) Winter, Kryptogamen Flora, 2: 165, 1887. Fig. 1 (8-13), Fig. 3 (1-2)**

Colonies on Czapek's Dox, yeast extract agar fast growing, attaining 4-5 cms in diameter after 7-8 days. Perithecia scattered or gregarious, semi-immersed, sometimes superficial, obpyriform, glabrous, black, 480-720 × 400-540 µm, basal part subglobose, narrowed towards the upper region into short conical neck, ostiolate, ostiole rounded, 100-120 µm wide. Neck conical, straight, cylindrical, dark brownish black, 120-130 × 160-170 µm. Peridium pseudoparenchymatous, thick walled, semitransparent. Asci octosporous, long cylindrical, clavate, hyaline, 70-170 × 10-30 µm, apical pore distinct, rounded, 5-10 µm wide, hyaline, stipe small, hyaline, 15-25 × 5-7.5 µm. Paraphyses filliform, cylindrical, hyaline 50-60 × 3-5.5 µm. Ascospores uniseriate to biseriate, sometimes biseriate in the middle and uniseriate towards apical and basal region, yellowish brown when young and brownish black at maturity. Spores obliquely or transversally arranged in an ascus, brownish black, obovate or subglobose, broadly rounded at both ends, with a single germ pore at the apical end, 20-35 × 12-21 µm.

The fungus was isolated from *Azadirachta* bark incubated in moist chambers. Perithecia appeared after 10 days of incubation. Part of this specimen has been deposited at CBS under reference number DU/KG 302.

The isolate resembles in all the characters with the type description. Earlier it was reported mostly on dung of different animals (coprophilous) (Agnihothrudu, 1958, Cailleux, 1921, Mahju, 1933).

Gnomonia fructicola Fall. Can. J. Bot. 29: 299-315, 1951 Fig. 2 (14-20)

Perithecia scattered to aggregated in groups of 4-9 (5), erumpent, glabrous, black, pyriform, 850-1070 μm long, basal part embedded in bark, globose to subglobose, thin walled, yellowish brown, 350-500 μm wide, neck long, cylindrical, thick walled, carbonaceous, black, protrude from the bark, 400-700 × 100-150 μm, ostiole hyaline to light yellowish brown, 60-90 × 70-110 μm, periphysis 3-3.5 μm wide. Peridium pseudoparenchymatous, cells thick walled. Asci octosporous, clavate, long, hyaline, 30-60 × 10-15 μm, apical pore indistinct, stipitate, stipe very small, hyaline, 5-7.5 × 4.5-5 μm. Paraphysis numerous, hyaline, filiform, 20-60 × 5-5.5 μm. Ascospores uniseriate to biseriate, hyaline, 2-4 guttulate when young, one septate, septa prominent at maturity, wall slightly constricted at the septum, bicelled with rounded ends, elliptic fusiform in shape, 10-17.5 × 5.75 μm.

This was isolated from bark of *Eucalyptus* incubated in moist chambers. Part of this specimen has been deposited at ITCC under reference number DU/KG 391. It shows resemblance with *Gnomonia fructicola* (type description) in most of its characters but differs in having smaller asci and ascospores.

Conicochaeta velutina (Fuckel) Munk, Danish-Pyrenomycetes. 90-92, 1957 Fig. 2 (1-7), Fig. 3 (5-6)

Perithecia scattered or gregarious, semi-immeresed, globose to ovoid, non stromatic, black, 100-270 × 200-250 μm, ostiolate, neck very small, covered with setae like hairs, 15-20 × 30-50 μm, slightly conical covered with periphysis, setae straight, simple, dark brown, aseptate, tip acute, 15-40 × 3-4.5 μm. Peridium very thick, pseudoparenchymatous, cells dark brownish black, 2.5-6 μm wide. Asci octosporous, clavate, cylindrical, hyaline, 73-93 × 4.5-10.5 μm, apical ring distinct, 3.5-4.5 μm wide, stipe very short, hyaline, 7.5-25 × 3-5 μm, paraphysis septate, filifoml, hyaline, 60-102 × 2-2.5 μm. Ascospores dark brown, 2-3 guttulate, ovoid to obovate, one celled, obliquely uiseriate, 6-11.5 × 5-9 μm.

This fungus was isolated from bark of *Ficus* incubated in moist chambers. Mycelium was found embedded in the cortical region of bark and fruiting bodies appeared after 8-10 days on the bark surface. Part of the living specimen has been deposited at CBS under reference number DU/KG338. This isolate resembles in all essential features with type description.

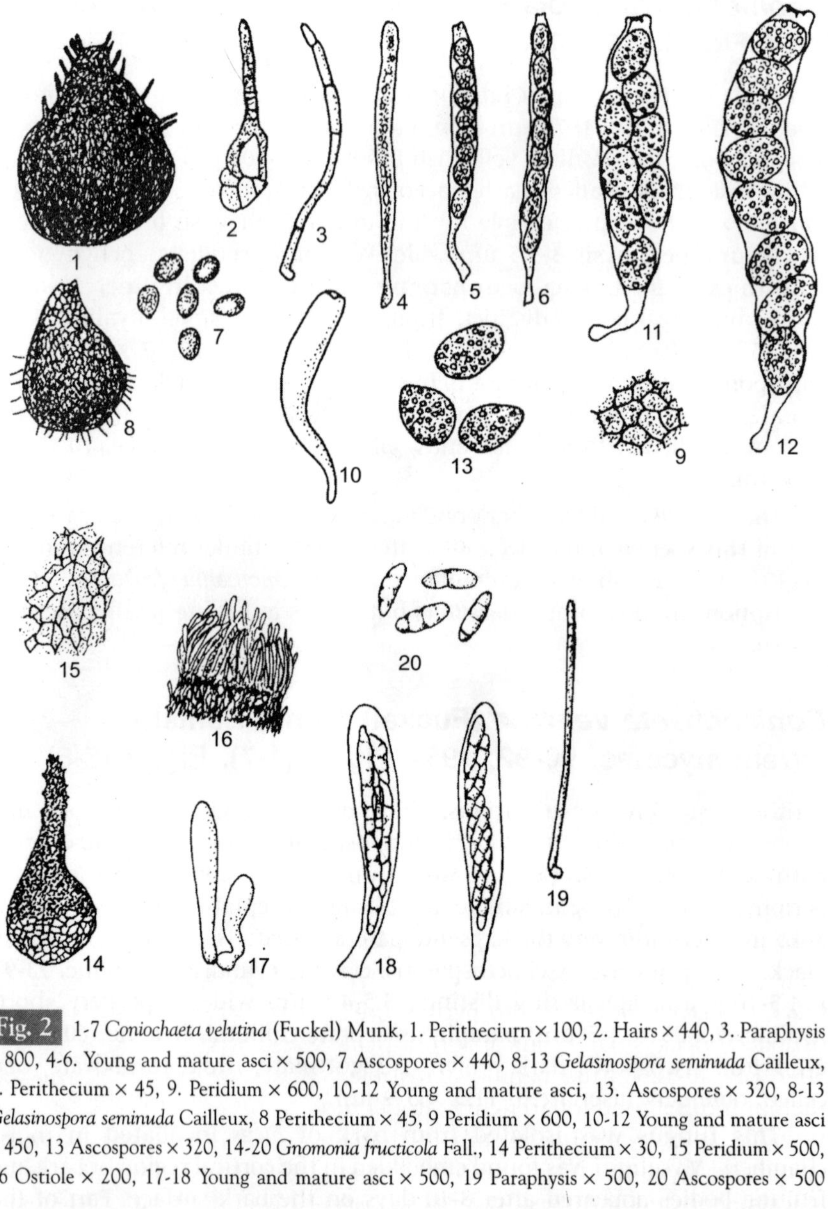

Fig. 2 1-7 *Coniochaeta velutina* (Fuckel) Munk, 1. Perithecium × 100, 2. Hairs × 440, 3. Paraphysis × 800, 4-6. Young and mature asci × 500, 7 Ascospores × 440, 8-13 *Gelasinospora seminuda* Cailleux, 8. Perithecium × 45, 9. Peridium × 600, 10-12 Young and mature asci, 13. Ascospores × 320, 8-13 *Gelasinospora seminuda* Cailleux, 8 Perithecium × 45, 9 Peridium × 600, 10-12 Young and mature asci × 450, 13 Ascospores × 320, 14-20 *Gnomonia fructicola* Fall., 14 Perithecium × 30, 15 Peridium × 500, 16 Ostiole × 200, 17-18 Young and mature asci × 500, 19 Paraphysis × 500, 20 Ascospores × 500

Gelasinospora seminuda Caileux, Bull. Trimest. De la Soc. Mycol. De Fr. 87: 461-626, 1971. Fig. 2 (8-13)

Vegetative hyphae light brown to dark brown, thick walled, septate, 4.5-5 μm wide. Perithecia sparse to gregarious, partly immersed, brownish

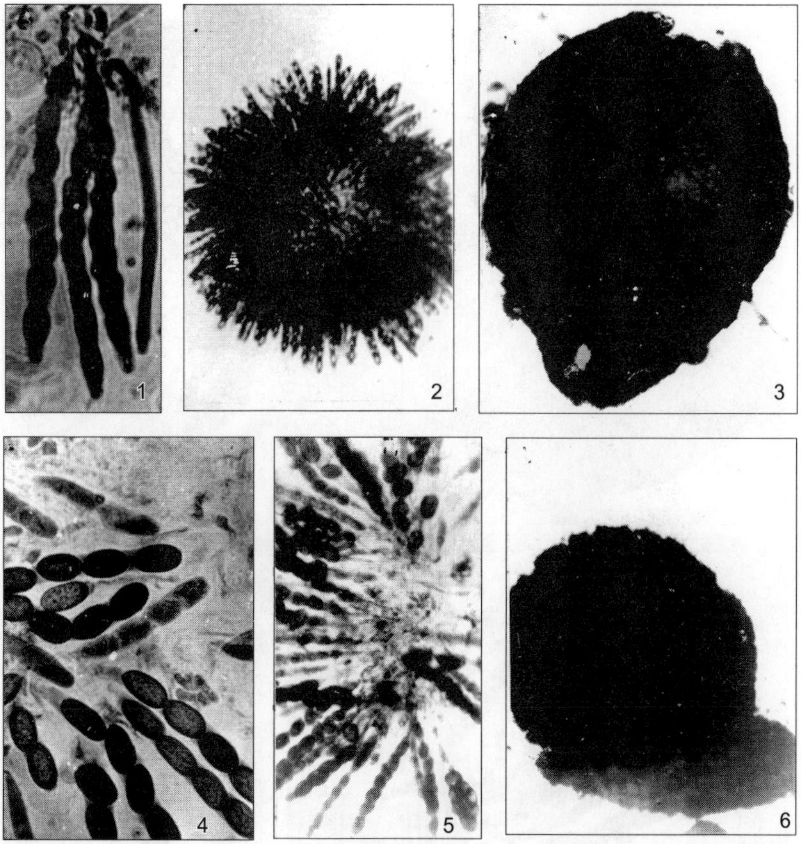

Fig. 3 1-2 *Sordaria macrospora* (Auersw) Winter, 1 Perithecium × 90, 2 Asci showing ascospores × 200, 3-4 *Sordaria tetraspora* Wint., 3 Perithecium × 240, 4 Asci and ascospores × 450, 5-6 *Coniochaeta velutina* (Fuckel) Munk., 5 Asci in group × 220, 6 Young and mature asci × 720

black to black, subglobose to pyrifoml, glabrous, 230-450 × 170-380 µm, ostiolate, neck small, conical, cylindrical, straight, black, thick walled, 30-70 × 40-120 µm, papillose, ostiole, 30-100 µm wide. Peridium membranous, pseudoparenchymatous, dark brown, cells angular to rounded thick walled, 5-10 µm in diameter. Asci octosporous, clavate, cylindrical, hyaline, 120-170 × 15-30 µm, stipate, persistent, broader at the apical region and narrowed towards base, apical opening 9.5-10.5 µm wide, stipe small, hyaline, 20-27.5 × 5-7.5 µm, pararaphysis lacking. Ascospores uniseriate to biseriate, particularly at the upper and the lower regions, uniseriate and in the middle biseriate, spores one celled, ellipsoid, ownish black, 25-35 × 15-25 µm, germ pores two, one at each end, reticulate at maturity, each part of reticulation 2.5-3 µm in diameter.

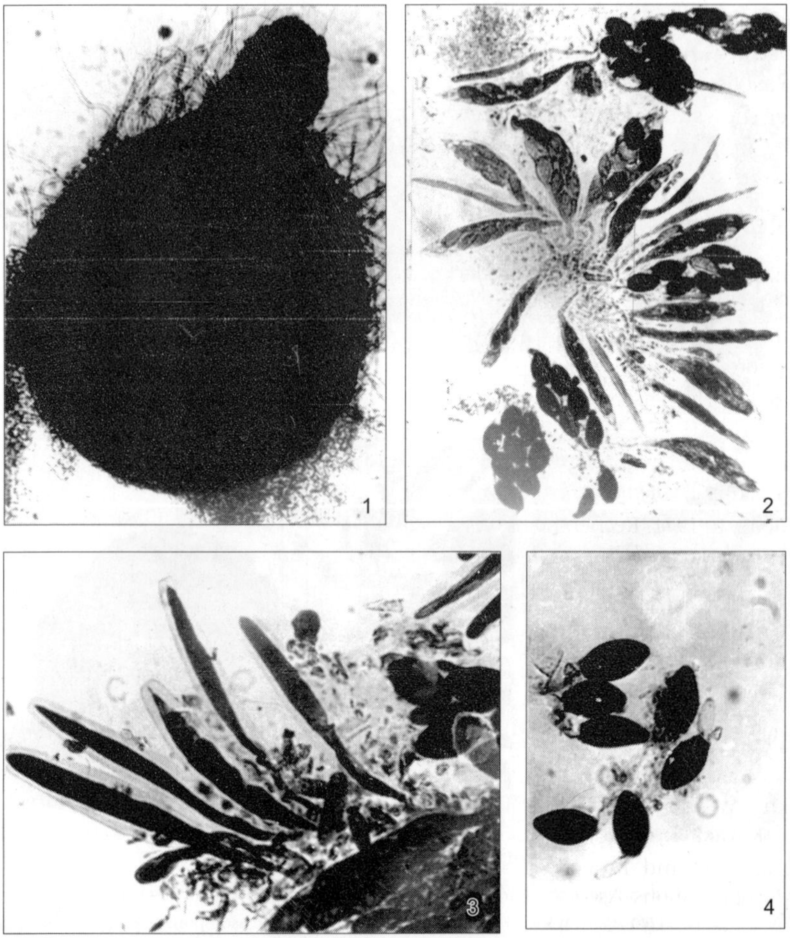

Fig. 4 1-3 *Podospora ellisiana* (Griff.) Mirza & Cain., 1. Perithecium × 120, 2. Group of asci showing ascospores with caudae × 140, 3. Young asci showing apical ring × 300, 4. Ascospores × 300

This form was isolated from bark and incubated in moist chamber at 25 + 1°C. Perithecia appeared after 12-15 days of incubation. Colonies on carrot and potato agar brownish black, 3-4 cm in diameter after 8-9 days of incubation and perithecia appeared after 17-18 days. Living cultures of the isolate have been deposited at CBS under reference number DU/KG 337. This isolate resembles with type description in most of its characters except smaller size.

Comments: In the present studies it is recorded as growing saprophytically on bark though typically coprophilous fungi (Cain, 1934,

1962, Cain & Farrow, 1956). However, it is possible that near the bark substrates the fungi might have been growing saprophytically on dung and the ascospores might have been ejected from these on to the bark and then after they started growing on the dead bark. This was also supported by Dr. B.J. Coppins, Royal Botanic Garden, Edinburgh.

References

Agnihothrudu, V. 1958. Fungi isolated from rhizosphere IV. J. Indian Bot. Soc. 37: 422-431.

Cain, R.F. 1934. Studies on Coprophilous Sphaeriales, in Ontario, University, Toronto. Biol. Series. 38: 1-126.

Cain, R.F. 1962. Studies of Coprophilous ascomycetes VIII. New species of *Podospora.* Can. J. Bot. 40: 447-490.

Cain, R.F. and Farrow W.M. 1956. Studies of Coprophilous ascomycetes III. The genus *Triangularia.* Can. J. Bot. 34: 689-697.

Clilleux, R. 1971. Recherches sur la mycoflora coprophile centra-fricaine les generes *Sordaria, Gelasinospora, Bombardia* (biologie Morphologie Systematique). Ecologie-Bull. Trimest de la Soc. Mycol. De. Fr. 87: 461-626.

Mahju, N.A. 1933. A contribution to our knowledge of Indian Coprophilous fungi. J. Indian Bot. Soc. 13 153-163.

Malhotra, G. and Mukerji, K.G. 1978. Fungi of Delhi XXVIII: Four Pyrenomycetes from bark. Trans. Mycol. Soc. Japan 19: 283-288.

Malhotra, G. and Mukerji, K.G. 1979. Fungi of Delhi XXX: A note on the genus *Sordariella.* Phytomorph. 32: 325-329.

Rikhy, M. and Mukerji, K.G. 1973. Fungi of Delhi XXVI. Three new Ascomycetes. Kavaka 1: 37-41.

Saxena, A. and Mukerji, K.G. 1973. Fungi of Delhi XVII. Three unrecorded coprophilous Ascomycetes. Ceska Mykologie 27(3): 165-168.

Frontiers in Plant Sciences (Pages 21-34)
Eds. K.G. Mukerji, K.V.B.R. Tilak, S.M. Reddy, L.V. Gangwane, P. Prakash, I.K. Kunwar

Family Xylariaceae–An Overview

Alaka Pande

ABSTRACT

The Family Xylariaceae is widely represented in tropics. Members of the family show variations in almost all morphological features used in taxonomy. Truly Xylariaceous features and variations seen are described. Recent trend is in selecting criteria from secondary metabolites and genetic finger printing techniques which are mentioned at the end.

Key Words: Xylariaceae, taxonomic criteria, recent trends

The family Xylariaceae is one of the oldest families. It exhibits a wide range of morphological variations. The members of this family are distributed in the tropics and temperate regions occurring mostly as saprophytes in shadowy moist places, in leaf litter and on decaying wood pieces or logs. However, some members are known to be pathogenic inciting wood rots and cankers or root rots. Some are known to occur on varied substrata such as termite hills, grassy lands, dung etc. These macroscopic forms belong to Class Ascomycotina Order Xylariales (or Sphaeriales) Family Xylariaceae; the family was established by Tulasne and Tulasne in 1863 which now has 40 genera (and also enlists 70 synonyms) with 600 species. Comprehensive monographs are available only for a few genera.

The members of the family Xylariaceae are most unattractive of the macro-fungi being dark black and have odd shapes. However, for a mycologist, they are very fascinating because of the variations shown by them, in every aspect- habit, habitat, size, shape, orientation and position of perithecia with reference to substratum, growing pattern, number of perithecia per stroma, ascospore etc.

Author address: Mycology & Plant Pathology Dept., Agharkar Research Institute, Pune 411 004
E-mail: alakapande@yahoo.com

Truly Xylaraiceous characters are: Generally saprophytes, or weak pathogens, many endophytic in wood or bark of angiosperms; stroma generally well developed, ascoma perithecial, black, globose, ostiole periphysate, inter-ascal paraphyses, unitunicate, asci cylindrical, wall persistent and with thick but inseparable layers, almost always with simple or complex J+ apical structure; spores usually dark brown to black, aseptate, with germ slit; sometimes with hyaline appendage (Miller 1928a,b, Dennis 1978).

Major genera are *Xylaria* Hill ex Schrank, *Hypoxylon* Bull., *Daldinia* Ces. & de Not., *Biscogniauxia* Kuntze, *Poronia* Willd, *Rosellinia* de Not.. In addition there are several small genera under this family.

Historial review

Saccardo (1882) had placed these fungi in Phaeosporae based entirely on the ascospore characters; However, Tulasnes established the name Xylariaceae in 1861. Lloyd (1917-1919) has drawn excellent figures of many *Xylaria* species, beautifully illustrating the morphology of stroma, other surface features and nature of perithecia. Lindau 1897 proposed Xylariaceae to be placed under Sphaeriales. Miller (1928 a,b) and Nannfeldt (1932) considered it a remarkably homogenous family under Sphaeriales, but Luttrell (1951). Arx & Muller (1954) recognised Xylariales as a separate order for pyrenomycetous fungi showing 'Xylaria type" of centrum and unitunicate asci provided with crown or apical pore. Miller(1928a,b, 1949) included genera like *Rosellinia, Anthostoma* and also *Sordaria* in this family. V. Arx & Muller (1954) employed microscopic characters alone such as cylindrical shape of asci, their unitunicate nature and germ pores or germ slits on the ascospores to delimit the family. Martin (1967) while accepting Xylariales as a separate order, included stromal features also for family characterization.

The floristics of F. Xylariaceae has never been studied in India and whatever little is known is found in scattered reports by different workers. The history of Xylariaceae in India is as old as history of Fungi in India. The earliest record of xylariaceous fungus in India seems to be by Klotzsch 1832 as *Sphaeria (Cordyceps) nigripes)* Klotzsch (= *Xylaria nigipes* (Klotzsch) Sacc.) on the basis of a collection by Mr. Wight from Belgam (Karnatak) (then Bombay Presidency). All earlier records are based on collections from India sent to I.M.I. and examined by British mycologists.

The true taxonomic treatment for F. Xylariaceae form India was first done by Mukerji et al. (1969) and Thind and Waraitch(1969). Lakanpal and Mukerji (1973) described from species of Xylaria. Dargan & Thind (1979) critically studied genus *Daldinia*. Based on anatomical studies of stroma, two subgenera were proposed by them as *Entodaldinia* and *Eudaldinia*. Dargan (1980) presented a review on Indian Xylariaceae. They have published 6 new species under *Rosellinia*. Agnihothrudu (1964, 1965) reported Xylariaceous fungi (*Hypoxylon, Rosellinia, Ustulina*) occurring in tea estates in North-East India, Rubber plantations in Southern India etc. Some other reports on the family are by Kar & Gupta from W. Bengal and by Tilak, S.B.Kale, V.S. Kale, U.K. Kulkarni, Thite, Srinivasulu, Ullasa, Mhaskar, Anahosur etc. from S. India.

Pande took up systematic survey for the first time from Peninsular India for F. Xylariaceae and listed/ published >80 names in a series of papers during 1970-80 on Xylariaceae of Western India: 14 papers species of *Xylaria. Hypoxylon, Rosellinia* etc. The work was further continued and other aspects like cytology, ascosporogenesis, spore discharge, anamorphs, cultural morphology, metabolites produced etc. were studied.

In other countries, the family is receiving much more attention in all aspects of studies and especially so in the recent past. The main contributors are from J.D. Rogers, S.C. Jong, O. Petrini, A.J.S. Whalley, P. Martin, J.H. Miller, R.W.G. Dennis, Yu Ming Ju, S.C. Teng, and several others. They have monographed several genera, have genus wise regional studies or reviewed the scenario in Xylariaceae (Petrini & Rogers,1986, Rogers,1979, 2000, Whalley, 1985, 1993, 1996, Vander Gucht & Whalley,1996). Cytological details and ascosporogenesis were also studied for *Hypoxylon* spp. by Rogers and others. Ecological considerations received attention for biodiversity aspects. Rogers, Ko are trying to get teleomorphic states in cultures with high percentage of success. Greenhalgh, Chesters and others are proving importance of anamorphic states in identification. Scientists like Marc, Edwards are paying attention to the family following modern concepts, by using analytical and biological tools like biochemical analyses with help of HPLC, GC-MS, molecular techniques etc. It is proved that members of Family Xylariaceae will be an important source for newer bio-active metabolites.

Taxonomic characters

Conventionally the criteria are drawn from the habitat and morphology of various structures including anamorphs, for taxonomic identification in fungi. Recently there is a search for supporting characters from other studies like extractable pigments, secondary metabolites and most recently from DNA mapping or finger printing etc.

Habitat/Habit

Xylariaceous fungi are known to colonize a variety of habitats ranging from dry and sun-exposed sites to shadowy and wet localities, in dense forests. Different genera show preferential habitats with a rough selection about the plant part to be colonized. (Hypoxylons mostly on wood, Xylarias with a variety of habitats- twigs, leaves, roots, soil, animal horn, fruit, seed etc. *Rosellinia* includes plant parasite forms; *Poronia* is generally coprophilous. *Xylaria nigripes* grows on termite nests; *Hypoxylon hydnicola* is reported as mycoparasite on *Irpex* on *Quercus* (Miller 1961); *Xylaria vaporaria* is an invader of mushroom beds(Hawksworth et al.,1995).

Endophytes: Over the past 20 years increasing number of investigations of endophytic fungi have drawn attention to the almost ubiquitous and sometimes dominant presence of members of Xylariaceae especially in tropical plants. A total of 11 genera of Xylariaceae have been recorded as endophytes. One *Xylaria* species is isolated, as an endophyte, from thallus of a liverwort *Bazzania*.

Although identification of Xylariaceous fungi isolated as endophytes, is difficult since they do not produce mature stromata in culture, Petrini and Petrini (1985) have developed keys for identification of endophytic Xylariaceae upto generic level for temperate isolates. The tropical endophytic Xylariaceae are more difficult to identify.

Morphological aspects

Stromata: The family is characterized by having hard stromatic tissue well developed, at times massive, may be crustose, applanate, pulvinate, globose, upright and branched or simple, turbinate, clavate etc. or may be reduced totally. The fertile region may be distinguishable from the sterile part. Internal tissue may be white or coloured. The stromal tissue types were used in taxonomy esp. in *Hypoxylon, Daldinia* (Martin 1961; Dargan & Thind 1984; Ju & Rogers 1996).

Size and Shape

The size and shape of the stroma can also be influenced to some extent by external/environmental conditions (part of substratum colonized, humidity, temperature etc.) and are not used in taxonomic identification as main characters. The size is variable from a few millimeter to 30 to 35 centimeters or even about half a meter in height. Breadth or diameter also ranges from 1 or 2 millimeter to upto 10 centimeter.

Shape is also variable in the family Xylariaceae. The shape varies with the number and manner of arrangement of perithecia within the stroma. With the development of single perithecium, the stroma may be globose to ovoid as in *Rosellinia*; multiperithciate stroma may also be globose as in *Hypoxylon, Daldinia* and even in some species of *Xylaria*. The multiperitheciate stroma with flattened surface is seen in some *Hypoxylon or Biscognauxia* species. The upright stroma may be pedicellate or sessile. It can be clavate, cylindrical, may be forked, branched. Branching may be at stipe region or in fertile part. But branching is not a stable feature. Stroma may be splitting vertically in center or not so.

Colour

The stroma is generally dark brown when mature but may show colour variations when young. All shades from bright red to dark brown, purple, black could be encountered determined by the stages of development e.g. *Hypoxylon rubiginosum*. Species like *H. haematostroma* may retain bright red, crimson red colour at maturity. *Daldinia* is characterized by reddish brown stromata.

KOH-extractable pigments are considered as a reliable supporting character by Yu Ming Ju & Rogers (1996). A small piece of stroma is crushed on microscope slide and a drop of KOH is added. The pigment diffusing out from the stroma ranges from green to brown to black. The shades of this pigment are considered as reliable diagnostic colours. Only eight species of *Hypoxylon* show purplish pigments in KOH, which helps in shortening the identification key.

Surface

Surface of stroma in F. Xylariaceae is another important diagnostic feature esp. in *Xylaria* where the genus is divided into section Xylorugosa, and Xyloglossa. The surface layer covering stroma, its splitting nature to expose ostioles can contribute towards identification. Surface can be smooth when perithecia are deeply embedded as in*X. allantoidea, H. nummularium*. It may be slightly roughened due to protruding ostioles or slightly projecting perithecia as in *H. rubiginosum, X. myosurus*. The stroma is described muriform when perithecia are strongly protruding over the surface e.g. *X. berkeleyii*. Surface of stroma is sometimes covered with fibrous layer which splits open around projecting ostioles e.g. *X. coccophora, X. stromatica*. Stromal surface in many Xylaria species is longitudinally wrinkled or furrowed or irregularly channeled e.g. *X. filiformis, X. melaneura*.

Fertility of stroma

In many members with applanate stromata the entire stroma is seldom fertile. In *Hypoxylon* -sect. applanata there may be only central fertile region surrounded by sterile peripheral margin (*H. sulcatum, H. nummularium*); or fertile and sterile patches may be intermixed (*H. flacculosum*). In cylindrical upright forms, such as *Xylaria, Poronia, Kretzschmaria* sterile stipes are produced. In some members of *Xylaria* the entire stroma is fertile till the growing tip while in some the tip is sterile and growing region can easily be distinguished.

Persistance of stroma

The carbonaceous stroma in *Hypoxylon* or giant *Xylaria* may last for one or two years or more as dead sterile crust or stroma. But normally all

Xylariaceae are annual. i.e. the stroma developed in one year does not produce fresh perithecia, asci or ascospores in the succeeding seasons.

Perithecia

The ascocarps in Xylariaceae are true perithecia with distinct perithecial wall typical, asci and paraphyses in wall layers and an apical ostiole lined internally with periphyses. These represent Luttrell's "Xylaria type of Centrum". The perithecia are generally globular to flask shaped, often conical or may be laterally compressed due to crowding appearing tubular (*H. haematostroma*). The perithecia are embedded in the stroma or slightly projecting, but sometimes with reduced stroma appear to be practically free on stroma and are connected by basal stroma only (*X. melaneura, X. schwackei, X. filiformis, H. oodes, H. rosellinoides*)

Ostiole

Ostioles are apical and central, predetermined, papillate, shining, dark coloured whereby they appear as darker spots on dull lighter stromata (*X. thyrsus, H. notatum*). Ostioles may be slightly conical, umbillicate. In some *Hypoxylon* species (Sect. annulata) the ostioles are surrounded by narrower or wider ostiolar discs. The ostiole characters were used with more weightage, by Miller to divide the genus *Hypoxylon* into subsections.

Asci, paraphyses

The asci are typically cylindrical, unitunicate, stipitate, provided with amyloid apical apparatus stainable with iodine. The apical structure may be simple, represented by rings or may be much complex. The spore bearing part of the ascus is longer that the stipe. The paraphyses are true, longer than asci simple, hyaline. Asci and paraphyses are produced in wall layers.

Ascospores

The ascospores are normally smooth, one celled, brown to black, opaque, monostichously arranged. The shape of ascospores generally is ellipsoidal-inequilateral. Navicular ascospores are seen in *H. notatum;* Hat shaped spores are seen in *H. coorgianum*; small ascospores with one side flat and the other conical appearing triangular in lateral view are seen in *X. nigripes var. microspora, X. berkeleyi.* Rectangular ascospores are produced by *H. rectangulosporium.* The species is also characterized by having 5-8 ribs or ridges on spore wall. reticulate spore surface is seen in *X. reticulospora.* Other types of ornamentation also are seen like spore wall pitted, furrowed, dotted etc., which are commonly seen in *Hypoxylon.*

Germ slit

The germ slit is mostly on convex surface in inequilateral spores. Sometimes two slits are seen. Three types of germ slits are observed in Xylariaceae (a) straight germ-slit, running from tip to tip on flat surface of the spores-elongated and straight (b) elongated and spiral or oblique running not restricted to flat surface only and one or two spirals may be seen on convex surface also (*Helicogermslita*). (c) poroid - equilaterally placed 4 or rarely 2 germ pores may be seen in some species.

Spore sheath

Ascospores are surrounded by a sheath in some species which may either be distinctly observed at either end (*X. azadirachtae, X. memecyli*) or may be of uniform thickness all round (*X. heloidea, X. hypoxylon*). The epispore is easily separable with application of 10% KOH where it splits horizontally at the central region and slips off from the ends where it may appear as appendages. The inequilateral spores are associated with dehiscent perispore. Spore appendage is a hyaline structure apparently continuous with spore sheath but is characteristic of the genus *Entosordaria*. The spore sheath and spore appendages are best seen in fresh materials. KOH treatment for epispore proved to be equally useful for herbarium specimens as for the fresh materials. This feature of epispore is of taxonomic value since it is species specific.

Ascospore discharge and spore germination

The spore dispersal through forcible ejection from the ostiole can be observed by keeping the fresh stromata on glass plates under humid chamber as black sooty ejected spores collected on the glass plate. *Daldinia, Hypoxylon* species exhibit such black rings of ejected ascospores around their stromata when incubated this way. The *Xylaria* spp. collected from field and brought into laboratory in paper bags also show spores collected on paper bags. Sometimes they are seen collected near the ostiole on the stromata held in masses waiting for crawling insects, or wind currents, to disperse them away from mother stroma. Ingold (1933) was probably the first to report forcible spore discharge in alternating patterns of light and dark responses in *Hypoxylon coccineum*. In this species delayed light positive, and delayed dark negative response was noticed. The spore discharge continued for sometime after the light conditions were changed. Kramer & Pady (1970) observed a rhythmic spore discharge in different species of *Hypoxylon* in response to light conditions. Some species were distinctively light positive where spore discharge began within half an hour of exposure to light. Some species showed delayed dark response or delayed light response; it took 2-4 hours for the beginning of spore discharge after exposure to particular light condition. Likewise endogenous rhythm of spore discharge is observed in *Daldinia concentrica*. Pande (1975) observed that asci grew towards the ostiole and eject the ascospores as propagules of different sizes-single spores or 2 to 8 spores together as a single propagule. The larger propagules travelled longer distances as against single spores.

After the spores are discharged they germinate by producing 1-2 germ tubes emerging through the germ slits. The epispore first splits and then the germ tubes emerge through the germ slit. On agar plates pure white or porcelain white colonies with radiating mycelium and rope like structure leading to conidial states are produced.

Anamorphs or conidial states

Anamorph characters are of taxonomic value. However, not all the Xylariaceae fungi have been associated with conidial state. In nature, the conidial states for several species of these fungi was first described by Tulasne and Tulasne (1863). The conidia develop freely on many young stromal surfaces. The conidiophores are simple or branched and conidia are one celled, hyaline oval or rounded to pyriform with truncate base. They are produced in large quantities and appear as white masses spread over the young stromata.

Practically every species of *Hypoxylon* has a conidial state. These are classified into types on the basis of conidiophore and conidial morphology as *Nodulisporium Acrostaphyllus, Sporothrix, Basidiobotrys, Geniculosporium,* or *Virgariella* type. (Jang and Rogers, 1972).

Development

Miller (1961) reviewed the development of *Hypoxylon sp.* in wood. Following ascospore germination the hyphae ramify between bark and cortex and xylem; the limits of the mycelium are visible as black lines in the host tissue. Eventually the hyphal network starts swelling in the bark region. The bark ruptures and fungal stroma becomes erumpent. It gets differentiated into entostroma and ectostroma. Once the ectostroma is defined and exposed it soon gets covered with conidial state. Simultaneously in the periphery of the entostroma directly below the conidial layer starts the development of perithecial initials as hyphal coils. In the center of each coil is a broad septate hypha–the ascogonium. This structure later gives rise to ascogenous hyphae; in turn they produce croziers asci and ascospores. This is suggestive of conidia functioning like spermatia.

Cytology

Cytological investigations were carried out as early as 1900, by Dawson in *Poronia*, by Brown (1913) in *Xylaria* and by Rhodes (1918) in *Daldinia*. Hayman (1964) investigated *Rosellinia limonispora* (n=6). Major contributions are from Rogers and his co-workers during 1964-1970, who studied cytology of ascus and developmental details in many Xylariaceous fungi. Rogers determined the haploid chromosome number to be in the range of n=4 (*H. fuscum, H. multiforma, H. punctatum*) to n = 8 (*X. polymorpha*). Both the species of *Poronia* studied had n=7.

All the species of Xylariaceae have 8 ascospores per ascus. Binucleate condition of spores was reported in *Poronia oedipus* while *P. punctata* had quadrinucleate spores. Initial uninucleate ascospores show further nuclear division and become binucleate but ultimately become uninucleate by loss of one nucleus.

Geographjic distribution and ecology

The observations on distribution patterns are based only on one aspect of their life style - the mature ascostromata. Other aspects of life cycle like conidial states, endophytic existence of several species, are not taken into consideration while generalizing distribution pattern. The term 'endemic to a region' may not remain valid for a particular species, with further surveys from different areas. Keeping this point in mind, still one can say that *Xylaria* has the widest range of distribution and is represented by maximum number of species. Next in order come *Hypoxylon, Rosellinia, Daldinia* etc. Other genera like *Kretzscmaria, Poronia, Ustulina* are small genera with only a limited number of species and restricted geographical distribution.

Cultures

Colony characters and anamorphic states of 5 species of *Xylaria* and one species of *Hypoxylon* in synthetic media were studied by Pande(1974). For most of the Xylariaceae it is difficult to obtain fertile stroma in culture. Some species can complete lifecycle on agar media and are said to be homothallic. E.g. *Poronia oedipus* (Jang & Rogers,1969), *Penzigia frustulosa* (Jang & Rogers,1970), *Penziggia spp., Daldinia vernicosa, Hypoxylon gwyneddii, H. atropurpureum var. brevistipitatum, H. serpens* etc. (Rogers 1969, Rogers & Berbee 1964, Jong & Rogers 1970, Whalley, Edwards & Francis 1983). However, a large number of species were induced to produce mature stroma by inoculating living trees with ascospores. e.g. *Hypoxylon atropunctatum, Ustulina deusta, Kretzschmaria clavus, Xylaria schweinitzii* (Thompson 1963, Ko 1979).

Modern concepts in Taxonomy

Classification of fungi is principally based on morphological features. Increasing interest is seen in using other aspects like biochemistry, molecular biology etc. for the purpose. In many fungi protein analysis using gel electrophoresis is employed to distinguish related species.

Computer application/numerical taxonomy

Attempt was made to apply numerical taxonomy to key out species of *Hypoxylon* (Whalley & Greenhalgh 1973) especially the British species. The features of *Hypoxylon* species in perfect states and in culture were coded and analyzed separately and in combination. They observed that combined analysis based on perfect state and cultural state characters clearly demonstrated the division between coloured and non-coloured species. The association between papillate and umbilicate species was also confirmed. **Pigment pattern**: Miller divided the genus *Hypoxylon* into 4 sections- Hypoxylon, papillata, annulata, primo-cinerea based on visible stromal colour and ostiole features. The stromal pigments as seen in their solvent extracts and analyzed by TLC of *Hypoxylon* speices provided useful taxonomic data for determination of the species and indicated that there is a major division in the genus *Hypoxylon* between the truly coloured and non-coloured species. (Greenhalgh & Whalley 1970, Whalley & Greenhalgh 1973, Pande & Dighe 1997). Therefore emphasis is to be given to pigments as visual character and also after analysis of solvent extracts. Pande & Dighe studied 36 collections of Xylariaceae members for their stromal pigments using TLC method. These fungi showed remarkable uniformity for a given species in TLC patterns irrespective of their substrata, geographical location or year of collection. Both fresh and old specimens preserved in herbarium for years showed uniform band patterns. The data from this technique proves to be useful as a supporting criteria in taxonomic identification when morphological characters are not well preserved.

Enzymes

All members of Xylariaceae have the capacity to degrade lignin and cellulose, i.e. to cause white rots. The enzymatic activity of these fungi is widely exploited except for the endophytic species or endophytic phases of the species.

Secondary metabolites

Secondary metabolites are restricted in distribution since these are synthesized by certain species or certain strains of the species only. Earlier, secondary metabolites were not considered as of taxonomic value for identification due to general lack of chemical investigations. Over the past 25 years the picture is changing. Xylariaceae is receiving more and more attention for the secondary metabolites produced by them. There is a promise concealed in studies on Xylariaceae for their secondary metabolites as useful taxonomic characters and for determining their natural relationships within genera and with other fungi. In Xylariaceae production of secondary metabolites is a constant and reliable feature even when species were grown on different media or under slightly varying conditions. These are produced in sufficiently large quantities and thus their chemical characterization can be attempted.

Whalley and his co-workers started work on secondary metabolites of Xylariaceae and obtained some interesting results. Whalley (1996) suggested that many of the secondary metabolites produced by Xylariaceae taxa have potential ecological role and also can be used for taxonomic considerations. Cytochalazins also are useful species markers which may be widely distributed in Xylariaceae. Individual species can often be recognized on the basis of exclusive metabolite produced. Presence of cytochalazin E (which is phytotoxic) as a secondary metabolite during growth of a root pathogen *Rosellina necatrix* and its absence in saprophytic species indicates that there are distinguishing differences (Whalley & Edwards 1987, 1994). Presence of punctoporin in *Poronia* punctata and its absence in other species of *Poronia* can be used as a marker character for species identification. Chemical data supports the possibility that *Ustulina* and *Kretzschmaria* have a strong affinity with some species of *Xylaria*.

Overall *Hypoxylon* and related genera have naphthalene derivatives and *Xylaria* with related genera shows succinic acid derivatives as secondary metabolites.

In spite of such examples, it should be clearly understood that no one class of chemical compound alone would provide sufficient or even suitable data for broad taxonomic grouping; it is the overall picture obtained from examination of significant chemical compounds and their derivatives.

Genetic fingerprints

Molecular genetics and secondary metabolite analyses are becoming increasingly important in applied research aiming at the discovery of new fungal metabolites with promising bioactivities and the concurrent characterization of producer strains. HPLC based secondary metabolite profiles and PCR techniques are already being combined for such purposes, such techniques are probably going to find broader application in the future as admissible data in patents (Satdler Marc et al 2001).

Type herbarium specimens, cultures of freshly collected materials and fresh specimens of *Daldinia* and allied Xylariaceae were studied by Stadler Marc & others (2001) for their morphology in both teleomorphic and anamorphic states. Their metabolites pattern in stromata and cultures were also studied by employing HPLC-UV-Vis and HPLC-ESI-MS. Further, PCR-amplified 18S rDNA restriction patterns (ARDRA) were investigated. These studies revealed that stromata of a given species never contained the same major metabolites as corresponding cultures. Most cultures of *Daldinia* spp. produced naphthalene and chromone derivatives and differed from allied genera by the absence of mellein. Stromata of *Daldinia* spp. generally contained binaphthys but no mitorubin. HPLC profiles were in accordance with colour of KOH-extractable stromal pigments. ARDRA resolved several species whose 18S rDNA contained specific insertions.

Genetic fingerprints and HPLC-based secondary metabolite profiles were used for the first time to study the taxonomy of the broad range of *Daldinia* spp. and other Xylariaceous fungi. Strong correlations were noted between secondary metabolite profiles and the taxonomically important colour of KOH-extractable stromal pigments. PCR based techniques proved helpful for discrimination of some taxa at both the genus and species level.

Economic significance

Role in wood decomposition: Xylariaceae can colonize a wide range of substrata, the vast majorities are wood inhabitants. Wood weight loss studies have proven the ability of Xylariaceae to participate in wood decomposition. Certain members have also shown to degrade lignin, others exhibit impressive production of cellulolytic enzymes.

Xylariaceae are not important plant pathogens but a growing numbers of species are being acknowledged to cause considerable economic losses e.g. canker disease in *Populus tremuloides* and other hosts in Temperate Zone caused by *Hypoxylon mammatum. Biscognauxia* causes die-back and canker. Root rot infections are commonly encountered with *Rosellina* - *R. necatrix* is a serious problem on many trees include grapes, walnut, apple, pear, sweet cherry, etc.; *R. bonodes* is a truly tropical species attacking roots of cocoa, coffee, rubber, tea and similar plantation crops; *R. herpotrichioides* is the casual agent for needle blight disease of conifers.

Acknowledgement

Grateful thanks are offered to the Director, Agharkar Research Institute, Pune 411 004 for facilities.

References

Agnihothrudu, V. (1964) Fungi from North East India XXII. *Mycopath. Mycol. Appl.* **23**: 111-115.

Agnihothrudu, V. (1965) *Hypoxlon truncatum (Schw. ex Fr.)* Miller var. *ponceanum (*Berk. Et Cooke) Miller from Mysore. *Curr. Sci.* **30**: 192.

Arx, v. J.A. & E. Muller (1954) *Die Gattungen der amerosporen Pyrenomycetes* Beitr. Kryptogamen Schweiz **11(1)**: 1-434.

Brown, H.B. (1913) Studies in the development of *Xylaria, Annl. Mycol.* **11**:1-13.

Dargan, J.S. (1980) The family Xylariaceae in India - a review. *Jour. Indian Bot. Soc.* **59**: 53-59.

Dargan, J.S. & K.S. Thind (1979) Xylariaceae of India – VII. the genus *Rosellina* in the North - West Himalayas. *Mycologia* **71**: 1010-1023.

Dargan, J.S. & K.S. Thind (1984) Xylariaceae of India VIII. Genus *Daldinia* Ces. & de Not. - a further segregation into two new subgenera. *Kavaka* **12**: 113-118.

Dawson, M. (1900) On the biology of *Poronia punctata* L. *Annl. Bot.* **14**:245-262.

Dennis R.W.G. (1956) Some Xylariaceae of tropical America *Kew Bull.* **1956**:401-444.

Greenhalgh. G.N. and A.J.S. Whalley (1970) Stromal pigments of some species of *Hypoxylon* Trans. *Br. Mycol. Soc.* **55**: 89-96.

Haymann, D.S. (1964) *Rosellina limoniispora* nuclear changes in developing asci *Can. J. Bot.* **42**: 13-21.

Hawksworth, D.L., P.M. Kirk, B.C. Sutton & D.N. Peglar (1995) *Ainswarth and Bisby's Dictionary of the Fungi* 8th Edition CAB International, IMI, U.K. pp. 616

Ingold, C.T. (1933) Spore discharge in Ascomycetes I. Pyrenomycetes. *New Phytologist* **32**: 175-196.

Jong, S.C. & J.D. Rogers (1969) *Poronia oedipus* in culture. *Mycologia* **61**: 853-862.

Jong, S.C. & J.D. Rogers (1970) *Penzigia frustulosa* in culture. *Mycologia* **62**: 851-855.

Jong, S.C. & J.D. Rogers (1972) Illustrations and descriptions of conidial states of same *Hypoxylon* species. *Washington Agri. Exp Str Tech. Bull. no.***71.**

Ju, Yu Ming & J.D. Rogers (1996) A revision of genus *Hypoxylon. Mycol. Memoire no.* **20** pp 365.

Ko, W.H. (1979) A method for reproduction of perithecial stromata by Xylariaceaeous fungi. *Mycologia* **71**: 441-443.

Kramer, C.L. & S.M. Pady (1970) Ascospore discharge in *Hypoxylon. Mycol.* **62**: 1170-1186.

Lakhanpal, T.N & Mukerji, K.G. 1973, Morphology of some Indian species of Xylariaceae and Clavicipitaceae, *Cěska Mykologie,* **27:** 169-173

Lindau, G. (1897) Pyrenomycetinae *in* Engler's *Naturlichen Pflanzenfamilien Lepizizo* pp 491.

Lloyd, G.C. (1917-19) Synopsis of some genera of the large Pyrenomycetes p.32

Luttrell, E.S. (1951) Taxonomy of the Pyrenomycetes *Univ. Missouri Stud. Sci Ser.* **24(3)**: 1-120.

Martin, P. (1967) Studies in the Xylariaceae: I. New and old concepts *J. South African Bot.* **33**: 205-240.

Miller, J.H. (1928 a & b) Biological studies in the Sphaeriales I & II *Mycologia* **20**: 187-213 & 305-339.

Miller, J.H. (1949) A revision of classification of the ascomycetes with special emphasis on the pyrenomycetes. *Mycologia* **41**: 99-127.

Miller, J.H. (1961) A monograph of the world species of *Hypoxylon*. Athens, *University of Georgia Press,* pp.158.

Mukerji, K.G., Bedi, K., Tiwari, J.P. & Tiwari, I. 1969, Morphology of Indian species of *Xylaria* and *Poronia, Phytomorphology,* 19: 219-224

Nannfeldt J.A. (1932) Studien uber die morphology and systematic der nicht-lichenisierten inoperculata discomycetes *Nova. Acta. Regiae Soc. Sci. Uppsala IV* **8(2)**: 1-368

Pande, A. (1974) Behaviour of some species of family Xylariaceae in artificial culture. *Jour. Univ. Poona Sci. Tech. Sect.* **46**: 69-72.

Pande, A. (1975) Ascospore discharge in *Anthostoma pimpriana* (Xylariaceae) *Biovigyanum* **1**: 41-46.

Pande, A. & Sangeeta Dighe (1997) Stromal pigments in *Hypoxylon* and allied genera collected from Peninsular India. *Jour. Plant Morph. Taxo.* **7**: 33-37.

Petrini, L.E. & O. Petrini (1985) Xylariaceous fungi as endophytes *Sydowia* **38**: 216-234.

Petrini, L.E. & J.D. Rogers (1986) A summary of the *Hypoxylon serpens* complex *Mycotaxon* **26**: 401-436.

Rhodes, A.S. (1918) *Daldinia vernicosa* a Pyroxylophilous fungus *Mycologia* **10**: 277-284.

Rogers J.D. & J.G. Berbee (1964) Developmental morphology of *Hypoxylon pruinatum* in bark of auaking asper. *Phytopath.* **54**: 154-162.

Rogers, J.D. (1979) Xylariaceae systematic, biological and evolutionary aspects *Mycologia* **71**: 1-42.

Rogers, J.D. (2000) Thoughts and musings in Xylariaceae *Mycol. Res.* **104:** 1412-1420.

Saccado, P.A. (1882) *Syll. Fung.* **I** : 1-815.

Stadler Marc & Hartmund Wollweber, Andrea, Muehlbaur, Thomas Henkel, Yu Ming Ju, J.D. Rogers, Hienz-George, Wetzstein & Hans Volker Tichy (2001) Secondary metabolite profiles, genetic fingerprints and taxonomy of *Daldinia* and allies *Mycotaxon* **77** : 379-429.

Thind , K.S. & K.S. Waraitch (1969) Xylariaceae of India I. *Proc. Indian Acad. Sci.* **B LXX**: 131-138

Thompson G.E. (1963) Decay of oaks caused by *Hypoxylon atropunctata Pl. Dis. Reptr.* **47**: 202-205

Tulasne, L. & C. Tulasne (1863) *Selecta fungorum carpologia* 319 Paris (English transl. By Grove 1931).

Van Der Gucht & A.J.S. Whalley (1996) Notes on ecology of Xylariaceae of Papua New Guinea. *Sydowia* **48**: 131-141.

Whalley, A.J.S. (1985) The Xylariaceae: Some ecological considerations *Sydowia* **38**:369-382.

Whalley, A.J.S. (1993) Tropical Xylariaceae: their distribution, ecological characteristics *in* Aspects of tropical Mycology Ed. S. Isaac *et al. Cambridge Univ. Press Cambridge,* pp 103-119.

Whalley, A.J.S. (1996) The Xylariaceous way of life *Mycol. Res.* **100**: 897-922.

Whalley, A.J.S. & R.L. Edwards (1994) Secondary metabolites and systematic arrangement within the Xylariaceae. *Can J. Bot.* **73 (suppl.)**: S802-S810.

Whalley, A.J.S., R.L. Edwards, & S.M. Francis (1983) *Hypoxylon gwynedii* sp. nov. from Wales Trans. *Br. Mycol. Soc.* **81**: 389-392.

Whalley, A.J.S. & R.L. Edwards (1987) Xylariaceous fungi-Use of Secondary metabolites *in* The evolutionary biology of fungi Ed. A.D.M. Rayner *et al.* pp. 423-434 *Cambridge Univ. Press, Cambridge, UK.*

Whalley, A.J.S. & G.N. Greenhalgh (1973) Numerical taxonomy of *Hypoxylon* Comparision of classification of the cultural and the perfect states. Trans. *Br. Mycol. Soc.* **61**: 435-454.

Frontiers in Plant Sciences (Pages 35-46)
Eds. K.G. Mukerji, K.V.B.R. Tilak, S.M. Reddy, L.V. Gangwane, P. Prakash, I.K. Kunwar

Marine Fungi from Orissa (India)-III: Mitosporic Fungi

K.N. Borse* and B.D. Borse**

ABSTACT

Biodiversity study of higher marine fungi from mangroves of Orissa coast yields 18 mitosporic fungi. *Coniothyrium obiones* Jaap, and *Trichocladium melhae* Jones, Abdel-Wahab & Vrijmoied are new records to the fungi of India. *Alternaria* sp., *Cirrenalia basiminuta* Raghukumar & Zainal, *C. pseudomacrocephala* Kohlmeyer, *C. pygmea* Kohlmeyer, *C. tropicalis* Kohlmeyer, *Exserohilum* sp., *Macrophoma* sp., *Monodictys pelagica* (Johnson) Jones, *Periconia prolifica* Anastasiou, *Phoma* sp., *Trichocladium achrasporum* (Meyers & Moore) Dixon, *T. alopallonellum* (Meyers & Moore) Kohlm. & Volkm. – Kohlm., *T. constrictum* Schmidt, *T. opacum* (Corda) Hughes, *Zalerion maritimum* (Linder) Anastasiou and *Z. varium* Anastasiou are new records for Orissa coast.

Key words: Marine fungi, mangroves, mitosporic fungi, Orissa.

INTRODUCTION

Investigations on marine fungi from Orissa coast are meager. In comparison to work carried out on marine fungi from west coast, less attention has been paid to the study of the marine fungi from East coast of India. However, to some extent marine fungi from east coast of India have been studied. (Raghukumar, 1973; Ravikumar and Vittal, 1987,1996; Sarma and Vittal, 2000, 2001; Borse 2000, Borse and Borse 2001; Borse and Pawar 2001, Borse, B.D. *et al.*2001,2002, Borse K.N. *et al.* 2001,2002). In the survey of

Author's address: *S.S.V.P.S's Arts, Commerce and Science college, Shindkheda, 425406 Dist. Dhule, Maharashtra.

**P.G.Department of Botany, S.S.V.P.S's L.K. Dr. P.R. Ghogrey Science College, Dhule- 424005, Maharashtra.

investigation of marine fungi along the Orissa coast (Mahanadi delta), mangrove marine fungi were collected from dead and decaying mangrove substrates like intertidal fronds of *Acrostichum aureum*, *Nypa fruticans*, intertidal stem of *Avicennia marina*, *Avicennia officinalis*, *Ceriops decandra*, *Kandelia candel*, *Heritiera fomes*, pneumatophores of *Sonneratia griffithii*. Eighteen mitosporic fungi were collected, illustrated and described.

MATERIALS AND METHODS

Intertidal wood samples from mangroves like *Acrostichum aureum*, *Nypa fruticans*, *Avicennia marina*, *Avicennia officinalis*, *Ceriops decandra*, *Kandelia candel*, *Heritiera fomes*, *Sonneratia griffithi* were collected along the coast of Orissa. The intertidal stem, leaf fronds, pneumatophores and roots along the mangrove creeks were collected at low tide, by cutting or sawing off pieces of the substrate. Collected samples were placed in polythene bags and sealed well to prevent loss of moisture. On return to the laboratory, specimens were examined for sporulating structures, then incubated in plastic boxes. Incubated material was periodically examined for the presence of fungi. Material was observed initially under an illuminated magnifying glass. Permanent slides were made by using double cover glass method (Kohlmeyer & Kohlmeyer, 1972) then necessary observations made under a research microscope and measurements were taken. Identification of the fungi and their reports from India, East coast of India and Orissa coast were confirmed with the help of Bilgrami *et al.*, (1979, 1981, 1991), Sarbhoy *et al.*, (1986, 1996), Kohlmeyer & Kohlmeyer (1979), Kohlmeyer & Volkmann–Kohlmeyer (1991), Hyde & Sarma (2000), Sarma & Hyde (2000) and other relevant literature. Photomicrographs of the material were taken by using Swiss research microscope (Agharkar Research Institute, Pune).

RESULTS AND DISCUSSION

Alternaria sp.

Conidiophores, cylindrical, septate, simple straight or curved, basal cell occasionally swollen, smooth, bearing conidia at the uniperforate apex, yellowish to brown. **Conidia,** 33-35 × 10-14 µm, ellipsoidal, tapering or not into an apical beak, muriform, constricted at the septa, smooth, olivaceous to brown.

Material examined: on intertidal fronds of *Acrostichum aureum*, Dungmal, 18/11/1999, Leg. Borse, K.N.

Remarks: The present fungus is rare in occurrence. No obligate marine species is known, but conidia of this genus are occasionally found on detritus along the shore as stated by Kohlmeyer & Kohlmeyer (1979). *Acrostichum aureum* is being reported as a new substrate record for the present fungus. Ananda & Sridhar (2001) reported this species on feathers from mangrove habitats of Karnataka.

Cirrenalia basiminuta Raghukumar & Zainal

Hyphae, 2.5–5 μm in diameter, septate, hyaline to pale brown. **Conidiophores:** terminal, integrated, monoblastic, determinate, 8-28 × 1 μm, conidia borne laterally and directly on conidiophore, solitary, helicoids, semi-contorted, 30-40 μm high and 20-30 μm in diameter. **Condia,** 3-4 septate, constricted at the septa, cells increasing in size from base to apex, apical cell 11-15 × 10-13 μm, subglobose, basal cell cylindrical and tapering, 5-15 μm long and 2-6 μm wide at the widest part. Pigmentation of cells increasing from base to apex, the apical cell light brown with a reddish tinge.

Material examined: on intertidal fronds of *Acrostichum aureum,* Dungmal, 11/11/2000, Leg. Borse, K.N.

Remarks: The present fungus is rare in occurrence. It has been reported for the first time from the Orissa coast. *Acrostichum aureum* is being reported as a new substrate/host record.

Cirrenalia pseudomacrocephala Kohlmeyer

Hyphae, 2-4.5 μm in diameter, septate, superficial or immersed, brown. **Conidiophores,** 22-28 μm long, 3-5 μm in diameter, cylindrical zero-two-septate, simple, straight or curved, acrogenous or laterally on hyphae, often remaining connected with detached conidia, sometimes obsolete, light brown. **Conidiogenous cells,** monoblastic, integrated, terminal, determinate. **Conidia,** acrogenous, solitary, helicoids, semicontorted to 1/4 times contorted, 3-5 septate, slightly or strongly constricted at the septa, fuscous to grayish-brown (upper three to five cells dark, lower one to three cells light-colored); cells increasing in diameter and pigmentation from base to apex, distinctly dissimilar; spirals 27-38 μm in diameter; terminal cell 15-18 μm high, 16-18 μm in diameter, subglobose to ellipsoidal, basally flattened, mostly largest and darkest of all cells; basal cells 5-9 μm high, 4-6 μm in diameter, obtusely conical or cylindrical , light brown; central cells subglobose of obtusely conical.

Material examined: on intertidal stem of *Avicennia marina,* Balramgadi, 25/5/2001, Leg. Borse, K.N.

Remarks: The present fungus is rare in occurrence. It forms a new record for the East coast of India.

Cirrenalia pygmea Kohlmeyer

Hyphae, 2.5-4 µm in diameter, septqte, ramose, fuscous. **Conidiophores,** obsolete. **Conidia,** acrogenous, solitary, helicoids, contorted ½ or 1 time, three or four septate, not or glightly constricted at the septa fist-shaped or reniform, black or fuscous, fulgent (upper three cells dark lower two or three cells light-coloured);) cells increasing in diameter), from base to apex distinctly dissimilar; spirals 24-33 × 28-32 µm; terminal cell 15-22 µm in diameter, central cells irregularly conical or almost wedge –shaped.

Material examined: on intertidal stem of *Avicennia marina,* Parki, 25/5/2001, Leg. Borse,K.N.

Remarks: The present fungus is rare in occurrence. It is an addition to the fungi of the Orissa state.

Cirrenalia tropicalis Kohlmeyer

Hyphae, 2-4.5 µm in diameter, septate, superficial or immersed, light brown. **Conidiophores,** 24-41 µm long, 2-4 µm in diameter, cylindrical, zero to four-septate, simple, acrongenous or laterally on hyphae, often remaining connected with detached conidia, sometimes obsolete, straight or curved, light brown. **Conidiogenous cells,** monoblastic, integrated, terminal, determinate. **Conidia,** acrogenous, solitary, regularly or irregularly helicoids, mostly 1 to 1 ½ times contorted. Rarely semicontorted, six to twelve-septate, not or slightly constricted at the septa, umber to reddish; spirals 21.5-36.5 µm in diameter; terminal cell 9-14 µm high, 11-19 µm in diameter, subglobose to ellipsoidal, basally flattened; basal cells 5-10 µm high, 4-6 µm in diameter; cylindrical; central cells subglobose, obtusely conical.

Material examined: on intertidal fronds of *Acrostichum aureum,* Dungmal, 11/11/2000, Leg. Borse, K.N.

Coniothyrium obiones Jaap

Pycnidia, 138-180 µm high, 150-219 µm in diameter, subglobose, at first immersed, finally erumpent, ostiolate, short papillate or coriaceous, thin-walled, brown, thick-walled cells that merge toward the venter with hyaline, thin-walled, flattened cells; cells of inner stratum bearing conidiophores; forming a *textura angularis.* **Papillae,** about 23 µm in diameter. **Conidiogenous cells,** up to 6 µm long, phialidic flask-shaped, tapering at the apex, one-celled, hyaline. **Conidia,** 4-6 µm long, 4-5 µm in diameter, ellipsoidal, ovoid or subglobose, one-celled, smooth and thick-walled, olivaceous to light brown, originating in basipetal succession on the apex of the conidiogenous cell; hyaline when released, maturing in the pycnidial venter.

Material examined: on pneumatophores of *Sonneratia griffithii*, Chandbali, 10/11/2000, Leg. Borse, K.N.

Remarks: The present fungus is occasional in distribution. The descriptions of the present collection well agree with the description of *C. obiones* given by Kohlmeyer & Kohlmeyer (1979). Therefore, it is refereed to that species. It has been first time reported from India.

Exserohilum sp.

Conidiophores, 62-140 μm long, 3-7 μm in diameter, single or paired, cylindrical, septate, often geniculate in the upper part, simple, straight or curved, smooth, brown. **Conidia,** 24-90 × 11-22 μm, simple, cylindrical to elipsoidal, slightly tapering toward the base, never rostrate, straight or somewhat curved, with a protuberant basal sear, with six to twelve pseudosepta, not constricted at the septa, smooth, thick-walled, brown; apical cells hyaline to light brown, separated from adjoining cells by distinct thick, dark-colored septa; central cells golden brown.

Material examined: on intertidal stem of *Avicennia marina,* Balramgadi, 30/5.2000, MFO, Leg. Borse, K.N.

Remarks: Traditionally, marine collection of fungi with conidia like the one described in the present study have been identified as *Drechslera halodes* (Drechsler) Subram. & Jain or *Exserohilum halodes* (Drechsler) Leonard & Suggs. Leonard (1976) consider *E. halodes* to be synonymous with *E. rostratum* (Drechsler) Leonard & Suggs. For the time being we prefer to list species similar to *E. halodes* from marine habitates as *Exerohilum* sp. until thorough studies clarify the identity of these collections as stated by Kohlmeyer & Kohlmeyer (1979). The present fungus is rare in distribution.

Macrophoma sp.

Conidiomata, Pycnidial, separate or gregarious, globose, subglobose, ampulliform, obpyriform or irregular, unilocular, immersed or erumpent, ostiolate, papillate or epapillate, thin-walled, dark-colored, glabrous or sometimes hairy or setose; occasionally with more than one ostiole. **Peridium,** pseudoparenchymatous, prosenchymatous, or rarely pseudosclerenchymatous; cells on the outside dark and thick-walled, toward the center hyaline and isodiametric. **Conidiogenous cells,** monophialidic, enteroblastic, annellidic, ampulliform, similar to the inner cells of the peridium, hyaline. **Conidia,** globose, obovoid, ellipsoidal or clavate, usually once or twice as long as wide, 22-24 × 9-11 μm, one-celled, but sometimes a secondary septation may result in two- or more-celled conidia, smooth-walled, hyaline, without appendages.

Material examined: on intertidal stem of *Ceriops decandra,* Jamboo, 28/5/2001, Leg. Borse, K.N.

Remarks: Coelomycetes resembling *Macrophoma* are occasionally encountered in marine and estuarine environment on variety of substrates. Some of these may be anamorphs of Ascomycetes, other may be spermogonia. No effort has been made so far to separate and describe the different species (Kohlm. & Kohlm., 1979). The present fungus is of occasional in occurrence.

Monodictys pelagica (Johnson) Jones

Hyphae, septate, rarely branched, dark brown or fuscous. **Conidiophores,** short or lacking, cylindrical, zero to two–septate, simple, lateral, dark brown. **Conidia,** 11-23 µm long, 11-35 µm in diameter, obpyriform, ovoid or rarely subglobose, muriform, not or slightly constricted at the septa, smooth, black, with one to three yellowish or light brown basal cells, solitary; the septation is only visible.in immature conidia.

Material examined: on intertidal stem of *Avicennia officinalis,* Ekakula, 12/11/2000, Leg. Borse, K.N.

Remarks: The present fungus is rare in occurrence. It has been collected from the Orissa coast for the first time.

Periconia prolifica Anastasiou

Condidiophores, 5-200 × 3 µm; cylindrical, septate, simple or branched, hyaline, often forming pustules on the surface of the substrates. **Conidiogenous cell,** ellipsoidal or ovoid, hyaline, produced acrogenously. **Conidia,** 6-9.5 µm in diameter, one-celled, subglobose or ovoid, smooth, thick-walled, light brown with a reddish tint or dark brown, developing basipetally, catenulate, cells finally separating.

Material examined: on intertidal stem of *Kandelia candel,* Balramgadi, 30/5/2000, Leg. Borse, K.N.

Remarks: The present fungus is very common in occurrence. It forms a new record for the Orissa state. *Acrostichum aureum* is a new substrate record of *P. prolifica.*

Phoma sp.

Pycnidia, 80-120 µm in diameter, separate or gregarious, globose, subglobose, ampulliform, obpyriform or irregular, unilocular, immersed or erumpent, ostiolate, papillate or epapillate, thin-walled, dark-colored, glabrous or sometimes hairy or setose. **Peridium,** pseudoparenchymatous, prosenchymatous, or rarely pseudosclerenchymatous; cells on the outside dark and thick-walled, toward the center hyaline and isodiametric. **Conidiogenous cells,** monophialidic, enteroblastic, annellidic, ampulliform, similar to the inner cells of the peridium, hyaline. **Conidia,** globose, obovoid, ellipsoidal or clavate, usually once or twice as long as wide, 8-10 µm, one-celled, but sometimes a secondary septation may result in two- or more-celled conidia, smooth-walled, hyaline, without appendages.

Material examined: on intertidal fronds of *Acrostichum aureum,* Dungmal, 2/6/2000, Leg. Borse, K.N.

Remarks: Coelomycetes resembling *Phoma* are occasionally encountered in marine and estuarine environment on variety of substrates. Some of these may be anamorphs of Ascomycetes, other may be spermogonia. No effort has been made so far to separate and describe the different species as stated by Kohlmeyer & Kohlmeyer (1979). The fungus is frequent in occurrence. *Acrostichum aureum* is a new substrate/host record.

Trichocladium achrasporum (Meyers &Moore) Dixon

Sporodochia, superficial, compact, fuscous to black. **Conidiophores,** absent or short, zero to four-septate, simple, formed laterally on hyphae, hyaline to light brown or fuscous. **Conidia,** (blastoconidia) 19-33 × 11-22 µm, clavate, ovoid or obpyriform, two-to five-septate, constricted at the septa, straight or slighty curved, increacing in diameter from base to apex, formed singly on the conidiophores; apical cells subglobose, dark brown; basal cells conical or subcylindrical, subhyaline to light brown or fuscous.

Material examined: on intertidal stem of *Avicennia marina,* Talsari, 8/11/2000, Leg. Borse, K.N.

Remarks: present fungus is occasional in occurrence. It has been first time reported from the Orissa coast.

Trichocladium alopallonellum (Meyers & Moore) Kohlm. & Volkm. – Kohlm.

Hyphae, septate, subhyaline to light brown. **Conidiophores,** 4-7 × 4-6 µm, macronematous, simple, one to two–celled, smooth, hyaline to light brown, lateral, short, sometimes indistinct. **Conidia:** 10-22 µm long, 17-20 µm in diameter, obpyriform, ovoidal or subglobose , one to two celled, fuscous, when two-celled apical cell larger (9-17× 7-11 µm), fuscous, basal cell, smaller obconical to cylindrical, light brown, conidiogenous cell usually remaining connected with the conidium.

Material examined: on intertidal stem of *Sonneratia griffithii,* Ekakula, 14/11/2000, Leg. Borse, K.N.

Remarks: The present fungus is of occasional occurrence and is an addition to the fungi of the Orissa state.

Trichocladium constrictum Schmidt

Hyphae, septate, subhyaline to light brown. **Conidiophores,** 4.-5 × 3.6 µm, macronematous, simple, one to two–celled, smooth, hyaline to light brown, lateral, short, sometimes indistinct. **Conidia,** light brown, 4-6 celled, 26-45- × 6 7-18 µm strongly constricted around septa, single cell nearly globular. Distal cell 19 µm in diameter.

Material examined: on intertidal leaf fronds of *Nypa fruticans,* Dungmal, 11/11/2000, Leg. Borse, K.N.

Remarks: The present fungus is of occasional occurrence. It forms a new report from the east coast of India. *Nypa fruticans* is a new substrate/ host record.

Trichocladium melhae Jones, Abdel-Wahab & Vrijmoed

Hyphae, sub-hyaline to light brown, septate and branched. **Conidiophores,** 5-18 µm long and 2-3 µm in diam., macronematous simple, one (two) celled, light brown, lateral (apical), and short. **Conidia,** dark brown to fuscous, and constricted at the septa. Apical cell 7-12 µm long and 4 -8 µm in diam., elongated to ellipsoidal in shape, middle cell 4-12 µm long and 3-6 µm in diam. While third cell is 5-6 µm long and 4-65 µm in diam. solitary conidia sometimes born directly from the mycelium.

Material examined: on intertidal stem of *Sonneratia griffithii,* Ekakula, 12/11/ 2000, Leg. Borse, K.N.

Remarks: Jones *et al.,* (2001) described *T. melhae* from Hong Kong, Malaysia, Singapur and Taiwan on exposed mangrove wood and drift wood buried in sand. The present fungus is reported second time in literature. It is rare in distribution and it has been reported from India for the first time. *Sonneratia griffithii* is being reported as a new substrate record.

Trichocladium opacum (Corda) Hughes

Colonies, effuse black. **Conidiophores,** 2.5-5 µm thick. **Conidia,** acropleurogenus, clavate, ellipsoidal, cyclindrical, rounded at the apex or pyriform 2-3 septate, smooth, proximal cell pale, other cells mid to dark brown, 26-38 × 11-17 µm.

Material examined: on intertidal stem of *Heritiera fomes,* Mahismara, 2/6/2000, Leg. Borse, K.N.

Remarks: The present fungus is rare in occurrence. It has been first time reported from the Orissa coast. *Heritiera fomes* is a new substrate record for the present fungus.

Zalerion maritimum (Linder) Anastasiou

There are about five synonyms to the species (Kohlmeyer & Kohlmeyer, 1979). **Hyphae,** 2-4 µm in diameter, septate, branched, hyaline to fuscous. **Conidiophores,** 12-98 µm long, 3-6 µm in diameter, macronematous or semimacronematous, slightly increasing in diameter from base to apex simple or rarely ramose, straight or curved, cylindrical, septate, rarely absent, superficial, light to dark fuscous. **Conidia,** 21-72 µm in diameter, acrogenous, solitary, blastic, coiled in tow or three dimensions. Forming 1-5 more or less irregular coils. Conidial filament 7-12 µm in diameter 5–28 septate, slightly or strongly constricted at the septa, fuscous or almost black.

Material examined: on intertidal stem of *Heritiera fomes,* Dungmal, 2/6/ 2000, Leg. Borse, K.N.

Remarks: The present fungus is very common in occurrence. It has been reported from the Orissa coast for the first time. *Acrostichum aureum* and *Heritiera fomes* are new substrate/host records for *Z. maritimum.*

Zalerion varium Anastasiou

Hyphae: septate, branched, immersed, hyaline, conidiophores up to 30 µm long, 2-3 µm in diameter, micronematous, simple, cylindrical, septate, sometimes absent, superficial, hyaline to light olive coloured. **Conidia:** 17-63 × 13-40 µm, solitary, irregularly helicoids or coiled in three planes, forming a knot or ball of about 9 to 27 cells, Conidial filament lateral, rarely branched or subtending an additional conidium, thick–walled, smooth, brown to dark brown, appearing black in mass; cells 6-14 × 3-10 µm.

Material examined: on intertidal stem of *Avicennia marina,* Ekakula, 12/11/2000, Leg. Borse, K.N.

Remarks: The present fungus is rare in distribution. It is being reported for first time from Orissa coast.

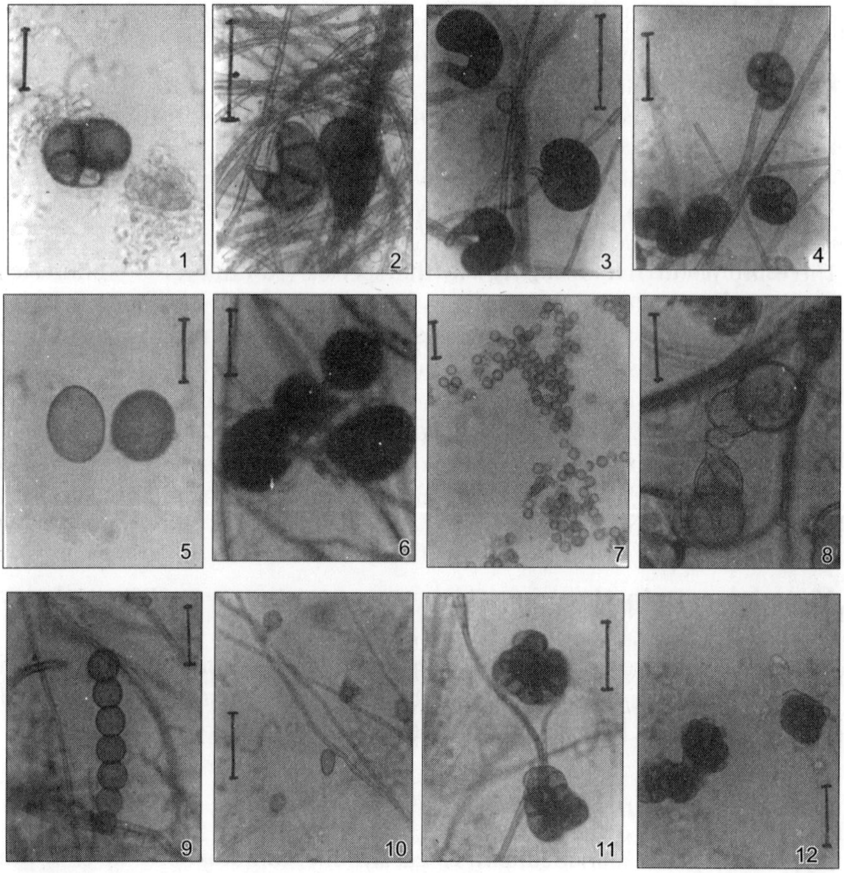

Fig. 1-2 Conidia of marine fungi 1. *Cirrenalia basiminuta*, 2. *C. pseudomacrocephala*, 3. *C. pygmea*, 4. *C. tropicalis*, 5. *Coniothyrium obiones* 6. *Monodictys pelagica*, 7. *Periconia prolifica*, 8. *Trichocladium achrasporum*, 9. *T. constrictum*, 10. *T. melhae*, 11. *Zalerion maritimum* 12. *Z. varium*
Scale bar = 10 μm except figs 5, 7, 8, 10 = 5 μm, fig 9 = 8 μm

Acknowledgements

Our sincere thanks are due to Dr. B.V. Kamble, Principal, S.S.V.PS's Arts, Commerce and Science College, Shindkheda, for facilities and encouragement. One of the authors (K.N.B.) is also grateful to the authorities of the University Grants Commission, W.R.O. Pune, for financial support.

References

Ananda, K. and Sridhar, K.R. 2001. Mycoflora on dead animal substrate of mangrove habitats of Karnataka coast, India. *Sri Lanka J. Aquat. Sci.* **6:** 85-93.

Bilgrami, K.S., Jamaluddin, S. and Rizvi, M.A. 1979. *Fungi of India* part I (List and references), Today and Tomorrow's printers and publishers, New Delhi pp. 467.

Bilgrami, K.S., Jamaluddin, S. and Rizvi, M.A. 1981. *Fungi of India part II.* (Host index and Addenda), *Today and Tomarrow's Printers and Publishers,* New Delhi, PP-268.

Bilgrami, K.S., Jamaluddin, S. and Rizwi, M.A. 1991. *Fungi of India* – List and References, *Today and Tomorrow's printers and Publishers,* New Delhi, pp. 798.

Borse, B.D. 2000. Marine fungi from India X. *Savoryella* Jones & Eaton (Ascomycotina). In: *Ecology of fungi* (Eds. Bhat D. J. & Raghukumar S.). *Goa University press,* pp.163-165.

Borse, B.D. and Borse, K.N. 2001. New reports of marine Ascomycetes from Orissa, India. *Geobios.* 28:62-64.

Borse, B.D. and Pawar, N.S. 2001. *Corbosphaerella* and *Dryospaera* two new generic records of marine Ascomycetes from West Bengal. *Geobios.* 28: 117-120.

Borse, B.D., Pawar, N.S. and Borse, K.N. 2001. Marine fungi from Sundarbans (India)-I sea foam spora. *J. Indian bot Soc. Vol.* 80, 237-240.

Borse, B.D., Pawar, N.S. and Borse, K.N. 2002. Marine fungi from Sunderbans (India)- II . The Genus *Arenariomyes* (Halosphaeriaceae). *J. Basic Appl. Mycol.* **1:** 8-10.

Borse, K.N., Pawar, N.S. and Borse, B.D. 2001. Marine fungi from Orissa. (India)-II. Arenicolous group. *BRI's JAST* 4 (1 & 11), 17-22.

Borse, K.N., Pawar, N.S. and Borse, B.D. 2002. Marine fungi from Orissa (India)– I The genus *corollospora* (Ascomycotina). *Geobios* **29 :** 258-263.

Hyde, K. D. & Sarma V.V. 2000. A Pictorial key to higher marine fungi In : *Marine Mycology – A Practical Approach* (eds. K.D. Hyde and S.B. Poinging). Fungal Diversity Research Series 1, *Fungal Diversity Press, Hong Kong*: 205-270.

Jones, E.B.G., Abdel - Wahab, M.A. and Vrijmoed L.L.P. 2001. *Trichlodadium melhae* sp. nov., A new marine fungus. *Fungal diversity*: 7: 49-52.

Kohlmeyer, J. and Kohlmeyer, Erika. 1972. Permanent microscopic mounts *Mycologia* 64: 666-669.

Kohlmeyer, J. and Kohlmeyer, E. 1979. *Marine mycology: The Higher Fungi.* Academic Press, New York, pp 698.

Kohlmeyer & Volkmann – Kohlmeyer, B. 1991. Illustrated key to the filamentous higher marine fungi. *Botanica Marina* Vol. 34: 1-61.

Raghukumar, S. 1973. Marine lignicolous fungi from India. *Kavaka.* 1: 73-85.

Ravikumar, D.R. and Vittal, B.P.R. 1987. Studies on mangrove fungi of India-I. notes on some Ascomycetes. *Kavaka.* 15: 99-103.

Ravikumar, D.R. and Vittal, B.P.R. 1996. Fungal diversity on decomposing of mangrove plant *Rhizophora* in Pichavrum estuary, East coast of India. *Indian Journal of marine Sci.* 25: 142-144.

Sarbhoy, A.K., Agrawal, D.K. and Varshey, J.L. 1986. *Fungi of India.* Associated publishing company, New Delhi, pp. 274

Sarbhoy A.K., Vershey J.L. and Agrawal D.K. 1996. *Fungi of India* (1982-1992). *S. K. Jain Publishers and Distributors,* New Delhi.

Sarma, V. V. and Hyde, K.D. 2000. Synoptic key to higher marine fungi In: *Marine Mycology – A Practical Approach* (eds. K.D. Hyde and S.B. Pointing) fungal Diversity Research Series 1, *Fungal Diversity Press, Hong Kong*: 271-307.

Sarma, V.V. and Vittal, B.P.R. 2000. Biodiversity of mangrove fungi on different substrata of *Rhizophora apiculata* and *Avicennia* sp. from Godavari and Krishna deltas, east coast of India. In : *Aquatic Mycology across the millennium* (Eds.. Hyde K.D., Ho W.H. and Pointing S.B.). *Fungal Diversity*. 5: 23-41.

Sarma, V.V. and Vittal, B.P.R. 2001. Biodiversity of mangrove fungi on selected plants in Godavari and Krishna deltas, East coast of India. *Fungal diversity*. 6: 115-130.

Frontiers in Plant Sciences (Pages 47-64)
Eds. K.G. Mukerji, K.V.B.R. Tilak, S.M. Reddy, L.V. Gangwane, P. Prakash, I.K. Kunwar

Notes on Some Species of *Agaricus* from N.W. Himalayas

Ashok Kumar* and T.N. Lakhanpal

ABSTRACT

Eleven species in the genus *Agaricus* have been reported from N.W. Himalaya out of a total of 35 recorded from all over the world. In the present study only four species in the genus have been recorded from the N.W. Himalaya. These are: *Agaricus campestris, A. angustus, A. placomyces* and *A. arvensis.* These four species have been repeatedly collected from Himachal Himalaya and are being described in detail. A key to all the eleven species has been provided.

Key words: Mushrooms, Agaricales, *Agaricus,* N.W. Himalaya.

INTRODUCTION

The genus *Agaricus* is considered almost cosmopolitan. Singer (1986) recognised 35 species in this genus. In this paper, 4 species have been described from the North-Western Himalayas, (marked with asterisk). Whereas a key is given to all the 11 species recorded so far, from the North-Western Himalayas.

Standard methodology was followed for collection, preservation and identification of the specimens as described earlier (Lakhanpal, 1996).

Key to the N.W. Himalayan species of *Agaricus*:

1. Pileus glabrous or innately squamose, lacking a woolly universal veil 3

1: Pileus and stipe base covered by a woolly universal veil often brightly

Author's address: Department of Biosciences, H.P. University, Shimla-171 005.
e-mail: tezlakhanpal@rediffmail.com
*Oilseed Research Station, C.S.K. Krishi Vishwavidyalaya, Palampur, H.P.

coloured 2

2. Pileus reddish orange .... *A. trisulphuratus*

2: Pileus yellowish orange .... *A. crocopeplus*

3. Pileus context becomes red on exposure and bruising 4

3: Basidiocarps surface yellowish on bruising, context not reddening 5

4. Annulus poorly developed, thin, fragile, spores ovoid-ellipsoid, cheilocystidia absent .... *A. campestris** sect. *Agaricus*

4: Annulus well developed, persistent, often with squamules on underside, pileus fibrillose—squamose ....*A. variegans* sect. *Sanguinolentae*

5. Basidiocarps fleshy, pileus 4.0-8.0 cm or more, annulus well developed, spores medium to large 6

Sect. *Majales*

5: Basidiocarps slender, pileus 3.0-5.0 cm diameter, pileus radially fibrillose, vinaceous .... *A. purpurellus*

6. Context deep yellow on exposure atleast at the stipe base 7

6: Context not deeply yellow 10

7. Stipe with swollen base or base atleast slightly swollen 8

7: Stipe cylindrical 9

8. Pileus with dark sepia brown squamules .... *A. xanthosarcus*

8: Pileus with reddish brown squamules .... *A. angustus**

9. Pileus with dark brown squamules .... *A. placomyces**

9: Pileus with pale grey fuscous squamules .... *A. iodolens*

10. Pileus white to pale white, yellow on drying, smooth .... *A. arvensis**

10: Pileus ochraceous, squamulose, stipe cylindric, ... *A. heterocystis*

1. *AGARICUS ANGUSTUS* FR. EPICRISIS, P. 212. 1838.

Pileus

5.0-2.0 cm diameter, fleshy, subglobosse to convex when young, broadly convex or flat with age, sometimes with a central depression, white to pale white when young, brownish orange to yellowish brown at the later stages, surface having very fine, brownish orange to reddish brown (8C6-8E6) fibrillose squamules, arranged in concentric rings, margin regular, sometimes appendiculate, incurved, becoming plane to decurved, non-striate.

Pileus Context

Up to 1.0 cm thick, white, changing to yellow on bruising and cutting after sometime, greyish orange (6B5) on drying, taste and odour pleasant. **Lamellae** 0.5–1.0 cm broad, moderately thick, finely adnexed, crowded, lamellulae of 5–6 lengths present, white to pinkish when young, changing to brownish orange or dark brown at the later stages, edges entire.

Stipe

Up to 15.0 cm long, 1.5–3.5 cm diameter, fleshy to tough, central, cylindrical, equal with a sub-bulbous base in the younger stages, white mycelial tomentum present at the base of the stipe, white to pale white when young, pale orange to yellowish brown at the later stages, surface having appressed, fibrillose, yellowish brown to reddish brown squamules, dry, solid when young, hollowing with age, flesh white, changing to yellow after sometime on cutting and bruising. **Partial veil** thick, membranous, white, covering the lamellae in the younger stages, leaves a pendant, white cottony, annular ring on the upper half of the stipe.

Spore print

Chocolate brown (6F4), spores 4.5–8.0 × 2.7–4.5 μm, broadly ellipsoid to ovate, thick walled, smooth, inconspicuously apiculate, yellowish brown in KOH, non-amyloid, mostly containing a small to large refractive guttule.

Basidia

16.2–22.5 × 6.3–8.0 μm, clavate, thin walled, mostly tetra sporic, sterigmata 1.8–3.6 μm.
Pleurocystidia absent

Cheilocystidia

9.0–27.0 × 9.0–18.0 μm, thin walled, clavate to pyriform, vesiculate, ovoid or globose, usually in chains but chains breaking and leaving the individual elements.

Subhymenium

Distinct, made up of thin walled, pseudoparenchymatous cells.

Hymenophoral Trama

Regular to subregular, consisting of thin walled, septate, branched, interwoven hyphae, 2.0–8.0 µm diameter.

Pileus Cuticle

A repent epicutis, made up of thin walled, septate, branched, interwoven hyphae, mostly 3.0–6.0 µm diameter.

Pileus Context

Consisting of thin walled, septate, branched, interwoven hyphae, 3.0–16.0 µm diameter

Stipe Cuticle

Made up of slightly thick walled, branched, interwoven hyphae, 2.0–15.0 µm diameter.

Veil Tissue

Made up of thin walled, septate, branched, interwoven hyphae, mostly 2.0–10.0 µm diameter, end cells inflated up to 20.0 µm diameter.

Macrochemical Tests

Reagent	Pileus cuticle	Pileus context
3% KOH	Light yellow	Golden yellow
30% NaOH	Negative	Light yellow
2% aq. Phenol	Negative	Reddish brown
Melzer's reagent	Negative	Reddish brown
30%KOH	Yellow	Reddish brown

Habit and Habitat

Solitary-scattered to gregarious, growing on soil rich in humus or on decaying tree foliage. The species is mycorrhizal with *Quercus incana*.

Associated Trees

Quercus incana, Pinus wallichiana, Rhododendron arboreum, Cedrus deodara

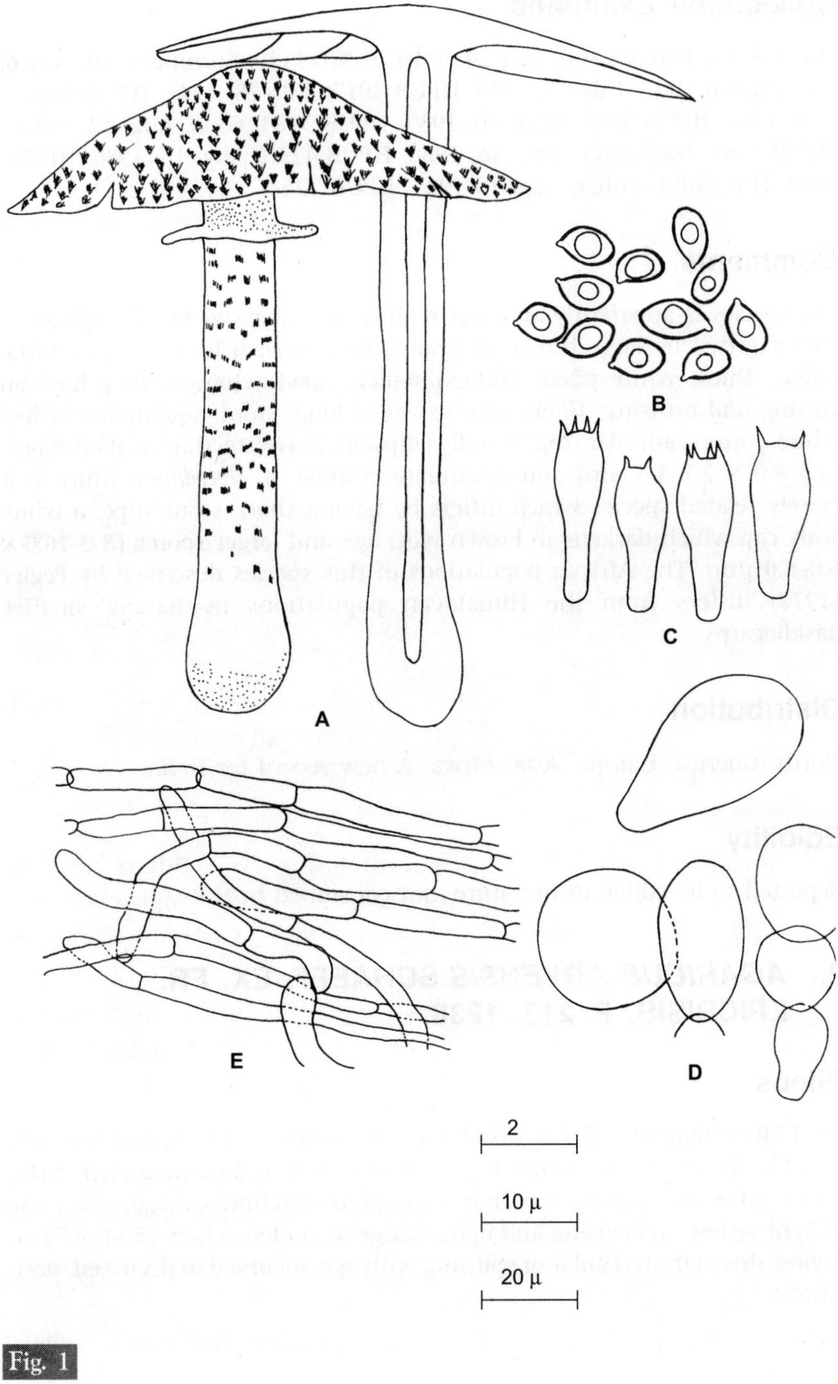

Fig. 1

Collections Examined

H.P. Shimla, Potter's Hill, July 24, 1984, HPUB 1119; Summer H ill, July 6, 1985, HPUB 3007; July 15, 1985, HPUB 3017; July 28, 1985, HPUB 3083; July, 1985, HPUB 3085; Sept. 10, 1985, HPUB 3384; Glen, Aug. 12, 1986, HPUB 4180; Narkanda, Aug. 16, 1986, HPUB 4236; Aug. 18, 1986, HPUB 4306; H.P. Kullu, Pulga, Aug. 13 1985, HPUB 3256.

Comments

The species is apparently an exact replica of *A. angustus* Fr. The species is characterized by large brownish orange to yellowish brown, squamulose pileus; thick, white pileus context which slowly changes to yellow on cutting and bruising; finely adnexed gills; long, stout, squamulose stipe, white, cottony annular ring; broadly ellipsoidal to ovate, thick walled spores (4.5–8.0 × 2.7–4.5 μm) and catenulate cystidia. *A. crocodilinus* Murr. is a closely related species which differs by having short, stout stipe, a white scaly cap which darkens to brown with age and larger spores (8.0–16.0 × 6.0–8.0 μm). The African populations of this species described by Pegler (1977) differs from the Himalayan populations by having smaller basidiocarps.

Distribution

North America, Europe, Asia, Africa. A new record for India.

Edibility

Reported to be edible in literature, not consumed in H.P.

2. *AGARICUS ARVENSIS* SCHAEFF. EX. FR. EPICRISIS, P. 213. 1838.

Pileus

6.0–12.0 cm diameter, fleshy, ovoid to convex when young, almost flat with age, faintly umbonate, white to pale white with yellowish brown (5D5) centre, smooth or having very fine yellowish brown fibrils, colour of the cap to light yellow on bruising and light orange to golden yellow (5A4-5b7) on drying, dry, margin regular or splitting with age, incurved to decurved, non-striate.

Pileus Context

Yellowish white, up to 0.5 cm thick in the centre, thinning towards the margin, slightly yellow on bruising, taste indistinct, odour fragrant.

Lamellae

0.3–0.6 cm broad, thin, free, crowded, lamellulae of atleast 3–4 lengths present, white to pinkish when young, dark brown at maturity and on drying, edges entire.

Stipe

6.5–12.0 cm long, 0.6–12.0 cm diameter, fleshy to subfleshy, central, cylindrical, almost equal or tapering from the base to the apex, white to pale white, surface smooth or very finely fibrillose, dry, stuffed when young, hollowing with age, context white, slightly changing ti yellow on bruising and on drying.

Partial Veil

Thick, membranous, white, smooth on the outer side, having white floccose patches on the inner side, leaving a persistent, pendant, annular ring on the upper half of the stipe.

Spore Print

Dark brown, spores 4.5–7.2 × 3.6–4.5 μm, ovoid to elliptical, thick walled, smooth, apiculate, lacking a germ pore, brown in KOH, non-amyloid, mostly containing a refractive guttule.

Basidia

16.2–21.6 × 6.3–8.1 μm, clavate, thin walled, bearing 2 or 4, sterigmata, sterigmata upto 2.5 μm long.

Pleurocystidia absent

Cheilocystidia

9.0–27.0 × 7.2–15.3 μm, thin walled, variable in shape, clavate, vesiculate, pyriform, ovate to sometimes globose.

Subhymenium

Distinct, made up of thin walled, pseudoparenchymatous cells.

Hymenophoral Trama

Irregular, consisting of thin walled, septate, branched, interwoven hyphae, 2.0–12.0 μm diameter, clamp connections present.

Pileus Cuticle

A repent epicutis, made up of thin walled, septate, branched, interwoven hyphae, sometimes projecting as erect or sub erect hyphae, 2.0–15.0 μm diameter, clamp connection not observed.

Pileus Context

Consisting of thin walled, septate, branched, interwoven hyphae, 2.0–15.0 μm diameter, clamp connection not observed.

Stipe Cuticle

Made up of thin walled, septate, branched, parallely to subparallely arranged hyphae, 2.0–8.0 μm diameter, inflated hyphae upto 2.0 μm diameter, clamp connection not observed.

Veil Tissue

Consisting of thin walled, septate, branched, interwoven hyphae, 2.0–8.0 μm diameter, end cells of the hyphae inflated up to 15.0 μm diameter, clamp connection not observed.

Macrochemical Tests

Reagent	Pileus cuticle	Pileus context
3% KOH	Negative	Yellow
Conc. HNO_3	Negative	Yellow
30% NaOH	Yellow	Yellow
40% NH_4 OH	Yellow	Yellow
30%KOH	Yellow	Yellow

Schaeffer's cross-reaction positive.

Habit and Habitat

Solitary-scattered, growing on soil rich in humus, in open areas, grasslands or mixed woods, mostly in the vicinity of oaks.

Associated Trees

Quercus incana, Cedrus deodara

Collections examined

H.P. Shimla, Glen, July 26, 1983, HPUB 35; Tara Devi, July 31, 1984, HPUB 1215; Jubbal, Aug. 4, 1985, HPUB 3096; Summer H ill, July 10, 1986, HPUB 4021; potter's Hill, July 11, 1986, HPUB 4029.

Comments

The present species resembles the description of *A. arvensis* Schaeff. ex. Fr. as given by (Freeman, 1979). This species belongs to section *Arvensis* Konr. and Maubl. of the genus *Agaricus* L. ex. Fr. sensu Singer (1986) because of it's sporophores tending to become yellow on cutting and bruising; surfaces showing a yellow colouration with alkaline reagents and positive Schaeffer's cross reaction.

A. macrosporus (Moller and Schaeffer) Pilat, a similar looking species, belonging to the same section differs in having buff coloured cap, scaly stipe below the ring, context of the stipe becoming flesh coloured when cut and large spores (9.0–12.0 × 5.5–7.7 μm).

A. xanthodermus Gen., another close species can be demarcated because it has usually a scaly cap, stipe base readily turns bright yellow on cutting and Schaeffer's macrochemical reaction is negative. *A. arvensis* var. *palustris* Smith described by Freeman (1979) also differs by having larger basidiocarps, stipe having bulbous base and larger basidia and cheilocystidia.

Distribution

North America, South America, and Europe. In India: U.P., W.B., Maharashtra. The species is being described fro the first time from the North-Western Himalayas.

Edibility: Reported to be edible in literature, not consumed in H.P.

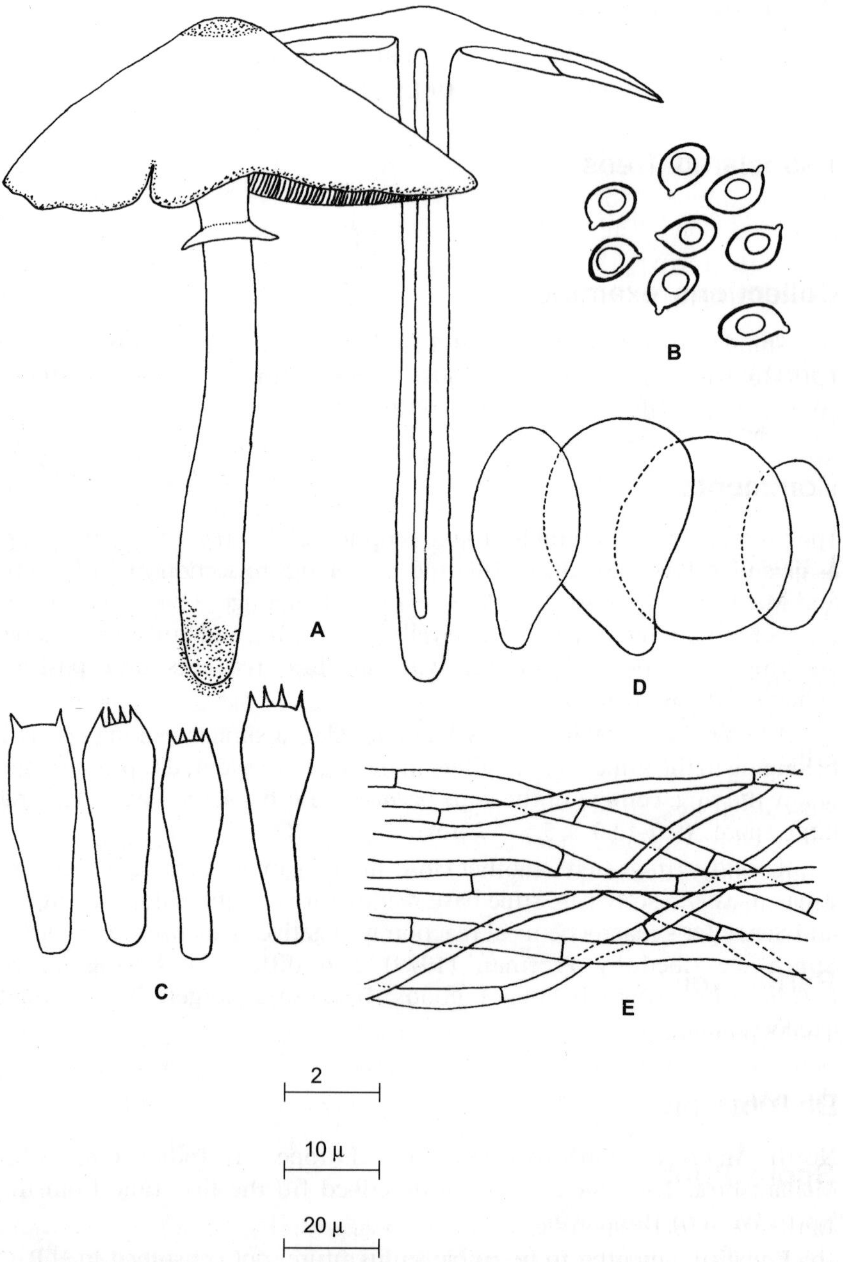

Fig. 2

3. *AGARICUS CAMPESTRIS* L. EX. FR. SYST. MYCOL., P. 281. 1821.

Pileus

1.5–8.0 cm diameter, fleshy, subglobosse to convex when young, broadly convex or almost flat with age, pure white to cream white, pale white to pale yellow on drying, smooth or having very fine silky fibrils which turn yellowish brown with age, dry or moist, margin regular, splitting with age, sometimes appendiculate, incurved, non-striate.

Pileus Context

Up to 1.0 cm thick in centre, evenly thinning towards the margin, white, unchanging or slightly reddening on cutting and bruising, taste and odour pleasant.

Lamellae

0.3–1.0 cm broad, moderately thick, free to finely adnexed, crowded, lamellulae of 3–4 lengths present, white to pink in younger stages, light brown to dark brown at maturity, edges entire.

Stipe

2.0–6.0 cm long, 0.5–1.2 cm diameter, fleshy, central, cylindrical, almost equal, white, surface smooth or having very fine, longitudinally appressed white fibrils, dry, solid when young, hollowing with age, flesh white, unchanging.

Partial Veil

Thick, membranous, white, covering the gills on younger stages leaves a pendant, friable annular ring on the upper half of the stipe, which soon disappears.

Spore Print

Dark brown (6F6), spores 6.3–9.5.0 × 4.5–6.3 μm, ovoid to broadly elliptical, thick walled, smooth, apiculate, apiculus hyaline, indistinct or papillate, brownish in KOH, non-amyloid, containing a small to large refractive guttule.

Basidia

21.6–28.8 × 7.2–9.9 µm, clavate, thin walled, bearing 4 sterigmata, sterigmata 2.0–4.0 µm long.

Pleurocystidia and Cheilocystidia absent

Subhymenium

Indistinct, made up of thin walled, pseudoparenchymatous cells.

Hymenophoral Trama

Almost regular, consisting of thin walled, septate, branched, regularly arranged hyphae, 2.0–10.0 µm diameter.

Pileus Cuticle

A repent epicutis, made up of thin walled, septate, branched, interwoven hyphae, mostly 3.0–6.0 µm diameter.

Pileus Context

Consisting of thin walled, septate, branched, interwoven hyphae, 2.0–25.0 µm diameter.

Stipe Cuticle

Made up of thick walled, septate, branched, subparallely arranged to interwoven hyphae, 2.0–15.0 µm diameter.

Stipe Context

Consisting of thin walled, septate, branched, interwoven hyphae, 2.0–20.00 µm diameter.

Veil Tissue

Made up of thin walled, septate, branched, interwoven hyphae, mostly 1.5–6.0 µm diameter, rarely up to 10.0 µm diameter. Clamp connections absent in all the hyphae.

Macrochemical Tests

Reagent	**Pileus cuticle**	**Pileus context**
Melzer's reagent	Yellowish brown	Yellowish brown

Habit and Habitat

Solitary-scattered to gregarious, growing on soil in meadows, grasslands, sometimes in open areas in forests or edges of forests.

Associated Trees

Pinus smithiana and *Cedrus deodara*

Collections Examined

H.P. Shimla, Narkanda, July 16, 1984, HPUB 1068, 1074; Aug. 16, 1986, HPUB 4222; Naldera, July 15, 1985, HPUB 3009, 3010; Baghi, Sept. July 26, 1985, HPUB 3526; Charabara, Oct. 2, 1985, HPUB 3762; Chhota Shimla, Aug. 8, 1986, HPUB 4137; H.P. Kullu, Manali, Aug. 11, 1985, HPUB 3200.

Comments

The is a meadow or pasture species, often called the 'blusher' or 'pink bottom' due to the pink colour of the gills in the immature specimens. The present species in its complete macro-and microscopic details fully agrees with the description of *A. campestris* L. ex. Fr. The species can be recognised by its white cap, white context, pink lamellae and white friable annular ring on the upper half of the stipe. The basidia are 4-sporedand cystidia are absent. *A. bitroquis* (Quel.) Sacc. is a related species which has a double ring and greyish pink gills. *A. arvensis* Schaeff. ex. Fr. is another similar mushroom but that bruises yellow.

Distribution

North America, Europe, Asia. In India: W.B., U.P., J & K. Maharashtra, South-West India. The species is being described for the first time from H.P. and all the developmental stages have been collected.

Edibility

Reported to be edible in literature. It is a prized edible mushroom and consumed in H.P. also.

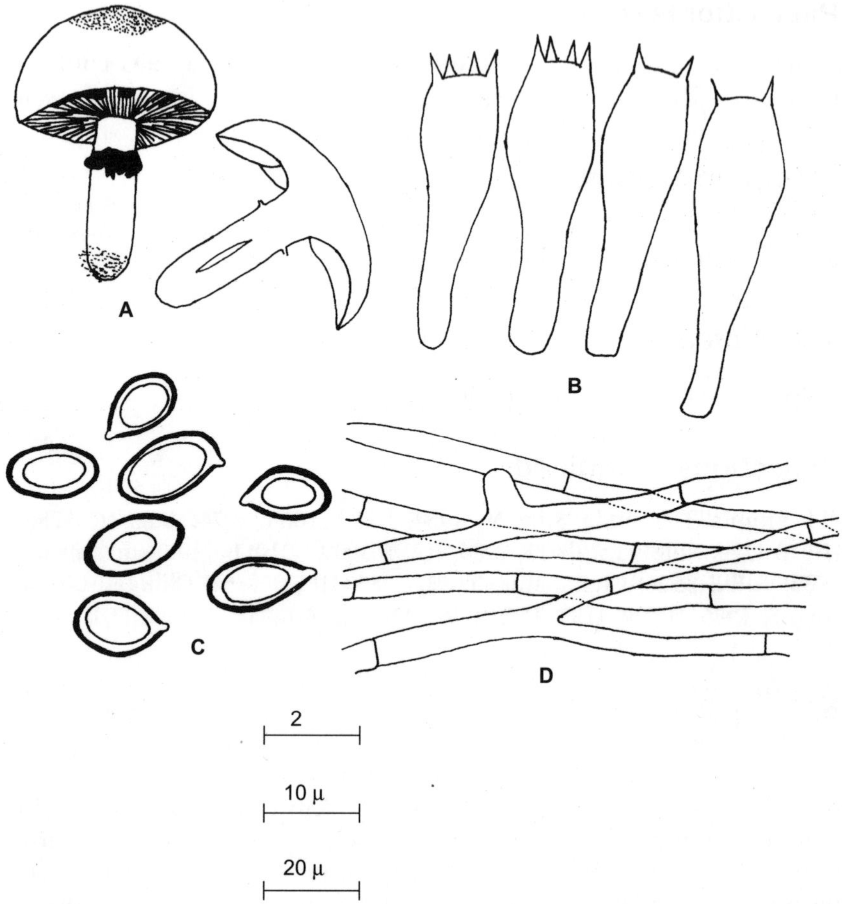

Fig. 3

4. *AGARICUS PLACOMYCES* PECK N.Y. STATE MUS. BULL., P. 40. 1878.

Pileus

4.0–12.0 cm diameter, fleshy, ovoid to convex when young, almost plane with age, faintly umbonate, dull white, covered with dark brown (6F4) radiating fibrillose squamules, squamules splitting with age, incurved to decurved, non-striate.

Pileus Context

Thin to moderately thick, up to 0.5 cm thick in the centre, thinning towards the margin, changing to yellow, then dull red on bruising after sometime, taste indistinct, odour unpleasant.

Lamellae

0.4–0.7 cm broad, fleshy to fragile, thin, free or finely adnexed, crowded, lamellulae of 3–4 lengths present, dull white to pinkish when young, chocolate brown (6F4) on maturity, edges entire.

Stipe

6.0–14.0 cm long, 1.0–2.5 cm diameter, fleshy, central, cylindrical, almost equal above, base bulbous, dull white to dingy brown, longitudinally appressed fibrils, dry, solid when young, hollowing with age, flesh white, slowly changing to yellow or reddish white on bruising. **Partial veil** thick, membranous, white to dull white, covering the gills in immature specimens, leaves a white to greyish white, persistent annular ring on the upper half of the stipe.

Spore Print

Chocolate brown, spores 4.5–6.3 × 2.7–3.6 μm, broadly elliptical to ovate, thick walled, smooth, apiculate, apiculus papilliate, hyaline, spores yellowish brown in KOH, non-amyloid, mostly containing a small to large refractive guttule. **Basidia** 13.5–22.5 × 4.5–7.2 μm, clavate, thin walled, mostly bearing 4 sterigmata, sterigmata 1.8–3.6 μm.
Pleurocystidia absent

Cheilocystidia

9.0–30.0 × 6.4–24.0 μm, thin walled, variable in shape, clavate, vesiculate, napiform, pyriform, sometimes ovate or globose.

Subhymenium

Distinct, made up of thin walled, pseudoparenchymatous cells.

Hymenophoral Trama

Irregular, made up of thin walled, septate, branched, interwoven hyphae, 2.0–16.0 μm diameter.

Pileus cuticle

A repent epicutis, made up of thin walled, septate, branched, interwoven hyphae, 2.0–9.0 μm diameter, hyphae tangled and fasciculated in the regions of squamules, having yellowish brown contents.

Pileus Context

Consisting of thin walled, septate, branched, interwoven hyphae, 2.0–18.0 μm diameter

Stipe Cuticle

Made up of slightly thin walled, branched, subparallely arranged to interwoven hyphae, 2.0–12.0 μm diameter.

Stipe Context

Consisting of thin walled, septate, branched, interwoven hyphae, 2.0–20.0 μm diameter.

Veil Tissue

Consisting of thin walled, septate, branched, interwoven hyphae, mostly 2.0–8.0 μm diameter, inflated hyphae up to 14.0 μm diameter. Clamp connection absent in all the hyphae.
Macrochemical tests All negative
Schaeffer's cross reaction negative

Habit and Habitat

Solitary-scattered to gregarious, growing on soil rich in humus in mixed coniferous forests.

Associated Trees

Abies pindrow, Picea smithiana, *Cedrus deodara, Pinus wallichiana.*

Collections Examined

H.P. Shimla, Narkanda, July 27, 1985, HPUB 3078; Aug. 23, 1985, HPUB 3380; Sept. 26, 1985, HPUB 3527; Aug. 16, 1986, HPUB 4286, 4291; Aug. 18, 1986, HPUB 4367; Jubbal, Aug. 4, 1985, HPUB 3110; Baghi, Sept. 27,

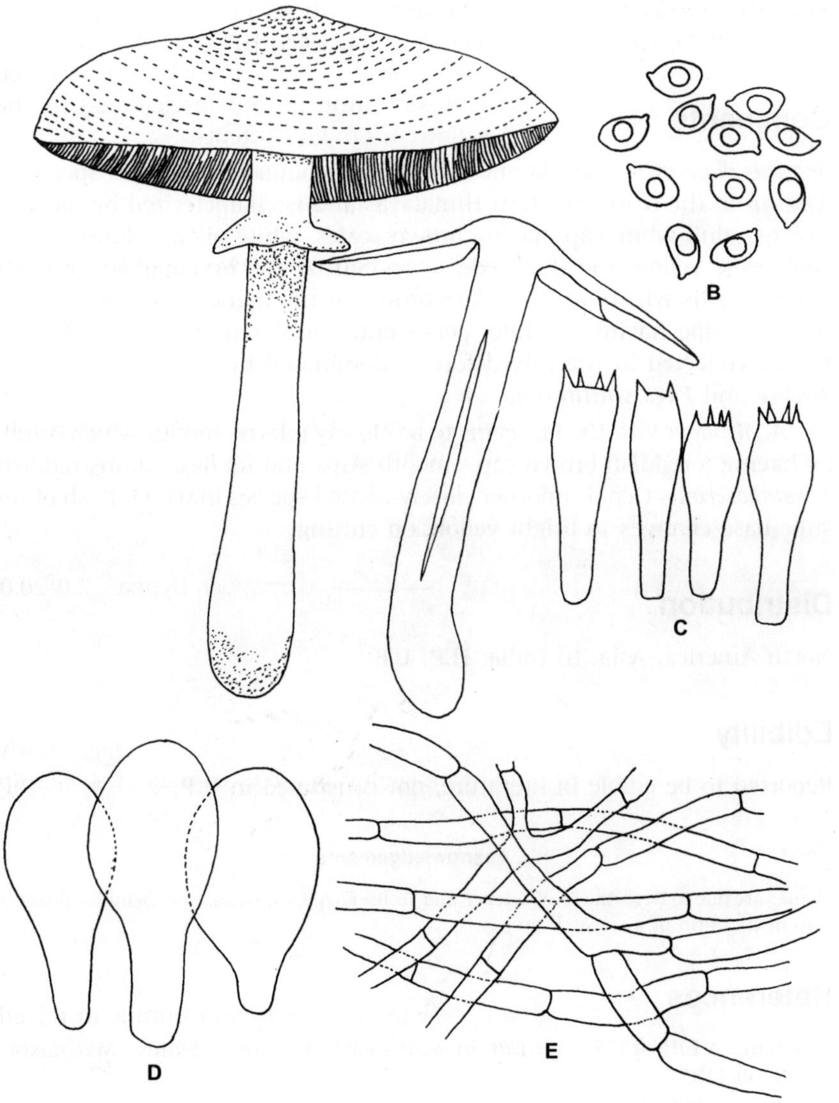

Fig. 4

1985, HPUB 3554; Chopla, Sept. 29, 1985, HPUB 3673;m H.P. Kullu, Manali, Aug. 11, 1985, HPUB 3210; Pulga, Aug. 14, 1985, HPUB 3274.

Comments

Agaricus placomyces Peck. Is one of the most common woodland species of *Agaricus* in the North-Western Himalayas and is characterized by medium to large, dull white cap, having brown scales, white pileus context which changes to yellow and then reddish on cutting and bruising; free ti finely adnexed gills which are chocolate brown in the mature specimens and a bulbous stipe having a white, persistent, annular ring. The species was mostly collected in the mixed forests dominated by *Abies pindrow, Cedrus deodara* and *Picea* **smithiana.**

A. silvaticus Vitt. Ex. Fr. seems to be closely related species which differs by having a reddish brown cap, smooth stipe and its flesh stains reddish. *A. xanthodermu*s Gen. is another closely related species in which flesh of the stipe base changes to bright yellow on cutting.

Distribution

North America, Asia. In India: H.P., U.P.

Edibility

Reported to be edible in literature, not consumed in H.P.

Acknowledgements

Thanks are due to Department of Science and Technology, GOI, India, New Delhi for financial help in the form of a research project.

References

Freeman, A.E.H. 1979. *Agaricus* in south-eastern United States. Mycotaxon. **8**: 50-118.

Lakhanpal, T.N. 1996. Mushrooms of India: Boletaceae Vol. I. A.P.H. Publishing Corporation. New Delhi, pp 170.

Pegler, D.N. 1977. A preliminary agaric flora of East Africa. Kew Bull. Additional series-VI. Royal Botanic gardens, Kew.

Singer, R. 1986. The Agaricales in Modern taxonomy. (4th ed.). Bishen Singh and Mahendra Pal Singh, Dehradun.

Frontiers in Plant Sciences (Pages 65-96)
Eds. K.G. Mukerji, K.V.B.R. Tilak, S.M. Reddy, L.V. Gangwane, P. Prakash, I.K. Kunwar

Variations in the Smut Fungi

R.V. Gandhe

ABSTRACT

The present paper deals with some morphological and physiological variations in the smut fungi, which are collected from different localities of Western Ghats of Maharashtra, India. Morphological variations observed in soral variations, spore morphology, spore-ball compositions are discussed with their importance and significance in smuts with examples. Host specialization has special importance and significance in identification of the species in smuts also discussed and stressed. Physiological variations such as mode of teliospore germination in Indian Smut Fungi are discussed in detail with illustrations. Teliospore germinations and their germination products were also studied with their variations and their importance in classification is emphasized.

Key words: Review, Ustilaginales, morphology, variations, spore germination,

INTRODUCTION

Smuts are parasitic on cereals, grasses, ornamentals and thousands of economic and noneconomic plants. They attack vast number of angiosperms in over 75 families.

Variation abounds in the smut fungi, in both morphology and physiology. Morphologically, many of the species vary widely in spore markings, culture characteristics, and gross symptoms expressed by the diseased hosts. Such characteristics as size, shape, colour, and surface ornamentations of the spores of many smut species are remarkably variable. Some others may exhibit much more variation in host range, symptomatology, and geographic distribution than in spore morphology. Very often smut species are notable for their variation in growth type and

Author address: Postgraduate Research Centre, Department of Botany, Modern College Pune 411 005.
E-mail: rvgandhe@hotmail.com

mutability of monosporidial colonies in culture; whereas in some species the gross morphological symptoms frequently display variation particularly in sorus characteristics, inflorescence modification, and plant height.

Physiological variation in the smut fungi involves all the phases of the life cycle process, including nutritional requirements and their composition, patterns of spore germination and their products, compatibility relationships, nuclear behavior, and pathogenicity. Apparently, variation in nutrient requirements in the haplophase is minimal, since most species up till reported grow well on standard composition of agar media. On the other hand, sharp variation in nutrient requirements of the dikaryophase is seemingly implicit in host specificity, which is common to most species investigated so far. Thus, in species parasitizing only one host, nutrient requirements for the dikaryophase are highly specific, while for those with a wide host range they are apparently much less so. Variation in teliospore germination may reflect differerential response to complex environmental factors, or inherent differences in longevity and germination rate, all of which may relate to variation in pathogenecity as exhibited by wide or restricted host range, degree of virulence, and specialization. Although the classification of families is still strictly dependent upon the mode of teliospore germination and their germination products, the classification of many genera and species is chiefly based upon the teliospore morphology, symptomology and host specificity. Except for the temporary impact of environment, variations in the smut fungi are genetically controlled through sexual processes involving either intra or inter species compatibility.

MATERIALS AND METHODS

During this extensive dissertation on smut fungi of the Western Ghats, several smut species from different localities were collected and identified with the monographs (Ainsworth and Sampson, 1950; Fischer, 1951; Mundkur and Thirumalchar, 1952; Fischer and Holton, 1953, 1957; Fischer and Shaw, 1953; Zundel, 1953; Duran and Fischer, 1961; Thrumalachar, 1966; Duran, 1972; 1973; Vanky, 1999a, 1999b, 2000, 2001a, 2001b, 2002) and related publications. The discussion in this paper is partially based upon the personal communications with Late Professor M.J. Thirumalachar, India and Kalman Vanky, Germany. New species were published and deposited in Herbarium Cryptogammae Orientalis, IARI, and New Delhi. Fresh collections of smuts were utilized for germination studies and the cultures. Exsiccate of the collected specimens is prepared and maintained. The data generated during his work is presented herewith (Tables 1) and utilized in the present paper for discussion. Teliospore suspension was

TABLE 1 *Showing the List of Smut Fungi in Maharshtra, India*

Genus	*Species*	*Host*	*Family of the host*	*Locality*	*Date of collection*	*Remarks*
1	2	3	4	5	6	7
1. *Ustilago* (Persoon) Roussel	1. *U. cynodontis* P. Hennings	*Cynodon dactylon* Pers.	Gramineae	Lonavala University Campus (Pune) Fruit Research Center (Pune)	8-9-73 9-9-73 12-9-73	In the inflorescence.
	2. *U. effusa* Sydow	*Chrysopogon aciculatus* (Retz.) Trin.	Gramineae	Ganeshkhind (Pune)	23-8-74	In the inner surface of leaf sheath. New record for Maharashtra. New host recorded.
	3. *U. neglecta* Niesal	*Setaria glauca* (L.) P. Beauv.	Gramineae	Pune Pune Karjat	25-9-75 20-8-76 5-9-76	In the ovaries. New record for Maharashtra.
	4. *U. pavgiensis* sp. nov.	*Eleusine coracana* Gaerth.	Gramineae	Kolhapur	15-10-76	In the ovaries. New species.
	5. *U. royleana* Sydow and Butler	*Digitaria royleana* (Nees) Prain.	Gramineae	Pune	21-10-76	In the spikelets. New record for Maharashtra.
	6. *U. scitaminea* Sydow	*Saccharum officinarum* L.	Gramineae	Vittalwadi (Pune)	4-10-74	In the inflorescence.
	7. *U. sparsa* Underwood	*Dactyloctenium aegyptiacum* Willd.	Gramineae	Pune	10-8-74 15-8-75	In the ovaries.

(Table 1 Contd.)

(*Table 1 Contd.*)

	8. *U. trichophora* (Link.) Kunze	*Paspalidium geminatum* Stapf. Echinochola colonum (L.) Limk.	Gramineae Gramineae	Khandala Khandala	4-8-74 16-8-74	In the ovaries. New host record in Maharashtra.
	9. *U. tripogonis* sp. nov. Syn. *Sporisorium tripogonis* Gandhe et al	*Tripogon lisboae* Stapf. *Tripogon jacquemontii* Stapf.	Gramineae Gramineae	Khandala Pune	5-7-73 5-10-75	In the ovaries. New species.
	10. *U. tritici* (Pers.) Rostrup	*Triticum vulgare* Vill.	Gramineae	Mahabaleshwar	27-1-75	In the spikelets.
	11. *U. utriculosa* (Nees) Tul.	*Polygonum barbatum* L.	Polygonaceae	Koyana	5-1-60	In the ovaries. New record for Maharashtra.
2. *Pericladium*	1. *P. tiliacearum* Passerini Thirum.	*Grewia tiliaefolia* Mundk. and	Tiliaceae Vahl.	Barsi	5-3-70	In the stems. New record for Maharashtra.
3. *Sphacelotheca* de Bary	1. *S. annulata* (Ell. And Evern.) Mundk.	*Dichanthium annulatum* (Forssk)	Gramineae	Vitthalwadi Nasik	4-10-74 24-8-74	In the ovaries.
	2. *S. bursa* (Berk.) Mundk. and Thirum.	*Themeda quadrivalvis* (L.) 0 Kuntze	Gramineae	Shirval	25-8-72	In the spikelets.
	3. *S. cornuta* (*Syd. and Butler*) Mundk.	*Ophiuros exaltatus* (L.) 0 Kuntze	Gramineae	Kapurwal (Bhor)	5-9-73	In the inflorescence. New to Maharashtra.
	4. *S. cymbopogonis* Yen.	*Cymbopogon martini* (Roxb.) Wats.	Gramineae	Pashan (Pune)	10-10-74	In the ovaries.

(*Table 1 Contd.*)

(Table 1 Contd.)

	5. S. *erythraeensis* (Syd.) Clinton	*Hackelochloa granularis* (L. F.) 0 Kuntze	Gramineae	Pune	5-9-75	In the ovaries.
	6. S. *iseilematis* (Sydow and Butler) Mundk. and Thirum.	*Iseilema laxum* Hack.	Gramineae	Pune	5-9-73	In the ovaries.
	7. S. *lohagadensis* sp. nov.	*Ischaemum tumidum* Stapf.	Gramineae	Lohagad	5-1-75	In the inflorescence. New species.
	8. S. *reiliana* (Huehn.) Clinton	*Sorghum vulgare* Pers.	Gramineae	Pune	10-10-75 5-11-76	In the inflorescence.
	9. S. *sinhagadensis* sp. nov.	*Coix lachryma-jobi* L.	Gramineae (Pune)	Sinhagad	1-12-75 5-11-76	In the ovaries, anthers and inflorescence. New species.
	10. S. *sorghi* (Link.) Clinton	*Sorghum vulgare* Pers.	Gramineae	Pune	25-9-74	In the ovaries.
	11. S. *tanglinesis* (Tracy and Earle) Zundle	*Sehima spathioflorum* (Hook F.) Blatt. Et McCann	Gramineae	East Thana (Bombay)	3-12-54	In the ovaries.
	12. S. *zilligii* Zundel	*Heteropogon ritchiei*	Gramineae (Hook F.)	Katraj (Pune)	6-9-75	In the inflorescence. New record for Maharashtra.
4. *Cintractia* Cornu	1. C. *axicola* (Berk.) Cornu	*Fimbristylis quinquangularis* Kunth	Cyperaceae	Khandala Khandala Pune	6-9-73 10-9-74 15-9-75	In the peduncles, pedicels and spikelets.
	2. C. *minor* (Clinton) Jackson	*Cyperus rotundus* L.	Cyperaceae	Lohagad Pune	20-9-73 29-10-74	In the peduncles, pedicels and inflorescence. New record for Maharashtra

(Table 1 Contd.)

(Table 1 Contd.)

5. *Farisia* Raciborski	1. *F. butleri* (Sydow) Sydow	*Scleria tessellata* Willd.	Cyperaceae	Pune	10-8-66	In the ovaries. New record for Maharashtra
6. *Liroa* Ciferri	1. *L. emodensis* (Berk.) Ciferri	*Polygonum chinense* L.	Polygonaceae	Mahabaleshwar	12-1-72	In the inflorescence ocrae, peduncles and stems.
7. *Melanopsichium* Beck	1. *M. pennsylvanicum* Hirschhorn	*Polygonum glabrum* Willd	Polygonaceae	Pune Khandala	25-11-74 6-1-74	In the inflorescence Stems and leaves.
8. *Sorosporium* Rudilphi	1. *S. apludae* Mishra	*Apluda aristata* L.	Gramineae	Sinhagad	1-10-75	In the ovaries. New record for Maharashtra.
	2. *S. bullatum* Pavgi and Thirum.	*Aristida adscensionis* L.	Gramineae	Pune	24-10-74	In the inflorescence leaf sheaths. New to Maharashtra.
	3. *S. contortum* Griffiths	*Heteropogon contortus* Beauv	Gramineae	Sinhagad	25-11-75	In the inflorescence New to Maharashtra
	4. *S. heteropogonicola* Mundk. and Thirum	*Heteropogon contortus* Beauv	Gramineae	Pune	28-11-74	In the ovaries. New record for Maharashtra.
	5. *S. kolhapurensis* sp. nov.	*Apuda mutica* L.	Gramineae	Kolhapur	7-10-75	In the ovaries. New species.
	6. *S. lophopogonis* Thirum. and Pavgi	*Lophopogon tridentatus* Hack.	Gramineae	Pune	18-9-73	In the inflorescence.
	7. *S. mahabaleshwarensis* sp. nov	*Ischaemum impressum* Hack.	Gramineae	Mahabaleshwar	12-10-76	In the inflorescence New species.
	8. *S. matheranensis* sp. nov.	*Spodiopogon rhizophorus* (Steud.)	Gramineae	Matheran	5-11-76	In the ovaries. New species.
	9. *S. punensis* sp. nov.	*Andropon pumilus* Roxb.	Gramineae	Pune	29-11-74	In the spikelets. New species.

(Table 1 Contd.)

(Table 1 Contd.)

	10. *S. semisagittatum* Thirum. and Pavgi	*Ischaemum semisagittatum* Roxb.	Gramineae	Khandala	12-11-75	In the ovaries.
	11. *S. tumefaciens* McAlp	*Chrysopogon montanus* Trin.	Gramineae	Katraj (Pune)	24-11-74	In the inflorescence New to Maharashtra.
9. Tolypospori-um Woronin	1. *T. andropogonis* Patel and Kulkarni	*Andropogon triticeus* R. Br.	Gramineae	Pune	Oct. 1950	In the inflorescence
	2. *T. ehrenbergii* (Kuehn) Pat.	*Sorghum vulgare* Pers.	Gramineae	Pune		In the ovaries.
	3. *T. penicillariae* Brefeld	*Pennisetum typhoides* Stapf ans Hubb.	Gramineae	Pune Nasik	10-10-75 15-10-76	In the ovaries.
Telletiaceae 10. *Telletia* Tulasne	1. *T. apludae* *Thirum. and Mishra. (Thirum. and Mishra.)*	*Apluda mutica var. varia* L.	Gramineae	Pune	15-10-76	In the ovaries. New record for Maharashtra.
	2. *T. fischeri* Patil & Gandhe	*Panicum psilopodium* Trin.	Gramineae	Khandala	10-9-75	In the ovaries. New species.
	3. *T. savilei* Gandhe et al.	*Tripogon jacquemontii Stapf.*	Gramineae	Pune	10-10-75	In the ovaries: New species.
	4. *T. thirumalachari* R.V. Gandhe & K.R.Gandhe	*Setaria verticillata* (L.) Beauv.	Gramineae	Pune	10-10-74	In the ovaries. New species.
	5. *T. vittata* (Berk.) Mundk. *var.* burmannii *Mishra*	*Oplismenus burmannii* Beauv.	Gramineae	Pune	20-10-74	In the ovaries. New record for Maharashtra.

(Table 1 Contd.)

(*Table 1 Contd.*)

11. *Neovossia* Kornicke	1. *N. punensis* Patil and Gandhe	*Polytoca barbata* (Roxb.) Stapf.	Gramineae	Pune	10-12-75	In the ovaries.
	2. *N. sumatii* Patil and Gandhe	*Coix lachryma jobi* L.	Gramineae	Sinhagad (Pune)	5-12-75	In the ovaries.
12. *Entyloma*	1. *E. bidentis* P. Hennings	*Bidens pilosa* L.	Compositeae	Sinhagad (Pune)	10-9-75	In the leaves.
	2. *E. chlorides* Thirum. and Pavgi	*Chloris virgata* Sw.	Gramineae	Pune	10-11-75	In the leaves and leaf sheaths.
	3. *E. dahliae* Sydow	*Dalhia coccinea* Des.	Compositeae	Victoria Gardens (Bombay)	- 1912	In the leaves
	4. *E. globigena* Thirum. and Safee.	*Blumea wightiana* DC	Compositeae	Khandala Lohagad Mahabaleshwar	11-9-72 13-9-72 15-8-74	In the leaves, stem, rarely in roots.
	5. *E. khandalensis* Patil and Gandhe.	*Centratherum* sp.	Compositeae	Khandala	1-10-63	In the leaves. New species.
	6. *E. mundkuri* Patil and Gandhe.	*Vicoa cernum* Dalz.	Compositeae	Khandala	17-9-74	In the leaves. New species.
	7. *E. nymphaeae* (Cunningham) Setchell	*Nymphaea stellata* Willd.	Nymphaeaceae	Kolhapur	5-9-76	In the leaves and petioles. New to Maharashtra.
	8. *E. oryzae* Sydow	*Oryza sativa* L.	Gramineae	Alibag		In the leaves and leaf sheaths.

(*Table 1 Contd.*)

(Table 1 Contd.)

13. *Melanotaenium* de Bary	1. M. *brachiariae* Viegas var. *paspalidii* *Shrinivasan and Thirum.*	*Paspalidium geminatum* Stapf.	Gramineae	Pune	20-9-72	In the leaves and leaf sheaths. New record for Maharashtra.
	2. M. *ischaemum* Thirum. and Pagvi	*Ischaemum semisaggitatum* Roxb.	Gramineae	Khandala	20-11-75	In the leaves and leaf sheaths.
	3. M. *oreophilum* *Sydow*	*Selaginella* sp.	Selaginellaceae	Khandala	10-9-60	In the leaves and the stem. New record for Maharashtra.
	4. M. *punensis* sp. nov.	*Themeda triandra* Forssk.	Gramineae	Pune	15-9-75	In the leaves and leaf sheaths. New species.
	5. M. *spermacoces* Thirum. et al	*Spermacose stricta* L.F.	Rubiaceae	Kolhapur	10-10-75	In the leaves. New host record.
14. *Doassansia* Cornu	1. *D. hygrophilae* Thirum	*Astercantha longifolia* Nees	Acanthaceae	Khandala	1-8-74	In the leaves.
	2. *D. khandalensis* Patil and Gandhe	*Hemiadelphis polysperma* Nees	Acanthaceae	Khandala	5-10-71	In the leaves.
15. *Dossansiopsis* Dietal	1. *D. nymphaeae* (Syd.) Thirum.	*Nymphaea stellata* Willd.	Nymphaeaceae	Kolhapur	9-9-76	In the petioles. New record for Maharashtra.
16. *Burrillia* Setchell	1. *B. kamatii* Thakur	*Hygrophila serphyllum* Anders	Acanthaceae	Khandala	7-9-74	In the leaves.
17. *Urocystis* Robenhorst	1. *U. hypoxyis* Thaxter	*Hypoxis aurea* Lour	Amaryllidaceae	Lonavala Khandala Lohagad	10-7-55 15-7-74 20-7-74	In the floral buds.

(Table 1 Contd.)

(Table 1 Contd.)

18. *Jamesdicksonia* Thirum. et al	1. *J. obesa* Thirum. et al	*Dichanthium annulatum* Stapf.	Gramineae	Pune	5-11-73	In the leaves and leaf sheaths.
19. *Georgefischeria* Thirum. and Naras	1. G. *narsimhanii* Patil and Gandhe	1) *Argyreia hookeri* Clke.	Convolvulaceae	Mahabaleshwar	7-2-69	In the leaves.
		2) *Argyreia elliptica* Chois.	Convolvulaceae	Khandala	6-9-70	In the leaves. New host record.
		3) *Lettosomia setosa* *Roxb.*	Convolvulaceae	Bhor	10-11-71	In the leaves. New host record.
	2. G. *riveae* *Thirum. and Naras. emend Ranganath*	*Rivea hypocrateriformis* Chois.	Convolvulaceae	Pune	1-11-74	In the axillary shoots forming witches' broom.
	3. G. *riveae* *Thirum. et al var.* argyreiae *Naras. et al*	*Argyreia cuneata* Ker	Convolvulaceae	Katraj (Pune) Bhor	1-2-74 12-2-75	In the axillary shoots forming witches' broom.

prepared by teasing the infected material in sterilized distilled water containing traces of tetracycline hydrochloride to avoid bacterial contamination. The suspension was then spread uniformly on glass slides with a clean blade or scalpel. These spores were germinated by the method suggested by Thirumalachar et al. (1950, 1953) with minor modifications. To study cytology of the germination products, the spores were germinated on 2% water agar and in sterile distilled water to the desired stages, killed and fixed in Randolph's modified Navaschin's fixative. The slides were stained with Heidenhain's hematoxylin-lightgreen combination as usual procedure (Johansen, 1940). Germinated spores were also killed and fixed in Cornoy's fixative and stained with Belling's 0.5% Aceto Carmine (Johansen,1940) for complementary observations.Monosporidial cultures were prepared and maintained on 2% PDA and HEA (host extract agar) by isolating basidiospores after teliospore germination on the medium.

RESULTS AND DISCUSSION

Host Specialization

Smuts are strictly host specific and often produce infections in the inflorescences or in the ovaries of the flowers so as convert them in to black powdery or agglutinated spore masses, causing tremendous direct economic loss. Realizing the importance of host specialization in the taxonomy of the smut fungi Fischer (1953), Fischer and Shaw (1953), proposed a species concept based partially on morphology and host-specialization at the family level. This is applicable as a partial solution to the species problem, particularly for taxa with a dearth of distinct morphological characters. Amongst species of *Ustilago, Sphacelotheca, Sorosporium* and *Entyloma* morphological similarities of sori make identification very difficult and impossible. Duran (1971) also strongly supported the concept of host specialization. He further added that morphology and host-specialization must both be stressed in smut taxonomy for correct identification. I have also noted the strict host specialization in my collections. Therefore in this work host specialization has given priority in species identification.

Variation in Morphology

Since morphology is the basic taxonomic criterion in the smut fungi, any variation in the primary morphological features becomes significant in relation to species delimitation. Conspicuous in this regard are variations in gross morphology of the diseased host and in spore morphology. Scarcely any of the morphological standards established for the various species have escaped the impact of variation and its consequences in species delimitation. For this reason, morphology as the basic taxonomic criterion in some genera has been supplemented by other criteria.

Gross Morphology and Soral Variation

For the most part, the gross morphology of the smuts reflects size and shape of the sori, both of these characteristics are determined largely by where the sori occur on the host. There are wide and remarkable variations in the morphology of sori and their development on the hosts. Variation often expected, therefore, other characters are considered as supporting characters for correct identification of the species. I feel, it is necessary to study soral development and composition for the classification of smut fungi. Moreover all these soral characters are often detectable and reflect salient features of different genera. There is a wide and notable variation in the gross morphology of inflorescence, stem, leaf and root smuts, examples of which are found in many smut genera particularly in the species of *Tilletia, Ustilago, Melanopschium, Tolyposporium, Soroporium, Sporisorium Urocystis* and *Sphacelotheca* (Figs. 1, 2, 8, 11, 12, 13, 14, 15, 24, 27, and 28). Majority of the species from the family Ustilaginaceae develop sori either in the inflorescence or in the reproductive structures such as ovaries, ovules, anthers and convert them into a black powdery mass/ agglutinated mass of spores. Whereas in the species of *Doassansia, Doassansiopsis, Burrillia, Entyloma, Melanotaenium, Jamesdicksonia* and *Georgefischeria* the sori are mostly restricted to the vegetative parts such as stem, leaves, roots, branches, axillary buds...etc. (Figs. 17,18,19,22,23,25,29 and 30). Number of species of *Ustilago, Sphacelotheca, Sorosporium, Sporisorium, Tolyposporium, Tilletia* and *Neovossia* develop sori in the inflorescences and destroy them completely or partially (Figs. 1-16, 24, 27 and 28). In such cases sori may develop limited or unlimited hypertrophy and deformity in the floral structures. The most striking morphological variation is observed in the species of *Ustilago*. The sori often remain inconspicuous in *Ustilago tripogonis* species.nov. (Fig. 3) and less hypertrophied but detectable in *Ustilago neglecta* Niessel as well as in. (Fig. 2). Because of this, it is very difficult to find out infected plants in the field. Similarly, sori are also inconspicuous in *Tilletia savilei* Gandhe et Vanky. and *Tilletia thirumalachari* R.V. Gandhe and K.R. Gandhe (Fig. 16). In contrast to this morphological feature, sori are distinct and much hypertrophied in *Ustilago trichophora* (Link) Kuntze (Fig. 4), *Liroa emodensis* (Ber.) Ciferri (Fig. 10), *Sorosporium bullatum* Pavgi and Thirum., *Tilletia fischeri* Patil and Gandhe. and in *Neovossia sumatii* Patil and Gandhe (Figs. 12). Moreover, elongated, large and often irregular hypertrophy is developed in

Figs. 1-14 Soral variation: (Figures not to the scale)

1 *Ustilago cynadontis* P. Hennings; 2 *Ustilago neglecta* Niessl; 3 and 3a *Ustilago tripogonis* species.nov.; 4 and 4a *Ustilago trichophora* Kunze.; 5 and 5a,b Sphacelotheca sinhagadansis species.nov.; 6 and 6a *Farysia buteri* Sydow; 7 *Pericladium tillecearum* Mundk. and Thirum.; 8 and 8a,b *Melanopschium pennsylvanicum* Hirachhorn; 9 and 9a *Cintractia axicola* Cornu; 10 Liroa emodensis Ciferri; 11 and 11a Sorosporium heteropogonicola Mundk. and Thirum.; 12 and 12a *Neovossia sumatii* Patil and Gandhe; 13 Sphacelotheca reiliana Clinton; 14 Ustilago scitaminea Sydow.

Melanospichium pennsylvanicum Hirschh. (Fig. 8). The sori in this species become erumpent and expose agglutinated spore mass at maturity. At this stage, it appears in the form of an irregular ball with thread like structures bursting out from the sorus. Most striking and very different type of soral development is observed in *Pericladium tilliacearum* Mundk. and Thirum., in which sori are developed in the form of blister- like small galls on the petiole and on the stem parts (Fig. 7). However, in the species of *Entyloma* remarkable variation is observed in soral morphology and sori are either inconspicuous or much distinctly hypertrophied. In *E. globigena* Thirum. and Safee. sori mostly develop in the leaves and stems and occasionally in the roots. In such cases the infected vegetative parts develop large sessile to sub sessile irregular galls (Fig. 26). On the other hand, sori are inconspicuous and rarely hypertrophied in *Entyloma mundkuri* Patil and Gandhe (Fig. 20). Most interesting soral development is observed in *Entyloma khandalensis* Patil and Gandhe in which sori show pustulate appearance similar to that of *Doassansia khandalensis* Patil and Gandhe. One of the most interesting and curious infection patterns was observed in *Georgefischeria riveae* Thirum. et al. wherein witches' broom – like malformations are developed in the axillary shoots (Fig. 29). However, other species in the genus show very distinct infection pattern than the type species. The species *Georgefischeria narasimhanii* Patil and Gandhe causes yellow patches in leaves and blackening of veins in the infected area (Fig. 30). In this species there is no malformation resulting in to witches broom of the axillary shoots as in the type species. The only species of *Doassansiopsis* namely, *D. nymphaeae* (Syd.) Thirum. develops distinct tumour–like growth in petioles and leaves of *Nymphaea* (Fig. 25). In *Urocystis hypoxyis* Thaxt. the sori are developed in floral buds resulting in to hypertrophy and destroy them completely the ovary and ovules (Fig. 24). However, in *Melanotaenium,* sori remain inconspicuous in *M. punensis* sp. nov. (Fig.). The sori are typically hypertrophied and evanescent in *Cintractia axicola* (Fig. 9). Whereas in *Farysia butlari* (Fig. 6). The monotypic genus *Jamesdicksonia* develops sori in the form of erumpent pustules similar to that of rust Puccinia graminis in the leaves and leaf sheaths (Fig. 18). More strikingly, the sori remain very inconspicuous but forming pustulate specks in *Burrillia kamatii* (Fig. 17). Whereas in *Ustilago cynodontis* entire inflorescence is infected and only crinkled rachis is observed after the dispersal of spores (Fig. 1) and often confused with the columella. More curious infection pattern was observed in *Sphacelotheca sinhagadensis* sp. nov. where the sori are developed in the anthers, ovaries and also in the inflorescence (Fig. 5). These sori then develop into distinct hypertrophied structures.

Figs. 15-30 Soral variation: (Not to the scale)

15 and 15a *Tilletia vittata* Mundkur var. *burmanii;* 16 and 16a,b *Tilletia thirumalachari* species nov.; 17 *Burrillia kamatii* Thakur; 18 and 18a *Jamesdicksonia obesa* Thirum.et al.; 19 and 19a *Melanotaenium oreophilum* Sydow; 20 *Entyloma mundkuri* species.nov.; 21 *Entyloma bidentis* P. Hennings; 22 and 22a *Doassansia khandalensis* Patil and Gandhe; 23 *Entyloma nymphaeae* Setchell; 24 *Urocystis hypoxyis* Thaxter; 25 *Doassansiopsis nymphaeae* Thirum.; 26 *Entyloma globigenum* Setchell; 27 *Tolyposporium penicillariae* Brefeld; 28 and 28a *Tolyposporium ehrenbergii* Pat.; 29 *Georgefischeria riveae* Thrum. and Naras.; 30 *Georgefischeria narasimhanii* Patil and Gandhe.

Spore Morphology

Size, Shape colour and exospore ornamentation are the principal and indispensable features in the species delimitation of smut fungi. These criteria are supplemented by the characters like sterile cells, gelatinous sheath, and pigmentation etc. of number of taxa.

Spore Size and Shape

There is a considerable variation in spore size in different genera and species of smut fungi. The variation in spore size, which often occurs in the same species, can be further demonstrated by comparing spore size in different species. Therefore spore size has greater importance and taxonomic significance both in small-spored species of *Ustilago, Sorosporium, Sporisorium,* and *Sphacelotheca* as well as relatively large-spored genera like *Cintractia, Tilletia,* and *Neovossia*. The smallest spores are observed in *Ustilago minima* in a range of 3.5 – 5.0 whereas largest spores are found in *Tilletia paradoxa* 57.0 μm, *Neovossia punensis* μm in a range of 30-39 μm (Fig. 56).

However, the character has some personal limitations and many times spore diameter for a species is expressed in the form range. In such cases spore size is expressed in the form of average mean value. The spores of *Ustilago cynadontis* (Fig. 31) are often cupped in dry condition but globose in turgid condition. The spores are small in *Ustilago tritici* (Pers.) Rostrup (4.8 μ in diameter), in contrast to the larger spores in *Cintractia axicola* (Berk.) Cornu (17.6 μ). Whether spore size alone in some instances can justify creation of new species must, in the final analysis, be left to the judgment of the individual.

Spore Shape

The spores of many species of smut fungi are commonly globose to subglobose or occasionally polyhedral. However, these descriptions of the spores are very loosely used by the Smutologist therefore this character of the spore has minor importance in the identification of species and often considered for species identification as the supporting character. The shapes are circular, globose, oval, elongated etc. and usually described in the range of globose to subglobose, circular to sub circular and so on. There is a notable variation in the spore shapes in the same sorus or different sori. It is necessary to observe shapes in several spores so that average spore shape can be described. In *Sorosporium, Tolyposporium* species spore ball configuration is curious where innermost spores of the ball sustain considerable internal pressure, and, hence, tend to assume a polygonal configuration visible when the spores are freed. The shape of the spore may change in the spore The shape of the spore may change in the spore The shape of the spore may change in the spore The shape of the spore may change in the spore The shape of the spore may change in the spore The shape of the spore may change in the spore The shape of the spore may change in the spore The shape of the spore may change in the spore The shape of the spore may change in the spore. In *Urocystis hypoxis,* outer few spores are wedge shaped, sometimes hyaline and sterile whereas inner darker spores are fertile and more or less globose. Various combinations of the components such as sterile cells, fertile pores, cortex, pseudoparenchyma and mycelial fragments are exhibited different shapes and positions in the spore balls. The spore ball forming genera *Doassansia, Burrillia, Doassansiopsis, Narasimhania, Tracya* showed remarkable compositions.

Spore Color or Pigmentation

Strictly speaking, colour is not a morphological character, but often included in all description of smut fungi. The spores are colored because of more or less pigmentation. This pigmentation may be evenly or unevenly distributed in the exospores of smuts. In various species of *Ustilago,* uneven pigmentation produces equatorial bands and lighter polar caps in *Planetella lironis.*

Variation in spore colour in the smut fungi is widely prevalent, expressed primarily in varying shades and tones of the basic colour for the species. For example, brown is the universal colour for spores of the species of Tilletia, which usually exhibit varying shades of this color in the range of light to dark, including shades on the light side that are yellowish, grayish, and violaceous and the dark side that are reddish, olivaceous, chestnut, and chocolate. Even within the same species variations in spore color are common, not only between spore masses from different sori but spores of single sori as well. Many species of *Tilletia* and *Neovossia* immature spores are hyaline and characteristically sculptured. There are intergradations in color ranging from light to heavily pigmented spores. There are several chances of wrong interpretation of colors in spores and depends upon individuality. Spore color, admittedly, is particularly subject to quantitative and qualitative interpretation. Nevertheless, because it is one of the few phenotypic characters available to ustilaginologist, no doubt it will continue to be extremely useful in smut systematics.

Spore Mass

Nearly in all the descriptions of the species the color of the spore-mass is described. It is generally black, with various shades of brown, fawn or pale

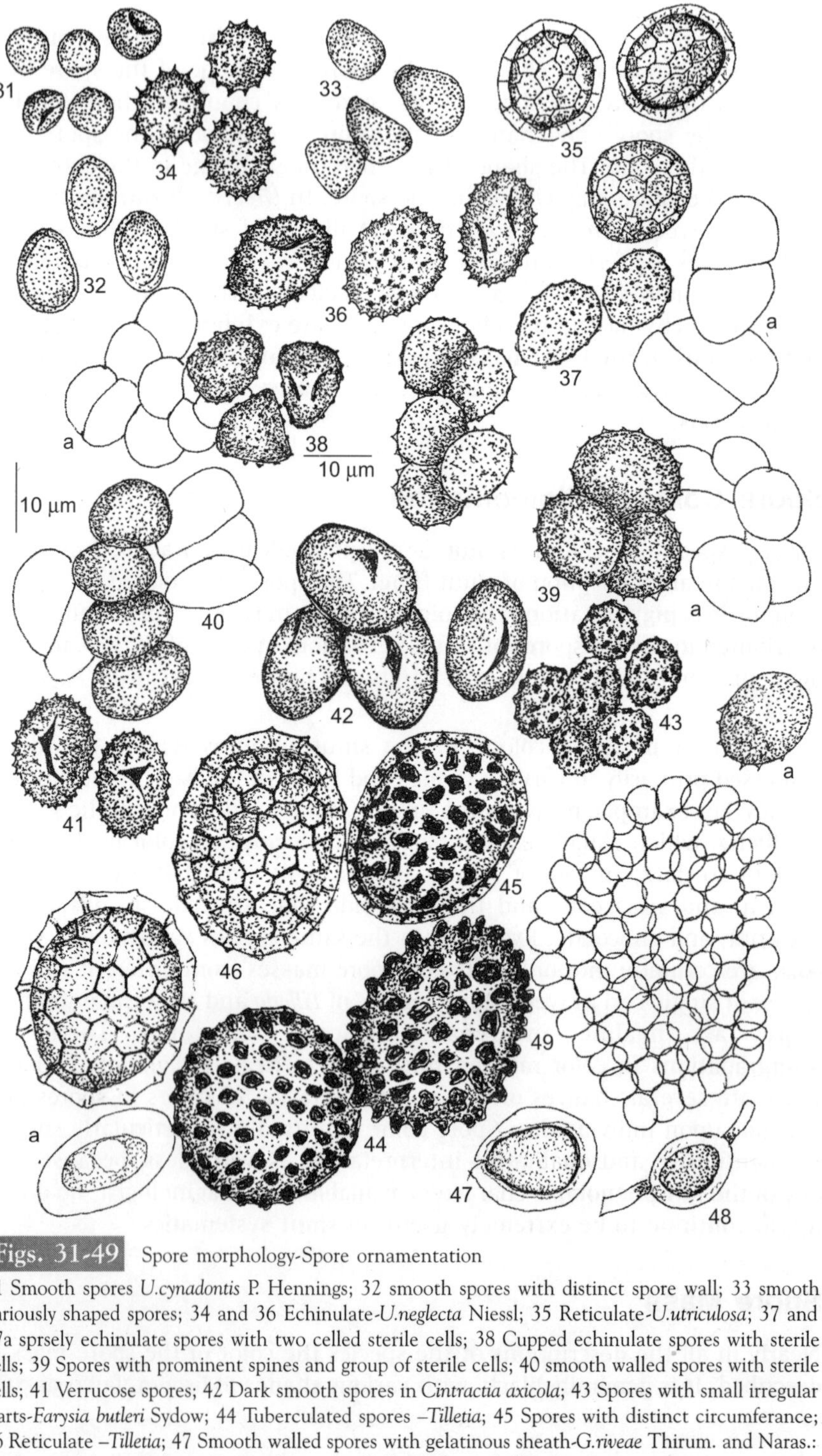

Figs. 31-49 Spore morphology-Spore ornamentation

31 Smooth spores *U.cynadontis* P. Hennings; 32 smooth spores with distinct spore wall; 33 smooth variously shaped spores; 34 and 36 Echinulate-*U.neglecta* Niessl; 35 Reticulate-*U.utriculosa*; 37 and 37a sprsely echinulate spores with two celled sterile cells; 38 Cupped echinulate spores with sterile cells; 39 Spores with prominent spines and group of sterile cells; 40 smooth walled spores with sterile cells; 41 Verrucose spores; 42 Dark smooth spores in *Cintractia axicola*; 43 Spores with small irregular warts-*Farysia butleri* Sydow; 44 Tuberculated spores –*Tilletia*; 45 Spores with distinct circumferance; 46 Reticulate –*Tilletia*; 47 Smooth walled spores with gelatinous sheath-*G.riveae* Thirum. and Naras.: 48 Spore with appendages-*E. nymphaeae* Setchell: 49 Spore-ball – *Sorosporium* species.

yellow. However, manner in which they are described are not standardized. Therefore in several species these descriptions are confusing for species identification. The colour of the individual spores does not necessarily correspond with that of spore-mass. In several species spore-mass colour differs as per the maturity of the sorus. However, in some smuts spore mass colour represents characteristic of the genus for example in *Liroa*, the spore-mass is typical chocolate brown. In the genus *Melanopschium* spore mass is comparatively hard, usually agglutinated and spores get separated only in water.

Spore Balls and their Composition

Nearly half of the genera produce their spores in clumps or aggregate to form spore balls. These spore balls may either consist only fertile spores or of a combination of fertile and sterile cells, Pseudoparenchymatous elements. The spore balls of *Sorosporium, Sporisorium* and *Tolyposporium* are made up of only fertile spores in which outer spores are globose to subglobose and sculptured (Fig. 49) while inner spores are mostly smooth and polygonal in shape. Moreover, the outer spores are rather thick walled as compare to the inner thin walled spores. In few species, the peridium of the sorus is made up of the fungal tissue; it flakes into sterile cells at maturity as in *Sphacelotheca*. One of the most curious spore ball is observed in *Tolyposporium* wherein spores are firmly fitted together by interconnected folds and thickenings on the spores. Whereas in *Urocystis* the central fertile spores are surrounded by the sterile wedge shaped cells (Fig. 55). However, the fertile spores are brownish whereas sterile cells are hyaline or slightly tinted. On the other hand, the spore balls of *Doassansia, Doassansiopsis* and *Burrillia* are permanent and remain embedded in the host tissues (Fig. 50-52). The composition of fertile and sterile cortex differs in spore balls. In *Doassansia* the central fertile spores are protected and surrounded by the sterile cortex (Fig. 50). In contrast to this peculiarity, the single outer layer in *Doassansiopsis* is made up of fertile spores and the central portion consists of parenchymatous cells (Fig. 52). In *Burrillia* the spore ball has no sterile cortex but the fertile spores remain intermixed with pseudoparenchymatous elements. But in *B. kamatii* Thakur there are no such Pseudoparenchymatous elements in the spore ball and all the spores are fertile (Fig. 51). In *Entyloma* species spores are embedded in a common gelatinous matrix (Fig. 54) whereas in *Georgefischeria* each spore develops individual gelatinous exine. In the genus *Melanotaenium* the spores are darker and remain aggregated mostly surrounding the vascular bundles (Fig. 53).

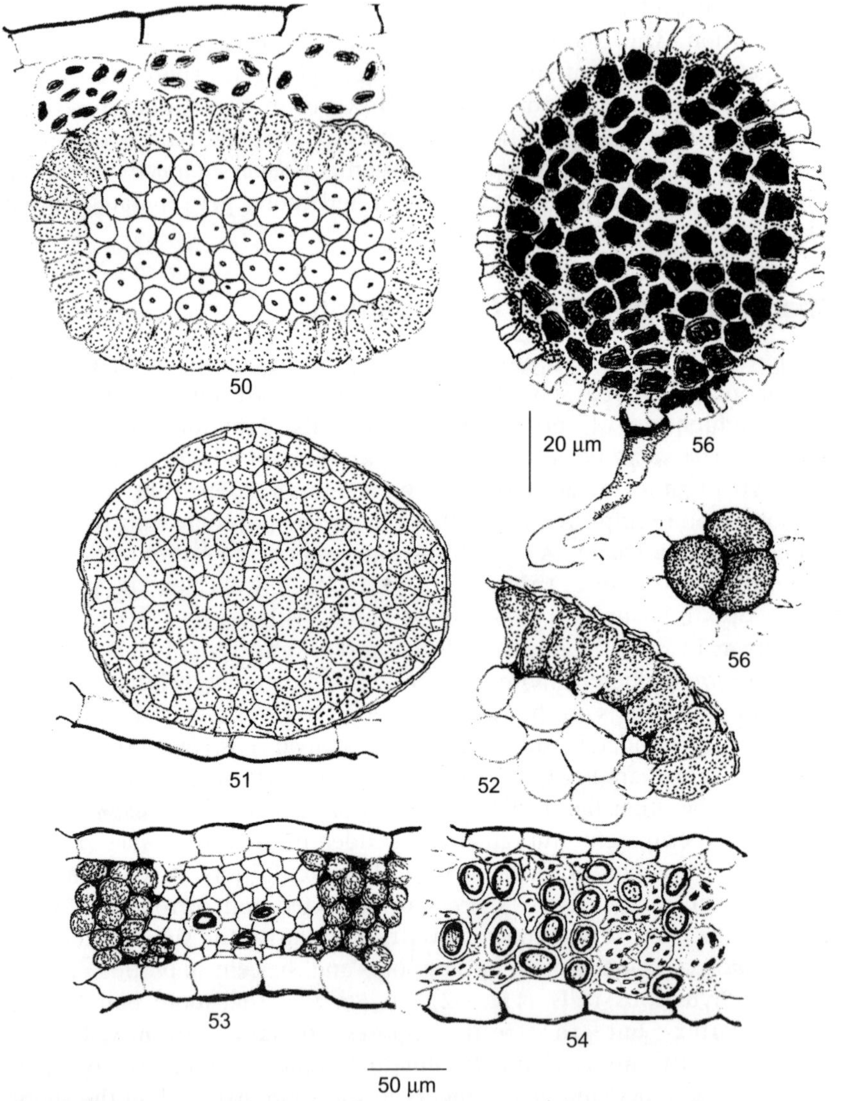

Figs. 50-56 Spore morphology and spore-balls

50 Permanently embedded spore-ball with central fertile and peripheral sterile spores- *D.khandalensis* Patil and Gandhe; 51 Permanently embedded spore-ball with all fertile spores- *Burrillia kumatii* Thakur; 52 Sectional view showing sterile core,peripheral fertile spores and fragmented mycelim; 53 Spore aggregation around the vascular tissue-*Melanotaenium*; 54 Spores remain embedded in the mucilage-*E. mundkuri* Patil and Gandhe; 55 Spore gall – *Urocystis hypoxysis* Thaxter; 56 – Spore appandage – *Neovossia sumatii* Patil and Gandhe.

Spore Ornamentation

Most of the authorities employed ornamentation as a criterion for the species delimitation in smut fungi. The spore ornamentations vary widely may be smooth, verrucose, echinulate, reticulate or frequently intergraded (Fig. 31-49). The spores are smooth in *Ustilago cynodontis* (Fig. 31) also in *Cintractia* (Fig. 42) *Urocystis, Entyloma, Melanotarnium Burrillia, Doassansia* and *Doassansiopsis* (Fig. 50-55). In most of the *Sorospoium* species the spore ornamentation ranges from smooth to verrucose, whereas in *Tilletia* species the spores either reticulate or warty, with rim or without rim (Fig. 44-46). In *T. tripogonis* sp. nov. the spores show tuberculated structures (Fig. 44). Apart from this character, in *Tilletia apludae* Mishra hyaline, mammilla like appendage is observed. Most interesting and curious hyphal appendage is observed in *Neovossia sumatii* Patil and Gandhe and in *N. punensis* Patil and Gandhe which clearly differs from the mammilla like structure in *T. apludae* (Fig. 56). Clearly reticulated spores are observed in *Ustlago utriculosa* (Fig. 35) and in *T. savilensis* sp. Nov. (Fig. 46). The spores are echinulate in *U. neglecta* (Fig. 34) whereas in *E. nymphaeae* they show a long hyphal attachment at one end and papillae at another end (Fig. 48).

Sterile Cells

To a certain extent, the characters of the so-called sterile cells, including their size, shape, and color have been used to characterize certain species or genera in smut fungi. Up to certain extent sterile cells play important role in the species identification. Unfortunately, the term does not refer to structures of common origin but used differently with respect to genera. However, the term is relative and differs in its application in various genera. For example, in some *Sphaceloheca* and *Sorosporium* species (Fig. 38-40) the cellular components of the sorus peridium, which tend to disintegrate into individual cells, in pairs or group of cells, are referred to as "sterile cells". In genera in which the spore balls consist of both fertile and sterile elements, the term "sterile cells" has been restricted to the sterile cellular components of the spore ball, as distinguished from the fertile spores for example *Urocystis, Thecaphora, Doassansia, Doassansiopsis.* In genera *Doassansia, Doassansiopsis, Burrillia* and *Urocystis* the spore balls are composed of both sterile and fertile components and the term sterile cells is applied for all the sterile components except the fertile spores (Fig. 50, 51 and 52). On the other hand, sterile cells have greater taxonomic importance among the species of *Tilletia* and sometimes referred as supporting character for identification of the species (Fig. 46). These sterile cells are intergraded in sculpture, hyaline to slightly tinted and often intermixed with the fertile spores. They are few to many in number but do not germinate like fertile spores. However, the exact function of the sterile cells not well understood.

Physiological Variations

Teliospore Germination and Germination Products

Many Smutologist used morphological features of teliospore germination as taxonomic criteria for classification, particularly at the family level.

Since the establishment of L.R. and C. Tulasne's (1847) two family classification system, the order Ustilaginales is basically classified into Ustilaginaceae, a phragmobasium with lateral and terminal basidiospores and Tilletiaceae, a holobasidium with only apical basidiospores. However, several other classificatory systems are also suggested mainly on the basis of teliospore germination. Some mycologists introduced a third family, Graphiolaceae, parasiting palms, in the Ustilaginales, but others included as imperfect fungi. Recently, Obrwinkler et al. (1982) showed that *Graphiola phonicis* is a basidiomycete, and described it, a new order, Graphiolales Donk ex Oberwinkler and Bandoni, with a single family.

To this two family classification system, Liou (1949) introduced a new family Yeniaceae, with a genus *Yenia,* which produces a rudimentary basidium, which produces progressively 1-4, pedicelled, usually septate " pseudo-promycelia" which in their turn produce either new pseudo-promycelia or basidiospores.

As time progressed increasing number of smuts were added from different countries and they exhibited very different type of germination from *Ustilago* and *Tilletia* types. The differences may be so great that sometimes it is very difficult to decide to which family the fungus in question belongs; as for instance is the case *Pericladium* or *Schroeteria* species.

Ciferri (1963) proposed four family classifications: Entylomellaceae, Tilletiaceae. Glomosoriaceae and Ustilaginaceae by introducing, Glomosporiaceae with a single genus, *Glomosporium,* on the basis of a branched promycelium, which partially splits up longitudinally forming two secondary branches in *G. amaranthi.* The splitting of the promycelium, however, was not observed in the type species *G. leptidium.*

During last two decades (Fischer and Holton 1957, Holton, Hoffmann and Duran, 1968) several germinated smuts showed problematic germinations than the typical germination for the families. The taxonomic status of the smuts has always been uncertain and repeatedly subjected to changes. During the last two decades, several yeasts and other saprophytic basidiomycetes fungi have been isolated and described. These fungi show Heterobasidiomycetes life cycle very similar to that of smut fungi (Bandoni and Johri, 1972,Fell 1970,1974,Fell et al. 1969,1973, Fell and Tallman, 1980, 1981, Moore, 1980, Nyland, 1950).

For a most unusual basidiomycete, *Filobasidium floriforme*, Olive (1968) introduced a new family: Filobasidiaceae, because of its superficial similarities with the smuts. Cox (1976) transferred Filobasidiaceae, to the Homobasidiomycetes,order Aphyllophorales, but Moore (1978) placed it in Order Tremellales, in Heterobasidiomycia.

Cunningham (1924), Fischer G. W. (1953), Fischer, G.W. and Holton, C.S. (1957), Linderberg (1959), and Vanky (1985) rejected the division in families and considered all smuts in only one family, Ustilaginaceae, because of this random variation in germination with certain limitations.

There is, however, great variation in basidial morphology within each family. In fact, many species do not germinate in accord with dictates of either family. Therefore, there is controversy regarding teliospore germination as the criterion of classification and some Smutologist still believed in the pattern of germination and germination products. My observations on teliospore germination, also in support of this opinion.

In few species of *Ustilago, Sphacelotheca* and *Cintractia* such as *U.neglecta* Niessl, *U.effusa* Sydow, *U.scitaminea* Sydow, *S. annulata* Mundkur, *S.sorghi* Clinton, *C.minor* Jackson, the basidia often developed long dikaryotic infection hyphae without developing basidiospores. (Fig. 56, 64, 65 and 76). Generally, all the genera from the Ustilaginaceae develop lateral and terminal sporidia on septate basidia such as *Ustilago, Sphacelotheca, Sorosporium, Cintractia, Melanopschium, Liroa,* and *Tolyposporium*. They are very frequently developed in *U.cynadontis* P.Hennings, *U.spersa* Underwood, *U.tripogonis* species. Nov., *U.trichophora* Kunze, *U.scitaminea*, *Sphacelotheca cornuta* Mundkur, *S.sinhgadansis* species. Nov., *C.axicola* Cornu, *L.emodensis* Ciferri, *Sorosporium heteropogonicola* Mundk.and Thirum. (Figs. 59-76)). No doubt there is a variation in number of septa in the basidium and this septation in the basidium differs species to species in the family Ustilaginaceae. The development of basidium in different species of Ustilaginaceae is critically discussed by C.T. Ingold with several illustrations.

On the other hand, the genera from the family Tilletiaceae such as *Tilletia, Neovossia, Doassansia, Doassansiopsis, Burrillia, Entyloma, Georgefischeria, Melanotaenium, Jamesdicksonia, and Urocystis* produce terminal basidiospores on aseptate holobasidia such as in the species *T. savilensis, T. fischeri, N. sumatii* Patil and Gandhe, *N. punensis* Patil and Gandhe, *E. bidentis* P. Hennings, *E. globigena* Thirum. and Safee., *E. mundkuri, M. punensis* Species. Nov., *D. khandalensis* Patil and Gandhe, *B. kamatii* Thakur, *U. hypoxis* Thaxter, *J.obesa* Thirum.et al., *G. narasimhanii* Patil and Gandhe (Fig. 78-92). The number of basidiospores differs in different genera with their germination

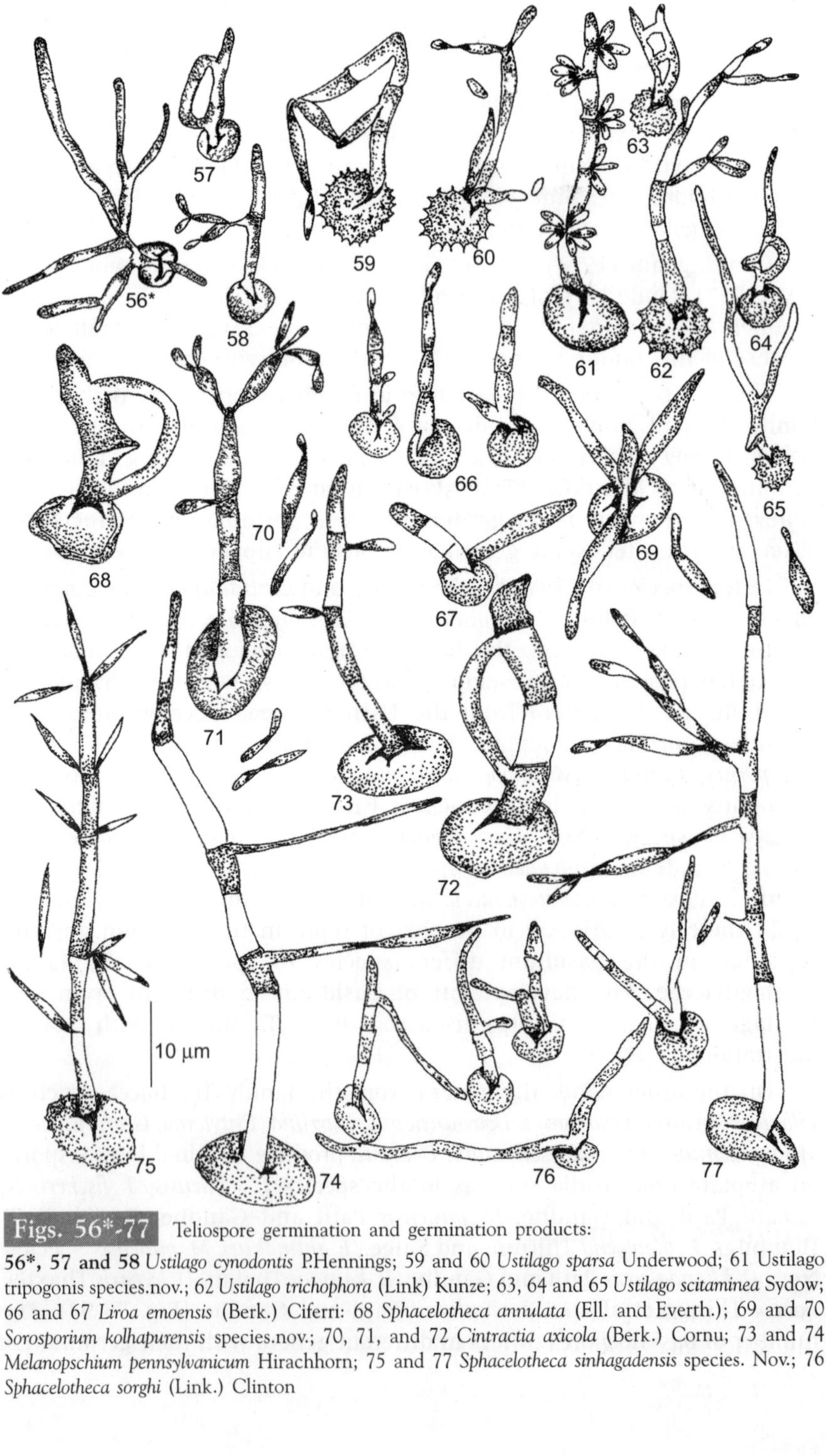

Figs. 56*-77 Teliospore germination and germination products:

56*, 57 and 58 *Ustilago cynodontis* P.Hennings; 59 and 60 *Ustilago sparsa* Underwood; 61 Ustilago tripogonis species.nov.; 62 *Ustilago trichophora* (Link) Kunze; 63, 64 and 65 *Ustilago scitaminea* Sydow; 66 and 67 *Liroa emoensis* (Berk.) Ciferri: 68 *Sphacelotheca annulata* (Ell. and Everth.); 69 and 70 *Sorosporium kolhapurensis* species.nov.; 70, 71, and 72 *Cintractia axicola* (Berk.) Cornu; 73 and 74 *Melanopschium pennsylvanicum* Hirachhorn; 75 and 77 *Sphacelotheca sinhagadensis* species. Nov.; 76 *Sphacelotheca sorghi* (Link.) Clinton

products. Many *Tilletia* species often develop limited preferably 8 apical basidiospores, which conjugate immediately or after their separation. They also develop ballistospores on the basidiospores. *Neovossia* species produce numerous filiform terminal basidiospores, which may or may not fuse *in situ* or after their separation from the basidia. The basidiospores which do not fuse, produce ballistoconidia, blastoconidia and/or secondary sporidia in *Tilletia* and *Neovossia*. In *N. sumatii* and *N. punensis* numerous apical sporidia are produced (Fig. 78 and 79) which do not fuse *in situ* or after their separation from the basidia. They develop ballistospores on the developing mycelium. There is a remarkable variation in the mode of germination in different genera and in many genera characteristic germination pattern is present.

Compatibility

Though **c**ompatibility of basidiospores differs species to species in smuts the genera from Ustilaginaceae showed more frequent fusions in situ or after their separations from the basidia. In many species of *Ustilago, Spacelotheca, Sorosporium, and Melanopschium* basidial cells act like basidiospores and undergo fusion so as to establish dikaryotic condition in the developing hypha. More strikingly, some species do not develop basidiospores but directly dikaryotic infection hypha. However basidiospores and the basidial cells develop blastospores. In all the genera of Tilletiaceae except the genus *Georgerfischeria,* basidiospores showed compatibility. Therefore terminally developed basidiospores on the holobasidia either conjugate in situ and/or after their separation. Almost all the species of *Entyloma, Melanotaenium, Tilletia, Doassansia, Doassansiopsis, Urocystis* and *Jamesdicksonia* showed conjugation of basidiospores either *in situ* or after their separation such as in *T. savilensis, T. fischeri, N. sumatii* Patil and Gandhe, *N. punensis* Patil and Gandhe, *E. bidentis* P. Hennings, *E. globigena* Thirum. and Safee., *E. mundkuri, M. punensis* Species. Nov., *D. khandalensis* Patil and Gandhe, *B. kamatii* Thakur, *U. hypoxis* Thaxter, *J. obesa* Thirum. et al.Therefore the conjugation of the compatible basidiospores resulted in to dikaryotic infection hyphae. However, in the genus *Georgefischeria, G. riveae* and *G. narasimhanii* basidiospores do not conjugate and develop ballistospores and blastospores. Dikaryotic phase in these species is resulted from the germination of dikaryotic basidiospores or the fusion of the monokaryotic hyphae.

Cultures of Smuts

Though smuts are obligate parasites can be cultured on different on solid or liquid media. However, their growth in culture is comparatively slow to the other fungi. They can be directly dusted on nutrient media such as PDA,

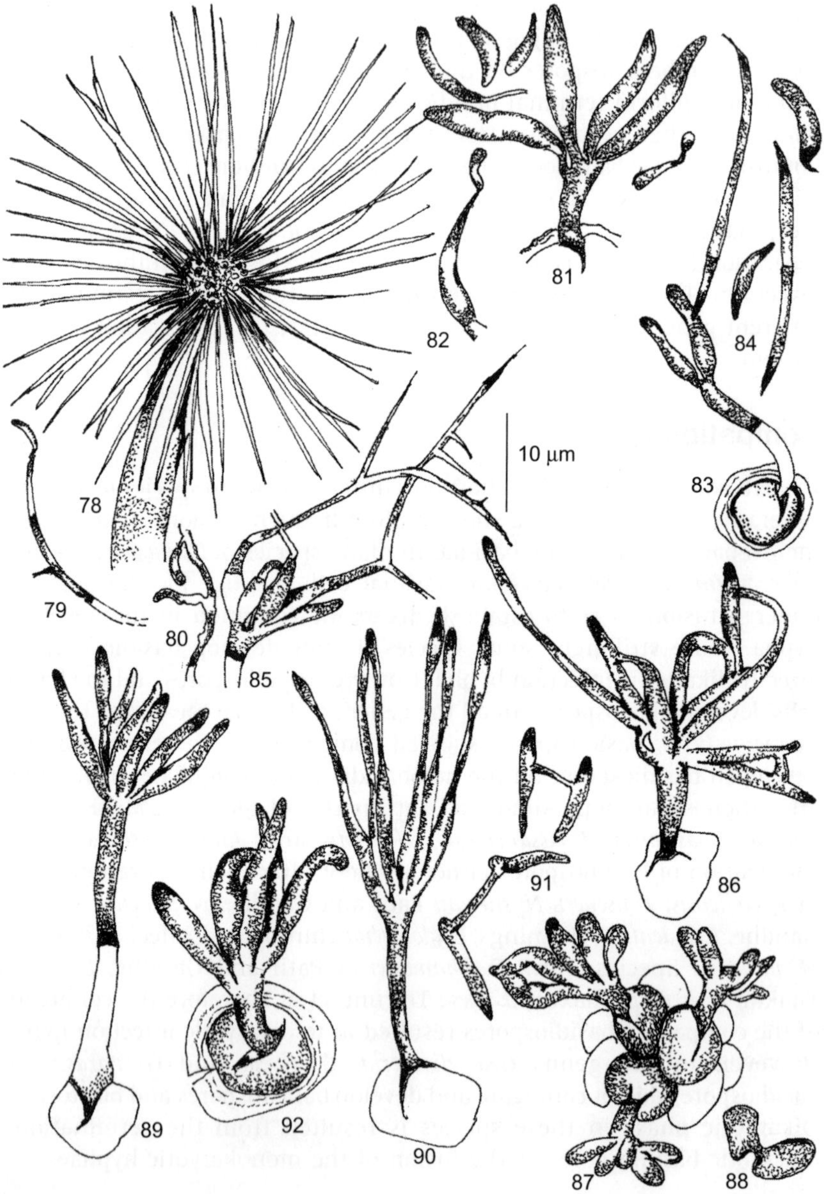

Figs. 78-92 Teliospore germination and germination products:

78-80 *Neovossia sumatii* Patil and Gandhe; 81 and 82 *Georgefischeria narasimhanii* Patil and Gandhe; 83 and 84 *Jamesdicksonia obesa* Thirum. et al.; 85 *Entyloma mundkuri* Patil and Gandhe; 86 *Burrillia kamatii* Thakur; 87 and 88 *Urocystis hypoxyis* Thaxter; 89, 90 and 91 *Doassansia khandalensis* Patil and Gandhe; 92 *Entyloma globigena* Thirum. and Safee.;

TABLE 2 *The following species are also recorded from Maharáshtra (Kamat et al 1971) but the author has not examined these species, hence, not included in the description.*

1. *Ustilago* (Persoon Roussel)	1. *U. avenae* (Pers.). Rostrup	*Avena sativa* L.	Gramineae	Pune
	2. *U. coicis* Brefeld	Polytoca barbata (Roxb.) Stapf	Gramineae	Bombay
		Coix lachryma jobi L.	Gramineae	Bombay
	3. *U. carmeri* Koren.	*Setaria italica* (L.) Beauv.	Gramineae	Pune
	4. *U. hordei* (Pers.) Lagerh	*Horedum vulgare* L.	Gramineae	General distribution
	5. *U. shirainna* P. Henn.	Bambuca sp.	Gramineae	Pune
2. *Pericladium* Passerini	1. *P. grewiae* Pass.	*Grewia* sp.	*Tiliaceae*	*Pune*
3. *Sphacelotheca* de Bary	1. *S.andropogonis – annulati* (Brefeld) Zundel	Dichanthium annulatum (Forssk.) Stapf	Gramineae	Pune
	2. *S. cruentiii* (Kuch.) Pottr.	*Andropogon halepensis* Brot. *Sorghum vulgare* Pers.	Gramineae	General Distribution
4. *Melanopsichium* Beck	1. M. *eleusinis* (Kulkarni) Mund. and Thirum.	*Eleusine coracana* Gaertn.	Gramineae	Kolhapur, Malkapur
5. *Sorosporium* Rudolphi	1. *S. formosanum (Sawada) Sawada*	*Panicum repens* L.	Gramineae	Pune

(Table 2 Contd.)

(Table 2 Contd.)

	2. *S. pseudanthistriae* Syd. And Butler	*Pseudanthistria hispida* Hook F.	Gramineae	Bassein
	3. *S.turneri* NcAlpine	*Eragrastis bifaria* (Vahl) Wight	Gramineae	Pune
6. *Tolyposporium* Woronin	1. *T.christensenii Raghunath*	*Cymbopogon flexuosus* (Ness) Wats	Gramineae	Pune
7. *Tilletia* Tulasne	1. *T.elusines* Syd.	*Dactyloctenium aegyptiacum* Willd	Gramineae	Lonavala
	2. *T. horrida Takahashi*	*Oryza sativa* L.	Gramineae	Ratnagiri
	3. *T. pulcherima* Ell. And Gall. Var. *brachiariae* Pavgi and Thirum.	*Brachiaria distachya* (L.) Stapf	Gramineae	Pune
	4. *T. tritici* (Brek.) Wint.	*Triticum vulgare* Vill.	Gramineae	General Distribution
8. *Melanotaenium* de Bary	1. M. *spermacoces Thirum. et al*	*Spermacoce hispida* L.	Rubiaceae	Pune

HA or 0.2% MA or first germinated on 2% plane Agar and then transferred to the nutrient media. Sometimes fragments of the host are inoculated with the smut spores for their rapid development in the culture. The greatest variation in the smut fungi is displayed by haploid cultures on agar media. Monosporidial cultures of many species usually vary widely in colony characteristics, some of which may represent temporary response to environment; the great majority, however, are inherently permanent. Smut cultures vary in type and rate of growth, topography, colour, pigmentation, and many combinations of these characteristics. The type of growth may be mycelial, Sporidial, or a combination of both; the color may range from white to black, with many intermediate shades. Some lines exhibit widely different rates of growth, especially those that are mycelial.

Acknowledgement

The author is grateful to Late Professor M.J. Thirumalachar, eminent mycologist for personal discussions at the laboratory in 1998. The author is also thankful to Professor Kalman Vanky, Smutologist, Germany for his suggestions.

References

Bandoni, R.J. and Johri, B.N. 1972. *Tilletiaria*: a new genus in the Ustilaginales. *Canad. J. Bot.* 50: 39-43.

Bauer, R., Begerow, D., Nagler, A. and Oberwinkler, F. 2001. The Georgefischeriales, A phylogenetic hypothesis. *Mycol. Res.* 105: 416-424.

Ciferri, R., 1924. Seconda contribuzione allo studio degli Ustilaginales. *Atti Ist Bot. Univ. Pavia, Ser.* 3: 1, 77-97.

Ciferri, R., 1963. The glomosporiaceae family (Ustilaginales). *Riv.Pat.Veg., Pavia, Ser.* 3, 3: 141-143.

Cox, D.E. 1976. A new homobasidiomycete with anomalous basidia. *Mycologia*, 68: 481-510.

Cunningham, G.H. 1924. *The Ustilagineae, or smuts, of New Zealand.* Trans. And Proc. New Zealand Institute, 55: 397-433.

Duran, R. and Fischer, G.W. 1961. *The genus Tilletia.* Washington State University, 138 pp.

Duran, R. 1971. Some host and distribution records of Mexican smut fungi. *Mycologia*, 62: 1094-1105.

Fell, J.W. 1970. Yeasts with Heterobasidiomycetes life cycles. In Ahearn, D.G.,(editor) .*Recent trends in yeast research.* Georgia State University, Atlanta, Georgia, 49-66pp.

Fell, J.W. 1974. Heterobasidiomycetes yeasts Leucosporidium and Rhodosporidium, Their systematics and sexual incompatibility systems. *Trans. Mycol. Soc. Japan.* 15: 316-323.

Fell, J.W., Statzell, A.C., Hunter, I.L. and Phaff, H.J., 1969. Leucosporidium gen. n., the Heterobasidiomycetes stage of several yeasts of the genus Candida. *Antonae van Leeuwenhoek*, 35: 433-462.

Fell, J.W., Hunter, I.L. and Tallman, A.S. 1973. Marine basidiomycetous yeasts (Rhodosporidium spp.n.) with tetrapolar and multiple allelic bipolar mating systems. *Canad. J Microbiol.* 19: 643-657.

Fell, J.W., and Tallman, A.S., 1980. Rhodosporidium paludigenum species nov., a basidiomycetous yeast from interdial waters of South Florida. *Internat. J Syst. Bacteriol.* 30: 658-659.

Fell, J.W., and Tallman, A.S., 1981. Heterothalism in the basidiomyceous yeast genus Sporidiobolus. *Syst. Bacteriol.* 5: 77-82.

Fischer G. W. 1953. *Manual of the North American smut fungi.* New York, Ronald Press Co. 343pp.

Fischer, G.W. and Holton, C.S. 1957. *Biology and control of Smut fungi.* New York, Ronald Press Co. 622pp.

Holton, C.S., Hoffmann, J.A. and Duran, R., 1968. Variations in the smut fungi. Ann. Rev. *Phytopathol.* 6: 213-242.

Fischer, G.W. and Shaw, C.G. 1953. a proposed species concept in the smut fungi, with application to North American species. *Phytopath,* 43: 181-188.

Johansen, D.A. 1940. *Plant Microtechnique.* Tata-McGraw Hill, New York and London, 523 pp.

Linderberg, B. 1959. Ustilaginales of Sweden. *Symb. Bot. Upsal.* 16(2): 1-175.

Liou, T.N., 1949. Sur I'existance d'une nouvelle famille des Ustilaginales. *Contr. Inst. Bot. Nat. Acad. Peiping* 6: 37-47.

Moore, R.T. 1978. Taxonomic significance of septal ultrastructure with particular reference to the jelly fungi. *Mycologia,* 70: 1007-1024.

Moore, R.T. 1980. Taxonomic proposals for the classification of marine yeasts and other yeast like fungi including the smuts. *Bot. Mar.*23: 361-373.

Mundkur, B.B. and Thirumalachar, M.J. 1952. *Ustilaginales of India.* C. M. I., Kew., 84pp.

Nyland, G. 1950. The genus Tilletiaopsis. *Mycologia,* 42: 487-496.

Oberwinkler, F., Bandoni, R.J., Blanz, P., Deml, G. and Kisimova-Horovitz, L., 1982. Graphiolales: Basidiomycetes parasitic on palms. – *Pl. Syst. Evol.* 140: 251-277.

Olive, L.S., 1968. An unusual new Heterobasidiomycete with Tilletia –like basidia. *J. Elisha Mitchell Sci. Soc.* 84: 261-266.

Tulasne, L.R. and. Tulasne, C., 1847. Memoire sur les Ustalaginees comparees Uredinees. *Ann. Sci. Nat. Bot.* Ser. 3, 7, 12-127.

Thirumalachar, M.J. 1966. Concepts of our knowledge in the genera of Ustilaginales.*Indian Phytopahol.* 19: 3-13.

Thirumalachar, M.J. and Narasimhan, M.J. 1950 Notes on some spore germination and mounting techniques. *Indian Phytopath* 3: 177-178.

Thirumalachar, M.J. and Narasimhan, M.J. 1953. Notes on some mycological methods. *Mycologia* 45:461-468.

Vanky, K. 1985. Carpathian Ustilaginales. *Symb. Bot. Upsal.* 24 (2) 1-309.

Vanky, K. 1999a. The new classificatory system for smut fungi and two new genera. *Mycotaxon,* 70: 35-49.

Vanky, K. 1999b. Three new genera of smut fungi. *Mycotaxon*, 71: 207-222.

Vanky, K. 2000. New taxa of Ustilaginomycetes. *Mycotaxon*, 74: 343-356.

Vanky, K. 2001a. The emended Ustilaginaceae of the modern classificatory system for smut fungi. *Fungal Diversity*, 6: 131-147.

Vanky, K. 2000b. *Lundquistia*, a new genus of Ustilaginomycetes. *Mycotaxon*,77: 371-374.

Vanky, K. 2002. *Illustrated Genera of Smut Fungi.* 2nd ed. APS Press, 245pp.

Vanky, K. 2002. The smut fungi of the world. *Acta Microbiologica et Immunologica Hungarica*, 49(2-3) 163-175.

Zundel, G.L. 1930. *The ustilaginales of the world.* – Pennsylvania State Coll. School Agric. Dept. Bot. Contrib. 176 : XI=1-410.

Frontiers in Plant Sciences (Pages 97-102)
Eds. K.G. Mukerji, K.V.B.R. Tilak, S.M. Reddy, L.V. Gangwane, P. Prakash, I.K. Kunwar

Biodiversity of Rust and Smut Fungi – India Perspective

D.K. Agarwal

ABSTRACT

Rusts are pleomorphic and obligate in nature (except a few). More than 160 genera of rusts are recorded, comprising of over 7 thousand species. *Puccinia* is the largest genus among the rusts. Smuts are parasitic mainly on monocots, having 1450 species distributed in 77 genera.

Key words: Uredinales, ustilaginales, distribution, heteroecious autoecious.

The abundant variation in the types of living organisms in any geophysical area is called biodiversity. With reference to large organisms (plants, animals etc.) the distinctive visible anatomical and morphological features enable us to identify the different genera and species and therefore, we can easily assess the extent of biodiversity. However, most fungi are not visible to the naked eyes and they need to be studied under the microscope. The culturable fungi can be identified provided they sporulate but the biotrophs which are not amenable to culture media go unnoticed. It is, therefore, imperative that there is a gap in knowledge of the biodiversity of all microbes including fungi. There are about one million named species of fungi and new species are being added at the rate of about 1500 each year. The total number of fungal species both known and unknown is estimated as 1.5 million.

The uredinales (rusts) are the largest group of plant parasitic fungi belonging to the class Basidiomycotina i.e. they produce basidia and basidiospores. They case severe diseases on economically important crop

Author address: Division of Mycology & Plant Pathology, IARI, New Delhi – 110 012.

plants like wheat, corn, cereals, legumes, bean and grasses. They are obligate in nature except a few and produce more than one spore forms in their life cycle ranging from monokaryotic to dikaryotic to diplophase. This phenomenon in rusts is known as Polymorphism or Pleomorphism. Many rusts complete their life cycle on two genetically different hosts and are known as heteroecious while few require only one host for completion of life cycle and are called autoecious. Polymorphism in rust is represented by five types of different spore forms (O – IV). O (Pycnia), I (Aecia), II (Uredinia), III (Telial or Teleuto) and IV (Basidial). In the literature there are more than one hundred and sixty four genera of rusts recorded out of which forty six are monotypic, comprising of a total of seven thousand species, world over. Out of which four thousand one hundred are only *Uredo, Puccinia,* and *Uromyces* only. Geographically rusts are distributed all over the world except the Antartica. The greatest number of species occur in temperate and near temperate regions. The most widely distributed species are of economic significance such as *Puccinia graminis tritici* Pers., *P. recondite* Rob. Ex Desm., *Uromyces phaseoli* (Rabenh.) Wint., and *Tranzschelia discolor* (Fckl.) Tranz. & Litv. An important pathogen which is found around the globe (circum global spread) is *Puccinia polysora* Underw. popularly known as Southern corn rust. Many rust fungi occur on dicots than on monocots. The dicots (Compositae & Leguminosae) provide support to genera like *Ravenelia, Uromyces, Dicheirinia,* and *Sphaerophragmium*. Similarly a series of genera thrive on Rosaceae e.g. *Phragmidium, Xenodochus, Frommea, Gerwasia* and *Hamaspora*.

The rust fungi parasitize ferns orchids, mints and composites. No rust fungus is known to parasitize the mosses or yet more primitive plants. Among the monocots, the grasses support as many species as all other families combined together. Species of *Puccinia* (Heteroecious species) are known to occur on both monocots and dicots and even on ferns. Example *P. pattersoniana* Arth. with telia on grasses and aecia on *Brodiaea* of Liliaceae. Other rusts produce telia on monocots and aecia on dicots e.g. *P. liatridis* (Arth. & Fromme) Bethal ex Arth. with telia on grasses and aecia on *Liatris* (Compositae). Few species of rust are host restricted which produce telia on ferns and aecia on first e.g. *Uredinopsis, Milesia, Hyalospora, Coleosporium* and *Cronartium* produces aecial stages on pine only whereas *Cronartium* produces them on stems, on cones but not on pine needles.

Puccinia is the largest genus of the rust fungi, with some 3000 to 4000 species recognized. Heteroecious and autoecious species occur as well as many types of life cycle variation. The genus has representatives in all regions of the world except the polar areas and contains parasites of most groups of vascular plants.

Many of the world's serious diseases are caused by species of *Puccinia,* e.g. black stem rust of small grains and grasses (*P. graminis* Pers.), leaf rust of small grains and grasses (*P. recondite* Rob. Ex Desm.), crown rust of oats and grasses (*P. purpurea* Cke.), sugar cane rust (*P. kuehnii* Butl.), sunflower rust (*P. helianthi* Schw.), safflower rust (*P. asparagi* DC.), hollyhock rust (*P. malvacearum* Bert. Ex Mont.), snapdragon rust (*P. antirrhini* Diet. & Holw.), and many others.

Some of the Important Rust Genera and their Species Causing Economically Important Plant Diseases

Puccinia

Puccinia graminis tritic	Black stem rust of wheat
P. graminis secale	Black stem rust of rye and barley
P. graminis agrostis	On grasses and poa
P. graminis avenae	On oats
P. recondite Rob. Ex Desm.	Orange or brown rust of wheat
P. striiformis	Yellow or stripe rust of wheat and barley
P. coronata Corda	Common rust of oats
P. sorghi Schw.	Rust of maize
P. helianthi Schw.	Sunflower rust
P. asparagi DC	Asparagus rust
P. kuehnii (Kruger) Butl.	Rust of sugarcane
P. erianthii Padwick & Khan	
P. polysora Underw.	Rust of maize
P. purpurea Cke.	Rust of sorghum
P. penicillariae Speg.	Rust of bazra
=*P. penniseti* Zimm.	
P.arachidis	Rust of grondnut

Uromyces

Uromyces pisi (Pers.) Wint.	Rust of pea
U. dianthi Niessl.	Carnation rust
U. betae	Beet rust
U. fabae (Pers.) DeBary	Pea rust
=*U. viciae-fabae* Pers.	
U. phaseoli Arth.	Rust of beans
U. ciceris-arietini (Gragun) Jacs.	Gram rust
U. striatus Schroet.	Alfa Alfa rust
U. trifolii (Hedwf. Ex DC.) Lev.	Clover rust

Hemileia

Hemileia vastatrix Rust of Coffee

H. coffeicola

Gymnosporangium

Gymnosporangium juniperi-virginianae Schw.

G. claviceps Cke. & Pk. Rust of juniperus and apple

Smuts (Ustilaginales) are the group of basidiomycetous fungi parasitizing higher plants and are of immense economic importance due to the serious losses they cause to numerous crop plants. Hence, they have been widely studied by various mycologists.

The role of smuts in nature is not under rated as many of them cause considerable economic loss to cultivated plants. The smut fungi are plant parasitic fungi mainly on angiosperms, specially on monocots. No smut fungus is known to occur on family Orchidiaceae having 1800 species. A saprobic smut e.g. *Telliticaria* is known only from culture. Three smut fungi that parasitise Pteridophyta are – a species of a *Exoteliospora* on Osmundaceae (*Osmunda*) and 2 species of *Melaniella* on Selaginellaceae (*Selaginella* sp.). Further more, two species *Uleiella* are known to occur on gymnosperms i.e. Araucariaceae (*Araucaria* sp.). The zoophilic *Melassezia* (Melasseziales, Exobasidiomycetidae), of which the sexual phase is unknown, produces the superficial skin lesions on humans, known as Pityraises versicolor.

The number of recognized smut species is 1450, distributed in 77 genera and about 3500 synonyms to the species and also 50 to the genera including 10 anamorphic genera. The number of host plant species is approximately 4100. About 800 species of smut fungi (55.2%) parasitise grasses (family gramineae) belonging to 19 genera. 220 species parasitise on Cyperaceae (15.2%), 108 on Compositae (7.4%), 50 on Polygonaceae and Ranunculaceae (3.4% each), 32 on Liliaceae (2.2%), 21 on Leguminoceae; 18 on both Juncaceae and Umbelliferae, 17 on Caryophyllaceae, 16 on Alismataceae, 15 on Scrophulariaceae (1.4 to 1.0%).

Teliospore forming smuts (Ustilaginomycetes) are parasites on Herbaceous, non woody plants, while those lacking teliospores (Microstromatales, Exobasidiales) mostly parasitise woody plants e.g. *Gemynago* (1 sp.) on *Triplochiton* (Sterculeaceae), *Mundkurella* (4 spp.) on 4 genera in the Araliaceae, *Pericladium* on *Grewia* (Tiliaceae; 3 species) and on Piper (Piperaceae; 1 spp.), and *Uteiella* (2 spp.) on Araucaria (Araucariaceae).

Reference

Arthur, J.C. 1925. Terminology Uredinales. *Bot. Ga.* 80: 219–223.

Arthur, J.C. 1934. *Manual of Rust in United States & Canada* . Purdue Foundation. Lafayette. IN. pp. 438.

Crowell, I.H. 1940. The geographical distribution of the genus *Gymnosporangium. Can. J. Res. C.* 18: 469–488.

Cummins, G.B. 1959. *Illustrated genera of rust fungi.* Burgess, Minneapolis.

Cunnigham, G.H. 1930. Terminology of sporeforms and associated spores in the rust fungi. *N.Z.J. Sci. Tech.* 128.

Faull, J.H. 1932. Taxonomy and geographical distribution of the genus *Milesia*. Contrib. Arnold Arbor. Harward Univ., No. 2, pp. 138.

Gopal Krishnan, K.S. 1951. Notes on the morphology of the genus *Hemileia*. *Mycologia* 43: 271–283.

Hiratsuka, N. 1958. Revision of taxonomy of the *Pucciniaceae*. Kasabi Publ. & Printing Co., Tokyo. Pp. 161.

Hiratsuka, Y. 1975. Recent Controversies on the terminology of the rust fungi. *Rept. Tottori. Mycol. Ist.* 12: 99–103.

Kern, F.D. 1911. A Biological & taxonomic study of genus *Gymnosporangium* . Bull. *N.Y. Bot. Gardl.* 7: 391–483.

Lappick, E. 1972. Evolutionary species of rust fungi (Uredinales) on the legume. *Anna. Fenn.* 9: 135–148.

Frontiers in Plant Sciences (Pages 103-126)
Eds. K.G. Mukerji, K.V.B.R. Tilak, S.M. Reddy, L.V. Gangwane, P. Prakash, I.K. Kunwar

A Century of Contribution Towards Indian Mycology

D.K. Agarwal

Carl von Linne or Linneaus in 1771 gave the first scientific description of a fungus from India after his "Species Plantarum" (1753) had appeared. The Gasteromycetous fungus was named as *Lycoperdon pistillare*. The fungus was collected from Tranquebar, near Thanjavur in Tamil Nadu by Koening, a Danish missonary (Payak, Sarbhoy and Lal, 1997).

The earliest records of Indian Fungi are difficult to trace because of the fact that the collections were sent abroad for identification. The regular collections were started by D.D. Cunningham (1871) and Barclay (1886) on Mucoraceae of Bengal and Uredineae of Shimla respectively. Cunningham ranks high as a pioneer in the field which came to be known as Aerobiology or study of airspora. He made the first attempt to "correlate the incidence of disease in man with variations in airspora in Calcutta during 1871-1873 (Ainsworth, 1979) for which he designed an apparatus called "aeroconicstope". He studied Mucorales and Uredinales and his name is commemorated for the genus *Cunninghamia* (*Choanephora*), Currey (1874) and later *Cunninghamella* (1903) of Mucorales. Barclay made a systematic study of Rust fungi of Shimla Hills. The life cycle studies and demonstration of atypical features of rust fungi like *Uromyces hobsonii* (the jasmine rust) and also the work on rusts affecting the Himalayan conifers and angiospermic plants is significant (Mundkur, 1952).

It is surprising that the coverage of Landmark during 19th Century, the contributions made by Indian Mycologists have been either over looked or ignored. Mundkur (1952) in his Presidential Address to the Botany Section of 38th Indian Science Congress, resurrected from the oblivion the name

Author address: Division of Plant Pathology, Indian Agricultural Research Institute, New Delhi-110 012

of nineteenth Century Indian Mycologist, Lt. Col. K.R. Kirtikar (1885) who collected and identified Agaricales and Gasteromycetes, coloured plates of which are still available at R.G. Medical College, Calcutta. A fungus collected at Thane was identified by C.G. Lloyd as *Calvatia lilicinia* and an agaric occurring on drumstick tree, *Moringa oleifera* Lamk. (= *M. pterygosperma*) was named as *Agaricus* (*Pleurotus*) *moringanus*. Fungi collected at Matheran were identified as *Agaricus campestris, Daedalea gibbora, D. versicolor* and *Polyporus giganteus*. These collections were made by Birdwood. Kirtikar subsequently brought out in collaboration with Basu, the famous work "Medicinal plants of India".

Systematic work of Mycology in India was started by Sir Edwin John Butler in 1901 and he is rightly regarded as "Father of Indian Mycology". Fries (1855) examined the collections made by Didrick Ferdinad Didrischen from the Nicobar Islands during 1845 to 1847. Perrottet in 1840, collected some fungi from the Nilgiri Hills which were worked out by Montagne. Later on a few fungal specimens found their way from India to Europe before 1850 and were studied by Corda (1850). Till 1850, the total number of Indian fungi known to science scarcely reached one hundred only.

Joseph D. Hooker and his co-worker Thomson made surveys of Sikkim Himalayas and Khasi Hills during 1849-50 and collected specimens of fungi belonging to fleshy fungi and also made their sketches. These specimens and drawings were sent to A. Barkeley. D.D. Cunningham (1871-1877) an Indian Medical Army Official published results of his studies with respect to Mucoraceae and Uredinales, collected from Calcutta and also from Eastern India. Curry (1874) collected several Burmese fungi along with many collections from Bengal. These samples were received and studied by McRae (1918). Rabenhorst, Brefeld, Dietel, Patouillard, Massee (1898-1912) and others. Later on A. Barclay (1886) and Sir George Watt made collections and observations on economically important fungi found growing on tea and other crops.

Hennings (1900-1901) started descriptions of "Fungi Indiae Orientalis" and Massee (1899-1912) "Fungi Exotici" in Kew Bulletins including descriptions of various Indian species. At the same time Thiessen (1912) reported "Fungi of Bombay". Bresadola and Saccardo also started examining Indian Fungi to which they had access. During the past quarter of century, H. & P. Sydow particularly the son H. Sydow examined hundreds of Indian collections of fungal origin, communicated to them from Pusa (Bihar) and Coimbatore (T.N.) which ultimately added in the list of publication.

The mediveal era started with E.J. Butler who visited India in Calcutta in 1901 as an Imperial Mycologist and was transferred to Dehradun in 1902 and finally to Pusa in 1905. He was officially associated with mycology in

India till 1920 especially with McRae, F.J.F. Shaw, Padwick, J.F. Dastur and others and was especially assigned to work on parasitic fungi. One of the important contribution of Butler in India was that he laid the foundation of Herbarium Cryptogammae Indiae Orientalis (HCIO) in 1905 at Pusa, Bihar. The other important contributions of Butler include the classical taxonomic work on the genus *Allomyces* (Butler, 1911), *Pythium* and certain other Chytridiaceous fungi such as *Sclerospora* and *Rhizophagus*. Nearly 200 diseases of field and plantation crops were described by Butler in his classical book "Fungi and diseases of plants" published in 1918 from Calcutta. Later on this book was enlarged and revised in the form "Plant Pathology" which was published in 1955 under the joint authorships of Butler and Jones. Apart from this he published series of papers in collaboration with, H. & P. Sydow in "Annals Mycologici" under the title "Fungi Indiae Orientalis" Part I-IV. In 1931 he also published "The Fungi of India" with G.R. Bisby which was later on revised and updated by R.S. Vasudeva (1954). In 1922 Butler succeeded in producing the first number of "Review of Applied Mycology" now known as "Review of Plant Pathology" from Commonwealth Mycological Institute, Kew, England.

The applied mycology aspect was taken up by S.R. Bose (Polyporaceae) who described for the first time from India an antibiotic polyporin. Mitter and Tandon (1930) published a series of papers on the fungal flora of Allahabad, Nainital and Mussoorie Hills (1938) as a result of which several serious diseases and new taxa were explored during 1930-38. Seven flourishing school of thoughts of Mycological excellence emerged under the able leadership of eminent mycologists viz., Thind (Chandigarh, Punjab), Subramanian (Madras, T.N.), Das Gupta (Lucknow, U.P.), Kamat (MACS, Pune, M.S.), Saksena (Allahabad, U.P.), Bose (Calcutta, W.B.) and Mundkur (IARI, Delhi). These schools made invaluable contributions in the development of mycology.

Dastur (1913) did monumental work on *Phytophthora parasitica* describing a new disease on castor. Later on he published several papers on *Hevea brasiliensis, Vinca rosea* and *Piper betle.*

Mhatre and Mundkur (1945) Studied the genus *Synchytrium* in India. Thomas and co-workers (1947-48) concluded that morphologically similar and sexually compatible *Phytophthora arecae, P. meadii, P. faberi* and *P. parasitica* var. *nicotianae* should be merged in *P. palmivora* & further that *P. palmivora* and *P. parasitica* with all their varieties should be merged into *P. colocaceae*. Ramakrishnan and Narasimhan (1941) reported *Darluca filum* on cereals and Kamat and Patil (1948) reported several new hosts for *Leveillula taurica*. Venkatarajan (1946d) recorded for the first time that *L. taurica* was parasitized by *Cicinobolus cesati.*

Mitra (1943, 1944) gave a full description of *Melanopsamma ranjanii* sp.nov., a parasite of *Selaginella chrysocaulis*. Thirumalachar (1945a) described *Strigula asteridiza* Vain. as a parasite of *Cephaleuros parasiticus* and *C. mycoidea*. Padamanabhan (1947) reported that occurrence of *Gibberella fujikuroi* var. *subglutinans* on the pupae of *Epipyros*. Sultan Ahmad (1949) at the same time contributed immensely to the fungus flora of Pakistan and India. Sen Gupta *et al.* (1950) described an *Actinomycete* as a new wood destroying fungus. Padwick and Azmatullah Khan (1943) recorded *Claviceps viridis* sp. nov. on *Ophismenus compositus* at Shimla. Banerjee (1944-1945) recorded and described *Mitrula agharkarii* sp. nov. from Darjeeling.

Considerable taxonomic work was done by Padwick and his associates (1940b, 1940a) on the genus *Fusarium*. Banerjee (1942b) worked on *F. equiseti* causing leaf spot of *Eichhornia crassipes*. Later on this work was further elaborated by Subramanian. Arya and Prasad (1952) reported Alternaria blight of linseed for the first time from India.

A systematic survey of the Indian rusts was made and four new genera viz., *Dasturella* (Mundkur and Kheswala, 1943), *Catenulospora, Mehtamyces* (Mundkur and Thirumalachar, 1943) and *Santapauella* (Mundkur, 1945) were proposed. The last two genera were been merged into synonymy of *Phragmidiella* by Thirumalachar and Mundkur (1949). They also wrote a monograph on the "Genera of Rusts". The genus *Mundkurella* is the most remarkable in having two kinds of chlamydospores and an unusual type of germination.

Genus *Vasudevella* for Sphaeropsidales was raised by Chona, Munjal and Bajaj (1956). The genus *Metachora* reported by Sydow and Butler (1911) has been redescribed as *Phyllachora* by Clement and Shear (1931). The genus *Hendersonia* was reported by Sydow and Butler (1916).

The genus *Catenulospora* was raised by Mundkur and Thirumalachar (1943) and *Kulkarniella* by Gokhale and Patel (1951). A monotypic genus *Saksenaea* was named after the late Prof. R.K. Saksena by Saksena, S.B. (1953) under whose guidance the work on Mucorales was initiated in India. *Franzpetrakia* was created by Pavgi and Thirumalachar (1957), *Koorchamella* by Chona, Munjal and Kapoor (1958), *Corynesporella* by Munjal and Gill (1961), *Mystrosporiella* by Munjal and Kulshrestha (1969), Rai *et al.* (1964a) described a new genes *Achaetomium* from Indian soils which now is known from all over the world. Von Arx *et al.* (1978a) described a new genes, *Leucosphaera* belonging to Pseudoeurotiateae from dear dung collected from Delhi Zoo. *Harikrishnaella* by Singh and Sarbhoy (1971), *Hemicarpentalis* by Sarbhoy and Elphick (1972), *Pseudographiella* by Sarbhoy and Iyer (1973), *Sclerotiella* by Sarbhoy and A. Sarbhoy (1974). Extensive study has been made in the taxonomy of Dowry and Podery mildews (Mukerji, 1968c, 1969, 1975). Singh (1979) created *Pseudoaristastoma*, a Coelomycetous genera undescribed to science. A new genus *Johnkarlingia* was proposed parasitizing the roots of cabbage and cauliflower (Pavgi and Singh, 1979). *Sarbhoyomyces* by Saikia (1981) and *Morrisiella* by Saikia and Sarboy (1985). Srivastava (1987) proposed *Aspleniumsporites trivedii* gen. et. sp. nov. isolated from Neyveli Lignite areas of South India. Sivanesan *et al.* (1988) proposed *Thaxterina*—a new tubeufiaceous genus with multi-spored asci from India, Sivanesan and Sinha (1989) also created *Aldonata,* a new genus of Ascomycetes from the family Parmulariaceae.

Mehrotra and Sarbhoy (1960) recorded *Chaetocladium hesseltineii* which was later on raised to a new genus by Benny and Benjamin as *Dichotomocladium* (Mehrotra & Sarbhoy, 1960). Rai *et al.* (1961) created a new *Helicostylum* species from Indian soils which was later found commonly from different parts of the world. Lingappa (1955) described several new species of *Physoderma* and *Synchytrium* from India.

Karling's (1942) monumental work on "Indian Chytridiales" described 64 species belonging to 24 genera and 11 families belonging to 6 orders compiled in the Monograph. Iyengar (1935) reported two species of *Coelomomyces* viz., *C. indiana* and *C. anophelesica* as an obligate parasite of *Anopheles* larvae.

Monographic treatment of "Cercosporae of India" was published by Vasudeva (1963) which comprised of more than 200 species recorded from India. Sarbhoy *et al.*, published several supplements to *"Fungi of India"* (1971, 1976, 1982, 1992) reviewing and updating the list of Indian Fungi recorded as new upto 1992. Bilgrami *et al.* (1979) also updated Fungi of India upto 2002 but did not include thirtyfive species which Sarbhoy *et al.* missed in their "Fungi of India".

Payak (1964) prepared a comprehensive account of spore morphology, physical properties and physiology, spore chemistry and fossil spores etc. listing more than 77 types of spores in fungi. Khanna *et al.* (1966, 1968. 1972. 1979) studied ultrastructure of spores of smut fungi by using TEM and Gum Acacia Replica Technique for the first time in India. Three species of *Neovossia* viz., *N. indica* (Karnal bunt), *N. horrida* (Bunt of rice) & *N. brachypodii* (Bunt of *Brachypodium*), their comprehensive studies were made with the help of optical and electron microscope.

During the period (1905-1992) 235 new genera have been proposed from India (Appendix-II). As the number of fungi reported from India from various ecosystems increased, it became necessary to have regional lists of fungi. Several such lists have been prepared and published (Appendix-I) from time to time.

Sarbhoy and Agarwal (1997) have given 102 authentic descriptions of "Tropical Plant Pathogenic Fungi" and their geographical distribution maps in India in the form of HCIO sets. They also listed more than 400 Meliolales from India (Hosagaudar *et al*.. 1996). Scanning electron microscopic studies of zygospores of the genus *Absidia* was worked out for the first time in India by Sarbhoy (1985) and later on he used freeze microtome technique, light microscopy and scanning electron microscopy to merge 16 perfect states of Aspergilli into eight genera of Subramanian.. Some of the important contributions made by various workers in different groups of fungi is described as under.

MYXOMYCOTINA

Wight collected *Physarum cinereum* (Batsch.) Pers. on grass in Madras before 1830 and is perhaps the first record of Myxomycetes from India. Arthur Lister and G. Lister (1925) in their monograph of the "Mycetozoa" have made mention of ten collections of different species of Myxomycetes. Lodha (1934) published "Indian Slime Moulds" and described 43 species of myxomycetes including five new of Bruhl and Gupta (1927). Thind (1977) published 25 papers in his monograph "Myxomycetes of India" comprising of 186 species, three varieties and six new forms. Lakhanpal and Mukerji (1975, 1976, 1977, 1979, 1981) and Nannenga-Bremekamp *et al.* (1979, 1984), Mukerji and Kaur (1996) described 400 different myxomycetes from India.

MASTIGOMYCOTINA

Chaudhuri (1931-1945) did notable work on aquatic phycomycetes and described *Myzocytium proliferum* Shenk. and *Rhizoidiomyces apophysatum*, the first Indian record of Mastigomycotina after *Allomyces arbuscula*. He also described *Hamidia* gen. nov. in 1942. Das. Gupta and John (1953) described a large number of aquatic fungi from Lucknow.

In Saprolegniales Chaudhuri and Kochhar (1935) discovered eight genera which included *Saprolegnia* which was earlier recorded by Butler (1907-1930) thus adding seven more genera to the list viz., *Achlya, Aplanes, Hamidia, Isoachlya Protoachlya, Pythiopsis* and *Thraustotheca*. Saksena and Tandon initiated researches on physiology of fungi and later on Ram Dayal and Tandon published many papers on the ecology of the aquatic moulds. Lacy (1949) Lucknow University, reproted four species of Chytridiales viz., *Chytrinmolla, Entophlyctis bulligera, Lagenidium entophytum* and *Aphanomyces apophysii*. Later, Lacy (1955) added seven more species to Chytridiales. Bhargava and Srivastava (1965) described two species of *Olpidiopsis viz., O. saprolegniae, O. lucurians* and *Aphanomyces laevis* new to India and later Bhargava and Srivastava (1977) added *Blastocladiella pringsheimii, B. globosa, B. ramosa* and *Allomyces arbuscula*. Das Gupta (1982), Das Gupta and Rachel

John (1953, 1988) described several genera of *Blastociadiella* from India *viz., Monophagus, Sparromyces* and *Indocytium* parasitizing on *Ulothrix* species. Das Gupta and John (1988) described 18 species of *Blastocladiella. B. glomerata, B. incrassata* and *B. angusta* are reported as new to India and the rest to the science. Das Gupta *et al.* (1990) reproted *Gonapodya terrestris* sp.nov. and *G. prolifera & G. polymorpha* as new records from India. Ram Dayal and his students developed a flourishing school in Varanasi and described 29 species and published a book on "Zoosporic Fungi of India" in 1988 along with Usha Kiran. Mer and Khulbe (1984) reported occurrence of 23 species of aquatic fungi belonging to Blastocladiales, Saprolegniales, Lagenidiales and Peronosporales in three lakes and their surroundings from Nainital.

Recently the group at Osmania University, Hyderabad became quite active under the stewardship of Manoharachary (1981, 1987, 1989) and has done excellent work on aquatic and soil fungi. They reported 125 species covering 64 genera although most of these are already described earlier from India. Chowdhury and Agarwal (1981) described two species of *Pythium* from India.

ZYGOMYCOTINA

Butler (1905) was the first to report *Chaonephora simsonii* as parasite on the flowers of *Zinnia elegans* from India, whereas Mitra (1929) reported *Rhizopus artocarpi* on *Artocarpus integrifolia* causing fruit rot in Andamans. Several Mucorales were reproted from soil, fruits, crops of various states by many workers (Bhattacharya and Barua, Rai and Mukerji, Choudhuri and Sacchar, Saksena *et al.* in the early fifties). By early sixties 135 species belonging to 25 genera were recorded from India (Tandon, 1968, Sarbhoy, 1967). Mehrotra and Mehrotra (1964) and Mehrotra and Baijal (1968) raised this number to 186 species comprising of 41 genera. Sarbhoy (1968) revised the Sphaerosporus group of *Mucor* and created a new species of *M. brunneogriseus* along with a key for their identification. He also created another species *Mucor variabilis* as well. Mukerji (1968b) described a interesting new species of genus *Piptocephalis* parasitic on several mucoraceous fungi. Behera and Mukerji (1974) created a interesting species *Chlamydoabsidia dasgupti* sp. nov. isolated from soils. Mehrotra *et al.* (1972) revised the above sections and added some important physiological differences between the species. Mehrotra *et al.* (1968) recorded 17 species of *Mortierella* including seven new species and three varieties from India whereas Prakash and Sarbhoy (1993) described *Mucor thermophilus*, as new species of *Mucor.*

The work on Mycorrhizal fungi started in India during 1980 and more than one hundred fungi were recorded by Khan and Sarbhoy, Bagyaraj, Mukerji and Tewari, (Bhattacharjee *et al.*, 1981; Manoharachari *et al.*, 2002;

Mukerji, 1996; Mukerji *et al.*, 1983; Neeraj *et al.* 1993). A new genus *Piriformospora* was described and patented in U.S.A by Verma *et al.* (1998). This fungus has ushered in a great potential in the world for increasing the yield of various crops.

ASCOMYCOTINA

In India, little attention was paid to investigations related to Ascomycotina, a highly heterogenous class of fungi because of the lapses by mycologists who showed apathy to fungi other than those responsible for destructive diseases of crops. This remark was made by Late Prof. Kamat in his Presidential Address (1976). Enormous amount of work has been done on different groups of Acomycetes by Mukerji and his coworkers isolated from varied substracts (Rai *et al.* 1963; Dakshini *et al.* 1970; Mukerji and Bhandari 1972; Saxena and Mukerji 1970a, c, 1972, Vön Arx *et al.* 1978a, b; Vön Arx and Mukerji 1981).Thind and his associates in Punjab, Kamal and his co-workers in Maharashtra, Kar and his collaborators in Eastern India, West Bengal and Tilak and his co-workers in Aurangabad described a large number of Ascomycetous fungi belonging to ascohymenial discomycetes and Ascoloculares. Around 60 ascomyceteous fungi are reported from Andhra Pradesh (Manoharachary and Rampulla Reddy 1991).

Notable contributions have also been made by Pavgi and his students from Benaras Hindu Univ. on Hemiascomycetes. Bose and Müller (Calcutta), Narayanaswamy and Ramakrishanan (Madras) contributed to the study of Powdery mildews. A monograph of this is in press and hope to be out this year. The family Cheatomiaceae has been studied intensively by Mukerji and his associates (Rai *et al.*, 1964a; Rai and Mukerji 1962a; Mukerji, 1968a; Saxena and Mukerji 1970c, 1972; Mukerji and Saxena, 1974; Mukerji and Singh 1974, Malhotra and Mukerji 1976, Mukerji and Khanna 1980) and Gymnoascaceae by G.R. Ghosh at Bhubaneshwar. Kamat and Anahousur proposed a new type of ascocarp development, the "Lecanidion type" and the family Clavicipataceae has received great attention by Govindu and Thirumalachar. Collections and descriptions of these fungi became an integral part of any floristic study necessitating a better understading of these organisms. Rai *et al.* (1964b) described a interesting perfect state of *Arpergillus* from soils.

Ullasa *et al.* (1973) established the "Clavicep type" of ascocarp development distinct from the "Xylaria type" to which this group was referred by Luttrell (1951). The genus *Phyllachora* was monographed by Kamat, Seshadri and Alka Pande who recognised 88 valid species with 19 synonyms. Hosagoudar (1985) described 17 species of *Phyllachora, P. pratapraji* Hosag. was created as a new species from Kerala. Later (1987, 1996) he listed 417 species of Meliolales comprising of 7 new species and

9 varieties including combinaiton novum and also prepared a CD. Marine and Coprophilous Ascomycetes have been studied by Lodha (1974) in Rajasthan. Saksena *et al.* (1967, 1968) published a series of papers on soil fungi including several ascomycetous species. So far 3800 ascomycetous fungi comprising of 248 genera have been recorded from India.

The family Gymnoascaceae was extensively studied by Ghosh. Ascocarp development of few genera was studied in *Glomerella, Chaetomium, Achaetomium* and *Sporormia* by Pande, Rao and Patil, Mukerji and Ricky, Rai, Tewari and Mukerji in the early eighties. In the Agharkar Herbarium, MACS, Pune some of these important ascomycetous fungi are maintained.

BASIDIOMYCOTINIA

Barclay (1856-1882) was the first to make notable contribution on mushrooms collected by him from Assam, Darjeeling, Sikkim (162 species) and Kashmir, Calcutta and Madhya Pradesh (10 species). He made a systematic study of rust fungi of Shimla Hills and also studied the life cycle of *Uromyces hobsonii* (Jasminum rust) and worked on rusts affecting the Himalayan conifers and angiospermic plants, Kirtikar during 1886 read papers before the Bombay Natural History Society reproting certain fungi belonging to Agaricales and Gasteromycetes, C.G. Lloyd (1898-1975) identified a fungus collected from Thane, M.S. as *Calvatia lilicina* and an agaric occurring on the drum stick tree, *Moringa oleifera* Lamk. (= *M. pterygosperma*) was named as *Agaricus* (*Pleurotus*) *moringamus*. Kirtikar subsequently brought out in collaboration with Major B.D. Basu, the famous work, "Medicinal Plants of India".. Ajrekar (1912) reported *Melampsorella ricini* on *Castor* from Bombay.

Sathe (1979), Massee (1898-1912) added another 32 species and Henning 72 species to the Indian list. Bose and his co-workers (1919-1951) recorded 34 mushrooms from W. Bengal followed by Patil, Singh and Mehrotra, Purukayastha, Bakshi, Jandaik, Kapoor, Munjal, Ghosh, Pathak, Koul, Sohi, Nair and Balakrishnan.

Thind and his students pioneered extensive exploration of mycoflora of Western Himalayas from 1952-1976 and Eastern Himalayas from 1977-1986. He studied encompass Xylariaceae, Pezizales, Helotiales, Clavariales and Hydnales, Thelephorales and Polyporoids among the Aphyllophorales and Gasteromycetes, His entire collection of 7000 specimens is housed at PAU, Chandigarh. He and his associates published two comprehensive accounts of Myxomycetes and Clavariacae of India. Bose (1919-1928) gave an account of Indian Polypores. Bakshi (1971) published a monograph on "Indian Polyporaceae".

Mundkur, Thirumalachar, Ramakrishnan, T.S. & K., Ramakrishnan, Payak and Joshi *et al.* (1986) studied extensively the rusts and smuts of

India. So far more than 180 genera of rusts have been recorded from India (List attached) comprising of 4958 species (Sarbhoy *et al.*, 1994). Karnal bunt of wheat (*Neovossia indica*) was recorded by Mitra (1931). It was considered a minor disease which has become a major one within a short span of few years. Hosagoudar (1985) studied Teliomycetes of South India and reported several Uredinial stages on different host plants from Coimbatore.

DEUTEROMYCOTINA (AGONOMYCETALES)

Several hyphomycetous fungi were described by Mitter and Tandon before 1930 from India but significant contributions were made by Cook (1825-1914), Hansford (1900-1960) Agnihothrudu and Subramanian (1956-1994) Bhat (1990-2004). Till 1930 only 461 species of Hyphomycetes were recorded from India out of these 66 belonged to the genus *Cercospora* (Appendix III). Mundkur (1938), Padwick (1939, 1940a) worked extensively on *Fusarium* diseases and their occurrence in India. Subramanian and Ramakrishnan (1956) reported several hyphomycetous fungi from various parts of India. Subramanian (1971) monographed more than 500 genera in his book "Hyphomycetes" including 31 new genera reported by him, Thirumalachar, Srinivasan, Agnihothrudu and Barua, Chona and Munjal. Saikia and Sarbhoy (1980, 1981) reported several Hyphomycetous fungi from North East India. Mukerji and coworkers described several new species of Deuterobmycetes (Rai and Mukerji; 1962b; Mukerji and Ranga Rao; 1968; Mukerji, 1969; Saxena and Mukerji; 1970b, 1973; Rikhy and Mukerji, 1973, Sharma *et al.*, 1973; Behera and Mukerji, 1974). Till date 75 new genera have been recorded representing more than 1000 new species (Sarbhoy *et al.*, 1994). Kumar and Sarbhoy (1988) and Subramanian and Sekar (1990) published monographic works on *Cladosporium* and Indian Coronophorales, respectively. Recently nine new genera of Hyphomycetes have been added to the Indian list viz. Synnemoellisia (Rao et al, 1988), Goosiomyces (Rao and Manoharachary 1989) Ramaraomyces (Rao et al 1989), *Mukerjiella* (Manoharachary and Krishna Roa, 1997), *Synnmukerjiomyces* (Aneja and Kumar, 1999), *Cylindromyces* (Manoharachary, Agarwal and Rao, 2004), *Beltraniomyces* (Manoharachary, Agarwal and Rao, 2003), *Hawksworthia* (Manoharachary, Agarwal and Rao, 2005, in press) and *Manoharacharyomyces*) (Agarwal and Rao, 2005, in press).

In India, of the 26000 fungi recorded, Deuteromycotina represents 45%, Basidiomycotina 25%, ascomycotina 20%, Zygomycotina and Mastigomycotina 10%. This clearly shows that maximum work has been done on Deuteromycotina especially on floristic fungi. During the last three decades, the number of Ascomycotina over shadowed the Deuteromycotina.

APPENDIX I *List of of Fungi of India Publications*

Authors	*Years of Publication*	*Period Considered*	*Number of Species*
Butler, E.J. and Bisby, G.R.	1931	1905-1930	2351
Mundkur, B.B.	1938	1930-1937	524
Ramarishnan, K. and Subramanian, C.V.	1952	1938-1951	812
Vasudeva, R.S.	1954	1905-1952	3895
Subramaniam, C.V. and Ramakrishna, K.	1956	1952-1956	765
Vasudeva, R.S.	1962	1953-1957	1006
Tandon, R.V. and Sudhir Chandra	1964	1957-1962	685
Tilak, S.T. and Ramchandra, R.	1970	1962-1967	655
Rangaswami, G., Sheshadri, V.S. and Lucy Chennamma, K.A.	1970	1964-1968	1983
Mukerji, K.G. and Juneja, L.C.	1974	1962-1972	3435
Sarbhoy, A.K., Girdhari Lal and Varshney, J.L.	1975	1967-1971	1050
Sarbhoy, A.K., Varshney, J.L. and Agarwal, D.K.	1980	1971-1976	2032
Sarbhoy, A.K., Agarwal, D.K. and Varshney, J.L.	1986	1977-1981	1792
Saikia, U.N.	1986	Upto 1980	130
Bhide, V.P., Pande, A., Sathe, A.V., Rao, V.G., and Patwardhan, P.G.	1987	1968-1985	3717
Sarbhoy, A.K., Varshney, J.L. and Agarwal, D.K.	1996	1982-1992	2566
Hosagoudar, V.B., Abraham, T.K. and Pushpanganathan, P.	1996	1981-1996	3251
Bilgrami, K.S. and Jamaluddin	–	1996-2003	-

APPENDIX II *List of New Genera Reported from India*

MASTIGOMYCOTINA

Allomyces Butler
Dicksonomyces Thirum., Rao & Salam
Hamidia Chaudhuri
Johnkarlingia Singh & Pavgi
Sclerophthora Thirum., Shaw & Naras.
Monophagus Das Gupta & John
Sparrowmyces Das Gupta & John
Indocytium Das Gupta & John

ZYGOMYCOTINA

Saksenea Saksena (Mehrotra & Sarbhoy)
Dichotomocladium Benny & Benjamin
Piriformospora Varma *et al.*

ASCOMYCOTINA

Achaetomium Rai, Tewari & Mukerji
Annajenkinsia Thirum. & Narasimhan
Arxonia Ullasa
Ascoglena Srivastava
Astomella Thirum.
Auriculoscypha Reid & Manimohan
Backusia Thirum.
Bitancourtia Thirum. & Jenkins
Bitrimonospora Sivanesan, Talde & Tilak
Boothiella Lodha & Mirza
Cerodothis Muthappa
Chaedefaudiomyces Ullasa, Rao & Patil
Chlorociboria Seaver ex Ramamurthi, Korf. & Batra
Deshpandiella Kamat & Rao
Dictyodochium Sivanesan
Discomycetella Sanwal
Eryachora Patil
Govindua Batista & Maia
Gymnoascoideus Orr, Roy & Ghosh
Helicogermslita Lodha & Hawksworth
Hemiascosporium Batra
Hemicarpenteles Sarbhoy & Elphick.
Hemisartorya Rai & Chowdhery
Kokkalera Ponnappa
Korfiella Pant & Tewari
Lambosina Thite & Kulkarni
Leucosphaeria v Arx, Mukerji & Singh
Narasimhella Thirum. & Mathur
Peethambra Subram. & Bhat
Petalosporus Ghosh, Orr & Kuhen
Phaeotricosphaeria Sivanesan
Plecaria Thind & Waraich
Pseudodelistschia Kapoor, Bahl & Lal
Pseudoeurotium Lodha
Raperia Subram. & Rajendren
Reticulosphaeria Sivanesan & Bahekar
Salmonomyces Chiddarwar
Sapucchaka K. Ramakrishnan
Sclerocleista Subram.
Sivanesniella Gawande & Agarwal
Sordariella Kapoor, Lal & Bahl
Spinulosphaeria Sivanesan
Thaxteriellopsis Sivanesan, Panwar & Kaur
Thaxterina Sivanesan, Rajak & Gupta
Thindia Korf & Waraitch
Tilakidium Vaidya, Naik & Rathod
Tilakiella Srinivasulu
Tripterosporella Subram. & Lodha
Tsypterospora Lodha
Tunstallia Agnihothrudu
Warcupiella Subram.

BASIDIOMYCOTINA

Ceraceosorus Bakshi
Chlorolepiota Sathe & Deshpande
Cytoliptoporus Dhana & Rajvardan
Eutypa Khurana, Thind & Berthier
Lyophyllopsis Sathe & Daniel
Muribasidiospora Kamat & Rajendren

UREDINALES

Acervulospora Thirum.
Catenulopsora Mundkur
Ceropsora Bakshi & Singh
Dasturella Mundkur & Kheswala
Didymopsorella Thirum.
Elateraecium Thirum.
Haplophragmiopsis Thirum.
Hiratsukamyces Thirum., Kern & Patil

(Appendix II Contd.)

(Appendix II Contd.)

Jacksoniella Kamat & Sathe
Kamatomyces Sathe
Karnkampella Rajendren
Kernella Thirum.
Kulkarniella Gokhale & Patel
Mehtamyces Mundkur & Thirum.
Peridiosora Kamat & Sathe
Ramakrishnania Ramachar & Bhagyanarayana
Stakmania Kamat & Sathe
Tunicopsora Singh & Pandey

USTILAGINALES

Franzpetrakia Thirum. & Pavgi
Georgefischeria Thirum. & Naras.
Jamesdicksonia Thirum., Pavgi & Payak
Narasimhania Thirum. & Pavgi
Mundkurella Thirum.
Zundelula Thirum. & Narasimhan

HYPHOMYCETES

Abgliophragma Roy and Gujarati
Acrostaphylus Arnaud ex Subram.
Adhogamina Subram. & Lodha
Agarwalia Tiwari & Agarwal, P.D.
Amerobotryum Subram. & Natrajan.
Angulimaya Subram. & Lodha.
Anelophorella Subram.
Anelophragmia Subram.
Ardhachandra Subram. & Sudha

Bahukalasa Subram. & Chadrashekhara
Bahupaathra Subram. & Lodha
Bahusakala Subram.
Bahusaganda Subram.
Bahusandhika Subram.
Bahusutraveeja Subram. & Bhat
Basifimbria Subram. & Lodha
Beejasamuha Subram. & Chandrashekhara
Beltraniella Subram.
Betraniomyces Manoharachary, Agarwal & Rao
Benjpalia Subram. & Bhat
Biharia Thirum. & Mishra
Bilgramia Panwar, Purohit & Chouhan

Cercocladospora Agrawal and Singh
Chaetendophragmia Rao & Mani Varghese
Chlamydorubra K.B. & K.S. Deshpande
Civisubramaniania Vittal & Dorai
Colemaniella Agnihothrudu
Cryptascoma Ananthapadmanaban
Cylindromyces Manoharachary, Rao & Agarwal

Dendrographiella Agnihothrudu
Drumopama Subram.
Duosporium Thind
Dwayabeeja Subram.
Dwayalome Subram.
Dwayamala Subram.

Edmundmasonia Subram.
Endosporostilbe Subram.

Gangliostilbe Subram. & Vittal
Gleichemia Sridhar & Kaveriappa
Gliocladiopsis Saksena, S.B.
Gliophragma Subram. & Lodha
Globuliroseum Sullia & Khan
Goosiomyces Krishna Rao & Manoharachary

Hawksworthia Manoharachary, Agarwal & Rao
Hyalostachybotrys Srinivasan

Koorchaloma Subram.
Koorchalomella Chona, Munjal & Kapoor
Kramasamuha Subram. & Vittal
Kumanasamuha P. Rao & D. Rao
Kutilakesa Subram.
Kutilkesopsis Agnihothrudu & Barua

Lacellinopsis Subram.
Laridospora Nawawi
Lecythispora Chowdhry
Lomachashaka Subram.
Lomaantha Subram.

Macraea Subram.
Manoharaharyomyces Agarwal & Rao
Menisporella Agnihothrudu
Monocillium Saksena, S.B.
Morrisiella Saikia & Sarbhoy

(Appendix II Contd.)

(Appendix II Contd.)

Moorella Rao, P.R. & Rao, D.
Mystrosporiella Munjal & Kulshreshtha

Nalanthamala Subram.

Panchanania Subram. & Nair*Parahelminthosporium* Subram. & Bhat
Parasympodiella Ponappa
Patharamaya Subram.
Petrakiopsis Subram. & Reddy
Phaeodactylium Agnihothrudu
Phaeotrichoconis Subram.
Phialographium Upadhyay & Kendrick
Phialomyces Mishra & Talbot
Phialotubus Roy & Leelavathy
Phragmospathula Subram. & Nair
Pirozynskia Subram.
Prathigada Subram.
Prathoda Subram.
Pseudotorula Subram.

Raizadenia Srivastava
Ramaraomyces Rao, Manoharachary & Goos

Sadasivanella Agnihothrudu
Sadasivania Subram.
Saporophragma K.B. & K.S. Desphande
Sarbhoyomces Saikia
Satwalekera D. Rao, V. Rao & P.R. Rao
Sclerotiella Sarbhoy & A. Sarbhoy
Searchomyces Mehrotra & Mehrotra
Sopagraha Subram. & Sudha
Sporidesmina Subram. & Bhat
Sporidesmiopsis Subram. & Bhat
Starkeyomyces Agnihothurudu
Subramania D. Rao & P.R. Rao
Subramaniomyces Mani Varghese & V.G. Rao
Sutravarana Subram. & Chandrashekhra
Syamithabeeja Subram. & Natarajan
Sympodina Subram. & Lodha
Synnemelliasia K. Rao, Manoharachary & Goos

Tandonella Prasad & Verma
Tandiella Despande & Desphande
Tharoopama Subram.
Thirumalacharia Rathaiah
Thozetellopsis Agnihothrudu
Tretophragmia Subram. & Natarajan
Trichoconiella Jain

Vakrabeeja Subram.
Vanbeverwijkia Agnihothrudu
Veronaella Subram. & Reddy
Vrikshopama D. Rao & R. Rao

COELOMYCETES

Ajrekarella Kamat & Kalani
Amphitiarospora Agnihothrudu
Bagcheea Mueller & Menon
Barnettella D. Rao & P.R. Rao
Bartaliniopsis Singh
Caudophoma Patil & Thirum.
Cyclodomella Mathur, Bhatt & Thirum.
Discosiellina Subram. & Chandra Reddy
Febulocada Nagaraj

Around 61 water-borne conidial fungi have been listed besides discussing various aspects and prospects of the above fungi from India (Manoharachary, 1985).

APPENDIX III Number of different groups of fungi recorded from India (1931-1992)

Fungal groups	*Butler & Bisbys 1931*	*Mundkur 1938*	*Ramakrishnan & Subramanian 1951*	*Butler & Bisby's Revised by R.S. Vasudeva 1960*	*R.S. Sarbhoy et al. Vasudeva 1962*	*1975*	*1988*	*1986*	*1992*
Myxomycotina	–	90	–	88	39	32	63	50	
Phycomycotina	75	54	61	203	98	80	140	118	
Ascomycotina	476	67	141	715	161	397	574	442	
Basidiomycotina	1339	135	357	1849	187	251	240	232	
Fungi Imperfecti (Agonomycetales)	461	178	253	1040	468	395	1043	963	
Total	2351	524	812	3875	953	1155	2060	1805	2566

References

Agarwal, D.K. and N.K. Rao 2004. *Manoharacharyomyces*—a new genus of Hyphomycetes from India. *J. Mycopathol. Res.* (In press).

Ahmad, S. 1949. A contribution to the fungus flora of Pakistan and India. *Indian Phytopath.* **2**: 11-16.

Ainsworth, G.C. 1979. Introduction to the History of Mycology. Cambridge Univ. Press, Cambridge, London, 351 pp.

Ajrekar, S.L. 1912. The castor rust (*Melampsporella ricini* de Toni). *J. Bombay Nat. Hist. Soc.*, **21**: 1092-1095.

Aneja, K.R. and Kumar, R. 1997. *Synnmukerjiomyces thermophile* a new Thermophilic Hyphomycete from India. In: *Advances in Microbial Biotechnology.* (Eds. J.P. Tewari et al.) APH Publishing Corp. New Delhi. pp. 1-5.

Arya, H.C. and Prasada, R. 1952. Alternaria blight of linseed. *Indian Phytopath.* **5**: 33-39.

vön Arx, J.A. and Mukerji, K.G. 1981. *Faurelina indica* sp. nov. *Sydowia,* 34: 39-41

vön Arx, J.A. and Mukerji, K.G. and Singh, N. 1978a. Leucosphera, a new genus of the Pseudoeurotiaceae. *Persoonia,* 10: 141-143.

vön Arx, J.A., Mukerji, K.G. and Singh, N. 1978b. A new coprophillous ascomycete from India. *Persoona,* 10: 144-146.

Bakshi, B.K. 1971. Indian Polyporaceae. ICAR Bulletin, New Delhi.

Banerjee, S.N. 1942b. *Fusarium equiseti* (Cda.) Sacc. (= *F. falcatum* App. et. Wr.) causing a leaf spot of *Eicchornia crassipes* Solms. Part I. *J. Deptt. Sci.* Calcutta Univ., N.S., **1**: 29-37.

Banerjee, S.N. 1944-45. A new species of discomycetes from Bengal, *Mitrula agharkarii* sp. nov. *Sci. & Cult.*, **10**: 308-309.

Barclay, A. 1886. On a Uredineae affecting the Himalayan spruce for *Abies smithiana* Forbes. *J. Asiatic. Soc.* **55**: 1-11.

Behera, N. and Mukerji, K.G.1974. Fungi of Delhi. XXV. *Chlamydoabsidia dasgupti* sp. nov. and *Polyschema indica.* Sharma, Behera and Mukerji, *Nor. J. Bot.*, 21: 1-3.

Bhargava, K.S. and Srivastava, Y.N. 1963. Observation on Indian aquatic fungi III. Some new species of Blastocladiales, Monoblepharidales and Peronosporales collected from Gorakhpur. *Proc. Nat. Acad. Sci.* **35**: 49-58.

Bhargava, K.S. and Srivastava, Y.N. 1977. Some fungi hitherto unrecorded from India. *Curr. Sci.* **46**: 46-54.

Bhattacharjee, M., Mukerji, K.G., Tewari, J.P. and Skoropad, W.P.1981. Structure and hyperparasitism of a new species of *Gigaopsra Trans. Br. Mycol. Soc.*, 77: 184-188.

Bilgrami, K.S., Jamaluddin and Rizvi, K.S. (1979). *Fungi of India.* Part I, Today and Tommorrow, New Delhi. 567 pp.

Bruhl, P. and Sen Gupta, J. (1927). Indian slime fungi (Myxomycetes or Mycetozoa). *J. Dept. Sci.*, Calcutta University,, **8**: 101-122.

Butler, E.J. 1905. Some Indian forest fungi. *Indian For.*, **31**: 487-496, 548, 556, 611-617, 670, 679.

Butler, E.J. 1907-1930. Reports of Imperial Mycologist, in the reports of Agricultural Research Institute & College (Pusa), pp. 34-80.

Butler, E.J. 1911. On *Allomyces*, a new aquatic fungus. *Ann. Bot.,* London, **25**: 1023-1034.

Butler, E.J. 1911. The leaf spot of turmeric (*Taphrina maculans* sp. nov.). *Ann. Myc.,* **9**: 36-39.

Butler, E.J. 1918. *Fungi and diseases in plants.* Thacker, Spink & Co., Calcutta, VI, 547 pp.

Butler, E.J. and Bisby, G.R. 1931. *Fungi of India.* ICAR. Sci. Monogr. I., XVIII, Calcutta, 237 pp.

Chaudhuri, H. 1942. Indian water moulds. V. A new genus of Saprolegniaceae. *Hamidia* gen. nov. *Proc. Indian Acad. Sci.* **15**: 225-230.

Chaudhuri, H. and Kochar, P.L. 1935. Indian Water mnoulds. I. *Proc. Ind. Acad. Sci.* **2**: 137-154.

Chona, B.L., Munjal, R.L. and Bajaj, B.S. 1956. *Vasudevella,* a new genus of the Sphaeropsidales. *Indian Phytopath.* **9**: 186-194.

Chona, B.L., Munjal, R.L. and Kapoor, J.N. 1958. *Koorchamella*: a new genus of Tuberculariaceae. *Indian Phytopath.* **11**: 130-132.

Choudhury, P.N. and Agarwal, G.P. 1981. Taxonomic studies on aquatic fungi from India. I. Pythiaceae. *Indian Phytopath.* **34**: 235-236.

Clements, F.E. and Shear, C.L. 1931. *The Genera of Fungi*-IV. 496. 588 plates. H.W. Wilson Comp., N.Y.

Cooke, M.C. 1880. Some exotic fungi. *Grev.* **8**: 93-96, 145-150, *Grev.* **9**: 10-15.

Corda, A.C.J. 1850. *Icones Fung.* Hubcusque Cognitorum. Prag., I-VI, Many plates, 1837-1854.

Cunningham, D.D. 1871. Seventh Annual Report. 1870. Appendix B, 230 pp.

Currey, F.J. 1874. *Trans. Linn. Soc.* London II, Ser., Bot., I. pp. 119-131.

Dakshim, K.M.M., Tandon, R.K. and Mukerji, K.G. 1970. A new species of *Phyllachora Mycologia.* 62: 296-300.

Das Gupta, S.N. 1982. Discourse on Aquatic Phycomycetes of India. *Indian Phytopath.* **35**: 193-216.

Das Gupta, S.N. and John Rachel. 1953. Studies on the Indian aquatic fungi. I. Some water moulds of Lucknow. *Proc. Indian Acad. Sci.,* Sec. B, **38**: 165-170.

Das Gupta, S.N. and John Rachel. 1988. A contribution to our knowledge of the genus *Blastocladia. Indian Phytopath.* **41**: 521-547.

Das Gupta, S.N., Sarbhoy, A.K. and Shome, S.K. 1990. *Gonapodya terrestris* sp. nov. *Indian Phytopath.* **43**: 218-222.

Dastur, J.F. 1935. Diseases of Pan (*Piper betle* L.) in the Central Provinces. *Proc. Indian Acad. Sci.,* Sec. B, **1**: 803.

Fries, E. 1855. Novae symbolae mycologicae. *Nova Acta Regiae Societitis Sci.,* Upsala, III, Ser. I, pp. 17-136.

Gawande, S. and Agarwal, D.K. 2004. *Sivanesaniella*-a new Ascomycetous genus from India. *Indian Phytopath.* **56**: 231-232.

Gokhale, V.P. and Patel, M.K. 1951. *Kulkarniella,* a new genus of Rusts. *Indian Phytopath.* **2**: 170-173.

Hennings, P. and Massee. 1900-1901. Fungi Indiae Orientalis.*Hedw*. XXXIX, pp. 150, 153, 1900.

Hosagouder, V.B. 1985. New and noteworthy species of *Phyllachora* from South India. *Indian Phytopath.* **38**: 447-451.

Hosagoudar, V.B. 1987. Meliolaceae of South India. II. *Sydowia* **60**: 113-121.

Hosagoudar, V.B. 1996. *Meliolales of India*. Published by Botanical Survey of India, Calcutta, pp. 362.

Iyengar, M.O.P. 1935. Two new fungi of genus *Coelomomyces* parasitic in larvae of *Anopheles sp. Parasitology* **27**: 440-449.

Iyer, Rohini, Ghosh, S.K. and Sarbhoy, A.K. 1973. Studies on Coprophilous Fungi-I. *Proc. Indian Natl. Sci. Acad.* **39**: 199-202.

Joshi, L.M., Singh, D.V. and Srivastava, K.D. 1986. Wheat and wheat diseases in India. In - *Problems and Progress of wheat pathology in South Asia,* Malhotra Publishing House, New Delhi, pp. 1-19.

Kamat, M.N. 1976. Twenty five years of Ascomycetology in India. Presidential Address. *Kavaka,* **4**: 1-13.

Kamat, M.N. and Patel, M.K. 1948. Some new hosts of *Oidiopsis taurica* (Lev.) Salmon in Bombay. *Indian Phytopath.* **1**: 153-158.

Karling, J.S. 1942. *The Plasmodiophorales.* New York, V, 144 pp.

Karling, J.S. 1942.*The simple holocarpic biflagellate Phycomycetes.* New York, IX, 123 pp.

Khanna, A. and Payak, M.M. 1972. Teliospore morphology of some smut fungi. IV. *Sphacelotheca reiliana. Nova Hedwigia* **26**: 907-913.

Khanna, A., Payak, M.M. and Mehta, P.C. 1966. Teliospore morphology of some smut fungi-I. Electron Microscopy. *Mycologia,* **58**: 562-569.

Khanna, A., Payak, M.M. and Mehta, S.C. 1968. Teliospore morphology of some smut fungi II. Light Microscopy. *Mycologia* **60**: 655-662.

Khanna, A., Payak, M.M. and Nam Prakash. 1971. Teliospore morphology of some smut fungi. III. *Ustilago nuda*. *Indian Phytopath.* **24**: 481-486.

Kumar, J. and Sarbhoy, A.K. 1988. Taxonomic studies on the genus *Cladosporium* Link. Fr. from India. *Advances in Mycology*, Botany Dept., PAU, Chandigarh. 262 pp.

Lacy, R.C. 1949. Studies in aquatic Phycomycetes. *Indian Phytopath.* **2**: 134-141.

Lacy, R.C. 1955. Studies in aquatic Phycomycetes. *Indian Phytopath.* **8**: 208-209.

Lakhanpal, T.N. and Mukerji, K.G. 1975. Taxonomic studies of Indian Myxomycetes-VII. The genus *Licea*. *Kavaka,* **3**: 101-105.

Lakhanpal, T.N. and Mukerji, K.G. 1976. Taxonomic studies on Indian Myxomycetes-V. Some new records of Trichiales.*Proc. Indian Natn. Sci. Acad., Sec. B,* **42**: 34-40.

Lakhanpal, T.N. and Mukerji, K.G. 1977. Taxonomic studies on Indian Myxomycetes. XII. Two new speices of *Comatrichia, Trans. Mycol. Soc.,* Japan, 18: 125-133.

Lakhanpal, T.N. and Mukerji, K.G. 1979. Taxonomic studies on Indian Myxomycetes. XVIII. Two new speices of *Didymium*. *Rev. de Mycol.* 43: 236-241.

Lakhanpal, T.N. and Mukerji, K.G. 1979a. Taxonomic studies of Indian Myxomycetes-XIV. *Kavaka,* **7**: 59-62.

Lakhanpal, T.N. and Mukerji, K.G. 1981. Taxonomy of Indian Myxomycetes. J. Cramer, Germany, p. 531.

Lingappa, B.T. 1955. Some new Indian species of Synchytrium.*Lloydia* **18**: 129-142.

Lingappa, B.T. 1955. Two new species of *Physoderma* from India. *Mycologia* **47**: 109-121.

Lister, A. and Lister, G.V. 1925.*A monograph of the Mycetozoa.* 3rd Edn., London.

Lloyd, C.G. 1898-1925. Mycological Notes, Nos. 1-75, pp. 1-1364, Plates & Figs. Cincinnati, Ohio.

Lodha, B.C. 1974. Studies on Coprophilous fungi-IV. Some clestothecial Ascomycetes. *J. Indian bot. Soc.* **50**: 196-208.

Lodha, S.A. 1934. Indian slime moulds. Univ. of Punjab, Lahore, 34 pp.

Luttrell, E.S. 1951. Taxonomy of the Pyrenomycetes. Univ. Monogr. Studies,**24**: 151.

Manoharachary C. 1981. The taxonomy and ecology of freshwater. Phycomycetes from India. Rev: Life. Sci. **1**: 3-21.

Manoharachary C. 1985. Aspects and prospects of water-borne conidied fungi from India. *Adv. Biol. Res* **4** (1 < 2) 160-183.

Manoharachary C. 1987. Aspects and prospects of freshwater fungi from India. Current Trends *Limnological Res.* 10-28. Narendra Publishers.

Manoharachary C. 1989. Perspectives in acquatic mycology from India. Economic plants & microbes (Ed. Purkayartha) Today & Tomorrow printers & publishers, New Delhi. 181-188.

Manoharachary C. and Rampulla Reddy 1991. Some considerations on the taxonomy and ecology of ascomycotina. Proc. Env. Sci. Marathwada University Aurangabad 180-201.

Malhotra, G. and Mukerji, K.G. 1976, Fungi of Delhi, XXIX. Three new species of *Chaetomium* from decaying wood. *Rev. de Mycol.* 40: 1979-1984.

Manoharachary, C., Agarwal, D.K. and Rao, N.K. 2003.*Beltraniomyces* a new genus of Dematiaceous hyphomycetes from India. *Indian Phytopath.* **56**: 418-421.

Manoharachary, C., Agarwal, D.K. and Rao, N.K. 2004. A new anamorphic genus from India. *Indian Phytopath.* **57**: 161-163.

Manoharachary, C., Agarwal, D.K. and Rao, N.K. 2005.*Hawksworthia* - genera nov. from India. *Indian Phytopath.* (in press).

Manoharachary, C., Kumar, I. and Mukerji, K.G. 2002. Arbuscular mycorrhizal fungi—identification, taxonomic criteria, classification, controversies and terminology. In: *Techniques in Mycorrhizal Studies* (Eds. Mukerji, K.G., Manoharachary, C. and Chamola, B.P.). Kluwer Academic Publishers, Netherlands. pp. 249-273.

Manoharachary, C. and Krishna Rao, N. 1999.*Mukerjiella* gen. et. sp. nov. from India. In: *New Approaches in Microbial Ecology* (Eds. Tewari, J.P. et al.) Aditya Books Pvt. Ltd., New Delhi. pp. 1-5.

Massee, G. 1898-1912. Fungi Exotici. *Kew Bull.* 113-135.

Mc Rae, W. 1918. *Phytophthora meadii* n. sp. on *Hevea brasiliensis. Mem. Dept. Agric. India,* Bot., Ser., **9**: 219-273.

Mehrotra, B.S. and Baijal, U. 1968. Is *Blakeslea* valid genus? *J. Elisha Mitchell Scient. Soc.* **84**: 207-210.

Mehrotra, B.S. and Mehrotra, B.R. 1964. Species of*Mortierella* from India - IV. *Zentbl. Bakt. Parasitkde Abt. Z.* **2**, 118: 171-177.

Mehrotra, B.S. and Sarbhoy, A.K. 1960. A new species of*Chaetocladium* from India. *Mycologia* **52**: 795-799.

Mehrotra, B.S., Baijal, U. and Mehrotra, B.R. 1968.*Mycopath. Mycol. appl.* **33**: 125-129.

Mehrotra, B.S., Singh, S.N. and Baijal, U. 1972. The section sphaer osporous group of Mucor, a reassessment. *Sydowia* **26**: 41-46.

Mer, G.S. and Khulbe, R.D. 1984. Water moulds of Sat Tal (Nainital), India*Sydowia* **37**: 195-207.

Mhatre, J.R. and Mundkur, B.B. 1945. The*Synchytria* of India. *Lloydia* **1**: 131-138.

Mitra, A.K. 1943. A new asomycetous fungus on *Selaginella*. *Curr. Sci*. **12**: 329.

Mitra, A.K. 1944. *Melanopsamma ranjanii* sp. nov. a new parasite of *Selaginella. J. Indian bot. Soc.* **23**: 158-163.

Mitra, M. 1929. *Phytophthora parasitica* Dast. causing "damping off" of disease of cotton seedlings and fruit rot of guava in India. *Trans. Brit. mycol. Soc.* **14**: 249-254.

Mitra, M. 1931. *Tilletia indica* Mitra sp. nov. A new bunt on wheat in India.*Ann. app. Biol.* **18**: 178-180.

Mitra, M. 1935. Stinking smut (Bunt) of wheat with special reference to *Tilletia indica* Mitra. *Indian J. Agric. Sci.* **5**: 51-74.

Mitra, M. and Mehta, P.R. 1938. Some leaf diseases of*Hevea brasiliensis* new to India. *Indian J. Agric. Sci.* **8**: 185-188.

Mitter, J.H,. and Tandon, R.N. 1930. The fungus flora of Allahabad.*J. Ind. bot. Soc.* **9**: 197.

Mitter, J.H. and Tandon, R.N. 1938. Fungi of Nainital-II*J. Ind. bot. Soc.* **17**: 177-182.

Mukerji, K.G. 1968a. The position of the genus *Achactomium* in Pyrenomycetes. *Proc. Natl. Inst. Sci.*, India, 34B: 288-292.

Mukerji, K.G., 1968b. Fungi of Delhi. IV. A new species of*Piptocephalis*. *Mycologia* 60: 326-336.

Mukerji, K.G., 1969. Fungi of Delhi. 1. A new species of the genus*Sepedomium* Links. *Mycopath. Mycol. Appl.*, 37: 304-312.

Mukerji, K.G., 1969b. On the nomenclature of the Podery mildew of*Dalbergia sisco* Roxb. *Mycolgia*. 61: 181-184.

Mukerji, K.G., 1968c. Descriptions of pathogenis bacteria and fungi. Set. 19 Nos. 181-190 (Emyphales) CMI, Kew, Surrey, England.

Mukerji, K.G., 1975. CMI Descriptions of pathogenic fungi and bacteria. Set 46; Nos. 451-460 (Revonorporales). CMI, Kew, Surrey, England.

Mukerji, K.G., 1996. Taxonomy of endomycorrhizal fungi. In: Advances in Botany (Eds. Mukerji, K.G., et al.). APH Publishing Corp, New Delhi, pp.213-221.

Mukerji, K.G., and Khanna, M. 1980. Fungi of Delhi, XII. *Chaetomium lawrence-amessi. Acta Mycologia.* 16: 255-256.

Mukerji, K.G., Bhattacharjee, M. and Tewari, J.P. 1983. New species of vesicular-arbuscular mycorrhizal fungi. *Trans. Br. Mycol. Soc.* 81: 641-643.

Mukerji, K.G. and Bhandari, N.N. 1972. Fungi of Delhi XI.*Sporormia cainia* sp. n. from India. *Trans. Mycol. Soc.*, Japan, 12: 99-103.

Mukerji, K.G., and Kaur, S. 1996. *Arcyria rufosa:* A new Indian myxomycete. Mycotaxon, 59: 479-481.

Mukerji, K.G. Ranga Rao, V. 1968.*Volutella lini* sp. nov. *Trans. Br. Mycol. Soc.* 51: 337-339.

Mukerji, K.G. and Saxena, A.S. 1974. Notes on*Achactomium, Anixiella, Boothiella, Chaetomium, Lophotriohus, Pseudoeurotium, Pycinidiophora* and the classification of the chaetomiales. *Beiheft. Nova Hedwigia,* 47: 373-404.

Mukerji, K.G. and Singh, N. 1974. Fungi of Delhi. XXI.*Chaetomium dehliamum* sp. *nov. Friesia,* 10: 265-269.

Mundkur, B.B. 1938. *Fungi of India,* supplement I. ICAR. Sci. Monogr. **12**: 54.

Mundkur, B.B. 1952. Ustilaginales of India.*Mycol. Pap (CMI)*, 84 pp.

Mundkur, B.B. and Kheswala, K.F. 1943.*Dasturella* - a new genus of Uredinales. *Mycologia* **35**: 201-246.

Mundkur, B.B. and Thirumalachar, M.J. 1943.*Catenulospora,* a new genus of Rusts. *Ann. Bot.* London, n.s. **7**: 213-220.

Mundkur, B.B. and Thirumalachar, M.J. 1945. Two new genera of rusts on Bignoniaceae. *Mycologia,* **37**: 619-628.

Munjal, R.L. and Gill, H.S. 1961. *Corynesporella*-a new genus of Hyphomycetes. *Indian Phytopath.* **14**: 6-9.

Munjal, R.L. and Kulshreshtha, D.D. 1969. *Mystrosporiella,* a new genus of Hyphomycetes. *Mycopath. Mycol. appl.* **39**: 355-357.

Nannenga-Bremekamp, N.E., Mukerji, K.G. and Singh, N. 1979. On two coprophillous Myxomycetes from India. *Proc. K. Ned. Akad. Wet. Ser. C,* 82: 217-221.

Nannenga-Bremekamp, N.E., Mukerji, K.G. and Pasricha, R. 1984. Note on Indian Myxomycetes-Three new species and comments on others. *Proc. K. Ned. Akad. Wet. Ser.* C. 87: 471-482.

Neeraj, Mukerji, K.G., Sharma, B.C. and Varma, A.K. 1993. A new species of *Gigaspora* from desert soils of Rajasthan, India. World J. Microbial. Biotech., 9: 291-294.

Padmanabhan, S.Y. 1947.*Fusarium* sp. parasitic on *Epipyros,* a Lepidopterous parasite of the sugarcane *Pyrilla. Proc. Indian Acad. Sci.* **26**: 77-92.

Padwick, G.W. 1939. The genus *Fusarium.* I. Known occurrence in India. *Indian J. agric. Sci.* **9**: 171-184.

Padwick, G.W. 1940a. A new disease of wheat in India. *Curr. Sci.* **9**: 179-180.

Padwick, G.W. 1940b. The genus*Fusarium* III. A critical study of the fungus causing wilt of gram (*Cicer arietinum* L.) and of the related species of the Sub-section Orthocera, with special relation to the variability of key characteristics.*Indian J. agric. Sci.* **10**: 241-248.

Padwick, G.W. and Azmatullah, M. 1943.*Claviceps purpurea* (Fr.) Tul. and a new species from Shimla. *Curr. Sci.* **12**: 257.

Pavgi, M.S. and Thirumalachar, M.J. 1957. *Franzpetrakia,* a new genus of Ustilaginales. *Sydowia* **9**: Suppl. I: 1-3.

Payak, M.M. 1964. *Palynology of Fungi*. Recent Advances in Palynology (Eds. P.K.K. Nair). Publ. by Indian Botanic Garden, Lucknow, 1964, pp. 27-55.

Payak, M.M., Sarbhoy, A.K. and Lal, S.P. 1997. History of Mycology in India. Internatl. Cong. on Integrated Plant Dis. Mgmt. Souvenir, IPSGJ, **97**, New Delhi, Nov. 10-13, pp. 24-33.

Prakash, R. and Sarbhoy, A.K. 1993. *Mucor thermophilus* sp. nov. L. The Mucorales from India. *Zentrabl. Mikrobiol.* **148**: 531-534.

Rai, J.N. and Mukerji, K.G. 1962a. A new species of *Chaetomium* from Indian soils. Can. J. Bot. 40: 857-860.

Rai, J.N. and Mukerji, K.G. 1962b. *Sporotrichum carthuiovirde* Rai and Mukerji, a new speices from Indian soils. Mycopath. Mycol. Appl. 18: 122-126.

Rai, J.N., Mukerji, K.G. and Tewari, J.P. 1963. *Tripterospora tetraspora* sp. nov., a new cleistothecial Ascomycetes. *Can. J. Bot.* 41: 327-329.

Rai, J.N., Tewari, J.P. and Mukerji, K.G., 1961. A new *Helicostylum* from Indian soils. *Can. J. Bot.,* 39: 1282-1284.

Rai, J.N., Tewari, J.P. and Mukerji, K.G. 1964a. *Achaetomium,* a new genus of Arcomycetes. *Can. J. Bot.,* 42: 693-697.

Rai, J.N., Tewari, J.P. and Mukerji, K.G. 1964b. A new *Aspergillus* from Indian soils—*A. striatus* sp. nov. *Can. J. Bot.* 42: 1521-1524.

Ramakrishnan, T.S. and Narasimhalu, I.L. 1941. The occurrence of *Darluca filum* (Biv.) Cast. on cereal rusts in South India. *Curr. Sci.* **10**: 290-291.

Ram Dayal and Kiran, U. 1988. *Zoosporic fungi of India*. Inter India Publ., New Delhi, 238 pp.

Rao, N.K. Manoharachary C. and Goos R.D. 1998. Forest litter Hyphomycetes from A.B, India III. A new synnematous Hyphomycete *mycologia*.

Rao, N.K. and Manoharachary 1989. *Goosiomyces,* A new genus of dematiaceous Hyphomycetes from Andhra Pradesh, India. *Mycol. Res.* 92(2), 249-251.

Rao, N.K. and Manoharachary C. and Goos, R.D. 1989. Forest litter Hyphomycetes from Andhra Pradesh India IV, A new genus of synnemetous Hyphomycetes. *Mycologa* 81(5): 790-793.

Rikhy, M. and Mukerji, K.G. 1973. Fungi of Delhi. XXIX *Bahusandhika compacta* sp. nov. and *Endorporostilibe minuta* sp. nov. Rev. *de Mycol.,* 38: 91-94.

Saikia, U.N. 1981. *Sarbhoyomyces* gen et. sp. nov. from India. *Curr. Sci.* **50**: 949.

Saikia, U.N. and Sarbhoy, A K. 1980. Hyphomycetes of North East India. I. *Indian Phytopath.* **33**: 459-461.

Saikia, U.N. and Sarbhoy, A.K. 1981. Hyphomycetes of North-East India. IV - The genus *Melarographium. Indian Phytopath.* **34**: 364-367.

Saikia, U.N. and Sarbhoy, A.K. 1985. *Morrisiella* - A new genus of Synnematous Hyphomycete. *Mycologia* **77**: 318-320.

Saksena, R.K., Krishnanand and Sarbhoy, A.K. 1968. Ecology of soil fungi of Uttar Pradesh-IV. *Proc. Nat. Inst. Sci.* **33**B: 298-306.

Saksena, R.K., Sarbhoy, A.K. and Krishnanand. 1967. Ecology of soil fungi of Uttar Pradesh. III. *Proc. Nat. Inst. Sci.* **33**B: 154-161.

Saksena, S.B. 1953. A new genus of Mucorales. *Mycologia* **45**: 426-436.

Sarbhoy, A.K. 1967. Species of the Mucorales recorded from India.*Mycol. Pap (CMI)* 108, Kew, England.

Sarbhoy, A.K. 1968. Revison of Sphaerosporus group of mucor.*Trans. Brit. mycol. Soc.* **61**: 25-32.

Sarbhoy, A.K. 1985. Cleistothecial states of*Aspergillus* and their taxonomic position. *Mycopathologia* **91**: 133-142.

Sarbhoy, A.K. 2002. Importance of taxonomy - A business Plan in 21stCentury. Tropical Mycology 2001, Proc. of IIIrd Natl. Symp. Organised by Indian Mycological Society, Calcutta, pp. 1-8.

Sarbhoy, A.K. and Agarwal, D.K. 1997. *HCIO Descriptions of Plant Pathogenic Fungi* (Set. Nos. 1-17). Malhotra Publishing House, Mayapuri, Delhi, pp. 1-102.

Sarbhoy, A.K. and Elphic, J. 1973. Hlemicarpentalis - a new genus of ascomycetes. *Curr. Sci.* **41**: 255-256.

Sarbhoy, A.K. and Iyer, Rohini. 1972. A new species of*Pseudographiella* from India. *Sydowia* **26**: 81-82.

Sarbhoy, A.K. and Sarbhoy, Asha (1974). *Sclerotiella indica* gen. and sp. nov. from Indian soil. *M.V.M. Patrika*, Poona **9**: 91-93.

Sarbhoy, A.K., Agarwal, D.K. and Varshney, J.L. 1976.*Fungi of India*. Suppl. V, Today and Tomorrow, Printers and Publishers, New Delhi, pp. 270.

Sarbhoy, A.K., Agarwal, D.K. and Varshney, J.L. 1994.*Fungi of India* (1982-1991). C.B.S. Publ., Daryaganj, New Delhi, pp.350.

Sarbhoy, A.K., Lal, G. and Varshney, J.L. 1974.*Fungi of India* (1977-84). Associated Publ. House, New Delhi. pp. 274.

Sarbhoy, A.K., Lal, G. and Varshney, J.L. 1975.*Fungi of India*. M/s. Navyug Traders, New Delhi, pp. 149.

Sathe, A.V. and Deshpande, S.D. 1979.*Chlorolepiota* - a new genus of Agaricales from India. *Curr. Sci.* **48**: 693-695.

Saxena, A.S. and Mukerji, K.G. 1970a.*Kernia bifurcotricha* sp. nov. *Trans. Br. Mycol. Soc.*, 54: 16-18.

Saxena, A.S. and Mukerji, K.G., 1970b.*Didymostilbe ellisi* sp. nov. *Trans. Br. Mycol. Soc.*, 55: 504-508.

Saxena, A.S. and Mukerji, K.G. 1970c. Fungi of Delhi, XV.*Lophotrichus indicus* sp. nov. *Acta Bot. Neerlandica*, 19: 722-726.

Saxena, A.S. and Mukerji, K.G. 1972. Fungi of Fungi, XVII. A new species of *Chaetomium Mycologia*, 64: 1326-1330.

Saxena, A.S. and Mukerji, K.G. 1973. Fungi of Delhi, XIX.*Bahupaathra minuta* sp. nov. *Mycopath. Mycol. Appl.*, 49: 201-204.

Sengupta, S.R., Nigam, S.S. and Tandon, R.N. 1950. A new wool-destroying fungus *Ctenomyces* species. *Text. Res. J.* **20**: 671-675.

Sharma, K.R., Behera, N. and Mukerji, K.G. 1973.*Podoxyphium indicum* sp. nov. *Polyschema indica* sp. nov. *Norw. J. Bot.*, 20: 27-29.

Singh, D.V. and Sarbhoy, A.K. 1971. *Harikrishnaella* - a new genus from India. *Sydowea* **26**: 66-69.

Pavgi, M.S. and Singh, S. 1979.*Johnkarlingia* - a new genus of Aynchytriaceae on *Brassica* roots.

Singh, S. 1979. *Pseudoaristastoma* - An undescribed genus of Coelomycetes from India. *Sydowia* **31**: 238-241.

Sivanesan, A. and Sinha, A.R.P. 1989. *Aldonata* - a new ascomycete genus in the Parmulariaceae. *Mycol. Res.* **92**: 246-247.

Sivanesan, A., Rajak, R.C. and Gupta, R.C. 1988. *Thaxeterina* - a new tubefiaceous genus with multispored asci from India. *Trans. Brit. mycol. Soc.* **90**: 662-665.

Srivastava, R.K. 1987. *Aspleniumsporites trivedii* gen. et. sp. nov. from Neyveli lignitie of south India. *Curr. Sci.* **56**: 305-306.

Subramanian, C.V. 1971. *Hyphomycetes.* ICAR Publ. New Delhi, 930 pp.

Subramanian, C.V. and Ramakrishnan, K. 1956. List of Indian Fungi (1952-1956). *J. Madras Univ.,* **26**: 327-421.

Subramanian, C.V. and Sekar, G. 1990. Coronophorales from India - A monograph. *Kavaka* **18**: 19-90.

Sydow, H. and Butler, E.J. 1911. Fungi Indiae Orientalis, Pars III. *Ann. Mycol.* **9**: 372-421.

Sydow, H. and Butler, E.J. 1916. Fungi Indiae Orientalis. Pars. V. *Ann. Mycol.* **14**: 177-220.

Tandon, R.N. 1968. *Mucorales of India.* ICAR Monogr. New Delhi, 120 pp.

Thiessen, F. 1912. Fungi aliquot Bombay euses a Rev. Ed. Blaller Colleeti *Ann. Myc.* **IX**, pp. 153-159, 1911.

Thirumalachar, M.J. 1945a. An ascomycetous parasite of *Cephaleuros. Proc. Indian Acad. Sci.* **22** (B): 374-377.

Thirumalachar, M.J. and Mundkur, B.B. 1949. Genera of rust-I. *Indian Phytopath.* **2**: 65-101.

Thirumalachar, M.J. and Mundkur, B.B. 1949. Genera of rust. II. *Indian Phytopath.* **2**: 193-244.

Thirumalachar, M.J., Gopal Krishnan, K.S. and Narasimhan, M.J. 1947. Morphology of spore forms and heterocism of the giant Bamboo rust, *Dasturella divina. Bot. Gaz.,* **108**: 371-379.

Thind, K.S. 1977. *Myxomycetes of India.* ICAR Publ. New Delhi. pp.

Thomas, K.M. and Ramakrishnan, T.S. 1948. Studies in the genus Phytophthora. II. *Proc. Indian Acad. Sci.* **27** (B): 55-73.

Ullasa, B.A., Rao, V.G. and Patil, A.S. 1973. *Chadefaudiomyces,* a new member of Diaporthaceae (Ascomycetes). *Rev. de mycol.* **38**: 19-22.

Verma, Savita, Varma, Ajit, Kail-Henir, Rexer, Annette, Hassel, Garlach Kost, Sarbhoy, A.K., Bisen Prakash, Butehori and Phillipp. Franken 1998. *Piriformospora indica* gen. et. sp. nov., a new root colonizing fungus. *Mycologia* **90**: 896.

Vasudeva, R.S. 1954. *Fungi of India.* Revised and updated version. ICAR Publ. pp. 554.

Vasudeva, R.S. 1960. (Butler, E.J. and Bisby, G.R.). Revised Edition *Fungi of India.* Printed and published by Govt. of India Press, Calcutta, pp. 552.

Vasudeva, R.S. 1962. *Fungi of India* (Supplement I). ICAR. New Delhi, III + 206 pp.

Vasudeva, R.S. 1963. *Indian Cercosporae.* ICAR Publ., pp. 245.

Venkatarajan, S.V. 1946d. A natural fungus parasite of powdery mildew on *Cyamopsis psoraloides* DC. *Curr. Sci.* **15**: 319.

SECTION 2

FUNGAL ECOLOGY

Frontiers in Plant Sciences (Pages 129-142)
Eds. K.G. Mukerji, K.V.B.R. Tilak, S.M. Reddy, L.V. Gangwane, P. Prakash, I.K. Kunwar

Occurrence and Phenology of the Oomycetes with Special Reference to Saprolegniaceae

S.K. Prabhuji

ABSTRACT

The occurrence and phenology of the members of Saprolegniaceae are reviewed with reference to the contributions by Indian mycologists. Observations on the distribution of Saprologniaceae in the water-bodies and soils of Padrauna (District: Kushinagar) have been presented. An attempt has been made to explain the difference in the peaks of occurrence of Saprolegniaceae in waterbodies and in soil. A comparison with the previous available literature indicated that eccentric oospore forms were more common in the warm temperate and tropical regions while the members with centric and sub-centric oospores were dominant in cold temperate regions. Distinct occurrence zones of the World may be identified for the dominance of eccentric-oospore and sub-centric or centric-oospore forms of Saprolegniaceae. An altitude-latitude relationship has also been observed in such distribution.

Key words: Phenology/Saprolegniaceae/Oospore type/Eccentric/Centric

INTRODUCTION

Since the inception of the studies on mycology the investigations on higher fungi have, undoubtedly, moved forward at a constant pace and is touching new heights. However, the graph of scientific explorations and advance studies in the field of water-moulds (particularly the members of Saprolegniaceae) is continuously going down as far as the Indian scenario

Author address: Department of Botany, U.N. Postgraduate College, Padrauna - 274304, U.P.
E-mail: shaktiprabhuji@rediffmail.com

is concerned and has arrived at the verge of extinction. Therefore, a special attention is required to revitalize it so that the studies in this branch of mycology may go on to regain its proper place.

The first references to any of the Saprolegniaceae appear to have been those of Ledermuller (1760), of Wrisberg (1765) and of Spallanzani (1777). For a long time they were regarded as algae and were described by most of the workers under the generic name - *Conferva,* which included, in its Linnaean application, the filamentous aquatic plants, generally. The earliest binomials appear to be those of Flora Danica (1780), *Byssus aquatica,* and of Schrank (1789), *Conferva piscium.* Previous workers had observed these fungi on flies in water, but as the name indicates, Schrank's observation was the first record of their occurrence on fishes. The earliest figures of these fungi are those of Flora Danica (1780), of Dillwyn (1809) and of Lyngbye (1819). Gruithuisen (1821) described a fungus on the remains of a dead snail, and for the first time figured escaping zoospores. Carus (1823) described a fungus on Salamander larvae with spores callected in the form of a globe at the mouth of a sporangium and called it - *Hydronema.* In an appendix to Carus' paper, Nees von Esenbeck (1823) established the genera *Saprolegenia* (Gruithuisen's fungus) and *Achlya* (Carus' fungus) on the distinctive difference in the escape of the zoospores which we recognize as their salient features even today. Then, the studies on Oomycetes went on step by step till the onset of 21st century (Agardh, 1824; Berkeley, 1833; Bennett, 1841; Naegeli, 1847; Pringsheim, 1851; DeBary, 1852; Hoffman, 1867; Cornu, 1872; Fischer, 1892; Humphrey, 1892; Petersen, 1909; 1910; Tiesenhausen, 1912; Kauffman, 1921; Coker, 1923; Apinis, 1930; Lund, 1934; Ito, 1936; J.R. Raper, 1936; Chaudhuri and Kochhar, 1935, Chaudhuri and Lotus, 1936; Hamid, 1942; Johnson, 1956; Scott, 1961; Srivastava, 1967; Milanez, 1968, 1970; Dayal and Thakurji, 1966, 1968; Dick, 1969; Lund, 1978; Prabhuji and Srivastava, 1977,1978; Prabhuji, 1984 a,b; Prabhuji and Sinha, 1993, 1994; and Prabhuji et. al., 1984).

Later, the Oomycetes were accepted as a class of heterotrophic organisms, probably derived from heterokont algal ancestry (Bessey, 1950; Copeland, 1956). Manton (1952, 1965), on the basis of his studies on the fine structure of flagella, confirmed the algal ancestry and Vogel (1960), too, confirmed it on the basis of modes of lysine synthesis. The chemical composition of the hyphal wall (Novaes- Ledieu et. al., 1967) and an assessment of the significance of microtubules versus flattened cristae within mitochondria (Turian, 1962) have also established the specific identity of these fungi.

Out of these the 'water-moulds', particularly the members of Saprolegniaceae, have been isolated, described and their phenology studied in different geographical regions of the World (Harvey, 1925; Lund, 1934; Forbes, 1935; Waterhouse, 1942; Chaudhuri et al., 1947; Rossy-

Valdderrama, 1956; Ziegler, 1958; Willoughby, 1962, 1965; Dayal and Tandon, 1962; Dick and Newby, 1961; Roberts, 1963; Srivastava, 1967 a,b; Milanez, 1967; 1968; 1970; Milanez and Beneke, 1968; Howard et a l., 1970; Alabi, 1971; Khulbe and Bhargava, 1977; Lund, 1978; Mer et al., 1980; Khulbe, 1980; Prabhuji, 1984 a,b; Prabhuji and Sinha, 1993; and Prabhuji et al., 1984; 1996; 2001).

Water-moulds, particularly the aquatic Oomycetes, have been isolated and their occurrence and phenology have been studied by various investigators under two different conditions, i.e., from water samples of different waterbodies and from soil samples near the waterbodies or away from it. Such variations in sampling sites caused difference in the results regarding periodicity of these fungi. Certain members which are very frequently present in waterbodies exhibit their absence in soil and *vice-versa.*

Coker (1923) was the earliest to point out that certain species of Saprolegniaceae show a seasonal periodicity and variation in abundance. Harvey (1925) was the first to make a study of the water-moulds occurring in soil. Later, Forbes (1935), Waterhouse (1942), Dayal and Tandon (1962) and Srivastava (1967b) have reported that winter is the most suitable period for the growth of water-moulds, while Chaudhuri et al. (1947), Naumov (1954), Perrott (1960) and Dick and Newby (1961) have observed two maxima, one in autumn and the other in spring. Khulbe and Bhargawa (1977) have also given two maxima for the periodicity of water-moulds, for temperate lakes during Summer and autumn and for sub-tropical lakes during spring and autumn. However, they have clearly indicated that the periodicity is governed by the temperature and not by the season. Following the observations of Dick and Newby (1961) in soil, Prabhuji (1979, 1984 a) has observed two peaks of occurrence of these fungi in soil - one during December-January and another in July-August. These observations were, later, confirmed by Prabhuji et al. (1996, 2001).

Hughes (1962) reviewed the literature on the distribution of watermoulds in water and put forward a theory that the members of Saprolegniaceae with eccentric oospores are found more commonly in tropical regions while the species with centric or sub-centric oospores occur in greater number in temperate regions. The observations of Hughes (1962) were supported by Srivastava (1967a, b) and Prabhuji (1979, 1984 a,b). Alabi (1971), however, reported that the eccentric species were found mainly in dry season (at high tenmperature) and centric and sub-centric species in the rainy season (at low temperature). Recently, Prabhuji et al. (1996, 2001) have also recorded two peaks of occurrence of Saprolegniaceae in soils and all the members, isolated, had eccentric oospores.

Besides the review of literature, the phenology (seasonal occurrence) of the members of Saprolegniaceae has been studied in the waterbodies and soils of Padrauna (Dist. Kushinagar), during the course of present investigations. Moreover, the distribution of Saprolegniaceae has also been studied with special reference to their oospore type, which is actually based on the present data together with the data obtained previously by various workers from different geographical regions of the World. Futhermore, an attempt has been made to explain the reason why the members of Oomycetes exhibit only one peak of occurrence in water bodies and two peaks in soils.

MATERIALS AND METHODS

To study the saproleginaceous flora of both aquatic and terrestrial habitats, two waterbodies (one near railway station-WB2 and another near sugar factory-WB1) and the environs of a waterbody- WB3 (for soil samples) near Chhawani area at Padrauna (Dist: Kushinagar) were surveyed. Water and Soil samplings were done at monthly interval for three years (April, 1999-March, 2002). Quadrat technique (Dick and Newby, 1961; Prabhu ji, 1984a) was employed for sampling and treatment of soil samples. At soil collection site four quadrat areas of one metre square were permanently marked around WB3. Each time from each quadrat area eighty samples (five samplings were done from one sub-division out of sixteen sub-divisions marked within each quadrat) were taken out (Prabhuji, 1984a, b). Every month the records of all the four sites were summed up to record monthly isolations. The methods given by Srivastava (1967a, b) were employed for the collection of water samples (from WB1 and WB2) and fungal isolations.

The isolates were identified using the keys provided by Coker (1923), Johnson (1956), Scott (1961) and Seymour (1970).

During isolations several members belonging to Pythiaceae, Blastocladiaceae and Chytridiaceae were also recorded, however, considering the specificity of the work only the members of Saprolegniaceae were taken into account and rest of the members were deleted from our present list of isolations.

RESULTS AND DISCUSSION

Ten species of Saprolegniaceae were recorded from two ponds (Table 1) and sixteen species were isolated from the soil samples (Table 2) at Padrauna. An examination of Tables 1 and 2 indicate that one or the other fungus was recorded every month from water as well as soil samples, although considerable variation in the number of species was apparent. A distinct high or low degree of fluctuation in 'Quadrat totals' (the sum of the counts for all species recorded from a given quadrat area) was indicated (Table 2)

TABLE 1 *Seasonal occurrence of ten species of Saprolegniaceae (considering two water bodies WB1 and WB2 at Padrauna together) between April, 1999 to March, 2002*

Species isolated	*Apr*	*May*	*Jun*	*Jul*	*Aug*	*Sep*	*Oct*	*Nov*	*Dec*	*Jan*	*Feb*	*Mar*
Achlya orion	+	+	+	+	+	+	+	+	+	+	+	–
A. diffusa	+	–	–	–	–	+	–	+	+	+	–	+
A. prolifera	+	–	–	–	–	–	+	+	+	+	+	+
A. flagellata	–	–	+	–	+	+	+	+	+	+	+	–
A. proliferoides	-	+	-	+	+	+	+	+	+	+	+	+
Dictyuchus monosporus	–	–	–	–	–	–	–	+	+	+	–	–
Saprolegnia sp.	–	–	–	+	+	–	–	–	+	+	–	–
S. ferax	–	–	–	–	–	–	–	+	+	+	+	–
S. diclina	–	–	–	–	–	–	–	–	+	+	+	–
Aphanomyces *laevis*	+	+	+	+	+	+	+	+	+	+	+	+
Number of species												
Isolated per month	4	3	3	4	5	5	5	8	10	10	7	4

– = absence of the species
+ = presence of the species

from mouth to mouth for the soil-sampling site. The data obtained from the two waterbodies exhibit, in general, a common trend in fluctauation of the number of species and there is only one 'peak' during December-January (Temp. 18-20°C), but, the data from the soil samples indicate two 'peaks' of occurrence- one during July-August (soil temp. 26-28°C) and another more pronounced during December-January (soil temp. 18-20°C) for the fluctuation in the number of species as well as for fluctuation in 'Qudrat totals' and 'troughs' during May-June (soil temp. 30-32°C) and September-October (soil temp. 24-26°C).

Forbes (1935), Waterhouse (1942) and Srivastava (1967 b) have shown that maximum yield of watermoulds from water samples was obtained during December-February (one peak of occurrence) and minimum in May-June. The results of the present study, obtained from the water bodies, very well agree with their findings. However, isolations from soil samples during the present study have indicated two 'peaks' of occurrence, similar to the observations of Dick and Newby (1961) and Prabhuji (1984 a).

Difference of Phenology in Water and Soil

Oomycetes have a definite affinity with water. Dick (1968), discussing the distribution of biflagellate members (aquatic oomycetes), has shown that

TABLE 2 *Number of isolations of species of Saprolegniaceae from soil samples in the environs of a water body in Padrauna between April, 1999 to March, 2002*

Species isolated	Apr	May	Jun	Jul	Aug	Sep	Oct	Nov	Dec	Jan	Feb	Mar
Achlya orion	–	–	–	8	12	10	4	5	10	13	–	7
A. diffusa	–	–	–	–	10	–	–	3	5	12	–	–
A. prolifera	–	–	–	–	3	–	–	5	6	10	–	3
A. flagellata	–	–	–	–	–	–	–	–	3	8	5	2
A. proliferoides	–	–	–	–	–	–	–	8	8	10	6	–
Brevilegnia diclina	–	–	–	–	–	–	–	–	2	8	3	–
B. subclavata	–	–	–	–	–	–	–	–	2	7	–	–
B. indica	–	–	–	–	–	–	–	2	4	8	–	–
Dictyuchus monosporus	–	–	–	–	–	–	–	–	2	4	–	–
D. sterile	–	–	–	–	–	–	–	–	6	10	5	3
Aphanomyces *laevis*	10	8	9	7	10	6	7	7	13	20	15	11
A. helicoids	8	8	10	12	11	5	6	8	12	18	16	10
Leptognia sp.	–	–	–	–	–	–	–	–	3	5	2	–
Saprolegnia ferax	–	–	–	–	–	–	4	4	6	8	–	–
S. diclina	–	–	–	–	–	–	3	4	3	5	–	–
Number of species	2	2	2	3	5	3	5	9	16	16	8	6
Ouadrat totals	18	16	19	27	46	21	24	46	88	154	54	36
Species totals for eccentric forms	18	16	19	27	46	21	17	38	76	136	52	36
Species totals for centric/sub-centric forms	–	–	–	–	–	–	7	8	12	18	2	–

– = absence of the species

these fungi could be affected by water in four ways- the effect of water on germination; requirement of water for vegetative growth influence of water on sporulation and the role of water in propagule dissemination.

It is a known fact, now, that the zoospores (evanescent propagules) provide the major inoculum for the isolation of water-moulds either from water or from soil. However, it is very difficult to identify the type of propagules present during the processing to isolate water-moulds, particularly from the soil. In a similar attempt Warcup (1955) has isolated *Thraustotheca* sp. (a member of Saprolegniaceae) from hyphal fragments present in soil.

The fundamental requirement of well-aerated water for asexual sporulation (formation of zoospores) and zoospore motility is a well established fact. These motile zoospores encyst and grow on the baits and result into isolation of these fungi from water and soil samples. In water samples the zoospores present there (released from the sporangia developed on floating baits or baits on the margin of the waterbody), easily encyst on the baits, *in vitro*. However, for soil samples the situation is a bit different. In this context I would like to draw attention towards the isolation procedure adopted for soil samples. The samples are first thoroughly mixed with water to prepare a dispersion liquid (mixture of soil and water) and soil particles etc. are, then, allowed to settle down. Afterwards, the baits (boiled hempseed cotyledons, parched rice grains, dead insects etc.) are floated over it. The resting propagules (gemmae, oospores etc.), present in the soil, germinate within 2-4 hrs duration on getting adequate moisture and the optimal temperature *in vitro*, which results into development of zoospores that, in turn, encyst on the baits. This contention has the support of Dick (1966) who has concluded that the resting propagules are initially responsible for almost all isolations of water-moulds from soil.

Now, I would like to discuss two completely different situations of isolation of water-moulds, i.e., from the waterbodies and from soil. Water bodies are supposed to be of two types- one, it may be large enough to have flowing water and the second, where it may be smaller to be in stagnant condition. In tropic regions the optimal temp. is reached only during December- January when asexual sporulation is at its peak resulting into abundance of zoospores. It causes maximum numbers of isolations (types and frequency) during this period. Later, following the high temp. period in May-June, a slightly low temp. condition is arrived in July-August, however, the temp. is not optimal like December-January. Therefore, only a few members (only constant species and moderate to high temp. species) sporulate and result into a few (not significant) isolations only. Furthermore, a high speed of water-flow in large waterbodies during rainy season and a high bacterial density in stagnant waterbodies, too, become the limiting factors. And thus, only one significant 'peak' (during December-January) develops.

In soil, the first situation (isolations during December-January) is the same as in waterbodies, but, in monsoon season (during July- August) it is not like that. The soil contains a variety of oomycetes in the form of resting propagules – the oospores. Dick (1968) has considered oospores (contained within oogonia) as non-dispersed, semi-permanent source of evanescent propagules- the zoospore and called it 'a zoospore bank' in soil. During rains the mature resting oospores get the germination-stimulus of rainwash and such temporary flushing would favour zoospore production and its release. Under soil environment the zoospore dispersion over less than a 5mm radius would be enough to provide an acceptable inoculum density (Dick, 1968). It, therefore, causes a sudden surge of zoospore production from the resting propagules (oospores) of a variety of

Oomycetes present within the soil and results into significant isolations from soil during July-August forming a second 'peak' of occurrence, however, less prominent than that of December-January. Being the thick-walled resting structures in soil, the oospores would not be affected by those limiting factors, which normally affect the thin-walled motile zoospores and the vegetative structures in waterbodies.

Oospore Type and Distribution

It is observed that the members of Saprolegniaceae with eccentric oospores and with centric or sub-centric oospores exhibit variability in occurrence in the soil. The eccentric forms of Saprolegniaceae occurred throughout the yaer and have wide range of fluctuations in 'species totals' (the sum of records for species for given quadrat areas), while centric and sub-centric forms have a restricted occurrence, i.e., only during September-March and have least fluctuation in 'species totals'. From tables 1 and 2 it is evident that the most common species is *Aphanomyces laevis* and *Dictyuchus monosporus* and *D. sterile* are the least represented forms.

It is also evident from tables 1 and 2 that out of ten (isolated from water samples) and sixteen (isolated from soil) members of Saprolegniaceae, seven and thirteen species had eccentric oospores respectively. This clearly shows the dominance of eccentric forms over centric and sub-centric forms during the present study.

Prabhuji (1984a) has raised a strong point about the dominance of the members of Saprolegniaceae with eccentric oospores in tropics. A perusal of the previous literature, particularly for the occurrence of eccentric and centric forms of Saprolegniaceae, reveals a very interesting and characteristic pattern of global distribution of Saprolegniaceae. Studies made in the Indian subcontinent (Srivastava, 1967a, b; and Prabhuji, 1984a) and Brazil (Milanez, 1967, 1968, 1970 and Milanez & Beneke, 1968), a part of tropical region of the globe, have shown a distinct dominance of eccentric forms over centric and sub- centric forms, whereas the investigations conducted in United Kingdom (Dick and Newby, 1961; Dick, 1966), Canada (Dick, 1971), Iceland (Howard et al., 1970), United States of America (Miller and Ristanovic, 1969) and Denmark (Lund, 1978), which represent the temperate region of the world, have indicated dominance of centric and sub-centric forms over eccentric forms of Saprolegniaceae. Based on such studies, a distinct eccentric and centric (including sub-centric forms) " Occurrence Zones" of the World may be identified for the members of Saprolegniaceae. It is hoped that further studies in the rest of the areas of the tropical and temperate regions may yield favourable results, as visualized, to confirm the afore-said contention.

There are some studies (Khulbe and Bhargava, 1977; Mer et al., 1980 and Khulbe, 1980), made in the hills of Nainital (India) about 1300-1700m above sea level, which have shown the dominance of centric and sub-centric forms of Saprolegniaceae over eccentric forms. By virtue of being in 'eccentric zone' of the World the investigations done in Indian subcontinent are supposed to exhibit the dominance of eccentric forms, and therefore, such observations raise a question on the validity of the present contention. However, if we consider climatic changes in accordance with the change in altitude, we may inunciate that at an altitude of 1,000-3,000 in above sea level the conditions would be identical to the temperate region, irrespective of the 'zone' in which it is situated. Thus, the observations in the 'centric zone' should follow at an altitude range of 1300-1700 m above sea level. This clearly indicates towards another curious pattern "the altitude-latitude relationship" in the distribution of Saprolegniaceae in relation to their oospore type. If such a relationship is correct, the present contention is very well valid.

Acknowledgements

Thanks are due to Dr. S. B. Singh, Principal, U.N. Postgraduate College, Padrauna (Dist. Kushinagar) for the facilities to work and encouragements.

References

Agardh, C.A. 1824. *Systema algarurn*, vol.1. Lund, Sweden.

Alabi, R.O. 1971. Seasonal periodicity of Saprolegniaceae at Ibadan, Nigeria. *Trans. Brit. Mycol. Soc.* 56: 337-341.

Apinis, A. 1930. Untersuchungen uber die in Lettland gefundenen Saprolegniaceen nebstr Bemmerkungen uber andere Wasserpilze, *Acta Horti. Bot. Univ. Latv.*, 4: 201-246.

Bennett, J. 1841. Parasitic fungi growing on living animals, *Aunals of Nat.Hist.*, 9: 66-67.

Berkeley, M.J. 1833. *Gleanings of British algae*, London.

Bessey, E.A. 1950. *Morphology and Taxonomy of Fungi*, Blakiston.

Carus, C.G. 1823. Beitrag zur Geschichte der unter Wasser an verwesenden Thierkorpern sich erzeugenden Schimmel-oder Algen Gattungen, *Nova Acta Acad* Bd.X, p.493.

Chandhuri, H. and Kochhar, P.L. 1935. Indian water-moulds. I. *Proc. Ind. Acad. Sci.*, Sect. B, 2:137-154.

Chandhuri, H. and Lotus, S.S. 1936. Indian water-moulds. II. *Proc. Ind. Acad. Sci.*, Sect. B, 2:137-154.

Chaudhuri, H., Kochhar, P.L., Lotus, S.S., Banerjee, M.L. and Khan, A.K. 1947. *A hand book of Indian water molds.* Punjab Univ. Bot. Publ., Lahore.

Coker, W.C. 1923. *The Saprolegniaceae, with notes on other water moulds,* Univ. North Carolina Press, Chapel Hill (U.S.A.), pp.201.

Copeland, H.F. 1956. *The classification of lower Organisms.* Palo Alto, California.

Cornu, M. 1872. Monographie des Saprolegniees: etude pysiologique et Systematique. *Ann.Sci.Nat.Bot.,* Ser.V, 15: 1-198.

Dayal, R and Thakurji. 1966. The occurrence and distribution of aquatic fungi in certain ponds of Varanasi. *Hydrobiol.,* 27:548-558.

Dayal, R and Thakurji. 1968. Studies on aquatic fungi of Varanasi. V. a taxonomic study. *Proc. Nat. Acad. Sci., India.* 38B: 32-38.

Dayal, R. and Tandon, R.N. 1962. Ecological studies of some aquatic Phycomycetes, *Hydrobiol.,* 20: 121-127.

De Bary, A. 1852. Beitrag zur Kenntniss der *Achlya prolifera,* Bot. Zeitung., Bd.X, p.473.

Dick, M.W. 1966. The Saprolegniaceae of the environs of Blelham Tran: sampling techniques and the estimation of propagule numbers. *J. gen Microbiol.,* 42: 257-282.

Dick, M.W. 1971. Oospore structure in *Aphanomyas Mycologia*- 63: 686-688.

Dick, M.W. and Newby, H.V. 1961. The occurrence and distribution of Saprolegniaceae, in certain soils of South-east England. I. Occurrence, *J. Ecol.* 49: 403-419.

Dick, M.W. 1968. Considerations of the role of water on the taxonomy and ecology of the filamentous biflagellate fungi in littoral zones. *Veroff. Des Inst. Fur Meeresf. Bremerhaven,* 3: 27-38.

Dick, M.W. 1969. Morphology and taxonomy of the Oomycetes, with special reference to Saprolegniaceae, Leptomitaceae and Pythiaceae. I. Sexual reproduction. *New Phytol.,* 68: 751-775.

Dillwyn, L.W. 1809. *British Confervae.* London.

Fischer, A. 1892. DiePilze Deutschlands, Oesterreichs und der Schweiz. Phycomycetes. In *Rabenhorst. Kryptogamen-Fl.,* 1 (4): 1- 505.

Flora Danica. 1780. Fasc. 14, Tom. V. *Havniae.*

Forbes, E.J. 1935. Water molds of the Manchester district. *Mem. Manchz. Lit. Phil. Soc.* 79: 1-11.

Gruithuisen, F.von P. 1821. Die Branchienschnecke und eine aus ihren Ueberresten hervorwschsende lebendig gebahrende Conferve. *Nova Acta Acad.,* Bd.X, p.437.

Hamid, A. 1942. Indian Water-moulds. III. *Proc. Ind. Acad. Sci.,* Sect. B, 15: 206-215.

Harvey, J.V. 1925. A study of the water molds and pythiums occurring in the soils of Chapelk Hill. *J. Elisha Mitchell Sci.* Soc. 41: 151-164.

Hoffmann, H. 1867. Ueber *Saprolegnia* und *Mucor, Bot. Zeitung,* Bd.XXV, p.345.

Howard, K.L., Seymour, R. and Johanson, Jr., T.W. 1970. Aquatic fungi of Iceland: Saprolegniaceae. *Jour. Elisha Mitchell Sci. Soc.,* 86: 63-79.

Hughes, G.C. 1962. Seasonal periodicity of the Saprolegniaceae. In the South-eastern United States. *Trans. Brit. mycol. Soc.,* 45: 519-531.

Humphrey, J.E. 1892. The Saprolegniaceae of the United States, with notes on other species, *Trans. Amer. Phil. Soc.*, Vol. 17: 63-148.

Ito, S. 1936. *Mycological flora of Japan* I. Phycomycetes, Tokyo, pp. 340.

Johnson, T.W., Jr. 1956. *The genus Achlya: Morphology and Taxonomy,* Univ. of Michigan Press, Ann Arbor, U.S.A.

Kauffman, C.H. 1921. *Isoachlya*, a new genus of the Saprolegniaceae, *Amer.J.Bot.*, 8: 231-237.

Khulbe, R.D. 1980. Occurrence of water molds is some lakes of Nainital, Kumaon Hills, India. *Hydrobiol.* 74:77-80.

Khulbe, R.D. and Bhargava, K.S. 1977. Distribution and seasonal periodicity of water molds in some lakes in Nainital Hills, India. *Hydrobiol.*, 54: 67-72.

Ledermiiller. 1760. Microscopic Augen. und Gemuths-Ergotzungen.

Lund, A. 1934. Studies on Danish freshwater Phycomycetes, and notes on their occurrence, particularly relative to the hydrogen-ion concentration of the water, *Kgl. Danske Vidensk. Selsk. Skrift.*, Naturv. Math., Afd. IX, 6 (1): 1-97.

Lund, A. 1978. Occurrence of Saprolegniaceae in Danish soils, *Nova Hedwigia*, 29: 577-592.

Lyngbye, H.C. 1819. Tentamen Hydrophytologiae Danicae, *Havuiae*. P. 248.

Manton, I. 1952. The fine structure of plant Cilia. *Symp. Soc. exp. Biol.*, 6: 306.

Manton, I. 1965. Some phyletic implications of flagellar structure in plants. In: *Advances in Botanical Research,* Vol. II. (Ed. R.D. Preston). London & New York.

Mer, G.S., Sati, S.C. and Khulbe, R.D. 1980. Occurrence, distribution and seasonal periodicity of some aquatic fungi of Sat-Tal (Nainital), India. *Hydrobiol.*, 76: 201-205.

Milanez, A.I. 1967. *A study of the Aquatic Phycomycetes, from the Gull Lake area in Michigan*. Ph.D. Thesis, Michigan State University.

Milanez, A.I. 1968. Aquatic fungi of the "Cerrado" region of Sao Paulo State. I. Frist Results. *Rickia*, 3: 97-109.

Milanez, A.I. 1970. Contribution to the knowledge of aquatic Phycomycetes of Sao Paulo State. I. Oomycetes from the West region. *Rickia*, 5: 23-43.

Milanez, A.I. and Beneke, E.S. 1968. New records of aquatic Phycomycetes from Michigan. *Pap. Mich. Acad. Sci.* Ann Arbor, 53: 11-22.

Miller, C.E. and Ristanovic, B. 1969. Studies on Saprolegniaceous filamentous fungi. *Ohio Jour. of Science,* 69: 105-109.

Naegeli, C. 1847. Zeitschrift fiir wiss., Botanik, Hefte 1, 3-4.

Naumov, N.A. 1954. *Flora gribov Leningradskoi oblasti, vypusk. I. Arikhimitsety Ficomitsety,* Acad. of Sci., USSR.

Nees von Esenbeck, C.G. 1823. Zusatz., (Appended to Carus, 1823, q.v.,) p.507.

Novaes-Ledieu, M., Jimenez-Martinez, A. and Villanueva, J.R. 1967. Chemical composition of hyphal wall of Phycomycetes. *J. gen. Microbiol.*, 47: 237.

Perrott, P.E. 1960. The ecology of some aquatic Phycomycetes, *Trans. Brit. mycol. Soc.*, 46: 213-224.

Petersen, H.E. 1909. Studier over Ferskvands Phycomyceter. Bidrag til Kundskaben om de submerse Phykomyceters Biologi og Systematik, samt om deres Udbredelse I Danmark. *Bot. Tidsskrift*, 29: 345-440.

Petersen, H.E. 1910. An account of Danish freshwater-Phycomycetes with biological and systematical remarks. *Aun. Myc.*, 8: 494-560 (same as paper above, in English).

Prabhuji, S.K. 1984a. Studies on some water moulds occurring in certain soils of Gorakhpur. *J. Indian bot. Soc.*, 63: 387-396.

Prabhuji, S.K. 1984b. Distribution of water moulds within quadrats in the soils of Gorakhpur. *Proc. Nat. Acad. Sci. India*, 54 (B): 21-32.

Prabhuji, S.K. and Sinha, S.K. 1993. Two new water moulds from India: *Allomyces recurvus* sp. nov. and *Brevilegnia indica* sp.nov., *Neo Botanica*, 1 (1& 2): 31-38.

Prabhuji, S.K. and Sinha, S.K. 1994. *Pythium multisporum* Poitras: morphology and phenology in the soils of Padrauna, India. *J. Living World*, 1 (2): 126-133.

Prabhuji, S.K. and Srivastava, G.C. 1977. Some members of Saprolegniaceae occurring in the soils of Gorakhpur. *Geobios*, 4: 154-155.

Prabhuji, S.K. and Srivastava, G.C. 1978. Addition to lower fungi of India. *Geobios*, 5: 137-138.

Prabhuji, S.K., Singh, S.B. and Sinha, S.K. 1996. Distribution of some aquatic fungi in the soils of Padrauna in relation to pH, moisture and organic contents. Proc. Nat. Symp. Plant Disease Management, Gorakhpur, India.

Prabhuji, S.K., Sinha, S.K. and Singh, S.B. 2001. Genus: Achlya (Saprolegniaceae) in the soils of Padrauna (U.P.) *Proc. Internat. Symp. on Frontiers of Fungal Diversity and Diseases in South-east Asia*, Gorakhpur.

Prabhuji, S.K., Srivastava, G.C. and Sinha, S.K. 1984. Aquatic fungi of India: IV. *Brevilegnia* Coker & Couch. Kavaka, 12 (2): 99-106.

Pringsheim, N. 1851. Entwickelungsgeschichte der *Achlya prolifera*, *Nova Acta Acad.*, Bd.XXIII, p.395.

Raper, J.P. 1936. Heterothallism and sterility in *Achyla* and observations on the cytology of *Achyla bisexualis*, *Journ. Elisha Mitchell. Sci. Soc.*, 52: 274-289.

Roberts, R.E. 1963. A study of the distribution of certain members of the Saprolegniales. *Trans. Brit. Mycol.* Soc., 46: 213-224.

Rossy-Valdderrama, C. 1956. Some water molds from Puerto Rico. *Jour. Elisha Mitchell Sci. Soc.*, 72: 129-137.

Schrank, F.von P. 1789. *Balersche Flora: Miinchen*, 2, p.553.

Scott, W.W. 1961. *A Monograph of the genus Aphanomyces. Virg. Agr. Expt. Stat. Tech. Bull.* No. 151.

Seymour, R.L. 1970. The genus Saprolegnia. *Nova Hedwigia*, 19: 1-122.

Spallanzani. 1777. Opuscules de physique. Geneve.

Srivastava, G.C. 1967 a. some species of Saprolegniaceae collected from Gorakhpur, India. *Hydrobiol.* 30: 281-292.

Srivastava, G.C. 1967 b. Ecological studies on some aquatic fungi of Gorakhpur, India. *Hydrobiol.* 30: 385-404.

Tiesenhausen, M.von. 1912. Beitrage zur Kenntnis der Wasserpilze der Schweiz., *Archiv f. Hydrobiol. und Planktonkunde,* 7: 261-308.

Turian, G. 1962. Cytoplasmic differentiation and dedifferentiation in the fungus *Allomyces. Protoplasma,* 54: 323.

Vogel, H.J. 1960. Two modes of lysine synthesis among lower fungi: evolutionary significance. *Biochim biophys. Acta,* 41: 172.

Warcup, J.H. 1955. Isolation of fungi from hyphae present in soil. *Nature,* London, 175: 953.

Waterhouse, G.M. 1942. Some water molds of the Hogsmill River collected from 1937-1939. *Trans. Brit. Mycol. Soc.* 25: 315-325.

Willoughby, L.G. 1962. The occurrence and distribution of reproductive spores of Saprolegniales in freshwater. *J. Ecol.,* 50: 733-759.

Willoughby, L.G. 1965. Some observations on the location of sites of fungal activity at Blelham Tarn. *Hydrobiol.* 25: 352-356.

Wrisberg. 1765. *Observations de animalcules infusorlis satura.* Gottingae.

Ziegler, A.W. 1958. The Saprolegniaceae of Florida. *Mycologia,* 50: 693-696.

Frontiers in Plant Sciences (Pages 143-162)
Eds. K.G. Mukerji, K.V.B.R. Tilak, S.M. Reddy, L.V. Gangwane, P. Prakash, I.K. Kunwar

Associative Effects of Beneficial Rhizosphere Microorganisms for Sustainable Agriculture

K.V.B.R. Tilak* and N. Ranganayaki

ABSTRACT

Soil microorganisms are very important in the biogeochemical cycles of both inorganic and organic nutrients in the soil and the maintenance of soil health and quality. It is, therefore, imperative to enhance the efficiency of external inputs by employing the best and suitable combination of beneficial microorganisms for sustainable agricultural production. Soil - plant –microbe interactions are complex and there are many ways in which the outcome can influence the plant health and productivity. The interaction may be harmful, beneficial and neutral. The present review focuses on the interaction of beneficial microorganisms in the rhizosphere for sustainable crop production.
Key words: Beneficial soil microbes, Rhizosphere, Soil health, Sustainable crop Productivity

INTRODUCTION

The vast majority of plant associated soil organisms inhabit the rhizosphere, defined as the zone around roots in which microbial growth is stimulated by the release of nutrients. Within the rhizosphere, there is a continuous interaction between plant roots and the microorganisms. These interactions can have an important influence on plant growth.

*Author's address: Department of Botany, Osmania University, Hyderabad-500 007, A.P., India
E-mail: tilakkvbr@yahoo.com

Although rhizosphere appears to be too complex to allow its manipulation, specific microorganisms (bacteria or fungi or actinomycetes) can be applied to seed or soil, which cause an alteration in the composition of the rhizosphere. Such manipulations have important and exciting implications. In addition to the manipulation of the microflora to discourage phytopathogens, it should be possible to promote the activity of beneficial ones, such as mycorrhizal fungi, nitrogen fixers, phosphate solubilizers, actinorhizae and several plant growth promoting microorganisms. Some recent investigations have brought to light instances where biological activities are markedly enhanced in two or three membered microbial associations (Saxena and Tilak, 1998, Gupta *et al.*, 1998, Saxena *et al.*, 2002; Dey *et al.*, 2004). Such syntropic associations are of ecological importance with implied agricultural significance.

Interaction among beneficial soil microorganisms has been reviewed earlier (Saxena and Tilak, 1994). In the present review, an attempt has been made to discuss interactions among beneficial rhizosphere microorganisms, which are of significance for soil health and crop yield.

Associative Effects Among Nitrogen Fixers

The beneficial effects of single inoculation of *Azotobacter, Azospirillum, Gluconoacetobacter diazotropicus (syn: Acetobacter diazotrophicus), Azoarcus, Rhizobium* and many other nitrogen fixers has been well documented and reviewed. (Paula *et al.*, 1992; Reinhold-Hurek and Hurek, 1998; Okon *et al.*, 1998; Saxena and Tilak, 1998; Bashan, 1999; Lata et a l., 2000; Tilak and Saxena, 2000; Rao and Pal, 2003). In recent years combined ability of these nitrogen fixers has been tested for enhanced growth and yield. The crop response to *Azospirillum* inoculation when applied in combination with other diazotrophs on the yield and nitrogen gains is well documented (Baltensperger *et al.*, 1978; Kapulinik *et al.*, 1987). Dual inoculation with *Rhizobium* and *Azospirillum* increased the nodulation and yield of clover, lucerne, chickpea and pigeonpea over rhizobial inoculation alone under field conditions (Tilak *et al.*, 1979; Kundu, 1988). Positive effects of combined inoculation with *Azospirillum* and *Rhizobium* have been reviewed earlier by Saxena and Tilak (1994). However, Plazinski and Rolfe (1985) reported inhibitory effect of such association on nodulation of different legumes, although there was stimulation in growth of plants.

Itzigsohn *et al* (1993) demonstrated with Fahraeus slides that *Azospirillum* increases the number of root hairs and their diameter although no increase in total number of infection threads was observed. Okon *et al.* (1998) concluded that *Azospirillum* exerts its effects through host plant, and not through direct interaction with *Rhizobium.*

In *Phaseolus vulgaris,* Inoculation with *Rhizobium etli* TAL 182 and *R.tropici* CIAT 899 increased seed yield (13%), and combined inoculation with *Rhizobium* and *Azospirillum* resulted in further increase (23%), whereas plants inoculated with *Azospirillum* alone did not differ in yield from uninoculated controls, despite a relative increase in shoot dry biomass (Burdman *et al.,* 1996). In experiments conducted with hydroponic solution, *Azospirillum* caused an increase in secretion of nod gene-inducing flavonoids, as observed by nod gene—inducing assays of root exudates fractionated by HPLC (Burdman *et al.,* 1996; Volpin *et al.,* 1996).

It was assumed that beneficial effect of dual inoculation of *Azospirillum brasilense* and *Bradyrhizobium* in soybean was attributed to increased number of effective nodules (Groppa *et al.,* 1998), greater proportion of nitrogen derived from fixation (Galal, 1997) or production of indole acetic acid by *Azospirillum* (Molla *et al.,* 2001). Steinberg *et al.* (1989) reported that co-inoculation of *Azospirillum lipoferum* and *Bradyrhizobium japonicum* (both are denitrifiers under saprophytic conditions) into soil produced higher denitrifying enzyme potential than inoculation with single strain although *B. japonicum* did not affect the *A. lipoferum* growth rate.

Seed inoculation with *Azospirillum brasilense* strain 4 and *Rhizobium* species specific to chickpea (*Cicer arietinum*) and cowpea (*Vigna unguiculata*) increased the dry matter production of shoots and N uptake of various minor millets (Agarwal and Tilak, 1989).

Beneficial effects of *Azospirillum* and *Azotobacter* coinoculation have been reported in rainfed crops like maize and sorghum (Tilak *et al.,* 1982), pearlmillet (Wani *et al,* 1985; Kharb and Kundu, 1986, Tilak and Subba Rao, 1986) and barley (Negi *et al.,* 1987). Combined inoculation of *Azospirillum lipoferum* and *Azotobacter chroococcum* was better than *A. lipoferum* alone in the presence 60 kg N ha^{-1} level (Subbian and Chamy, 1984).

Associative effect of combined inoculation *Azotobacter* and *Rhizobium* has been reported in several legumes (Singh *et al.,* 1981; El-Baharawy, 1983; Kothari and Saraf, 1986; Ram *et al.,* 1988; El-Mokadem *et al.,* 1989 El-Gamai, 1992).

Associative Effects of Phosphate-solubilizing and Nitrogen-fixing Microorganisms

Though, a majority of phosphate solubilizing microorganisms (PSM) has been isolated from the rhizosphere of leguminous and non-leguminous crops, a few have also been reported from non-rhizosphere soils. Some are known to occur in soils of rock phosphate deposit areas, marine environment and phyllosphere of crop plants. The presence of root exudates which contain carbohydrates, protein, amino acids and certain growth promoting substances in the rhizosphere zone may be responsible for harbouring majority of PSM in their vicinity.

In vitro solubilization studies led to the investigation of PSM as P releaser from native soil P pool as well as increasing the availability of phosphorus from added chemical fertilizers. PSM's like species of *Aspergillus, Bacillus, Escherichia, Arthrobacter* and *Pseudomonas* are used as inoculants for various crops in our country (Gaur, 1990; Gaind *et al.* 2000).

The synergistic effects of PSM and N_2 fixers on the growth and yield of crops have been reported by several workers (Smith *et al.*, 1961; Kundu and Gaur, 1984; Alagawadi and Gaur, 1988). Co-inoculation studies using *Rhizobium* and phosphate solubilizing bacteria (PSB) have been carried out by many workers (Dudeja *et al.*, 1981; Kucey, 1987; Downey and van Kessel, 1990).

Soybean seed inoculated with *Bradyrhizobium japonicum* grown in a plot inoculated with 4 kg ha^{-1} PSB showed significantly higher nodulation and yield (Chandra *et al.*, 1995). Increased nodulation, yield attributes, seed index and seed yield have also been reported due to combined inoculation of *Pseudomonas striata* and *B. japonicum* (Dubey, 1997; Kumrawat *et al.*, 1997). Combined inoculation of *Rhizobium* and PSM along with 50% N and P doses resulted in increased plant height, number of branches, leaves per plant, number of pods per plant, grains per pod and pod yield in pea as compared to 20 kg N + 80 kg P2O5 and 40 kg K2O ha $^{-1}$ applied through chemical fertilizers (Patel *et al.*, 1993). A significant increase in nitrogenase activity, growth and grain yield of pea due to dual inoculation of *Rhizobium leguminosarum* and PSB has been reported by Srivastava *et al.* (1998). In field trials, Jeyabal *et al.* (1992) observed that *B. japonicum* and PSB inoculation resulted in increased seed yield by 37% under varying soil-textured conditions. In a long-term experiment on inoculated soybeans grown in P-deficient soils in Madya Pradesh, Dubey (1998) found that combined use of *Bradyrhizobium* and PSB increased soybean yield from 0.95 t/ha in control to 1.21 t/ha.

Inoculation of a mixed cultures of rhizobia (Tt 9 or TNAU 14) along with *Azospirillum brasilense* strain SP7 and PSB (Bacillus *megaterium var. phosphaticum*) PB1 in groundnut cv. CO1, produced maximum nodulation, plant nutrient content and yield compared to individual inoculation or dual inoculation of *Rhizobium* with *Azospirillum* or phosphobacteria (Balamurugan and Gunasekaran, 1996).

Associative Effects of Arbuscular Mycorrhizae (AM) with Nitrogen Fixers and Phosphate Solubilizing Microorganisms

The tripartite symbiosis between legumes, *Rhizobium* and AM fungi has been the subject of intensive research (Tilak, 1993; Saxena and Tilak, 1994;

Srinivasa *et al.*, 1995; Ibijbijen, *et al.*, 1996; Saxena *et al.*, 1997; Zhao *et al.*, 1997; Olesniewicz and Thomas, 1999; Saxena *et al.*, 2002) Generally AM fungi improve phosphate nutrition of legume, which in turn enhances plant growth and nitrogen fixation. The effects of AM fungi cannot be duplicated by increasing P availability to non-AM plants.

Associative effects of AM with rhizobia have been reported for legumes like soybean (Xie *et al*, 1995; Srinivasa *et al.*, 1995; Soliman *et al.*, 1996; Cobera and Hernandez, 1997; Zhao *et al.*, 1997); clovers (Lin *et al.*, 1993; Moawad *et al.*, 1996); chickpea (Subba Rao *et al.*, 1986; Singh and Tilak, 1989; Chaturvedi and Kumar, 1991, Rawat *et al.*, 1991); faba beans (Ishac *et al.*, 1994); Medicago (Azcon *et al.*, 1991); greengram (Baruah *et al.*, 1996, Saxena *et al.*, 1997; Das *et al.*, 1997); cowpea (Sodi *et al.*, 1994; Ahiabor and Hirata, 1994); groundnut (Shasidhara *et al.*, 1994; Mandhare *et al.*, 1995); pigeonpea (Costa *et al.*, 1994; Ahiabor and Hirata, 1994; Olsen and Habte, 1995; Singh, 1996) and legumes like *Anthyllis cystisoides* (Requena *et al.*, 1997; *Leucaena leucocephala* (Mallesh and Bagyaraj, 1990; Dixon., 1992); *Albizzia procera* (Muhadiono *et al.*, 1990); *Sesbania* (Sengupta and Choudhuri, 1995; Tilak, 2004) and *Prosopis* (Dixon ,1992).

Specific compatibility between rhizobia and AM fungi can exist in terms of competitive ability of introduced rhizobia (Thiagarajan*et al.*, 1992; Saxena *et al.*, 1997) and root colonization by introduced AM fungi (Saxena *et al.*, 1997). AM colonized roots exhibit a different physiology and alter root exudation (Schwab *et al.*, 1983). Seasonal response of groundnut genotypes to AM fungus, *Glomus fasciculatum*, and *Bradyrhizobium* inoculation was studied by Ibrahim *et al.* (1995). They have proposed a basis for explaining yield sustenance by dual inoculation with both endosymbionts by significant correlations between mycorrhizal colonization and root phosphorus content and of yield with them in addition to the desirable correlation between CO_2 exchange rate, biomass and root P. The results provide ample reason for integrating AM schedules into breeding and agronomic approaches in programmes on stabilizing groundnut yields (Ibrahim et al., 1995). Dual inoculation of *Glomus fasciculatum* along with rhizobia significantly improved nodulation, nitrogen fixation, nutrient content in shoot and yield of soybean (Kumutha and Santhanakrishnan, 1994). Rhizobial inoculated soybean responds to AM inoculation due to increased indole acetic acid and abscisic acid contents in plants and nodules (Murakami *et al.*, 1991), increased size and activity of nodules along with higher photosynthetic efficiency in leaves (Pacovsky, 1986) and increased resistance to infection by the pathogen, *Pseudomonas syringae* (Shalaby and Hanna, 2000).

Interaction between AM fungi and *Azospirillum* has been reported for crops like wheat (Panwar, 1991); barley (Subba Rao *et al.*, 1985, Negi et al., 190); maize (Barea *et al.*, 1983); sorghum (Pacovsky *et al.*, 1985; Sreekumar and Tilak, 1988; Subba Rao *et al*, 1989; pearl millet (Subba Rao, 2000); rye grass (Barea *et al.*, 1983); sweet potato (Kandaswamy *et al.*, 1988; Rasco *et al.*, 1992); jute (Bali and Mukerji, 1991) and coffee (Kumari and Balasubramanian, 1993). Specific compatibility between AM fungi and *Azospirillum* in barley has been reported by Subba Rao *et al.* (1985).

Occurrence of nitrogen-fixing azospirilla within the surface sterilized spores of various AM fungi has been reported by Tilak *et al.* (1986). Also, the enhancement of AM spore germination with cell-free extracts of different nitrogen-fixing bacteria has also reported (Tilak *et al.*, 1990).

Associative effects of *Azotobacter* and AM fungi are more prominent in soils of low fertility than inoculation with individual organisms (Azcon *et al.*, 1978; Bagyaraj and Menge, 1978; Brown and Carr, 1984; Ramachandra and Rai, 1987; Mohandas, 1987; Ho, 1988; Kshirsagar *et al.*, 1994). Gerdemann and Trappe (1974) reported *Azotobacter* cells encrustated on spores of AM fungi.

Synergistic effects of AM fungi and *Glucanoacetobacter* (syn. *Acetobacter*) *diazotrophicus*, a saccharophilic diazotroph associated with sugar/starch rich plants, on sugarcane and sweet potato (Paula *et al.*, 1991, 1992), sweet sorghum (Paula *et al.*, 1991; Isopi *et al.*, 1995) have been reported. AM fungal hypae may promote the penetration and spread of *G. diazotrophicus* in cortical regions in mycorrhizal plants (Paula *et al.*, 1991).

Beneficial responses due to interaction between PSM and AM fungi have been obtained in soils with low concentrations of calcium. Nevertheless, Azcon-Aguilar *et al.*. (1986) observed that PSM did not improve the utilization by mycorrhizal plants of a labeled source of insoluble phosphate added to the soil (pH 7.4). Perhaps, the high concentration of exchangeable calcium in the soil precluded any potential bacterial solubilization of labeled phosphate sources. Singh annd Kapoor (1998) reported increase in straw and grain yield of mungbean due to PSM and *Glomus fasciculatum*. PSM population was more in mycorrhizal plants. Similarly, Ghosh and Poi (1998) reported improved nodulation, plant growth, P uptake and PSM population due to combined inoculation of *Bacillus polymyxa* and *G. fasciculatum* in soybean, groundnut, mungbean and lentil. Singh and Kapoor (1998) observed improved root colonization by AM fungi in wheat. Higher grain and straw yield and nitrogen and phosphorus uptake were recorded with combined inoculation of *Bacillus circulans*, *Cladosporium herbarum* and *Glomus* spp with Mussorie rock phosphate amendment. However, Singh (1994) reported that dual inoculation of VAM and *Bradyrhizobium japonicum* or PSM did not cause significant increase in nodulation, plant growth and seed yield of soybean. Tilak *et al.* (1995) reported that dual inoculation with *Pseudomonas striata* and AM fungi (*Glomus fasciculatum* or *G.mosseae*) significantly increased the bean yield, root biomass and total phosphorus uptake by soybean plant over uninoculated control in alluvial soils.

Dual inoculation of *Frankia*, an actinorhizal symbiont, and AM fungi resulted in increased nodule weight and nitrogenase activity of *Alnus acuminata* (Russo, 1989) and higher nutrient content (N,P, micronutrients) in *Casuarina equisetifolia* (Karthikeyan and Raj, 1991). Ravichandran and Balasubramanian (1999) studied the inoculation effects of four *Frankia* strains in combination with *Glomus fasciculatum*. Strains CC13 and ORS 020607 were more compatible with AM fungus than ORS 021001 and 022601. Dual inoculation of *Gigaspora margarita* and *Frankia* was more beneficial and compatible than other AM fungi (Udaiyan and Sugavanam, 1995). In sesame, soil inoculation with AM fungi and seed inoculation with PSM (*Pseudomonas striata, Bacillus polymyxa* and *Aspergillus awamori*) resulted in significant increase in dry biomass and phosphorus uptake by plants (Tandon and Prakash, 1998).

The symbiotic establishment of mycorrhizae on plant roots is influenced by bacteria of the plant rhizosphere. Some of these bacteria promote consistently the colonization and establishment mycorrhiza in plants. This notion has led to the concept of mycorrhization helper bacteria (MHB). The use of such MHB's can improve the effect of mycorrhizal inoculum (Garbaye, 1994).

Associative Effects of Plant Growth Promoting Rhizobacteria (PGPR) with Nitrogen Fixers, Phosphate Solubilizers and AM Fungi

Plant growth promoting rhizobacteria (PGPR) are a group of beneficial bacteria inhabiting the soil ecosystem (Kloepper *et al.*, 1989). The effects of PGPR on plant growth can be mediated by direct or indirect mechanisms (Glick, 1995). The direct effects are attributed to production of phytohormones, solubilization of insoluble phosphates or the supply of biologically fixed nitrogen. The indirect mechanisms are suppression of phytopathogens through the production of siderophores, HCN, ammonia, antibiotics, volatile metabolites, chitinases etc. (Glick, 1995; Glick and Pasternak, 2003). PGPR include bacteria like *Arthrobacter, Bacillus, Burkholderia, Enterobacter, Klebsiella, Pseudomonas, Xanthomonas, Serratia* and several other bacterial genera. The effects of PGPR on crop growth have been evaluated by a number of workers (Malik *et al.*, 1995; Dasti *et al.*, 1997; Sturz *et al.*, 1997; Kozyrovska *et al.*, 1997, Chiarini *et al.*, 1998, Gupta *et al.*, 1998; Tilak *et al.*, 2003; Dey *et al.*, 2004).

Inoculation of PGPR isolates namely *Proteus vulgaris, Klebsiella planticola* and *Bacillus subtilis* increased seed yield of sunflower cv. OSH 1007 from

49-62% over uninoculated controls under field conditions (Malik *et al.*, 1995). Increase in plant paramerters of wheat was reported following inoculation with 12 different strains belonging to *Pseudomonas aeruginosa, P. cepacia, P. fluorescens* and *P.putida* (De Freitas and Germida, 1990). Seed bacterization with fluorescent pseudomonads resulted in yield increase of 27% in field-grown wheat (Weller and Cook, 1986). Three fluorescent pseudomonads (PGPR 1, PGPR 2 and PGPR 4) isolated from groundnut rhizosphere, have been identified as effective biofertilizers for groundnut cultivation under rainfed system (Pal *et al.*, 2000). These strains gave higher pod and haulm yield, nodule number and weight, plant biomass and nutrient (N, P) contents in soil, shoot and kernels under varying soil-agroclimatic conditions of India.

Recent research suggests that PGPR's in combination are more effective than single inoculations. Auxin producing *Bacillus* strains promoted root growth and nodulation when coinoculated with *Rhizobium etli* on *Phaseolus vulgaris* cv. Contender in growth chambers (Srinivasan *et al.*, 1996). Likewise, Coinoculation of soybean with *Bradyrhizobium japonicum* and *Serratia liquefaciens* 2-68 or *S. proteamaclans* 1-102 increased grain and protein yield of soybean over uninoculated control (Dashti *et al.*, 1997). Gupta *et al.*, (1998) observed that *Enterobacter* isolate EG-ER-2 increased the nodule occupancy of *Bradyrhizobium* spp.strain S 24 on mungbean and another isolate KG-ER-1 did the same for rhizobial strain COG 15 Developments in the area of engineering of the rhizosphere that specifically enhance the growth of plant growth promoting bacteria, to improve the sustainability of agriculture has been reviewed by Nautiyal (2000).

While assessing the effect of different PGPR's (*Azotobacter chroococcum, Azospirillum brasilense, Pseudomonas fluorescens, P.putida* and *Bacillus cereus*) on pigeonpea (*Cajanus cajan)* cv. P-921 inoculated with *Rhizobium* sp. (AR-2-2k), Tilak *et al.* (2004) observed that these PGPR's did not antagonize the *Rhizobium*. The Dual inoculation of *P.putida, P.fluorescens* and *B.cereus* registered a significant response on growth, nodulation and nitrogen fixation in pigeonpea over uninoculated control and seed inoculation with *Rhizobium* culture alone. Further, they reported an increase in nodule occupancy of the introduced strain from 50% (with *Rhizobium* alone) to 85% in the presence of *P.putida*. These results suggest that PGPR's affect differentially the *in situ* growth of *Rhizobium* and additionally raise the possibility of altering nodulation and competition among strains through associated root microflora (Tilak *et al.*, 2004).

The interactions of PGPR and AM fungi have been recently reviewed by Dey *et al.* (2004). Some rely on interactions with potential phytopathogens such that 'bioprotection' results (Azcon-aguilar and Barea,1996) while others produce compounds that can directly stimulate plant growth (Barea, 1997; 2000). Microbial populations in the Rhizosphere can either interfere with or benefit establishment of AM fungi (Vosatka and Gryndler, 1999). Mycoparasitic relationships (Jeffries, 1997) have been found to interfere with AM development, while many microorganisms can stimulate AM formation and /or functioning (Gryndler, 2000; Barea *et al.*, 2002). The microbial interactions in the mycorrhizosphere may involve a variety of microorganisms with specific functional capabilities, which may influence plant growth (Linderman, 1992).

CONCLUSIONS

Enhancement of plant growth depends on the combinations of different beneficial rhizosphere microorganisms. The efficient strains should be identified according to the host, soil and environmental conditions and their compatibility and associative effects tested. Selection of suitable combinations to fit the growth situations will increase the chances of exploitation of rhizosphere effect which in turn will help in improved soil health and increased crop productivity.

With the introduction of new techniques such as stable-isotope probing method, fluorescent *in situ* hybridization combined with micro-autoradiography and tracking of labeled substrate uptake, it becomes reality to study the interaction of beneficial microorganisms in the rhizosphere. These studies will have an impact in tailoring the beneficial microorganisms in the rhizosphere for improved environmental quality, microbial diversity and sustainable agriculture.

References

Agarwal, R. and Tilak, K.V.B.R. 1989. *Azospirillum brasilense* and *Rhizobium* sp. (cowpea misc.) inoculum effect on yield and nitrogen uptake of various minor millets. *Indian Jour. Microbiology.* 29: 229-232.

Ahiabor, B.D. and Hirata, H. 1994. Characteristic responses of three tropical legumes to the inoculation of two species of VAM fungi in Andosol soils with different fertilities. *Mycorrhiza.* 5: 63-70.

Alagawadi, A.R. and Gaur, A.C. 1988.Associative effect of *Rhizobium* and phosphate solubilizing bacteria on the yield and nutrient uptake of chickpea. *Plant and Soil.* 105: 241-246.

Azcon,R., Azcon, G., De Aguilar, C. and Barea, J.M. 1978. Effect of plant hormones present in bacterial cultures on the formation and response to VA endomycorrhiza. *New Phytologist.* 80: 334-336.

Azcon, R., Rubio, R. and Barea, J.M. 1991. Selective interactions between different species of mycorrhizal fungi and *Rhizobium meliloti* strains, and their effects on growth, N_2 fixation (^{15}N) and nutrition of *Medicago sativa* L. *New Phytologist.* 117: 399-404.

Azcon-Aguilar, C. and Barea, J.M. 1996. Arbuscular mycorrhizas and biological control of soil-borne plant pathogens: an overview of the mechanisms involved. *Mycorrhiza*. 6: 457-464.

Azcon-Aguilar, C., Diaz-Rodriguez, R.M. and Barea, J.M. 1986. Effect of soil microorganisms on spore germination of vesicular-arbuscular mycorrhizal fungus *Glomus mosseae*. *Trans. Brit. Mycol. Soc.* 86: 337-340.

Bagyaraj, D.J. and Menge, J.A. 1978. Interaction between a VA mycorrhiza and *Azotobacter* and their effects on rhizosphere microflora and plant growth. *New Phytologist*. 80: 567-573.

Balamurugan, S. and Gunasekaran, S. 1996. Effect of combined inoculation of *Rhizobium* sp., *Azospirillum* and phosphobacteria in groundnut. *Madras Journal of Agriculture*. 83: 506-508.

Bali, M. and Mukerji, K.G. 1991. Interaction between VA mycorrhizal fungi and root microflora of jute. In: *Plant Roots and Their Environment*. (Eds. McMichael, B.L. and Persson, H.) pp. 396-401. Elsevier Science Publishers, The Netherlands.

Baltensperger, A.A., Schank, S.C., Smith, R.L., Little, R.C., Bouton, J.H. and Dudeck, A.E. 1978. Effect of inoculation with *Azospirillum* and *Azotobacter* on turf type bermuda genotypes, *Crop Science*. 18: 1043-1045.

Barea, J.M. 1997. Mycorrhiza/bacteria interactions on plant growth promotion. In: *Plant Growth-Promoting Rhizobacteria, Present Status and Future Prospects.* (Eds. Ogoshi *et al.*) pp. 150-158. OECD, Paris.

Barea, J.M. 2000. Rhizosphere and mycorrhiza of field crops. In: *Biological Resource Management: Connecting Science and Policy.* (Eds. Touant *et al.*) pp. 110-125. OECD, INRA Editions and Springer, Berlin, Heidelberg, New York.

Barea, J.M., Bonis, A.F. and Olivares, J. 1983. Interaction between *Azospirillum* and VA mycorrhiza and their effects on growth and nutrition of maize and rye grass. *Soil Biol. and Biochem*. 15: 705-709.

Barea, J.M. Toro, M., Orozco, M.O., Campos, E. and Azcon, R. 2002. The application isotopic (^{32}P and ^{15}N) dilution techniques to evaluate the interactive effect of phosphate solubilizing rhizobacteria, mycorrhizal fungi and *Rhizobium* to improve the agronomic efficiency of rock phosphate for legume crops. *Nutritional Cycling for Agro-ecosystems.* 63: 35-42.

Baruah, R., Barthakur, H.P. and Thakur, A.C. 1996. Effect of indigenous VA-mycorrhizal and rhizobial inoculation on growth and nodulation of greengram (*Vigna radiata* L.) as affected by sources of phosphorus. *Proceedings of Seminar on Problems and Prospects of Agricultural Research and Development in North East India,* pp. 213-219., Assam Agricultural University, Jorhat, India.

Bashan, Y. 1998. Inoculants of plant growth-promoting bacteria for use in agriculture. *Biotechnology Advances*. 16: 729-770.

Brown, M.E. and Carr, G.R. 1984. Interactions between *Azotobacter chroococcum* and VAM and their effect on plant growth. *Journal of Applied Bacteriology*. 56: 429-437.

Burdman, S., Volpin, H. Kigel, J. and Okon, Y. 1996. Promotion of *nod*-gene inducers and nodulation in common bean *(Phaseolus vulgaris)* roots inoculated with *Azospirillum brasilense*. *Appl. Envi. Microbiology*. 62: 3030.

Chandra, K., Mukherjee, P.K., Karmakar, J.B. and Sharma, B.K. 1995. Effect of phosphate solubilizing bacteria on rhizobial symbiosis in soybean at rainfed conditions of Manipur. *Environment and Ecology*. 13: 436-438.

Chaturvedi,C. and Kumar, A. 1991. Nodulation and nitrogenase activity in gram (*Cicer arietinum* L.) as influenced by *Rhizobium* and VA mycorrhiza interaction. *Natl. Acad. Sci. Letter*. 14: 289-292.

Chiarini, L., Bevivino, A., Tabacchioni, S. and Dalmastri, C. 1998. Inoculation of *Burkholderia cepacia*, *Pseudomonas fluorescens* and *Enterobacter* sp. on *Sorghum bicolor*: Root colonization and plant growth promotion of dual strain inocula. *Soil Biol. and Biochem*. 30: 81-87.

Cobera, J. and Hernandez, A. 1997. Assessment of the effects of the rhizobia-vesicular arbuscular mycorrhiza association on growth and development of soybeans (*Glycine max* (L.) Merrill). *Cultivos-Tropicales*. 18: 9-12.

Costa, J.M. and Loper, J.E. 1994. Characterization of siderophore production by the biological control agent *Enterobacter cloacae*. *Mol. Plant Microbe Interactions*. 7: 440-448.

Das, P.K., Sahoo, P.N. and Jena, M.K. 1997. Effect of VA-mycorrhiza and *Rhizobium* Inoculation on nutrient uptake, growth attributes and yield of green gram (*Vigna radiata* L.). *Environment and Ecology*. 15: 830-833.

Dasti, N., Zhang, F., Hynes, R. and Smith, D.L. 1998. Plant growth promoting rhizobacteria accelerate nodulation and increase nitrogen fixation activity by field grown soybean (*Glycine max* (L) Merr.) under short season conditions. *Plant and Soil*. 200: 205-213.

DeFreitas, J.R. and Germida, J.J. 1990.Plant growth promoting rhizobacteria for winter wheat. *Canadian Jour. Microbiology*. 36: 265-272.

Dey, R., Pal, K.K. and Tilak, K.V.B.R. 2004. Interaction of plant growth promoting rhizobacteria and AM fungi: Present status and future prospects. In: *Basic Research and Applications of Mycorrhizae* (Eds. Gopi K. Podila and Ajit K. Varma). pp. 83-110. I.K. International Pvt. Ltd., New Delhi, India.

Dixon, R. 1992. Inoculation of *Leucaena* and *Prosopis* seedlings with *Glomus* and *Rhizobium* species in saline soils: rhizosphere relations and seedling growth. *Arid Soil Research Rehabilitation*. 7: 133-140.

Dubey, S.K. 1998. Response of soybean to biofertilizers with and without nitrogen, phosphorus and potassium on swell shrink soil. *Indian Jour. Agronomy*. 43: 546-549.

Dubey, S.K. 2001.Associative effect of nitrogen fixing and phosphate solubilizing bacteria in rainfed soybean (*Glycine max*) and its residual effect on succeeding wheat (*Triticum aestivum*) crop. *Indian Jour. Agricultural Sci*. 61: 769-771.

Dudeja, S.S., Khurana, A.L. and Kundu, B.S. 1981. Effect of *Rhizobium* and phosphomicroorganism on yield and nutrient uptake in chickpea. *Curr. Sci*. 50: 503.

Downey, J. and van Kessel, C. 1990. Dual inoculation of *Pisum sativum* with *Rhizobium leguminosarum* and *Penicillium bilaji*. *Biol. and Fert. Soils*. 10: 194-196.

El-Bahrawy, S.A. 1983. Associative effect of mixed culture of *Azotobacter* and different rhizosphere fungi with *Rhizobium japonicum* on nodulation and symbiotic nitrogen fixation in soybean. *Zentralblatt fur Mikrobiologie.* 138: 443-449.

El-Gamai, M.S. 1992. Interactions between *Azotobacter* spp., *Rhizobium sesbani* in the rhizosphere of *Sesbania sesban* (L.) Merrill plants and its efficiency on growth and symbiotic nitrogen fixation. *Zentralblatt fur Mirobiologie.* 147: 112-118.

El-Mokadem, M.T., Helemish, F.A., Abou-Bakr, Z.Y.M. and Shetewas, S.A. 1989. Associative effect of *A. lipoferum* and *Azotobacter chroococcum* with *Rhizobium* sp.on mineral composition and growth of chickpea (*Cicer arietinum*) on sandy soils. *Zentralblatt fur Mikrobiologie.* 144: 255-265.

Gaind, S., Mathur, R.S. and Tilak, K.V.B.R. 2000. Phosphate solubilizing microorganisms. In: *Recent Advances in Biofertilizer Technology* (Eds. Yadav,Singh, Raychoudhury) pp. 191-206.

Galal, Y.G.M. 1997. Dual inoculation with strains of *Bradyrhizobium japonicum* and *Azospirillum brasilense* to improve growth and biological nitrogen fixation of soybean (*Glycine max* L.). *Biol. and Fert. Soils.* 24: 317-322.

Garbaye, J. 1994. Bacteres Auxiliaries dela mycorrhization:nouvelle dimension de la symbiose mycorrhizienne (Mycorrhizae helper bacteria: a new dimension of the mycorrhizal symbiosis). *Acta Botanica Gallica.*141: 52-57.

Gaur, A.C. 1990. *Phosphate Solubilizing Microorganisms as Biofertilizers.* Pp. 176. Omega Scientific Publishers, New Delhi, India.

Gerdemann, J. M. and Trappe, J.M. 1974. *Mycologia Memoir #5.* pp. 76. New York Botanical Garden, Mycological Society of America.

Ghosh, G. and Poi, S.C. 1998. Response of *Rhizobium*,phosphate solubilizing bacteria and mycorrhizal organisms on legume crops. *Environment and Ecology.* 16: 607-610.

Glick, B.R. 1995. The enhancement of plant growth promotion by free living bacteria. *Canadian Jour. Microbiology.* 41: 109-117.

Glick, B.R. and Pasternak, J.J. 2003. Plant growth promoting bacteria. In: *Molecular Biotechnology-Principles and Applications of Recombinant DNA* (Eds. Glick, B.R.and Pasternak, J.J.) pp. 436-454. ASM Press, Washington D.C., U.S.A.

Groppa, M.D., Zawoznik, M.S. and Tomaro, M.L.1998. Effect of co-inoculation with *Bradyrhizobium japonicum* and *Azospirillum brasilense* on soybean plants. *European Journal of Soil Biology.* 34: 75-80.

Gupta, A., Saxena, A.K., Gopal, M. and Tilak, K.V.B.R. 1998. Effect of plant growth promoting rhizobacteria oncompetitive ability of introduced *Bradyrhizobium* sp. (*Vigna*) for nodulation. *Microbiological Research.* 153: 113-117.

Gryndler, M. 2000. Interactions of arbuscular mycorrhizal fungi with other soil microorganisms. In: *Arbuscular Mycorrhizas: Physiology* and *Function.* (Eds. Kapulinik, Y. and Douds D.D. Jr.) pp. 239262. Kluwer Academic, Dordrecht, The Netherlands.

Ho, I. 1988. Interaction between VA mycorrhiza fungus and *Azotobacter* and their effects on growth of tall fescue. *Plant and Soil.* 105: 291-293.

Ibjibijen, J., Uruquiaga, S., Ismail, M., Alves, B.J.R. and Boddey, R.M. 1996. Effect of arbuscular mycorrhizal fungi on growth, mineral nutrition and nitrogen fixation of three varieties of common beans (*Phaseolus vulgaris*). *New Phytologist.* 134: 353-360.

Ishac, Y.Z., Angle, J.S., El-Borollosy, M.A., El-damerdash, M.E., Mostafa, M.I. and Fares, C.N. 1994. Growth of *Vicia faba* as affected by inoculation with vesicular arbuscular mycorrhizae and *Rhizobium leguminosarum bv.viceae* in two soils. *Biol. and Fert. Soils.* 17: 27-31.

Isopi, R., Fabbri, P. Gallo, M., Dell and Puppi, G. 1995. Dual inoculation of *Sorghum bicolor* (L) Monech spp.*bicolor* with vesicular arbuscular mycorrhizas and *Acetobacter diazotrophicus.* Symbiosis. 18: 43-45.

Ibrahim, K.K., Arunachalam, V., Kesava Rao, P.S. and Tilak, K.V.B.R. 1995. Seasonal response of groundnut genotypes to arbuscular mycorrhiza-*Bradyrhizobium* inoculation. *Microbiological Research.* 150: 218-224.

Itzigsohn, R. Kapulnik, Y., Okon, Y. and Dovrat, A. 1993. Physiological and morphological aspects of interactions between *Rhizobium meliloti* and alfalfa (*Medicago sativa*) in association with *Azospirillum brasilense. Canadian Jour. Microbiology.* 39: 610.

Jeffries, P. 1997. Mycoparasitism. In: *The Mycota IV. Environmental and Microbial Relationship.* pp. 95-113. *Springer verlag,* Berlin/Hydelberg/New York.

Jeyabal, A., Kuppuswamy, G. and Lakshmanan, A. 1992.Effect of seed coating on yield attributes and yield of soybean(*Glycine max* L.). *Jour.Agronomy and Crop Science.* 169: 145-150.

Kandaswamy, D., Palaniswamy, D. and Oblisami, G.1988. Screening of germplasm of sweet potato for VA mycorrhiza fungal occurrence and response of the crop to inoculation of VAM fungi and *Azospirillum. Jour. Root Crops.* 14: 37-42.

Kapulnik, Y., Okon, Y. and Henis, Y. 1987. Yield response of spring wheat cultivars to inoculation with *Azospirillum brasilense* under field conditions. *Biol. and Fert. Soil.* 4: 27

Karthikeyan, V. and Raj, J. 1991. Interaction effect of *Frankia* sp., and VA-mycorrhizal fungus on growth response in *Casuarina equisetifolia. Abs.XXXI Annual Conference of association of Microbioogists of Indi.* Coimbatore, India, pp. 79.

Kharb, P. and Kundu, B.S. 1986. Response of pearlmillet to associative diazotroph inoculation. *Environment and Ecology.* 4: 12.

Kloepper, J.W., Lifshitz, R. and Zablotowicz, R.M. 1989. Free-living bacterial inocula for enhancing crop productivity. *Trends in Biotechnology.* 7: 39-93.

Kothari, S,K, and Saraf, C.S. 1986. response of greengram (*Vigna radiata* Wilczek) to bacterial seed inoculation and application of P fertilizer. *Jour. Agricultural Sciences.* 107: 463-466.

Kozyrovska, N., Kovtunovych, G., Giomodova, E., Kuharchuck, P. and Kordyum, V. 1997. Novel inoculanmts for an environmentally-frienly crop production. In: *Resources, Conservation and Recycling.* (Ed. Wise, D.L.), pp. 79-85. Global Environmental Biotechnology. Proceedings of the 3rd International Symposium of International Society for Biotechnology. Boston, USA.

Kshirsagar, C.R., Mandhare, V.K., Kalbhor, H.B. and Patil, P.L. 1994. Response of onion to *Azotobacter* and VA-mycorrhizal inoculation alongwith phosphorus levels. *Jour. Maharastra Agricultural University.* 19: 476-477.

Kucey, R.M.N., Janzen, H.H. and Leggett, M.E. 1989. Microbially mediated increases in plant available phosphorus. *Advances in Agronomy*. 42: 199-228.

Kumari, S.M.P. and Balasubramanian, A. 1993. Effect of combined inoculation of VAM and *Azospirillum* on the growth and nutrient uptake by coffee seedlings. *Indian Coffee.* 57: 5-11.

Kumatha, K. and Santhanakrishnan, P. 1994. Enhancement of nitrogen fixation in soybean by the dual inoculation of VA-mycorrhiza and *Bradyrhizobium*. *Madras Agricultural Journal.* 81: 497-499.

Kumrawat, B., Dighe, J.M., Sharma, R.A. and Katti, G.V. 1997. Response of soybean to biofertilizer in black clay soils. *Crop Research.* 14: 209-214.

Kundu, B.S. 1988.Effect of dual inoculation with *Azospirillum* and *Rhizobium* on nodulation and nitrogen fixation in pigeonpea. *Indian Jour. Microbiology.* 28: 233.

Kundu, B.S. and Gaur, A.C. 1984.Rice response to inoculation with N_2-fixing and P-solubilizing microorganisms. *Plant and Soil,* 79: 227-234.

Lata, Saxena, A.K. and Tilak, K.V.B.R. 2000. Biofertilizers to augment soil fertility and crop production. In: *Soil Fertility and Crop Production* (Ed. Krishna, K.R.). *Science Publishers Inc.* Enfield (NH) USA/Plymouth (UK), pp. 279-312.

Lin, X.G., Hao, W.Y., Shi, Y.Q. and Wu, T.H. 1993. Response of white clover to *Rhizobium*-VA mycorrhiza dual inoculation. *Grassland of China.* 3: 20-25.

Linderman, R.G. 1992. Vesicular-arbuscular mycorrhizae and soil microbial interactions. In: *Mycorrhizae in Sustainable Agriculture. American Society of Agronomy* Special publication. ASA, Madison, Wisconsin.USA, pp. 1-26.,

Malik, R.D.S., Gaur, A.C. and Kavimandan, S.K. 1995. Growth promoting bacteria associated with sunflower. *Indian Jour. Expl. Biology.* 33: 311-312.

Mallesh, B.C. and Bagyaraj, D.J. 1990. Reaction of *Leucaena* to single inoculation versus dual inoculation with *Glomus mosseae* and/or *Gigaspora margarita. Leucaena Research Reports.* 11: 56-57.

Mandhare, V.K., Kalbhor, H.B. and Patil, P.L. 1995. Effects of VA-mycorrhiza, *Rhizobium* and phosphorus on summer groundnut. *Jour. Maharastra Agricultural University.* 20: 261-262.

Moawad,H., El-Maksoud, H.K.A. and Saad, R.N. 1996. Response of berseem clover (*Trifolium alexandrinum*)to rock phosphate fertilization and dual inoculation with *Rhizobium* and vesicular arbuscular mycorrhiza (VAM). *Egyptian Jour. Microbiology.* 31: 13-23.

Mohandas, S. 1987. Field response of tomato (*Lycopersicon esculentum* Mill "*Pusa Ruby*") to inoculation with VA mycorrhizal fungus *Glomus fasciculatum* and *Azotobacter vinelandii. Plant and Soil.* 98: 295-297.

Molla, A.H., Shamsuddin, Z.H. and Halimi Mohd-Saud. 2001. Mechanism of root growth and promotion of nodulation in vegetable soybean by *Azospirillum brasilense. Commu. Soil Science and Plant Analysis.* 32: 2177-2187.

Muhadiono, I., Supiandi, S., Mansjoer, I., Garcia, M.U. , Kartasubrata, J. and Ummaly, R.C. 1990. Agroforestry technology: *Rhizobium* and endomycorrhizal infection in roots of *Albizzia procera* (Roxb.) Benth as biofertilizer in the future. In: *Proceedings of Symposium on Agroforestry systems and Technologies, Bogor, Indonesia* (Eds. Katrasubrata, J. and Tjitrosoma), *Biotrop-Special Publication.* 39: 107-114.

Murakami-Mizukami, Y., Yamamoto, Y. and Yamaki, S. 1991. Analyses of Indole acetic acid and abscisic acid contents in nodules of soybean plants bearing VA mycorrhizas. *Soil Science and Plant Nutrition.* 37-291-298.

Nautiyal, C.S. 2000. Plant beneficial rhizosphere competent bacteria. *Proc. Natl. Acad. Sciences, India.* 70(B): 107-123.

Negi, M., Sachdev, M.S. and Tilak, K.V.B.R. 1990. Influence of soluble phosphorus fertilizer on the interaction between a vesicular-arbuscular mycorrhizal fungus and *Azospirillum brasilense* in barley (*Hordeum vulgare* L.) *Biol. and Fert. Soils.* 10: 57-60.

Negi, M., Sadasivam, K. and Tilak, K.V.B.R. 1987. Establishment of *Azotobacter* and *Azospirillum* in the rhizosphere of barley in organic amended soils. *Zentralblatt fur Mikrobiologie.* 142: 149.

Okon, Y., Bloemberg, G.V. and Lugtenberg, B.J.J. 1998. Biotechnology of biofertilization and phytostimulation. In: *Agricultural Biotechnology* (Ed. Altman, A.), *Marcel Dekker, Inc.* New York/Basel/Hongkong, pp. 327-349.

Olesniewicz, K.S. and Thomas, R.B. 1999. Effects of mycorrhizal colonization on biomass production and nitrogen fixation of black locust (*Robinia pseudoacacia*) seedlings grown under elevated atmospheric carbon dioxide. *New Phytologist.* 142: 133-140.

Olsen, T. and Habte, M. 1995. Mycorrhizal inoculation effect on nodulation and N accumulation in *Cajanus cajan* at soil P concentrations sufficient or inadequate for mycorrhiza free growth. *Mycorrhiza.* 5: 395-399.

Pacovsky, R.S. 1986. Micronutrient uptake and distribution in mycorrhizal and phosphorus fertilized soybeans. *Plant and Soil.* 95: 379-385.

Pacovsky, R.S., Fuller, G. and Paul, F.A. 1985. Influence of soil on the interaction between endomycorrhizae and *Azospirillum* in sorghum. *Soil Biol and Biochem.* 17: 525-531.

Pal, K.K., Dey, R., Bhatt, D.M. and Chauhan, S.M. 2000. Plant growth promoting fluorescent pseudomonads enhanced peanut growth,yield and nutrient uptake, http//www.ag.auburn.edu/argentina/pdfmanuscript/pal.pdf pp. 12.

Panwar, J.D.S. 1991. Effect of VAM and *Azospirillum brasilense* on photosynthesis, nitrogen metabolism and grain yield in wheat. *Indian Jour. Plant Physiology.* 35: 357-361.

Patel, R.M., Patel, N.M. and Patel, Z.G. 1993. Role of biofertilizer in productivity of fingermillet (*Eleusine coracana*). *Indian Jour. Agronomy.* 38: 119-121.

Paula, M.A., Reis, V.M. and Dobereiner, J. 1991. Interaction of *Glomus clarum* with *Acetobacter diazotrophicus* in infection of sweet potato (*Ipomoea batata*), sugarcane (*Saccharum spp.*) and sweet sorghum (*Sorghum vulgare*). *Biol. and Fert. Soils.* 11: 111-115.

Paula, M.A., Urguiaga, S., Siqueira, J.O. and Dobereiner, J. 1992. Synergistic effects of vesicular arbuscular mycorrhizal fungi and diazotrophic bacteria on nutrition and growth of sweet potato. *Biol. and Fert. Soils.* 14: 61-66.

Plazinski, J. and Rolfe, B.G. 1985. *Azospirillum-Rhizobium* interaction leading to a growth stimulation without nodule formation. *Canadian Jour. Microbiology.* 31: 1026-1030.

Ram, G., Rawat, A.K. and Joshi, B.S. 1988. Effect of *Rhizobium* and *Azotobacter* on *Cicer arietinum* in Chattisgarh of M.P. *Legume Research.* 11: 150-152.:

Ramachandra, G.S. and Rai, P.V. 1987. Effect of *Azotobacter chroococcum* and *Glomus fasciculatum* on growth and yield of brinjal (*Solanum melongena* L.). *Indian Jour. Microbiology.* 27: 78-80.

Rao. D.L.N. and Pal, K.K. 2003. Biofertilizers in oilseeds production-Status and future strategies. *National Seminar on Stress Management in Oilseeds for Attaining Self-Reliance in Vegetable Oils.* pp. 195-220. Directorate of Oilseeds Research (ICAR), Hyderabad, A.P., India.

Rasco, E.T. (Jr), Amante, V.D.R. and Lopez, P.J.S. 1992. effect of genotype and water regime on sweet potato response to biofertilizers. In: *Annual Report. South Asian Prtogramme for Potato Research and Development* (Eds. Rasco, E.T. (Jr), Aromin, F.B., Amante, V.D.R. and Lopez, P.J.S.), pp. 239-250. Manila, Phillipines.

Ravichandran, V.K. and Balasubramanian, T.N. 1999. Growth response of *Casuarina equisetifolia* seedlings to dual inoculation of *Frankia* and VAM (*Glomus fasciculatum*). *Jour. Ecobiology.* 11: 175-179.

Rawat, A.K., Khare, A.K. and Vaishya, V.K. 1991. Dual effect of *Rhizobium* and mycorrhiza on nodulation and yield of chickpea. *Legume Research.* 14: 150-152.

Reinhold-Hurek, B. and Hurek, T. 1998. Interaction of graminaceous plants with *Azoarcus* spp. and other diazotrophs: Identification, localization and perspectives to study their function. *Critical Review of Plant Science.* 17: 29-54.

Requena, N., Jimenez, I., Toro, M. and Barea, J.M. 1997. Interactions between plant-growth-promoting rhizobacteria (PGPR), arbuscular mycorrhizal fungi and *Rhizobium* spp. in the rhizosphere of *Anthyllis cytisoides,* a model legume for revegetation in mediteranean semi-arid ecosystems. *New Phytologist.* 136: 667-677.

Russo, R.O. 1989. Evaluating alder-endophyte (*Alnus accuminata-Frankia*-mycorrhizal) interactions: I. Acetylene reduction in seedlings inoculated with *Frankia* Ar 134 and *Glomus intraradices,* under three phosphorus levels. *Plant and Soil.* 118: 151-155.

Saxena, A.K., Rathi, S.K. and Tilak, K.V.B.R. 1997. Differential effect of various endomycorrhizal fungi on nodulating ability of greengram by *Bradyrhizobium* sp. (*Vigna*) strain S 24. *Biol. and Fert. Soils.* 24: 175-178.

Saxena, A.K., Shende, R. and Tilak, K.V.B.R. 2002.Interaction of arbuscular mycorrhiza with nitrogen-fixing bacteria. In: *Arbuscular Mycorrhizae: Interactions in Plants, Rhizosphere and Soils* (Ed. Johri, B.N.) *Science Publishers, Inc.* pp. 47-68. Enfield (NH) USA/Plymouth (UK).

Saxena, A.K. and Tilak, K.V.B.R. 1994. Interaction among beneficial soil microorganisms. *Indian Jour. Microbiology.* 34: 91-106.

Saxena, A.K. and Tilak, K.V.B.R. 1998. Free living nitrogen fixers: Their role in crop productivity. In: *Microbes for Health, Wealth and Sustainable Environment.* (Ed.Varma, A.K.) pp. 25-64. *Malhotra Publishing House,* New Delhi, India.

Schwab, S.M., Menge, J.A. and Leonard, R.T. 1983. Quantitative and qualitative effects of phosphorus on extracts and exudates of sudan grass roots in relation to vesicular arbuscular mycorrhiza formation. *Plant Physiology.* 73: 761-765.

Sengupta, A. and Choudhury, S. 1995. Effect of dual inoculation of *Rhizobium* and mycorrhiza on growth response of *Sesbania grandiflora* L. in coastal saline and sand dune soil. *Indian Jour. Forestry.* 18: 35-37.

Shalaby, A.M. and Hanna, M.M. 2000. Interactions between VA mycorrhizal fungus *Glomus mosseae, Bradyrhizobium japonicum* and *Pseudomonas syringae* in soybean plants. *Egyptian Jour. Microiology.* 35: 199-209.

Shasidhara, G.B., Srinivasa, M.N., Bidari, B.I. and Manjunathaiah, H.M. 1994. Response of groundnut to inoculation of VA mycorrhizal fungi and/or *Rhizobium* in vertisols. *Jour. Maharastra Agricultural University.* 19: 464-465.

Singh, C.S. 1996. Arbuscular mycorrhiza (AM) in association with *Rhizobium* sp. improves nodulation, N_2 fixation and N utilization of pigeonpea (*Cajanus cajan*) as assessed with a N_2 technique in pots. *Microbiological Research.* 151: 87-92.

Singh, H.P. 1994. Response to inoculation with *Bradyrhizobium,* VAM and PSM on soybean in mollisol. *Indian Jour. Microbiology.* 34: 27-31.

Singh, M. and Tilak, K.V.B.R. 1989. Field response of chickpea to inoculation with *Glomus versiforme. Plant and Soil.* 119: 281-284.

Singh, R.P., Singh, L. and Teotia, R.S. 1981. Effect of symbiotic nitrogen-fixing bacteria on growth and seed compositionof soybean. *Acta Botanica Indica.* 9: 104-109.

Singh, S. and Kapoor, K.K. 1998. Effect of inoculation of phosphate solubilizing microorganisms and vesicular arbuscular mycorrhizal fungus on mungbean under natural soil conditions. *Mycorrhiza.* 75: 249-253.

Smith, J.H., Allioson, F.E. and Soulides, D.A. 1961. Evaluation of phosphobacterin as a soil inoculant. *Proc. Soil Sci. Soc. America.* 25: 109-111.

Sodi, R.R., Srinivasa, M.N., Chittapur, B.M. and Babalad, H.B. 1994. Field response of cowpea to dual inoculation of VA-mycorrhiza and *Rhizobium* at different fertilizer levels. *Jour. Maharastra Agricultural University.* 19: 459-460.

Soliman, S., El-Ghandour, I.A. and Khadra-Abbady, A. 1996. Effects of nitrogen and phosphorus supply and of *Rhizobium* and VAM fungus inoculants on dinitrogen fixation in soybean. *Folia Microbiologia.* 41: 197-200.

Sreekumar, K.R. and Tilak, K.V.B.R. 1988. Nitrogen fixation in forage plants. In: *Pasture and Forage Crops Research-A State of Knowledge Report.* pp. 248-260. (Ed. Panjab Singh), New Delhi, India.

Srinivasa, M.N., Basavarga, G.T. and Kulkarni, J.M. 1995. Vesicular arbuscular mycorrhiza assist in nodulation and N_2 fixation in soybean. *Jour. Maharastra Agricultural University.* 20: 292-293.

Srinivasan, M., Petersen, D.J. and Holl, F.B. 1996. Influence of Indole-acetic-acid-producing *Bacillus* isolates on the nodulation of *Phaseolus vulgaris* by *Rhizobium etli* under genobiotic conditions. *Canadian Jour. Microbiology.* 42: 1006-1014.

Srivastava, T.K., Ahlawat, I.P.S. and Panwar, J.D.S. 1998. Effect of phosphorus, molybdenum and biofertilizers on productivity of pea. *Indian Jour. Plant Physiology*. 3: 237-239.

Steinberg, C., Gamard, P., Faurie, G. and Lensi, R. 1989. Survival and potential denitrifying activity of *Azospirillum lipoferum* and *Bradyrhizobium japonicum* inoculated into sterilized soil. *Biol. and Fert. Soils*. 7: 101-107.

Sturz, A.V., Christie, B.R., Matheson, B.G. and Nowak, J. 1997. Biodiversity of endophytic bacteria which colonize red clover nodules,roots,stems and foliage and their influence on host growth. *Biol. and Fert. Soils*. 25: 13-19.

Subba Rao, N.S. 2000. *Soil Microbiology*. Pp.407. 4th Edition. *Oxford and IBH Publishing Co. Pvt. Ltd.*, New Delhi, India,

Subba Rao, N.S., Tilak, K.V.B.R. and Singh, C.S. 1985. Synergistic effect of vesicular-arbuscular mycorrhizasand *Azospirillum brasilense* on the growth of barley in pots. *Soil Biology and Biochemistry*. 17: 119-121.

Subba Rao, N.S. , Tilak, K.V.B.R. and Singh, C.S. 1986. Yield and nitrogen gains of sorghum as influenced by *Azospirillum brasilense*. In: *Cereal Nitrogen Fixation: Proceedings of the Working Group Meeting*. (ICRISAT), Patancheru, A.P., India, pp. 69-74.

Subbian, P. and Chamy, A. 1984. Effect of *Azotobacter* and *Azospirillum* on the growth and yield of sesame. *Madras Agricultural Journal*. 71: 615-617.

Tandon, V. and Prakash, A. 1998. Influence of soil inoculation with VA-mycorrhiza and phosphorus solubilizing microorganisms on growth and phosphorus uptake in *Sesamum indicum*. *Intl. Jour. Tropical Agriculture*. 16: 201-209.

Thiagarajan, T.R., Ames, R.N., and Ahmad, M.H. 1992. Response of cowpea (*Vigna unguiculata*) to inoculation with co-selected vesicular-arbuscular mycorrhizal fungi and *Rhizobium* strains in field trials. *Canadian Jour. Microbiology*. 38: 573-576.

Tilak, K.V.B.R. 1993. Associative effects of vesicular-arbuscular mycorrhizae with nitrogen fixers. *Proc. Indian Natl Sci. Acad.* B59: 325-332.

Tilak, K.V.B.R., Li, C.Y. and Dwivedi, A. 1990. Enhancement of spore germination of *Glomus fasciculatum* by bacterial cell-free extracts. *Indian Jour. Expl. Biology*. 28: 373-375.

Tilak, K.V.B.R., Li, C.Y. and Ho, I. 1989. Occurrence of nitrogen fixing *Azospirillum* within the spores of vesicular arbuscular mycorrhizal fungi. *Plant and Soil*. 116: 286-288.

Tilak, K.V.B.R., Ranganayaki, N. and Manoharachari, C. 2003. Contribution of plant growth promoting rhizobacteria in crop production—Indian scenario. Session II: Overview of PGPR work in India. pp. 247-248. Sixth International PGPR Workshop, Calicut, India.

Tilak, K.V.B.R., Ranganayaki, N. and Manoharachari, C. 2004. Synergistic effects of plant growth promoting rhizobacteria and *Rhizobium* on nodulation and nitrogen fixation by pigeonpea (*Cajanus cajan* (L) Milsp.). Abstract, Poster Session 12, International Congress-Rhizosphere 2004, Munich,Germany.

Tilak, K.V.B.R. and Saxena, A.K. 2000. *Azospirillum*-its impact on crop production. In: *Recent Advances in Biofertilizer Technology*. pp. 177-189. (Eds. Yadav, Singh and Raychoudhury).

Tilak, K.V.B.R., Saxena, A.K. and Sadasivam, K.V. 1995. Synergistic effects of phosphate solubilizing bacterium (*Pseudomonas striata*) and arbuscular mycorrhizae on soybean. In: *Proceedings of Third National Conference on Mycorrhiza* (Ed. Adholeya, A. and Singh, S.), Tata Energy Research Institute, New Delhi, India, pp. 224-226.

Tilak, K.V.B.R., Singh, C.S. and Rana, J.P.S. 1981. Effect of combined inoculation of *Azospirillum brasilense* with *Rhizobium trifolii, R. meliloti* and *Rhizobium* sp. (cowpea misc.) on nodulation and yield of clover (*Trifolium repens*), lucerne (*Medicago sativa*) and chickpea (*Cicer arietinum*). *Zentralblatt fur Bakteriologie*. 136: 117-120.

Tilak, K.V.B.R., Singh, C.S., Roy, N.K. and Subba Rao, N.S. 1982. *Azospirillum* and *Azotobacter* inoculation. Effect on yield of maize and sorghum. *Soil Biol. and Biochem*. 14: 417-418.

Tilak, K.V.B.R. and Subba Rao,N.S. 1987. Association of *Azospirillum brasilense* with pearl millet (*Pennisetum americanum* (L.) Leeke). *Biol. and Fert.Soils*. 4: 97-102.

Udaiyan, K. and Sugavanam, V. 1995. Interaction of vesicular-arbuscular mycorrhizal fungi ad *Frankia* sp. On growth and nutrient uptake in *Casuarina equisetifolia*. *Forestry Journal*. 11: 231-237.

Volpin, H., Burdman, S., Castro-Sowinski, S., Kapulnik, Y. and Okon, Y. 1996. Inoculation with *Azospirillum* increased exudation of rhizobial nod-gene inducers by alfalfa roots. *Mol. Plant–Microbes Inter.* 9: 388.

Vosatka, M. and Gryndler, M. 1999. Treatment with culture fractions from *Pseudomonas putida* modifies the development of *Glomus fistulosum* mycorrhiza and the response of potato and maize plants to inoculation. *Appl. Soil Ecol.* 11: 245-251.

Wani, S.P., Chandrapalaiah, S. and Dart, P.J. 1985. Response of pearlmillet cultivars to inoculation with nitrogen fixing bacteria. *Expl Agriculture*. 21: 175-182.

Weller, D.M. and Cook, R.J. 1986. Increased growth of wheat by seed treatments with fluorescent pseudomonads and implications of *Pythium* control. *Canadian Jour. Plant Pathology*. 8: 328-334.

Xie, Z., Staehelin, C., Vierheili, H., Wiemken, A., Jabbouri, S., Brouhton, W.J., Vogeli-Lange, R., Boller, T. and Xie, Z.P. 1995. Rhizobial factors stimulate mycorrhizal colonization of nodulating and non-nodulating soybeans. *Plant Physiology*. 108: 1519-1525.

Zhao, B., Trouvelot, A., Gianinazzi, S. and Gianinazzi-Pearson, V. 1997. Influence of two legume species on hyphal production and activity of two arbuscular mycorrhizal fungi. *Mycorrhiza*. 7: 179-185.

Frontiers in Plant Sciences (Pages 163-218)
Eds. K.G. Mukerji, K.V.B.R. Tilak, S.M. Reddy, L.V. Gangwane, P. Prakash, I.K. Kunwar

Interactions between Rhizosphere microorganisms and Plant Pathogens

C.S. Reddy, G.S. Laha and K. Muralidharan

ABSTRACT

The rhizosphere is the first-line defense for roots against attack by pathogenic microorganisms. The rhizosphere region of plants is heavily inhabited by complex and dynamic communities of microorganisms, including both beneficial and deleterious ones. Of particular interest are the rhizosphere microorganisms with ability as antagonists to provide defense against soil borne plant pathogens. Fluorescent pseudomonads constitute a large population of the rhizosphere microflora because of their nutritional versatility, being able to utilize a large number of organic substrata. Specific strains of *Pseudomonas fluorescens, P. putida* readily colonize the growing roots and enhance the plant growth by reduction of root zone population of deleterious microflora and pathogens. These beneficial strains are also potential biocontrol agents of several soil-borne plant pathogens. The mechanisms by which these microorganisms provide defense include competition for nutrition and space sites, forms of hyperparasitism, induced resistance and by production of a variety of metabolites that include antibiotics, siderophores and other substances such as cyanide.

Key words: Interactions, rhizosphere microflora, pathogens

INTRODUCTION

Soil is the natural habitat of countless hordes of microorganisms, consisting of many kinds and types, living in a constantly changing environment. It is an important medium in nature for the growth, multiplication, survival and other activities of microbes which live in soil not in pure culture but

Author's address: Directorate of Rice Research, Rajendranagar, Hyderabad-500030, India.

as complex populations wherein various forms of antagonistic relationships are operative. Usually their existence produces a healthy soil and a more favourable environment for plant growth. The role of microorganisms in natural processes and in maintaining the biological balance in nature is among the most significant ones. The scope of this review is restricted to rhizosphere bacteria only and their interactions with plant pathogens, though a mention has also been made about the rhizosphere fungi, actinomycetes, and microfauna at appropriate places of the article.

The health of a flowering plant is determined by countless biotic and abiotic interactions in the soil and on the root surfaces. As early as in 1687, Hooke first suggested the existence of a relationship between root exudates and the soil environment (Curl and Truelove 1996). Later, the activity of microorganisms was observed to be stimulated in the soil volume close to roots of a plant and this part of soil volume was denoted as rhizosphere (Hiltnner 1904). Since then, some remarkable advances have been made on the importance of the rhizosphere microorganisms on the plant growth. The rhizosphere is heavily populated by different microorganisms that exert both beneficial and deleterious effects on plant growth. Many strains of plant growth promoting rhizobacteria (PGPR) were discovered in the rhizosphere. These bacteria were shown to survive in the rhizosphere. The new opportunities to apply biotechnological developments provided the impetus to exploit commercially the key beneficial microorganisms that inhabit plant roots (Schroth and Hancock, 1982). Seed bacterization which introduced and established beneficial microorganisms in the rhizosphere was identified as an excellent way for improving plant growth. The natural defense provided by microorganisms in the rhizosphere deserves much more scientific attention, both because of its importance in plant ecology and because of its potential as a practical means to improve the health and productivity of crop (Cook et al., 1995).

THE NATURE AND DIVERSITY OF RHIZOSPHERE MICROBIAL POPULATION

The rhizosphere is a dynamic environment that differs from the bulk soil both in physical and chemical properties. The extent of "rhizosphere effect" in the soil will depend on the diffusion away from the root of many compounds released from the root, their diffusion characteristics and water status of the soil (Bowen and Rovira, 1999). The extremely complex physical environment of the rhizosphere is the region of intense biological activity (Lynch 1982; Schippers et al., 1987). Microorganisms inhabit surfaces of roots and soil particles metabolizing nutrients exuded by roots (Lynch 1990; Lynch and Whipps 1991). Much of our knowledge of the physical state of the rhizosphere and the rhizoplane has been derived from transmission and scanning electron microscopy (Foster et al. 1983; Foster 1986). These studies have shown that the epidermal cells are coated with polysaccharides of both plant and animal origin and of variable thickness in which microbial colonies are embedded. Other microbial colonies lie outside of this mucilage among the mineral components of the soil and are nourished by soluble root exudates. The quality and quantity of rhizodeposition varies with plant species, root age, growing conditions and prevailing soil environment (Hale and Moore 1979; Curl and Truelove 1986). This includes sloughed root cap cells (Hawes and Pueppke 1986), mucilages, exudates and lysates, which contain amino acids, proteins, sugars, complex carbohydrates, alcohols, vitamins and hormones (Rovira et al., 1979; Bowen 1980; Lynch 1982; Chaboud 1983; Newman 1985; Vancura 1988). In addition, the rhizosphere bacterial population can induce the production and release of various growth regulators (Barber and Martin 1976; Prikryl and Vancura 1980; Vancura 1988).

The resource base in the rhizosphere is highly complex (Loper et al. 1984; Lynch and Whipps 1991) and accordingly, the composition of the rhizosphere microorganisms undergoes a dynamic change. Among the microbial population in rhizosphere, the bacteria have received most attention. However, a marked rhizosphere effect also occurs with actinomycetes, fungi, nematodes, protozoa and other microflora and fauna (Bowen and Rovira 1999). The rhizosphere effect is generally measured by comparison of microbial populations in bulk soil (S) with those from soil intimately associated with the root (R) and is expressed as the R:S ratio (Katznelson 1946). The R:S ratio of different plants range from 23 to 58 for bacteria and from 1.3 to 12 for fungi (Ivarson and Katznelson 1960; Rouatt et al., 1960). Root exudates can also have a strong influence on soil pathogens. The nutrients in exudates release dormant spores and sclerotia from the grip of soil fungistasis, a natural phenomenon imposed by chemical inhibitors and microbial competition for nutrients. Propagules freed from fungistasis germinate and establish infection if they are not first consumed by mycophagous animals. Root exudates are also important in the activation of quiescent nematode stages and in the movement of plant parasitic forms to infection sites.

Available evidences indicate that the rhizosphere microflora have the potential to influence the plant growth and crop yield significantly (Schippers et al. 1987). Some microorganisms colomis plant roots in nature such as rhizobia, mycorrhiza and fungal root pathogens. Others such as plant growth promoting rhizobacteria (PGPR), harmful rhizosphere microorganisms are largely restricted to rhizosphere. Beneficial plant associated microorganisms in the rhizosphere are part of a vast and largely untapped natural biological resources that interact with crop plants and with deleterious microorganisms (Cook 1991). Fluorescent pseudomonads were identified as most effective rhizosphere bacteria as they suppressed harmful microorganisms in addition to promoting plant growth. Specific strains of *Pseudomonas fluorescens—P. putida* group were shown to readily colonize the growing roots and cause statistically significant yield increases (Schroth and Hancock 1982). Enhanced plant growth induced by these bacteria was often associated with reduction in the population of deleterious microorganisms and pathogens in the rhizosphere (Suslow, 1982; Geels and Schippers 1983; Gaskins et al., 1985). Studies have shown that gram negative bacteria predominate in the rhizosphere (Clark 1940; Rouatt and Katznelson 1961) and among these fluorescent pseudomonads are especially abundant and constitute a proportionately greater fraction of the microflora. Lambert et al. (1987) reported that flouorescent pseudomonads represented 35% of the total bacterial population in the maize rhizosphere. Kloepper et al. (1992) found that the number of identifiable species of *Pseudomonas* to be greater than other genera in the rhizosphere. Other genera often found in the rhizosphere include *Achromobacter, Alkaligenes, Enterobacter, Flavobacterium,* and *Serratia* (Brown 1972; Suslow 1982; Ordentlich et al. 1987).

Weller (1985) listed several compelling reasons why flouorescent pseudomonads have received the most attention among the rhizosphere microorganisms. First, the natural habitat for the flouorescent pseudomonads is soil and more specifically rhizosphere. They readily colonize the roots in nature (Burr and Caesor 1984; Schippers et al. 1987; Weller 1988). Thus when flouorescent pseudomonads are introduced on roots through seed bacterization, they are placed in an environment suitable for their growth and survival. Second, fluorescent pseudomonads are nutritionally versatile, being able to utilize a large number of organic substrates that exude from the roots (Weller 1988; Mazzola and Cook 1991; Chanway and Holl 1992). Third, the pseudomonads are fast growers relative to other bacteria in the rhizosphere. The generation time on the roots of *Pinus radiata* was 5.2 h for pseudomonads, 7.2 h for viable bacteria and 39 h for *Bacillus* spp. (Bowen and Rovira 1976). Numerous studies have shown that the population of pseudomonads grows exponentially, for the first several days after inoculation (Bahme and Schroth 1987; Kluepfel et al. 1991; Baath et al. 1992; Kloepper and Beauchamp 1992). Further, the flouorescent pseudomonads produce a large number of secondary metablolites which include siderophores, the microbial iron chelating agents (Kloepper et al. 1980; Misaghi et al. 1982; Leong 1986; Schippers et al. 1987), antifungal compounds (Dowling and O'Gara 1994), antibiotics (Fravel 1988; Thomashow and Weller 1990; Keel et al. 1992; Mazolla et al. 1992; Harrison et al. 1993; Schimidli – Sacheria et al. 1997), and other substances like hydrogen cyanide (Voisard et al. 1989). Thus, if antibiosis is one of the mechanisms of biological control, the ability to produce antibiotics and other antimicrobial secondary metabolites could be useful for an antagonist in root colonization and pathogen suppression. In addition, the demonstrated antagonistic effects of fluorescent pseudomonads include competition for nutrients (O'sulliran and O'Gara 1992), production of lytic enzymes (chitinases and glucanases) that act on fungal cell wall components (Singh et al. 1999) and more importantly Induced systemic resistance (Maurhoper et al. 1998; Ramamurthy et al. 2001) and through induction of plant chitinases and peroxidases (Dalisay and Kuc, 2001; Nandkumar et al. 2001). There are enormous reports (Table 1) that flouorescent pseudomonads are potential biocontrol agents and many of these have plant growth promoting ability. Flouorescent pseudomonads have been clearly demonstrated as an integral part of the microbial population in disease suppressive soils (Cook and Rovira, 1976; Scher and Baker, 1982; Weller, 1983). Scher and Baker (1980, 1982) isolated flouorescent pseudomonads from the *Fusarium* wilt suppressive soils and induced supressiveness by adding the bacteria into soils conducive for the manifestation of wilt. Flouorescent pseudomonads are amenable to genetic manipulation. Chet et al. (1993) reported the transfer of chitinase genes from *Serratia*, *Aeromonas* and *Trichoderma* into *Pseudomonas putida* resulting in the increased antagonistic activity of the transformant against *Sclerotium rolfsii* and *Rhizoctonia solani*. Using this advantage, fluorescent pseudomonads have been used to deliver anti-microbial and insecticidal compounds into the rhizosphere (Mazzola et al. 1992; Obukowicz et al. 1986, 1987).

Another group of microorganism that is abundant in the rhizosphere is fungi. Among the fungal antagonists, *Trichoderma* and *Gliocladium* are the most important. As expected with the soil fungi, species of *Trichoderma* use a wide range of compounds (Danielson and Davey 1973; Arrieta-Escobar and Belin 1982). Suggested mechnisms for biocontrol by the two genera are antibiosis, lysis, competition and mycoparasitism (Ayers and Adams 1981; Cook and Baker 1983; Papavizas and Lumsden 1980). Weindling and Emerson (1963) isolated an organic metabolite in crystalline form, very toxic at high dilution to *Rhizoctonia solani* which was later designated as gliotoxin. Brian and McGowan (1945) described a second antibiotic viridin from *Trichoderma viride*. Later, Webster and Lomas (1964) showed that these two antibiotics are actually produced by *Gliocladium*. Howell and Stipanovic (1983) reported isolation and characterization of a new toxic metabolite from *Gliocladium virens*, a diketopiperazine designated as gliovirin. Godtfredson and Vangedal (1965) described a sesquiterpenoid metabolite designated as trichodermin from *Trichoderma*. *Trichoderma* and *Gliocladium* are not only good sources of various toxic metabolites and antibiotics, but also various enzymes such as exo- and endo-glucanases, cellobiase and chitinase (Papavizas 1985; Chet and Baker 1980, 1981) and degrades the glucans in the walls of *Pythium* spp., and chitin and glucan in the cell walls of *Rhizoctonia solani.* Other examples of fungal biocontrol agents encountered in the rhizosphere are *Sporidesmium scleritiovorum, Coniothyrum minitans* etc. (Adams and Ayers, 1982). Another fungal antagonist which has shown promise against take-all disease is *Phialophora* sp. I-52.

Actinomycetes, another group of microorganisms, as resident antagonists are important in maintaining satisfactory biological balance in soil because of their ability to produce antibiotics. It is well known that antagonistic actinomycetes are widely distributed in nature and soil is the store house of inexhaustible stock of such potent antagonists. The abundance of antagonistic actinomycetes in soils of different countries and the antagonistic properties of actinomycetes effective against various plant pathogenic fungi have been studied by Waksman et al. (1942), Krasilnikov et al. (1953), Mukherjee and Nandi (1955), Shete et al. (1960), Rangaswami et al. (1967), Broadbent et al. (1971) and, Singh and Reddy (1979). Gasperini (1890) is credited with being the first to demonstrate that actinomycetes are potential destroyers of fungi and bacteria. Subsequently, the work of Muller (1908), Lieske (1921), Porter (1924) and Gratia (1934) confirmed that *Streptomyces* spp. have marked inhibitory effect on fungi and bacteria. Since antagonistic microorganisms have ecological significance in soil, emphasis has been placed on soil actinomycetes as agents for suppression of root diseases. Hartley (1921) was one of the first to investigate the use of antagonistic organisms in the control of soil borne plant diseases. Severity of soil borne plant diseases, viz., damping-off of tomato, seedling blight of chick pea and sugarbeet has been reduced by addition of pure cultures of antagonistic strains of *Streptomyces diastaticus* to previously autoclaved or natural, artificially infested soil (Singh and Reddy, 1979).

The diverse group of microfauna includes the protozoa, nematodes, and some arthropods, primarily mites and springtails (collembola). There is a relationship between flora and fauna in organic matter decomposition and nutrient cycling. As nematodes and protozoans feed upon microbial populations, they consequently affect organic matter decomposition. It is likely that such feeding ultimately liberates nutrients immobilized in microbial cells or reduces competition between microorganisms so that mineralization is actually accelerated. These activities not only influence the general nutrition, health, and vigor of higher plants (which relates to disease susceptibility), but also determine the competitive behavior of root-infecting fungi and their microbial antagonists. These indirect and direct impacts are important ones for further study. The microarthropods like mites and collembola are known to fragment organic matter as they feed on soil microflora. This fragmentation to finer particles creates new surface areas for microbial colonization and consequently speeds the decomposition and mineralization processes. Several other interesting features of the microbiotic community include increased nitrogen fixation by nitrogen-fixing bacteria *(Azotobacter* spp.*)* where protozoa are present, and the behavior of bacterial and fungal-feeding nematodes (Curl, 1988). In the final analysis, efforts to manipulate and exploit the friendly fauna populations for crop benefit must be compatible with microbial symbionts, and other plant-growth promoting rhizosphere organisms, and with fungi and bacteria that are being promoted for biological control of diseases. This is clearly an area with great opportunities for further research (Curl, 1988).

Deleterious Rhizosphere Microorganisms

A group of rhizosphere microorganisms which was not widely considered as plant pathogens has been reported by several workers as a major contributing factor for plant growth retardation and yield decrease (Colhoun, 1979; Suslow and Schroth, 1982). According to Schippers et al. (1987), the microorganisms that affect the plant health by their metabolites without actually parasitizing the plant's tissue are termed as deleterious rhizosphere microorganisms (DRMO). The DRMO include both deleterious rhizobacteria (DRB) and deleterious rhizo-fungi (Suslow and Schroth, 1982). In experiments performed under controlled greenhouse conditions, several DRBs have been shown to be associated with the reduced germination rates, root lesions, seed decay, reduced radicle elongation, increased colonization by pathogenic fungi and stunting of seedlings (Suslow and Schroth, 1982). Most of the symptoms are subtle and are not obvious under field condition but the overall effects on early growth and final yield can be substantial.

Although most field soils contain populations of DRMO, their effect on plant health and crop yields seems to become more pronounced with certain cropping practices. Schippers et al. (1987) reported that potatoes grown in the same plots every third or fourth year have yielded 10-15% less, than the yield from plots grown every sixth year. Yields was much lower (30% less) in fields where potatoes were grown every year or every second year. This reduction in yield due to repeated cultivation was not due to increased accumulation of pathogens like *Verticillium dahliae, Rhizoctonia solani,* and *Streptomyces scabies* or nutrient deficiencies but due to the increased accumulation of deleterious rhizosphere microorganisms or the factors in the soil that stimulate the production of their toxic metabolites (Schippers et al. 1987). The effect is obviously host specific, as the yields of other crops in rotations e.g., wheat and sugar beet are not affected by the frequencies of the potato cropping (Hoekstra, 1981). Host specificity of deleterious rhizosphere pseudomonads has been demonstrated for wheat and citrus at the cultivar level (Frederickson and Elliott 1985; Gardner et al. 1984). The DRMO probably survive on the root residues that provide a source of inoculum for root colonization of the next potato crop, as stated for the other rhizosphere microorganisms (Bowen, 1979). Alstrom (1991) found that the growth retardation caused by two pseudomonads (*Pseudomonas fluorescens* and a non-fluorescent pseudomonad) varied greatly between wheat, *Brassica* spp., spinach and lettuce with large differences between cultivars within each plant species. The differences in response were attributed to difference in level of root colonization. Frederickson et al. (1987), however, considered that differences between cultivars in response to DRMO were due to different cultivars varying in their susceptibility. Van Peer et al. (1990) reported that fluorescent pseudomonads isolated from inside the roots and the root surface differed in their response to plant growth; some 60% of the internal colonists reduced the plant growth, while rhizosphere isolates either had no effect (55%) or stimulated the plant growth (35%).

Reduction in yield due to repeated cultivation of the same crop has been reported in other crops also. This phenomenon has been observed in wheat, barley, bean and maize (Schippers et al. 1985). Wheat roots in the soil with large amount of crop residues in no-till sowing systems carry higher populations of root growth inhibiting pseudomonads than those in the conventional tillage systems. These pseudomonads were suggested to be responsible for decreased crop yields frequently observed in US-Pacific Northwest and England (Elliott and Lynch, 1984; Frederickson and Elliott, 1985). The involvement of deleterious rhizobacteria in the replant diseases of a range of trees e.g., cherries, pears, apples and citrus has been well documented (Catska et al. 1982; Jaffee et al. 1982; Utkhede and Smith, 1994).

Different mechanisms have been suggested by which DRMO exert their effect on the plant growth. Increased fungal colonization of the roots from bacterially treated plants indicates that DRB may either increase the susceptibility of roots to colonization by certain pathogenic fungi or stimulate the fungi to colonize the roots. Fluorescent pseudomonads, in particular, have been much studied with regard to plant growth inhibition. The other genera that have been identified to be deleterious to plant growth are *Enterobacter, Klebsiella, Citrobacter, Flavobacterium, Achromobacter,* and *Arthrobacter*. Production of hydrogen cyanide by the rhizosphere pseudomonads is suspected to be partly responsible for the inhibition of plant growth and for the reduction of yields in high frequency potato cropping (Bakker and Schippers, 1987; Schippers et al. 1987). This suspicion originates from the observed HCN-induced inhibition of potato root cytochrome oxidase respiration (Bakker and Schippers, 1987), from the high percentage of HCN-producing pseudomonads in the rhizosphere, from strong dependence of pseudomonas-HCN production on the availability of iron, and from the known presence of particular amino acids, such as glycine in the potato root exudates. Deleterious *Pseudomonas* spp. have been suspected to inhibit potato root development by their production of hydrogen cyanide (Bakker and Schippers, 1987; Bakker et al. 1989). Cyanide is an inhibitor of the terminal step in respiration and therefore extremely harmful to living cells (Knowles, 1976). Cyanide has also been reported to be responsible in the deleterious activity of *Pseudomonas* spp. causing growth inhibition in lettuce and bean (Alstrom and Burns, 1989). Cyanide production by deleterious rhizosphere pseudomonads thus is postulated to have a detrimental effect on plant establishment in several crops (Schippers et al. 1987). Phytotoxin production by deleterious *Pseudomonas* spp. has been implicated as a factor in the growth retardation of wheat seedlings (Frederickson and Elliott 1985). Similarly, Campbell et al. (1987) reported the involvement of polygalacturonic acid hydrolase in the inhibition of canola seedlings by deleterious *Pseudomonas* spp.

Plant Growth Promoting Rhizobacteria

Root colonizing bacteria especially in the genus *Pseudomonas* have captured special attention in the last two decades for their beneficial and deleterious effects on plant development. Since 1978, there has been increasing interest in plant growth promoting rhizobacteria (PGPR), a term first proposed by Kloepper and Schroth (1978, 1981), when they reported that fluorescent pseudomonads promoted early growth of radish and potatoes. The term plant growth promoting rhizobacteria (PGPR) denotes the group of bacteria which, when introduced to roots through seed bacterization cause significant increase in plant growth and yield. Specific strains of *Pseudomonas fluorescens—Pseudomonas putida* group rapidly colonize plant roots of several crops and cause statistically significant yield increases (Schroth and Hancock 1982). In addition to *Pseudomonas*, a large array of other bacteria viz., *Azospirilum, Azotobacter, Bacillus, Klebsiella, Hafnia, Enterobacter, Alcaligenes, Arthrobacter* and *Serratia* have also been shown to enhance plant growth in many instances.

The concept of PGPR is now well established (Bashan, 1998; Shishido and Chanway 1999). Enhanced plant growth caused by these bacteria is often accompanied with the suppression of root zone population of deleterious fungi and bacteria (Suslow, 1982; Geels and Schippers 1983; Gaskins et al. 1985). These beneficial strains are also potential biocontrol agents of several soil borne plant pathogens (Schroth and Hancock, 1982; Leong 1986; Whipps 2001). Kloepper et al. (1980) first conclusively proved that PGPR exert their plant growth promoting activity by depriving native microflora of iron. PGPR produce extracellular low molecular weight microbial iron transport agents called siderophores which effectively sequester available iron in the root environment making it unavailable to the native microflora. The hypothesis that the main mechanism of plant growth promotion by PGPR is competition with DRMO for iron is based on the observation that siderophore defective mutant of wild PGPR strains obtained by ultra-violet light or mutagenic chemicals or transposon mutagenesis also lost their plant growth promoting property (Kloepper and Schroth 1981; Marugg et al. 1985). Also, addition of ferric iron to the iron-limiting environment abolishes in vitro antagonism and also prohibits the plant growth promotion by PGPR. It has been also hypothesized that siderophores of PGPR have a higher affinity for iron than those of DRMO and pathogens (Leong and Neilands 1982; Buyer and Sikora 1990). Thus PGPR produce siderophores and sequester iron more efficiently and also they have the ability to use the ferrisiderophore complex produced by DRMO or the pathogens. On the contrary, the DRMO lack the receptor protein for the uptake of PGPR's ferrisiderophore complex. This mechanism is widely operative in the rhizosphere. Another means of plant growth promotion by PGPR is by suppression of plant diseases by induction of systemic resistance which has been termed as induced systemic resistance or ISR (Van Loon et al. 1998).

Alternatively, PGPR may increase plant growth by other, more direct means also e.g. by associative N_2 fixation (Hong et al. 1991), production of IAA and other harmones (Prikryl et al. 1985; Loper and Schroth 1986; Costacurta and Vanderleyden 1995), and enhancement of phosphate uptake (Lifshitz et al. 1987), solubilizing nutrients such as phosphorus (Whitelaw 2000), promoting mycorrhizal functions (Garbaye 1994), regulation of ethylene production in roots (Glick 1995), releasing phytoharmones (Arshad and Frankenberger, 1991; Beyeler et al. 1999), and decreasing heavy metal toxicity (Burd et al. 1998). Therefore, Bashan and Holguin (1998) suggested that these bacteria can be grouped into biocontrol plant growth promoting bacteria (biocontrol PGPB) and PGPB.

Plant growth promotion and yield increases by seed and seed piece bacterization with fluorescent pseudomonads and other bacteria have been well demonstrated in many crops. Burr et al. (1978) first demonstrated that strains of fluorescent *Pseudomonas* spp. applied to potato seed pieces increased the yield in field tests. These results were confirmed and growth promotion was also demonstrated for sugarbeet (Suslow and Schroth 1982), radish (Kloepper and Schroth 1978) and other crops. Potato yields increased from 5 to 33 % in field plots of California and Idaho (Kloepper et al. 1978). The most striking results were obtained with a 30 day old radish crop. Treatment with various strains of fluorescent pseudomonads resulted in increased root weight ranging from 60 to 144% in all field trials (Kloepper and Schroth 1978). Kloepper et al. (1988) reported significant yield increases in canola by the application of PGPR. The best commercial success story of PGPR was observed in China. When commercial application of PGPR began in China, PGPR was changed to yield increasing bacteria (YIB) because this term was more readily accepted by growers (Chen et al. 1996). The first YIBs were used in production agriculture in China in 1985 on turnip and rapeseed. Within a decade, commercial application of YIBs spread in China to cover 3.35 million ha with 1,132 tons of inoculants of YIB on 48 different crop species (Utkhede 2000). The research conducted in Australia by Ryder et al. (1996) and Yan et al. (1996) confirmed that some *Bacillus* spp. isolated in China and used as YIB gave both a PGPR effect and biocontrol of root diseases in many plants.

Root Colonization Potential of Plant Growth Promoting Rhizobacteria

The ultimate success of plant growth promotion by introduction of PGPR is largely dependent on the establishment of a significant population size along an elongated root system and their persistence throughout the growing season at sites where DRMO may become active. Many studies have addressed the competitive ability of specific rhizobacteria on roots under natural systems (Schroth and Hancock 1982). The manipulation of the root environment to favor the introduced organisms depends on the inoculum delivery system and the root colonizing ability of the particular organism. Delivery system of PGPR or biocontrol agents have mainly been effected by applying them to seeds or tubers, from which it is believed that PGPR or biocontrol agents will move and establish on and around the growing roots. Root colonizing ability, which encompasses the movement and proliferation in competition with indigenous microorganisms, appears to be the most important pre-requisite for subsequent effects on plant growth promotion and disease control (Parke 1990). Ahmad and Baker (1987) used the term rhizosphere competence to describe the ability of biocontrol agents to grow and function in the rhizosphere. Rhizosphere competence might also be thought of as the relative root colonizing ability of a rhizobacterial strain. Thus rhizosphere competence varies among bacteria, with strains unable to colonize roots being rhizosphere incompetent. Howie et al. (1987) hypothesized that colonization of wheat roots by *P. fluorescens* occurs in two phases. In first phase, the bacteria attach to roots, which are then transported on the elongated root tips, as the seed germinates. In second phase, the bacteria spread locally and proliferate. The ultimate fate of introduced bacteria is governed by its ability to compete. Nutrients rather than space are thought to be the limiting factor in competition among bacteria during rhizosphere colonization. The first step in such bacterial colonization is the rapid colonization of the germinating seed which profusely exudes a wide range of amino acids, carbohydrates and organic acids during germination (Hayman 1969; Stanghellini and Hancock 1970; Balasubramanian and Rangaswami 1978; Lynch 1978). Chemotaxis towards rich seed exudates containing nutrient may represent a competitive advantage for certain bacteria in early establishment on seed and radicle. Scher et al. (1985) demonstrated the chemotactic attraction of *Pseudomonas putida* strain RW1 to asparagine, threonine and valine at c+oncentrations present in seed exudates and these compounds induced the effects comparable to those of the exudates itself. Similarly, chemotaxis of fluorescent pseudomonads towards seed exudates and germinating tomato seeds was demonstrated by Gamliel and Katan (1992). The role of flagella in root colonization is controversial. The non-motile mutants of *Pseudomonas* were not impaired in root colonization of wheat and soyabean (Howie et al. 1987 and Scher et al. 1988). de Weger et al. (1987), however, found that several non-motile mutants of *P. fluorescens* strain WCS 374 were severely impaired in potato root colonization. Several strains of biocontrol agent *P. fluorescens* carrying mutation in the general chemotaxis gene *che A* were highly defective in tomato root tip colonization as non-motile strains indicating that chemotaxis rather than random motility is involved in root colonization (Lugtenberg et al. 2001). Vesper (1987) found that take-all suppressive strains of *P. fluorescens* Z-79 produce fimbriae that mediate attachment to corn roots.

Surface charged properties of bacteria and plant cells have been suggested to be involved in the next event in colonization process (James et al. 1985). Interaction of polysaccharides and proteins at the substratum surface probably influence the subsequent bacterial adhesive interactions (Pringle and Fletcher 1986). Dekkers et al. (1998) revealed that O-antigen of lipopolysaccharide of *Pseudomonas fluorescens* is required for competitive colonization. However, the exact mechanism by which this part of lipopolysaccharide plays a role in colonization is still not clear. Another factor that affects the movement and distribution of the introduced bacteria along the root system is water infiltration or downward flow of water. Parke et al. (1986) experimentally showed a major effect of downward movement of water along the roots in distributing the microorganisms to the sites for subsequent growth. Liddel and Parke (1989) demonstrated that infiltrating water contributed to a downward movement of *Pseudomonas fluorescens* strains PRA 25 along the root system which probably redistributed the bacteria from seed or seed-piece to lower parts of the root system. Competition for nutrients supplied by exudates from roots and seed is an important factor in successful root colonization. Generally, aminoacids, monosaccharides and organic acids are the major exudate compounds. A mutant of *P. fluorescens* strain WCS 365, designated as PCL 1085 which was defective in utilization of malic acid and succinic acid was found to be poor root colonizer indicating that utilization of exudate organic acids is the nutritional basis of tomato root tip colonization. Dekkers et al. (1998) reported that membrane bound NADH dehydrogenase I (NDH-I) plays an important role in root colonization as mutation in the *nuo D* gene (involved in NDH-I production) impaired the competitive root colonization. Fluorescent pseudomonads are better root colonizer because they are nutritionally versatile and can utilize a wide range of organic substrates (Stolp and Gadkari 1981) and are well adapted to root exudates. In addition, they have a very fast growth rate relative to other bacteria in the rhizosphere, resulting in their better establishment in the rhizosphere (Rouatt and Katznelson 1961).

Colonization may involve simply root surface development. However, endophytic colonization of the root is also known. Endophytic colonization has been observed with *P. fluorescens* Sm 3-RN on spruce (Shishido et al. 1999), *P. fluorescens* 2-79RN$_{10}$ on wheat (Kim et al. 1997), *Bacillus pumilus* Meyer & Gottheil SE 34 and *P. fluorescens* 63-28 on pea (Benhamou et al. 1996; M' Piga et al. 1997), *P. fluorescens* CHA0 on tobacco (Troxler et al. 1997), and *P. fluorescens* WCS 4176 on tomato (Duijff et al. 1997). These endophytic bacteria are particularly important because they are able to grow and compete in rhizosphere as well as within roots. Indeed, many seeds, roots and tubers are normally colonized by endophytic bacteria (McInroy and Kloepper 1995; Sturz et al. 1999).

Biological Control by Rhizobacteria

Interest in biological control of plant diseases has increased recently fuelled by the public concerns over the use of chemicals in the environment. To achieve successful and reproducible biological control, the knowledge on ecological interactions in soil and root environment is the most essential. Significant disease suppression can be achieved by manipulation of physicochemical and microbiological environment through management practices such as soil amendments, soil solarization and crop rotations. In some soils, described as microbiologically suppressive to pathogens (Schneider 1982), microbiological antagonism of the pathogens is especially great, leading to substantial disease control. However, at present the prime interest is in the development and application of specific biocontrol agents for the control of plant diseases on seeds and roots. Biological control has thus become an important input in the integrated disease management.

Microorganisms that can grow in the rhizosphere are ideal for use as biocontrol agents since the rhizosphere provides the front line defense for roots against attack by the plant pathogens. Among the various antagonists used for the management of plant diseases, *Trichoderma* and *Pseudomonas* play a vital role. The isolates of *Trichoderma, T. viride, T. harzianum, T .virens* and *T. hamatum* are used against the management of various diseases of crop plants especially with dreaded soil-borne pathogens. It has many advantages as a bio-control agent owing to its high rhizosphere competence, ability to synthesize polysaccharide–degrading enzymes, amenability for mass multiplication, broad spectrum of action against various pathogens and above all its environmental friendliness. Fluorescent pseudomonads which are important part of the rhizosphere microflora play a significant role in the suppression of plant diseases. Fluorescent pseudomonads antagonize the plant pathogens by aggressive root colonization and by producing one or more of a variety of metabolites that include antibiotics (Fravel 1988), siderophores (Schippers et al. 1987) and other substances like cyanide (Voisard et al. 1989). The different species of *Bacillus* form the second most important group of rhizobacteria that have the potential to act as biocontrol agents. Apart from these two groups of bacteria plant disease control has been reported by a number of other rhizosphere bacteria.

Good soil biocontrol systems have been reported for species of *Fusarium, Sclerotium, Sclerotinia, Pythium* and *Rhizoctonia*. The following biocontrol agents have already been registered; *Agrobacterium radiobactor* against crown gall (USA, Australia, NZ); *Bacillus subtilis* for growth enhancement (USA); *Pseudomonas fluorescens* against bacterial blotch (Australia); *Pseudomonas fluorescens* for seedling diseases (USA); *Peniophora gigantea* against *Fommes annosus* (UK); *Pythium oligandrum* against *Pythium* spp. (USSR); *Trichoderma viride* against timber pathogens (Europe); *Trichoderma* spp. for root diseases (USSR); *Fusarium oxysporium* against *Fusarium oxysporum* (Japan); *Trichoderma harzianum* against root diseases (USA); *Gliocadium virens* for seedling diseases (USA) and *Trichoderma harzianum* or *Polysporum* against wood decay (USA).

Biological control of different plant diseases by rhizobacteria is summarized in Table 1.

BIOCONTROL OF SOIL-BORNE DISEASES

Chemical control of soil borne plant diseases is frequently ineffective because of the physical and chemical heterogeneity of the soil, which may prevent effective concentrations of the chemical from reaching the pathogen. Biological control agents colonize the rhizosphere, the site requiring protection and leave no toxic residues, as opposed to chemicals.

Microorganisms have been used extensively for the biological control of soil borne plant diseases as well as for promoting plant growth. Fluoroscent pseudomonas are the most frequently used bacteria for biological control and plant growth promotion, but *Bacillus* and *Streptomyces*

TABLE 1 *A list of plant disease control by biocontrol bacteria:*

Antagonist	Pathogen	Disease	Reference
Rice			
Pseudomonas fluorescens	*Rhizoctonia solani*	Sheath blight	Devi et al. (1989), Sivakumar and Narayanaswamy (1998), Nagaraju et al. (2002)
P. fluorescens str. Pf ALR 2	*R. solani*	Sheath blight	Rabindran and Vidhyasekaran (1996)
P. fluorescens str. Pf 7-14 *P.putida* V 14 i	*R. solani*	Sheath blight	Krishnamurthy and Gnanamanickem (1998)
P. fluorescens str. Pf1	*R. solani*	Sheath blight	Vidhyasekaran and Muthamilan (1999)
Fluorescent *Pseudomonas sp.*	*R. solani*	Sheath blight	Laha and Venkataraman (2001)
P. fluorescens (Strain Pf1 & FP7)	*R. solani*	Sheath blight	Commare et al. (2002)
P fluorescens Strain P, FP7 and PB2	*R. solani*	Sheath blight	Nandakumar et al. (2002)
P. fluorescens str. Pf 7-14	*R. solani* *Pyricularia grisea*	Sheath blight Blast	Chatterjee et al. (1996)
P. fluorescens str. Pf 7-14	*Magnaporthe grisea*	Blast	Krishnamurthy and Gnanamanickem (1998)
P. fluorescens str. Xh 216	*M. grisea*	Blast	Lin et al. (1998)
P. fluorescens (ABPf-1)	*M. grisea*	Blast	Kumar et al. (2000)
P. fluorescens	*M. grisea*	Blast	Karpagavalli et al. (2001)
P. fluorescens	*M. grisea*	Blast	Mosa (2002)
P. fluorescens	*Fusariem moniliforme*	Foot rot	Ahamad and Srivastava, (2002)
P. fluorescens Strain S3	*Achlya klebsiana* *Pythium spinosum*	Seedling disease	Adhikari et al. (2001)
Enterobacter cloacae MR 12 *Alcaligenes paradoxus* R4	*Xanthomonas oryzae* pv. *oryzae*	Bacterial blight	Yang et al. (1999)
P. fluorescens Strain Pf1	*X. oryzae pv. oryzae*	Bacterial blight	Vidhyasekaran et al. (2001)
P. fluorescens	*X. oryzae* pv. *oryzae*	Bacterial blight	Manav and Thind (2002)
P. fluorescens Strain Pf1	*Hirschmanniella gracilis*	Rice root nematode	Ramakrishnan et al. (1999)

(Table 1 Contd.)

(Table 1 Contd.)

Antagonist	*Pathogen*	*Disease*	*Reference*
Wheat			
Pseudomonas spp. B-10	*Gaeumannomyces graminis* var. *tritici*	Take all	Kloepper et al. (1980)
P. fluorescens strain 2-79	*G. graminis* var. *tritici*	Take all	Weller and Cook (1983), Thomashow and Weller (1988)
P. fluorescens strains 2-79 RNo, R1a-80R, R7z-80R	*G. graminis* var. *tritici*	Take all	Weller et al. (1988)
P. aureofaciens	*G. graminis* var. *tritici*	Take all	Vincent et al. (1991)
P. fluorescens	*G. graminis* var *tritici*	Take all	Wuthrich and Defago (1991)
P. fluorescens strain Q 8r 1-96	*G. graminis* var. *tritici*	Take all	Raaijmakers and Weller (1998)
P. corrugata strain 2140	*G. graminis* var. *tritici*	Take all	Choi et al. (2003)
P. fluorescens Strain 2-79	*G. graminis* var *tritici* and *Rhizoctonia spp.*	Take all and Rhizoctonia root rot	Duffy (2000)
Bacillus spp. strain L 324-92R12	*Rhizoctonia solani* *Pythium* AG 8 spp.	Root rot	Kim et al. (1997)
P. fluorescens strain 2-79B *Burkholderia cepacia* strain 1-23 *Pseudomonas* spp. strain 1-30	*Pythium irregulare* *P. torulosum* *P.graminicola*	Root rot	Milus and Rothrock (1997)
Maize			
Pseudomonas fluorescens str. AB 254	*P. ultimum*	Pre-emergence damping off	Callan et al. (1990)
P. corrugata strain 1 & 7	*Pythium ultimum* *P. arrhenomanes* *Fusarium graminearum*	Seedling damping off	Pandey et al. (2001)
P. fluorescens Strain PF1	*Rhizoctonia solani f.sp. sasaki*	Banded leaf and sheath blight	Sivakumar et al. (2000)
Bacillus mojavensis	*F. moniliforme*	Wilt	Bacon and Hinton (2002)
P. fluorescens str. PJO 210	*Bipolaris maydis*	Leaf spot	Mohamed and Caunter (1995)
Barley			
Pseudomonas chlororaphis MA 342	*Drechslera graminea* *D. teres*	Seed infection	Johnsson et al. (1998)
Pantoea agglomerans	*Pseudomonas syringae* pv. *syringae*	Basal kernel blight	Braun et al. (2000)

(Table 1 Contd.)

(Table 1 Contd.)

Antagonist	Pathogen	Disease	Reference
Pearl millet/Fingermillet			
Pseudomonas fluorescens	*Sclerospora graminicola*	Downy mildew of pearl millet	Umesha et al. (1998)
P. fluorescens str. 7-14	*Magnoporthe grisea*	Blast	Sittar and Gnanamanickam (1996)
Potato			
Pseudomonas putida strain M 17	*Erwinia carotovora sub sp. carorovora*		Soft rot Colyer and Mount (1984)
P. putida strain pp 22	*E. carotovora* sub sp. *carotovora*	Soft rot	Liao (1989)
P. fluorescens strain F 113	*E. carotovora* sub sp. *carotovora* *E. carotovora* sub sp. *atroseptica*	Soft rot	Cronin et al. (1997)
Pseudomonas sp. *Enterobacter cloacae*	*Giberella pulicaris*	Dry rot	Schisler et al. (2000)
P. aureofaciens IS 1	*Clavibacter michiganensis* sub sp. *sepedoniurs*	Ring rot	De la Cruz et al. (1992)
Bacillus spp. isolate S 1	*Ralstonia solanacearum*	Bacterial wilt	Shekhawat et al. (1993)
P. fluorescens strain BC 8	*R. solanacearum*	Bacterial wilt	Ciampi et al. (1997)
Cellulomonas fimi	*H. solani*	Potato silver scurf	Martinez et al. (2002)
Alcaligenes piechaudii *Arthrobacter oxydans* *Bacillus mycoides*	*Helminthosporium solani*	Potato silver scurf	Michaud (2002)
Tomato			
Streptomyces diastaticus MN-196	*Pythium aphanidermatum*	Damping-off	Singh and Reddy (1979)
Pseudomonas fluorescens Strain Pf1	*P. aphanidermatum*	Damping-off	Ramamoorthy et al. (2002)
P. aureofaciens strain AB 244	*P. ultimum*	Damping off	Warren and Bennet (1999)
P. fluorescens 5014 & 5-2/4	*P.ultimum* var. *ultimum*	Damping-off	Hultberg et al. (2000)
Enterobacter cloacae	*P. ultimum*	Damping-off	Kageyama and Nelson (2003)

(Table 1 Contd.)

(Table 1 Contd.)

Antagonist	*Pathogen*	*Disease*	*Reference*
Streptomyces sp. Di-944	*R. solani*	Rhizoctonia damping-off	Sabaratnam and Traquair (2002)
P. fluorescens str. CHA0	*Fusarium oxysporum f.sp. radicis-lycopersici*	Wilt	Duffy and Defago (1997)
P. chlororaphis str. PCL 1391	*F. oxysporum f. sp. radicis-lycopersici*	Wilt	Chin-A-Woeng et al. (1998)
P. fluorescens Strain Pf1	*F.oxysporum* f.sp. *lycopersici*	Wilt	Ramamoorthy et al. (2002)
P. fluorescens str. Pfcp	*Ralstonia solanacearum*	Wilt	Anuratha and Gnanamanickam (1990)
P. fluorescens Strain WCS356	*F.oxysporum* f.sp. *lycopersici*	Foot and root rot	Dekkers et al. (2000)
P. fluorescens Strain CHA0 *P. aeruginosa* Strain IE-6S+	*Macrophomina phaseolina* *F. solani* and *Rhizoctonia solani*	Root infecting fungi	Siddiqui and Shaukat (2002a)
Bacillus subtilis Isolate 728	*Botrytis cineria*	Grey mold	Zhao et al. (2003)
P. fluorescens 89B61 *Bacillus pumilis* SE 34	*Phytophthora infestans*	Late blight	Yan et al. (2002)
P. syringae Strain Cif 7	*Pseudomonas syringae* pv. *syringae*	Bacterial speck	Wilson et al. (2002)
P. fluorescens Strain CHA0 *P. aeruginosa* Strain IE-6S+	*Meloidogyne javanica*	Root knot nematode	Siddiqui and Shaukat (2002b)
Radish/Carrot			
Pseudomonas fluorescens strain A 12	*Fusarium oxysporum f. sp. conglutinans*	Wilt	Scher and Baker (1982)
P. fluorescens strain WCS 374	*F. oxysporum f. sp. conglutinans*	Wilt	Leeman et al. (1991)
P. fluorescens strain WCS 374	*F. oxysporum f. sp. raphani*	Wilt	Leeman et al. (1995)
P. *fluorescens* WCS 417	*F. oxysporum f. sp. raphani*	Wilt	Duijff et al., (1998)
P. putida	*F. oxysporum f. sp. raphani*	Wilt	De Boer et al. (1999)
P. putida WCS358	*Fusarium* sp.	Wilt	De-Boer et al. (2003)
Enterobacter cloacae	*Pythium ultimum*	Damping-off	Kageyama and Nelson (2003)

(Table 1 Contd.)

(Table 1 Contd.)

Antagonist	Pathogen	Disease	Reference
Cucumber/Muskmelon/ Watermelon			
Pseudmonas aureofaciens 63-28 *P. corrugata* 13	*Pythium aphanidermatum*	Cucumber damping off	Chen et al. (1998)
Bacillus subtilis Strain BACT-0	*P. aphanidermatum*	Cucumber damping-off	Utkhede et al. (1999)
Serratia plymuthica	*P. ultimum*	Cucumber damping off	Benhamou et al. (2000)
P. fluorescens *P. corrugata*	*P. ultimum*	Cucumber damping- off	Georgapoulos et al. (2002)
P. aeruginosa Strain CR56	*P. ultimum*	Cucumber damping-off	Lou et al. (2002)
Enterobacter cloacae	*P. ultimum*	Cucumber damping-off	Kageyama and Nelson (2003)
P. fluorescens str. BTP 7 *P. putida* str. BTP 1	*P. aphanidermatum*	Cucumber root rot	Ongena et al. (1999)
P. fluorescens strain A 12 & N1R	*Fusarium oxysporum f. sp. cucumerium*	Cucumber wilt	Parke et al. (1988)
Streptomyces spp. 385 *Paenibacillus* spp. 300	*F. oxysporum f. sp. cucumerium*	Cucumber wilt	Singh et al. (1999)
B. subtilis Isolate 1 &2	*P. ultimum* *F. oxysporum* f.sp. *cucumerinum*	Basal stem rot and wilt of cucumber	Hamed (1999)
E. agglomerans (B8Fr) *B. subtilis* (AGS-4)	*Didymella bryoniae*	Gummy stem blight of green house cucumber	Utkhede and Koch (2002)
Serratia plymuthica Strain IC14	*Botrytis cineria, Sclerotinia sclerotiorum*	Grey mold and white mold of cucumber	Kamensky et al. (2003)
P. fluorescens	*F. oxysporum* f.sp. *melonis*	Muskmelon wilt	Ozaktan and Bora (2000)
Cauliflower/Cabbage/Rape/ Broccoli/Lettuce			
Bacillus polymyxa, BP1	*Xanthomonas campestris* pv. *campestris* *Alternaria brassicicola*	Black rot and damping-off of cauliflower	Pichard and Thouvenot (1999)
B. subtilis Strain -BB	*X. campestris* pv. *campestris*	Black rot of brassicas	Wulff et al. (2002)
Serratia marcescens *Streptomyces viridodiasticus* *Micromonospora carbonacea*	*Sclerotinia minor*	Root disease of lettuce	El-Tarabily et al. (2000)

(Table 1 Contd.)

(Table 1 Contd.)

Antagonist	*Pathogen*	*Disease*	*Reference*
Chilli/Black pepper			
Pseudomonas fluorescens	*Pythium aphanidermatum*	Damping –off of chilli	Manoranjitham et al. (1999)
Alcaligenes sp. Starin AMB 8	*Phytophthora capsici*	Nursery rot of black pepper	Anith and Manomohandas (2001)
Serratia plymuthica A21-4	*P. capsici*	Phytophthora blight of pepper	Shen et al. (2002)
Peas and Beans			
Pseudomonas cepacia str. AMMD *P. fluorescens* str. PRA 25	*Aphanomyces euteiches f.sp.pisi* and *Pythium spp.*	Root rot and Damping-off of Pea	Parke et al. (1991) Bowers and Parke (1993)
P. fluorescens str. PRA 25	*Pythium ultimum Rhizoctonia solani*	Pea root rot	Xie et al. (1996)
P. fluorescens Strains SBW25 and CHA0	*P. ultimum*	Pea damping-off	Naseby et al. (2001)
Bacillus subtilis Strain GBO3	*Fusarium solani* f.sp. *solani* R. *solani* *F. oxysporum*	Dry bean root rot	De-Jensen et al. (2002)
P. putida	*F. oxysporum f. sp. phaseoli*	Bean wilt	Anderson and Guerra (1985)
P. fluorescens str. S 97	*P. springae pr. phaseolicola*	Bean halo blight	Alstrom (1991)
Chick pea			
Streptomyces diastaticus FA-217	*Sclerotium rolfsii*	Seedling blight	Singh and Reddy 1979
Pseudomonas fluorescens Strain Pfs3	S. *rolfsii*	Seedling mortality	Sarma et al. (2002)
P. aeruginosa Strain Pag	S. *rolfsii*	Root rot	Singh et al. (2003)
P. fluorescens Isolate 4-92	*Macrophomina phaseolina*	Charcoal rot	Srivastava et al. (2001)
P. fluorescens	*Rhizoctonia bataticola*	Dry root rot	Sindhan et al. (2002)
P. fluorescens NBRI 1303	*Fusarium oxysporum* f. sp. *ciceris* M. phaseolina *Pythium spp.*	Root rot & wilt	Nautiyal (1997)
P. fluorescens (Pf1, Pf27 & Pf2)	*F. oxysporum* f. sp. *ciceris*	Wilt	Vidhyasekaran and Muthamilan (1995)
P. fluorescens strs. Q29Z-80 & M8Z-80	*Pythium ultimum*	Pre-emergence damping off	Kaiser and Hannan (1989)

(Table 1 Contd.)

(Table 1 Contd.)

Antagonist	*Pathogen*	*Disease*	*Reference*
P. P *fluorescens* str. Q29Z-80	*Pythium* spp.	Pre-emergence damping off	Trapero-Casas et al. (1990)
Pigeon pea and Lentil			
Pseudomonas fluorescens	*Fusarium udum*	Pigeon pea wilt	Vidhyasekaran et al. (1997)
P. fluorescens	*F. udum*	Pigeon pea wilt	Siddiqui et al. (1998)
P. fluorescens strs. GRP 3 & PRS 9	*F. oxysporum* f. sp. *lini*	Wilt of lentil	Rao et al. (1999)
Ground nut/Pea nut			
Pseudomonas fluorescens	*Sclerotium rolfsii*	Collar rot	Ganesan and Gnanamanickam (1987)
P. fluorescens FDP-15	*S. rolfsii*	Collar rot	Patil et al. (1998)
P. fluorescens Pf 1	*Macrophomina phaseolina*	Root rot	Shanmugam et al. (2002)
P. fluorescens	*Cercosporidium personatum*	Late leaf spot	Meena et al. (2002)
Sunflower			
Pseudomonas fluorescens isolate IV	*Alternaria helianthi*	Seedling blight	Prasad and Kulshrestha (2002)
P fluorescens str. MCO 7	*Pythium ultimum*	Damping-off	Kim et al. (1998)
P. fluorescens str. GL-92 *P. putida* str. TI-5	*Sclerotina sclerotiorum*	Wilt	Expert and Digat (1995)
Safflower			
Pseudomonas fluorescens	*Rhizoctonia bataticola*	Root rot	Prashanthi et al. (2000)
Sesame			
Bacillus polymyxa KB-8	*Fusarium oxysporum* f. sp. *sesami*	Wilt	Hyun et al. (1999)
Flax			
Pseudomonas spp. B-10	*Fusarium oxysporum* f. sp. *lini*	Wilt	Kloepper et al. (1980)
P. putida strain A 12	*F. oxysporum* f. sp. *lini*	Wilt	Scher and Baker (1982)
P. fluorescens strain WCS 358	*F. oxysporum* f. sp. *lini*	Wilt	Duijff et al. (1998)
Cotton			
Pseudomonas fluorescens strain 3551	*Pythium ultimum*	Pre-emergence damping off	Loper (1988)

(Table 1 Contd.)

(Table 1 Contd.)

Antagonist	*Pathogen*	*Disease*	*Reference*
P. fluorescens Hr 37a R2	*P. ultimum*	Damping off	Howie and Suslow (1991)
Enterobacter cloacae	*P ultimum*	Damping-off	Kageyama and Nelson (2003)
P. fluorescens	*Rhizoctonia solani*	Root rot	Howel and Stipanovic (1979)
P. fluorescens Pf1	*R. bataticola*	Root rot	Marimuthu et al. (2002)
P. fluorescens	*Pythium carolinianum*	Root rot	Abdelzaher and Elnaghy (1998)
P. fluorescens strain EG 1053	*P. ultimum* *R. solani*	Seedling disease complex	Hagedorn et al. (1990)
P. fluorescens (CRb-26)	*Xanthomonas axonopodis* pv. *malvacearum*	Bacterial blight	Mondal et al. (1999)
P. fluorescens Strain Crb-26	*X. axonopodis* pv. *malvacearum*	Bacterial blight	Monda, et al., (2001)
P. fluorescens Strain PR8	*X. axonopodis* pv. *malvacearum*	Bacterial blight	Bhowmik et al. (2002)
Tobacco			
Pseudomonas fluorescens str. CHA0	*Thielaviopsis basicola*	Black root rot	Keel et al. (1992)
P. fluorescens str. CHA0	*T. basicola*	Black root rot	Laville et al. (1992)
P. fluorescens strain CHA0	Tobacco necrosis virus	Tobacco necrosis virus disease	Maurhofer et al. (1995)
Sugarcane			
Pseudomonas fluorescens Starin EP1 and Pf1	*Colletotrichum falcatum*	Red rot	Viswanathan and Samiyappan (2002)
Sugar beet			
Streptomyces diastaticus FA-217	*Sclerotium rolfsii*	Seedling blight	Singh and Reddy (1979)
Pseudomonas fluorescens *P. corrugata*	*Pythium ultimum*	Damping off	Georgapoulos et al. (2002)
P. fluorescens DR54	*P. ultimum*	Damping-off	Thrane et al. (2000)
Bacillus subtilis Isolate Bac-B	*Cercospora beticola*	Cercospora leaf spot	Collins and Jacobsen (2003)
Fruits			
Pseudomonas fluorescens	*Fusarium oxysporum f. sp.cubense*	*Fusarium* wilt of banana	Raguchander et al. (1997)

(Table 1 Contd.)

(Table 1 Contd.)

Antagonist	Pathogen	Disease	Reference
P. fluorescens Starin Pf1	*F. oxysporum* f. sp. *cubense*	Banana wilt	Rajappan et al. (2002)
Bacillus subtilis Isolate-01	*Lasiodiplodia theobromae*	*Lasiodiplodia* rot of banana	Mortuza and Ilag (2000)
Pantoea agglomerans Strain C9-1 & E 325	*Erwinia amylovora*	Fire blight of apple and pear	Pusey et al. (2002)
Pantoea agglomerans Strain Eh252	*E. amylovora*	Fire blight of apple and pears	Stockwell et al. (2002)
P. fluorescens Strain A506	*E. amylovora*	Fire blight of pear	Lindow and Suslow (2003)
B. subtilis Strain AB27	*E. amylovora*	Fire blight of pear	Basim et al. (2002)
B. subtilis *Erwinia herbicola* Strain JII/E2	*Eutypha lata*	Die back of grapevine	Schmidt et al. (2001)
B. subtilis Isolate 25021	*Botrytis cineria*	Grey mold of strawberry	Helbig and Bochow, (2001)
P. cepacia	*Penicillium digitatum*	Green mould	Smilanick and Denis-Arrue, (1992)
Serratia plymuthica HRO-C48	Verticillium dahliae *Phytophthora cactorum*	*Verticillium* wilt and *Phytophthora* root rot	Kurze et al. (2001)
B. subtilis Strain EBW-4	*Phytophthora cactorum*	Crown and root rot	Utkhede et al. (2001)
Flowers			
Bacillus polymyxa	*Fusarium oxysporum* f.sp. *tulipae*	wilt	Saniewska (1999)

species have also been commonly used. *Trichoderma, Gliocladium,* and *Coniothyrium* are the most commonly used fungal biocontrol agents. Perhaps the most successful biocontrol agent of a soilborne pathogen is *Agrobacterium radiobactor* strain K84, used against crown gall disease caused by *A.tumefaciens* (Kerr, 1980).

Competition as a mechanism of biological control has been exploited with soil borne plant pathogens as with the pathogens on the phylloplane. Naturally occurring, nonpathogenic strains of *Fusarium oxysporium* have been used to control wilt diseases caused by pathogenic *Fusarium* spp. Molecular techniques have also facilitated the introduction of beneficial traits into rhizosphere competent organisms to produce potential biocontrol agents (Chet 1993). Chitin and β – (1,3) – glucan are the two major structural components of many plant pathogenic fungi, except oomycetes, which contain cellulose in their cell wall and no appreciable levels of chitin. Biological control of some soil borne fungal diseases has been correlated with chitinase production. Bacteria producing chitinases or glucanases exhibit antagonism *in vitro* against fungi (Ordentlich et al. 1987). A recombinant *Escherichia coli* expressing the *chi A* gene from *Serratia marcescens* was effective in reducing disease incidence caused by *Sclerotium rolfsii* and *Rhizoctonia solani* (Gal et al. 1997). In other studies, chitinase genes from *S. marcescens* have been expressed in *Pseudomonas* spp. and the plant symbiont *Rhizobium meliloti*. The modified *Pseudomonas* strain controlled the pathogen *F. oxysporium* f. sp. *rodelens* and *Gauemannomyces graminis* var. *tritici*.

BIOCONTROL STRATEGIES FOR DISEASE ON GREENHOUSE PLANTS

The greenhouse environment presents a unique situation that may make conditions more favourable for plant diseases (Jarvish 1982). Floriculture, ornamental, aromatic and medicinal plants are grown in greenhouses or polyhouses. The competitive nature of the floriculture industry in the world trade and the stiff competition from foreign firms has forced many growers to become more efficient in production. This means that adequate control measures must be found to prevent losses due to diseases caused by recurring pathogens like *Botrytis, Pythium*, and *Rhizoctonia*. As a part of developing improved disease control strategies, studying new technologies for the use of biological control is under way in the USA and elsewhere. Biocontrol takes advantage of environmentally-safe microorganisms, predators, or parasites that are introduced during crop production for disease or pest control. Several microorganisms like *Burkholderia cepacia* and binucleate *Rhizoctonia* have been identified as effective biocontrol agents previously. The technology must now be developed to extend their range of effectiveness against multiple pathogens as well as to develop formulations that provide suitable shelf life, efficacy, and user-friendliness in the greenhouse environment. Greenhouse crops have a high economic value, and therefore can absorb higher cost input to control plant diseases. The biocontrol agents can be directly applied to growing mix, in the fertigation system, sprayed on plants, or applied to high value hybrid seed. Because of the reduced area and high density planting, less inoculum is needed than in treating a field. The use of chemical fungicides may result in pathogen resistance over time, but long-term use of bicontrol is not an issue because plant pathogens do not develop resistance to biocontrol agents. The first objective of research should be to understand the mechanisms associated with biocontrol of several diseases on greenhouse plants (*Rhizoctonia* stem rot, *Pythium* dampimg-off, *Sclerotium* rot, *Fusarium* wilt and *Botrytis* gray mold). Understanding mechanisms involved in biocontrol is essential for developing strategies to control plant diseases as well as assessing possible resistance development of pathogens to biocontrol agents. In some cases, it is hard to control plant diseases during the entire crop production period by applying single biocontrol agent. Additional application of the same biocontrol agent or integrated use of two or more biocontrol agents needs to be considered (Cook 1993; Van Loon et al. 1998). This approach may provide benefits to ornamentals industry by extending crop protection period and controlling wide range of plant diseases.

(i) The *Trichoderma* System as a Model of Biocontrol of Plant Pathogens

Trichoderma spp. act against a range of economically important aerial and soil borne plant pathogens (Samuels, 1996). They have been used in the field and greenhouse against silver leaf on plum, peach and nectarine; Dutch elm disease on elm's honey fungus (*Armillaria mellea*) on a range of tree species; and against rots on a wide range of crops, caused by *Fusarium, Rhizoctonia, Pythium* and *Verticillum,* and sclerotium forming pathogens such as *Sclerotinia* and *Sclerotium* (Weindling 1932). In many, experiments', showing successful biological control, the antagonistic *Trichoderma* was mycoparasite.

Mycoparasitism is a complex process, including several successive steps. The first detectable interaction shows that the hyphae of the mycoparasite grow directly towards its host. This phenomenon appears a chemotropic growth of *Trichoderma* in response to some stimuli in the host or growth towards a gradient of chemicals produces by the host (Inbar and Chet 1995). When the mycoparasite reaches the host, its hyphae often coil around it or are attached to it by forming hook like structures. In this respect, production of appressoria at the tips of short branches has been described for *T. hamatum* and *T. harzianum*. The interaction of *Trichoderma* with its host is specific. The possible role of agglutinins in the recognition process determining the fungal specificity has been recently examined. Indeed, recognition between *T. harzianum* and two of its major hosts, *R. solani* and *S. rolfsii,* was controlled by two different lectins present on the host hyphae. *R. solani* carries a lectin that binds to galactose and fructose residues on the *Trichoderma* cell walls. This lectin agglutinates conidia of a mycoparasitic strain of *T. harzianum,* but did not agglutinate two non-parasitic strains (Stasz et al. 1996). This agglutinin may play a role in prey recognition by the predator. Moreover, because it does not distinguish among biological variants of the pathogen, it enables the *Trichoderma* species to attack different *R. solani* isolates. D-glucose or D-mannose residues apparently present on the cell walls of *T. harzianum*, inhibited the activity of a second lectin isolated from *S. rolfsii* (De la Cruz, et al. 1995) Following these interactions the mycoparasite sometime penetrates the host mycelium, apparently by partially degrading its cell wall. Microscopic observations led to the suggestion that *Trichoderma* spp. produced and secreted mycolytic enzymes responsible for the partial degradation of the host's cell wall. The complexity and diversity of the chitinolitic system of *T.harzianum* involves the complementary modes of action of six enzymes, all of which might be required for maximum efficiency against a broad spectrum of chitin-containing plant pathogenic fungi. The level of hydrolytic enzymes produced differs from host-parasite interaction analysed. This phenomenon correlates with the ability of each *Trichoderma* isolate to control a specific pathogen. It is considered that mycoparasitism is one of the main mechanisms involved in the antagonism of *Trichoderma* as a biocontrol agent. The process apparently includes chemotropic growth of *Trichoderma*, recognition of the host by the mycoparasite, secretion of extracellular enzymes, hyphae penetration, and lysis of the host (Sivan and Harman 1991).

The involvement of volatile and nonvolatile antibiotics in the antagonism by *Trichoderma* has been proposed. Indeed some isolates of *Trichoderma* excrete growth-inhibitory substances. Thus, the biocontrol ability of *Trichoderma* strains is most likely conferred by more than one exclusive mechanism. In fact, it seems advantageous for a biocontrol agent to suppress a plant pathogen using multiple mechanisms (Goldman et al. 1994)

(ii) Mechanism of Antagonism and Disease Suppression

For achieving biological control, antagonists should have the ability to colonize the root system effectively (Weller 1988; Parke 1990) and to produce certain antagonistic secondary metabolites (Defago and Haas 1990). These beneficial bacteria suppress the plant pathogens by producing one or more of a variety of mechanisms and metabolites that include antibiotics (Fravel 1988; Thamashow and Weller 1990; Keel et al.1992), siderophores (Kloepper *et al.* 1980; Leong, 1986; Schippers et al. 1987), parasitism and lysis (Ordentich et al. 1987; Campbell and Ephgrave 1983), volatile substances such as cyanide (Voisard et al. 1989) or induced systemic resistance or ISR (van Loon et al. 1998). Activity is not restricted to only one of these mechanisms, and indeed an efficient biocontrol agent may affect pathogens by a combination of mechanisms (Chet, 1987). The improvement and eventual commercialization of biocontrol bacteria depends in part on understanding and exploiting the mechanisms involved in these anatagonistic interactions. For example, Chet et al. (1993) reported the transfer of chitinase genes from *Serratia*, *Aeromonas* and *Trichoderma* into *Pseudomonas putida* resulting in the increased antagonistic activity of the transformant against *Sclerotium rolfsii* and *Rhizoctonia solani*. Some specific examples of modes of pathogen suppression by bacterial biocontrol agents are discussed below.

(iii) Competition

Competition between the biocontrol bacteria and the pathogen can result in the significant displacement of the pathogen. Microorganisms compete for nutrients, space and other essential elements in the soil and around plant rhizosphere. For example, Elad and Chet (1987) found that various strains of *Arthrobacter* sp. compete for various nutrients with germinating oospores of *Pythium aphanidermatum* and results in the pathogen suppression. However, the greatest interest has been developed in the study of competition for iron. Kloepper et al. (1980) were the first to demonstrate the importance of siderophores {a low molecular weight, high affinity iron (ferric) chelating agent that transport iron into the cells of microorganisms} in the mechanisms of biological control by fluorescent pseudomonads. Siderophore mediated competition is considered as one of the most important mechanisms involved in the biocontrol of plant pathogens. Under iron limiting conditions these iron chelators sequester the limited iron supply in the rhizosphere making it unavailable to the pathogen and thereby restricting their growth (Kloepper et al. 1980; Loper and Henkels 1999). Competition for iron has been extensively studied in the genus *Pseudomonas* and the role of siderophore pyoverdine produced by many fluorescent pseudomonads has been clearly demonstrated in the control of *Pythium* and *Fusarium* species either by studying the purified pyoverdine with synthetic iron chelators or through the use of pyoverdine minus mutants (Loper and Buyer 1991; Duijff et al. 1993;). Buysens et al. (1996) reported the involvement of another siderophore pyochelin in the protection of tomato plants from *Pythium* damping-off by *Pseudomonas aeruginosa*. Kloepper et al. (1980) reported the involvement of pseudobactin (a siderophore from *Pseudomonas* sp. B-10) in the suppression of *Gauemannomyces graminis* var. *tritici* in wheat and *Fusarium oxysporium* f.sp. *lini* in flax. Similarly, Scher and Baker (1982) reported the involvement of siderophore by *Pseudomonas putida* strain A-12 in the suppression of Fusarium wilt of flax and radish induced by *F. oxysporium* f.spp. *lini* and *conglutinans* respectively. Loper et al. (1988) showed the involvement of a fluorescent siderophore in the biological control of *Pythium ultimum* in cotton by *Pseudomonas fluorescens* strain 3551. An important attribute of a successful rhizosphere biocontrol agent would be the ability to remain at high population density on the root surface, providing protection of the whole root for the duration of its life. Mycorrhizal fungi can also be considered to act as a sophisticated form of competition or cross-protection, decreasing the incidence of root disease.

(iv) Antibiosis

Several studies have demonstrated that antibiosis plays an important role in the disease suppression by certain bacteria. The production of antibiotics by actinomycetes, bacteria and fungi is very simply demonstrated *in vivo*. Numerous agar plate tests have been developed to detect volatile and non-volatile antibiotic production by putative biocontrol agents and to quantity their effects on pathogens. Howell and Stipanovic (1979, 1980) were among the first to demonstrate the production of antibiotics by biocontrol fluorescent pseudomnads; they identified pyrrolnitrin and pyoluteorin, as active compounds against *Rhizoctonia solani* root rot and *Pythium* induced damping off of cotton, respectively. Treatment of seed with either the cultures of *Pseudomonas fluorescens* or the two antibiotics protected the seedlings from root disease. Thomashow and Weller (1988 1990) conclusively proved that the antibiotic Phenazine-1-carboxylic acid (PCA) produced by *Pseudomonas fluorescens* strain 2-79 (NRRL B-15132) and its rifampicin–resistant mutant 2-79 RN_{10} was the major factor in the take-all suppression. Phenazine deficient (Phz-) mutants produced by Tn5 mutagenesis failed to inhibit *G. graminis* var. *tritici* in vitro and were coordinatedly reduced in their ability to suppress take all disease. However, when these mutants were restored for the ability to produce PCA by complementation with homologous DNA from a genomic library of strain 2-79, the complemented strains were also fully restored for the ability both to inhibit *G.graminis* var.*tritici* in vitro and to suppress take-all disease. This suggested the importance of phenazine antibiotic in take-all suppression. Similar results have been obtained with *Pseudomonas chlororaphis* strain 30-80 which produces three phenazines. While it is not difficult to demonstrate antibiotic production by bacteria in nutrient-rich culture media; it is rather very important to detect and quantify antibiotic production in the root environment. Thomashow et al. (1990) extracted 28-133 ng of PCA/gm root and adhering soil when wheat was grown in unsterilzed soil with *Pseudomonas fluorescens* 2-79 and *P.auriofaciens*. When steam sterilized soils were used, the amount of PCA increased to 265-578 ng/gm root plus adhering soil. No PCA was detected in the rhizosphere of wheat when treated with PCA-negative mutants. Mazolla et al. (1992) demonstrated the importance of antibiotic producing ability for ecological competence using phz- and phz+ wild type and complemented strains of 2-79. Phz- strains declined significantly faster than phz+ strains in natural soil.

The broad spectrum antibiotic 2, 4-di-acetyl phluroglucinol (Phl or DAPG) is one of the most important factors in the biological control of soil borne pathogens by fluorescent pseudomonads. Phl production by *Pseudomonas fluorescens* strain CHA0 has been shown to play an important role in the suppression of black root rot of tobacco caused by *Thielaviopsis basicola* (Keel et al. 1992). The same antibiotic produced by *Pseudomonas fluorescens* strain PFM2 has been found to be inhibitory against *Septoria tritici* (Levy et al. 1992). Cronin et al. (1997) reported that production of DAPG by strain F113 of *Pseudomonas fluorescens* is the key factor in the suppression of potato soft rot pathogen *Erwinia carotovora* sub sp. *atroseptica*. Vincent et al. (1991) reported the involvement of Phl in the suppression of take all by *Pseudomonas fluorescens* strain Q 2-87. Kraus and Loper (1992) reported involvement of phl in the suppression of *Pythium ultimum* and *R. solani* by *Pseudomonas fluorescens* strain pf-5. Raaijmakers et al. (1999) demonstrated that Phl production in the rhizosphere of wheat was strongly related to the density of the bacterial population present and the ability to colonize roots.

A novel antibiotic, zwittermicin A has been reported to contribute to disease suppression by *Bacillus cereus* UW 45 (He et al. 1994). The reduction in root infection of cotton by *Pythium ultimum* and increase in seedling emergence by *Pseudomonas fluorescens* str. Hv37a R2 was attributed to the biosynthesis of an antibiotic oomycin A (Howie and Suslow 1991). Similarly, production of antibiotic kanosamine by strain UW 85 of *Bacillus cereus* was demonstrated by Milner et al. (1996). A novel furanone, 3-(1-Hexenyl)-5-methyl-2-(5H)-furanone isolated from the biocontrol agent *Pseudomonas aureofaciens* (= *P. chlororaphis*) was found to have antifungal activity against *Pythium ultimum, Fusarium solani, Fusarium oxysporum* and *Thielaviopsis basicola* (Paulitz et al. 2000). *Paenibacillus polymyxa*, the type species of the genus, is known to produce two types of peptide antibiotics (Beatty and Jensen, 2002). One group is active against bacteria and includes the polymyxin-colistin-circulin family, polypeptins, jolipeptin, gavaserin, and saltavalin, which contain 2, 4-diaminobutyric acid. The other is active against fungi and actinomycetes and includes gatavalin and fusaricidins. The species also synthesize plant hormones auxin (Lebuhn et al. 1997) and cytokinin (Timmusk et al. 1999). The complexity of the mechanisms involved in the biological control are highlighted in the works of Schmidli-Sachera et al. (1997), who showed that biocontrol against *P. fluorescens* strain CHA0 produces antimicrobial agents 2,4-diacetyl phluroglucinol, pyoluteorin, HCN and siderophores (Pyoverdine, salicylic acid, pyochelin). Mutants of *P.fluorescens* CHA0, which lacked global regulator GacA-, did not have the ability to produce Phl, Plt and HCN and did not protect dicotyledonous plants against *Pythium* damping-off, but could still protect monocotyledonous plants, wheat and maize against the same pathogen.

(v) Parasitism and Lysis

Microorganisms capable of lysing other microorganisms are wide spread in nature and play an important role in microbiological equilibrium in soil. Mycoparasitism occurs when one fungus exists in intimate association with another from which it derives some or all its nutrients while conferring no benefit in return. Biotrophic mycoparasites have a persistant contact with or occupation of living cells, whereas necrotrophic mycoparasites kill the host cells, often in advance of contact and penetration. Species of *Gliocladium* and *Trichoderma* are well-known biological control agents that produce a range of antibiotics that are active against pathogens *in vitro* and, consequently, antibiotic production has commonly been suggested as a mode of action for these fungi. Mitchell and Hurwitz (1965) protected tomato plants against damping-off caused by *Pythium aphanidermatum* with lytic *Arthrobacter* sp. Campbell and Ephgrave (1983) showed that hyphal lysis, induced by *Bacillus* sp. in soil, was involved in the biological control of *Gaeumannomyces graminis* var. *trtici*. Considerable research has been done in identifying the key enzymes involved in the lysis. Ordentlich et al. (1988) found the involvement of chitinase in the suppression of *Sclerotium rolfsii* by *Serratia marcescens*. Chitinolytic enzymes produced by *Bacillus cereus* also appears to be involved in biocontrol of *Rhizoctonia solani* (Pleban et al. 1997). Biocontrol of *Phytopthora cinnamomi* induced root rot of *Banksia grandis* was obtained by a cellulase producing isolate of *Micromonospora carbonacea* (El-Tarabily et al. 1996). Singh et al. (1999) demonstrated that production of chitinolytic enzymes were involved in the suppression of *F.o* .f.sp. *cucumerinum* in cucumber by *Paenibacillus* sp. 300. *Bdellovibrio bacteriovorus* is a wide-spread bacterium that can attack several bacteria especially gram-negative ones. Sherff (1973) demonstrated that an isolate of *B. bacteriovorus* obtained from rhizosphere of soybean prevented the symptoms of soybean blight caused by *Pseudomonas savastonoi* pv. *glycinea* in green house condition.

(vi) Hydrogen Cyanide

The production of HCN is another factor involved in the activity of biological control agents as many of these organisms produce sufficient HCN to inhibit fungal growth *in vitro*. The ability to produce HCN is most common among the rhizosphere pseudomonads. Bakker and Schippers (1981) found that approximately 50% of the pseudomonads isolated from potato and wheat rhizosphere were able to produce HCN in vitro. Voisard et al. (1989) stressed the importance of cyanide production by *Pseudomonas fluorescens* CHA0 in the suppression of black root rot of tobacco. Cyanide also seems to play a role in the suppression of take all disease in wheat and *Rhizoctonia solani* on cotton by *P.fluorescens* CHA0. Strain CHA0 reduced the number of running hyphae of *G. graminis* var. *tritici* while the HCN—negative mutant CHA5 had no such effect (Defago et al. 1990). It was postulated that bacterial HCN modified the metabolism of tobacco in a way that induces some plant defense mechanism against the pathogen (Defago et al. 1990; Voisard et al. 1989).

(vii) Induced Systemic Resistance

Substantial evidences showed that beside a direct antagonistic effect on soil-borne plant pathogens, a different form of systemic resistance is elicited by specific strains of root colonizing plant growth promoting rhizobacteria, which is referred to as induced systemic resistance (ISR). Phenotypically, rhizobacteria-mediated ISR resembles classic pathogen-induced resistance, in which non-infected parts of a plant previously infected by a necrotizing pathogen becomes more resistant to further infection, a phenomenon which is called systemic acquired resistance (SAR) (Ross 1961). The state of SAR is associated with local and systemic increase in the level of salicylic acid (SA) (Malamy et al. 1990) and expression of a large set of genes including pathogenesis related (PR) proteins (van Loon and van Strien 1998). Compelling evidence for an essential role of SA in SAR came from experiments with plants transformed with the bacterial *nahG* gene, which encodes salicylate hydroxylase. This enzyme catalyzes the conversion of SA to the inactive compound catechol. Plants expressing *nahG* gene failed to accumulate SA following pathogen attack, and they were unable to activate SAR genes or to develop resistance against pathogens (Gaffney et al. 1993). Further studies on the mechanisms of SAR signalling pathway revealed that the gene *npr1* (for non-expresser of PR genes) acts as a master gene in the SAR signaling pathway and *npr1* mutant plants are able to accumulate SA in response to pathogen infection but appear to be defective in SA perception or in subsequent SA-sensing events (Cao et al. 1994; Delany et al. 1995).

The rhizobacteria mediated ISR is independent of SA (Pieterse et al. 2000). This was substantiated by the observation that strain WCS358 of *Pseudomonas putida*, that does not produce SA in vitro was able to induce resistance and the strain WCS374 of *Pseudomonas fluorescens* which produces largest amount of SA in vitro, could not induce resistance indicating that the bacterially produced SA is not involved in this ISR (van Wees et al. 1997). To investigate whether the bacteria induce the production of SA by the plant, experiments were conducted with *Arabidopsis* plants transformed *nahG* gene. Such transformants cannot accumulate SA and do not express SAR. However, starin WCS of *Pseudomonas putida* could induce ISR indicating that ISR is independent of SA. In addition, the biocontrol agent *Serratia marcescens* 90-166 has been shown to induce protection in both wild-type and transgenic *nahG* tobacco plants against *Pseudomonas syringae* pv. *tabaci* (Press et al. 1997), indicating that the ability to trigger SA-independent pathway controlling systemic resistance is not uncommon among rhizobacteria.

Further studies on the molecular mechanisms of rhizobacteria mediated ISR revealed that there was no increased accumulation of PR proteins. Radish plants of which the roots were treated with ISR-inducing *P. fluorescens* WCS 417r did not accumulate PR proteins, although these plants clearly showed enhanced resistance against fusarium wilt disease (Hoffland et al. 1995). Similarly *Arabidopsis* plants expressing *P. fluorescens* WCS417r mediated ISR showed enhanced resistance against *F. oxysporum* f.sp. *raphani* and *Pseudomonas syringae* pv. *tomato*, but did not coincide with the activation of the SAR marker genes PR-1, PR-2 and PR-5 (Pieterse et al. 1996; van Wees et al. 1997). Further genetic dissection of rhizobacteria mediated ISR revealed that the *Arabidopsis* jasmonic acid (JA) response mutant *jar1* as well as the ethylene-insensitive mutants *etr1* and *ein2*, were impaired in the expression of ISR, whereas they were fully capable of expressing SAR in response to infection with an avirulent pathogen. Thus, in contrast to SAR, ISR requires components of JA and ethylene signalling pathways (Pieterse et al. 1998). However, ISR was found to share with SAR the requirement for *npr1*. Thus SAR and ISR are regulated differently but converge at the level of *npr1*. Yet, in the SAR response the involvement of *NPR1* is associated with the expression of PR-genes, whereas in the ISR response it is not.

CONCLUSIONS

Despite the many achievements of modern agriculture, certain cultural practices have actually enhanced the destructive potential of plant diseases. These practices include use of genetically similar crop varieties in continuous monoculture, use of plant cultivars susceptible to pathogens, and use of nitrogenous fertilizers at concentrations that enhance disease susceptibility. Plant disease control, therefore, has now become heavily dependent on fungicides to combat the wide variety of fungal diseases that threaten agricultural crops. The governments of many countries are increasingly aware of the drawbacks of many chemical pesticides, in terms of their effect on the environment, as well as on the growers and consumers of agricultural products. Studies aimed at replacing pesticides with environmentally safer methods are currently being conducted at many research centres. The heightened scientific interest in biological control of plant pathogens is partly a response to growing public concerns over chemical pesticides. Biological control is a potent means of reducing the damage caused by plant pathogens.

It is evident from the innumerable research reports over the years that several rhizosphere bacteria have the potential to suppress several plant pathogens and to promote plant growth. However, in spite of these extensive research efforts, the use of biocontrol in commercial agriculture is still limited. The inconsistency in performance of these biocontrol agents over time and space in comparison to chemical pesticides is a major impediment to their large scale use in commercial agriculture. Thus the use of biocontrol agents is being tried in a system where chemical fungicides provide a better and more economical result i.e., the biocontrol agents are tried to fit in to a chemical paradigm (Harman 2000). Therefore, efforts are needed to develop widely adapted biocontrol agents with broad-spectrum activity. Clear understanding of different molecular mechanisms of antagonism and root colonization are essential so that desirable traits can be pulled together genetically in one biocontrol agent. Improving known biological agents, with genetic manipulation, can enhance their biological control attributes and expand their spectrum of activity. Cook (1993) advocated an approach to biological control that uses combinations of biocontrol agents that suppresses diseases by complementary mechanisms. This may also allow long term control since individual components colonize different crops, are adapted to different environments. Similarly, timely and augmented releases of biocontrol agents have been suggested to make biocontrol more effective. Another aspect which needs attention is the wide spread and multilocation testing of the biocontrol agents against specific diseases at the respective hot spot locations. Research should also be directed to focus on shelf life of the biocontrol products. Thus it is hoped that biocontrol will form an important component in the integrated disease management in near future.

References

Abdelzaher HMA, Elnaghy MA (1998) Identification of *Pythium carolinianum* causing 'root rot' of cotton in Egypt and its possible biological control by *Pseudomonas fluorescens*. *Mycopathologia* 142: 143-151.

Adams PB, Ayers WA (1982) Biological control of *Sclerotinia lettuce* drop in the field by *Sporidesmium sclerotiovorum*. *Phytopathology* 72: 485-488.

Adhikari TB, Joseph CM Yang G, Phillips DA, Nelson LM (2001) Evaluation of bacteria isolated from rice for plant growth promotion and biological control of seedling disease of rice. Can. J. Microbiol. 47 916-924.

Ahamad S, Srivastava M (2002) Evaluation of fungal and bacterial antagonists against *Fusarium moniliforme* causing foot rot disease of rice. Ann Agric Res 23: 319-321.

Ahmad JS, Baker R (1987) Rhizosphere competence of *Trichoderma larizianum*. *Phytopathology* 77: 182-189.

Alstrom, S. (1991). Induction of disease resistance in common bean susceptible to halo blight bacterial pathogen after seed bacterization with rhizopshere pseudomonads. *J. Gen. Microbiol.* 37: 495-501.

Alstrom, S. and Burns R.G. (1989) Cyanide production by rhizobacteria as a possible mechanism of plant growth inhibition. *Bio Fert Soils* 7: 232-238.

Anith, K.N. and Manomohandas, T.P. (2001). Combined application of *Trichoderma harzianum* and *Alcaligenes* sp. strain AMB 8 for controlling nursery rot disease of black pepper. *Indian Phytopath.* 54(3): 335-339.

Anuratha, C.S. and Gnanamanickam, S.S. (1990). Control of bacterial wilt caused by *Pseudomonas solanacearum* in India with antagonistic bacteria. *Plant Soil* 124: 109-116.

Arrieta-Escobar, A. and Belin, J.M. (1982). Effects of polyphenolic compounds on the growth and cellulolytic activity of a strain of *Trichoderma viride. Biotechnol. Bioeng.* 24: 983-989.

Arshad, M. and Frankenberger, Jr., W.T. (1991). Microbial production of plant hormones. *In*: *The Rhizosphere and Plant Growth.* 327-334.

Ayers, W.A. and Adams, P.B. (1981). Mycoparasitism and its application to biological control of plant diseases. In: G.C. Papavizas (ed.). *Biological Control in Crp Protection.* Totowa, NJ, Allanheld, Osmun, pp 91-103.

Bacon, C.W. and Hinton, D.M. (2002) Endophytic and biological control potential of *Bacillus mojavensis* and related species. *Biological Control* 23: 274-284.

Bahme, J.B. and Schroth, M.N. (1987). Spatial temporal colonization patterns of a rhizobacterium on under ground organs of potato. *Phytopathology.* 77: 1093-1100.

Bakker, A.W. and Schippers, B. (1987). Microbial cyanide production in the rhizosphere in relation to potato yield reduction and *Pseudomonas* spp.–mediated plant growth stimulation. *Soil Biol Biochem.* 19: 249-256.

Bakker, A.W., Bakker, P.A.H.M. and Schippers, B. (1989). Deleterious cynamide–producing rhizosphere pseudomonads as a factor limiting potato root growth and tuber yield in high frequency potato cropping soil. *In*: J. vos, C.D. van Loon and G.J. Bollen (eds.) *Effects of crop rotation on potato production in the temperate zones.* Kluwer Academic Publishers, Dordrecht pp 153-162.

Balasubramanian, A. and Rangaswami, G. (1978) Influence of seed and rot exudations on the rhizosphere effect in *Sorghum vulgare* and *Crotolaria juncea.* Folia Microbiol. 23: 418-488.

Barber, D.A. and Martin, J.K. (1976) The release of organic substances by cereal roots into the soil. New Phytol 76: 69-80

Bashan, Y. and Holguin, G. (1998) Proposal for the division of plant growth promoting rhizobacteria into two classifications: biocontrol PGPB (plant growth promoting bacteria) and PGPB. *Soil Biol Biochem* 30: 1225-1228

Bath, E., Frostegard, A. and Fritze, H. (1992) Soil bacterial biomass, activity, phospholipid fatty acid pattern, and pH tolerance on an area polluted with alkaline dust deposition. *Appl. Environ. Microbiol.* 58: 4026-4031.

Beatty P.H. and Jensen S.E. (2002) *Paenibacillus polymyxa* produces fusaricidin-type antifungal antibiotics active against *Leptosphaeria maculans,* the causative agent of blackleg disease of canola. *Can. J. Microbiol.* 48: 159-169

Benhamou, N., Kloepper, J.W., Quadt-Hallman, A. and Tuzun, S. (1996). Induction of defense related ultrastructural modifications in pea root tissues inoculated with endophytic bacteria. *Plant Physiol* 112: 919-929

Beyeler, M., Keel, C., Michaux, D. and Haas, D. (1999). Enhanced production of indole-3-acitic acid by a genetically modified strain of *Pseudomonas fluorescens* CHAO affects root growth of cucumber, but does not improve protection of the plant against *Pythium* root rot. *FEMS Microbial Ecology* 28: 225-233.

Bhowmik, B., Singh, R.P. Jayaraman, J. and Verma, J.P. (2002). Population dynamics of cotton endophytic *Pseudomonas*, their antagonism and protective action against the major pathogens of cotton. *Indian Phytopath*. 55 (2): 124-132.

Bowen, G.D. (1979). Integrated and experimental approaches to the study of growth of organisms around roots. *In*: B. Schippers and W. Gams (eds) *Soil borne Plant Pathogens* Academic Press, London pp. 209-227

Bowen, G.D. (1980). Misconceptions, concepts and approaches in rhizosphere biology. *In* : D.C. Ellwood, J.N. Hedger, M.J. Latham, J.M. Lynch, J.H. Slater (Eds) *Contemporary Microbial Ecology* Academic Press, London, pp 283-304

Bowen, G.D. and Rovira, A.D. (1976). Microbial colonization of plant roots. *Annu. Rev. Phytopathol*. 14: 121-144.

Bowen, G.D. and Rovira, A.D. (1999) The rhizosphere and its management to improvement plant growth. *Advan Agron* 66: 1-102

Bowers, J.H. and Parke, J.L. (1993) Epidemiology of *Pythium* damping-off and *Aphanomyces* root rot of peas after seed treatment with bacterial agents for biological control. *Phytopathology* 83: 1466-1473

Braun, K. A., Jacobsen, B.J. and Sands, D.C. (2000) Biological control of *Pseudomonas syringae* pv. *syringae*, the causal agent of basal kernel blight of barley, by antagonistic *Pantoea agglomerans. Phytopathology* 90: 368-375

Brian, P.W. and McGowan, J.C. (1945). Viridin: A highly fungistatic substance produced by *Trichoderma viride* . *Nature* 156: 144-145.

Broadbent, P., Baker, K.F. and Waterworth, Y. (1971). Bacteria and actinomycetes antagonistic to root pathogens in Australian soils. *Aust. J. Biol. Sci.* 24: 925-944.

Brown, M.E. (1972). Plant growth substances produced by micro organisms of soil and rhizosphere. *J. Appl Bacteriol* 35: 443-451

Burd, G.I., Dixon, D.G. and Glick, B.R. (1998). A plant growth promoting bacterium that decreases nickel toxicity in seedlings.*Appl Environ Microbiol* 64: 3663-3668.

Burr, T.J. and Caesor, A. (1984). Beneficial plant bacteria. *CRC Crit. Rev. Plant Sci*. 2: 1-20.

Burr, T.J., Schroth, M.N. and Suslow, T.V. (1978). Increased potato yields by treatments of seed species with specific strains of *Pseudomonas fluorescens* and *P. putida. Phytopathology* 68: 1377-1383.

Buyer, J.S. and Sikora, L.J. (1990). Rhizosphere interactions and siderophorus. *Plant Soil* 129: 101-107.

Buysens, S., Heungens, K., Poppe, J. and Hofte, M. (1996). Involvement of pyrochelin and pyoverdino in suppression of *Pythium*–induced damping off of tomato by *Pseudomonas aeruginosa* 7 NSK 2. *Appl Environ Microbiol* 62: 865-871.

Callan, N.W., Mathre, D.E. and Miller, J.B. (1990). Bio-priming seed treatment for biological control of *Pythium ultimum* pre-emergence damping-off in sh2 sweet corn. *Pl Dise* 74: 368-372

Campbell J.N., Cass, D.D. and Peteya, D.J. (1987) Colonization and penetration of intact conola seedling roots by an opportunistic fluorescent *Pseudomonas* spp., and the response of host tissue. *Phytopathology* 77: 1166-1173.

Campbell, R and Ephgrave, JM (1983). Effect of bentonite clay on the growth of *Gaeumannomyces graminis* var. *tritici* and on the interaction with antagonistic bacteria. *J. Gen. Microbiol.* 129: 771-777.

Cao, H., Bowling, S.A., Gordon, A.S. and Dong, X. (1994). Characterization of an *Arabidopsis* mutant that is non responsive to inducers of systemic acquired resistance. *Plant Cell* 6: 1583-1592.

Catska, V., Vancura, V., Hudsla, G. and Prikryl, Z. (1982) Rhizosphere micro organisms in relation to the apple replant problem. *Plant Soil* 69: 187-197.

Chaboud, A. (1983) Isolation, purification and chemical composition of maize root cap slime. *Plant Soil* 73: 395-402

Chanway, C.P. and Holl, F.B. (1992) Influence of soil biota on Douglas-fir (*Pseudotseuga menziesii*) seedling growth: the role of rhizosphere bacteria. *Can. J. Bot* 70: 1025-1031

Chatterjee, A., Valasubramanian, R., Ma, W.L., Vachhani, A.K., Gnanamanickam, S. and Chatterjee, A.K. (1996). Isolation of ant mutants of *Pseudomonas fluorescens* strain Pf 7-14 altered in antibiotic production, cloning of ant DNA, and evaluation of the role of antibiotic production in the control of blast and sheath blight of rice. *Biological Control* 7(2): 185-195.

Chen, C., Belanger, R.R., Benhamou, N. and Paulitz, T.C. (1998). Induced systemic resistance (ISR) by *Pseudomonas* spp. impairs pre- and post-infection development of *Pythium aphanidermatum* on cucumber roots. *European J. Plant Pathology* 104: 877-886.

Chet, I. (1987). *Trichoderma* – application, mode of action, and potential as a biocontrol agent of soil borne plant pathogenic fungi. In: *Innovative Approaches to Plant Disease Control.* I. Chet, (Ed.). pp 137-160, John Wiley and Sons, New York.

Chet, I. And Baker, R. (1980). Induction of suppressiveness to *Rhizoctonia solani* in soil. *Phytopathology* 70: 994-998.

Chet, I. And Baker, R. (1981). Isolation and biocontrol potential of *Trichoderma hamatum* from soil naturally suppressive to *Rhizoctonia solani. Phytopathology* 71: 286-290.

Chet, I., Barak, Z. and Oppenheim, A. (1993). Genetic engineering of micro organisms for improved biocontrol activity. *In*: *Biotechnology in Plant Disease* Control. I. Chet, (Ed.) Wiley-Liss Inc., New York.

Chin-A-Woeng, T.F.C., Bloemberg, G.V., van der Bij, A.J. (1998). Biocontrol by phenazine-1-carboxamide producing *Pseudomonas chlororaphis* PCL 1391 of tomato root rot caused by *Fusarium oxysporum* f.sp. *radicis-lycopersici. Molecular Plant Microbe Interactions* 11: 1069-1077.

Choi, H.Y, Ryder, M.H., Gillings, M.R., Stokes, H.W., Ophel Keller, K.M., Veal, D.A. (2003). Survival of a lacZY-marked strain of *Pseudomonas corrugata* following a field release. *FEMS Microbiology Ecology* 43(3): 367-374.

Ciampi, L., Burzio, L.O. and Burzio, L.A. (1997). Carriers for *Pseudomonas fluorescens*, antagonistic to *Pseudomonas* (Ralstonia) *solanacearum*, causal agent of bacterial wilt. *Fitopatologia* 32 (1): 64-70.

Clark, F.E. (1940). Notes on types of bacteria associated with plant roots. *Trans. Kansas Acad. Science* 43: 75–84.

Colhoun, J. (1979). Minor pathogens and complex root disease: concluding remarks. *In*: *Soil borne plant pathogens*. N. Schippers, W. Gams, (Eds.) pp. 351-354, Academic Press, New York.

Collins, D.P. and Jacobsen, B.J. (2003). Optimizing a *Bacillus subtilis* isolate for biological control of sugar beet *Cercospora* leaf spot. *Biological Control* 26: 153-161.

Commare, R.R., Nandakumar, R., Kandan, A., Suresh, S., Bharathi, M., Raguchander, T. and Samiyappan, R. (2002). *Pseudomonas fluorescens* based bio-formulation for the management of sheath blight disease and leaf folder insect in rice. *Crop Protection* 21(8): 671-677.

Cook R.J. and Baker, K.F. (1983). *The Nature and Practice of Biological Control of Plant Pathogens*. St. Paul, Minn: APS, 539 pp.

Cook R.J. and Rovira, A.D. (1976). The role of bacteria in the biological control of *Gacumannomyces graminis* by suppressive soils. *Soil Biol. and Biochem*. 8: 269-273.

Cook, R.J. (1991). Twenty five years of progress towards biological control. *In*: *Biological control of soil-borne Plant Pathogens.* pp 1-14. D. Hornby (Ed.) CAB International, UK, 479 pp.

Cook, R.J. (1993). Making greater use of introduced microorganisms for biological control of plant pathogens. *Annu. Rev. Phytopath*. 31: 53-80

Cook, R.J., Thomashow, L.S., Weller, D.M., Fuzimoto, D., Mazzola, M., Bangera, G. and Kim, D.S. (1995). Molecular mechanisms of defense by rhizobacteria against root disease. *Proc. National Acad. Sci.*, USA 92: 4197-4201.

Costacurta, A. and Vader Leyden, J. (1995). Synthesis of phytonormones by plant associated bacteria. *Cirt. Rev. Microbiol*. 21: 1.

Cronin, D., Moenne-Loccoz, Y., Fenton, A., Dunne, C., Dowling, D. N. and O' Gara, F. (1997). Ecological interaction of a biocontrol *Pseudomonas fluorescens* strain producing 2, 4-diacetylphloroglucinol with the soft rot potato pathogen *Erwinia carotovora* subsp. *atroseptica*. *FEMS Microbiology Ecology* 23(2): 95-106.

Curl, E.A. and Truelove, B.(1986). *The rhizosphere* (Advanced series in Agric. Science 15). Springer Verlag, Berlin, Heidelberg.

Dalisay, R.F. and Kuc, J.A. (1995). Persistence of reduced penetration by *Colletotrichum lagenarium* into cucumber leaves with induced systemic resistance and its relation to enhanced peroxidase and chitinase activity. *Physiol. Mol. Plant Pathol.* 47: 329-338.

Danielson, R.M. and Davey, C.B. (1973). Nonnutritional factors affecting the growth of *Trichoderma* in culture. *Soil Biol. Biochem*. 5: 495-504.

De Boer, M., Van der sluis, I., Van Loon, L.C. and Bakker, P.A.H.M. (1999). Combining fluorescent *Pseudomonas* spp. to enhance suppression of *Fusarium* wilt of radish. *European J. Plant Pathology* 105: 201.210.

De Jensen, C.E., Percich, J.A. and Graham, P.H. (2002). Integrated management strategies of bean root rot with *Bacillus subtilis* and *Rhizobium* in Minnesota. *Field Crops Research* 74: 107-115.

De la Cruz, A.R., Poplawsky, A. R. and Wiese, M.V. (1992). Biological suppression of potato ring rot by fluorescent pseudomonads. *Applied and Environmental Microbiology* 58(6): 1986-1991.

De la Cruz, J., Pintor Toro, J.A., Benitez, T. and Llobell, A. (1995). Purification and characterization of an endo-b-1,6-glucanase from *Trichoderma harzianum* that is related to its mycoparasitism. *J. Bacteriol.* 177: 1864-1895.

De Boer, M., Bom, P., Kindt, F., Keurentjes, J.J.B., van der Sluis, I.,van Loon, L.C. and Bakker, P.A.H.M. (2003). Control of *Fusarium* wilt of radish by combining *Pseudomonas putida* strains that have different disease suppressive mechanisms. *Phytopathology* 93(5): 626-632.

Defago, G. and Haas, D. (1990). Pseudomonads as antagonists of soil borne plant pathogens: mode of action and genetic analysis. *Soil Biol. and Biochem.* 6: 249-291.

Defago, G., Berling, C.H., Burger, U., Haas, D., Kahr, G., Jeel, C., Voisard, C., Wirthner, Ph. and Wutrich, B. (1990). Suppression of black root rot of tobacco by *Pseudomonas* strain: potential applications and mechanisms. *In*: *Biological control of soil-borne Plant Pathogens*. D. Hornby (Ed.) 93-108 pp., CAB International.

Dekkers, L.C., Mulders, I.H.M., Phoelich, C.C., Chin-A-Woeng, T.F.C., Wijfjes, A.H.M. and Lugtenberg, B.J.J. (2000). The sss colonization gene of the tomato-*Fusarium oxysporum* f.sp. *radicis-lycopersici* biocontrol strain *Pseudomonas fluorescens* WCS365 can improve root colonization of other wild-type *Pseudomonas* spp. bacteria. *Mol. Plant Microbe Interact.* 13(11): 1177-1183.

Dekkers, L.C., Van der Bij, A.J., Mulders, I.H.M., Phoelich, C.C. Wentwoord, R.A.R. et al. (1998). Role of the O-antigen of lipopolysaccharide, and possible roles of growth rate and NADH: ubiquinone oxidoreductase (*nuo*) in competitive tomato root-tip colonization by *Pseudomonas fluorescens* WCS365. *Mol. Plant Microbe Interact.* 11: 763-771.

Delany, T.P., Friedrich, L. and Ruals, J.A. (1995). *Arabidopsis* signal transduction mutant defective in chemically and biologically induced disease resistance. *Proc. Natl. Acad. Sci.* USA. 92: 6602-6606.

Devi, T.V., Vizhi, R.M. Sakthivel, N. and Gnanamanickam, S.S. (1989). Biological control of sheath blight of rice in India with antagonistic bacteria. *Plant Soil* 119(2): 325-330.

De-Weger, C.A., Vander Hugt, C.I.M., Wifjfijes, A.H.M. and Bakker, P.A.H.M., Schippers, B. and Lugtenberg, B. (1987). Flagella of a plant growth promoting *Pseudomonas fluorescens* strains are required for colonization of potato roots. *J. Bacteriol.* 169: 2769-2773.

Dowling, D. N. and O'Gara, F. (1994). Metabolites of *Pseudomonas* involved in the biocontrol of plant disease. *Trends in Biotechnology* 12: 133-141.

Duffy, B. (2000). Combination of pencycuron and *Pseudomonas fluorescens* strain 2-79 for integrated control of *Rhizoctonia* root rot and take-all of spring wheat. *Crop Protection* 19(1): 21-25.

Duijff, B. J.,Pouhair, D., Olivain, C., Alabouvette, C. and Lemanceau, P. (1998). Implication of systemic induced resistance in the suppression of *Fusarium* wilt of tomato by *Pseudomonas fluorescens* WCS417r and non-pathogenic *Fusarium oxysporum* F047. *European J. Plant Pathology* 104: 903-910.

Duijff, B.J., Gianinazzi–Pearson, V. and Lemanceau, P. (1997). Involvement of the outer membrance lopopolysaccharides in the endophytic colonization of tomato roots by biocontrol *Pseudomonas fluorescens* strain WCS 417r *New Phytologist* 135: 325-334.

Duijff, B.J., Meijer, J.W., Bakker, P.A.H.M. and Schippers, B. (1993). Siderophore–mediated competition for iron and induced resistance in the suppression of *Fusarium* wilt of carnation by fluorescent *Pseudomonas* spp. *Netherlands J. Plant Pathology* 99: 277-289.

El Tarabily, K.A., Soliman, M.H., Nassar, A.H., Al Hassani, H.A., Sivasithamparam, K., McKenna, F. and Hardy GE St, J. (2000). Biological control of *Sclerotinia* minor using a chitinolytic bacterium and actinomycetes. *Plant Pathology* 49(5): 573-583.

Elad, Y. and Chet, I. (1987). Possible role of competition for nutrients in biocontrol of *Pythium* dumping off by bacteria. *Phytopathology* 77: 190-195.

Elliott, L.F. and Lynch, J.M. (1984). Pseudomonads as a factor in the growth of winter wheat (*Triticum aestirum* L.). *Soil Biol. and Biochem.* 16: 69-71.

El-Tarahbily, K.A., Sykes, M.L., Kurtboke, H.D., Hardy, G.E, St. J, Barbosa, A.M. and Dekker, R.F.H. (1996). Synergistic effects of a cellulose producing *Micromonospora carbonacea* and abtibiotic-producing *Streptomyces violascens* on the suppression of *Phytophthora cinnamomi* root rot of *Bauksia grandis*. *Can J. Bot* 74: 618-624.

Expert, J.M. and Digat, B. (1995). Biocontrol of *Sclerotinia* wilt of sunflower by *Pseudomonas fluorescens* and *Pseudomonas putida* strains. *Can J. Microbiol* 41(8): 685-691.

Foster, R.C. (1986). The ultra structure of the rhizoplane and the rhizosphere.*Annu. Rev. Phytopath*. 24: 211-234.

Foster, R.C., Rovira, A.D. and Cock, T.W. (1983). *Ultra structure of the root soil interface* p. 157. The American Phytopathological Society, St. Paul, MN.

Fravel, D.R. (1988). Role of antibiosis in the biocontrol of plant diseases.*Annu. Rev. Phytopath*. 26: 75-91.

Frederickson, J.K. and Elliott, L.F. (1985). Effects on winter wheat seedling growth by toxin producing rhizobacteria. *Plant Soil* 83: 399-409.

Frederickson, J.K., Elliott, L.F. and Engibaus, J.C. (1987). Crop residues as substrate for host-specific pseudomonads. *Soil Biol. and Biochem*. 19: 127-134.

Gaffney, T., Friedrich, L., Vernooij, B., Negrotto, D., Nye, G., Ukens, S., Ward, E., Kessmann, H. and Ryals, J. (1993). Requirement of salicylic acid for the induction of systemic acquired resistance. *Science* 261: 754-756.

Gal, S.W., Choi, J.Y., Kim, C.Y. et al. (1997). Isolation and characterization of 54 kDa and 22kDa chitinase gene of *Serratia marcescens* KCTC2172. *FEMS Microbiol. Lett.* 151: 197-204.

Ganesan, P. and Gnanamanickam, S.S. (1987). Biological control of *Sclerotium rolfsii* Sacc. In peanut by inoculation with *Pseudomonas fluorescens. Soil Biol. Biochem.* 19:35-38.

Garbaye, J. (1994). Helper bacterial: a new dimension to the mycorrhizal symbiosis. *New Phytologist* 128: 197-210.

Gardner, J.M., Chandler, J.L. and Feldman, A.W. (1984). Growth promotion and inhibition by antibiotic producing florescent pseudomonads on citrus roots. *Plant Soil* 77: 103-113.

Gaskins, M.H., Albrecht, S.L. and Hubbel, D.H. (1985). Rhizosphere bacteria and their use to increase plant productivity: A review. *Agric. Ecosyst. Environ.* 12: 99-116.

Gasperini, G. (1890). Recherches morphologiques et biolpgiques sur un microorganisme del' atmosphere, le *Streptothrix foersteri* Cohn. *Ann. Microg*. 10: 449-474.

Gaunliel, A. and Katan, J. (1992). Chemotaxis of fluorescent pseudomonads towards seed exudates and germinating seeds in solarized soil. *Phytopathology* 82: 328-332.

Geels, F.P. and Schippers, B. (1983). Reduction of yield depression in high frequency potato cropping soil after seed tuber treatments with antagonistic fluorescent *Pseudomonas* spp. *Phytopathol.* Z. 108: 207-214.

Georgakopoulos, D.G., Fiddaman, P., Leifert, C. and Malathrakis, N.E. (2002). Biological control of cucumber and sugarbeet damping-off caused by *Pythium ultimum* with bacterial and fungal antagonists. *J. Applied Microbiology* 92(6): 1078-1086.

Godtfedsen, W.O. and Vangedal, S. (1965). Trichodermin, a new sesquiterpine antibiotic. *Acta Chem. Scand.* 19: 1088-102.

Goldman, G.H., Hayes, C. and Harman, G.E. (1994). Molecular and cellular biology of biocontrol by *Trichoderma* spp. *Trends Biotechnol.* 12: 478-486.

Gratia, A. (1934). La dissolution des bactéries et ses applications thérapeutiques. *Bull. acad. roy. med. belg.* 14: 285-300.

Hagedorn, C., Nelson, N. and Skwara, J. E. (1990). Evaluation of a *Pseudomonas fluorescens* strain for repression of seedling disease in cotton. *Virginia J. Science* 41: 492-500.

Hale, M.G. and Moore, L.D. (1979). Factors affecting root exudation. II: 1970-1978. *Advances in Agronomy* 31: 93-124.

Hamed, H.A. (1999). Biological control of basal stem rot and wilt of cucumber caused by *Pythium ultimum* and *Fusarium oxysporum* f.sp. *cucumerinum*. *African J. Mycology and Biotechnology* 7(1): 81-91.

Harrison, L.A., Letendre, I., Kovacevich, P., Pierson, E. and Weller, D.M. (1993). Purification of an antibiotic effective against *Gaurmannomyces* graminis var. *tritici* produced by a biocontrol agent, *Pseudomonas aureofaciens. Soil Biol. Biochem.* 25: 215-221.

Hartley, C. (1921). *Damping-off in forest nurseries*. Dept. Bull., U.S.D.A., No. 934.

Hawes, M.C. and Pueppke, S.G. (1986). Isolated peripheral root cap cells: yield from different plants and callus formation from single cells. *American J. Botany* 73: 1466-73.

Hayman, D.S. (1969). The influence of temperature on the exudation of nutrients from cotton seeds and on pre-emergence damping-off of *Rhizoctonia solani*. *Can. J. Bot.* 47: 1663-1669.

He, H., Silo Suh, L.A., Clardy, J. and Handelsman, J. (1994). Zwittermicin A, an antifungal and plant protection agent from *Bacillus* cereus. *Tetrahedron Let.* 35: 2499-2502.

Helbig, J. and Bochow, H. (2001). Effectiveness of *Bacillus subtilis* (isolate 25021) in controlling *Botrytis cinerea* in strawberry. *Zeitschrift fur Pflanzenkrankheiten und Pflanzenschutz* 108(6): 545-559.

Hiltner, L. (1904). Uber neure Ertahrunger and problemse auf dem Gebeit der bodenbak teriologileund unter besonderer Beruck sichtigung der Grundung ung and Vrache, *Arb. Dtsch. Landw. Ges. Berl.* 98: 59-78.

Hockstra, O. (1981). 15 Jaar "de Schreef" Resultaten van 15 jaar vrunch twisselingson- derzock op net bouwplannenproefveld "de Schreef", Proefstn. *Akkerbouw Groenteteelt Vollegrond* 11: 1-93.

Hoffland, E., Pieterse, C.M.J., Bik, L. and Van Pelt, J.A. (1995). Induced systemic resistance in radish is not associated with accumuation of pathogenesis – related proteins. *Physiol. Mol. Plant Pathol.* 46: 309-320.

Hong, Y., Pasturnak, J.J. and Glick, B.R. (1991). Biological consequences of plasmid transformation of the plant growth promoting rhizobacterium *Pseudomonas putida* GR 12-2, *Can. J. Microbiol.* 37.

Howell, C.R. and Stipanovic, R.D. (1979). Control of *Rhizoctonia solani* on cotton seedlings with *Pseudomonas fluorescens* and with an antibiotic produced by the bacterium. *Phytopathology*. 69: 480-482.

Howell, C.R. and Stipanovic, R.D. (1980). Suppression of *Pythium ultimum* induced damping off of cotton seedlings by *Pseudomonas fluorescens* and its antibiotic, Pyoluteorin. *Phytopathology* 70: 712-715.

Howell, C.R. and Stipanovic, R.D. (1983). Gliovirin, a new antibiotic from *Gliocladium virens*, and its role in the biological control of *Pythium ultimum*. *Can. J. Microbiol.* 29: 321-324.

Howie, W.J. and Suslow, T. V. (1991). Role of antibiotic biosynthesis in the inhibition of *Pythium ultimum* in the cotton spermosphere and rhizosphere by *Pseudomonas fluorescens*. *Molecular Plant Microbe Interactions* 4(4): 393-399.

Howie, W.J., Cook, R.J. and Weller, D.M. (1987). Effects of soil matric potential and cell motility on wheat root colonization by fluorescent *Pseudomonas* suppressive to take all. *Phytopathology* 77: 286-292.

Hultberg, M., Alsanius, B. and Sundin, P. (2000). In vivo and in vitro interactions between *Pseudomonas fluorescens* and *Pythium ultimum* in the suppression of damping-off in tomato seedlings. *Biological Control* 19: 1-8.

Hyun, J.W., Kim, Y.H., Lee, Y.S. and Park, W.M. (1999). Isolation and evaluation of protective effect against *Fusarium* wilt of sesame plants of antibiotic substance from *Bacillus polymyxa* KB-8. *Plant Pathology Journal* 15(3): 152-157.

Inbar, J. and Chet, I. 1995. The role of recognition in the induction of specific chitinases during mycoparasitism by *Trichoderma harzianum*. *Microbiology* 117: 187-199.

Ivarson, K.S. and Katznelson, H. (1960). Studies on the rhizosphere microflora of yellow birch seedlings. *Plant and Soil* 12: 30-40.

Jaffre, B.A., Abawi, G.S. and Mai, W.F. (1982). Role of soil microflora and *Pratylenchus penetrans* in an apple replant disease. *Phytopathology* 72: 247-251.

James, D.W., Suslow, D. and Steinback, K.E. (1985). Relationship between rapid, "firm adhesion", and long term colonization of roots by bacteria. *Appl. Environ. Microbiol.* 50: 392-397.

Jarvis, W.R. (1992). *Managing diseases in green house crops.* St. Paul. MN. APS, 288 pp.

Johnssin, L., Hokeberg, M. and Gerhardson, B. (1998). Performance of the *Pseudomonas chlororaphis* biocontrol agent MA 342 against cereal seed-borne diseases in field experiments. *European J. Plant Pathology* 104: 701-711.

Kageyama, K. and Nelson, E. B. (2003). Differential inactivation of seed exudate stimulation of *Pythium ultimum* sporangium germination by *Enterobacter cloacae* influences biological control efficacy on different plant species. *Applied and Environmental Microbiology* 69(2): 1114-1120.

Kaiser, W.J. and Hannan, R.M. (1989). Biological control of seed rot and pre-emergence damping-off of chickpea with fluorescent pseudomonads. *Soil Biol. and Biochem.* 21(2): 269-273.

Kamensky M., Ovadis, M., Chet, I. and Chernin, L. (2003). Soil-borne strain IC14 of *Serratia plymuthica* with multiple mechanisms of antifungal activity provides biocontrol of *Botrytis cinerea* and *Sclerotinia sclerotiorum* diseases. *Soil Biol. and Biochem.* 35(2): 323-331.

Karpagavalli, S., Marimuthu, T., Jayaraj, J. and Ramabadran, R. (2001). An integrated approach to control rice blast through nutrients and biocontrol agent. *Research on Crops* 2(2): 197-202.

Katznelson, H. (1946). The rhizosphere effect of mangels on certain groups of micro organisms. *Soil Science* 62: 343-354.

Keel, C., Schnider, U., Maurhofer, M., Voisard, C., Laville, J., Burger, U., Wirthner, P., Haas, D. and Defago, G. (1992). Suppression of root diseases by *Pseudomonas fluorescens* CHA0: importance of the bacterial secondary metabolite 2, 4-diacetylphloroglucinol. *Molecular Plant Microbe Interactions* 5(1): 4-13.

Kerr, A. (1980). Biological control of crown gall through production of agrocin 84. *Plant Disease* 64: 25-30.

Kim, D.S., Weller, D.M. and Cook, R.J. (1997). Population dynamics of *Bacillus* sp. L 324-92^{R} 12 and *Pseudomonas fluorescens* 2-79 R10 in the rhizosphere of wheat. *Phytopathology* 87: 559-564.

Kim, J.W., Park, B.K., Hwang, I.Y. and Park, C.S. (1998). Antifungal activity of root colonizing *Pseudomonas fluorescens* MC07 is responsible for its disease suppression ability. *Korean J. Plant Pathology* 14(6): 606-611.

Kloepper, J.N. and Schroth, M.N. (1978). Plant growth promoting rhizobacteria on radishes. *Proc. IV Int. Conf. Plant Pathogen Bacter.* 2: 879-882.

Kloepper, J.N. and Schroth, M.N. (1981). Plant growth promoting rhizobacteria and plant growth under gnotobiotic conditions. *Phytopathology* 7: 642-644.

Kloepper, J.W. and Beau Champ, C.J. (1992). A review of issues related to measuring colonization of plant roots by bacteria. *Can. J. Microbiol.* 38: 1219-1232.

Kloepper, J.W., Hume, D.J., Scher, F.M., Singleton., Tipping, B., Laliberte, M., Frauley, K., Kutchaw, T., Simonson, C., Lifshitz., R., Saleska, I. and Lee, L. (1988). Plant growth promoting rhizobacteria on canola (rape seed). *Plant Disease* 72: 42-46.

Kloepper, J.W., Leong, J., Teintze, M. and Schroth, M.N. (1980). Enhanced plant growth by siderophores produced by plant growth promoting rhizobacteria. *Nature* 286: 885-886.

Kloepper, J.W., Mc Inroy, J.A. and Brown, K.L. (1992). Comparative identification of fatty acid analysis of soil, rhizosphere and geocarposphere bacteria of peanut (*Arachis hypogaea, L.). Plant Soil* 139: 85-90.

Kluepfel, D.A., Kline, E.L., Skipper, H.D., Hughes, T.A., Gooden, D.T. (1991). Release and tracking of genetically engineered bacteria in the environment. *Phytopathology*. 81: 328-352.

Knowler, C.J. (1976). Microorganisms and cyanide. *Bacterial Rev.* 40: 652-680.

Krasilnikov, N.A., Koreniako, A.I. and Artamonava, O.I. (1953). The distribution of actinomcete antagonists in soils. *Microbiologiya* 22: 3-10.

Kraus, J. and Loper, J.E. (1992). Lack of evidence for a role of antifungal metabolite production by *Pseudomonas fluorescens* Pf-5 in biological control of *Pythium* damping-off of cucumber. *Phytopathology* 82: 264-271.

Krishnamurthy, K. and Gnanamanickam, S.S. (1997). Biological control of sheath blight of rice: induction of systemic resistance in rice by plant-associated *Pseudomonas* spp. *Current Science* 72: 331-334.

Krishnamurthy, K. and Gnanamanickam, S.S. (1998). Biological control of rice blast by *Pseudomonas fluorescens* strain Pf7-14: evaluation of a marker gene and formulations. *Biological Control* 13(3): 158-165.

Kumar, L.P., Niranjana, S.R., Prakash, H.S. and Shetty, H.S. (2000). Effect of *Pseudomonas fluorescens* formulation against *Pyricularia grisea* in rice. *Crop Improvement* 27(2): 159-166.

Kurze, S., Bahl, H, Dahl, R. and Berg, G. (2001). Biological control of fungal strawberry diseases by Serratia plymuthica HRO-C48. *Plant Disease* 85(5): 529-534.

Laha, G. S. and Venkataraman, S. (2001). Sheath blight management in rice with biocontrol agents. *Indian phytopath.* 54(4): 461-464.

Lambert, B., Layns, F., Rooyen, L., Van Gossele, F., Papon, Y. and Swings, J. (1987). Rhizobacteria of maize and their fungal activities. *Applied and Environmental Microbiology* 53 (8): 1866-1871.

Laville, J., Voisard, C., Keel, C., Maurhofer, M., Defago, G. and Haas, D. (1992). Global control in *Pseudomonas fluorescens* mediating antibiotic synthesis and suppression of black root rot of tobacco. *Proc. National Academy of Sciences, U.S.A.* 89(5): 1562-1566.

Lebuhn M, Heulin T. and Hartmann, A. (1997). Production of auxin and other indolic and phenolic compounds by *Paenibacillus polymyxa* strains isolated from different proximity to plant roots. *FEMS Microbiol.* Ecol. 22:325-334.

Leeman, M., Scheffer, R. J., Van Pelt, J.A., Bakker, P.A.H.M. and Schippers, B. (1991). Control of *Fusarium* wilt of radish by *Pseudomonas fluorescens* WCS374, in greenhouse trials. *Bulletin SROP* 14(8):34-38.

Leeman, M., Van Pelt J. A., Den Ouden, F.M., Heinsbroek, M., Bakker, P.A.H.M. Schippers, B. (1995). Induction of systemic resistance against *Fusarium* wilt of radish by lipopolysaccharides of *Pseudomonas fluorescens*. *Phytopathology* 85(9):1021-1027.

Leong, J. (1986). Siderophores : their biochemistry and possible role in the biocontrol of plant pathogens. *Annu. Rev. Phytopathol.* 24: 187-209.

Leong, S.A. and Neilands, J.B. (1982). Siderophore production by phytopathogenic microbes. *Biophysics* 218 (2): 351-359.

Levy, E., Gough, F.J., Berlin, K.D., Guiana, P.W. and Smith, J.T. (1992). Inhibition of *Septoria tritici* and other phytopathogenic fungi and bacteria by *Pseudomonas fluorescens* and its antibiotics. Plant Pathology 41: 335-341.

Liao, C. H. (1989). Antagonism of *Pseudomonas putida* strain PP22 to phytopathogenic bacteria and its potential use as a biocontrol agent. Plant Disease 73(3): 223-226.

Liddell, C.M. and Parke, J.L. (1989). Enhanced colonization of pea tap roots by afluorescent pseudomonad biocontrol agent by water infiltration into soil. Phytopathology 79: 1327-1332.

Lieske, R. (1921). Morphologie und Biologia dar Strahlenpilze. Leipzig. 9:138-143.

Lifshitz, R., kloepper, J.W., Kozlowski, M., Simonson, C., Carloso, J., Tipping, E.M. and Zaleska, I. (1987). Growth promotion of canola (rapeseed) seedlings by a strain of *Pseudomonas putida* under gnotobiotic conditions. Can. J. Microbiol. 33:390.

Lin, F. C., Hong, X., Lin, W.T., Chen, W.L. and Li, D.B. (1998). Screening of antagonistic bacteria to *Magnaporthe grisea*. J. Zhejiang Agricultural University 24(6): 591-594.

Lindow, S.E. and Suslow, T.V. (2003). Temporal dynamics of the biocontrol agent *Pseudomonas fluorescens* strain A506 in flowers in inoculated pear trees. Phytopathology 93(6): 727-737.

Loper, J.E. (1988). Role of fluorescent siderophore production in biological control of *Pythium ultimum* by a *Pseudomonas fluorescens* strain. Phytopathology 78: 166-172.

Loper, J.E. and Buyer, J.S. (1991). Siderophores in microbial interactions on plant surfaces. Molecular Plant Microbe Interactions 4: 5-13.

Loper, J.E. and Henkels, M.D. (1999). Utilization of heterologous siderophores enhances levels of iron available to *Pseudomonas putida* in the rhizosphere. Applied and Environmental Microbiology 65: 5357-5363.

Loper, J.E. and Schroth, M.N. (1986). Influence of bacterial sources of indole-3-acetic acid on root elongation of sugar but. Phytopathology 76: 386.

Loper, J.E., Suslow, T.V. and Schroth, M.N. (1984). Lognormal distribution of bacterial populations in the rhizosphere. Phytopathology 74: 1454 – 1460.

Lou, B.G., Zhang, B.X., Ryder, M., Warren, R., Harvey, P. (2002). Biological control of seedling damping-off of cucumber. Acta Phytophylacica Sinica 29 (2): 109-113.

Lugtenberg, B.J.J., Dekkers, L. C. and Bloemberg, G.W. (2001). Molecular determinanats of rhizosphere colonization by *Pseudomonas*. Annu. Rev. Phytopathol. 39: 461-490.

Lynch, J.M. (1978). Microbial interactions around imbibed seeds. Ann. Appl. Biol. 89: 165-167.

Lynch, J.M. (1982). Interactions between bacteria and plants in the root environment. *In*: *Bacteria and Plants*, M.E. Rhodes-Roberts and F.A. Skinner (Eds.) pp. 1-23, London, Academic, 264 pp.

Lynch, J.M. (1990). *Microbial* metabolites. *In*: *The Rhizosphere*, JM Lynch (Ed.) pp. 177-206., Chichester; Wiley.

Lynch, J.M. and Whipps, J.M. (1991). Substrate flow in the rhizosphere. In : *The rhizosphere and Plant growth*. D.L. Keister., P.B. Cregan (Eds.) pp.15-24, Dordrecht, The Netherlands, Kluwer, 386 pp.

M'Piga, P., Belanger, R.R., Paulitz, T.C. and Benhamou, N. (1997). Increased resistance to *Fusarium oxysporum* f.sp. radicis-lycopersici in tomato plants treated with the endophytic bacterium, *Pseudomonas fluorescens* strain 63-28. *Physiological and Molecular Plant Pathology* 50: 301-320.

Malamy, J., Carr, J.P., Klessig, D.F. and Raskin, I. (1990). Salicylic acid: a likely endogenous signal in the resistance response of tobacco to viral infection. Science 250: 1002-1004.

Manav, M. and Thind, B.S. (2002). Management of bacterial blight of rice with bioagents. *Plant Disease Research* 17(1): 21-28.

Manoranjitham, S.K., Prakasam, V. and Rajappan, K. (1999). Effect of antagonists on Pythium aphanidermatum (Edson) Fitz and the growth of chilli seedlings. *J. Biological Control* 13(1-2): 101-106.

Marimuthu, S., Subbian, P., Ramamoorthy, V. and Samiyappan, R. (2002). Synergistic effect of combined application of *Azospirillum* and *Pseudomonas fluorescens* with inorganic fertilizers on root rot incidence and yield of cotton. *Zeitschrift fur Pflanzenkrankheiten und Pflanzenschutz* 109(6): 569-577.

Martinez, C., Michaud, M., Belanger, R. R. and Tweddell, R.J. (2002). Identification of soils suppressive against *Helminthosporium solani*, the causal agent of potato silver scurf. *Soil Biol. and Biochem*. 34 (12): 1861-1868.

Marugg, J.D., Van Spanje, M., Hockstra, W.P.M., Schippers, B., Weisbeek, P.J. (1985). Isolation and analysis of genes involved in siderophore – biosynthesis in plant growth – stimulating *Pseudomonas putida* WCS 358. *J. Bacteriol*. 174: 563-570.

Mathre, D.E., Cook, R.J. and Callan, N.W. (1999). From discovery to use: traversing the world of commercializing biocontrol agents for plant disease control. *Plant Disease* 83:972-983.

Maurhofer, M., Hase, C., Meuwly, P., Metraux, J.P. and Defago, G. (1994). Induction of systemic resistance of tobacco to tobacco necrosis virus by the root-colonizing *Pseudomonas fluorescens* strain CHA0: influence of the gacA gene and of pyoverdine production. *Phytopathology* 84(2): 139-146.

Maurhofer, M., Reismman, C., Scherer, S.P., Heebs, S., Haas, D. and Defago, G. (1998). Salicylic acid biosysthetic genes expressed in *Pseudomonas fluorescens* starin P3 improve the induction of systemic resistance in tomato against tobacco necrosis virus. *Phytopathology* 88:678-684.

Mazzola, M. and Cook, R.J. (1991). Effects of fungal root pathogens on the populations dynamics of biocontrol strains of fluorescent pseudomonads in the wheat rhizosphere. *Appl. Environ. Microbial.* 57: 2171-2178.

Mazzola, M., Cook, R.J., Thomashow, L.S., Weller, D.M. and Pierson, L.S. III, (1992). Contribution of phenazine antibiotic biosynthesis to the ecological competence of fluorescent pseudomonads in soil habitats. *Appl. Environ. Microbial.* 58: 2616-2624.

Mc In roy, J.A. and Kloepper, J.W. (1995). Population dynamics of endophytic bacteria in field grown sweet corn and cotton. *Can J. Microbiol* 41: 895-901.

Meena, B., Radhajeyalakshmi, R., Marimuthu, T. , Vidhyasekaran, P and Velazhahan, R. (2002). Biological control of groundnut late leaf spot and rust by seed and foliar applications of a powder formulation of *Pseudomonas fluorescens. Biocontrol Science and Technology* 12(2): 195-204.

Michaud, M., Martinez, C., Simao-Beaunoir, A.M., Belanger, R. R. and Tweddell, R.J. (2002). Selection of antagonist microorganisms against *Helminthosporium solani,* causal agent of potato silver scurf. *Plant Disease* 86 (7): 717-720.

Milner, J.L., Silo-Suh, L, Lee, J.C., He, H., Clardy, J. and Handelsman, J. (1996). Production of kanosamine by *Bacillus cereus* UW 85. *Applied and Environmental Microbiology* 62: 3061-3065.

Milus,E.A, Rothrock, C.S. (1997). Efficacy of bacterial seed treatments for controlling *Pythium* root rot of winter wheat. *Plant Disease* 81(2): 180-184.

Misaghi, I.J., Stowell, L.J., Grogan, R.G., and Spearman, L.C. (1982). Fungistatic activity of water soluble fluorescent pigment of fluorescent pseudomonads. *Phytopathology* 72: 33-36.

Mitchell, R. and Hurwitz, R. (1965). Suppression of *Pythium debaryanum* by lytic rhizosphere bacteria. *Phytopathology* 55: 156-158.

Mohamed, S. and Caunter, I.G. (1995). Isolation and characterization of a *Pseudomonas fluorescens* strain suppressive to *Bipolaris maydis. J. Phytopathology* 143 (2): 111-114.

Mondal, K.K., Dureja, P. and Verma, J. P. (2001). Management of *Xanthomonas camprestris* pv. *Malvacearum* induced blight of cotton through phenolics of cotton rhizobacterium. *Current Microbiology* 43(5): 336-339.

Mondal, K.K., Singh, R. P. and Verma, J.P. (1999). Beneficial effects of indigenous cotton rhizobacteria on seed germinability, growth promotion and suppression of bacterial blight disease. *Indian Phytopath.* 52 (3):228-235.

Mortuza, M.G. and Ilag, L.L. (2000). Biological control of Lasiodiplodia rot in banana fruits by *Bacillus* species. Bangladesh *J. Plant Pathology* 16(1-2): 47-50.

Mosa, A.A. (2002). Induced resistance in rice against blast disease using abiotic and biotic agents. *Annals of Agricultural Science,Cairo* 47(3): 993-1008.

Mukherjee, S.K. and Nandi, P.N. (1955). Studies on the antagonistic actinomycetes from the soils of West Bengal. *Indian Phytopath*. 8:64-71.

Müller, R. (1908). Eine Diphtheridee und eine Streptothrix mit gleichem blauen Farbstoff, sowie Untersuchungen über Streptothrix-arten imallgemeinen. *Centr. Bakteriol. Parasitenk*. Abt. I. Orig. 46:195-212.

Nagaraju, P., Dronavalli, N. and Biradar, D.P. (2002). Biological control of sheath blight (*Rhizoctonia solani*) in transplanted rice (*Oryza sativa*). *Indian J. Agricultural Sciences* 72(5): 306-307.

Nandakumar, R., Babu, S., Radjacommare, R., Raguchander, T. and Samiyappan, R. (2002). *Pseudomonas fluorescens* mediated antifungal activity against *Rhizoctonia solani* causing sheath blight in rice. *Phytopathologia Mediterranea* 41(2): 109-119.

Nandkumar, R., Babu, S., Viswanathan, R., Raghuchander, T. and Samiyappan, R. (2001). Induction of systemic resistance in rice against sheath blight disease by *Pseudomonas fluorescens*. *Soil biol. Biochem*. 33: 603-612.

Naseby, D.C., Way, J.A., Bainton, N.J. and Lynch, J.M. (2001). Biocontrol of *Pythium* in the pea rhizosphere by antifungal metabolite producing and non-producing Pseudomonas strains. *J. Applied Microbiology* 90 (3): 421-429.

Nautiyal, C.S. (1997). Selection of chickpea-rhizosphere-competent *Pseudomonas fluorescens* NBRI1303 antagonistic to *Fusarium oxysporum* f.sp. *ciceris*, *Rhizoctonia bataticola* and *Pythium* sp. *Current Microbiology* 35(1): 52-58.

Newman, E.I. (1985). The rhizosphere, carbon sources and microbial populations. In: A.H. Fritter (ed.) *Ecological interactions in soil.* Boston, Blackwell. Sci, pp 107-121.

Obukowicz, M.G., Perlak, F.J., Bolton, S.L., Kusano-Kretzmer, K., Mayer, E.J., *et al* (1987). Is 50L as a non-self transposable rector used to integrate the *Bacillus thuringiensis* delta endotoxin gene into the chromosome of root colonizing pseudomonads. *Gene* 51: 91-96.

Obukowicz, M.G., Perlak, F.J., Jusano-Kretzmer, K., Mayer, E.J., Bolton, S.L., *et al* (1986). In 5 – mediated integration of the delta endotoxin gene from *Bacillus thuringiensis* into the chromosome of root colonizing pseudomonads. *J. Bacteriol.* 168: 982-989.

Ongena M, Daayf F, Jacques P, Thomart P, Benhamou N, Paulitz TC, Cornelis P, Koedam N, Belanger RR (1999) Protection of cucumber against *Pythium* root rot by fluorescent pseudomonads: predominant role of induced resistance over siderophores and antibiotics. *Plant Pathology* 58: 66-76.

Ordentlich A, Elad Y, Chet F (1987) Rhizosphere colonization by *Serratia marcescens* for the control of *Sclerotium rolfisii. Soil Biol. and Biochem.* 19: 747-751.

Ordentlich, A., Elad, Y. and Chet, I. (1988). The role of chitinase of *Serratia marcescens* in biocontrol of *Sclerotium rolfisii. Phytopathology* 78: 84-88.

O'Sullivan, D. J. and O'Gara, F. (1992). Traits of fluorescent *Pseudomonas* spp. involved in suppressing plant root pathogens. *Microbial Rev*. 56: 662-676.

Ozaktan, H. and Bora, T. (2000). Biological control of *Fusarium oxysporum* f. sp. *melonis* by the formulations of fluorescent pseudomonads. *J.Turkish Phytopathology* 29(2-3): 133-149.

Pandey, A., Palni, L.M.S. and Hebbar, K.P. (2001). Suppression of damping-off in maize seedlings by *Pseudomonas corrugata*. *Microbiological Research* 156: 191-194.

Papavizas GC and lumsden, R.D. (1980). Biological control of soil-borne fungal propagules. *Annu. Rev. Phytopathol.* 18: 389-413.

Papavizas, G.C. (1985). Trichoderma and Gliocladium: Biology, ecology, and potential for biocontrol. *Annu. Rev. Phytopathol.* 23:23-54.

Parke, J.L. (1990). Root colonization by indigenous and introduced microorganisms. In: *The Rhizosphere and Plant Growth*. D.L. Keister, and P.B. Gregan (Eds.) pp. 33-42, Kluwer Academic Publishers, Dordrecht, The Netherlands.

Parke, J.L. (1991). Efficacy of *Pseudomonas cepacia* AMMD and *Pseudomonas fluorescens* PRA25 in biocontrol of *Pythium* damping-off and *Aphanomyces* root rot of pea. *Bulletin SROP*. 14(8): 30-33.

Parke, J.L., Moen, R., Rovira, A.D. and Bowen G.D. (1986). Soil water flow affects the rhizosphere distribution of a seed-borne biological control agent, *Pseudomonas fluorescens*. *Soil Biol. Biochem.* 18: 583-588.

Patil, R.P., Jagadeesh, K.S., Krishnaraj, P.U. and Kulkarni, J.H. (1998). Bacterization of groundnut with *Pseudomonas fluorescens* for the control of collar rot caused by *Sclerotium rolfsii* Sacc. *Karnataka J. Agricultural Sciences* 11 (2): 423-425.

Paulitz, T., Nowak, T.B., Gamard, P., Tsang, E. and Loper, J. (2000). A novel antifungal furanome from *Pseudomonas aureofacieus*, a biocontrol agent of fungal plant pathogens. *J. Chemical Ecology* 26 (6): 1515-1524.

Pichard, B. and Thouvenot, D. (1999). Effect of *Bacillus polymyxa* seed treatments on control of black-rot and damping-off of cauliflower. *Seed Science and Technology* 27(2): 455-465.

Pieterse, C.M.J, Van Wees, S.C.M., Hoffland, E., Van Pelt, J.A. and Van Loon, L.C. (1996). Systemic resistance in *Arabidopsis* induced by biocontrol bacterial is independent of salicylic acid accumulation and pathogenesis related gene expression. *Plant Cell* 8: 1225-1237.

Pieterse, C.M.J, Van Wees, S.C.M., Van Pelt, J.A., Knoester, M., Laan, R., Gerrits, H., Weisbeek, P.J. and Van Loon, L.C. (1998). A novel signaling pathway controlling induced systemic resistance in *Arabidopsis*. *Plant Cell* 10: 1571-1580.

Pieterse, C.M.J., Van Pelt, J.A., Ton, J., Parchmann, S., Mueller, M.J., Buchala, A.J, Metraux, J.P. and Van Loon, L.C. (2000). Rhizobacteria mediated induced systemic resistance (ISR) in *Arabidopsis* requires sensitivity to jasmonate and ethylene but is not accompanied by an increase in their production. Physiol. Mol. Plant Pathol. 57: 123-134.

Pleban, S., Chernin, L. and Chet, I. (1997). Chitinolytic activity of an endophytic strain of *Bacillus cereus*. Letters in Applied Microbiology 25: 284-288.

Porter, C.L. (1924). Concerning the characters of certain fungi as exhibited by their growth in the presence of other fungi. Am.J.Bot.11:168-187.

Prasad, R.D.and Kulshrestha, D.D. (2002). Efficacy of some bacterial antagonists against seedling blight of sunflower caused by *Alternaria helianthi*. Plant Disease Research 17(1): 65-66.

Prashanthi, S.K., Kulkarni, S. and Anahosur, K.H. (2000). Management of safflower root rot caused by *Rhizoctonia bataticola* by antagonistic microorganisms. Plant Disease Research 15(2): 146-150.

Press, C.M., Wilson, M., Tuzun, S. and Kloepper, J.W. (1997). Salicylic acid produced by *Serratia marcescens* 91-166 is not the primary determinant of induced systemic resistance in cucumber or tobacco. Mol. Plant Microbe Interact. 10: 761-768.

Prikryl, Z. and Vancura, V. (1980). Root exudates of plants VI. Wheat root exudation as dependent on growth, concentration gradient of exudates and the presence of bacteria. Plant and Soil 57: 69-83.

Pringle, J.H. and Fletcher, M. (1986). Absorption of bacterial surface polymers to attachment substrate. J. Gen. Microbiol. 132: 743-749.

Prinkryl, Z., Vancura, V. and Wurst, M. (1985). Auxin formation by rhizosphere bacteria as a factor of root growth. Biologia Plantarum. 27: 159.

Raaijmakers, J. M. and Weller, D. M. (1998). Natural plant protection by 2, 4-diacetylphloroglucinol-producing *Pseudomona* spp. in take-all decline soils. Molecular Plant Microbe Interactions 11: 144-152.

Raaijmakers, J.M., Bonsall, R.F. and Weler, D.M. (1999). Effect of population density of *Pseudomonas fluorescens* on production of 2, 4- diacetylphloroglucinol in the rhizosphere of wheat. Phytopathology 89: 470-475.

Rabindran, R. and Vidhyasekaran, P. (1996). Development of a formulation of *Pseudomonas fluorescens* PfALR2 for management of rice sheath blight. Crop Protection 15(8): 715-721.

Raguchander, T., Jayashree, K. and Samiyappan, R. (1997). Management of Fusarium wilt of banana using antagonistic microorganisms. J. Biological Control 11(1-2): 101-105.

Rajappan, K., Vidhyasekaran, P., Sethuraman, K. and Baskaran, T.L. (2002). Development of powder and capsule formulations of Pseudomonas fluorescens strain Pf-1 for control of banana wilt. Zeitschrift fur Pflanzenkrankheiten und Pflanzenschutz 109(1): 80-87.

Ramakrishnan, S., Sivakumar, C.V. (1999). Biological control of rice root nematode with *Pseudomonas fluorescens*. *In*: Proc. National symposium on rational approaches in nematode management for sustainable agriculture, Dhawan, S.C. ,Kaushal, K.K. (ed.). Anand, India, 23-25 November, 1998. pp. 43-46, Nematological Society of India; New Delhi; India.

Ramamoorthy, V., Raguchander, T. and Samiyappan, R. (2002). Enhancing resistance of tomato and hot pepper to *Pythium* diseases by seed treatment with fluorescent pseudomonads. European J. Plant Pathology 108 (5): 429-441.

Ramamoorthy, V., Raguchander, T. and Samiyappan, R. (2002). Induction of defense-related proteins in tomato roots treated with *Pseudomonas fluorescens* Pf1 and *Fusarium oxysporum* f. sp. *lycopersici*. Plant and Soil 239(1):55-68.

Ramamurthy, V., Viswanathan, R., Raghuchnder, T., Prakasam, V. and Samiyappan, R. (2001). Induction of systemic resistance by plant growth promoting rhizobacteria in crop plants against pest and diseases. Crop Protection 20: 1-11.

Rangaswami,G., Oblisami, G. and Swaminathan, R. (1967). Antagonistic actinomycetes in the soils of South India. U.A.S., Bangalore, India and U.S.D.A., P.L.480-FG-in-129. pp.156.

Rangeshwaran, R., Prasad, R.D. and Anuroop, C.P. (2001). Field evaluation of two bacterial antagonists, *Pseudomonas putida* (PDBCAB 19) and *P. fluorescens* (PDBCAB 2) against wilt and root-rot of chickpea. J. Biological Control 15 (2): 165-170.

Rao, C.V.S., Sachan, I.P. and Johri, B.N. (1999). Influence of fluorescent pseudomonads on growth and nodulation of lentil (*Lens esculentus*) in *Fusarium* infested soil. Indian J. Microbiology 39(1): 23-29.

Ross, A.F. (1961). Systemic acquired resistance induced by localized virus infections in plants. Virology 14: 340 – 358.

Rouatt, J.W. and Katznelson, H. (1961). A study of the bacteria on the root surface and in the rhizosphere soil of crop plants. J. Applied Bacteriology 24 (2): 164-171.

Rouatt, J.W., Katznelson, H. and Payne, T.M.B. (1960). Soil Sci. Soc. Amer. Proc. 24: 271-273.

Rovira, A.D., Foster R.C. and Martin J.K. (1979). Note on terminology: Origin, nature and nomenclature of the organic materials in the rhizosphere. *In*: The soil-root interface, J.L. Harley and R.S. Russel (Eds.) pp 1-4, London, Academic, 448 pp.

Ryder, M., Yan, Z., Terrace, T., Rovira, A.D. and Tang, W. (1996). Control of take all and *Rhizoctonia* root rot on wheat and seedling growth promotion by *Bacillus* and other root associated bacteria. *In* : Advances in Biological control of Plant Diseases, W. Tang, R. Cook and A. Rovira (Eds.) pp. 94-98, China Agricultural University Press, Beijing.

Sabaratnam, S. and Traquair, J.A. (2002). Formulation of a *Streptomyces* biocontrol agent for the suppression of *Rhizoctonia* damping-off in tomato transplants. Biological Control 23: 245-253.

Samuels, G.J. (1996). *Trichoderma*: A review of biology and systematics of the genus. Mycol. Res. 100:923-934.

Saniewska, A. (1999). Influence of Bacillus polymyxa on the growth and development of *Fusarium oxysporum* f.sp. *tulipae*. Acta Agrobotanica 52(1-2): 5-10.

Sarma, B.K., Singh, D.P., Mehta, S., Singh, H.B. and Singh, U.P. (2002). Plant growth-promoting rhizobacteria-elicited alterations in phenolic profile of chickpea (*Cicer arietinum*) infected by *Sclerotium rolfsii*. J. Phytopathology 150 (4-5): 277-282.

Scher, F.M. and Baker, R. (1980). Mechanism of biological control in a *Fusarium suppressive* soil. Phytopathology 70: 412-417.

Scher, F.M. and Baker, R. (1982). Effect of *Pseudomonas putida* and a synthetic iron chelator on induction of soil suppressiveness to *Fusarium* wilt pathogens. Phytopathology 72: 1567-1573.

Scher, F.M., Kloepper, J.W. and Singreton, C.A. (1985). Chemotaxis of fluorescent *Pseudomonas* spp. to soybean seed exudates *in vitro* and in soil. Can. J. Microbiol. 31: 570-574.

Scher, F.M., Kloepper, J.W., Singleton, C., Zaleska, I. and Lalibeite, M. (1988). Colonization of soybean roots by *Pseudomonas* and *Serratia* species. Relationship to bacterial mobility, chemotaxis and germination time. Phytopathology 78: 1055-1059.

Scherff, R.H. (1973). Control of bacterial blight of soybean by *Bdellovibrio bacteriovorus*. Phytopathology 63 : 400-402.

Schippers, B., Bakker, A.W. and Bakker, P.A.H.M. (1987). Interactions of deleterious and beneficial rhizosphere microorganisms and the effect of cropping practices. Annual Review of phytopathology 25: 339-358.

Schippers, B., Geels, F.P., Hoekstra, O., Lamers, J.G., Maenhout, C.A.A.A. (1985). Yield depressions in narrow rotations caused by unknown microbial factors and their suppressions by selected pseudomonads. *In*: Ecology and Management of Soil-Borne Plant Pathogens. C.A .Parker, A.D. Rovira, K.J. Moore, P.T. Wong, J.F. Kollmorgen (Eds.). pp. 127-130. St. Paul, Minnesota, APS.

Schisler, D. A., Slininger, P. J., Kleinkopf, G., Bothast, R. J. and Ostrowski, R.C. (2000). Biological control of *Fusarium* dry rot of potato tubers under commercial storage conditions. American J. Potato Research 77(1): 29-40.

Schmidi Sacheria, P., Keel, C. and Defago, G. (1997). The global rgulator GacA of Pseudomonas *fluorescens* CHAO is required for suppression of root diseases in dicotyledons but not in graminaeae. Plant Pathology 46: 80-90.

Schmidt, C.S., Lorenz, D., Wolf, G.A. and Jager, J. (2001). Biological control of the grapevine dieback fungus *Eutypa lata* II: influence of formulation additives and transposon mutagenesis on the antagonistic activity of *Bacillus subtilis* and *Erwinia herbicola*. J. Phytopathology 149(7-8): 437-445.

Schroth, M.N. and Hancoc, J.G. (1982). Disease suppressive soils and root colonizing bacteria. Science 216:1376-81.

Shanmugam, V., Senthil, N., Raguchander, T., Ramanathan, A. and Samiyappan, R. (2002). Interaction of *Pseudomonas fluorescens* with *Rhizobium* for their effect on the management of peanut root rot. Phytoparasitica 30(2): 169-176.

Shekhawat, G.S., Singh, R. and Chakrabarti, S.K. (1993). Possibility of biological management of potato bacterial wilt with microbial antagonists and latent potato viruses. J. Indian Potato Assoc. 20 (3-4): 219-222.

Shen, S.S., Kim, J.W. and Park, C.S. (2002). *Serratia plymuthica* strain A21-4: a potential biocontrol agent against *Phytophthora* blight of pepper. Plant Pathology Journal 18(3): 138-141.

Shete, K., Srivastava, G.P. and Vora,V.C. (1960). Screening of actinomycetes to antifungal and antibacterial activity. Antib. and Chemo.11:438-440.

Shishido, M., Brevil, C. and Chanway, C.P. (1999). Endophytic colonization of spruce by plant growth promoting rhizobacteria. FEMS Microbioogy Ecology. 29: 191-196.

Siddiqui, I.A. and Shaukat, S.S. (2002 a). Mixtures of plant disease suppressive bacteria enhance biological control of multiple tomato pathogens. Biology and Fertility of Soils 36 (4): 260-268.

Siddiqui, I.A. and Shaukat, S.S. (2002 b). Rhizobacteria-mediated induction of systemic resistance (ISR) in tomato against *Meloidogyne javanica*. J. Phytopathology 150 (8-9): 469-473.

Siddiqui, Z.A., Mahmood, I. And Hayat, S. (1998). Biocontrol of *Heterodera cajani* and *Fusarium udum* on pigeonpea using *Glomus mosseae, Paecilomyces lilacinus* and *Pseudomonas fluorescens.* Thai J. Agricultural Science 31(3): 310-321.

Sindhan, G.S., Hooda, I. and Karwasra, S.S. (2002). Biological control of dry root rots of chickpea caused by *Rhizoctonia bataticola.* Plant Disease Research 17(1): 68-71.

Singh, P.P., Shin, Y.C., Park, C.S. and Chung, Y.R. (1999), Biological control of Fusarium wilt of cucumber by chitinolytic bacteria. Phytopathology 89: 92-99.

Singh, R.S. and Reddy, C.S. (1979). Suppression of damping-off of tomato and seedling blight of chick pea and sugarbeet by strains of *Streptomyces diastaticus.* Indian Phytopath. 32:374-377.

Singh, U.P., Sarma, B.K. and Singh, D.P. (2003). Effect of plant growth-promoting rhizobacteria and culture filtrate of *Sclerotium rolfsii* on phenolic and salicylic acid contents in chickpea (*Cicer arietinum*). Current Microbiology 46 (2): 131-140.

Sislow, T.V. and Schroth, M.N. (1982). Role of deleterious rhizobacteria as minor pathogens in reducing crop growth. Phytopathology 72: 111-115.

Sitther, V. and Gnanamanickam, S.S. (1996). Biological control of blast disease of fingermillet (*Eleusine coracana* L.) and an analysis of fertility of *Magnaporthe grisea*. Current Science 71(2): 144-147.

Sivakumar, G. and Narayanaswamy, N.T. (1998). Biological control of sheath blight of rice with *Pseudomonas fluorescens.* Oryza 35 (1): 57-60.

Sivakumar, G., Sharma, R.C. and Rai, S.N. (2000). Biocontrol of banded leaf and sheath blight of maize by peat based *Pseudomonas fluorescens* formulation. Indian Phytopath. 53(2): 190-192.

Sivan, A. and Harman, G.E. (1991). Improved rhizosphere competence in a protoplast fusion progeny of *Trichoderma harzianum*. J.Gen.Microbiol. 137:23-32.

Smilanick, J.L. and Denis Arrue, R. (1992). Control of green mold of lemons with *Pseudomonas* species. Plant Disease 76 (5): 481-485.

Srivastava, A.K., Singh, T., Jana, T.K.and Arora, D.K. (2001). Induced resistance and control of charcoal rot in *Cicer arietinum* (chickpea) by *Pseudomonas fluorescens.* Canadian J. Botany 79(7): 787-795.

Stanghellini, M.E. and Hancock, J.G. (1970). Radial extent of the bean spermosphere and its relation to the behaviour of *Pythium ultimum.* Phytopathology 61: 165-168.

Stasz, T.E., Harman, G.E. and Weeden, N.F. (1996). Protoplast preparation and fusion in two biocontrol strains of *Trichoderma harzianum.* Mycologia 80: 141-148.

Stockwell, V.O., Johnson, K.B., Sugar, D. and Loper, J.E. (2002). Antibiosis contributes to biological control of fire blight by *Pantoea agglomerans* strain Eh252 in orchards. Phytopathology 92(11): 1202-1209.

Stolp, H. and Gadkari, D. (1981). Non-pathogenic members of the genes *Pseudomonas. In*: The Prokaryotes : A Handbook on Habitats, isolation and identification of bacteria, Vol. 1, MP Starr, H. Stolp, HG Truper, A Balows and HG Schlegel (Eds.) pp. 719-741, Springer-Verlag, Berlin.

Sturz, A.V., Christie, B.R., Matheson, B.G., Arsenault, W.J. and Buchanan, N.A. (1999). Endophytic bacterial communities in the periderm of potato tubers and their potential to improve resistance to soil-borne plant pathogens. Plant Pathology 48: 360-369.

Suslow, T.V. (1982). Role of root colonizing bacteria in plant growth. *In*: Phytopathogenic Prokaryotes. MS. Mount., GH Lacy (Eds.) Vol. 1. pp 187-223. Academic Press, New York/London.

Thomashow, L.S. and Weller, D.M. (1988). Role of a phenazine antibiotic from *Pseudomonas fluorescens* in biological control of *Gacumannomyces graminis* var. *tritici*. J. Bacteriol. 170: 3499-3508.

Thomashow, L.S. and Weller, D.M. (1990). Role of antibiotics and siderophores in biocontrol of take all disease of wheat. Plant and Soil 129: 93-99.

Thomashow, L.S. and Weller, D.M., Bonsall, R.F. and Pierson, C.S. III (1990). Production of the antibiotic phenazine – 1- carboxylic acid by fluorescent *Pseudomonas* species in the rhizosphere of wheat. Appl. Environ. Microbiol. 56: 908-912.

Thrane, C., Nielsen, T.H., Nielsen, M.N., Sorensen, J. and Olsson, S. (2000). Viscosinamide-producing *Pseudomonas fluorescens* DR54 exerts a biocontrol effect on *Pythium ultimum* in sugar beet rhizosphere. FEMS Microbiology Ecology 33: 139-146.

Timmusk S, Nicander B., Granhall, U. and Tillberg E. (1999). Cytokinin production by *Paenibacillus polymyxa*. Soil Biol. Biochem. 31:1847-1852.

Trapero Casas, A., Kaiser, W.J. and Ingram, D.M. (1990). Control of *Pythium* seed rot and pre-emergence damping-off of chickpea in the U.S. Pacific Northwest and Spain. Plant Disease 74(8): 563-569.

Umesha, S., Dharmesh, S.M., Shetty, S.A., Krishnappa, M. and Shetty, H.S. (1998). Biocontrol of downy mildew disease of pearl millet using *Pseudomonas fluorescens*. Crop Protection 17(5): 387-392.

Utkhede, R.S. (2000). Success and future of plant growth promoting rhizobacteria and biocontrol agents. In: Proc. International Conference on Integrated Plant Disease Management for Sustainable Agriculture, Indian Phytopathological Society, pp. 1273-1278.

Utkhede, R.S. and Koch, C.A. (2002). Chemical and biological treatments for control of gummy stem blight of greenhouse cucumbers. European J. Plant Pathology 108 (5): 443-448.

Utkhede, R.S. and Smith, E.M. (1994). Biotic and abiotic causes of plant problems of fruit trees. Acta Horticod 363: 25-32.

Utkhede, R.S., Sholberg, P.L. and Smirle, M.J. (2001). Effects of chemical and biological treatments on growth and yield of apple trees planted in *Phytophthora cactorum* infested soil. Canadian J. Plant Pathology 23(2): 163-167.

Utkhede, R.S., Koch, C.A. and Menzies, J.G. (1999). Rhizobacterial growth and yield promotion of cucumber plants inoculated with *Pythium aphanidermatum*. Canadian J. Plant Pathology 21(3): 265-271.

Van Loon, L.C. and Van Strien, E.A. (1999). The families of pathogenesis – related proteins, their activities and comparative analysis of PR-1 type proteins. Physiol. MO1. Plant Pathol. 55: 85-97.

Van Loon, L.C., Bakker, D.A.H.M. and Pietrse, C.M.J. (1998). Systemic resistance induced by rhizosphere bacteria. Annu. Rev. Phytopathol. 36: 453-483.

Van Peer, R., Punte, H.C.M., de Weger, L.A. and Schippers, B. (1990). Characterization of root surface and endorhizosphere pseudomondas in relation to their colonization of roots. Applied and Environmental Microbiology 56: 2462-2470.

Van Wers, S.C.M., Pieterse, C.M.J., Trijssenaar, A., Van't Westende, Y., Hartog, F. and Van Loon, L.C. (1997). Differential induction of systemic resistance in *Arabidopsis* by biocontrol bacterial. Mol. Plant Microbe Interact. 10: 716-724.

Vancura, V. (1988). Plant metabolites in soil. *In*: soil microbial associations: control of structures and functions. V. Vanicura and F. Kune (Eds.) pp 57 – 144, Amsterdam/New York, Elsevier, 498 pp.

Vesper, S.J., Malik, N.S.A. and Bauer, W.D. (1987). Transposon mutants of *Bradyrhizobium japonicum* altered in attachment to host roots. Appl. Environ. Microbiol. 53: 1959-1961.

Vidhyasekaran, P. and Muthamilan, M. (1995). Development of formulations of *Pseudomonas fluorescens* for control of chickpea wilt. Plant Disease 79(8): 782-786.

Vidhyasekaran, P. and Muthamilan, M. (1999). Evaluation of a powder formulation of *Pseudomonas fluorescens* Pf1 for control of rice sheath blight. Biocontrol Science and Technology 9(1): 67-74.

Vidhyasekaran, P., Kamala, N., Ramanathan, A., Rajappan, K., Paranidharan, V. and Velazhahan, R. (2001). Induction of systemic resistance by *Pseudomonas fluorescens* Pf1 against *Xanthomonas oryzae* pv. *oryzae* in rice leaves. Phytoparasitica 29(2): 155-166.

Vidhyasekaran, P., Sethuraman, K.,Rajappan, K. and Vasumathi, K. (1997). Powder formulations of *Pseudomonas fluorescens* to control pigeonpea wilt. Biological Control 8(3): 166-171.

Viswanathan, R. and Samiyappan, R. (2002). Induced systemic resistance by fluorescent pseudomonads against red rot disease of sugarcane caused by *Colletotrichum falcatum*. Crop Protection 21 (1): 1-10.

Voisard, C., Keel, C., Haas, D. and Defago, G. (1989). Cynamide production by *Pseudomonas fluorescens* helps suppress black root rot of tobacco under gnotobiotic conditions. EMBO J. 8: 351-359.

Waksman, S.A., Horning, E.S., Welsch, M. and Woodruff, H.B. (1942). Distribution of antagonistic actinomycetes in nature. Soil Sci. 54:281-296.

Warren, J.E and Bennet, M.A. (1999). Bio-osmopriming tomato (*Lycopersicon esculentum* Mill.) seeds for improved stand establishment. Seed Science and Technology 27: 489-499.

Webster, J. and Lomas, N. (1964). Does *Trichoderma viride* produce gliotoxin and viridian. Trans. Br. Mycol. Soc. 47: 535-540.

Weindling, R. (1932). *Trichoderma lignorum* as a parasite of other fungi. Phytopathology 22: 837.

Weindling, R. and emerson, O.H. (1936). The isolation of a toxic substance from the culture filtrate of *Trichoderma.* Phytopathology 26:1068-1070.

Weller, D. M. and Cook, R.J. (1983). Suppression of take-all of wheat by seed treatments with fluorescent pseudomonads. Phytopathology 73: 463-469.

Weller, D.M. (1983). Colonization of wheat roots by a fluorescent pseudomonad suppressive to take all. Phytopathology 73: 1548-1553.

Weller, D.M. (1985). Application of fluorescent pseudomonads to control root diseases. *In*: Ecology and Management of Soil borne plant pathogens. CA Parker, A.D. Rovira, K.J. Moores, P.T.W. Wong, J.F. Kollmorgen (Eds.). The American Phytopathological Society. St. Paul, Minnesota, USA.

Weller, D.M. (1988). Biological control of soil borne plant pathogens in the rhizosphere with bacteria. Annu. Rev. Phytopathol. 26: 379-407.

Weller, D.M., Howie, W.J. and Cook, R.J. (1988). Relationship between *in vitro* inhibition of *Gaeumannomyces graminis* var. *tritici* and suppression of take-all by fluorescent pseudomonads. Phytopathology 78: 1094-1100.

Whipps, J.M. (2001). Microbial interactions and biocontrol in the rhizosphere. J. Experimental Botany 52: 487-511.

Whitelaw, M.A. (2000). Growth promotion of plants inoculated with phosphate – solubilizing fungi. Advances in Agronomy 69:99-151.

Wilson, M., Campbell, H.L., Ji, P., Jones, J.B. and Cuppels, D.A. (2002). Biological control of bacterial speck of tomato under field conditions at several locations in North America. Phytopathology 92(12): 1284-1292.

Wulff, E.G., Mguni, C.M., Mortensen, C.N., Keswani, C.L. and Hockenhull, J. (2002). Biological control of black rot (*Xanthomonas campestris* pv. *campestris*) of brassicas with an antagonistic strain of *Bacillus subtilis* in Zimbabwe. European J. Plant Pathology 108(4): 317-325.

Wuthrich, B. and Defago, G. (1991). Suppression of wheat take-all and black root rot of tobacco by *Pseudomonas fluorescens* strain CHA0: results of field and pot experiments. Bulletin SROP 14(8): 17-22.

Yan, Z., Harris, A., Adkins, P., Ryder, M., Rovira, A. and Tang, W. (1996). Growth promotion of Chinese cabbage, capsicum and rape seed by yield increasing bacteria (YIB's) under Australian conditions. *In* : Advances in Biological control of Plant Diseases, W. Tang, R. Cook and A. Rovira (eds.) pp. 136-138, China Agricultural University Press, Beijing.

Yan, Z.N., Reddy, M.S., Ryu, C.M., McInroy, J.A., Wilson, M. and Kloepper, J.W. (2002). Induced systemic protection against tomato late blight elicited by plant growth-promoting rhizobacteria. Phytopathology 92 (12):1329-1333.

Yang, H.L., Sun, X.L., Song, W., Wang, Y.S. (1999). Studies on disease resistance induced in rice by plant growth promoting bacteria. Acta Phytopathologica Sinica 29(3): 286-287.

Zhao, J.H., Li, J.Z. and Kong, F.Z. (2003). Biocontrol activity against *Botrytis cinerea* by *Bacillus subtilis* 728 isolated from marine environment. Annals of Microbiology 53: 29-35.

Frontiers in Plant Sciences (Pages 219-286)
Eds. K.G. Mukerji, K.V.B.R. Tilak, S.M. Reddy, L.V. Gangwane, P. Prakash, I.K. Kunwar

The Ecology of Fungi on Plant Remains above the Soil from Dharwad

Ch. Ramesh and R.S. Chavan

ABSTRACT

The present investigation deals with the study of seasonal periodicity and percentage occurrence of fungi on different substrates from Dharwad region of western ghat forest area for a period of two years (viz: October 1998 to September 2000). The study was initially taken up mainly to understand the mycoflora of the area and the variation occurring on different types of substrates.

Key words: ecology, fungi, plant remains, substrates.

INTRODUCTION

The most significant factor affecting the distribution and life styles of fungi is their heterotophy, so that although they have some ability to fix atmospheric carbon dioxide, previously elaborated organic compounds, which they absorb from their immediate environment, must meet virtually all their growth needs. Fungi are confined to habitats in which a suitable supply of such compounds is available.

Ecology has been defined by Daubenmire (1974) as the study of the reciprocal relations between organisms and their environment" Fungi are the heterotrophic organisms which cannot manufacture their basic food requirements and so are dependent on food materials produced by other organisms either as saprobes or parasites.

Author's address: Department of Botany, Karnatak University, Dharwad–580003. Karnataka, India.

Most of the fungi have a multicellular thallus or mycelium composed of many fine threads called hyphae. In some groups, the thallus is not divided into cells, but it is elongate branched and coenocytic, in other groups, the thallus is unicellular, and each is a minute single entity.

Fungi may occur as mycelia, spores. Only mycelia decompose substrates. Fungal ecology is still dominated by a process-oriented point of view and it therefore seems desirable to identify those fungi that are present in an active mycelial state. Spore germination, mycelial development, fruiting body production, spore discharge and dissemination, and the development of rhizomorphs, sclerotia and resting cells, all fall within the province of fungus ecology.

That the environment in which a fungus grows is exceedingly complex (Billings 1952) has been demonstrated in a series of proposed outlines to be used in the study of ecological life history of the fungi (Cooke 1951) As a specific habitat for specific association of fungi, the litter is becoming important in ecological and other studies (Dickinson and Pugh, 1974) the ability of certain known and unknown organisms to decay portions of the litter has been studied (Broad foot and Pierre, 1939) and the ability of certain species of litter fungi to use lignin and cellulose has been considered (Lindeberg 1944.) there is a large population of fungi, which grows on the totally unrecognizable remains of plants in the fermentation, duff and humus layers under litter. In the development of our knowledge of fungi, both morphologically and ecologically, it is important that the taxonomy of the group be understandable to all. The present investigation reports on the seasonal periodicity and percentage occurrence of fungi on different substrates from Dharwad region of Western Ghats forest area over period of two years.

Patil and Ramesh (1983,1984,1986,1987) and Ramesh (1986, 1987 a, b, c, d, e, 1988 a, b, c, 1990 a, b, c 1992, 1993 a ,b, c, d, e) reported Ascomycetes from Maharashtra especially Loculoascomycetes. Studies on Marine Loculoascomycetes were under taken from Maharashtra by Borse (1983) and Ramesh (1985). Studies on water borne fungi occurring on different substrate in various water bodies were investigated by Ramesh and Vijayakumar (2001, 2003)

MATERIALS AND METHODS

All the specimens described in the present investigation has been collected by the authors. Frequent visits were paid to the different areas in the locality during different seasons from October 1998 to September 2000.

The forms collected were growing either saprophytically or parasitically or were present as the prologues on varied hosts/substrate like living leaves, stem and on the dead fallen leaves, twigs, stem, bark, wood etc. the saprophytic forms were found abundant in areas where accumulations of humus was more. They were growing either on the dead leaves,dead stems, twigs or parasitic forms were collected mainly from infected plant parts, especially leaves and stems. The isolated fungal prologues present on all the collected host/substrata were also considered for the present investigations.

a. Study Area and Environmental Data

Dharwad is selected as the study area located in Karnataka state, India (map. 1) at 15°30′ N latitude and 75°30′ E longitude, because of its rich vegetation especially of Angiosperms with other groups of plants harboring different groups of Mycoflora. The site chosen for the study in Dharwad were viz: (1) Karnatak university campus (2) Mango orchard near Haliyal road (3) Tapovan garden (4) University of Agricultural Sciences (5) Rayapur (map 2). There is lot of variation as far as vegetation is concerned this is due to hilly and plateau region. Due to change in vegetation there is likely that Mycoflora varies.

Data on temperature, relative humidity and rainfall recorded at Meteorological department, University of Agricultural sciences, Dharwad. Karnataka, India was considered for ecological studies of fungi.

b. Field Procedure

The samples were collected twice in a month, randomly from all the chosen sites. The collections included infected living leaves, living stems (scrapped or cut) leaf litter (dead leaves), stem litter (dead stems). The samples were brought to the laboratory in a sterile polythene bags/paper covers.

RESULTS

A total of 213 species belonging to 92 genera have been isolated in the present investigation. Out of 213 species, 12 species belonged to Zygomycotina, 28 species belonged to Ascomycotina and 173 species belonged to Deuteromycotina. Among Deuteromycotina 167 species belonged to Hyphomycetes where as 6 species belonged to Coelomycetes.

1.a. Percentage Occurrence of Zygomycotina on Leaf.

Among Zygomycotina a total of six species belonging to 3 genera have been recorded (Table No. 1). All the species were abundant except *Circinella simplex* which was rare in its percentage occurrence.

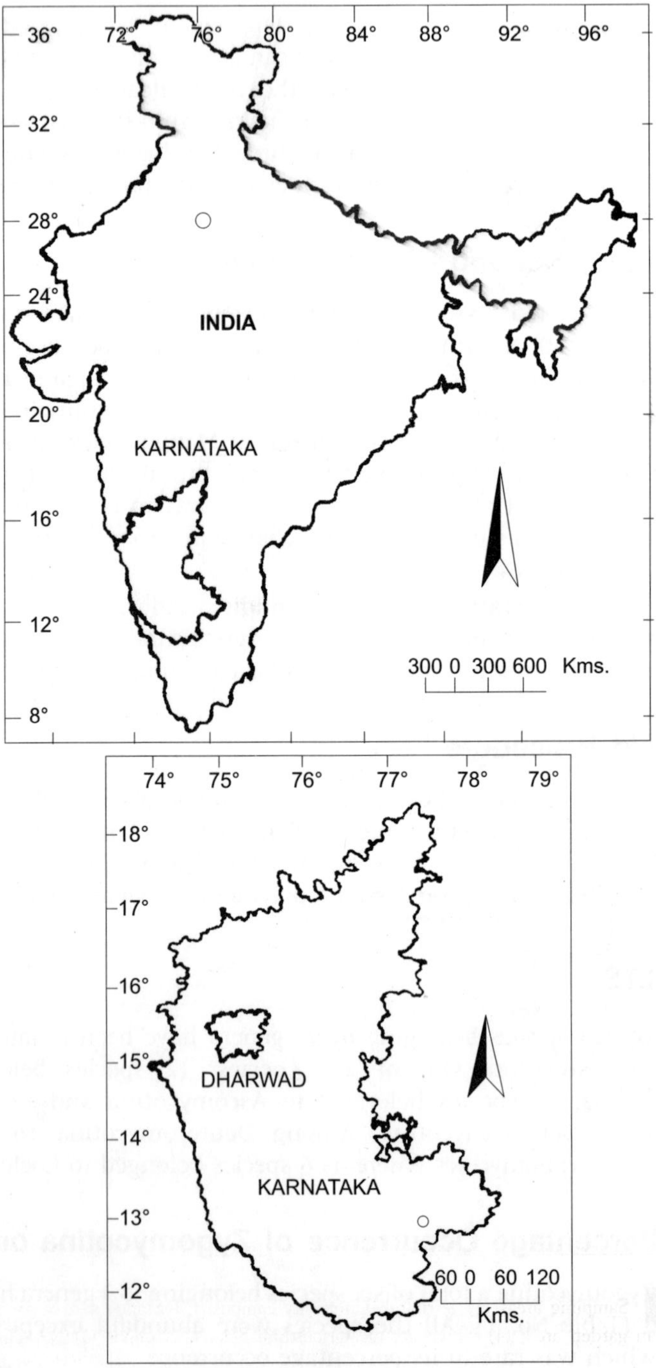
36°
72°
76°
80°
84°
88°
92°
96°
32°
28°
24°
20°
16°
12°
8°
INDIA
KARNATAKA
300 0 300 600 Kms.
74°
75°
76°
77°
78°
79°
18°
17°
16°
15°
14°
13°
12°
DHARWAD
KARNATAKA
60 0 60 120
Kms.

Map 1

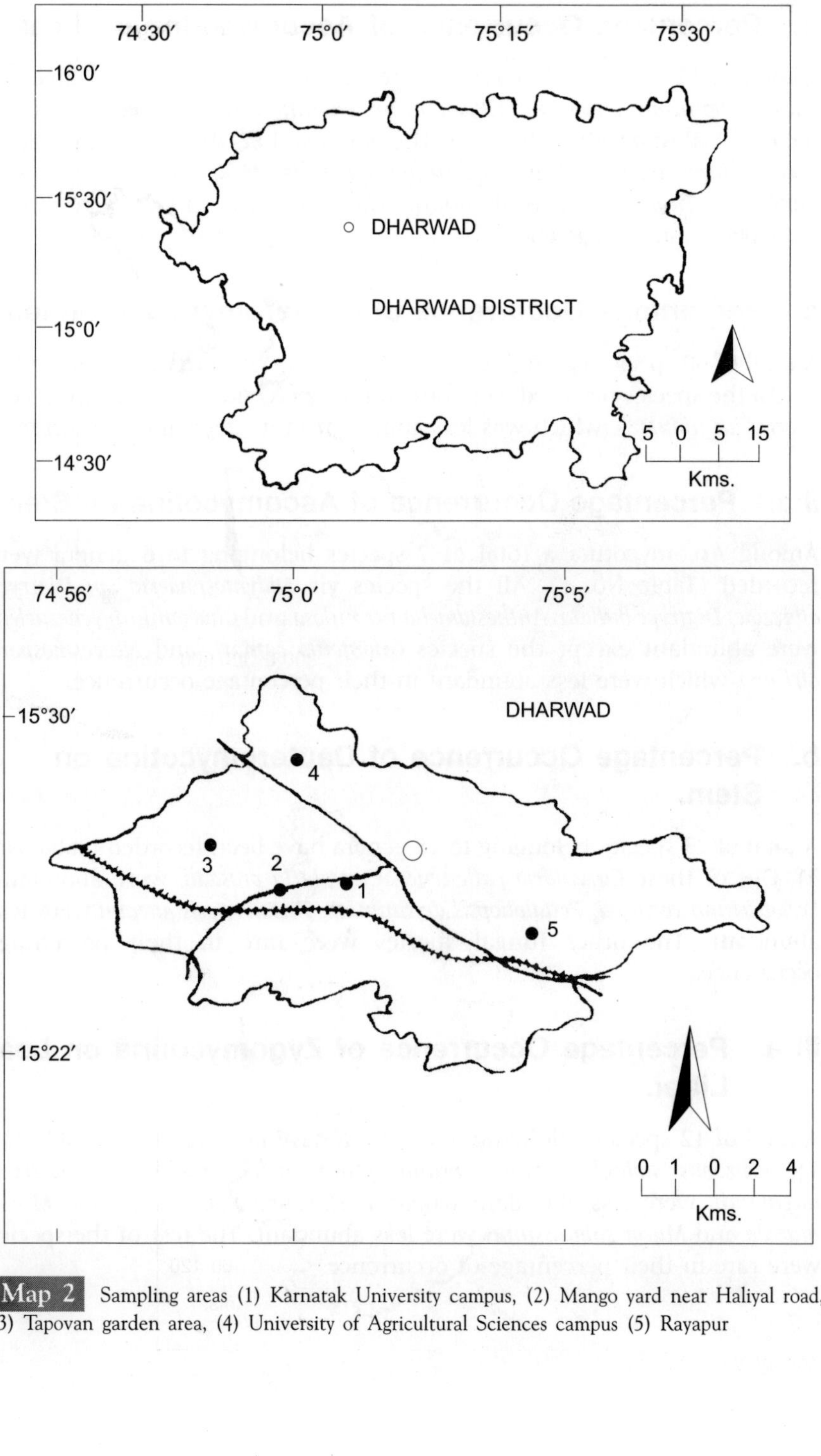

Map 2 Sampling areas (1) Karnatak University campus, (2) Mango yard near Haliyal road, (3) Tapovan garden area, (4) University of Agricultural Sciences campus (5) Rayapur

b. Percentage Occurrence of Ascomycotina on Leaf

A total of 12 species belonging to 8 genera have been isolated (Table No. 1) *Achaetomium globosum, Chaetomium elatum* and *Emericella sp.* Were recorded abundantly where as the species like *Ascotricha chartarum, Chaetomium abuense, Leptosphaerulina australis, Melanospora zamiae* and *Thielavia terricola* were less abundant. The rest of the species were rare in their percentage occurrence.

c. Percentage Occurrence of Deuteromycotina on leaf.

A total of 60 species belonging to 34 genera have been recorded (Table No. 1). All the species recorded were rare in their percentage occurrence except *Aspergillus nidulans,* which was less abundant in its percentage occurrence.

II a. Percentage Occurrence of Ascomycotina on Stem.

Among Ascomycotina a total of 7 species belonging to 6 genera were recorded (Table No. 1). All the species viz *Didymosphaeria sp: Diatrype albizziae, Diatrype bullata, Anthostomella turmulosa* and *Chaetomium venezuelens* were abundant except the species *Gnomonia comari,* and *Neocosmospora africana* which were less abundant in their percentage occurrence.

b. Percentage Occurrence of Deuteromycotina on Stem.

A total of 23 species belonging to 18 genera have been recorded (Table No. 1). Out of these *Curvularia pallescence, Curvularia prasadii,* were abundant. *Peziotrichum corticola, Pestalotiopsis dessiminata, Pestalotiopsis guepini* were less abundant. The other fungal species were rare in their percentage occurrence.

III a. Percentage Occurrence of Zygomycotina on Leaf Litter.

A total of 12 species belonging to 8 genera have been recorded (Table No. 1). *Rhizopus stolonifer, Mucor praini, Mucor racemosus, Syncephalastrum racemosum* were less abundant where as *Haplosporangium bisporale, Mucor mucedo* and *Mucor subtilissimus* were less abundant. The rest of the species were rare in their percentage of occurrence.

TABLE 1 Percentage Occurrence of Zygomycotina on Leaf

S.No.	NAME OF THE FUNGUS	*Period of study, Seasons, No. of Fungal Occurrences*						*Total No. of Fungal Occurrences*	*% of Fungal Occurrence*
		1998–1999			*1999–2000*				
		Winter	*Summer*	*Rainy*	*Winter*	*Summer*	*Rainy*		
		Oct–Jan	*Feb–May*	*Jun–Sept*	*Oct–Jan*	*Feb–May*	*Jun–Sept*		
	ZVGOMYCOTINA								
1	*Circinella simplex*	3	–	–	2	3	–	8	2.63
2	*Mucor mucedo*	9	11	7	4	16	–	47	15.46
3	*Mucor praini*	13	20	–	24	13	14	84	27.63
4	*Mucor racemosus*	20	6	22	26	–	17	91	29.93
5	*Mucor subtilissimus*	7	9	6	12	4	5	43	14.14
6	*Rhizopus arrhizus*	2	7	5	6	11	–	31	10.19

Total No. of Occurrence of Fungi = 304

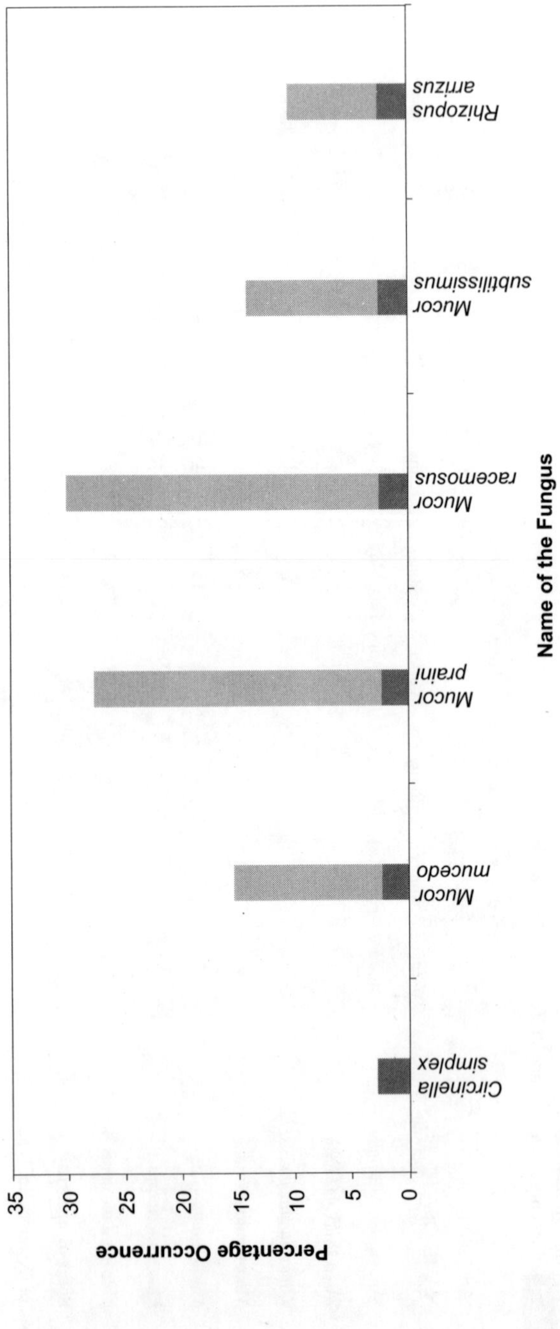

Fig. 1 Percentage Occurrence of Zygomycotina on Leaf

TABLE 2 Percentage Occurrence of Ascomycotina on Leaf

S.No.	NAME OF THE FUNGUS	*Period of study, Seasons, No. of Fungal Occurrences*						*Total No. of Fungal Occurrences*	*% of Fungal Occurrence*
		1998–1999			*1999–2000*				
		Winter	*Summer*	*Rainy*	*Winter*	*Summer*	*Rainy*		
		Oct–Jan	*Feb–May*	*Jun–Sept*	*Oct–Jan*	*Feb–May*	*Jun–Sept*		
	ASCOMYCOTINA								
1	*Achaetomium globosum*	21	3	13	21	17	8	83	21.89
2	*Anthostomella tumlosa*	3	–	4	2	–	7	16	4.22
3	*Ascotricha chartarum*	4	7	6	11	3	4	35	9.23
4	*Chaetomium abuense*	5	–	4	10	4	3	26	6.86
5	*Chaetomium aurangabadense*	–	3	4	–	2	7	16	4.22
6	*Chaetomium elatum*	7	–	9	7	–	18	41	10.81
7	*Chaetomium globosum*	1	3	2	1	1	5	13	3.43
8	*Chaetomium seminudum*	–	–	6	–	–	7	13	3.43
9	*Emericella* sp.	13	8	9	16	12	8	66	17.41
10	*Leptosphaerulina australis*	9	–	7	5	4	2	27	7.12
11	*Melanospora zamiae*	3	3	3	4	–	9	22	5.80
12	*Thielavia terricola*	–	–	8	3	–	10	21	5.54

Total No. of Occurrence of Fungi = 379

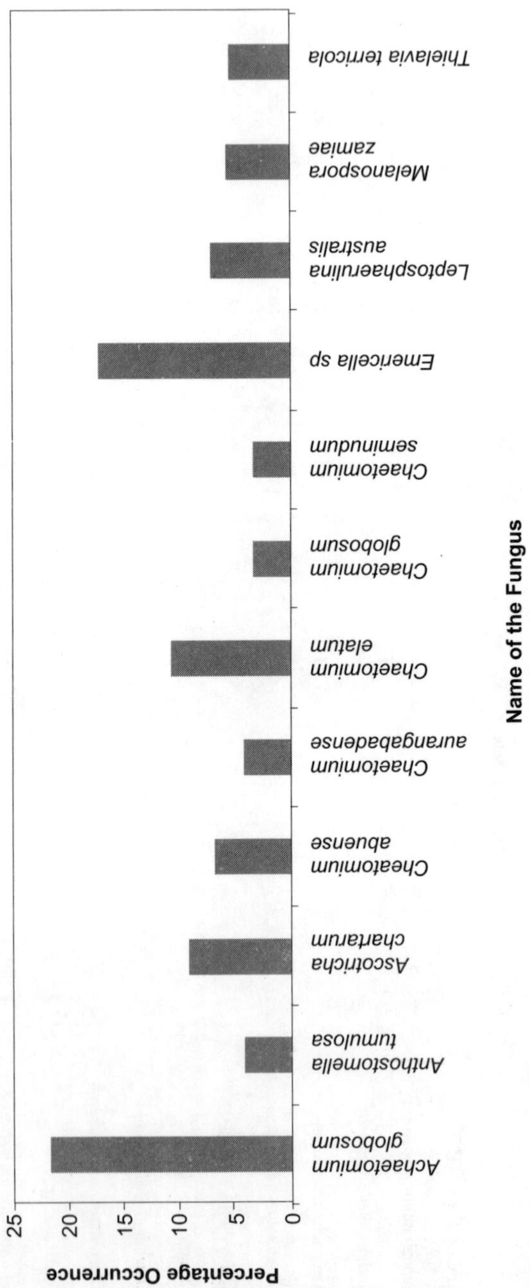

Fig. 2 Percentage Occurrence of Ascomycotina on Leaf

TABLE 3 Percentage Occurrence of Deuteronycotina on Leaf

S.No.	NAME OF THE FUNGUS	*Period of study, Seasons, No. of Fungal Occurrences*						*Total No. of Fungal Occurrences*	*% of Fungal Occurrence*
		1998–1999			1999–2000				
		Winter	*Summer*	*Rainy*	*Winter*	*Summer*	*Rainy*		
		Oct–Jan	Feb–May	Jun–Sept	Oct–Jan	Feb–May	Jun–Sept		
	DEUTEROMYCOTINA								
	HYPHOMYCETES								
1	*Acremonium killiense*	4	–	7	4	–	3	18	0.85
2	*Alternaria burnsi*	5	2	4	3	4	1	19	0.90
3	*Alternaria cucumerina*	2	–	5	5	–	6	18	0.85
4	*Alternaria dennisii*	5	–	7	4	–	9	25	1.18
5	*Alternaria pluriseptata*	20	–	12	8	–	40	80	3.79
6	*Alternaria sonchi*	2	2	2	5	1	4	16	0.75
7	*Alternaria triticina*	7	–	7	7	–	4	25	1.18
8	*Aspergllus erythrocephalus*	11	7	2	9	5	8	42	1.99
9	*Aspergillus nidulans*	53	39	–	51	36	–	179	8.49
10	*Aspergillus ustus*	15	–	23	10	–	17	65	3.08
11	*Bahusaganda indica*	–	–	16	–	–	21	37	1.75
12	*Beltraniella portoriscensis*	9	–	10	9	–	8	36	1.70

(Table 3 Contd.)

(Table 3 Contd.)

13	*Bipolarisspicifera*	–	5	8	–	2	8	23	1.09
14	*Bloxomia truncata*	–	–	16	–	–	28	44	2.08
15	*Capnobotrys neesii*	7	–	1	4	–	1	13	0.61
16	*Cladosporium cladosporloides*	3	4	12	7	2	5	33	1.56
17	*Cladosporium cucumerinum*	4	–	3	7	–	4	18	0.85
18	*Cladosporium elatum*	2	–	7	7	–	9	25	1.18
19	*Cladosporium musae*	5	–	1	3	–	2	11	0.52
20	*Curvularia brachyspora*	6	–	4	3	–	8	21	0.99
21	*Curvularia oryzae*	6	–	7	11	–	11	35	1.66
22	*Curvularia penniseti*	9	–	4	17	–	4	34	1.61
23	*Cylindrocladium floridanum*	7	–	2	11	–	2	22	1.04
24	*Cylindrocladium parvum*	9	3	2	7	6	5	32	1.51
25	*Drechslera neergardi*	9	–	4	6	–	4	23	1.09
26	*Drechslera rostrate*	18	–	21	9	–	7	55	2.61
27	*Drechslera spicifer*	5	–	3	4	–	2	14	0.66
28	*Fusariam concinna*	–	–	27	–	–	13	40	1.89
29	*Fusarium nivale*	6	5	2	6	2	4	25	1.18
30	*Fusarium oxysporum*	7	–	8	13	–	7	35	1.66

(Table 3 Contd.)

(Table 3 Contd.)

31	*Geotrichum candidum*	2	3	1	5	1	4	16	0.75
32	*Gliomastix murorum*	3	–	8	6	–	14	31	1.47
33	*Gyrothrix circinata*	–	–	17	–	–	11	28	1.32
34	*Monodictys asperospora*	4	4	7	5	4	4	28	1.32
35	*Myrothecium advena*	2	–	5	5	–	5	17	0.8
36	*Nigrospora oryzae*	7	11	21	13	13	14	79	3.75
37	*Nigrospora sphaerica*	16	–	8	13	–	5	42	1.99
38	*Paecilomyces carneus*	6	4	2	9	1	4	26	1.23
39	*Paecilmyces varioti*	11	–	16	12	–	5	44	2.08
40	*Penicillium citrinum*	10	14	4	2	17	6	53	2.51
41	*Penicillium minio–luteum*	18	–	8	3	–	15	44	2.08
42	*Penicillium oxalicum*	37	–	15	32	–	6	90	4.27
43	*Penicillium rubrum*	16	11	32	12	13	17	101	4.79
44	*Penicillium simplicissimum*	12	–	9	17	–	7	45	2.13
45	*Penicillium steckii*	7	–	4	4	–	5	20	0.94
46	*Rhizoctonia sp.*	2	5	1	4	8	6	26	1.23
47	*Schizotrichella lunata*	4	–	4	4	–	3	15	0.71
48	*Scolecobasidium variable*	7	–	9	13	–	8	37	1.75

(Table 3 Contd.)

(Table 3 Contd.)

49	*Scolecobasidium verruculosum*	4	–	7	5	–	4	20	0.94
50	*Scopulariopsis brevicaulis*	2	–	6	4	–	7	19	0.9
51	*Scytalidium lingnicola*	5	2	–	6	5	3	21	0.99
52	*Scytalidium thermophilum*	5	4	–	8	10	–	27	1.28
53	*Tetracoccosporium paxianum*	6	–	3	9	–	2	20	0.94
54	*Trichoderma lignorum*	–	–	16	–	–	14	30	1.42
55	*Veronaea coprophila*	3	–	4	7	–	4	18	0.85
56	*Verticillium tenerum*	–	–	6	–	–	9	15	0.71
57	*Verticillium terrestre*	4	–	4	3	–	4	15	0.71
58	*Volutina concentrica*	12	–	28	12	–	21	73	3.46
59	*Wardomyces anomalous*	9	–	5	7	–	9	30	1.42
	COELOMYCETES								–
60	*Chaetomella atra*	5	–	3	3	–	2	13	0.61

Total No. of Occurrence of Fungi = 2106

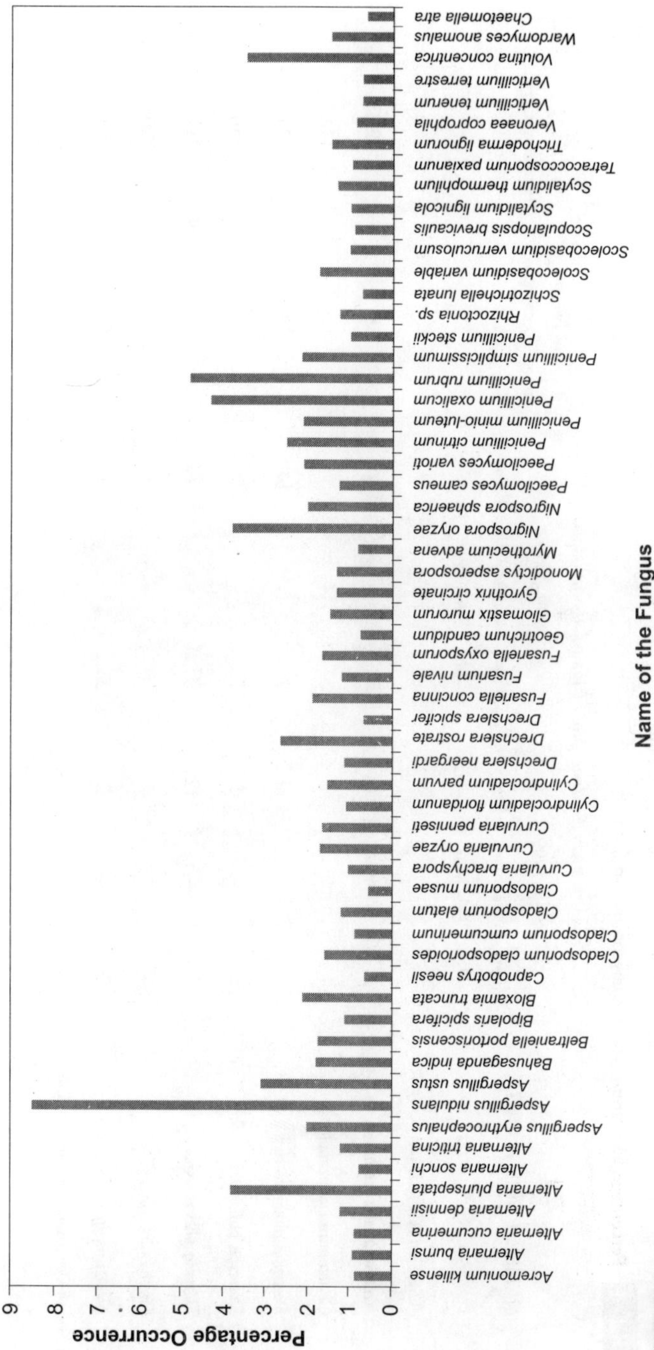

Fig. 3 Percentage Occurrence of Deuteromycotina on Leaf

TABLE 4 Percentage Occurrence of Ascomycontina on Stem

S.No.	NAME OF THE FUNGUS	*Period of study, Seasons, No. of Fungal Occurrences*						*Total No. of Fungal Occurrences*	*% of Fungal Occurrence*
		1998–1999			1999–2000				
		Winter	*Summer*	*Rainy*	*Winter*	*Summer*	*Rainy*		
		Oct–Jan	*Feb–May*	*Jun–Sept*	*Oct–Jan*	*Feb–May*	*Jun–Sept*		
	ASCOMYCOTINA								
1	*Anthostomella tumulosa*	2	–	5	5	–	6	18	13.23
2	*Chaetomium venezuelens*	1	3	2	2	4	3	15	11.02
3	*Diatrype albizziae*	2	–	12	3	–	10	27	19.85
4	*Diatrype bullata*	5	–	5	7	–	5	22	16.17
5	*Didymosphaeria* sp.	7	–	12	7	–	7	33	24.26
6	*Gnomonia comari*	–	3	2	–	4	2	11	8.08
7	*Neocosmospora africana*	2	–	2	3	–	3	10	7.35

Total No. of Occurrence of Fungi = 136

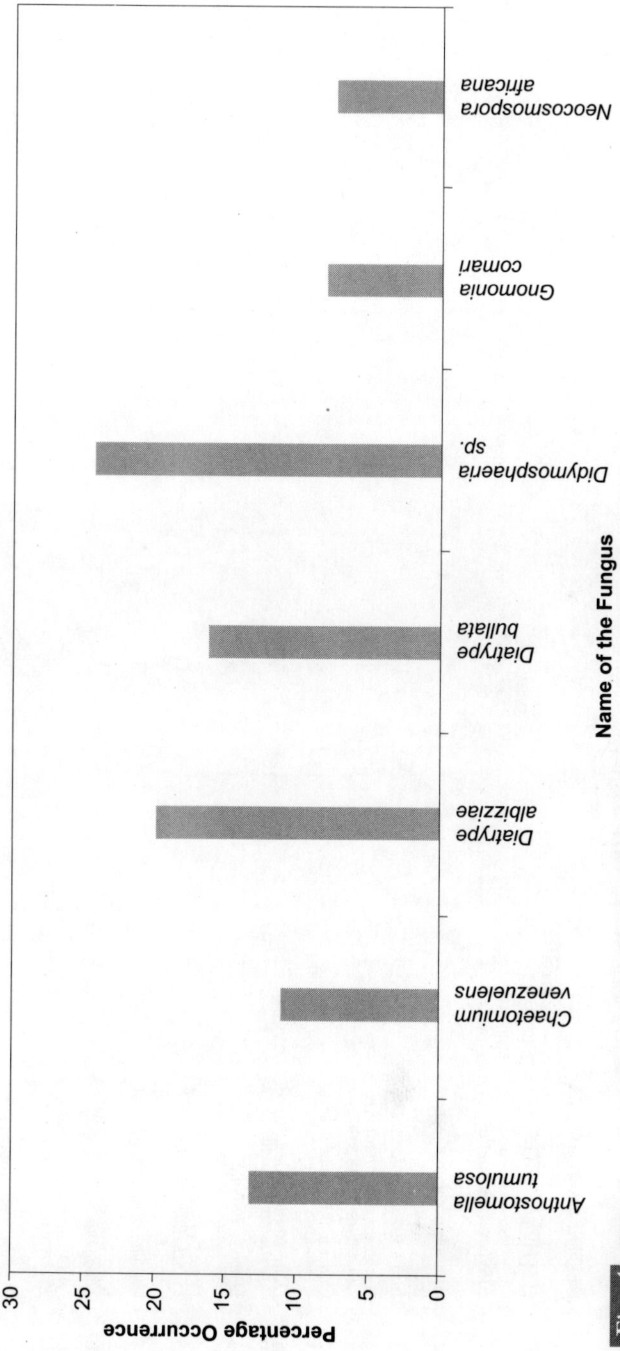

Fig. 4 Percentage Occurrence of Ascomycotina on Stem

TABLE 5 Percentage Occurrence of Deuteromycotina on Stem

S.No.	NAME OF THE FUNGUS	*Period of study, Seasons, No. of Fungal Occurrences*						*Total No. of Fungal Occurrences*	*% of Fungal Occurrence*
		1998–1999			*1999–2000*				
		Winter	*Summer*	*Rainy*	*Winter*	*Summer*	*Rainy*		
		Oct–Jan	*Feb–May*	*Jun–Sept*	*Oct–Jan*	*Feb–May*	*Jun–Sept*		
	DEUTEROMYCOTINA								
	HYPHOMYCETES								
1	*Alternaria cheiranthi*	5	2	4	4	1	1	17	3.26
2	*Alternaria chlamydospora*	2	6	–	2	11	–	21	4.03
3	*Botryosporium madrasense*	4	–	–	10	–	–	14	2.68
4	*Chloridium schulzerii*	6	–	2	4	–	1	13	2.49
5	*Curvularia pallescens*	14	–	13	19	–	7	53	10.17
6	*Curvularia prasadii*	7	–	16	24	–	14	61	11.70
7	*Exosporium stilbaceum*	–	–	11	4	–	7	22	4.22
8	*Gliocladium roseum*	7	–	–	11	–	–	18	3.45
9	*Graphium samaragdinum*	2	1	3	1	1	1	13	2.49
10	*Memnoniella levispora*	–	–	9	–	–	3	12	2.30
11	*Periconia byssoides*	4	–	9	6	–	7	16	4.99
12	*Peziotrichum corticola*	11	5	–	6	5	–	27	5.18

(Table 5 Contd.)

(Table 5 Contd.)

13	*Pithomyces atro–olivaceous*	4	–	12	3	–	6	25	4.79
14	*Pithomyces graminicola*	–	–	11	–	–	10	21	4.03
15	*Sporoschisma saccardoi*	–	2	2	–	2	4	10	1.91
16	*Stachybotrys nilagirica*	3	–	3	6	–	2	14	2.68
17	*Stachybotry spulchra*	3	–	2	2	–	1	8	1.53
18	*Taeniolella atta*	2	2	2	2	1	3	12	2.30
19	*Torula herbarum f. quaternella*	3	–	5	5	–	4	17	3.26
20	*Volutella fructi*	4	–	2	5	–	7	18	3.45
	COELOMYCETES								
21	*Lasiodiplodio threobromae*	2	–	2	3	–	3	10	1.91
22	*Pestalotiopsis disseminata*	10	7	6	9	4	12	48	9.21
23	*Pestalotiopsis guepini*	7	6	5	2	7	14	41	7.86

Total No. of Occurrence of Fungi = 521

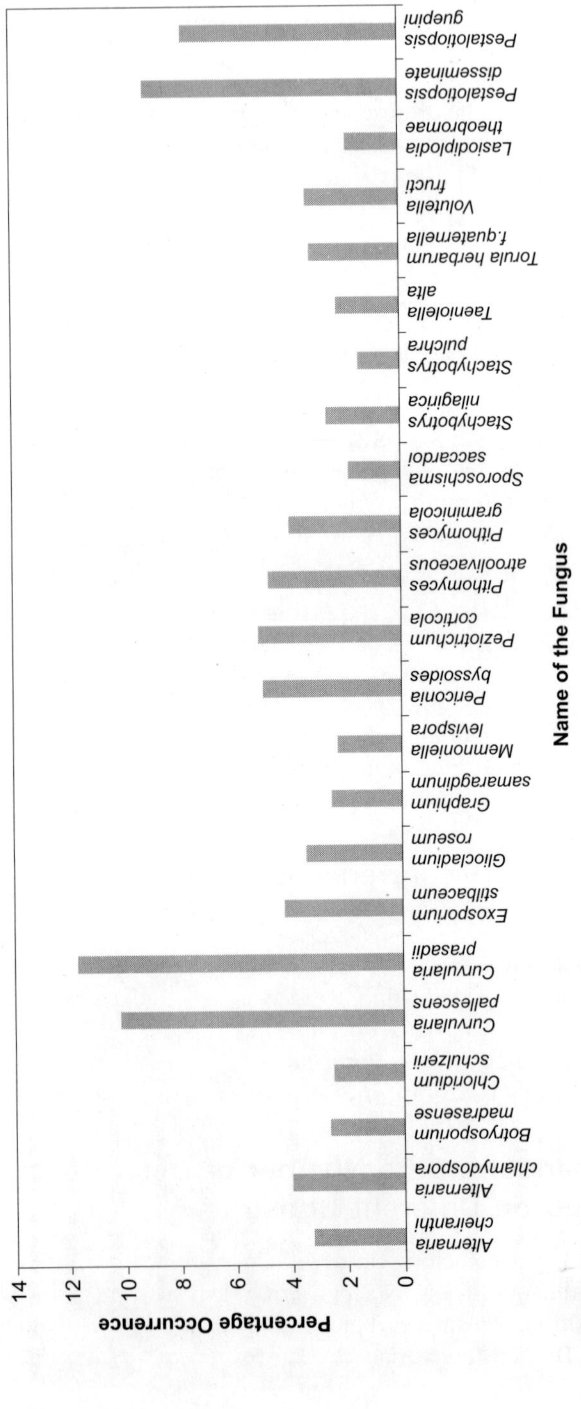

Fig. 5 Percentage Occurrence of Deuteromycotina on Stem

b. Percentage Occurrence of Ascomycotina on Leaf Litter

A total of 14 species belonging to 7 genera have been recorded (Table No. 1) among these *Neurospora sitophila* and *Chaetomium trilaterale* were recorded abundantly where as the species like *Anthostoma amoenum, Ascotricha chartarum, Chaetomium elatum, Chaetomium spirale* and *Lophotrichous bartlettii* were less abundant. The rest of the organisms were rare in their percentage occurrence.

c. Percentage Occurrence of Deuteromycotina on Leaf Litter

Among Deuteromycotina a total of 76 species belonging to 27 genera have been recorded (Table No. 1). Among these *Aspergillus candidus* and *Aspergillus japonicus were* less abundant. Where as others were rare in their percentage occurrence. No species was recorded as abundant.

IV a. Percentage Occurrence of Ascomycotina on Wood litter

A total of 5 species belonging to 5 genera have been recorded (Table No. 1). All the species were abundant in their percentage of occurrence in the order *Hypoxylon fraxinophilum, Herpotrichia rhodosticta, Neocosmospora africana, Trichothyrina pinophylla* and *Gnomonia comari*

b. Percentage Occurrence of Deuteromycotina on Wood Litter.

A total of 25 species belonging to 21 genera have been recorded (Table No. 1). Out of these *Aspergillus fumigatus* was abundant in its percentage of occurrence where as the species like *Fusariella obstipa, Penicillium atromentosum, Periconia cooke, Pestalotiopsis guepini* were less abundant. The rest of the species were rare in their percentage of occurrence.

V. Comparison in the Number of Occurrence of Fungi Isolated on Different Substrata.

A total of 213 species belonging to 92 genera have been isolated on all the substrata (Table no. 11, Fig. No. 11). Out of 213 species, 12 species belonged to Zygomycotina, 28 species belonged to Ascomycotina and 173 species belonged to Deuteromycotina.

TABLE 6 Percentage Occurrence of Zygomycotina on Leaf Litter

S.No.	NAME OF THE FUNGUS	*Period of study, Seasons, No. of Fungal Occurrences*						*Total No. of Fungal Occurrences*	*% of Fungal Occurrence*
		1998–1999			1999–2000				
		Winter	*Summer*	*Rainy*	*Winter*	*Summer*	*Rainy*		
		Oct–Jan	*Feb–May*	*Jun–Sept*	*Oct–Jan*	*Feb–May*	*Jun–Sept*		
	ZYGOMYCOTINA								
1	*Absidia blakesleana*	9	–	–	9	–	–	18	3.08
2	*Circinella simplex*	2	–	–	1	1	–	4	0.68
3	*Cunninghamella echinulata*	8	5	–	14	–	–	27	4.63
4	*Haplosporangium bisporale*	–	–	16	–	–	17	33	5.66
5	*Helicostylum pyriformi*	2	3	–	7	2	–	14	2.40
6	*Mucor mucedo*	7	16	4	4	10	–	41	7.03
7	*Mucor praini*	15	12	–	20	10	7	64	10.97
8	*Mucor racemosus*	17	4	14	22	–	10	67	11.49
9	*Mucor subtilissimus*	8	8	5	14	3	4	42	7.20
10	*Rhizopus arrhizus*	1	9	7	4	6	–	27	4.63
11	*Rhizopus stolonifer*	48	22	20	54	18	24	186	31.90
12	*Syncephalastrum racemosum*	15	–	17	16	–	12	60	10.29

Total No. of Occurrence of Fungi = 583

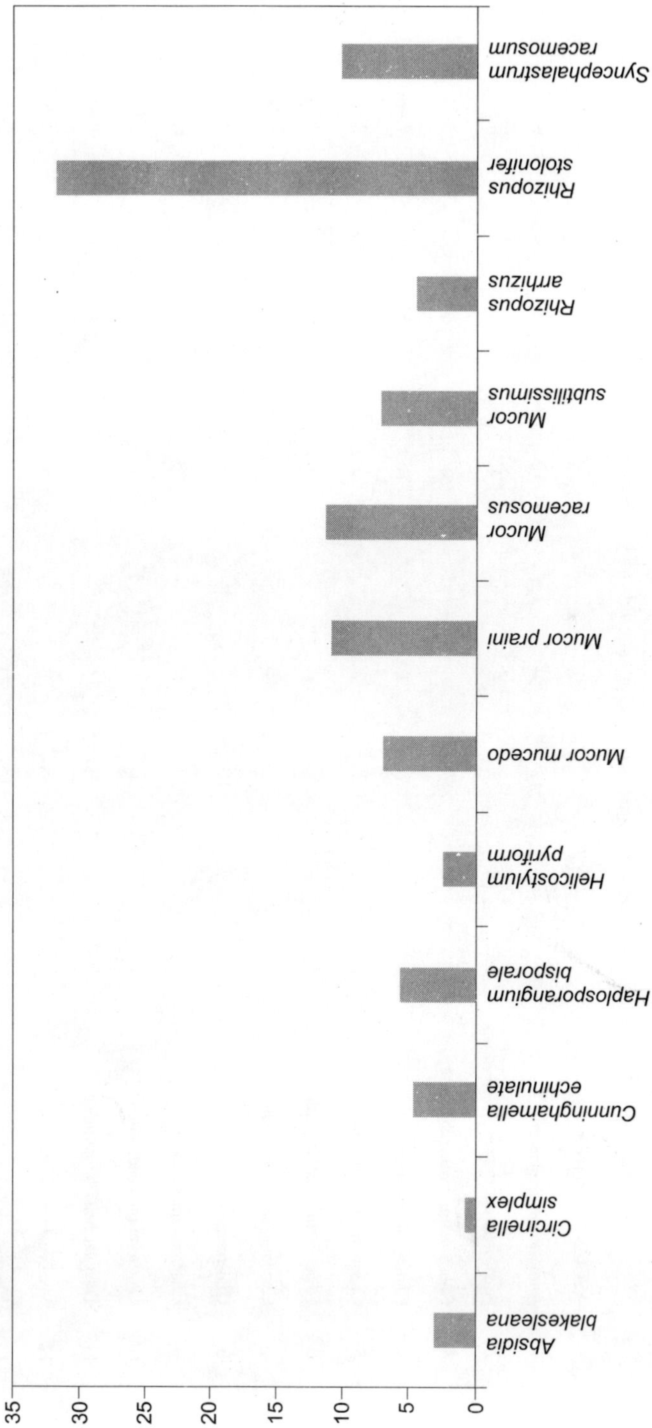

Fig. 6 Percentage Occurrence of Zygomycotina on Leaf-Litter

TABLE 7 Percentage Occurrence of Ascomycotina on Leaf Litter

S.No.	NAME OF THE FUNGUS	*Period of study, Seasons, No. of Fungal Occurrences*						*Total No. of Fungal Occurrences*	*% of Fungal Occurrence*
		1998–1999			*1999–2000*				
		Winter	*Summer*	*Rainy*	*Winter*	*Summer*	*Rainy*		
		Oct–Jan	*Feb–May*	*Jun–Sept*	*Oct–Jan*	*Feb–May*	*Jun–Sept*		
	ASCOMYCOTINA								
1	*Anthostoma amoenum*	–	–	12	–	7	10	29	6.38
2	*Ascotricha chartarum*	3	4	7	8	4	4	30	6.60
3	*Chaetomium aurangabadense*	–	2	7	–	2	6	17	3.74
4	*Chaetomium elatum*	7	–	7	4	–	16	34	7.48
5	*Chaetomium globosum*	1	2	2	3	2	3	13	2.86
6	*Chaetomium microcephalum*	4	–	5	10	–	3	22	4.84
7	*Chaetomium seminudum*	–	–	4	–	–	5	9	1.98
8	*Chaetomium spirale*	–	5	18	–	–	19	42	9.25
9	*Chaetomium trilaterale*	10	–	17	11	–	18	56	12.33
10	*Chaetomium venezuelens*	1	2	1	1	3	1	9	1.98
11	*Lophotrichous barllettii*	7	–	14	8	–	11	40	8.81
12	*Myxotrichum chartarum*	–	–	7	–	–	8	15	3.30
13	*Neurospora sitophila*	31	11	20	21	25	9	117	25.77
14	*Trichothyrina pinophylla*	3	–	6	2	–	10	21	4.62

Total No. of Occurrence of Fungi = 583

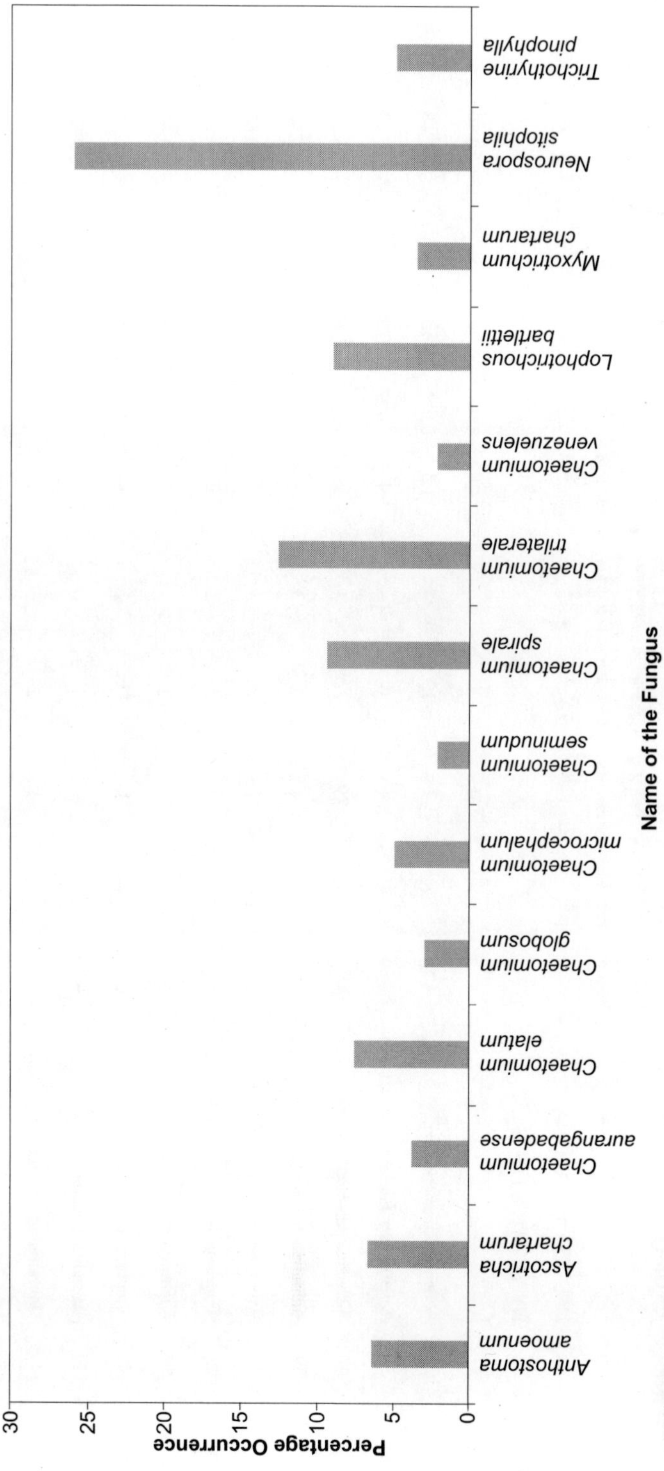

Fig. 7 Percentage Occurrence of Ascomycotina on Leaf-Litter

TABLE 8 Percentage Occurrence of Deuteromycotina on Leaf Litter

S.No.	NAME OF THE FUNGUS	*Period of study, Seasons, No. of Fungal Occurrences*						*Total No. of Fungal Occurrences*	*% of Fungal Occurrence*
		1998–1999			*1999–2000*				
		Winter	*Summer*	*Rainy*	*Winter*	*Summer*	*Rainy*		
		Oct–Jan	*Feb–May*	*Jun–Sept*	*Oct–Jan*	*Feb–May*	*Jun–Sept*		
	DEUTEROMYCOTINA								
	HYPHOMYCETES								
1	*Acremonium* sp.	4	–	9	5	–	2	20	0.59
2	*Acremonium alternatum*	4	7	7	10	–	12	40	1.18
3	*Acremonium kiliense*	4	–	4	6	–	4	18	0.53
4	*Alternaria alternata*	20	–	29	22	–	38	109	3.23
5	*Alternaria burnsi*	4	3	4	2	3	1	17	0.50
6	*Alternaria citri*	5	–	7	2	–	7	21	0.62
7	*Alternaria crassa*	9	–	9	6	–	9	33	0.98
8	*Alternaria longipes*	2	–	7	2	–	8	19	0.56
9	*Alternaria longissima*	6	4	3	11	2	8	34	1.00
10	*Alternaria palandui*	9	–	5	7	–	6	27	0.80
11	*Alternaria tenuissima*	2	3	3	2	6	3	19	0.56
12	*Aspergillus argentums*	7	2	2	5	4	7	27	0.80

(Table 8 Contd.)

(Table 8 Contd.)

13	*Aspergillus awamori*	6	–	6	9	–	4	25	0.74
14	*Aspergillus candidus*	29	26	42	39	23	43	202	5.99
15	*Aspergillus ellipsoideus*	6	–	–	4	–	–	1	0.29
16	*Penicillium atromentosum*	9	–	9	10	–	17	45	6.69
17	*Periconia byssoides*	3	–	7	4	–	7	21	3.21
18	*Periconia cookie*	4	–	12	3	–	16	35	5.20
19	*Stachybotrys atra*	5	–	3	5	–	6	19	2.82
20	*Stachybotrys pulchra*	2	–	1	2	–	1	6	0.89
21	*Torula caligans*	5	1	1	9	2	3	21	3.12
22	*Zygosporium masoni*	–	–	2	–	–	4	6	0.89
	COELOMYCETES								
23	*Diplodia elastica*	2	–	2	4	–	3	11	1.63
24	*Pestalotiopsis disseminata*	9	7	13	5	1	7	42	6.25
25	*Pestalotiopsis guepini*	8	7	11	1	11	7	45	6.69

Total No. of Occurrence of Fungi = 672

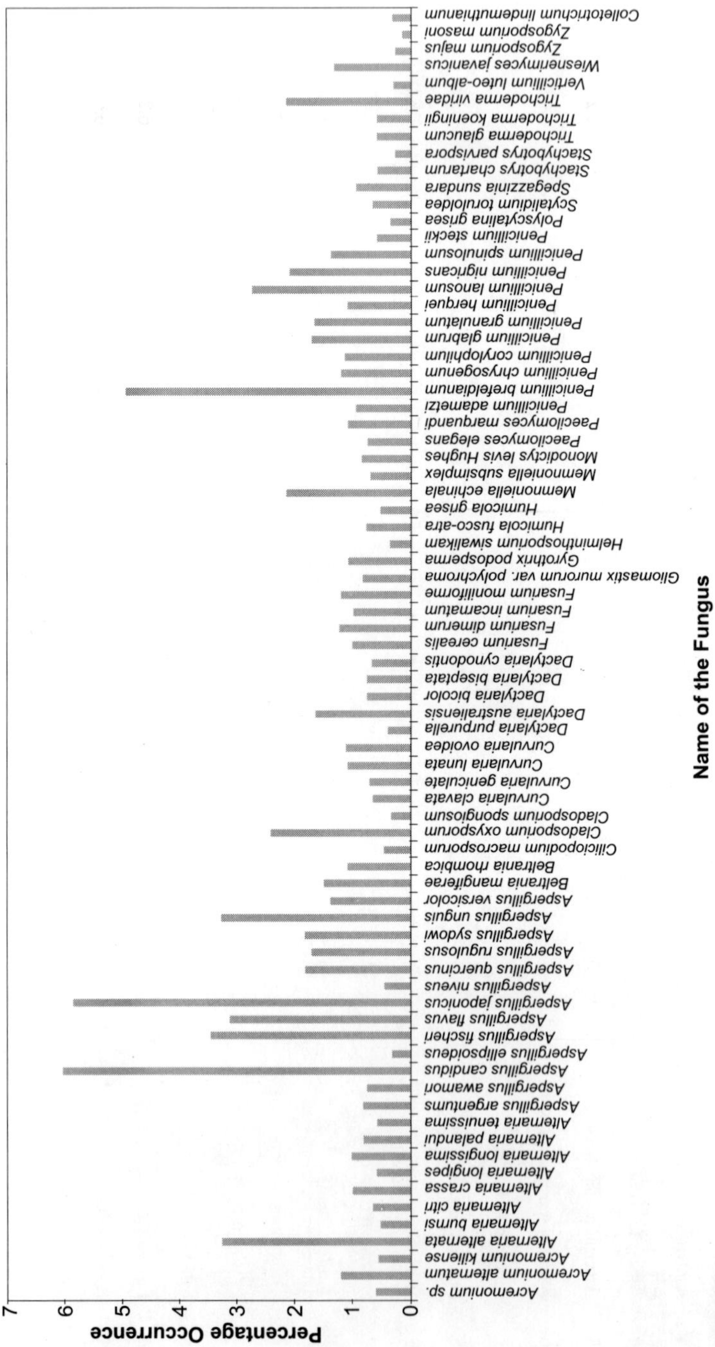

Fig. 8 Percentage Occurrence of Deuteromycotina on Leaf-litter

TABLE 9 Percentage Occurrence of Ascomycotina on Wood Litter

S.No.	NAME OF THE FUNGUS	*Period of study, Seasons, No. of Fungal Occurrences*						*Total No. of Fungal Occureences*	*% of Fungal Occurrence*
		1998–1999			*1999–2000*				
		Winter	*Summer*	*Rainy*	*Winter*	*Summer*	*Rainy*		
		Oct–Jan	*Feb–May*	*Jun–Sept*	*Oct–Jan*	*Feb–May*	*Jun–Sept*		
	ASCOMYCOTINA								
1	*Gnomonia comari*	–	2	1	–	3	2	8	10.12
2	*Herpotrichia rhodosticta*	–	–	11	–	–	9	20	25.31
3	*Hypoxylon fraxinophilum*	–	–	17	–	–	11	28	35.44
4	*Neocosmospora africana*	3	–	2	3	–	4	12	15.18
5	*Trichothyrina pinophylla*	1	–	3	2	–	5	11	13.92

Total No. of Occurrence of Fungi = 79

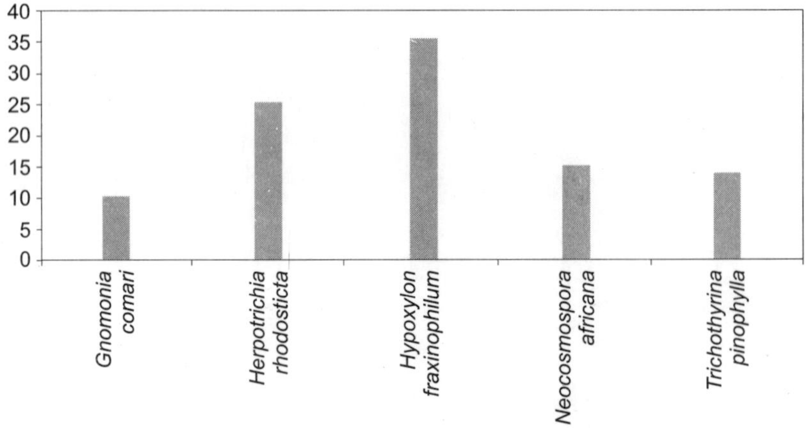

Fig. 9 Percentage Occurrence of Ascomycotina on Wood-Litter

Among 173 species of Deuteromycotina, 167 species belong to Hyphomycetes where as 6 species belonged to Coelomycetes.

On leaf, a total of 6 species belonging to 3 genera of Zygomycotina, 12 species belonging to 8 genera of Ascomycotina and 60 species belonging to 34 genera of Deuteromycotina have been recorded.

Among 12 species of Zygomycotina 6 species belonging to 3 genera on leaf and 12 species belonging to 8 genera on leaf litter have been recorded.

Circinella simplex, Mucor mucedo, Mucor praini, Mucor racemosus, Mucor subtillissimus, Rhizopus arrhizus were recorded both on leaf as well as on leaf litter. No Zygomycotina species was observed on stem and wood litter. *Absidia blakesleana, Cunnigahamella echinulata, Haplosporangium bisporale, Helicostylum pyriformi, Rhizopus stolonifer,* and *Syncephalastrum racemosum* were observed only on leaf litter but absent in all other substrates.

Where as in case of Ascomycotina more number of species were observed on leaf litter followed by leaf, stem and wood litter.

Among 27 species of Ascomycotina, 12 species belonging to 8 genera on leaf, 7 species belonging to 6 genera on stem, 14 species belonging to 7 genera on leaf litter and 5 species belonging to 5 genera on wood litter have been recorded.

Achaetomium globosum, Chaetomium abuense, Emericella sp, Leptosphaerulina australis, Melanospora zamiae, and *Thielavia terricola* were isolated only on leaf where as *Diatrype albizziae, Diatrype bullata, Didymosphaeria sp,* were isolated only on stem. In case of leaf litter

TABLE 10 Percentage Occurrence of Deuteronycotina on Wood Litter

S.No.	NAME OF THE FUNGUS	*Period of study, Seasons, No. of Fungal Occurrences*						*Total No. of Fungal Occureences*	*% of Fungal Occurrence*
		1998–1999			1999–2000				
		Winter	*Summer*	*Rainy*	*Winter*	*Summer*	*Rainy*		
		Oct–Jan	Feb–May	Jun–Sept	Oct–Jan	Feb–May	Jun–Sept		
	DEUTEROMYCOTINA								
	HYPHOMYCETES								
1	*Alysidium resinae*	2	5	–	5	7	–	19	2.81
2	*Aspergillus fumigatus*	37	17	22	31	55	18	180	26.78
3	*Botryosporium madrasense*	3	–	–	4	–	–	7	1.04
4	*Cladosporium chlorocephalum*	2	1	5	4	–	8	20	2.97
5	*Cladosporium tenuissimum*	5	–	4	4	–	4	17	2.52
6	*Cordana musae*	–	1	3	–	3	9	16	2.38
7	*Curvularia trifolii*	3	1	2	5	–	1	12	1.78
8	*Excipularia narsapurensis*	6	–	2	3	–	2	13	1.93
9	*Exosporium stilbaceum*	–	–	9	3	–	5	17	2.52
10	*Fusariella obstipa*	16	–	10	7	–	12	55	8.18
11	*Gonytrichum macrociadum*	3	–	2	2	–	2	9	1.33
12	*Graphium putredinis*	2	–	4	2	–	3	11	1.63

(Table 10 Contd.)

(Table 10 Contd.)

13	*Memnoniella levispora*	–	–	10	–	–	3	13	1.93
14	*Monodictys castanae*	5	6	–	6	3	–	20	2.97
15	*Myrothecium roridum*	5	–	2	2	–	3	12	1.78
16	*Aspergillus fischeri*	20	30	–	25	41	–	116	3.44
17	*Aspergillus flavus*	11	–	32	30	–	32	105	3.11
18	*Aspergillus japonicus*	57	–	48	4	–	49	198	5.88
19	*Aspergillus niveus*	47	–	39	18	–	46	150	0.44
20	*Aspergillus quercinus*	5	16	–	21	19	–	61	1.81
21	*Aspergillus rugulosus*	9	22	–	17	9	–	57	1.69
22	*Aspergillus sydowi*	15	–	24	6	–	16	61	1.81
23	*Aspergillus unguis*	19	14	17	29	21	10	110	3.26
24	*Aspergillus versicolor*	9	–	16	5	–	17	47	1.39
25	*Beltrania mangiferae*	–	–	27	–	–	24	51	1.51
26	*Beltrania rhombica*	–	9	11	–	4	13	37	1.09
27	*Ciliciopodium macrosporum*	1	–	4	4	–	6	15	0.44
28	*Cladosporium oxysporum*	13	–	26	26	–	17	82	2.43
29	*Cladosporium spongiosum*	5	–	1	3	–	2	11	0.32
30	*Curvularia clavata*	7	–	3	5	–	7	22	0.65

(Table 10 Contd.)

(Table 10 Contd.)

31	*Curvularia geniculate*	4	2	5	5	4	4	24	0.71
32	*Curvalaria lunata*	4	–	11	16	1	5	37	1.09
33	*Curvalaria ovoidea*	12	–	6	11	–	9	38	1.12
34	*Dactylaria purpurella*	6	–	3	2	–	2	13	0.38
35	*Drechslera australiensis*	20	–	20	10	–	6	56	1.66
36	*Drechslera bicolor*	9	–	6	7	–	4	26	0.77
37	*Drechslera biseptata*	11	–	7	6	–	2	26	0.77
38	*Drechslera cynodontis*	11	5	–	4	3	–	23	0.68
39	*Fusarium cerealis*	7	–	15	6	–	6	34	1.00
40	*Fusarium dimerum*	10	–	23	7	–	2	42	1.24
41	*Fusarium incarnatum*	9	–	6	15	–	3	33	0.98
42	*Fusarium moniliforme*	13	–	14	2	–	12	41	1.21
43	*Gliomastix murorum var. polychroma*	7	–	10	7	–	5	29	0.85
44	*Gyrothrix podosperma*	4	–	19	3	–	11	37	1.09
45	*Helminthosporium siwalikam*	4	–	2	1	–	5	12	0.35
46	*Humicola fusco–atra*	8	–	3	9	–	6	26	0.77
47	*Humicola grisea*	–	4	6	–	4	4	18	0.53
48	*Memnoniella echinata*	–	–	39	–	–	34	73	2.16

(Table 10 Contd.)

(Table 10 Contd.)

49	*Memnoniella subsimplex*	8	–	7	5	–	4	24	0.71
50	*Monodictys levis Hughes*	6	–	14	5	–	4	29	0.86
51	*Paecilomyces elegans*	5	–	7	5	–	9	26	0.77
52	*Paecilomyces marquandi*	9	–	10	9	–	9	37	1.09
53	*Penicillium adametzi*	14	–	4	10	–	4	32	0.95
54	*Penicillium brefeldianum*	40	36	27	27	23	14	167	4.95
55	*Penicillium chrysogenum*	1	–	6	7	–	17	41	1.21
56	*Penicillium corylophilum*	12	–	7	8	–	12	39	1.15
57	*Penicillium glabrum*	24	–	6	14	–	14	58	1.72
58	*Penicillium granulatum*	13	–	9	2	–	12	56	1.66
59	*Penicillium herquei*	9	–	36	7	–	8	37	1.09
60	*Penicillium lanosum*	23	–	16	13	–	21	93	2.76
61	*Penicillium nigricans*	23	–	27	10	–	11	71	2.1
62	*Penicillium spinulosum*	13	–	14	14	–	6	47	1.39
63	*Penicillium steckii*	4	–	3	6	–	7	20	0.59
64	*Polyscytalina grisea*	3	–	2	2	–	5	12	0.35
65	*Scytalidium toruloidea*	6	–	2	4	–	10	22	0.65
66	*Spegazzinia sundara*	2	–	7	7	–	16	32	0.95

(Table 10 Contd.)

(Table 10 Contd.)

67	*Stachybotrys chartarum*	–	–	6	–	–	13	19	0.56
68	*Stachybotrys parvispora*	3	–	2	2	–	2	9	0.26
69	*Trichoderma glaucum*	4	2	5	3	2	4	20	0.59
70	*Trichoderma koningii*	4	–	9	3	–	4	20	0.59
71	*Trichoderma viridae*	19	–	20	15	–	19	73	2.16
72	*Verticillium luteo–album*	4	–	1	3	–	2	10	0.29
73	*Wiesnerinomyces Javanicus*	9	7	12	6	3	8	45	1.33
74	*Zygosporium majus*	3	–	2	2	–	2	9	0.26
75	*Zygosporium masoni*	–	–	3	–	–	2	5	0.14
	COELOMYCETES								
76	*Colletotrichum lindemuthianum*	3	–	2	3	–	3	11	0.32

Total No. of Occurrence of Fungi =3396

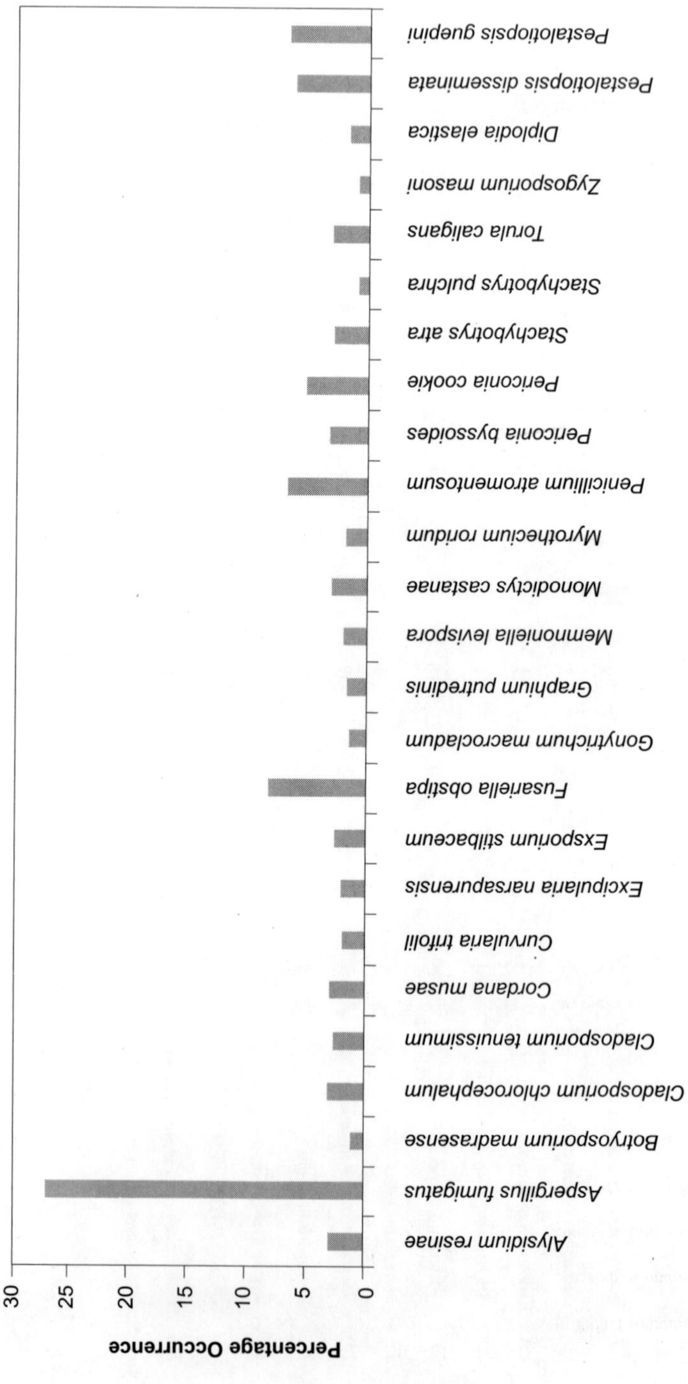

Fig. 10 Percentage Occurrence of Deuteromycotina on Wood-Litter

TABLE 11 Comparison in the Occurrence of Fungi on Different Substrates (Leaf/Stem/Leaf–Litter/Wood–Litter)

S.No.	NAME OF THE FUNGUS	*Host*		*Litter*	
		Leaf	*Stem*	*Leaf*	*Wood*
	ZYGOMYCOTINA				
1	*Absidia blakesleana*	–	–	+	–
2	*Circinella simolex*	+	–	+	–
3	*Cunninghamella echinulata*	–	–	+	–
4	*Haplosporangium bisporale*	–	–	+	–
5	*Helicostylum pyriformi*	–	–	+	–
6	*Mucor mucedo*	+	–	+	–
7	*Mucor praini*	+	–	+	–
8	*Mucor racemosus,*	+	–	+	
9	*Mucor subtilissimus*	+	–	+	–
10	*Rhizopus arrhizus*	+	–	+	–
11	*Rhizopus stolonifer*	–	–	+	–
12	*Syncephalastrum racemosum*	–	–	+	–
	ASCOMYCOTINA				
13	*Achaetomium globosum*	+	–	–	–
14	*Anthostoma amoenum*	–	–	+	–
15	*Anthostomella tumulosa*	+	+	–	–
16	*Ascotricha chartarum*	+	–	+	–
17	*Chaetomium abuense*	+	–	–	–
18	*Chaetomium aurangabadense*	+	–	+	–
19	*Chaetomium elatum*	+	–	+	–
20	*Chaetomium globosum*	+	–	+	–
21	*Chaetomium microcephalum*	–	–	+	–
22	*Chaetomium seminudum*	+	–	+	–
23	*Chaetomium spirale*	–	–	+	–
24	*Chaetomium trilaterale*	–	–	+	–

(Table 11 Contd.)

(Table 11 Contd.)

25	*Chaetomium venezuelens*			–	+	+	–
26	*Diatrype albizziae*			–	+	–	–
27	*Diatrype bullata*			–	+	–	–
28	*Didymosohaeria* sp.			–	+	–	–
29	*Emericella* sp.			+	–	–	–
30	*Gnomonia comari*			–	+	–	+
31	*Herpotrichia rhodosticta*			–	–	–	+
32	*Hypoxylon fraxinophilum*			–	–	–	+
33	*Leptosphaerulina australis*			+	–	–	–
34	*Lophotrichous bartlettii*			–	–	+	–
35	*Melanosopra zamiae*			+	–	–	–
36	*Myxotrichum chartarum*			–	–	+	–
37	*Neocosmospora africana*			–	+	–	+
38	*Neurospora sitophila*			–	–	+	–
39	*Thielavia terricola*			+	–	–	–
40	*Trichothyrina pinophylla*			–	–	+	+
	DEUTEROMYCOTINA						
	HYPHOMYCETES						
41	*Acremonium* sp.	+	–	+	+	–	+
42	*Acremonium alternatum*	+	+	+	+	–	+
43	*Acremonium kiliense*	+	–	+	+	–	+
44	*Alternaria alternata*	+	–	+	+	–	+
45	*Alternaria burnsi*	+	+	+	+	+	+
46	*Alternaria cheiranthi*	+	+	+	+	+	+
47	*Alternaria chlamydospora*	+	+	–	+	+	–
48	*Alternaria citri*	+	–	+	+	+	
49	*Alternaria crassa*	+	–	+	+	–	+
50	*Alternaria cucumerina*	+	–	+	+	–	+
51	*Alternaria dennisii*	+	–	+	+	–	+

(Table 11 Contd.)

(Table 11 Contd.)

52	*Alternaria longipes*	+	–	+	+	–	+
53	*Alternaria longissima*	+	+	+	+	+	+
54	*Alternaria palandui*	+	–	+	+	–	+
55	*Alternaria pluriseptata*	+	–	+	+	–	+
56	*Alternaria sonchi*	+	+	+	+	+	+
57	*Alternaria tenuissima*	+	+	+	+	+	+
58	*Alternaria triticina*	+	–	+	+	–	+
59	*Alysidium resinae*	+	+	–	+	+	–
60	*Aspergillus argentum*	+	+	+	+	+	+
61	*Aspergillus awamori*	+	–	+	+	–	+
62	*Aspergillus candidus*	+	+	+	+	+	+
63	*Aspergillus ellipsoideus*	+	–	–	+	–	–
64	*Aspergillus erythrocephallus*	+	+	+	+	+	+
65	*Aspergillus fischeri*	+	+	–	+	+	–
66	*Aspergillus flavus*	+	–	+	+	–	+
67	*Aspergillus fumigatus*	+	+	+	+	+	+
68	*Aspergillus japonicus*	+	–	+	+	–	+
69	*Aspergillus nidulans*	+	+	–	+	+	–
70	*Aspergillus niveus*	+	–	+	+	–	+
71	*Aspergillus quercinus*	+	+	–	+	+	–
72	*Aspergillus rugulosus*	+	+	–	+	+	–
73	*Aspergillus sydowi*	+	–	+	+	–	+
74	*Aspergillus unguis*	+	+	+	+	+	+
75	*Aspergillus ustus*	+	–	+	+	–	+
76	*Aspergillus versicolor*	+	–	+	+	–	+
77	*Bahusaganda indica*	–	–	+	–	–	+
78	*Beltrania mangiferae*	–	–	+	–	–	+
79	*Beltrania rhombica*	–	+	+	–	+	+
80	*Beltraniella portoriscensis*	+	–	+	+	–	+

(Table 11 Contd.)

(Table 11 Contd.)

81	*Bipolaris spicifera*	–	+	+	–	+	+
82	*Bloxamia truncata*	–	–	+	–	–	+
83	*Botryosporium madrasense*	+	–	–	+	–	–
84	*Capnobotrys neesii*	+	–	+	+	–	+
85	*Chloridium schuizerii*	+	–	+	+	–	+
86	*Ciliciopodium macrosporum*	+	–	+	+	–	+
87	*Cladosporium chlorocephalum*	+	+	+	+	–	+
88	*Cladosporium cladosporioides*	+	+	+	+	+	+
89	*Clildosporium cucumerinum*	+	–	+	+	–	+
90	*Cladosporium elatum*	+	–	+	+	–	+
91	*Cladosporium musae*	+	–	+	+	–	+
92	*Clodosporium oxysporum*	+	–	+	+	–	+
93	*Cladosporium spongiosum*	+	–	+	+	–	+
94	*Cladosporium tenuissimum*	+	–	+	+	–	+
95	*Cordana musae*	–	+	+	–	+	+
96	*Curvularia brachyspora*	+	–	+	+	–	+
97	*Curvularia clavata*	+	–	+	+	–	+
98	*Curvularia geniculata*	+	+	+	+	+	+
99	*Curvularia lunata*	+	–	+	+	+	+
100	*Curvularia oryzae*	+	–	+	+	–	+
101	*Curvularia ovoidea*	+	–	+	+	–	+
102	*Curvularia pallescens*	+	–	+	+	–	+
103	*Curvularia penniseti*	+	–	+	+	–	+
104	*Curvularia prasadii*	+	–	+	+	–	+
105	*Curvularia trifolii*	+	+	+	+	–	+
106	*Cylindrocladium floridtmum*	+	–	+	+	–	+
107	*Cylindrocladium parvum*	+	+	+	+	+	+
108	*Dactylaria purpurella*	+	–	+	+	–	+
109	*Drechslera australiensis*	+	–	+	+	–	+
110	*Drechslera bicolor*	+	–	+	+	–	+

(Table 11 Contd.)

(Table 11 Contd.)

111	*Drechslera biseptata*	+	–	+	+	–	+
112	*Drechslera cynodontis*	+	+	–	+	+	–
113	*Drechslera neergardi*	+	–	+	+	–	+
114	*Drechslera rostrate*	+	–	+	+	–	+
115	*Drechslera spicifer*	+	–	+	+	–	+
116	*Excipularia narsapurensis*	+	–	+	+	–	+
117	*Exosporium stilbaceum*	–	–	+	+	–	+
118	*Fusariella concinna*	–	–	+	–	–	+
119	*Fusariella obstipa*	+	–	+	+	–	+
120	*Fusarium cerealis*	+	–	+	+	–	+
121	*Fusarium dimerum*	+	–	+	+	–	+
122	*Fusarium incarnatum*	+	–	+	+	–	+
123	*Fusarium moniliforme*	+	–	+	+	–	+
124	*Fusarium nivale*	+	+	+	+	+	+
125	*Fusarium oxysporum*	+	–	+	+	–	+
126	*Geotrichum candidum*	+	+	+	+	+	+
127	*Gliocladium roseum*	+	–	–	+	–	–
128	*Gliomastix murorum*	+	–	+	+	–	+
129	*Gliomastix murorum* var. *polychroma*	+	–	+	+	–	+
130	*Gonytrichum macrocladum*	+	–	+	+	–	+
131	*Graphium putredinis*	+	–	+	+	–	+
132	*Graphium samaragdinum*	+	+	+	+	+	–
133	*Gyrothrix circinata*	–	–	+	–	–	+
134	*Gyrothrix podosperma*	+	–	+	+	–	+
135	*Helminthosporium siwalikam*	+	–	+	+	–	+
136	*Humicola fusco–atra*	+	–	+	+	–	+
137	*Humicola grisea*	–	+	+	–	+	+
138	*Memnoniella echinata*	–	–	+	–	–	+
139	*Memnoniella levispora*	–	–	+	–	–	+

(Table 11 Contd.)

(Table 11 Contd.)

140	*Memnoniella subsimplex*	+	–	+	+	–	+
141	*Monodictys asperospora*	+	+	+	+	+	+
142	*Monodictys castanae*	+	+	–	+	+	–
143	*Monodictys levis Hughes*	+	–	+	+	–	+
144	*Myrothecium advena*	+	–	+	+	–	+
145	*Myrothecium roridum*	+	–	+	+	–	+
146	*Nlgrospora oryzae*	+	+	+	+	+	+
147	*Nigrospora sphaerica*	+	–	+	+	–	+
148	*Paecilomyces carneus*	+	+	+	+	+	+
149	*Paecilomyces elegans*	+	–	+	+	–	+
150	*Paecilomyces marquandi*	+	–	+	+	–	+
151	*Paecilomyces varioti*	+	–	+	+	–	+
152	*Penicillium adametzi*	+	–	+	+	–	+
153	*Penicillium atromentosum*	+	–	+	+	–	+
154	*Penicillium brefeldianum*	+	+	+	+	+	+
155	*Penicillium chrysogenum*	+	–	+	+	–	+
156	*Penicillium citrinum*	+	+	+	+	+	+
157	*Penicillium corylophilum*	+	–	+	+	–	+
158	*Penicillium glabrum*	+	–	+	+	–	+
159	*Penicillium granulatum*	+	–	+	+	–	+
160	*Penicillium herquei*	+	–	+	+	–	+
161	*Penicillium lanosum*	+	–	+	+	–	+
162	*Penicillium minio–luteum*	+	–	+	+	–	+
163	*Penicillium nigricans*	+	–	+	+	–	+
164	*Penicillium oxalicum*	+	–	+	+	–	+
165	*Penicillium rubrum*	+	+	+	+	+	+
166	*Penicillium simplicissimum*	+	–	+	+	–	+
167	*Penicillium spinu1osum*	+	–	+	+	–	+
168	*Penicillium steckii*	+	–	+	+	–	+

(Table 11 Contd.)

(Table 11 Contd.)

169	*Periconia byssoides*	+	–	+	+	–	+
170	*Periconia cookie*	+	–	+	+	–	+
171	*Peziotrichum corticola*	+	+	–	+	+	–
172	*Pithomyces atro–olivaceous*	+	–	+	+	–	+
173	*Pithomyces graminicola*	–	–	+	–	–	+
174	*Polyscytalina grisea*	+	–	+	+	–	+
175	*Rhizoctonia* sp.	+	+	+	+	+	+
176	*Schizotrichella lunata*	+	–	+	+	–	+
177	*Scolecobasidium variable*	+	–	+	+	–	+
178	*Scolecobasidium verruculosum*	+	–	+	+	–	+
179	*Scopulariopsis brevicaulis*	+	–	+	+	–	+
180	*Scytalidium lignicola*	+	+	–	+	+	+
181	*Scytalidium thernophilum*	+	+	–	+	+	–
182	*Scytalidium toruloidea*	+	–	+	+	–	+
183	*Spegazzinia sundara*	+	–	+	+	–	+
184	*Sporoschisma saccardoi*	–	+	+	–	+	+
185	*Stachybotrys atra*	+	–	+	+	–	+
186	*Stachybotrys chartarum*	–	–	+	–	–	+
187	*Stachybotrys nilagirica*	+	–	+	+	–	+
188	*Stachybotrys parvispora*	+	–	+	+	–	+
189	*Stachybotrys pulchra*	+	–	+	+	–	+
190	*Taeniolella alta*	+	+	+	+	+	+
191	*Tetracoccosporium paxianum*	+	–	+	+	–	+
192	*Torula caligans*	+	+	+	+	+	+
193	*Torula herbarum f.quatemella*	+	–	+	+	–	+
194	*Trichoderma glaucum*	+	+	+	+	+	+
195	*Trichoderma koningii*	+	–	+	+	–	+
196	*Trichoderma lignorum*	–	–	+	–	–	+
197	*Trichoderma viridae*	+	–	+	+	–	+

(Table 11 Contd.)

Contd.

198	*Veronaea coprophila*	+	–	+	+	–	+
199	*Verticillium luteo–album*	+	–	+	+	–	+
200	*Verticillium tenerum*	–	–	+	–	–	+
201	*Verticillium terrestre*	+	–	+	+	–	+
202	*Vobltella fructi*	+	–	+	+	–	+
203	*Volutina concentrica*	+	–	+	+	–	+
204	*Wardomyces anomalous*	+	–	+	+	–	+
205	*Wiemeriomyces javanicus*	+	+	+	+	+	+
206	*Zygosporium majus*	+	–	+	+	–	+
207	*Zygosporium masoni*	–	–	+	–	–	+
	COELOMYCETES						
208	*Chaetomella atra*	+	–	+	+	–	+
209	*Colletotrichum lindemuthianum*	+	–	+	+	–	+
210	*Diplodia elastica*	+	–	+	+	–	+
211	*Lasiodiplodia theobromae*	+	–	+	+	–	+
212	*Pestalotiopsis disseminata*	+	+	+	+	+	+
213	*Pestalotiopsis guepini*	+	+	+	+	+	+

Note: + = Present
– = Absent

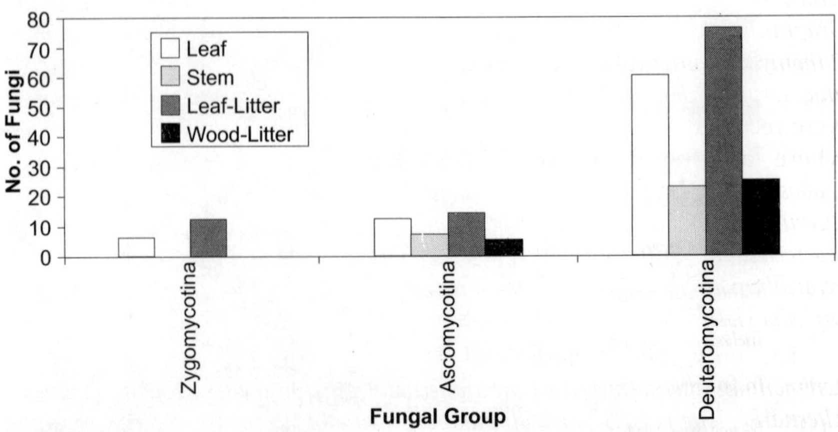

Fig. 11 Comparison in the Occurrence of No. of Fungi on Different Substrata

Anthostoma amoneum, Chaetomium microcephalum, Lopotrichous bartlettli, Myxotrichum chartarum and *Neurospora sitophila* were present but absent in other substrates. *Herpotrichia rhodosticta* and *Hypoxylon fraxinophilum* were isolated only on wood litter.

Among Deuteromycotina 60 species belonging to 34 genera on leaf, 23 species belonging to 18 genera on stem, 76 species belonging to 27 genera on leaf litter and 25 species belonging to 21 genera on wood litter have been recorded.

Alternaria cucumerina, Alternaria dennissi, Alternaria pluriseptata Alternaria sonchi, Alternaria triticina Aspergillus erythrocephalus, Aspergillus nidulans, Aspergillus ustus, Bahusaganda indica, Beltraniella portoriscensis, Bipolaris spicifera, Bloxamina truncata, Caprobotrys neessi, Cladosporium cladosporides, Cladosporium cucumerinum, Cladosporium elatum, Cladosporium musae, Curvularia brachyspora, Curvularia oryzae, Curvularia penniseti, Cylindrocladium folridanum, Cylindrocladium parvum, Drechslera neeragaradi, Drechslera rostrate, Drechslera spicifer, Fusariella concinna, Fusarium nivale, Fusarium oxysporum, Geotrichum candidum, Gliomastrix murorum, Gyrothrix circinata, Monodictys asperospora, Myrothecium advena, Nigrospora oryzae, Nigrospora sphaerica, Paecilomyces carneus, Paecilomyces varioti,Penicillium citrinium, Penicillum minio-luteum, Penicillium, oxalicum, Penicillium simplicissimum, Penicillium rubrum, Rhizoctina sp, Schizotrichella lunata, Scolecobasidium variable, Scolecobasidium verruculosum, Scopulariopsis brevicaulis, Scytalidium lignicola, Scytalidium thermophilum, Tetracoccosporium paxianum, Trichoderma lignorum, Veronaea coprophila, Verticillum tenerum, verticellum terrestre, Volutina concentrica, Wardomyces anomalous and *Chaetomella atra* were isolated only on leaf but absent in other substrata.

The species like *Alternaria cheiranthi, Alternaria chlamydospora, Chloridium schulzerri, Curvularia pallescence, Curvularia prasadii, Gliocladium roseum, Graphium samargdinum, Peziotrichum corticola, Pithomyces atro-olivaceous, Pithomyces graminicola, Sporoschisma saccardi, Stachybotrys nilagirica, Taeniolella alta, Torula herbarum f quaternella, Volutella fructi,* and *Lasiodiplodia theobrome,* were recorded only on the stem but were not recorded on other substrata where as the species like Alysidium *resinae, Aspergillus fumigatus, Cladosporium chlorocephalum, Cladosporium tenuissimum, Cordana musae, Curvularia trifoli, Excipularia narsapurensis, Fusariella obstipa, Gonytrichum macrocladium, Graphium putredinis Monodictys castanae, Myrothecium roridum, Penicillium atromentosum, Periconia cookie, Stachybotrys atra, Torula caligens,* and *Diplodia elastica* were recorded only on wood litter.

The only species harboured by leaf – litter were *Acremonium sp. Acremonium alternatum, Alternaria alternata, Alternaria citri , Alternaria crassa, Alternaria longipes, Alternaria palandus, Alternaria tennissima, Aspergillus candidus, Aspergillus ellipsoideus, Aspergillus fischeri, Aspergillus flavus, Aspergillus japonicus, Aspegillus niveus, Aspergillus quericinus, Aspergillus rugulosus, Aspergillus sydowii, Aspergillus unguis, Aspergillus versicolor, Beltrania mangiferae, Belternia rhombica, Cilicopodium macrosporum, Cladosporium oxisporum, Cladosporium spongiosum, Curvularia clavate, Curvularia seniculate, Curvularia lunata, Curvularia ovoidea, Dactylaria purpurella, Drechslera australiensis, Drechslera bicolor, Drechslera biseptata, Drechslera cynodontis, Fusarium dimerum, Fusarium incarnatum, Fusarium monilformae, Gliomastrics murorumyar. Polychoma, Gyrothrix podosperma, Helminthosporium siwalikam, Humicola fuscoatra, Humicola grisea, Memnoniella echinata, Memnoniella subsimplex, Monodictys levis, Paecilomyces eligans, Paecilomyces marqandi, Penicillium adametzi, Penicillium brefeldianum, Penicillium chrysogenum, Penicillium corylophillum, Penicillium glabrum, Penicillium granulatum, Penicillium nigricans, Penicillium spinulosum, Polyscytalina grisea, Scytalidium toruloidea, Spegazzinia sundara, Stachybotrys chartarum, Stachybotrys paroispora, Trichoderma glaucum, Trichoderma koningii, Trichoderma viride, Verticillium luteoalbum, Wiesneriomyces javanicus, Zygosporium majus* and *colletotrichum lindemuthianum*. The species like *Botryosporium madrasense, Exosporium stilbaceum, Memnoniella levispora, Pericoria byssoides, Stachybotrys pulchra, Pestalotiopsis* disseminata and *Pestalotiopsis guepini* were recorded on stem as well as wood litter. The rest of the species showed difference in their occurrence on different substrata not even single species was recorded on all the substrata.

Seasonal Occurrence of Isolated Fungi

Among Zygomycotina except *Haplosporangium bisporale* all other species were recorded in winter seasons of both the years where as the former was recorded only in rainy seasons of both the years of study (Fig no. 2). This may be due to the low temperature (Fig no. 13, 14). The species *Absidia blakesleana* was recorded only in winter season. *Mucor mucedo* and *Rhizopus arrizus* were recorded in all the season of the 1998-99 year of the study where as they were not recorded in rainy season of the year 1999-2000. this may be due to heavy rainfall in the second year of the study (Fig no. 14). This shows the occurrence of the species is not consistent but may change in its occurrence from year to year and season to season depending on the environmental factors. The species like *Mucor subtilissimus* and *Rhizopus stolonifer* were consistent in their occurrence in all the seasons and all the years of study. The rest of the species showed difference in their occurrence in different seasons.

Among Ascomycotina all the species were recorded in a rainy seasons of both the years of study (Fig no. 12). This may be due to heavy rainfall,

TABLE 12 Percentage Occurrence of Ascomycotina on Wood Litter

S.No.	NAME OF THE FUNGUS	*Period of study, Seasons, No. of Funga1 Occurrences*					
		1998–1999			*1999–2000*		
		Winter	*Summer*	*Rainy*	*Winter*	*Summer*	*Rainy*
		Oct–Jan	*Feb–May*	*Jun–Sept*	*Oct–Jan*	*Feb–May*	*Jun–Sept*
	ZYGOMYCOTINA						
1	*Absidia blakesleana*	+	–	–	+	–	–
2	*Circinella simplex*	+	–	+	+	–	
3	*Cunninghamella echinulata*	+	+	–	+	–	–
4	*Haplosporangium bisporale*	–	–	+	–	–	+
5	*He licostylum pyriformi*	+	+	–	+	+	–
6	*Mucor mucedo*	+	+	+	+	+	–
7	*Mucor praini*	+	+	–	+	+	+
8	*Mucor racemosus*	+	+	+	+	–	+
9	*Mucor subtilissimus*	+	+	+	+	+	+
10	*Rhizopus arrhizus*	+	+	+	+	+	–
11	*Rhizopus stolonifer*	+	+	+	+	+	+
12	*Syncephalastrum racemosum*	+	–	+	+	–	+
	ASCOMYCOTINA						
13	*Achaetomium globosum*	+	+	+	+	–	+
14	*Anthostoma amoenum*	–	–	+	–	+	+
15	*Anthostomella tumulosa*	+	–	+	+	–	+
16	*Ascotricha chartarum*	+	+	+	+	+	+
17	*Chaetomium abuense*	+	–	+	+	+	+
18	*Chaetomium aurangabadens*	–	+	+	–	+	+
19	*Chaetomium elatum*	+	–	+	+	–	+
20	*Chaetomium globosum*	+	+	+	+	+	+
21	*Chaetomium microcephalum*	+	–	+	+	–	+
22	*Chaetomium seminudum*	–	–	+	–	–	+
23	*Chaetomium spirale*	–	+	+	–	–	+

(Table 12 Contd.)

(Table 12 Contd.)

24	*Chaetomium trilaterale*	+	–	+	+	–	+
25	*Chaetomium venezuelens*	+	+	+	+	+	+
26	*Diatrype albizziae*	+	–	+	+	–	+
27	*Diatrype bullata*	+	–	+	+	–	+
28	*Didymosphaeria* sp.	+	–	+	+	–	+
29	*Emericella* sp.	+	+	+	+	+	+
30	*Gnomonia comari*	–	+	+	–	+	+
31	*Herpotrichia rhodosticta*	–	–	+	–	–	+
32	*Hypoxylon fraxinophilum*	–	–	+	–	–	+
33	*Leptosphaerulina australis*	+	–	+	+	+	+
34	*Lophotrichous bartlettii*	+	–	+	+	–	+
35	*Melanospora zamiae*	+	+	+	+	–	+
36	*Myxotrichum chartarum*	–	–	+	–	–	+
37	*Neocosmospora africana*	+	–	+	+	–	+
38	*Neurospora sitophila*	+	+	+	+	+	+
39	*Thielavia terricola*	–	–	+	+	–	+
40	*Trichothyrina pinophylla*	+	–	+	+	–	+
	DEUTEROMYCOTINA						
	HYPHOMYCETES						
41	*Acremonium* sp.			–	–	+	–
42	*Acremonium alternatum*			–	–	+	–
43	*Acremonium kiliense*			+	–	+	–
44	*Alternaria alternata*			–	–	+	–
45	*Alternaria burnsi*			+	–	+	–
46	*Alternaria cheiranthi*			–	+	–	–
47	*Alternaria chlamydospora*			–	+	–	–
48	*Alternaria citri*			–	–	+	–
49	*Alternaria crassa*			–	–	+	–
50	*Alternaria cucumerina*			+	–	–	–

(Table 12 Contd.)

(Table 12 Contd.)

51	*Alternaria dennisii*	+	–	–	–
52	*Alternaria longipes*	–	–	+	–
53	*Alternaria longissima*	–	–	+	–
54	*Alternaria palandui*	–	–	+	–
55	*Alternaria pluriseptata*	+	–	–	–
56	*Alternaria sonchi*	+	–	–	–
57	*Alternaria tenuissima*	–	–	+	–
58	*Alternaria triticina*	+	–	–	–
59	*Alysidium resinae*	–	–	–	+
60	*Aspergillus argentum*	–	–	+	–
61	*Aspergillus awainori*	–	–	+	–
62	*Aspergillus candidus*	–	–	+	–
63	*Aspergillus ellipsoideus*	–	–	+	–
64	*Aspergillus erythrocephalus*	+	–	–	–
65	*Aspergillus fischeri*	–	–	+	–
66	*Aspergillus flavus*	–	–	+	–
67	*Aspergillus fumigatus*	–	–	–	+
68	*Aspergiillus japonicus*	–	–	+	–
69	*Aspergillus nidulans*	+	–	–	–
70	*Aspergillus niveus*	–	–	+	–
71	*Aspergillus quercinus*	–	–	+	–
72	*Aspergillus rugulosus*	–	–	+	–
73	*Aspergillus sydowi*	–	–	+	–
74	*Aspergillus unguis*	–	–	+	–
75	*Aspergillus ustus*	+	–	–	–
76	*Aspergillus versicolor*	–	–	+	–
77	*Bahusaganda indica*	+	–	–	–
78	*Beltrania mangiferae*	–	–	+	–
79	*Beltrania rhombica*	–	–	+	–

(Table 12 Contd.)

(Table 12 Contd.)

80	*Beltraniella portoriscensis*	+	–	–	–
81	*Bipolaris spicifera*	+	–	–	–
82	*Bloxamia truncata*	+	–	–	–
83	*Botryosporium madrasense*	–	+	–	+
84	*Capnobotrys neesii*	+	–	–	–
85	*Chloridium schulzerii*	–	+	–	–
86	*Ciliciopodium macrosporum*	–	–	+	–
87	*Cladosporium chlorocephalum*	–	–	–	+
88	*Cladosporium cladosporioides*	+	–	–	–
89	*Cladosporium cucumerinum*	+	–	–	–
90	*Cladosporium elatum*	+	–	–	–
91	*Cladosporium musae*	+	–	–	–
92	*Cladosporium oxysporum*	–	–	+	–
93	*Cladosporium spongiosum*	–	–	+	–
94	*Cladosporium tenuissimum*	–	–	–	+
95	*Cordana musae*	–	–	–	+
96	*Curvularia brachyspora*	+	–	–	–
97	*Curvularia clavate*	–	–	+	–
98	*Curvularia geniculata*	–	–	+	–
99	*Curvularia lunata*	–	–	+	–
100	*Curvularia oryzae*	+	–	–	–
101	*Curvularia ovoidea*	–	–	+	–
102	*Curvularia pallescens*	–	+	–	–
103	*Curvularia penniseti*	+	–	–	–
104	*Curvularia prasadii*	–	+	–	–
105	*Curvularia trifolii*	–	–	–	+
106	*Cylindrocladium floridanum*	+	–	–	–
107	*Cylindrocladium parvum*	+	–	–	–
108	*Dactylaria purpurella*	–	–	+	–

(Table 12 Contd.)

(Table 12 Contd.)

109	*Drechslera australiensis*	–	–	+	–
110	*Drechslera bicolor*	–	–	+	–
111	*Drechslera biseptata*	–	–	+	–
112	*Drechslera cynodontis*	–	–	+	–
113	*Drechslera neergardi*	+	–	–	–
114	*Drechslera rostrate*	+	–	–	–
115	*Drechslera spicifer*	+	–	–	–
116	*Excipularia narsapurensis*	–	–	–	+
117	*Exosporium stilbaceum*	–	+	–	+
118	*Fusariella concinna*	+	–	–	–
119	*Fusariella obstipa*	–	–	–	+
120	*Fusarium cerealis*	–	–	+	–
121	*Fusarium dimerum*	–	–	+	–
122	*Fusarium incarnatum*	–	–	+	–
123	*Fusarium moniliformae*	–	–	+	–
124	*Fusarium nivale*	+	–	–	–
125	*Fusarium oxysporum*	+	–	–	–
126	*Geotrichum candidum*	+	–	–	–
127	*Gliocladium roseum*	–	+	–	–
128	*Gliomastix murorum*	+	–	–	–
129	*Gliomastix murorum* var. *polychroma*	–	–	+	–
130	*Gonytrichum macrocladum*	–	–	–	+
131	*Graphium putredinis*	–	–	–	+
132	*Graphium samaragdinum*	–	+	–	–
133	*Gyrothrix circinata*	+	–	–	–
134	*Gyrothrix podosperma*	–	–	+	–
135	*Helminthosporium siwalikam*	–	–	+	–
136	*Humicola fusco–atra*	–	–	+	–
137	*Humicola grisea*	–	–	+	–

(Table 12 Contd.)

(Table 12 Contd.)

138	*Memnoniella echinata*	–	–	+	–
139	*Memnoniella levispora*	–	+	–	+
140	*Memnoniella subsimplex*	–	–	+	–
141	*Monodictys asperospora*	+	–	–	–
142	*Monodictys castanae*	–	–	–	+
143	*Monodictys levis*	–	–	+	–
144	*Myrothecium advena*	+	–	–	–
145	*Myrothecium roridum*	–	–	–	+
146	*Nigrospora oryzae*	+	–	–	–
147	*Nigrospora sphaerica*	+	–	–	–
148	*Paecilomyces carneus*	+	–	–	–
149	*Paecilomyces elegans*	–	–	+	–
150	*Paecilomyces marquandi*	–	–	+	–
151	*Paecilomyces varioti*	+	–	–	–
152	*Penicillium adametzi*	–	–	+	–
153	*Penicillium atromentosum*	–	–	–	+
154	*Penicillium brefeldianum*	–	–	+	–
155	*Penicillium chrysogenum*	–	–	+	–
156	*Penicillium citrinum*	+	–	–	–
157	*Penicillium corylophilum*	–	–	+	–
158	*Penicillium glabrum*	–	–	+	–
159	*Penicillium ganulatum*	–	–	+	–
160	*Penicillium herquei*	–	–	+	–
161	*Penicillium lanosum*	–	–	+	–
162	*Penicillium minio–luteum*	+	–	–	–
163	*Penicillium nigricans*	–	–	+	–
164	*Penicillium oxalicum*	+	–	–	–
165	*Penicillium rubrum*	+	–	–	–
166	*Penicillium simplicissimum*	+	–	–	–

(Table 12 Contd.)

(Table 12 Contd.)

167	*Penicillium spinulosum*	–	–	+	–
168	*Penicillium steckii*	+	–	+	–
169	*Periconia byssoides*	–	+	–	+
170	*Periconia cookie*	–	–	–	+
171	*Peziotrichum corticola*	–	+	–	–
172	*Pithomyces atro–olivaceous*	–	+	–	–
173	*Pithomyces graminicola*	–	+	–	–
174	*Polyscytalina grisea*	–	–	+	–
175	*Rhizoctonia* sp.	+	–	–	–
176	*Schiwtrichella lunata*	+	–	–	–
177	*Scolecobasidium variable*	+	–	–	–
178	*Scolecobasidium verruculosum*	+	–	–	–
179	*Scopulariopsis brevicaulis*	+	–	–	–
180	*Scytalidium lignicola*	+	–	–	–
181	*Scytalidium thermophilum*	+	–	–	–
182	*Scytalidium toruloidea*	–	–	+	–
183	*Spegazzinia sundara*	–	–	+	–
184	*Sporoschisma saccardoi*	–	+	–	–
185	*Stachvbotrys atra*	–	–	–	+
186	*Stachybotrys chartarum*	–	–	+	–
187	*Stachybotrys nilagirica*	–	+	–	–
188	*Stachybotrys parvispora*	–	–	+	–
189	*Stachybotrys pulchra*	–	+	–	+
190	*Taeniolella alta*	–	+	–	–
191	*Tetracoccosporium paxianum*	+	–	–	–
192	*Torula caligans*	–	–	–	+
193	*Torula herbarum £ quaternella*	–	+	–	–
194	*Trichoderma glaucum*	–	–	+	–
195	*Trichoderma koningii*	–	–	+	–

(Table 12 Contd.)

(Table 12 Contd.)

196	*Trichoderma lignorum*	+	–	–	–
197	*Trichoderma viride*	–	–	+	–
198	*Veronaea coprophila*	+	–	–	–
199	*Verticillium luteo–album*	–	–	+	–
200	*Verticillium tenerum*	+	–	–	–
201	*Verticillium terrestre*	+	–	–	–
202	*Volutella fructi*	–	+	–	–
203	*Volutina concentrica*	–	+	–	–
204	*Wardomyces anomalous*	+	–	–	–
205	*Wiesneriomvces javanicus*	–	–	+	–
206	*Zvgosponum majus*	–	–	+	–
207	*Zvgosporium masoni*	–	–	+	+
	COELOMYCETES				
208	*Chaetomella atra*	+	–	–	–
209	*Colletotnchum lindemuthianum*	–	–	+	–
210	*Diplodia elastica*	–	–	–	+
211	*Lasiodiplodia theobromae*	–	+	–	–
212	*Pestalotiopsis disseminata*	–	+	–	+
213	*Pestalotiopsis guepini*	–	+	–	+

Note: + = Present
– = Absent

more relative humidity low temperature of the environment both the years (Fig no. 13, 14). The species like *Achaetomium globosum, Ascotricha chartarum, Chaetomium globosum, Chaetomium venezuelens, Emericella sp: and Neurospora sitophila* were recorded in all the seasons and all the years of study. The species like *Anthostomella tumulosa, Chaetomium microcephalum, Chaetomium trilaterale, Diatrype albizziae, Diatrype bullata, Didymospheria sp: Loptotrichous bartlettii, Neocosmospora africana* and *Trictothyrina pinophylla*. Viz: winter and summer of both the years of the study.

The species like *Chaetomium seminudum, Herpotrichia rhodosticta Hypoxylon fraxinophilum* and *Myxotrichum chartarum* were recorded only in

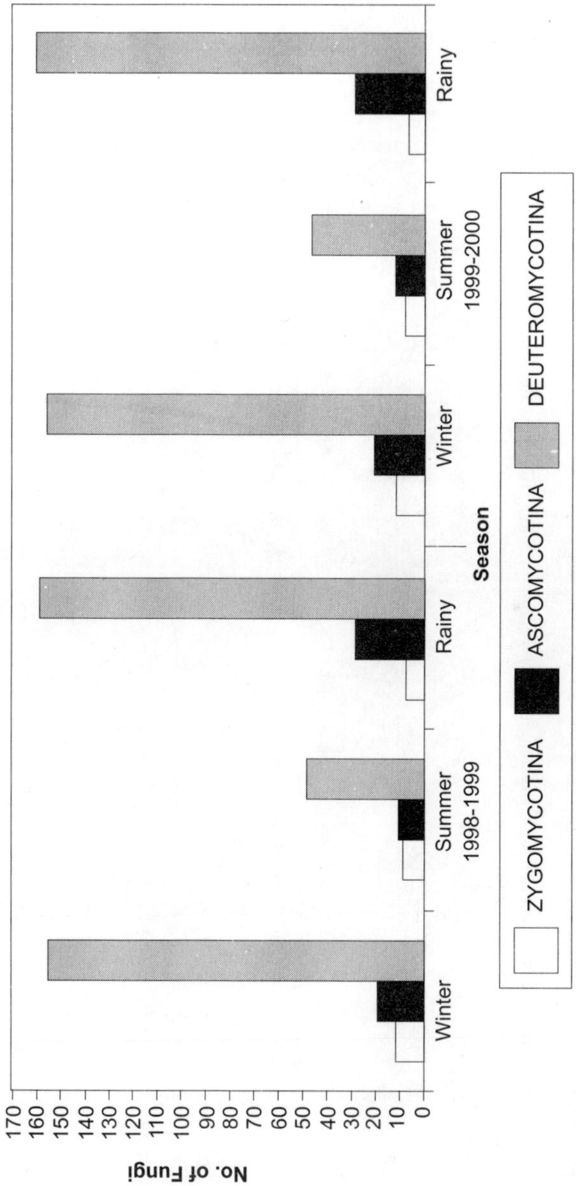

Fig. 12 Comparison in the Occurrence of No. of Fungi in Different Seasons

TABLE 13 Record of Rainfall(R.F.), Temperature(Temp.)–Maximum(Max.)–Minimum(Min.), Relative Humidity(R.H.) from October 1998 to September 1999.

S.No.	*Environmental Factors*	*Years*											
		1998			*1999*								
		Months											
		Oct	*Nov*	*Dec*	*Jan*	*Feb*	*Mar*	*Apr*	*May*	*Jun*	*Jul*	*Aug*	*Sept*
1	*Rainfall (mm)*	98.1	40.2	0	0	0	0	14.7	32.8	71.8	113.9	19.5	88.0
2	*Temp. Max. (°C)*	28.7	29.4	28.7	29.4	32.8	36.2	36.6	32.2	28.0	26.4	27.1	28.3
3	*Temp. Min. (°C)*	20.2	17.1	14.1	12.4	16.9	20.2	21.1	21.3	21.0	20.8	20.4	20.0
4	*Relative humidity (%)*	84.0	77.0	73.0	75.0	63.0	63.0	65.0	75.0	85.0	89.0	85.0	83.0

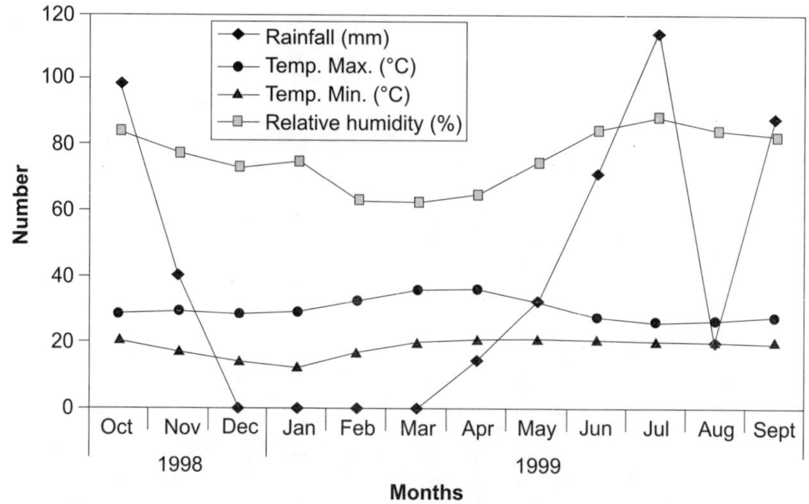

Fig. 13 Variations in the Environmental Factors

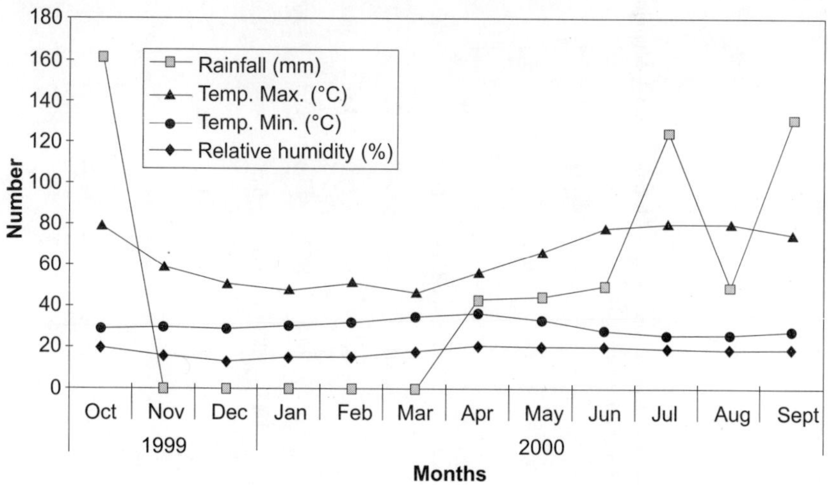

Fig. 14 Variations in the Environmental Factors

rainy seasons of both the years where as these were not recorded in other seasons. The rest of the species showed variations in their occurrence in all the seasons and all the years of the study.

Among the Deuteromycotina the species like *Alternaria burnsi, Alternaria cheiranthi, Alternaria longissima, Alternaria sonchi, Alternaria tenuissima,*

TABLE 14 Record of Rainfall(R.F.), Temperature(Temp.)–Maximum(Max.)–Minimum(Min.), Relative Humidity(R.H.) from October 1999 to September 2000.

S.No.	*Environmental Factors*	*Years*											
		1999			*2000*								
		Months											
		Oct	*Nov*	*Dec*	*Jan*	*Feb*	*Mar*	*Apr*	*May*	*Jun*	*Jul*	*Aug*	*Sept*
1	Rainfall (mm)	161.1	0.0	0.0	0.0	0.0	0.0	44.1	45.4	50.7	125.2	50.4	132.1
2	Temp. Max. (°C)	28.9	29.6	28.9	30.6	32.2	35.2	37.3	33.9	29.0	26.8	27.2	29.0
3	Temp. Min. (°C)	19.8	16.0	13.4	15.1	15.7	18.5	21.5	21.1	21.3	20.4	20.2	20.4
4	Relative humidity (%)	79.0	59.0	51.0	48.0	52.0	47.0	57.0	67.0	79.0	81.0	81.0	76.0

Aspergillus argentums, Aspergillus candidus, Aspergillus erythrocephalous, Aspergillus unguis, Cladosporium cladosporides, Curvularia geniculata, Cylidrocladium parvum, Fusarium navale, Geotrichum candidum, Graphium samaragdinum, Monodictys asperospora, Nigrospora oryzae, Paecilomyces carneus, Penicillium brefeldianum, Penicillium citrunum, Penicillium rubrum, Rhizoctinia sp. Taeniolella alta, Torula caligans, Trichoderma glaucum, Wiesneriomyces javanicus, Pestalotiopsis disseminata and *Pestalotiopsis guepini* were recorded consistently in all the seasons and all the years of the study (Fig no. 12).

The species like *Bahusuganda indica, Beltrania mangifera, Bloxamia truncata, Fusariella concinna, Gyrothrix circinata, Memnoniella echinata, Memnoniella levispora, Pithomyces graminicola. Stachybotrys chartarum, Trichoderma lignorum, Verticillium tenerum,* and *Zygosporium masonii* were recorded only in the rainy seasons of both the years of study where as the same were not recorded in the other seasons. This may be due to the low temperature of the environment and heavy rainfall. The rest of the species differed in their occurrence in different seasons where as few occurred in three seasons of the first year and the same were differed in their second year of the study. This shows that the seasonal variation in the occurrence of the fungi differed from season to season and year to year depending on the environmental factors.

DISCUSSION

An important first step towards understanding fungal ecology is to delineate the nature of the substrate relationships. Fungi may absorb substrates either directly from their substrata, if the molecules are capable of passing through the cell wall and membrane, or they may first break them down in to more assailable units by means of extra cellular enzymes. Fungi a are heterotrophic organisms which can not manufacture their basic food requirements and so are dependent on food materials produced by other organisms either as saprobes or parasites.

The results of these experiments demonstrates that physical factors such as substratum type and temperature can influence the fungal interactions using method similar those in the present study, Widen and Hsu (1987) likewise found that substratum and temperature were major determinants in the expressed competitive performance of several isolated species.

Occurrence of Fungi on Different Plant Parts/substrata

In the present investigation different parts and substrata of the plants have been tested to estimate the Mycoflora of Dharwad District. These plant parts and substrata are categorized in to Host and litter. Host is nothing but a living plant material like leaf and stem which congregate fungal growth and multiplication.

Living plant affords an extensive variety fungal habitats and there is bulky amount of comprehensive information about fungi that occur on, in and around healthy plant organs. The assemblage associated with the plant organs are affected by a wide range of interacting environmental factors.

The Leaves

The leaf is a flattened lateral out growth of stem or branch, developed from a node and having a bud in its axil. These are exogenous in origin and they are mainly concerned with preparation of food material.

As a fungal habitat leaves has been intensively examined and although investigation have been over whelming taxonomic or floristic, some notable efforts have been made to define ecological relationships. Broadly, the aim of ecological studies has been to characterize the Mycoflora at the stages during leaf development and relate its composition to constantly changing capacity of the leaf to support fungal growth. The importance of major climatic and nutrient factors as determinants of microbial behaviour on the leaf are widely appreciated.

Phylloplane climate reflects general climate, seasonal climate as determined by regional climatic zone, has an important broad, long term influence. Assuming a nutrient supply, then the activity of fungi on phylloplane is determined by its thermal and radiation climate and the availability of the water. These are obviously interlinked and are fundamentally governed by constantly shifting interaction between various atmospheric factors, which act directly at both the leaf atmospheric interface and indirectly through their effects on leaf function. With respect to their temperature requirements for generation and growth there is no evidence to suggest that phylloplane fungi are anything but mesophilic. It would therefore seem that expect in the most stable and most favorable of climates, temperature fluctuation and associated changes in water availability limit their activity to relatively brief periods, fungal propagules may have sufficient internal reserves of food materials to support their germination or development.

The Stem

Very little is known of the ecology of fungi on living stems with respect to green stems, it might be expected that colonization would be parallel to that of the phylloplane, but there may be important differences between recruitment patterns for stem and leaves. For instance, contamination of the lower parts of the stems by rain splash from soil may play an important part in the establishment of fungi on them. Although relatively few studies have been made, a larger number of species have been found on woody stems, but despite these floristic data ecological relationships. Remain obscure although some fungi can degrade bark, lack of easily available substrate, presence of fungistatic resins all combine to create stress conditions and restrict access of saprotrophs because of these reasons in the present study the spectrum of mycobiota is not that much to the expected level as it was in the leaves.

The highest level of the fungal activity possibly occur at the nodes, lenticels or other irregularities or in slime fluxes where stress may be somewhat reduced.

The Leaf Litter

Considering that as a precise habitat for the specific association of fungi, the leaf litter is becoming important in ecological and in other studies. Although it is well established that the leaf litter offers an appropriate substratum for many saprophytic fungi, most studies of these were confined to isolation and distribution, especially of new genera and innovative species.

Ecological studies of these leaf litter fungi and the pattern of the colonization on naturally occurring plant debris above the soil received little attention till the middle of the 20th century (Dorai 1988) Pugh 1958 investigated a fungal succession on leaf litter of *Crex paniculata* other workers reported on the Mycoflora of various types of plant leaf litter at one stage in the decomposition process, and many thousands of isolated observations of fungi on decaying plant debris have been recorded. Viz: Rabenhort's kryptogamen flora, but careful sequential observations are almost invariably lacking.

Leaf litter decay is an important component of the nutrient cycling process in the forest ecosystem. Slow rates of leaf litter decomposition can result in the accumulation of large nutrient stocks in the soil surface horizons and nutrient limitation for primary producers. The fungi perform an initial softing and predigestion of the leaf litter and animals complete the process of chemical and physical reduction to humus.

The decomposed leaf litter is the energy source for microorganisms and acts as a carrier of nutrients to soil because of its importance in nutrient cycling and supporting biotic components of ecosystems.

The Wood-litter

The woods accounts for a larger proportion of annual net primary production within the biosphere, and hence represent an immense energy store for potential exploitation by heterotrophs wood and bark form over 90% of the above ground biomass in all the forest types, and over 80% in savannah of the annual above ground litter input in forest ecosystems most estimate indicate that an average above 25% is woody and the remainder is composed of leaves fruits and other structures.

Wood become available for colonization in a diverse array of forms and conditions, ranging from living, standing tree to felled and processed timber. Associated with this, pattern of population and community development are extremely varied, so that value for overall decomposition rates, such as those quoted above the poor indicators of performance in individual substrata. All fungi, which habitat and derive their nutrition from woody tissue, will be regarded as essentially saprophytic. It has been observed that woody plant materials including logs decompose at varying rates of speed, depending on species of tree, location, fungi involved and other factors.

Comparison in the Occurrence of Fungi on Different Plant Parts and Substrata

As the aforementioned conservation various type of plant parts and substrata have different rhythms of fungi when compared in qualitative and quantitative manners. Coming to percentage occurrence of Zygomycoytina on leaf *Mucor racemosus* got high percentage occurrence with 29.93% followed by *Mucor priani, Mucor mucedo, Mucor subtilissimus, Rhizopus arrhizus,* and *Circinella simplex* where as on stem no Zygomycotina species was isolated. But on leaf litter maximum number of Zygomycotina species was isolated. *Rhizopus stoloniferous* obtained high percentage of occurrence with 31.90% followed by *Mucor racemosus, Mucor priani, Syncephalastrum racemosum* etc. No Zygomycotina species was isolated on wood litter. All fungi which habitat and derive their nutrition from woody tissues and stem, will be regarded as essentially saprophytic and parasitic. Most of the species of Zygomycotina are not that much saprophytic in nature and they are sugar loving fungi, possibly for the reason that, in the present study the researcher was unable to isolate Zygomycotina species on stem and wood litter.

Among Ascomycotina species like *Achaetomium globosum, Emercella sp.* and *Chaetomium elatum* got high percentage of occurrence on leaf i.e 21.89%, 17.41% and 10.81% respectively when compared with other species of same group. Where as all the species isolated, except *Gnomonia comari,* and *Neurospora africana* on stem were with high percentage of occurrence. In case of leaf litter the species like *Neurospora sitophila* with 25.77% and *Chaetomium trilaterale* with 12.33% were highest in their percentage occurrence. When compared with other species in the same group and same substratum. The wood litter harbored 5 species of Ascomycotina, which were with high percentage of occurrence.

When compared to Ascomycotina species on different plant parts and substrata *Achaetomium globosum, Ascotricha chartarum, Chaetomium abuense, Emercella sp. Leptosphaeria australis, Melanospora zamiae, Thielavia terricola* were found on leaf but absent in all tested plant parts and substrata. The species like *Diatyrpe albizziae, Diatyrpe bullulata, Didymosphaeria sp: Gnomonia comari,* were isolated only on stem. The species like *Anthostoma amoenum, Chaetomium microcephalum, Chaetomium spirale, Chaetomium trilaterale, Lophotrichus bartlettii, Myxotrichum chartarum, Neurospora sitophila, Trichothyrina pinophylla* were isolated only on leaf litter but absent in all other tested materials and the species like *Herpotrichia rhodostica, Hypoxylon fraxinophilum* were isolated only on wood litter. These domino effect shows that there must be the effect of different plant parts or substrata on the occurrence of different species of Ascomycotina.

All the above information clearly shows that the high percentage occurrence of Ascomycotina species was on the wood-littter when compared with other plant parts/substrata but as far the qualitative and quantitative occurrence of Ascomycotina species is concerned the leaf litter followed by leaf is the best substrata for them to grow.

The Deuteromycotina, which contributed larger magnitude in their occurrence, showed significant change in their occurrence while screening the plant parts and substrata. The species *Aspergillus nidulans* that contributed maximum in its percentage occurrence on leaf is absent in all tested materials. The species like *Curvularia pallescence* (10.17%) and *Curvularia prasadii,* (11.70%) were with high percentage of occurrence when compared with other species of the same group, on stem. no species of Deuteromycotina were with high percentage occurrence on leaf litter. similar results with slight variation in the quantitative aspect of Mycoflora was recorded by Bills and Polishook (1994). In case of wood- litter the *Aspergillus fumigates,* (26.78%) alone was with high percentage occurrence when compared with other species of same group and same substrata.

Maximum number of Deuteromycotina species was occurred on leaf litter and leaf followed by wood–litter and stem. This may be the reason that leaves and their surfaces accumulate a variety of deposited air borne contaminants many of which are potential nutrient sources. In many situations input of such exogenous nutrients in to the phylloplane is considerable and may assume a greater role in promoting fungal development than nutrients supplied by the leaf.

The Environmental Influence and Seasonal Occurrence

The interlocking effects of the abiotic environment are many and complex. The physical environment will always exert an influence directly on the organisms in a community. An influence of season would appear to be reflected in isolation seasonal fluctuations in the fungal population showed most of the members of Deuteromycotina were isolated during all the seasons from different localities of the Dharwad district on different plant parts and substrata.

Maximum temperature during the course of study in Dharwad was 37.3°C in the month of April of the year 2000 (Fig. No. 14). Followed by 36.6°C in the month of April of the year 1999 (Fig. No. 13). Where as the minimum temperature recorded was 12.4°C in the month of January of the year 1999 (Fig. no. 13) followed by 13.4°C in the month of December of the year 2000. (Fig. no. 14).

Minimum number of species was recorded in rainy season followed by winter and summer (Fig no. 12).

Among Zygomycotina species like *Mucor subtillissimus* and *Rhizopus stolonifer* occurred rather consistently throughout the year. However the species like *Absidia blakeesleeana* were isolated only during winter season. But in Ascomycotina species like *Achaetomium globosum, Ascotricha chartarum, Chaetomium globosum, Chaetomium venezuelans, Emericella* sp: occurred rather consistently throughout the year. Species like *Chaetomium seminudum, Herpotrichia rhodosticta, Hypoxylon fraxinophilum* and *Myxotrichum chartarum* were isolated only in rainy season. This may be because of the reason that the area was with consistent relative humidity and received rather consistent rainfall in winter and rainy seasons.

In Deuteromycotina many species of *Alternaria, Aspegillus, Cladosporium, Fusarium, Geotrichum, Graphium, Monodictys, Nigrospora, Paecilomyces, Penicillium, Rhizoctinia, Wiesneriomyces javanicus, Pestolotiopsis disseminata* were observed to adaptable to the various temperature fluctuations and hence, are found in all the seasons forms like *Acremonium sp. Alternaria citri, Alternaria crassa, Alternaria cucumerina, Alternaria dennissi, Alternaria longipes, Alternaria palandui, Alternaria pluriseptata, Aspegillus niveus, Aspergillus sydowii, Aspergillus ystus, Cladosporium cucumerinum, Cladosporium elatum, Cladosporium musae, Curvularia pallescens, Curvularia parasadii, Drechslera bicolor, Drechslera biseptata, Drechslera rostrate, Fusarium incarnatum, Fusarium monilformae, Gliomastrics murorum. Graphium putredinis, Humicola fuscotra, Memnoniella subsimplex, Myrothecium roridum, Penicillium adametzi, etc* were not isolated during summer season but they were observed to be adaptable to winter and rainy seasons.

Fungi have little or no control over the temperature of the environment and consequently over the thermodynamics of their metabolism. Enzyme activity and structure are also affected by temperature so is membrane fluidity. So, temperature is an extremely important factor regulating fungal life. During the course of study, it was observed that the total number of fungi decrease with the increase in the temperature of the environment.

In all 213 species belonging to 92 genera have been isolated. This data indicates that with the changes in environment moisture and fluctuations in temperature during different seasons fungal population also varies.

Of course the enumeration of fungi in the environment remains always hypothetical, so one has to be very critical while using a particular method of isolation, in order to obtain a reliable picture of fungal population in the environment.

SUMMARY AND CONCLUSION

The Mycoflora associated with leaf, stem, leaf litter and wood litter were investigated for two years in different vegetation areas of Dharwad District. A total of 213 species of 92 genera were isolated from all the above parts of the plant and substrata. As a specific habitat for specific association of the fungi, the leaf litter is becoming a important in ecological and other studies and a maximum number of species were isolated from leaf litter followed by leaf.

Group	Genera	Species
Zygomycotina	8	12
Ascomycotina	19	28
Deuteromycotina	65	173
Total	92	213

Currently, there is awareness about biodiversity almost all around the world. It is the total variability within all the living organisms and the ecological complexes they inhabit and forms basis of human survival for the reason that ecology is defined as the study of organism in relation to the surrounding in which they live. These surroundings are called the environment of the organisms. This environment is madeup of many components, including other organisms and their effects, and purely physical features such as temperature, relative humidity, Rainfall and pH etc.

Altered environmental circumstances due to geographical variation are known to affect the association between plant pathogens and their exotic hosts/ this may be also be exact with the associated saprophytic fungi which ultimately bring out the decomposition of leaves, stems leaf litter and wood litter.

Many more fungi, which are hidden underneath, the ecological niches to be surveyed, cultured, classified, conserved and utilized for human welfare. Fungi are sophisticated, omnipresent, beautiful economically useful used for human welfare. Micro–macroscopic and the natural scavengers hence deserve promotion. Our dedicated approach is essential in studying them, classifying them, growing and multiplying them.

The area under investigation showed that the Mycoflora is very rich and represented by various types of fungi. This is because of very rich and varied type of vegetations.

References

Bansal, M. and Mukerji, K.G. 1996. The role of endomycorrhiza in fine root, litter decomposition. In: Root demographics and their efficiences in sustainable agriculture. Grass and forest Ecosystem (ed. Box, J.G. Jr) Kluwer Academic Press, Dordrecht, Netherlands, pp. 339-401.

Bills, G.F. and Polishook, J.D. 1994. Abundance and diversity of microfungi in leaf litter of low land rain forest in Cost Rica. *Mycologia* **86**: 187-198.

Billings, W.D. 1952. The environmental complex in relation to plant growth and distribution. *Q. Rev. Biol.* **27**: 251

Broadfoot, W.M. and Pierre, W.H. 1939. Forest soil studies-I, Relation of rate of decomposition of tree leaves to their acid base balance and other chemical properties. *Soil Sci.* **48**: 329-348.

Cooke, W.B. 1951. Ecological life history out lines for fungi. *Ecology* **32**: 736

Daubenmire, C.H. and Pugh. G.J.F. 1974. *Plants and Environment* 3rd ed. John wiley and Sons, New York.

Dickinson, C.H and Pugh. G.J.F. 1974. *Biology of plant litter decomposition.* Vol. I and II. *Academic press*, London.

Dorai, M. 1988. *Taxonomic and Ecological studies of fungi colonizing leaves and litter of Eucalyptus sp. in South India.* Ph.d thesis, Univ. of Madras, Chennai.

Gupta, R., Mukerji, K.G., Kuhad, R.C. and Saxena, R.K. 1996. Plant surface mycoflora its role in decomposition and soil fertility. In: *Concepts in Applied Microbiology and Biotechnology* (eds. Mukerji, K.G. et al. 1996). Aditya Books Pvt. Ltd., New Delhi. pp 120-377.

Linderberg, G. 1944. Uber die Phyiologische Ligninannuenden Bodenhymenomy. *Symb. Bot. Ups.* **8** (2): 1

Patil, S.D and Ramesh, Ch. 1983. Loculoascomycetes from Maharashtra. *Life Science and Advances* **2**: 215-217

Patil, S.D. and Ramesh, Ch. 1984. Some Pleospora species from Maharashtra. *Indian Bot. Reptr*. **3**(1): 50

Patil, S.D and Ramesh, Ch. 1986. Some Loculoascomycetes from Maharashtra. *Ind. Bot. Reptr*. **5**(1): 27-31

Patil, S.D and Ramesh, Ch. 1987. Notes on some fungi of Pleosporaceae (Loculoascomycetes) from Maharashtra (India). *Trans. Mycol. Soc*. Japan. **22**(3): 229-236.

Pugh, G.J.F. 1958 Leaf litter fungi found on *Carex paniculata. Trans. Br. Mycol. Soc.* **41:** 185-195.

Ramesh, Ch. 1985. *Studies in the Ascomycetes of Maharashtra with special emphasis on Loculoascomycetes*, Ph.D thesis, University of Poona, Poona.

Ramesh, Ch. 1986. Some Loculoascomycetes from Maharashtra, Genus-*Mycosphaerella. Curr. Sci.* **55**(7): 369-371.

Ramesh, Ch. 1987 a. Some Loculoascomycetes from Maharashtra. *Acta Botanica Indica*. **15**(1): 89.

Ramesh, Ch. 1987 b. The Genus *Sporormia* from Maharashtra, *India. Bot. Reptr*, 5(2): 148.

Ramesh, Ch. 1987 c. Loculoascomycetes from Maharashtra, Genus- Bagnisiella, *India. Bot. Reptr.* **5**(2): 203.

Ramesh, Ch. 1987 d. Frequency of occurrence of Loculoascomycetous fungi from Maharashtra. *Proc. Nat. Symp. Fungal Ecol.* 79-85.

Ramesh, Ch. 1987 e. Three Loculoascomycetes new to India. *Acta Bot. Ind.* **15**(1): 106.

Ramesh, Ch. 1988 a. Loculoascomycetes from Maharashtra. *Ind. J. Bot.* **12**(1): 78-80

Ramesh, Ch. 1988 b. Loculoascomycetes from Maharashtra. (India). *Ind. Bot. Reptr.* **4**(142): 54.

Ramesh, Ch. 1988 c. A new species of *Vestergrenia, V. ixorae* from Maharashtra. *Ind. Bot. Reptr.* **4**(142): 105.

Ramesh, Ch. 1990 a. Some Loculoascomycetes from Maharashtra. *J. Ind. Bot. Soc.* **69**: 165.

Ramesh, Ch. 1990 b. Some Loculoscomycetes from India, Genus- *Botryosphaeria, Indian Phytopath.* **40**(3): 295- 300.

Ramesh, Ch. 1990 c. Studies on Loculoascomycetes, Genus- *Didymosphaeria. Indian Phytopath*. **43**(3): 361-367.

Ramesh, Ch. 1992. The Genus–*Massarina* from Maharashtra (India). *Ind. Bot. Reptr.* **11**: 1-5.

Ramesh, Ch. 1993 a. The Genus–*Melanomma* from Maharashtra. *Asian J. Pl. Sci.* **5**(1): 33-37.

Ramesh, Ch. 1993 b. The Genus–*Trematosphaeria* from Maharashtra. *Ind. J. Bot.* **17**: 56-58.

Ramesh, Ch. 1993 c. Some Loculoascomycetes from Maharashtra, Genus-*Keisseriella*. *Ind. Bot. Reptr.* **22**: 23-27.

Ramesh, Ch. 1993 d. The Genus–*Pleospora* from India. *J. Kar. Univ. (Sci.)* **37**: 17-32.

Ramesh, Ch. 1993 e. The Genus *Leptosphaeria* from India. *J. Kar. Univ. (Sci.)* **37**: 45-61.

Ramesh, Ch. and Vijaykumar, S. 2001. Seasonal occurrence of water borne fungi in Panda stream (Uttara Kannada region) Karnataka State, India. *Ecol. of Fungi.* Vol. 21-27pp.

Ramesh, Ch. and Vijaykumar, S. 2003. Seasonal occurrence of water borne fungi in different water bodies of Uttara Kannada region. *Frontiers in fungal Biotechnology and Plant pathogen relations.* Vol: 116-122pp.

Widden, P. and Hsu, D. 1987. Competition between *Trichoderma* species: Effect of temperature and litter type. *Soil Biology and Biochemistry.* **19**: 89-93.

Frontiers in Plant Sciences (Pages 287-300)
Eds. K.G. Mukerji, K.V.B.R. Tilak, S.M. Reddy, L.V. Gangwane, P. Prakash, I.K. Kunwar

Fungal Diversity and Pattern of Colonization on Leaf Litter of *Anacardium occidentale* L.

S. Shanthi and B.P.R. Vittal

ABSTRACT

The biodiversity of microfungi associated with decomposing leaf litter of *Anacardium occidentale* L. and the pattern of fungal colonization was investigated by litter bag technique. The fungi colonizing decomposing litter were analysed by moist chamber incubation technique. A total number of 77 species belonging to 57 genera were recorded. Depending on the first appearance of the fungi and their further continuance, three main groups of fungal colonizers were identified. Fungi which occurred initially during early stages of decomposition disappeared as decomposition progressed; some which showed less activity initially showed gradual increase in their activity and declined in the final stages; a few fungi appeared in the late stages of decomposition.

Key words: *Anacardium occidentale*, fungal diversity, leaf litter, pattern of colonization.

INTRODUCTION

Since 1950 when Chesters described the succession of ascomycetes on decaying deciduous logs and branches, a number of studies have attempted to describe the fungal successional patterns found on various types of decomposing plant debris on the soil surface. Hudson (1968) published a detailed review of all the major work investigating these fungal successional patterns. Subsequently many papers have dealt with decomposition of

Author's address: Centre for Advanced Studies in Botany, University of Madras, Guindy Campus, Chennai – 600 025.

plant matrices in different ecosystems and the changes in fungal saprophytic communities in the litter layers over a period of time (Visser & Parkinson, 1975; Subramanian & Vittal, 1979,1980; Chasseur & Beguin, 1990; Kjoller & Struwe, 1990; Aoki, Tokumasu & Tubaki, 1990; Aoki, Tokumasu & Oberwinkler, 1992; Sieber-canavesi & Sieber, 1993; Tokumasu, Aoki & Oberwinkler, 1994; Bills & Polishook, 1994; Aoki & Tokumasu, 1995; Tokumasu, 1996; Kshattriya *et al.*, 1994; Pasqualetti *et al.*, 1999). The microfungal succession associated with decomposing leaf litter of *Anacardium occidentale,* a tree growing abundantly along coastal areas, is reported in this paper.

MATERIALS AND METHODS

Study Site

The present study was carried out within the compound of Theosophical Society, located at Adayar, South-East of Chennai city along the coast of Bay of Bengal. The Society is spread over an area of 266 acres with scattered buildings surrounded by luxuriant growth of vegetation. The vegetation includes three distinct types namely, mangroves, psammophytes and tropical dry evergreen elements. The soil is sandy loam with high porosity. The vegetation is mostly dominated by *Anacardium occidentale* occupying large areas. In addition, plant taxa such as *Berrrya cordifolia, Cassia fistula, Dillenia indica, Swietenia mahogoni, Terminalia catappa, Tectona grandis, Syzygium cumini, Mimusops elengi, Adansonia digitata, Sterculia foetida, Adenanthera pavonina, Couropita gigantea* are found scattered in the campus. The climate is hot and humid for 8 months of the year. Rainfall is high during June and November due to the onset of the Southwest and Northeast monsoons respectively. The annual precipitation during the study period was 903.3 mm. with the average maximum and minimum temperatures being 33.77°C and 24.58°C respectively.

Sampling Procedure

The fungal diversity and pattern of colonization was studied by litter bag technique (Kshattriya *et al*, 1994) over a period of two years (Jan. 02 to Nov. 03). Forty eight nylon bags, each measuring 25 × 20 cms. and with mesh size of 2 mm. were used. Twenty air dried senescent leaves of *Anacardium occidentale* collected directly from the tree were packed into each bag and the bags were placed randomly on the forest floor underneath the trees. A second batch of another 12 bags, each containing 5 gm. of the senescent leaves were also placed on the floor underneath the trees to monitor the weight loss during the course of decomposition which was measured by oven drying at 80°C for 48 hours (Lousier and Parkinson, 1976).

Four randomly selected litter bags were removed from the forest floor at bimonthly intervals, brought to the laboratory for the enumeration of fungal diversity. The first set of bags were removed during the last week of January 2002. The leaf litter samples were incubated in moist chambers for 48 hours and then examined under stereomicroscope. The fungi in sporulating condition were recorded and identified down to species level. Further, the 'relative frequency' (RF) and 'relative abundance' (RA) for each fungal species was calculated. The RF denotes the number of samplings in which a taxa was recorded as against total samplings (12); the RA is calculated on the basis of the number of leaves in which a particular taxa was recorded as against the total number of leaves examined (80). The data is expressed as percent frequency and percent abundance. Based on RF, the fungi were classified as 'Most common – 76-100%'; 'Common – 51-75%'; 'Occasional – 26-50%'; and 'Sporadic – 1-25%'.

RESULTS AND DISCUSSION

Altogether 960 leaves were examined during the 2-year period of study and it yielded 77 species belonging to 57 genera which included seven taxa belonging to Myxomycota, 7 of Zygomycota, 3 of Ascomycota and the remaining 66 were Mitosporic fungi. The latter included 6 Coelomycetes and 60 Hyphomycetes. The number of fungal taxa recorded per sampling varied from 8-27 (Fig. 1) with greater numbers being recorded in the initial stages. The number of fungi recorded decreased with progress of decomposition.

The estimated dry weight values showed rapid weight loss during the initial eight-months period of incubation, with gradual loss thereafter (Fig. 2). The weight loss was almost 93% in the final set of bags collected during Nov. 2003 and the leaves were fragmentary suggesting that they are in advanced state of decomposition.

Of the 77 species recorded, four taxa, viz. *Beltrania rhombica, Beltraniella portoricensis, Gyrothrix circinata* and *Subramaniomyces fusisaprophyticus* were most common and recorded in all the 12 samplings. Among the six species which were common, *Acremonium stictum, Circinotrichum falcatisporum* and *Parasympodiella laxa* showed high relative frequency. Similarly, *Pestalotia theae* and *Cladosporium cladosporioides* occurred with high frequency among the 11 taxa recorded occasionally. Fifty-six species comprising 73% of the total species recorded were 'sporadic' in occurrence and their frequency is confined to only 1-3 samplings.

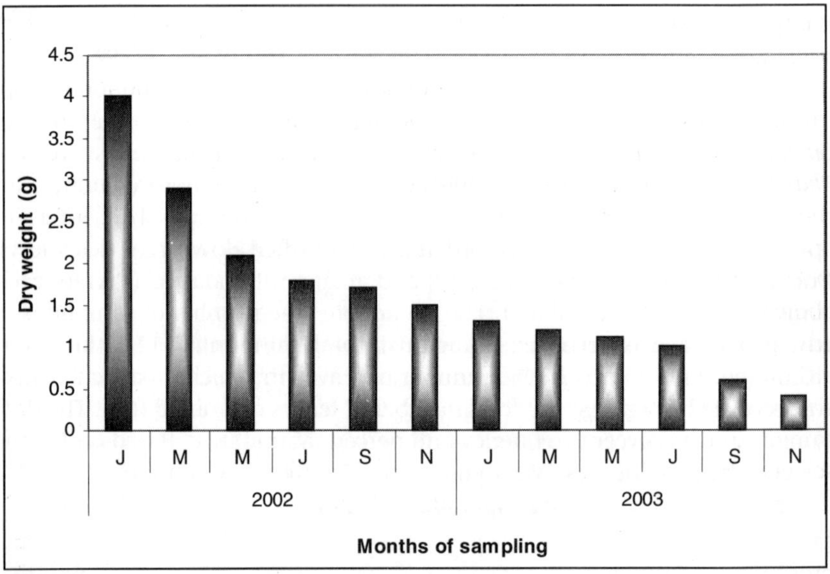

Fig. 1 Leaf litter dry weight recorded (g)

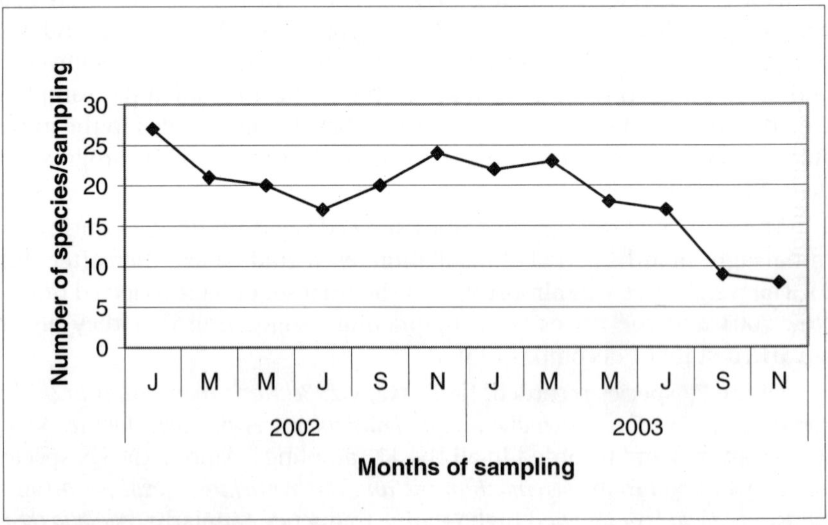

Fig. 2 Number of species recorded in different samplings

The data recorded on the relative abundance of individual taxa over the two-year period did not show any specific pattern. However, a critical analysis of the data reveals that fungi such as *Beltrania rhombica* and *Beltraniella portoricensis* although present in all the samplings, they were more abundant in the later stages of decomposition when compared with initial stages of decay (Fig. 4). On the contrary, the abundance of *Circinotrichum falcatisporum*, *Gyrothrix circinata*, *Parasympodiella laxa*, *Phaeotrichoconis crotalariae* and *Zygosporium oscheoides*, although low during the initial stages it gradually increased as the decomposition progressed and once again declined in the final stages of decomposition. But *Thozetella tocklaiensis* made its appearance during mid stages of decay and its abundance increased as decomposition progressed. No clear cut pattern emerged in the relative abundance of *Subramaniomyces fusisaprophyticus* although it was one of the most dominant colonizer of *Anacardium* litter. It showed consistently high abundance except during the mid–stage samplings. The average relative abundance (average of 12 samplings) of selected fungi is compared in Fig. 3.

Taking the average relative abundance of each of the species (computed as an average of the number of samplings in which a particular fungus was present) as an index of colonizing efficiency, it is observed that the most active colonizers of *Anacardium* litter were *Beltrania rhombica* (56.25%), *Subramaniomyces fusisaprophyticus* (55.42%), *Beltraniella portoricensis* (42.92%), *Gyrothrix circinata* (32.92%), *Circinotrichum falcatisporum* (25%),

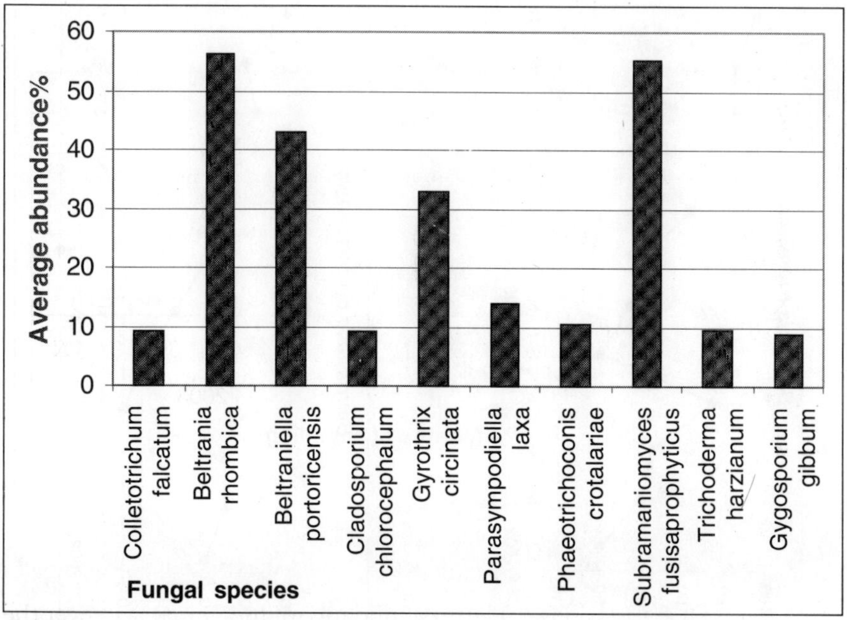

Fig. 3 Comparison of average abundance % of some of the selected species

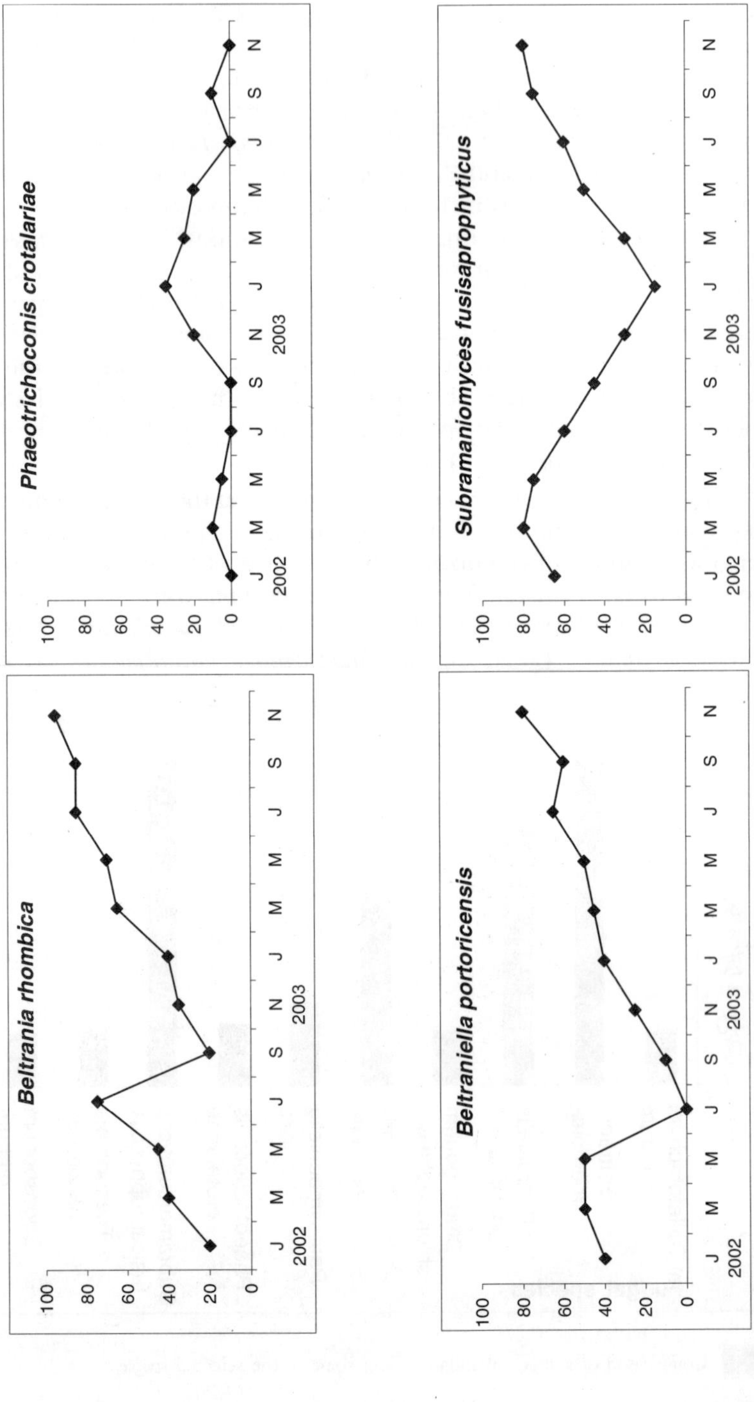
Beltrania rhombica
Phaeotrichoconis crotalariae
Beltraniella portoricensis
Subramaniomyces fusisaprophyticus
0
20
40
60
80
100
J M M J S N J M M J S N
2002
2003

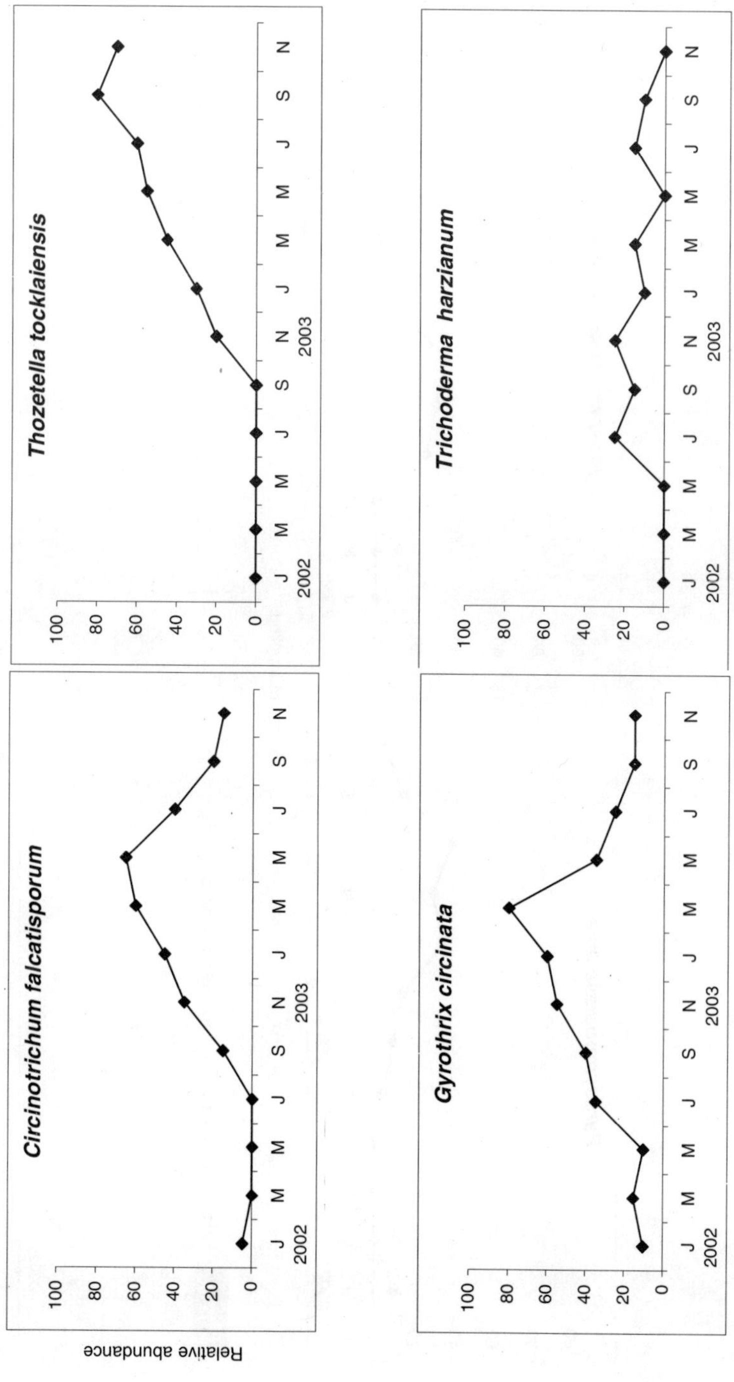
Circinotrichum falcatisporum
Thozetella tocklaiensis
Gyrothrix circinata
Trichoderma harzianum
Relative abundance
100
80
60
40
20
0
J M M J S N J M M J S N
2002
2003

(Fig. 4 Contd.)

(*Fig. 4 Contd.*)

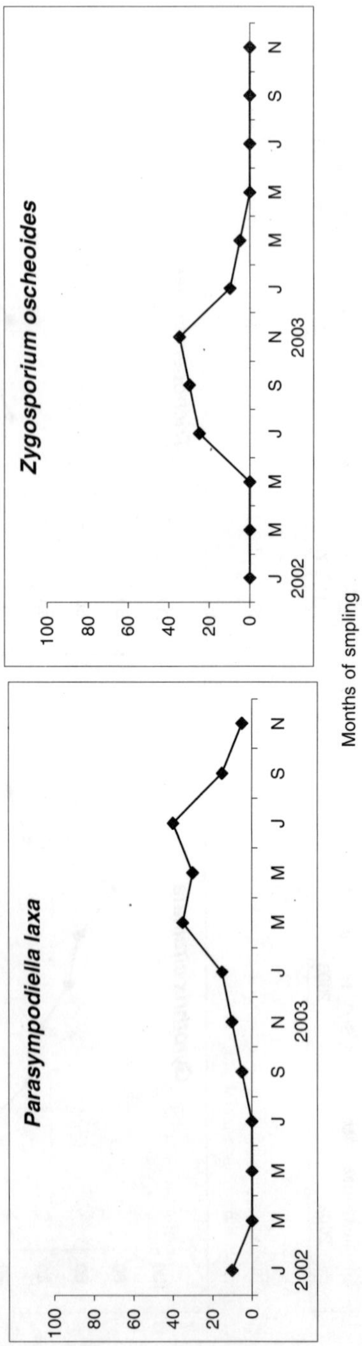

Fig. 4 Relative abundance of some selected fungal taxa

TABLE 1 *List of fungi recorded and their relative frequency in different samplings*

S.No.	Species	Months of sampling												Average abundance %	Relative frequency
		2002						2003							
		J	M	M	J	S	N	J	M	M	J	S	N		
	MYXOMYCOTA														
1	Comatricha laxa	10	-	-	-	-	-	-	-	-	-	-	-	0.83	S
	ZYGOMYCOTA														
2	Absidia corymbifera	-	-	-	-	5	15	25	15	-	10	-	-	5.83	O
3	Choanephora cucurbitarum	-	-	-	15	-	-	-	-	-	-	-	-	1.25	S
4	Cunninghamella echinata	-	-	-	-	-	10	-	-	-	-	-	-	0.83	S
5	Mucor mucedo	-	-	-	5	-	-	-	-	-	10	-	-	1.25	S
6	M. racemosus	-	-	5	30	15	-	20	-	-	-	-	-	5.83	O
7	Rhizopus stolonifer	-	-	-	-	-	15	25	5	-	-	-	-	3.75	S
8	Syncephalastrum racemosum	-	-	-	-	-	-	-	10	-	-	-	-	0.83	S
	ASCOMYCOTA														
9	Chaetomium seminudam	-	-	-	-	-	-	-	-	25	-	-	-	2.08	S
10	Euantennaria sp.	5	10	-	-	-	-	-	-	-	-	-	-	1.25	S
11	Meliola sp.	5	-	-	-	-	-	-	-	-	-	-	-	0.42	S
	MITOSPORIC FUNGI														
	COELOMYCETES														
12	Bartalinia robillardoides	-	-	-	-	-	-	20	25	-	-	-	-	7.5	S
13	Ciliochorella sp.	5	-	-	-	-	-	-	-	-	-	-	-	0.42	S
14	Colletotrichum falcatum	-	-	-	-	20	15	30	25	20	-	-	-	9.17	O
15	Gloeosporium sp.	-	-	-	-	-	-	-	-	15	-	-	-	1.25	S
16	Lasiodiplodia theobromae	-	-	-	-	15	-	-	-	-	-	-	-	1.25	S
17	Pestalotia theae	-	20	20	-	-	-	15	10	25	-	-	-	7.5	O

(Table 1 Contd.)

(Table 1 Contd.)

	HYPHOMYCETES														
18	Acremonium strictum	5	-	15	15	25	30	15	10	5	5	-	-	10.42	C
19	Alternaria alternata	5	-	10	-	15	20	-	15	-	-	-	-	5.42	O
20	Aspergillus candidus	-	-	5	-	-	-	-	-	-	-	-	-	0.42	S
21	A. flavipes	-	-	-	10	-	-	-	-	-	-	-	-	0.83	S
22	A. japonicus	-	5	-	25	5	10	-	-	-	-	-	-	3.33	O
23	A. niger	-	-	10	-	10	-	-	-	-	-	-	-	1.67	S
24	A. terreus	-	-		10	-	-	-	-	-	-	-	-	0.83	S
25	Aureobasidium pullulans	-	-	5	-	-	-	-	-	-	-	-	-	0.42	S
26	Beltrania rhombica	20	40	45	75	20	35	40	65	70	85	85	95	56.25	MC
27	Beltraniella portoricensis	40	50	50	-	10	25	40	45	50	65	60	80	42.92	MC
28	Camposporium antennatum	-	15	-	-	-	-	10	25	-	30	-	-	6.67	O
29	Circinotrichum maculiforme	15	-	5	35	-	-	-	-	-	-	-	-	4.58	S
30	C. falcatisporum	5	-	-	-	15	35	45	60	65	40	20	15	25	C
31	Cladosporium chlorocephalum	5	-	-	-	-	-	-	-	-	-	-	-	0.42	S
32	C. cladosporioides	10	-	-	-	20	10	-	20	30	-	-	20	9.17	O
33	C. inaequiseptatum	5	-	-	-	-	-	-	-	-	-	-	-	0.42	S
34	C. oxysporum	10	-	-	-	-	-	-	-	-	-	-	-	0.83	S
35	Curvularia lunata	-	-	10	-	15	-	-	-	-	-	-	-	2.08	S
36	C. pallescens	-	-	15	-	-	-	-	-	-	-	-	-	1.25	S
37	Dactylaria sp.	-	-	10	10	-	10	15	-	-	20	-	-	5.42	O
38	Dicyma vesiculifera	-	-	-	-	-	-	10	5	-	-	-	-	1.25	S
39	Drechslera australiensis	-	-	-	-	10	-	-	-	-	-	-	-	0.83	S
40	D. hawaiiensis	-	5	-	-	-	-	-	-	-	-	-	-	0.42	S
41	Fusarium dimerum	-	15	-	-	-	-	-	-	-	25	-	-	3.33	S

(Table 1 Contd.)

(Table 1 Contd.)

42	F. lateritium	-	-	-	-	-	5	15	-	-	-	-	-	1.67	S
43	Gyrothrix circinata	10	15	10	35	40	55	60	80	35	25	15	15	32.92	MC
44	G. podosperma	-	-	-	-	-	-	-	-	-	15	-	-	1.25	S
45	Hansfordia pulvinata	-	-	-	-	-	-	-	15	20	-	-	-	2.92	S
46	Helicoma sp.	15	-	-	-	-	-	-	-	-	-	-	-	1.25	S
47	Helicosporium serpentinum	-	-	-	45	-	-	-	-	-	-	-	-	3.75	S
48	Helicoubisia coronata	-	10	-	-	-	-	-	-	-	-	-	-	0.83	S
49	Memnoniella echinata	-	15	-	-	-	10	-	-	-	-	-	-	2.08	S
50	Monodictys putredinis	5	-	-	-	-	-	-	-	-	-	-	-	0.42	S
51	Myrothecium advense	-	-	-	5	-	15	-	-	-	-	-	-	1.67	S
52	Nigrospora sphaerica	5	-	-	-	-	-	-	-	-	-	-	-	0.42	S
53	Parasympodiella laxa	10	-	-	-	5	10	15	35	30	40	15	5	13.75	C
54	Penicillium citrinum	-	10	-	15	10	-	-	-	-	15	-	-	4.17	O
55	P. funiculosum	-	5	5	-	-	-	-	-	-	-	-	-	0.83	S
56	Phaeotrichoconis crotalariae	-	10	5	-	-	20	35	25	20	-	10	-	10.42	C
57	Pithomyces cupaniae	5	-	-	-	-	-	-	-	-	-	-	-	0.42	S
58	P. maydicus	20	-	-	-	-	-	-	-	-	-	-	-	1.67	S
59	P. sacchari	10	-	-	-	-	-	-	-	-	-	-	-	0.83	S
60	Polyscytalum	-	15	-	-	-	-	-	-	-	-	-	-	1.25	S
61	Pseudobeltrania penizigii	5	-	-	-	-	-	-	-	-	-	-	-	0.42	S
62	Scolecobasidium variabile	-	-	5	-	-	-	-	-	25	25	-	-	4.58	S
63	Selenosporella curvispora	-	5	-	-	-	-	-	-	30	-	-	-	2.92	S
64	Spegazzinia tessarthra	10	-	-	-	-	-	-	-	-	-	-	-	0.83	S
65	Sporidesmium brachypus	-	5		-	-	-	-	-	-	-	-	-	0.42	S
66	Stachybotrys parvispora	-		20	-	-	-	-	-	-	-	-	-	1.67	S

(Table 1 Contd.)

(Table 1 Contd.)

67	Subramaniomyces fusisaprophyticus	65	80	75	60	45	30	15	30	50	60	75	80	55.42	MC
68	Tetraploa aristata	5	5	-	-	-	-	-	-	-	-	-	-	0.83	S
69	Thozetella tocklaiensis	-	-	-	-	-	20	30	45	55	60	80	70	30	C
70	Thozetella sp.	-	30	35	-	-	-	-	-	-	-	-	-	5.42	S
71	Trichoderma harzianum	-	-	-	25	15	25	10	15	-	15	10	-	9.58	C
72	Torula herbarum	-	-	-	-	-	-	-	10	-	-	-	-	0.83	S
73	Volutella ciliata	-	-	-	-	-	10	-	-	-	-	-	-	0.83	S
74	Wiesneriomyces javanicus	5	-	-	-	-	-	-	-	-	-	-	-	0.42	S
75	Zygosporium gibbum	-	-	-	-	-	10	5	-	15	-	-	-	2.5	S
76	Z. echinosporum	-	25	-	-	-	-	-	-	-	-	-	-	2.08	S
77	Z. oscheoides	-	-	-	25	30	35	10	5	-	-	-	-	8.75	O

MC - 'Most Common'; **C** - 'Common'; **O** - 'Occasional; **S** - 'Sporadic'

Parasympodiella laxa (13.75%), *Acremonium strictum* (10.42%), and *Phaeotrichoconis crotalariae* (10.42%) in that order.

It is evident from the data that some of the fungi belonging to Zygomycota, and mitosporic fungi such as *Cladosporium, Curvularia, Drechslera, Pithomyces* and *Nigrospora* were recorded only during initial one or two samplings and this suggests that they may be carry over species from the phylloplane. Their abundance on surfaces of green foliage has been reported by several earlier workers (Subramanian & Vittal, 1979, 1980; Sudha, 1978; Dorai, 1988) and they are considered as primary saprophytes which are ubiquitous inhabitants of aerial plant surfaces (Hudson, 1968). The activity of these fungi was limited to freshly fallen litter and any records of these species in later stages of decomposition would be sporadic. Based on the hierarchial clustering it is found that *Beltrania rhombica, Beltraniella portoricensis, Subramaniomyces fusisaprophyticus* and *Gyrothrix circinata* are the typical litter colonizers in the initial stages of decomposition. These are considered as true 'litter fungi'. As the decomposition progressed these taxa were joined by a second group of colonizers, such as *Acremonium strictum, Parasympodiella laxa* and *Trichoderma harzianum.* Finally, in the late stages of decomposition a third group of fungi consisting of *Phaeotrichoconis crotalariae* and *Thozetella tocklaiensis* appear and continue. The significant features of colonization of *Anacardium* litter by microfungi as elucidated in the present study revealed that the primary colonizers are the 'foliicolous fungi' which arrive on host surface as airborne spores and act as initial colonizers after abscission. These are soon joined by 'true litter fungi', which are perhaps those that colonize leaves after they are shed and show activity for varying periods thereafter.

Acknowledgements

We are pleased to contribute this paper in honour of Prof. C. Manoharachary who made significant contributions to Mycology in India. The second author (BPRV) first came in contact with Prof. Manoharachary in 1978 and their friendship and dialogue on litter microfungi continued during the last three decades. Dr.BPRV thanks Prof. K.V.B.R. Tilak for his invitation to contribute an article for the Festchrift Volume of Prof. Manoharachary.

References

Aoki, T. and Tokumasu, S. 1995. Dominance and diversity of the fungal communities on fir needles. *Mycological Research*. **99**: 1439-1449.

Aoki, T., Tokumasu, S. and Oberwinker, F. 1992. Fungal succession on fir needles in Germany. *Transactions of the Mycological Society of Japan*. **33**: 359-374.

Aoki, T., Tokumasu, S. and Tubaki, K. 1990. Fungal succession on momi fir needles. *Transactions of the Mycological Society of Japan.* **31**: 355-374.

Bills, G.F. and Polishook, J.D. 1994. Abundance and diversity of microfungi in leaf litter of a lowland rain forest in Costa Rica. *Mycologia*. **86**: 187-198.

Chasseur, C. and Beguin, H. 1990. Etude de la microfonge dans la litiere de *Fagus sylvatica*. *Belgian Journal of Botany*. **123**: 117-133.

Dorai, M. 1988. Taxonomic and ecological studies of fungi colonizing leaves and litter of *Eucalyptus* species, in South India. Ph.D. Thesis, Univ. of Madras.

Hudson, H.J. 1968. The ecology of fungi on plant remains above the soil. *New Phytologist*. **67**: 837-874.

Kjoller, A. and Struwe, S. 1990. Decomposition of beech litter. A comparison of fungi isolated on nutrient rich and poor media. *Transactions of the Mycological Society of Japan*. **31**: 5-16.

Kshattriya, S., Sharma, G. D. and Mishra, R. R. 1994. Fungal succession and microbes on leaf litters in two degraded tropical forests of North-East India. *Pedobiologia*. **38(2)**: 125-137.

Lousier, J.D. and Parkinson, D. 1976. Litter decomposition in a cool temperate deciduous forest. *Can. J. bot.* **54**: 419-436.

Takayuki Aoki and Seiji Tokumasu. 1995. Dominance and diversity of the fungal communities on fir needles. *Mycol. Res.* **99(12)**: 1439-1449.

Pasqualetti. M., Mulas, B, Zucconi, L. and Rambelli, A. 1999. Succession of microfungal communities on *Myrtus communis* leaf litter in a Sardinian Mediterranean maquis ecosystem. *Mycol. Res.* **103(6)**: 724-728.

Sieber-canavesi, F. and Sieber, T. N. 1993. Successional patterns of fungal communities in needles of European silver fir (*Abies alba* Mill.). *New Phytologist*. **125**: 149-161.

Subramanian, C.V. and Vittal, B.P.R. 1979. Studies on litter fungi II. Fungal colonization of *Atlantia monophylla* Corr. leaves and litter. *Nova Hedwigia* Beiheft. **63**:361-369.

Subramainan, C.V. and Vittal, B.P.R. 1980. Studies on litter fungi IV. Fungal colonization of *Gymnosporia emarginata* leaves and litter. *Trans. Mycol. Soc. Japan*. **21**: 339-344.

Sudha, K. 1978. A study of mycoflora of leaves and litter. Ph.D. Thesis, Univ. of Madras.

Tokumasu, S. 1996. Mycofloral succession on *Pinus densiflora* needles on a moder site. *Mycoscience*. **37**: 313-321.

Tokumasu, S., Aoki, T. and Oberwinker, F. 1994. Fungal succession on pine needles in Germany. *Mycoscience*. **35**: 29-37.

Vardavakis, E. 1988. The mycoflora, production and decomposition of leaf litter in an Orno-Quercetum ilicis association. *Pedobiologia*. **32**: 167-176.

Visser, S. and Parkinson, D. 1975. Fungal succession on aspen poplar leaf litter. *Can. J. Bot.* **53**: 1641-1651.

Frontiers in Plant Sciences (Pages 301-312)
Eds. K.G. Mukerji, K.V.B.R. Tilak, S.M. Reddy, L.V. Gangwane, P. Prakash, I.K. Kunwar

Aeropalynology and Allergic Diseases

B.N. Pande

ABSTRACT

Air carries innumerable fungal spores, hyphal fragments, pollen grains etc. The aeromicrobiota of a locality is both transient and variable. Pollen monitoring of the environment has been extensively carried out in various regions of the world. With the increasing awareness of pollen as aeroallergen clinical investigations have now been formed a routine follow up method associated with aeropalynological surveys. Vast geographic and climatic zones which remained unexplored would provide useful information. Aeropalynological studies conducted suggested that an estimated 30% of Indian population suffers from one or the other type of allergy. Studies also suggest that meterological factors also determine the aerial incidence of pollen grains. In aeropalynological investigation carried out at Aurangabad during 1996 – 1997 & 1997 – 1998 a total of 15 types of pollen grains recorded were known to be potentially allergenic.

Clinical Data suggests that grass pollen was found to be the major component in the air having contributed 35.35% followed by *Amaranthus* 23.44%, *Parthenium* 3.24%, *Dodonia* 2.54%, *Xanthium* 0.98%, *Azadirachta* 0.39% and *Cassia* 0.22%. Among 150 patients tested 34 exhibited manifestation of allergic and 64 exhibited bronchial asthma. Seasonal asthma was more prevalent during summer and winter seasons. The maximum number of patients had the onset of allergic symptoms between 21 – 30 years and 31 – 40 years age.

This work would stimulate further investigation in exploring aeropore allergenic pollen and also the prevalence of indoor airborne pollen and fungal spores of different regions and its effect on human health.

Key words: Aeriomicrobiota, aeroallergen, asthma, pollen grains.

Author's address: Department of Environmental Science, Dr. Babasaheb Ambedkar Marathwada University, Aurangabad 431004 (M.S.) India.

INTRODUCTION

Air carries innumerable fungal spores, hyphal fragments, pollen grains etc. The aeromicrobiota of a locality is both transient and variable. The bioparticles remain suspended in air for some time and may be transported from one locality to another through wind currents.

Pollen monitoring of the environment has been extensively carried out in various regions of the world. Earlier work was essentially oriented towards qualitative and quantitative estimation of the pollen load in the atmosphere, seasonal variation, pollination biology, preparation of pollen calendar and pollen spectra. With the increasing awareness of pollen as aeroallergen clinical investigations have now been formed a routine follow up method associated with aeropalynological surveys. Extensive aeropalynological work in India has been carried out in almost all parts and it is hoped that the vast geographic and climatic zones which remained unexplored would provide useful information.

The term "palynology" was coined by Hyde and Williams (1943) as the substitute for the "science of pollen grains and spores". It was derived from the Greek verb "Palynin", i.e. scatter pollen grains and spores which are often disseminated by the wind. Erdtman (1952) coined the term "Aeropalynololgy" to cover the study of airborne pollen, spores and associated plant materials (palynomixtum). Pollen grains found in the air of a geographical area were widely known to be the cause of various allergic complaints like hay-fever, eczema, asthma and urticaria (Mandal and Chanda, 1979; Nair, 1963). Allergic response to hay-fever or pollinosis was generally recognized as the most prevalent and important of all allergies (Stanley and Linskens, 1974).

Pollen grains are among the earliest known allergens and are the major source of morbidity among atopic individuals (Kjelman, 1994). Jhon Bostoc (1819) was the first to suspect pollen as the cause of hay fever. Half a century later Blackley (1873) established that grasses were responsible for his allergic symptoms. Subsequent studies carried out during 20th century established that pollen are the chief causative agents of respiratory allergy (Naranjo, 1958; Shivpuri, 1980; Kantoor *et al.*, 1966; Lewis, 1984; Singh and Singh, 1994; Tilak, 1988). Over the years there has been a significant global progress in understanding the role of palynology in allergic diseases.

Allergies of Pollen

The allergens of pollen grains are located in extracellular wall sites of the pollen i.e. in the intine and exine. This can be established by using fluorescent antibody methods, taking great care to prevent diffusion from the original sites during processing. The prevention can be achieved by the use of freeze substitutions methods, because allergens get differed from their wall sites almost. The diffusates of the pollen contain a high number of protein fractions. Some carry diploid specified determinates being originated from the exine cavities, an exudates from pollen protoplasts carrying haploid specified determinates. Thus making it possible that different forms of allergens could be present in two sites.

PROGRESS OF THE AEROPALYNOLOGICAL STUDIES IN INDIA

Till late fifties and early sixties research relating to aeropalynology and pollen allergy was not given due importance largely due to the belief that allergy is a malady confined mainly to temperate countries. However, surveys conducted Aeropalynological studies suggested that an estimated 30% of Indian population suffers from one or the other type of allergy (Agarwal, 1988).

During the period of late fifties and early sixties, preliminary studies in this directions were undertaken by Kasliwal *et al.* (1955, 1958, 1959), Sanghavi *et al.* (1957), Kalra and Dumbrey (1957) at Pune and Shivpuri and his associates (1963) at Delhi, constitute the pioneering research work in India. Subsequent work carried out at various places in India has contributed significantly to our knowledge of aerial pollen and pollen allergy. This was then continued and further expanded by Singh and his co-workers (1993) at Delhi, Chanda and his associates (1971, 1978, 1979). Nair (1963) described the incidence of atmospheric pollen grains at Vellore and Tripathi *et al.* (1983) at Bombay, Lakhanpal and Nair (1958), Khandelwal (1973), Khandelwal and Mittre (1975) at Lucknow; Chitaley (1976) at Nagpur; Kumar (1984) at Bareilly; Lakhanpal (1960) at Almora; Tilak and Srinivasulu (1967) and Tilak and Vishwe (1980) at Aurangabad; Rai (1994) at Bhopal; Agashe and Vinay (1980), Agashe *et al.* (1980 and 1982) at Bangalore; Maribhat and Rajasab (1988) at Gulbarga; Ramalingam (1971) at Mysore; Ravindran *et al.* (1988) at Trivandrum; Jain and Das (1981) at Gwalior; Reddi (1976) at Visakhapatnam; Janakibai (1979) at Visakhapatnam; Janakibai and Reddi (1980) at Visakhapatnam; Gupta and Singh (1965) at Bikaner; Singh *et al.* (1987) at Dehradun etc.

Manake (1991) clearly brought out the correlation between meteorological factors and the aerial incidence of pollen grains. Temperature and relative humidity could be correlated with aerial incidence of pollen however with temperature as parameter. Significant correlation was negative while the humidity it may be positive or negative.

The size and shape of the pollen grain is also determining factor. Most of the allergenic pollen grains are minute or medium in size and mostly sphaeroidal or subsphaeroidal with 3 pores. The low incidence of pollen in air is typical of entomophilous or anemophilous like *Adhatoda, Clerodendron* etc. Their low incidence may be due to the medium or large size or clustering of pollen grains.

AEROPALYNOLOGICAL WORK AT AURANGABAD

Tilak and Patil (1983) carried out extensive work on *Parthenium* pollen. The maximum concentration was recorded in the month of July (655/m^3) and minimum in December (35/m^3). The airborne pollen was present almost throughout the year (Table 1). Indoor air monitoring for pollen indicated

TABLE 1 *Monthly average % contribution of Parthenium pollen to total airborne components from 1st July 1952 to 30th June 1983*

July	Aug.	Sept.	Oct.	Nov.	Dec.	Jan.	Feb.	Mar.	Apr.	May	Jun
26.67	8.15	12.42	11.81	10.39	1.43	1.83	6.52	6.72	7.13	3.67	3.26

TABLE 2 *Monthwise average % of potentially allergenic pollen grains*

Sr. No.	Pollen types	Nov. %	Dec. %	Jan. %	Feb. %	Total
1.	Grass pollen	25.13	18.99	53.56	43.70	35.35
2.	*Amaranthus* sp.	31.96	22.94	21.96	16.92	23.44
3.	*Parthenium* sp.	2.20	7.44	0.96	2.36	3.22
4.	*Dodonia* sp.	14.40	5.77	-	-	2.54
5.	*Xanthium* sp.	0.11	0.15	3.66	-	0.98
6.	*Azadirachta* sp.	-	-	-	1.57	0.39
7.	*Cassia* sp.	-	0.30	0.19	0.39	0.22

the presence of 43 different types of which 19 types were allergic. Some of the important types have been included in Table 2.

MATERIALS AND METHODS

During the investigation period Boehm's individual (personal) pollen collector and nasal filters were used. Both have been considered of utmost use to the patients suffering from nasobronchial allergy due to pollen grains.

Boehm's individual pollen collector provides more accurate information of the personal exposure of any individual to various allergens. This Boehm's pollen collector (Leuschner and Boehm 1977) is very light and can be easily carried from one place to other. The device is tagged to the patients clothing during the day time and kept horizontally left on the bed side overnight in the room. A slide smeared with white petroleum jelly is inserted in a plexiglass holder of the trap. The white petroleum jelly serves to trap the airborne components. The slide is changed every 24 hrs. Mounting and scanning is done as suggested by Tilak and Srinivasulu (1967). For estimating the airborne components of the pollen grains, slides are then scanned for showing their percentage contribution.

Nasal filters have minute pores which prevent entry of pollen grains into nostrils and their ultimate travel to the nasobronchial system of human beings. The pollen grains are considerably big in size and heavier in weight as compared to fungal spores. Therefore, it becomes difficult for them to enter the nostrils through the microspores of the nasal filters.

RESULTS AND DISCUSSION

During the investigation period i.e. in 1996-97 and 1997-98, a total of 15 types of pollen grains were recorded. Almost all the pollen types were known to be potentially allergenic. The highest pollen catch was that of

TABLE 3 *Monthwise prevalence of pollen types and their % contribution to the total airspora components*

Pollen type	*July*	*Aug.*	*Sept.*	*Oct.*	*% Contribution to total airborne components*
Acalypha sp.	–	–	7	3	0.22
Amaranthus sp.	–	15	25	3	0.96
Asteraceae	1	2	–	1	0.08
Cyperaceae	3	3	3	1	0.22
Eucalyptus sp.	–	–	2	3	0.11
Feronia sp.	–	–	9	2	0.24
Grass pollen	24	270	80	20	8.79
Parthenium sp.	49	289	231	2	12.74
Tephrosia sp.	–	–	3	2	0.11
Lantana sp.	–	–	3	–	0.06
Bougainvillea sp.	–	5	1	–	0.13

Parthenium (12.74), followed by grass pollen (8.79%). The other pollen grain types and their concentration and percentage contribution of different months is expressed in Table 3.

In order to understand the allergenic pollen grains of Aurangabad, prick tests and intra dermal skin tests were performed in the allergy clinic, Government Medical Hospital, Aurangabad with different available pollen antigens. Before carrying out the skin tests on the patients, a comprehensive case history of patients was recorded. The physical examination such as urine, blood, stool, nasal smear, chest, X-ray, etc. was made before carrying out the skin tests of the patients, all antihistamines and tranquilizers were discontinued for 72 hrs, prior to the skin tests.

After 20 minutes interval, the results of the skin tests were recorded and were graded according to the criteria of Shivpuri (1962) and Shivpuri and Agarwal (1969). The reactions ranging from 2^+ to 4^+ were termed as "markedly positive reactions".

Clinical Data, July 1996 to June 1997

The importance of inhalant allergens for respiratory allergy, skin diseases and gastrointestinal tracts are well known. The estimation of potentially aeroallergenic pollen grains was carried out in the Chest Disease Ward of Government Medical College Hospital, Aurangabad. The results based on experiments are as under: Out of 28 pollen grains, 19 pollen types were known to be potentially allergenic, of which antigens of 7 available pollen types were used for skin test. Grass pollen. During the investigation period, grass pollen was found to be the major component in the air having contributed 35.35% followed by *Amaranths* 23.44%, *Parthenium* 3.24%, *Dodonia* 2.54%, *Xanthium* 0.98%, *Azadirachta* 0.39% and *Cassia* 0.22% (Table 2).

During the period of investigation from July 1996 to June 1997 as many as 150 patients were tested and out of 150, 52 patients exhibited the symptoms of bronchial asthma and allergic rhinitis. 34 patients exhibited manifestation of allergy or allergic disorders may be due to inheritance and 64 patients exhibited bronchial asthma (Table 4.).

TABLE 4 *Disease wise distribution of patients during 1996-1997 and 1997-1998*

Disease	*Male*	*Female*	*Total*	*% contribution*
Bronchial asthma	32	32	64	44.38
Allergenic rhinitis	24	10	34	22.53
Allergenic rhinitis with bronchial asthma	32	20	52	33.09
Total	88	62	150	

The seasonal asthma was more prevalent during summer and winter seasons (Table 5) and at this time there was significant rise in the pollen count in the air.

TABLE 5 *Seasonal prevalence of asthma*

No. of cases	*Perennial*	*Non seasonal*	*Seasonal*		
			Summer	*Railway*	*Winter*
150	31	32	41	17	29

There was a correlationship between the seasonal and perennial incidence of allergic symptoms and the atmospheric pollen concentration.

However, antigens of eleven pollen types were tested for their intradermal test on the patients who showed seasonal symptoms of allergic disorders like bronchial asthma and allergic rhinitis (Tables 6 and 7). From the antigen tests it was clearly evident that the pollen grains of *Cassia* (37.0%), *Sorghum* (33.3%), *Amaranthus* (30.0%), *Acacia* (29.00%) etc. showed significantly higher percentage of reactivity respectively.

TABLE 6 *Skin testing of 150 patients with the antigen of pollen of various taxa*

Pollen type	*Total number of cases tested*	*No. of patients showing +ve, ID + test*	*% of reactivity*
Cynodon sp.	31	8	23.3
Cyperus sp.	20	5	20.0
Acacia sp.	17	5	29.4
Lantana sp.	25	8	32.0
Azadirachta sp.	34	16	32.3
Cassia sp.	27	10	37.0
Sorghum sp.	12	4	33.3
Parthenium sp.	21	6	28.6
Moringa sp.	17	3	17.6
Bougainvillea sp.	20	5	25.0
Amaranthus sp.	70	21	30.0

Clinical Data, July 1997 to June 1998

The intradermal skin tests were performed on 160 patients during this period, where in 58.75% of the patients were male and 41.25% female. Out

TABLE 7 *Results of skin tests with various pollen antigens*

Pollen antigen tested	*Total test done*	*Skin response %*					*Total % of*
		1^+	2^+	3^+	4^+	*Total*	2^+, 3^+ *and* 4^+
Amaranthus sp.	176	13.06	3.97	1.7	–	18.73	5.67
Argamone sp.	138	3.62	2.17	–	7.26	3.64	–
Azadirachta sp.	131	11.36	3.78	1.51	–	16.65	5.29
Brassica sp.	107	9.34	7.47	1.86	–	18.67	9.33
Cassia sp.	270	8.88	8.14	2.22	–	19.24	7.79
Dodonia sp.	134	2.98	2.23	1.49	0.74	7.44	4.46
Grass	567	11.28	5.64	1.94	–	18.86	7.58
Parthenium sp.	135	12.59	8.14	2.22	1.48	24.43	11.84
Prosopis sp.	118	11.01	3.38	–	–	14.39	2.7
Xanthium sp.	113	6.19	6.19	2.65	–	15.03	8.84
Eucalyptus sp.	17	17.64	–	–	–	17.64	–

of the total cases tested, 142 patients showed positive response to one or more pollen antigens, 18 showed negative response to all the antigens. However, 95 patients, out of the 142 patients tested, showed positive reactions i.e. 2^+ to 4^+. It is quite evident from table 6 that 2^+ and 3^+ reactions were commonly found in allergic patients with *Dodonia* and *Parthenium* pollen antigens. However, *Eucalyptus* pollen was found to be an exceptional one.

The assessment of the reactivity to the various pollen antigens was made as per the different age and sex groups and as it is evident from the table 4. The maximum number of patients had the onset of allergic symptoms between 21-30 years and 31-40 years age.

During aeropalynological survey in the years 1996-1997 and 1997-1998, a total of 28 pollen types were identified. Grass pollen with 47.2% were dominant in both the years. Hyde and Adams (1958) also reported the highest incidence of grass pollen at open places with cultivated lands in Great Britain. During this study the grass pollen was trapped throughout.

The pollen grains of *Parthenium* with 9% also contributed in highest concentration followed by *Lantana* with 3.9% and *Amaranthus* with 1.9%. The other contributors of pollen grains being *Hibiscus, Nerium* and *Tephrosia*. *Azadirachta* pollen contributed 9.7% to the total airspora and *Moringa* pollen 6.1%. Investigations carried out for two consecutive years i.e. 1996-97 and 1997-98 clearly revealed the fluctuations in the pollen concentration from year to year as it was also reported earlier at Cardif U.K. by Hyde (1956), Dua and Shivpuri (1962) at Delhi. Reddi (1970) at Visakhapatnam, Leuschner and Boehm (1977) at Basel, Tilak (1998) at Aurangabad, Pande (1994) at Aurangabad.

The general frequency of airborne pollen showed two peak periods, one during period of September to November of herb pollen and second during February to April of tree pollen. There was synchronization between these two periods and the incidence of various allergenic disorders recorded during this period in city. The pollen concentration in the atmosphere is closely related with the flowering periods.

The role of weather conditions over the peak incidence in the atmosphere was also well established. High concentration of pollen grain was recorded when the temperature range was between 27°C and 31°C with high wind velocity, whereas increase in relative humidity or dew formation decreased the concentration. Similar observations were also reported earlier by Hyde and Williams (1943), Hirst (1953), Sreeramulu and Ramalingam (1966), Leuschner (1974), Pande (1994) and Tilak (1998). The rain had profound effect on the pollen in the air. Either it cleared the pollen number by inhibiting anthesis or by washing of effect. Similar findings were also reported by Hyde and Williams (1943), Gregory (1973). The antigens of the eleven pollen types were tested on the 150 patients, the intradermal tests showed *Cassia, Sorghum, Azadirachta, Lantana, Amaranthus, Acacia* and *Parthenium* to be the most potentially allergenic.

It is hoped that this article would stimulate further investigations to work with keen interest in exploring aeroborne allergenic pollen. There is a great demand for understanding the prevalence of indoor airborne pollen and fungal spores of different regions and its effect on human health.

In views of the above, a long term investigations of indoor survey of different places is absolutely necessary in order to understand the clear spectrum of indoor airborne allergenic pollen and fungal spores for their effective control.

Acknowledgement

The author is grateful to Prof. S.T. Tilak for giving valuable suggestions.

References

Agarwal, M.K. 1988. Abs. XXII Ann. Conv. *The Ind. College of Allergy & Applied Immunol.* Delhi.

Agashe, S.N. and Vinay, P. 1980. *Adv. In Pollen Spore Res.*, **V-VII**: 185.

Agashe, S.N. and Anand, P. 1982. *Asp. Allergy & Applied Immunology*, **15**: 49.

Blackley, C.H., 1973. Experimental researches on the case of nature of *Cartarrahus aestivus* London.

Chanda, S. and Nandi, N.C. 1971. *Grana* **5**: 128.

Chanda, S., Mandal, S. and S.C. Lahiri, 1978. Occurrence of allergenic pollen in the atmosphere of Calcutta and another cause for organic environmental pollution. *Sci. and Cult*. **44**: 175-177.

Chanda, S. and Mandal, **S.**1979. *Ind. College of Allergy & Appl. Immunol*. Ab.7.

Chanda, S. and Mandal, **S.**1976. *4th Inter. Palynol. Conf*. Lucknow, Ab. 29.

Dua, K.L. and Shivpuri, D.N. 1962. Atmospheric pollen studies in Delhi area in 1958-59. *J. Allergy* **33**: 507-512.

Erdtman, G. 1952. Pollen morphology and plant taxonomy angiosperms (An Introduction to Palynology-I). *Almquist, Stockholm*, pp. 539.

Gregory, P.H. 1973. Microbiology of Atmosphere. 2nd Ed. Leonard Hill, *Aylesbury*, pp. 377.

Gupta, K.D. and Singh, J. 1965. *J. Ass. Phys. (India)* **13**: 919.

Hirst, J.M. 1953. Changes in the atmospheric pollen and pollinosis in West Japan. Proc. IX Int. Congr. Allergology Buenos Aires, pp. 1-2.

Hyde, H.A. 1956. the tree pollen in Great Britain. *Acta Allergol.* **10**: 224-245.

Hyde, H.A. and Williams, D.A. 1943. A census of atmospheric pollen. *Nature* **151**: 82-83.

Hyde, H.A. and Adams, K.F. 1958. An atlas of airborne pollen grains. Macmillan, London, pp. 112.

Janakibai, A., 1979. The airspora of Visakhapatnam. Ph.D. Thesis, Andhra University, Waltair.

Janakibai, A. and Reddy, S. 1980. *Adv. in Pollen Spore Res*. **V-VIII**: 209.

Jain, A.K. and Das, R.R. 1981. *J. Ind. Bot. Soc*. **60**: 344.

Kalra, S.L. and Dumbrey, D.G. 1957. *Armed Force Med. J*. (India), **13**: 3.

Kantoor, S.R., Frank, M., Kantoor, H.D., Golan, B.R., Marian, D., Scachnner, E., Kessler, A. and De Varies, A. 1966. Airborne allergens and clinical response of asthmatic in Arad, a new town in desert area in Aisrael. *J. Allergy* **32**: 65-74.

Kasliwal, R.M., Sanghvi, L.M. and Gupta, K.D. 1955. *J. Ass. Phy*. (India), **3**: 184.

Kasliwal, R.M., Gupta, K.D., Sethi, J.P., Solomon, S.K. and Sogani, I.C. 1958. Report submitted to ICMR.

Kasliwal, R.M., Sethi, J.P. and Sogani, C. 1959. Studies in atmospheric pollen. A daily census of pollen at Jaipur 1957-1959. *Ind. J. Med. Res*. **47**: 515-521.

Khandelwal, A. 1973. Studies in airborne pollen and fungal spores at Lucknow. Ph.D. Thesis, Lucknow University, Lucknow.

Khandelwal, A. and Mittre, V. 1975. *Paleobot*. **22(3)**: 178.

Kjellman, N.I.M., 1994. Prediction and prevention of allergy in infants and children. *Allergy Clin. Immunol. New,* **5**: 131-134.

Kumar, S. 1984. *J. Palynol.*, **20(1)**: 81.

Lakhanpal, R.N.N. and Nair, P.K.K. 1958. *Industr. Res*. **17C**: 80.

Lakhanpal, R.N.N. 1960. *Industr. Res*. **19C**: 51.

Leuschner, R.M., 1974. Pollen calendars for Switzerland. In: Atlas of European allergenic pollen (ed. Charpin *et al.*) pp. 193-199.

Leuschner, R.M. and Boehm, G. 1977. Ubescin pollen fonggerat fur Eingel person, *Schwciz. Med. Wescher* **107**: 57-59.

Lewis, W.H., 1984. Pollen allergy. In Allergy: Theory and Practice, Eds. Korenblat, P.E. and Wedner, H.J., Grune and Startlon Orlando, pp. 353-369.

Manake, S., 1991. Studies on certain aspects of palynology at Bhopal with reference to aerial dispersal productivity and allergenic significance. *Ph.D. Thesis, Bhopal University*, Bhopal.

Mandal, S. and Chanda, S. 1979. Studies on the allergenic pollen of Kalyani, West Bengal, Botanical Aspect. *Asp. Allergy & Appl. Immunol*. **12**: 101-106.

Maribhat, M. and Rajasab, A.H. 1988. *Ind. J. Aerobiol*. **1(2)**: 59.

Nair, P.K.K., 1983. An analysis of atmospheric pollen, fungal spores and other vegetable matter at Vellore, Madras State (India). *Ind. J. Med. Res.*, **51**: 447-452.

Naranjo, P., 1958. Etiological agents of respiratory allergy in tropical countries of Central and South America. *J. Allergy*, **29**: 363-374.

Pande, B.N., 1994. Indoor aeroallergens of hospital ward. Curr. Trends in Life Sciences. *Adv. in Mycol. and Aerobiol.* (edt. U.K. Talde), **20**: 301-313.

Rai, P.K., 1994. Studies on pollen allergy and allied aspects in India with special reference to Bhopal – A review. Curr. Trends in Life Sciences. *Adv. in Mycol. and Aerobiol.* (edt. U.K. Talde), **20**: 301-313.

Ramalingam, A., 1971. *Proc. Ind. Acad. Sci*. **74(B)**: 227.

Ravindran, P., Joshi, M., Sundaram, P., Ghogh, S., Gopi, T.V. and Kumar, R.P. 1988. *And. J. Aerobiol*. **1(2)**: 71.

Reddi, C.S., 1970. A comparative survey of atmospheric pollen and fungal spores at two places twenty miles apart. *Acta Allergol.* **25**: 189-215.

Reddi, C.S., 1976. *J. Palynol.*, **10**: 155.

Sanghavi, L.M., Sethi, J.P. and Kasliwal, R.M. 1957. Pollen allergy in Rajasthan. A preliminary study of botanical flora and aerial pollen. *Ind. J. Med. Res.*, **29**: 1-13.

Shivpuri, D.N. 1980. Clinically important pollen, fungal and insect allergens for nasobronchial allergy patients in India. *Asp. of Allergy Appl. Immunol.* **13**: 19-23.

Shivpuri, D.N., Dua, K.L. and Subramanian, T.A.V. 1963. *J. Med. Res.*, **51**: 64.

Shivpuri, D.N. and Agarwal, M.K. 1969. Studies on the allergenic fungal spores of Delhi, India metropolitan area clinical aspects. *J. Allergy*, **44**: 204-213.

Singh, A.B. and Singh, A. 1994. Pollen allergy – A global scenario. In recent trends in Aerobiology and Immunology (Ed. S.N. Agashe), Pub. *Oxford IBH*, pp. 143-173.

Singh, B.P., Singh, A.B., Nair, P.K.K. and Gangal, S.V. 1987. *Ann. Allergy*, **59**: 229.

Singh, A.B., Malik, P., Prakash, D. and Gangal, S.V. 1993. Identification of specific IgE binding fractions in castor beam (*Ricinis communis*) pollen obtained from different source materials. *Grana* **32**: 376-380.

Sreeramulu, T. and Ramalingam, A. 1966. A two year study of airspora of a paddy field near Visakhapatnam. *Indian J. Agric. Sci*., **36**: 111-132.

Stanley, R.G. and Linskens, H.F. 1974. Pollen biology, biochemistry management. Springer-Verlag Berlin, Heidelberg, New York, pp. 307.

Tilak, S.T. 1988. Aeropalynology of Maharashtra. J. of Aeropalynology. G. Thankmani Comm. Vol. *Today and Tomorrows Printers*, New Delhi, pp. 101-106.

Tilak, S.T. 1998. Aerobiology, Biology of Living Organisms Series-II. Satyajeet Prakashan (edt. J.G. Vaidya), Pune, pp. 504.

Tilak, S.T. and Vishwe, D.B. 1980. Aeroallergenic pollen at Aurangabad. *Maharashtra Medical Journal*, **26**: 291-294.

Tilak, S.T. and Vishwe, D.B. 1980. Aeroallergenic pollen at Aurangabad. *Adv.* Pollen *Spore Res*., **V-VIII**: 145-149.

Tilak, S.T. and Patil, B.Y. 1983. Status of *Parthenium* pollen as aeroallergen. *J. Bio. Res.* **3**: 12-15.

Tilak, S.T. and Srinivasulu, B.V. 1967. Airspora of Aurangabad. *Ind. J. Microbiol.* **7**: 167-170.

Tripathi, D.M. and Parikh, K.M. 1983. *Med. & Surg*., **23(7)**: 17.

Frontiers in Plant Sciences (Pages 313-330)
Eds. K.G. Mukerji, K.V.B.R. Tilak, S.M. Reddy, L.V. Gangwane, P. Prakash, I.K. Kunwar

A Study of the Aerial Mycoflora of Mauritius in Relation to Allergy and Asthma

A. Peerally and Vasant Rao*

ABSTRACT

Aerial fungal isolates from different parts of Mauritius were collected and their identification to generic and/or specific level determined. This study was initiated in view of the increasing importance of fungi in various human disorders particularly in asthma, allergic reactions, allergic sinusitis and lung diseases. In Mauritius the incidence of asthma is in the range of 5 to 8% while for sinusitis it is 10 to 15%.

In the past 100 years changes have occurred in the outdoor environment and in houses which have contributed to increased prevalence of allergy and asthma. This is creating greater awareness of the need, in the management of such problems, to seek additional causes of allergy and asthma, and in particular the precise identity of allergy causing fungi and the actual mechanism of these allergic reactions.

A review of the literature and the survey of fungal isolates from the air in Mauritius indicate that several genera of aerial mycoflora in Mauritius are similar to those known to cause various allergic reactions in humans elsewhere. The common genera are *Cladosporium, Aspergillus, Penicillium, Alternaria* and others. Some 200 isolates were studied and species were identified. This probably represents about 20% of the total fungal flora of Mauritius. Further studies are underway to identify the others.

Author's address: University of Mauritius, Reduit, Mauritius
E-mail: apeeraly@uom.ac.mu
*Presently at Department of Botany, O.U., Hyderabad 500 007 A.P. India.

This initial study will pave the way for various tests to establish the role of these species causing or implicated in allergy and asthma in children and adults and subsequently to enable further biochemical studies in collaborative projects with specialized laboratories overseas.

Key words: Aerial flora, allergy, Asthma

INTRODUCTION

In the past 100 years, changes have occurred in the outdoor and indoor environments which have contributed to the increased prevalence of hay fever allergy and asthma. Dotterud *et. al.* (1995) sampled fungal aerospores in the homes of 19 HDM – sensitized children. They found *Penicillium* as the most prevalent fungal isolate followed by yeasts, *Aspergillus, Cladosporium* and *Mucor.* High aerospore count was related to high indoor humidity. Aerospore counts were higher in the HDM-sensitised group than in the control group.

Similarly, Dill & Niggemann (1996) studied the homes of twenty children in order to determine a possible relationship between domestic fungal exposure and sensitization to fungi. Twelve of the nineteen homes (63%) revealed a considerable amount of viable fungal propagules and/or visible fungal growth. *Penicillium, Cladosporium, Aspergillus,* Mycelia *Sterilia* and yeasts were the most prevalent.

In India Verma et al (1995) reported fungal allergens as the most prevalent aeroallergens. Their most important view was that there were common antigenic/allergenic components among different fungi, a matter needing to be known for the proper diagnosis and treatment of patients allergic to fungi. They studied *Alternaria alternata, Aspergillus flavus, A. fumigatus, A. niger, A. tamarii, A. versicolor, Cladosporium herbarum, Curvularia lunata, Mucor hiemalis, Penicillium citrinum,* and *Fusariam solani.*

Li and Hsu (1997) found, in a study of airborne fungi in Taipei, the existence of an association between the occurrence of *Cladosporium* and the history of asthmatic children. According to Akiyama (1997) the prevalence of allergic diseases in Japan has been increasing in the last 30 years. About 3-5% of adults and children are suffering from bronchial asthma and was attributable to increased exposure to the allergens of mites and fungi.

The aerial fungi in Sagamihara (Japan) were studied for a periods of ten years from 1983 to 1992 in connection with allergy. The most predominant fungi found were *Epicoccum, Aureobasidium, Curvularia, Ulocladium. Pestalotia, Penicillium, Arthrinium, Nigrospora, Fusarium* and *Trichoderma* (Takatori et al, 1994), similarly in a study of the aeromycoflora of Cadiz in Spain the authors Gonzalez et al, (1992) reported *Cladosporium, Chaetomium* and *Ustilago* to be the most frequent, followed by *Alternaria, Ascochyta* and *Vernturia.*

In China, Xia (1992) collected 47 species of airborne fungi. Skin tests were made with 13 species for 114 cases of asthmatic patients and the positive ratio was 73.7%.

Various other studies on fungal allergens have been undertaken in specific places such as bakery, home environment, library and others, in a study of 383 bakery workers Smith et al (1997) found that 16% of them showed positive skin prick tests to fungal amylase. At the University of Sao Paulo, Brazil, a total of 31 librarians were questioned and 49% of them reported asthmatic or rhinitic symptoms. Tests for allergy towards 20 fungi most frequently present were also conducted. Gambale et al (1993) in his study concluded that since airborne fungi isolated in the libraries were likely to be present anywhere in the city of Sao Paulo, they were more likely to be present in high concentration in libraries, subsequently producing respiratory allergies.

Another interesting case of allergic reaction to fungal protein is the report by Tee et al (1993) that the fungal food "Quorn" prepared from *Fusarium graminearum* shared multiple common allergenic determinants with *Aspergillus fumigatus, Cladosporium herbarum* and *Alternaria alternata.*

Studies have also focused on the allergens, the allergies caused and on the control of the fungal agents responsible for the asthma. Kim et al (1996) from Japan encountered nasal allergy due to *Candida, Aspergillus* and *Alternaria.* Their studies revealed *Alternaria* as showing the highest ratio of strong positivity upon both RAST and skin test. They, therefore, concluded that *Alternaria* was an important fungal allergen. It is interesting to also mention the work of Granel et al (1995) who investigated allergy to *Alternaria.* Of the 662 patients allergic to *Alternaria* 40% had rhinitis and 60% suffered from asthma.

Lim et al (1995) from Singapore found that spores of *Bipolaris* contained at least four allergenic components, while Szantho et al (1992) found an interesting case of cross reactions with *Alternaria alternata* and *Phoma betae* in allergic children in Hungary. Their results suggest the presence of a common allergen in these two fungi.

Backman et al (1995) reported *Fusarium vasinfectum* as causing allergic bronchopulmonary fusarioris. On the other hand *Aspergillus spp* have been found to be responsible for allergic bronchopulmonary aspergillosis in 1-2% of patients with asthma, in the USA (Krasnick et al, 1995).

Species of *Penicillium* have also been reported as important in extrinsic bronchial asthma (Shen et al, 1996). The literature, also shows that allergic fungal sinusitis has been consistently and frequently reported as a cause of chronic sinusitis in immuno competent individuals. Different fungi, all of which are members of the Dematiaceae, have been reported as contributing to AFS as follows: *Aspergillus* and *Curvularia* (Quaraishi & Ramadan, 1997);

Aspergillus and *Penicillium* (Perez – Jaffe et al, 1997); *Aspergillus flavus, A. fumigatus* and *A. niger* (Chhabra et al, 1996); *Aspergillus fumigatus, Mucor and Cladosporium* (Deshpande et al, 1995), dematiaceous fungi (Kinsella et al, 1996).

Fugimoto et al (1994) found a case of allergic bronchopulmonary fungosis associated with *Aspergillus fumigatus, Penicilum luteum* and *Cladosporium cladosporioides*. Investigation of the allergy showed elevated specific IgE level, positive immediate and late skin reactions, positive dual asthmatic responses and positive precipitating antibodies for these three fungal species.

In an attempt to control the incidence of allergenic fungi in dwellings, Chaumont and Leger (1992) observed that citral vapours were effective against *Cladosporium* and some *Aspergillus* and *Penicillium* which Mucorales and *Geotrichum* strongly resisted.

According to Bent & Kuhn (1996), twenty-two fungal cultures grown from 15 AFS patients, were studied for *in-vitro* susceptibility to five common antifungal drugs; ketoconazole, amphotericin B, itraconarole, nystatin and fluconazole. The results obtained showed that ketoconazole and amphotericin B were the most effective. Tsushima et al (1996), however, reported amphotericin B as more effective than flucconazole in the treatment of a case of allergic bronchopulmonry aspergillosis.

AERIAL MYCOFLORA IN MAURITIUS

Aerial mycopropagules were allowed to produce colonies on potato-dexture agar (PDA) and tap water agar in petri plates. Individual colonies were transplanted to fresh plates of PDA and the new colonies were then used for microscopic studies. Indentification was attempted as far as possible to specific level but some isolates could be identified for the time being to generic level only. The account given provides fairly short descriptions of genera and species of the isolates obtained. Most identifications were done or confirmed with the help of the classical publications of Ellis (1971, 1976), and others.

1. *Pestalotia* de Not, *Memo. R. Acad.* (Sci. Torino 293): 80 (1839), Sutton, *Coelomycetes*, 163 (1980)

Sutton (1980) had earlier fully reviewed this genus and compared its similarities and differences with other genera. Presently, *Pestalotia* is reduced to a single species. The conidiomata eustromatic, cupulate, black or dark drown; conidiophores hyaline, formed on pseudoprenchymatous stroma.

Conidiogenous cells enteroblastic, annellidic. Conidia fusiform 5-6 celled, basal and apical cells hyaline to subhyaline, truncate at base, apical cell conic with 2-3 apical hyaline appendages, euseptate.

Pestalotia Pezizoides de Not, *Memo. R. Acad,* (Sci. Torino*: 80, (1839)*

Coniddiomata in culture blackish brown with brownish exudation, 1-2 mm across. Mycelium septate, branched, subhyaline to dark pigmented, thin to thick walled, pseudoparenchymatous, forming fruit body. Conidiophores hyaline to subhyaline, thin walled 10-20 µm long, 2-3 µm broad. Conidiogenous cells enteroblastic, usually conidia rectangular, annellidic with the conidiophores, 5-6 celled, 25-30 × 5-7 µm broad in the broadest region, basal cell and apical cell hyaline to subhyaline, truncate and conic respectively, middle cell thick-walled, pigmented. Apical and basal cells with appendages.

Isolated from air in Mauritius.

2. *Colletotrichum* Corda, Sturm *Deutschlands Flora* 3:41 91831-1832)

It is the conidial state of *Glomerella* Spauld & V Schrenck. Sometimes *Gloeosporium* species are also described under this genus.

It produces acervular conidiomata. Sclerotia are present or absent. Setae present in natural collection but not in culture or they may be rare. Conidiophores macronematous forming upper layer of conidiomata. Conidiogenous cells phialidic. Conidia aseptate, variable in shape, smooth hyaline, appresoria present.

Colletotirctum lindemuthianum (Sacc. & Magn.) Br. & Cav., Fungi Parass. 50 (1889).

Colonies slow–growing, dark brown to black. Sclerotia not observed. Setae absent, pale brown to dark brown. Conidiophores compact, cylindrical. Conidiogenous cells phialidic. Conidia straight, cylindrical, 7-10 µm long 2-3 µm broad, smooth, rounded or pointed at apex, base truncate, hyaline to subhyaline.

Isolated from air in Mauritius.

3. *Rhizopus* Eherenb. ex Corda, *Icones Fung.* 2; 20, 1838

Rhizopus occurs mostly on rotting bread, wet leather products, rotting fruits, decaying plant materials and in air. Cosmopotalitan in distribution. Fruiting bodies visible to naked eyes. Sporangiophores solitary or in clusters connected with hyphae and rhizoidal at the base.

Rhizopus stolonifer (Ehrenb. ex. Fr.) Lind. Danish Fungi: 72, 1913, Punithalingam CMI, Descrip. Path. Fungi. & Bac. No 524

Colonies cottony blackish brown with profuse cottony mycelium, hyphae coenocytic, aseptate, brown in colour, 3-5 µm broad. Sporanga spherical or globular, 200-300 µm across. Spores brownish grey, striate, spherical and the wall of columella collapsing back like umbrella.

Isolated from air in Mauritius.

4. *Gliocladium* Corda, *Icones Fung* 4:30 –31, 1840

It is a conidial state of *Nectria* vide Booth. It produces whitish cottony colonies. Conidiophores erect, branched, terminal phialide aseptate, cylindrical, narrow at apex, hyaline to subhyaline conidia, phialidic, aseptate, variable in shape, rounded at base and apex.

Gliocladium penicilloides Corda, *Icones Fung* 4: 30–31, 1840

Colonies cottony whitish effuse with penicilloid apices. Mycelium abundant, septate, branched, hyaline to subhyaline, smooth, 2-4 µm broad. Conidiophores macronematous, branched 2-4 times, septate. Conidiogenous cells phialidic, terminal, cylindrical, slightly narrower at base and apex, smooth, one-celled, hyaline to subhyaline 10-25 µm long up to 4ìm broad in the broadest and 1-2 µm at the pex. Conidia phialidic, smooth, allantoid, elongated, straight or curved, 5-15 µm long, 2-3 µm broad.

Isolated from air in Mauritius.

5. *Epicoccum* Link ex Fr., *System Mycol.* 3:466 (1832)

Sporodochia pulvinate black. Stroma reddish to blackish brown. Conidiophores clustered, simple. Conidia muriform. Very common on decaying plants and in air.

Epicoccum purpurascens Ehreng. ex Schlecht *Synop. Pl. crypt.*: 136 (1824)

Colonies sporodochial, up to 1 mm, coalescent, black brown, mild yellowish or reddish, exudates. Conidiophores 5-10 μm long. Conidiogenous cells monoblastic. Conidia muriform, subspherical to pyriform, dark golden brown, rough to verrucose, opaque with subhyaline to light brown basal cell.

Isolated from air in Mauritius.

6. *Papulaspora* Preuss, *Linnacea*, 24:112, 1851

It is a member of the Mycelia Sterilia belonging to Agonomycetales. It produces mycelium with bulbous propagules which are carried away by wind. It may be parasitise on plants and cause human diseases.

Papulaspora sepedonoides Preuss, Linnaca, 24:113, 1851

Propogules solitary multi cellular oval or globular.

Isolated from air in Mauritius.

7. *Fusarium* Link, *Mag. Ges. Naturff. Freude*, Berlin, 3:10, 1809

It produces cottony whitish colony in culture. Mycelium hyaline to subhyaline, septate, branched, with or without chlamydospores. Conidiophores producing phialides, branched. Conidiogenous cells phialidic or of two kinds giving rise to microconidia and macroconidia in the same culture or only one type of conidia. Microconidia aseptate, spherical. Macroconidia crescent shaped, hyaline or opaque in septal region.

Fusarium Moniliformae Shelect ex Fries, *Syst, Mycol,* 13:471, 1832.

Colonies cottony, whitish milky or cream coloured, reverse rosy. Mycelium creeping or submerged, 2-4 μm broad, septate, hyaline to subhyaline, branched, smooth chlamydospores visible. Conidiophores macronematous, mononematous, branched, hyaline to subhyaline. Conidiogenous cells phialidic producing macroconidia, microconidia not seen. Macroconidia crescent shaped, pointed at both ends but a flat notch at the base or slightly truncate, smooth.

Islolated from air in Mauritius.

8. *Nigrospora* Zimmermann, *Bakt. Parasitkde, Abt.* 2, 8:220, 1902

Colonies white with small shining black conidia, later brown to black. Mycelium creeping, anastomosing. Conidiophores macronematous or micronematous. Conidia solitary, spherical or broadly ellipsoidal, compressed dorsiventrally, black aseptate.

Nigrospora oryzae State of *Khuskia oryzae* Hudson, *Trans. Br. Mycol. Soc.* 46:355-360 =*N. oryzae* (Berk & Br.) Petch., *J. Indian Bot. Soc.*, 4:24, 1924

Colonies white with small shining black conidia, later brown to black, reverse brown with exudation. Mycelium septate, branched, superficial, anastomosing 3-5 µm broad. Conidiophores semi-macronematous, 2-8 µm long and 2-3 µm broad, subhyaline, conidiogenous cells monoblastic and up to 7 µm broad, subhyaline, smooth. Conidia solitary, spherical or broadly ellipsoidal, compressed dorsiventrally, black shining, 10-15 µm in diam, smooth.

Isolated from air in Mauritius.

9. Cladosporium Link ex Fries, Link *Magazin Ges Naturf. Freunder,* Berlin, 7: 37-38, 1815, Fries, *Syst. Mycol.*, 1: XLVI, Ellis, *Dema. Hypho.* 304-319, 1971.

Colonies effuse or punctiform, olivaceous grey, brown to blackish brown, floccose or hairy. Mycelium septate. Conidiophores macronematous or micronematous, straight, erect, branched or unbranched. Ramoconidia often present. Conidiogenous cells polyblastic, nodose, integrated terminal or intercalary, cicatrised scars prominent. Conidia blastic, catenate or solitary or branched in chains, variable in shape, usually cylindrical, fusiform, ovoid or spherical or subspherical with protuberant scar at both ends or at base, pale to olive brown or brown, smooth or ornamented, aseptate or septate.

Cladosporium herbarum (Pers.) Link ex S F Gray *Nat. Arr. Br.* Pl. 1:556, 1821

Colonies effuse, olive green or brown, velvety, reverse greenish black. Stroma not developed. Mycelium septate, branched, smooth, thin or thick walled, 2-4 µm broad, sometimes swollen thick cells are produced with olive green exudates. Conidiophores macronematous, mononematous, erect, straight or flexuous, geniculate, nodose, pale to olivaceous brown, smooth 200-250 µm, 3-5 µm thick, terminal or intercalary vesicular swellings 5-7 µm diam. With prominent scars. Conidia in branched chains, ellipsoidal or oblong rounded at the ends, pale to mid-olivaceous brown, smooth or finely verrucose, 5-20 µm long and 3-5 µm broad with protuberant scars at the ends.

Isolated from the air in Mauritius.

Cladosporium cladosporioides (Fresen.) de Vries, *Contribution to the Knowledge of the Genus Cladosporium* Link ex Fr. Baarn,: 57, 1952

Colonies effuse, olive green or brown, velvety; reverse greenish black. Mycelium septate, branched thin or thick-walled, smooth, 3-5 µm broad. Conidiophores macronematous, 150-300 µm long or shorter sometimes, septate, branched. Conidiogenous cells terminal, ramoconidia often present, 0-1 septate, 20-30 µm long, 2-4 µm thick, smooth or verrucose. Conidia formed in chains, aseptate, ellipsoidal or limoniform 3-10 µm long, 2-4 µm broad, pale olivaceous brown, smooth or very finely verrucose.

Isolated from air in Mauritius.

Cladosporium sphaerospermum Penz., *Michelia*, 2:473, 1882

Colonies olive green to olivaccous brown, often furrowed, reverse greenish black. Mycelium septate 2-3 µm branched, subhyaline to brown. Conidiophores macronematous, mononematous 150-300 µm long, 3-6 µm broad, septate, pale to dark brown, smooth. Conidiogenous cells terminal or intercalary with prominent scars. Ramoconidia present septa 0-2,10-25 µm long, smooth, pale brown, cylindrical, slightly broad, at the apical end with protuberant scars. Conidia globose or subglobose, 3-5 µm diam, olivaceouis brown smooth or verrucose.

Isolated from air in Mauritius.

Cladosporium oxysporum Brek. & Curt, *J. Linn. Soc.*, 10 (46): 362, 1868

Colonies effuse pale to grayish brown, cottony or loosely felted. Mycelium septate, branched 2-3 μm broad, smooth, subhyaline to light brown. Conidiophores macronematous, erect, nodose pale to mid brown, smooth 150-250 μm long, 3-5 μm broad, with terminal or intercalary nodes or vesicular swellings 5-8 μm diam. Ramoconidia present, elongated or cylindrical, 5-15 μm long 2-5 μm broad. Conidia ellipsoidal, limoniform or subspherical, subhyaline to light olivaceous brown, smooth 3-5 μm long. 2-3 μm broad 0-1 septate.

Isolated from air in Mauritius.

10. *Alternaria* Nees ex Fr. Nees, *Syst. Pil. Schwam.* 72:1816, Fries, *Syst. Mycol.*, 1:XLVI, 1821

Colonies effuse dark blackish brown to black. Mycelium septate, branched, pale to brown. Conidiophores macronematous, mononematous, simple or branched, solitary or in fascicles. Conidiogenous cells integrated, terminal or intercalary, polytrelic, sympodial or cicatrised. Conidia muriform with beak at apex and flat or rostrate at base, light to dark brown or variously pigmented in simple or branched chains.

Alternaria alternata (Fr.) Keisler, *Beih. Bot.* Zbl., 29:434, 1912 =*Torula alternata* Fr. *Syst. Mycol*, 3:500, 1821

Colonies black or grey. Mycelium septate, 2-5 μm broad, branched, subhyaline to brown. Conidiophores simple or branched 20-30 μm long, 2-5 μm broad, septate, brown. Conidiogenous cells polytretic terminal or intercalary cicatrised, sympodial. Conidia solitary or in chains often with a short conical or cylindrical beak, rostrate base, transversely or longitudinally septate, pale to brown, smooth or verrucose 15-45 μm long, 7-5 μm broad in the broadest region, beak pale to light brown.

Isolated from air in Mauritius.

11. Mucor Mich. ex Fr, Micheli *Nova pl. Gen.;* 1729, Fries, *Syst. Mycol*, 317; 1832

Mucor occurs as a saprophyte on rotting fruits, cosmopolitan in distribution, sporangiophores in clusters or solitary. Mycelium coenocytic. Sporangia with columella, containing spores.

Mucor hiemalis Wehner *Annls. Mycol.* 1:27, 1903

Colonies cottony, light brown to grayish. Mycelium coenocytic, 3-7 μm broad, aseptate, branched. Sporangiophores pin head like, very long, aseptate. Sporangia spherical to globular, light grayish 250 μm across. Spores light greenish, spherical or slightly irregular, 3-5 μm diam, smooth.

Isolated from air in Mauritius.

Mucor mucedo Mich. Ex Fries, Syst. Mycol. 3:317, 1832

Colonies cottony, brown to grey. Mycelium coenocytic 3-5 μm broad, aseptate, branched. Sporangiophores pin-head like, long aseptate. Sporangia spherical to globlular, light brown, 100-300 μm across. Spores spherical light grey to brown, 2-4ìm across, smooth.

Isolated from air in Mauritius.

12. Rhizoctonia dc ex Fr. *Syst. Mycol* 2:265, 1823

It is a sterilie stage but its relation to a basidiomycetous fungus has been established. Colonies cottony, abundant growth. Sterile mycelium, septate, subhyaline to pale brown, mycelium on maturity breaking up into pieces (arthro-conidia) or chlamydospores or elongated mycelial ropes or clumps.

Rhizoctonia crocorum (Persoon) De Candolle (Vide Carmichael et al, 1980)

Colonies cottony with abundant floppy growth, fast growing. Mycelium hyaline to subhyaline or brown 2-5 μm broad, sometimes thick walled, toruloid or forming clusters or ropes, septate, branched. No conidiophores and conidia. The hyphae breaking into arthrospores, toruloid, irregularly swollen, pale to dark brown, variable in shape and size.

Isolated from air in Mauritius.

13. Aspergillus Micheli ex Fries, Micheli, Nova Pl. Gen.,: 1729; Fries *Syst. Mycol.* 1: XLV, 1821, Subramanian, *Hypho*, PP. 506–554, 1971 Thom & Raper, *A Manual of Aspergilli*, PP. 1-373, 1945

Colonies effuse, variously coloured. Mycelium septate, branched. Sclerotia may be produced by some species. Conidiophores macronematous, mononematous with upper part swollen into vesicle of various shapes, surface of which is usually covered by phialides or short branches (metulae) in 1 or several series, terminal ones, always bear phialides. Conidiogenous cells monophialidic, variable in shape. Conidia phialidic, catenate variable in shape, always in short or long unbranched chains, aseptate.

The genus *Aspergillus* is found in all habitats. Its species are responsible for airborne diseases in human beings, rotting of foodstuffs, grains and fruits. Some species cause allergy to man, containing toxic substances and some are employed in industries for useful by-products. Many species occur in air in Mauritius.

Aspergillus niger Van Tiegh, *Annls. Sci. Nat. (Bot.), Ser.* 5, 8:240, 1867; Subramanian, *Hypho.* 534-536, 1971

Colonies effuse, blackish brown to black. Mycelium septate, branched 2-4 µm broad. Conidiophores macronematous, mononematous, very long with terminal spherical vesicle, subhyaline to brown, 30-60 µm diam. Conidiogenous cells, monophialidic, flask shaped 5-8 µm long, 2-3 µm thick, 1-1.4 µm broad at the open end. Conidia spinulose catenate, globose, brown to black, 3-5 µm in diam.

Isolated from air in Mauritius.

Aspergillus oryzae (Ahlburg in Korschelt) Cohn, *Jahreb. Schles. Gesell. Vaterl. Cultur.* (1183) 61:226, 1884; Subramanian, *Hypho.* Pp. 549-550,1971

Colonies whitish to grey, tough, felty mass developing pale yellow shades with the production of conidial areas. Mycelium submerged. Conidiophores, very long 10-15 µm broad, thin, smooth or rough, subhyaline. Conidial heads, globose to subglobose. Phialides borne directly on the visicle or on metulae, 10-12 µm long and 3-5 µm broad. Conidia in chains, pale to yellow or dark brown, spiny, 5-7 µm in diam.

Isolated from air in Mauritius.

Aspergillus flavus Link ex Fries *Syst. Mycol.* 3:386, 1832

Colonies floccose, yellowish in colour, reverse yellow to brown. Mycelium, septate, branched, 2-3 µm broad, smooth, yellowish submerged. Conidiophores, very long arising from submerged hyphae, rough walled, yellowish in colour with flask shape heads. Phialides directly borne on the head or on branches (metulae) in one series. Variable in size, 7-10 μm long and 2-3 μm broad. Conidia pyriform to globose in chains, yellowish green, 3-5 μm in diam, aseptate, spiny or ridges walled.

Isolated from air in Mauritius.

14. *Sagramala* Subramanian, *Curr. Sci.* 41: 755-761, 1972

While studying the various perfect states of *Aspergillus* and the conidial chains Subramanian (1972) created a genus viz., *Sagramala* which has bead–like conidia in chains closely held together.

Isolated from air in Mauritius.

15. Penicillium Link, *Mag. Ges. Naturf.* Freunde, Berlin, 3:16, 1809; Raper & Thom, *A Manual of Penicillia,* 875-1949; Samson et al, *Studies in mycol.* 11; 1-47,. 1976; Subramanian, *Hypho.*, pp. 555-600, 1971.

Colonies effuse, penicilloid, variously coloured, usually greenish, reverse yellowish green to different shades. Mycelium septate, branched hyaline to subhyaline, coloured or colourless. Conidiophores, erect simple or branched, penicilloid. Branches symmetric or asymmetric in one or more than one series. Phialides terminal on the branches, variable in shape and size. Conidia phialidic in simple chains.

This genus is cosmopolitan in distribution. It contains more than 200 species. Its various species have been considered in other anamorphic genera and it has teleomorph forms like *Eupenicillium, Talaromyces* etc.

Penicillium chrysogenum Thom, *U.S. Dept. Agri. Bon Anim Ind.* 118:58–60, 1910

Colonies effuse, granular, dark greenish to brown. Condidiophores penicilloid, erect branched in various orders, smooth or verrucose. Penicilli symmetric phialidic botryose or cylindrical or ampuliform with narrow neck. Conidia in chains, limoniform or spherical, greenish brown in colour. Isolated from air in Mauritius.

Penicillium terrestre **Jensen,** ***Cornell Univ. Exp. Stn. Bull. 315; 486-87, 1912***

Colonies effuse, velvety, zonate often deeply furrowed, floccose, fumiculose, yellowish grey in colour, reverse, uncoloured or pale tan. Conidiophores, erect long up to 400 µm along and 3-5 µm broad, branched, walls smooth or rough. Pencilli large, asymmetric, branches bearing phialides. Phialides compact botryose with narrow neck, 4-5 µm long and 2-3 µm broad. Conidial chains in well defined columns. Conidia aseptate, elliptical, rarely globose, 3-4 µm in diam, smooth in yellowish green mass.

Isolated from air in Mauritius.

Penicillium expansum **Link Ex Fries** ***Syst. Mycol.*** **3:407, 1832; Subramanian,** ***Hypho. pp.*** **585-586, 1971**

Colonies effuse, cottony surface, granular, yellowish green, Mycelium, septate, branched. Conidiophores, erect 100-300 µm long, 3-4 µm broad, smooth or finely rough, terminating in large penicillia. Penicilli asymmetric, fasciculate, branched once or twice. Metulae borne in verticals, conidia, aseptate, elliptical and then globose, 2-3 µm in diam, walls smooth, dull yellowish green.

Isolated from air in Mauritius.

16. ***Curvularia*****, Boedijn,** ***Bull. Jard. bot. Buitenz.,*** **III, 13(10; 120-234, 1933.** ***Curvularia lunata*** **(Wakker) Boedijn,** ***Bull. Jard. bot. Buitenz.,*** **III, 13(10; 120-234, 1933.**

Colonies effuse grey to black. Mycelium 3-5 µm broad, septate, branched. No stromata, condiosphores erect, macronematous, mononematous, pale brown, 150-250 µm, 3-5 µm broad, septate, geniculate, cicatrised. Conidiogenous cells polytrlic, integrated, terminal or intercalary. Conidia 20-40 µm long and 10-15 µm broad in the broadest region, middle cells darker than end cells which are lighter in colour than the other cells, smooth 2-3 µm septate, clavate, curved, brodly fusiform smooth with basal hilum.

Isolated from air in Mauritius.

17. Sclerotinia. Fuckel, *Symb. Mycol.* PP 330, 1870

Mycelia sterilial state. No conidial stage was observed in culture.

Isolated from air in Mauritius.

18. Unidentified Yeast–Like Cells

They are identical to across *Saccharomyces* and *Schizosaccharomyces* spp. In shape and process of fun duding.

DISCUSSION AND CONCLUSION

This paper reports on a sampling of aerial mycoflora in Mauritius. Some 200 isolates were screened and species are described. From the survey it was very clear that *Cladosporium* spp were by far the commonest. This is a pertinent observation since *Cladosporium* is one of the most potent fungus reported to cause allergy and asthma in children and adults. The species described here probably represent about 20-25% of the approximate total fungal aeroflora of Mauritius. Further studies are being effected to identify the remainder.

A special effort has been made in this paper to give the main diagnostic characteristics of each isolate. A comparison of aerial fungal isolates from Mauritius with those reported in other countries indicates the occurrence of similar or related fungi in Mauritius. This is an interesting observation since Mauritius is an outlandish island far from the other continents. There is also the interesting possibility that many different fungi from different regions might harbour the same major fungal allergens

Lack of knowledge of the identity of fungal allergens still constitutes a major obstacle for the improvement of diagnosis and therapy of allergies to mould. In view of this constraint Breitenbach et al (1997) analysed the allergens of two moulds, *Alternaria alternata* and *Cladosporim herbarum* and found that enolases were major allergens. These enolases were then cloned from cDNA libraries constructed from the vegetative cells of these two species of fungi. Sequencing of these two enolases showed very close relationship with other known fungal enolases. What is interesting is that a patient reactive to *Cladosporium* enolase reacted equally well with the enolases of *Alternaria, Saccharomyces* and *Candida.* About 50% of the sera from the patients reactive to *Cladosporium* and *Alternaria* were strongly reactive to the recombinant enolases. Similar trends have been noted by Tee et al (1994) and Szantho et al (1992). Such studies would go a long way towards elucidating the nature of fungal allergens and the therapy for their allergic reactions. A good inventory of fungi causing allergies from different parts of world is a necessary first step towards elucidating the ensuing biochemical and therapeutic studies.

Acknowlegements

The authors are grateful to Govt. of Mauritius and Dr. Shyam Bhagwant for the kind help rendered in this study. Further they take pleasure in presenting this paper in honour of Prof. C. Manoharachary, Osmania University, Hyderabad on the occasion of his getting superannuation.

References

Akiyama, K. 1997. Environmental allergens and allergic diseases. *Rinsho Byori* 45(1). 13-18.

Alonso, A, Rodriguez, S.M., Mouchian, K., albonico, J.F. and Iraneta, S.G. 1997. Interstitial pneumonitis in guinea pigs by the antigens of *Rhizopus nigricans.* I Investing Allergol. *Clin. Immunology*, 7(2), 102-109.

Backman, K.S., Roberts, M. and Patterson, R. 1995. Allergic bronchopulmonary mycosis caused by *Fusarium vasinfectum: Am. J. Respir. Crit. Case Med*., 152(4 pts), 1379-1381.

Bent, J.P. and Kuhn, F.A. 1996. Antifungal activity against fungal sinusitis organisms. *Laryngoscope* 106(11), 1331-1334.

Carmichael, J.W., Kendrick, W.B., Corners, I.L., and Sigler, L. 1980. Genera of Hyphomycetes. Univ. of Alberta pr. Edmonton, Alberta, Canada.

Chaumont, J.P. and Leger, D. 1992. Campaign against allergenic moulds in dwellings. Inhibitor properties of essentital oil of Geranium "Bourdon", citronellol, geranion and citral, *Ann. Pharm. Fr.* 50(3), 156-166.

Chhabra, A., Handa, K.K. Chakrabarti, A., Mann, S.B. and Panda, N. 1996. Allergic fungal sinusitis: clinicopathological characteristics. *Mycoses*, 39(11-12), 437-441.

Dill, I. and Niggermann, B. 1996. Domestic fungal viable propagules and sensitization in children with IgE mediated allergic diseases. Pediater. Allergy Immunol 7(31), 151-155.

Dotterud, L.K., Vorland, L.H. and Falk, E.S. 1995. Viable fungi in indoor air in homes and schools in the Sor–Varanger community during winter. *Pediatr. Allergy Immunol.* 6(4), 181-186.

Ellis, M.B. 1971. Dematiaceous Hyphomycetes, *CAB Int. U.K.*

Ellis, M.B. 1976. MoreDematiaceous Hyphomycetes, *CAB, Int. U.K.*

Fujimoto, K., Sato, A., Genma, H., Taniguchi, M., Okano, H. and Iwata, M. 1994. A case of allergic bronchopulmonary fungusis caused by several kinds of fungi. *Nippon Kyobu Shikkan Gakkai Zasshi* 32(1), 48-54.

Gambale, W, Croce, J., Costa – Manso, E., Croce, M. and Sales, M. 1993. Library fungi at the University of Sao Paulo and their relationship with respiratory allergy. *J. Investing. Allegrol. Clin. Immunol.*, 3(1), 45-50.

Gonzalez, G.M.F.J., Candau, P., Gonzlez, G.R.M.L. and Romero, F. 1992. A study of aeromycoflora of Cadiz relationship to anthropogenic activity. *J. Investig. Allergol. Clin. Immunol.* 2(4), 211-215.

Granel, C., Tapias, G., Valencia, M., Randazzo, L., Anglada, E. and Olivia, A. 1993. Allergy to *Alternaria.* Allergol, Immunopatho. (Madr), 21(1), 15-19.

Kim, Y., Suzuki, K., Kitao, S., Takagi, I. and Baba, S. 1996. Clinical review of nasal allergy due to fungus. *Acta Lorlaryngol. Suppl.* (Stockh.) 525, 105-107.

Kinsella, J.B., Bradfied, J.J., Gourley, W.K., Calhoun, K.H. and Rassekh, C.H. 1996. Allergic fungal sinusitis. *Clin. Otolaryngol* 21(5), 389-392.

Krasnick, J. Greenberge,r P.A., Roberts, M. and Patterson, R. 1995. Allergic bronchopulmonary aspergillosis serologic update for 1995. *J. Clin. Lab. Immunol.* 46(3), 137-142.

Li, C.S. and Hsu, L.Y 1997. Airborne fungus allergen in association with residential characteristics in artopic and control children in a subtropic region. Arch. Environ. Health 52 (1), 72-79.

Lim, S.H., Chew, F.T., Sim, S.M., Huang, Y.T. Goh, D.Y., Tan, H.T. and Lee, B.W. 1995. Allergens of Bipolaris species. *Asian Pac. J. Allergy Immunol.* 13(2), 101-105.

Perez – Jaffe, L.A., Lanza, D.C., Loeurner, L.A., Kennedy, D.W and Motone, K.F. 1997. *In situ* hybridization of *Aspergillus* and *Penicillium* in allergic fungal sinusitis : a rapid means of speciating fungal pathogens in tissues. *Laryngoscope* 107(2), 233-240.

Matsushima, T. 1975. Icones Fungorum Matsushima Lectarum, Kobe, Japan.

Quraishi, H.A. & Ramadan H.H. 1997. Endoscopic treatment of allergic fungal sinusitis. Otolaryngol. *Head Neck Surg.*, 117 (1), 29-34.

Shen, H.D., Lin W.L., Tsai, J.J., Liaw S.F. and Han, S.H. 1996. Allergenic components in thee different species of *Penicillium* cross reactivity among major allergens. *Clin. Exp. Allergy* 26(4), 444-451.

Smith, T.A., Lumely, K.P. and Hui, E.H. 1997. Allergy to flour and fungal amylase in bakery workers. *Occup. Med.* (Oct) 47(1), 21-24.

Subramanian, C.V. 1991 Hyphomycetes, ICAR, New Delhi.

Sutton, B.C. (1980). Coelomycetes, CAB, U.K.

Szantho, A., Osvath, P., Horvath, Z., Novak, E.K. and Kujalek, E. 1992. Study of mould allergy in asthmatic children in Hungary. *J. Investig. Allergol. Clin. Immunol* 2(2), 84-90.

Takatori, M., Shida, T., Akiyama, K. and Takatori, K. 1994. Airborne fungi during the last ten years in Sagamihara, *Arerugi* 43(1), 1-8.

Tee, R.D., Gordon, D.J., Welch, J.A. and Newman Taylor, A.J. 1993. Investigation of possible adverse allergic reactions to mycoprotein "Quorn". Clin. *Exp. Allergy,* 23(4), 257-260.

Tsushima, K., Fujimoto, K., Kubo, K. and Sekiguchi, M. 1996. Successful treatment of fungus ball in a patient with allergic bronchopulmonary aspergillosis: continuous percutaneous instillation of antifungal agents into the cavity. *Intern. Med.,* 35(9), 736-741.

Von Arx, J.A. 1974. The genera of fungi sporulating in pure culture. J. Cramer, Germany.

Verma, J., Sridhara, S., Singh, B.P. and Gangal, S.V. 1995. Studies on shared antigenic/allergic components among fungi. *Allergy,* 50(10), 811-816.

Xia, F.Y. 1992. Observations on the airborne fungi and their relationship to asthmatic patients in Shenyang Malukan area. Chung Hua Yu Fang I Hsueh Tsa Chih, 26(6), 353-356.

Frontiers in Plant Sciences (Pages 331-344)
Eds. K.G. Mukerji, K.V.B.R. Tilak, S.M. Reddy, L.V. Gangwane, P. Prakash, I.K. Kunwar

Interactions Among *Chrysosporium*, Keratinophilic and Nondermatophytic Fungi

R.K.S. Kushwaha

ABSTRACT

Seventy five keratinophilic fungi including 58 *Chrysosporium* isolates and 14 nondermatophytic fungi were screened for their antagonistic properties against *Microsporum gypseum, Aspergillus niger, Fusarium oxysporum* and *Penicillium citrinum*. Hyphal and colony interactions, antagonistic potential and influence of volatile and nonvolatile metabolites on these fungi revealed that most of the screened out fungi have inhibitory potential while few promoted the growth of test organisms.

Key words: Interactions, keratinophiles, nondermatophytes

INTRODUCTION

The global distribution of keratinophilic fungi and related dermatophytes in soil is now very well documented. These fungi mainly survive on keratinous substrate in soil of all the habitats studied so far. Development of teleomorph in soil is an additional feature for their survival through ascospore for longer period. During this period keratinophilic fungi have to tolerate influence of other neighboring geophilic fungi. Some of their antagonistic potentialities has been discussed earlier (Kushwaha, 2000).

The keratinophilic fungi inhabiting soil are affected by microbial flora of surrounding environment and also indicates antagonistic and hypo parasitic activity. Dwivedi and Garrett (1968) noticed a changed fungal

Author address: DST-FIST Sponsored Department of Botany, Christ Church College, Kanpur 208 001
E-mail: rkskushwaha@vsnl.net

species spectrum with the degree of staling caused by earlier established fungal colonies. The involvement of number of toxic metabolites has been frequently suggested to explain inhibition in growth of an organism involved in interaction in cultural conditions (Park,1964; Robinson, 1969; Upadhaya and Arora, 1980).

There are very few records of antagonistic studies between keratinophilic fungi and dermatophytes. Nigam (1987) studied the antagonistic potential of some *Chrysosporium* species. Dixit (1991) made some observations on *M. gypseum, M. fulvum, M. racemosum* and *M. vanbreuseghemii.* An attempt has been made here to study the effect of staling substances of keratinophilic fungi on some common dermatophytes and nondermatophytes

MATERIALS AND METHODS

The fungi used in this study were isolated from different soil samples by using hair bait technique. The isolated fungi were cultured on Sabourauds dextrose agar. The number shown against the isolates are their accession numbers of Germplasm Centre for Keratinophilic Fungi of the Department of Botany, Christ Church College, Kanpur.

The method described by Nigam and Kushwaha (1990) was exactly followed for studying interactions. The record of inhibition in radial growth for each of the colony was resulted after Fokkema (1973). Qualitative assessment of the antagonistic interaction for assessing interaction for assigning 'type' to the mode of interaction and reaction type was made by the criteria proposed by Johnson and Curl (1978) as follows: A-mutual intermingling; B-mutual inhibition on contact or space between colonies smaller< 2mm; C-mutual inhibition at distance of greater >2mm; D-inhibition of some organism on contact the inhibitor species continuous to grow unchanged or at a reduced rate through the inhibited colony; E-inhibition of one fungal partner at a distance the inhibitor species then continuing to grow through resulting clear zone at the inhibited colony perhaps at reduced rate.

RESULTS AND DISCUSSION

Sixty one keratinophilic fungi and 14 nondermatophytic fungi were studied for their inhibitory ability against *Microsporum gypseum, Aspergillus niger, Fusarium oxysporum* and *Penicillium citrinum.* Some staled Petridishes inoculated with *M. gypseum, A. niger, F. oxysporum* and *P. citrinum* were totally suppressed due to high antifungal activity of screened out keratinophilic fungi.

Most of the staled agar of keratinophilic fungi revealed a decrease in colonization of a dermatophyte and soil mycoflora but few keratinophilic fungi exhibited promotion of *M. gypseum* and other fungi. Maximum promotion in the growth of *F. oxysporum* was caused by *Chrysosporium* 395 (Table 1). This promotion was 3.80%, while minimum promotion in *F. oxysporum* (1.83%) was caused by *Chrysosporium* 402. *C. evolceauni* 363 showed 4.26% promotion in the growth of *M. gypseum*. *C. indicum* 360 caused maximum promotion in the growth of *F. oxysporum* and minimum by *C. indicum* 411. *C. tropicum* 352 induced maximum growth promotion (9.21%) in *M. gypseum*. Rest keratinophilic fungi did not promote the growth of *M. gypseum, A. niger, F. oxysporum* and *P. citrinum* but they depressed the growth of these fungi as compared to control.

TABLE 1 *Effect of staling products of keratinophilic fungi on the inhibition [%] of a dermatophyte and nondermatophytic fungi*

Staling contributor fungus		*M.gypseum*	*A.niger*	*F.oxysporum*	*P.citinum*
Chrysosporium sp. GPCK	349	58.55	34.59	15.24	28.16
Chrysosporium sp. GPCK	347	31.16	74.13	5.49	62.03
Chrysosporium sp. GPCK	348	81.26	71.22	16.16	52.77
Chrysosporium sp. GPCK	361	66.79	73.55	8.23	34.48
Chrysosporium sp. GPCK	374	42.1	100	10.08	49.55
Chrysosporium sp. GPCK	340	68.42	100	8.39	100
Chrysosporium sp. GPCK	355	20.7	70.35	100	23.15
Chrysosporium sp. GPCK	368	46.39	14.24	2.29	100
Chrysosporium sp. GPCK	381	100	20.64	19.21	12.14
Chrysosporium sp. GPCK	395	0.66	1.16	3.80*	35.19
Chrysosporium sp. GPCK	402	57.24	1.16	1.83*	64.81
Chrysosporium sp. GPCK	435	69.42	45.21	1.32	20.37
Chrysosporium sp. GPCK	379	26.32	61.92	11.32	100
Chrysosporium sp. GPCK	359	78.95	100	9.61	65.74
Chrysosporium sp. GPCK	391	55.61	61.63	39.02	58.81
Chrysosporium sp. GPCK	399	67.11	52.3	57.48	70.37
Chrysosporium sp. GPCK	440	66.45.	61.05	27.74	47.7
Chrysosporium sp. GPCK	429	42.11	38.66	50	60.67
Chrysosporium sp. GPCK	419	85.53	60.62	54.42	100
Chrysosporium sp. GPCK	409	100	49.42	100	57.4
Chrysosporium sp. GPCK	415	62.53	60.87	51.22	64.81
Chrysosporium sp. GPCK	387	97.36	90.12	7.93	44.81
Chrysosporium sp. GPCK	372	44.75	27.03	17.68	45.37
Chrysosporium sp. GPCK	367	75	1.16	30.34	46.78
Chrysosporium sp. GPCK	389	61.18	6.39	33.23	57.41
Chrysosporium sp. GPCK	418	42.76	46.8	12.5	25.93

(Table 1 Contd.)

(Table 1 Contd.)

Chrysosporium sp. GPCK	403	51.32	28.49	39.02	25.43
Chrysosporium sp. GPCK	390	51.32	12.21	20.12	16.67
Chrysosporium sp. GPCK	356	59.21	9.88	0.16	37.04
Chrysosporium sp. GPCK	353	37.5	5.09	0.61*	22.37
C.evolceanui GPCK	363	4.26*	5.09	0.61*	22.37
C.evolceanui GPCK	433	14.67	6.39	32.32	8.81
C.indicum GPCK	341	64.47	15.12	38.41	33.33
C.indicum GPCK	427	39.47	20.35	33.23	18.52
C.indicum GPCK	357	64.81	28.35	18.59	47.22
C.indicum GPCK	407	36.18	6.98	3.19*	23.63
C.indicum GPCK	411	17.76	4.51	8.84	21.29
C.indicum GPCK	358	57.89	7.71	3.35	50
C.indicum GPCK	360	34.21	14.48	3.35*	10.19
C.indicum GPCK	428	6.58	66.89	2.44	22.22
C.keratinophilum GPCK	342	36.18	19.92	40.55	45.37
C.lobatum GPCK	343	100	40.69	100	100
C.lobatum GPCK	416	100	100	22.57	100
C.lobatum GPCK	406	71.71	9.01	6.56	55.55
C.lobatum GPCK	398	65.13	17.01	62.8	58.33
C.queenslandicum GPCK	438	71.05	68.17	24.69	44.44
C.queenslandicum GPCK	423	64.47	18.31	4.87	72.22
C.tropicum GPCK	420	37.84	2.33	42.07	23.15
C.tropicum GPCK	346	45.74	31.39	22.87	44.44
C.tropicum GPCK	413	42.45	0.15	74.39	35.19
C.tropicum GPCK	344	100	72.67	75.46	85.19
C.tropicum GPCK	412	42.11	49.42	6.09	557.4
C.tropicum GPCK	345	72.37	52.91	3.66*	100
C.tropicum GPCK	352	9.21	33.14	1.68	29.63
C.tropicum GPCK	439	18.1	0.58	1.52	22.22
C.tropicum GPCK	392	7.75	16.43	1.83*	46.29
C.tropicum GPCK	404	13.16	13.08	1.07	59.26
C.tropicum GPCK	350	40.13	8.43	6.56	35.18
Malbranchea pulchella GPCK	351	76.32	84.3	57.93	42.59
Myceliophthora sp. GPCK	380	81.26	12.21	70.43	52.33
Trichophyton mentagrophytes GPCK	354	68.42	50.87	3.96	37.04

*Promotion in fungal growth.

M. *gypseum* = *Microsporum gypseum*; A. *niger* = *Aspergillus niger*; F. *oxysporum* = *Fusarium oxysporum*; P. *citinum* = *Pencillium citinum*

The effect of staled agars of 10 strains of *Acremonium, Cladosporium* sp., *Gliomastix* sp., *Sporotrichum* 1 and *Sporotrichum* 2 against the growth of *M. gypseum, A. niger, F. oxysporum* and *P. citrinum* was recorded (Table 2). The effect of staled agars of non keratinophilic fungi revealed less inhibition in the growth of *M. gypseum, A. niger, F. oxysporum* and *P. citrinum* as compared to control. In comparison to keratinophilic fungi, nondermatophytic fungi were not found to much inhibitory for the growth of a dermatophyte and other fungi. Total inhibition in the growth of *M. gypseum, F. oxysporum* and *P. citrinum* was caused by *Cladosporium* sp. while *Sporotrichum* 1 caused 100% inhibition in *F. oxysporum*, 39.47% in *M. gypseum*, 41.28% in *A. niger* and minimum 34.36% in *P. citrinum*. *Gliomastix* sp. caused 37.5% inhibition in *M. gypseum*, 68.6% in *A. niger*, 86.59% in *F. oxysporum* and 67.59% in *P. citrinum*. *Sporotrichum* 2 was seen to induce maximum inhibition in *M. gypseum* (76.97%), while minimum was in *F. oxysporum* (13.42%). The 81.39% and 3.78% inhibition in the growth of *A. niger* was caused by the activity of *Acremonium* 9 and 4 and 83.54 and 6.71% inhibition respectively in *F. oxysporum* was caused by *Acremonium* 6 and 4 respectively. *Acremonium* 6 and 1 caused 62.93% and 22.22% inhibition in the growth of *P. citrinum*. None of these nondermatophytic fungi showed promotion in the growth of *M. gypseum, A. niger, F. oxysporum* and *P. citrinum*.

TABLE 2 *Effect of staling products of some nondermatophytic fungi on dermatophyte and nondermatophytic fungi*

Staling contributor fungus	*M.gypseum*	*A.niger*	*F.oxysporum*	*P.citrinum*
Acremonium sp.1	47.37	6.98	8.54	22.22
Acremonium sp.2	37.5	39.98	38.11	55.11
Acremonium sp.3	34.87	29.65	12.19	34.74
Acremonium sp.4	57.58	3.78	6.71	37.52
Acremonium sp.5	51.32	62.21	26.83	25.92
Acremonium sp.6	71.05	63.37	83.54	62.93
Acremonium sp.7	78.08	13.23	48.94	62.04
Acremonium sp.8	63.82	79.94	65.85	62.04
Acremonium sp.9	59.87	81.39	57.62	60.18
Acremonium sp.10	52.95	75.87	21.65	42.59
Cladosporium sp.	100	76.74	100	100
Gliomastix sp.	37.5	68.6	86.59	67.59
Sporotrichum sp.1	39.47	41.28	100	34.26
Sporotrichum sp.2	76.97	64.48	13.43	31.03

A total inhibition in growth of some fungi on staled agar plates of *Chrysosporium* 381,409 was recorded. Staled agar plates of *C. lobatum* 343 and 416 caused hundred per cent inhibition in growth of *M. gypseum, F.*

oxysporum and *P. citrinum*. Some keratinophilic fungi were found to yield toxin or antibiotic like substances. The antagonistic capabilities of *C. tropicum* has also been demonstrated by Nigam (1987). Dixit (1991) also recorded the effect of staled substances of *Microsporum* spp. against the growth of *P. citrinum* and showed their inhibitory nature.

Promotion in growth of *F. oxysporum* was caused by staled agar of two different strains of *Chrysosporium* spp. 395 and 402, while promotion in *M. gypseum* was caused by *C. evoleceanui* 363. *C. evolceanui* 363, *C. indicum* 411 caused promotion in growth of *F. oxysporum*. Promotion in growth of *M. gypseum* and *F. oxysprum* was caused by *C. tropicum* 352 and 392 respectively. No promotion was found in the staled agar substances of nondermatophytic fungi when compared to control.

The present investigation has revealed that the effect of staled substances grown on agar medium showed inhibitory nature of most keratinophilic and nondermatophytic fungi against a dermatophyte, *A. niger, F. oxysporum* and *P. citrinum*.

Colony and Hyphal Interactions

The result presented here showed that different kind of interactions may be possible between keratinophilic and related dermatophytes on agar. The variable antagonistic inhibitory zone developed by eight test fungi against *C. indicum* 400 on agar may be due to different diffusible mycostatic staling substances produced by them. The antagonistic factor present may be volatile or non volatile metabolites (Webber and Hedger, 1986), antifungal toxins (Pachneri and Dix, 1980), substances altering in the pH of the medium (Bartman et al 1981).

The antagonistic potential of *C. indicum* 427 against 8 *Chrysosporium* species was shown in Table (3) where it could not show more than 10 AP against *C. tropicum* 344 and it also exhibited mutual inhibition in 5 cases. *C. tropicum* 352 could show positive AP against 5 *Chrysosporium* species while it was inhibited by *C. indicum* 341 and *C. tropicum* 344 and mutual inhibition was in 5 cases (Table 4). *M. fulvum* 414 when tested against 8 test fungi, it was inhibited by all except *C. tropicum* 345 and 346. No inhibition zone was recorded in 3 cases (Table 5). *M. gypseum* 385 was also inhibited by all except *C. tropicum* 344 and *Chrysosporium* species 340 developing inhibition zone of 1.5-3 mm and mutual inhibition in 6 cases was seen (Table 6). *T. rubrum* 400 could induce AP against *C. tropicum* 344 and 404 and rest 6 fungi inhibited this fungus (Table 7). *T. mentagrophytes* 398 was inhibited by *C. tropicum* 404 only and it could show AP against 6 test fungi. During this interference several micromorphological changes in test as well as experimental fungi were observed.

TABLE 3 *Interactions between Chrysosporium spp (T) and C. indicum 427 (E)*

Test Fungus	*Quantitative Result*				*Qualitative Result*		
	Inhibition %		*AP*	*Inhibition Zone*	*Inhibition Type*	*Micromorphological changes %*	
	E	T				E	T
C.indicum 341	30.76	4.45	IN	3.0	C	SW,C,D,P	C,SW,CO
C.keratinophilum 342	28.8	23.18	IN	2.0	B	P,SW,L,G,D	P,D,L
C.lobatum 343	27.77	18.18	IN	2.5	C	D,SW,P,L	C,D
C.tropicum 344	14.51	24.59	10.08	1.5	B	D,P,C,SW	D,SW,CH,CO,C
C.tropicum 345	27.14	13.63	IN	1.8	B	SW,CO,CH,G	P,C
C.tropicum 346	43.47	11.11	IN	3.0	C	CO,L,D,P,SW	D,C
C.tropicum 404	39.68	10.41	IN	1.5	B	P,SW,D,L,C	SW,C
Chrysosporium sp 340	21.31	16.66	IN	2.0	B	CH,SW,L	P,CO

TABLE 4 *Interactions between Chrysosporium spp (T) and C. tropicum 352 (E)*

Test Fungus	*Quantitative Result*				*Qualitative Result*		
	Inhibition %		*AP*	*Inhibition Zone*	*Inhibition Type*	*Micromorphological changes %*	
	E	T				E	T
C.indicum 341	28.82	15.68	IN	2.5	C	P,C,D,SW	CH,CO,C
C.keratinophilum 342	5.5	18.03	12.63	1.8	B	SW,C,CH,P	G,CO,P
C.lobatum 343	3.03	18.18	15.15	2.5	C	C,CH,D,P,SW	C,CH,SW
C.tropicum 344	6.25	3.03	IN	3	C	CH,SW,C,P,D	C,CO,P,SW
C.tropicum 345	5.71	5.71	0	1.6	B	D,SW,CH,CO	C
C.tropicum 346	2.7	13.33	10.63	2	B	C,CH,D,SW	C,SW,CH
C.tropicum 404	7.31	13.04	5.73	0	A	SW,CO,D,P	C,SW
Chrysosporium sp 340	6.97	10.63	3.66	2	B	D,P,G,CO	C,P,CO

TABLE 5 *Interactions between Chrysosporium spp (T) and M.fulvum 414 (E)*

Test Fungus	Quantitative Result			Qualitative Result			
	Inhibition %		AP	Inhibition Zone	Inhibition Type	Micromorphological changes %	
	E	T				E	T
C.indicum 341	21.17	12.12	IN	0	D	P,D,L,C	C,G,SW
C.keratinophilum 342	20	2.08	IN	0	D	P,CO,SW,L,D,G	D,CO,SW,C
C.lobatum 343	47.82	33.33	IN	2	B	P,CH,L,G,SW,C	SW,D
C.tropicum 344	32.78	11.59	IN	1.3	B	LL,D,CH,CO	D,P,SW
C.tropicum 345	15.25	20.45	5.28	1.5	B	P,L,C,SW,D	D
C.tropicum 346	7.69	23.07	15.38	2	B	P,CH,C,L	SW
C.tropicum 404	23.43	21.31	IN	2	B	P,CH,L,C,SW	C,SW
Chrysosporium sp 340	16.17	16	IN	0	D	SW,CO,D,P,CH	CH,CO

TABLE 6 *Interactions between Chrysosporium spp (T) and M.gypseum 385 (E)*

Test Fungus	Quantitative Result			Qualitative Result			
	Inhibition %		AP	Inhibition Zone	Inhibition Type	Micromorphological changes %	
	E	T				E	T
C.indicum 341	22.02	8.5	IN	3	C	SW,CH,G,D,P	SW,P,F,CH
C.keratinophilum 342	25.8	6	IN	1.8	B	C,SW,D,CH	SW,D
C.lobatum 343	16.52	9.52	IN	2	B	C,P,G,SW	C,SW
C.tropicum 344	4.76	17.64	12.88	3	C	L,P,CO,D,CH,G	CH,C,G
C.tropicum 345	11.32	5.88	IN	1.7	B	C,P,D,SW	C,SW
C.tropicum 346	22.22	2.94	IN	1.5	B	L,SWD,P,CH,C	SW,C,CH
C.tropicum 404	10.52	7.64	IN	2	B	D,CO,P	SW,CH,C
Chrysosporium sp 340	6.25	12.19	5.94	2	B	C,P,SW,D,CH	C,SW,P

TABLE 7 *Interactions between Chrysosporium spp (T) and T.rubrum 400 (E)*

Test Fungus	Quantitative Result			Qualitative Result			
	Inhibition %		AP	Inhibition Zone	Inhibition Type	Micromorphological changes %	
	E	T				E	T
C.indicum 341	19.67	9.52	IN	3	C	C,D,CO,CH	C
C.keratinophilum 342	16.66	13.14	IN	1.5	B	G,P,C,D	D,SW,C
C.lobatum 343	24.19	14.92	IN	0	A	L,SW,C,CO	C,CH
C.tropicum 344	3.92	12.76	8.84	2	B	L,CO,P,D,SW,CH	SW,CH,D,C
C.tropicum 345	33.87	8.7	IN	1.8	B	L,CO,D,SW	SW,CH
C.tropicum 346	19.23	3.55	IN	2	B	SW,G,L,D,CH	CH,C
C.tropicum 404	24.56	29.78	5.22	1.5	B	L,CH,C,SW	D,P,SW
Chrysosporium sp 340	19.23	15.55	IN	0	A	L,CO,D	C

TABLE 8 *Interactions between Chrysosporium spp (T) and T.mentagrophytes 398 (E)*

Test Fungus	*Quantitative Result*				*Qualitative Result*		
	Inhibition %		*AP*	*Inhibition Zone*	*Inhibition Type*	*Micromorphological changes %*	
	E	T				E	T
C.indicum 341	43.13	43.83	0.7	1.5	B	P,C,L,CH,D,G	C,CH
C.keratinophilum 342	11.8	31.99	17.11	2	B	SW,CH,G,D,P	SW
C.lobatum 343	16.32	30.33	13.98	0	D	L,P,SW,D	D,SW
C.tropicum 344	15.55	43.83	28.28	3	C	G	C,CH
C.tropicum 345	30.95	37.77	6.82	2	B	P,G	L,CH
C.tropicum 346	19.14	38.57	19.43	0	D	C,L,SW,D	SW,D
C.tropicum 404	21.42	20.45	IN	1.7	B	CH,C,SW,G	CH
Chrysosporium sp 340	13.33	27.27	13.94	0	D	L,CO,G	CO

E=Experimental fungus; T=Tst fungus; C=Curling;CH=Chlamydospore formation;SW=Swelling; IN=Inhibition of experimental fungus;
Type of inhibition=Detail of types are given in the text.Data are means of three replicates.

The overall sequence of antagonistic potentialities of these fungi are as follows. *T. mentagrophytes*>*C. tropicum*> *M. fulvum*> *M. gypseum*>*C. indicum*>*T. rubrum*. The inhibitory substances produced by *T. mentagrophytes* seems to be most important as *C. tropicum* a common keratin colonizer overgrown and inhibited by this fungus.

Nonvolatile Metabolites

The involvement of toxic metabolites or staling products or antibiotics has been frequently suggested to explain inhibition in radial growth of organism involved in the interaction in cultural conditions. Direct evidence in support of antibiotics production is often lacking, but supporting informations on their in vitro production has led many workers to believe this as the main cause of antagonism. A number of evidences are available that support secretion of antibiotics and toxic antimicrobial non volatile products by number of fungi in laboratory conditions (Brian, 1960; Park, 1964; Robinson, 1969; Upadhya and Arora, 1980).

The data presented here showed that culture filtrate of 8 keratinophilic fungi tested caused a vide range of inhibition in keratinophilic fungi and dermatophytes. Filtrate of all the 8 keratinophililc fungi tested strongly inhibited *C. tropicum, M. fulvum, M. gypseum* and *T . rubrum* (Table 9). A slight inhibition in *C. indicum* was recorded by antifungal substances in the filtrates of *C. keratinophilum* 342 and *C. indicum* 341. Similarly, *T mentagrophytes* was slightly inhibited by *C. indicum* 341, *C. keratinophilum* 342 and *C. tropicum* 344. It is most probably due to poor secretion of antifungal substances in filtrates of keratinophilic fungi.

Promotion in growth of *C. indicum* 427 due to six keratinophilic fungi. *T. mentagrophytes* 398 and *T. rubrum* 400 by *C. tropicum* 345 is noteworthy. Promotion of dermatophytes and keratinophilic fungi during antagonism has been observed by Nigam (1987) and Dixit (1991).

The survival of different fungi in microbial culture filtrates also depends on the concentration of unused residual nutrients and nature of toxic mycostatic products produced by microorganisms (Dwivedi and Garrett, 1968). Park also supported the view that competitive survival of any organism depends on their tolerance to antagonistic substance produced during microbial interactions. However, no report on micro morphological changes in these fungi due to the effect of nonvolatile metabolites has been given earlier. The changes like penetration, lyses, chlamydospore formation, swelling of hyphae were found frequently in *M. gypseum* with antagonistic effect of filtrates of most of the keratinophilic fungi but some other changes such as zonation, raised centre and secretion in *C. tropicum* and *M. fulvum* by some keratinophilic fungi are noteworthy as these appear to change the

TABLE 9 *Effect of cullture filtrates produced by Chrysosporium spp on some other keratinophilic and dermatophytic fungi*

Metabolite contributor	*C.indicum 427*		*C.tropicum 352*		*M.fulvum 414*		*M.gypseum 385*		*T.mentagrophytes 398*		*T.rubrum 400*	
	IN	*PN*	*IN*	*PN*	*IN*	*PN*	*IN*	*PM*	*IN*	*PN*	*IN*	*PN*
C.indicum 341	10.34	..	13.27	..	34.93	..	56.7	..	26.93	..	20.34	..
C.keratinophilum 342	10.02	..	9.15	..	32.75	..	19.26	..	33.86	..	30.5	..
C.lobatum 343	..	7.97	57.08	..	13.52	..	46.06	..	..	3.07	..	..
C.tropicum 344	..	0.71	23	..	0.86	..	44	..	..	..	..	..
C.tropicum 345	..	5.54	17.12	..	6.97	..	35.05	..	26.93	6.16	32.8	1.5
C.tropicum 346	..	1.39	58.41	..	7.86	..	29.22	..	..	..	..	..
C.tropicum 404	..	2.08	61.66	..	3.03	..	57.4	..	..	4.62	..	..
Chrysosporium sp 340	..	1.39	5.75	..	17.03	..	18.56	..	..	6.53	10.4	..

Inhibition, Promotion in %; Data are means of three replicates.

TABLE 10 *Effect of volatiles metabolites produced by Chrysosporium spp on some other keratinophilic and dermatophytic fungi*

Metabolite contributor	*C.indicum 427*		*C.tropicum 352*		*M.fulvum 414*		*M.gypseum 385*		*T.mentagrophytes 398*		*T.rubrum 400*	
	IN	*PN*	*IN*	*PN*	*IN*	*PN*	*IN*	*PM*	*IN*	*PN*	*IN*	*PN*
C.indicum 341	5.08	..	2.87	..	5.17	..	4.88	..	..	5.59	5.78	..
C.keratinophilum 342	1.96	..	5.74	..	4.99	..	10.11	..	..	9.19	..	1.24
C.lobatum 343	2.11	..	1.07	..	8.8	..	7.48	..	..	6.79	0.39	..
C.tropicum 344	10.58	..	1.58	..	1.54	..	.5.24	..	11.99	..	..	1.67
C.tropicum 345	14.82	..	2.33	..	5.36	..	7.48	..	31.19	..	1.59	..
C.tropicum 346	12.71	..	4.32	..	..	1.15	2.99	..	..	11.18	..	1.23
C.tropicum 404	11.65	..	1.7	..	4.99	..	9.73	..	40	..	9.46	..
Chrysosporium sp 340	13.12	..	2.87	..	3.45	..	9.37	..	..	0.79	..	1.24

Inhibition, Promotion in %; Data are means of three replicates.

colony morphology. Thus this study provides an understanding of the fungal succession and demonstrates their competitive ability.

Volatile Metabolites

The data presented here have shown depression and acceleration in growth of keratinophilic fungi and dermatophytes. The volatiles of *C. indicum* 427, *C. tropicum* 352 and *M. gypseum* 385 were inhibitory for all the fungi. Volatiles of *M. fulvum* 414 were inhibitory for 7, of *T. mentagrophytes* 398 for 3, of *T. rubrum* for 4 fungi (Table 10). *T. mentagrophytes* 398 and *T. rubrum* 400 promoted growth of 5 and 4 fungi respectively. Inhibition and promotion in growth of a range of fungi by fungal volatile metabolites have been presented in several other studies but their effects on keratinophilic fungi and dermatophytes have been reported in only studies of Nigam (1987) and Dixit (1991).

The stimulation and inhibition of fungus probably depend on the assay fungi being used as each fungus did not necessarily act in the identical manner in depressing or promoting test fungi. The antagonistic nature of *M. fulvum, T. rubrum* and *T. mentagrophytes* against some *Chrysosporium* is reported here.

It was concluded that all the test fungi caused inhibition in growth of *C. indicum, C. tropicum, M. gypseum. M. fulvum* 414 was inhibited by all the fungi except *C. tropicum* 346 which promoted the growth of this fungus. *C. tropicum* 344, 345 and 404 were inhibitory for *T. mentagrophytes* while *C. indicum* 341, *C. keratinophilum* 342, *C. lobatum* 343 and *C. tropicum* 346 and 340 promoted the growth.

Acknowledgements

This work was financially supported by UGC. Special thanks are due to Professor C. Manoharachary for the constant encouragement and appreciation.

References

Bertman, C.D., Doerfler, D.L., Bird, B.A., Peace, A.T. and Campbell, I.M. 1981. Mycophenolic acid production by *Penicillium brevicompaetum* on solid media. *Appl. Environ. Microbiol.* **41**: 729-736.

Brian, P.W. 1960. Antagonistic and competitive mechanism limiting survival and activity of fungi in soil. In: *Ecology of soil fungi.* (Eds. Parkinson, D. and Waid, J.S.). 115-129. Liverpool University.

Dixit, A.K. 1991. Occurrence and antagonistic studies of keratinophilic fungi with special reference to *Microsporum* species. *Ph.D. Thesis*. Kanpur University.

Dwivedi, R.S. and Garrett, S.D. 1968. Fungal competition on agar plate colonization from soil inocula. *Trans. Br. Mycol. Soc.* **55**: 999-1000.

Fokkema, N.J. 1973. The role of saprophytic fungi against *Drechslera sporokiniana* and on agar plates and on rye leaves with pollen. *Physiol. Pl. PathoL.* **31**: 195-205.

Johnson, L.F. and Curl, E.A. 1978. Methods of research on the ecology of soil borne plant pathogens. Bugess Publishing Co. Minnesota U.S.A.

Kushwaha, R.K.S. 2000. The genus *Chrysosporium*, its physiology and biotechnological potential. In: *Biology of dermatophytes and other keratinophilic fungi*. [Eds. Kushwaha, R.K.S. and Guarro, J.].66-76. Revista Ibero Americana Micologia, Spain.

Nigam, **N.** 1987. Studies on *Chrysosporium* and allied fungi from soil. *Ph.D. Thesis*. Kanpur University.

Nigam, N. and Kushwaha, R.K.S 1990. Hyphal interference among *Chrysosporium tropicum*, keratinophilic fungi and saprophytic fungi. *Trans. Br. Mycol. Soc.* **31**: 399-404.

Pachneri, A. and Dix, N.J. 1980. Production of toxins and wall degrading enzymes by Gliocladium roseum. *Trans. Br. Mycol. Soc.* **74**: 561-566.

Park, D. 1964. Some properties of a staling substance from *Fusarium oxysporium. Trans. Br. Mycol. Soc.* **47**: 71-79.

Robinson, P.M. and Garrett, M.K. 1969. Identification of volatile sporostatic factors from cultures of *Fusarium oxysporum. Trans. Br. Mycol. Soc.* **52**: 293-299.

Upadhaya, R.K. and Arora, D.K. 1980. Role of fungal staling products in interspecific competition among phylloplane fungi. *Experientia* **36**: 185-186.

Webber, J.F. and Hedger, J.N. 1986. Comparision of interactions between *Ceratocystis ulmi* and Elm bark saprobes in vitro. *Trans. Br. Mycol. Soc.* **86**: 93-101.

SECTION 3

MICROBIAL PHYSIOLOGY

Frontiers in Plant Sciences (Pages 347-354)
Eds. K.G. Mukerji, K.V.B.R. Tilak, S.M. Reddy, L.V. Gangwane, P. Prakash, I.K. Kunwar

Role of Type III Protein Secretion in *Rhizobium*-legume *Symbiosis*

K. Annapurna[1] and P. Prakash[2]

ABSTRACT

Type III secretion systems present in plant and animal pathogenic bacteria have found their homologues in symbiotic bacterial members of Rhizobiaceae. Considered as a machine used by bacterial pathogens to deliver virulence factors into their hosts on contact, this elegant protein secretion system may play a role in symbiotic interactions between *Rhizobium* and their legume hosts.

INTRODUCTION

Soil bacteria of the genera *Azorhizobium, Bradyrhizobium, Mesorhizobium* and *Rhizobium* (collectively called rhizobia) are prokaryotic symbionts that induce the formation of nodules on leguminous plant roots. Within the nodules bacteria differentiate into bacteroids that reduce atmospheric nitrogen to ammonia. In exchange for this assimilated nitrogen, plants supply carbohydrates to the bacterium. Nodule formation requires extensive molecular communication between the prokaryotic and eukaryotic partners. Plant root-produced flavonoids, *via* bacterial activator NodD proteins, induce the transcription of nodulation *(nod)* genes. Enzymes encoded by these genes are involved in the synthesis of lipochitooligo-saccharides (Nod factors) that initiate nodule development and bacterial entry (Broughton *et al* 2000; Downie and Walker 1999; Long, 2001; Perret *et al.,*2000). Besides these two key components for nodule formation, there are many other determinants that influence the extent of

Author's address: [1]Division of Microbiology, IARI, New Delhi 110 012, India
[2]UGC, Bahadur Shah Zafar Marg, New Delhi 110 002, India

symbiosis. Type III secretion systems (TTSSs) are one such element in rhizobia.

A number of Gram-negative bacterial plant and animal pathogens use type III secretion systems to inject virulence effector proteins into host cells (Hueck 1998; Galan and Collmer, 1999). These pathogens include the animal pathogens *Yersinia* spp. *Salmonella typhimurium, Shigella flexneri, Escherichia coli* and *Pseudomonas aeruginosa* and the plant pathogens *Pseudomonas syringae, Erwinia amylovora, Xanthomonas campestris* and *Ralstonia soldancearum,* TTSSs have also been recently identified in insect pathogens and eukaryotic-associated bacteria (Dale *et al.,* 2001; Marie *et al.,* 2001). Type III protein secreting systems which is employed to deliver bacterial effector proteins into the host cells has several distinguishing features: (1) the TTSS is activated by contact with the host cells *in vivo,* (2) the secretion signal is localized in the 5′ region of the mRNA or at the N-terminal region of the secreted protein, (3) the secretion of TTSS proteins requires the involvement of chaperones and (4) the association of filamentous appendages with TTSS. Interestingly certain animal pathogens have two distinct TTSSs, whereas plant pathogens, excluding the flagellar TTSS, appear to have one (Shea *et al.,* 1996; Haller *et al.,* 2000). TTSSs are composed of several proteins that are associated with the inner and outer bacterial membranes.

Upon contact with host cells, proteins are secreted via TTSS membrane channels either into the extra-cellular medium or into the eukaryotic cytoplasm where they subvert the functioning of the aggressed cell (Cernelis, 2000; He SY, 1998). Since TTSSs was thought to be an exclusive phenomena of pathogens, its discovery in *Rhizobium* came as a surprise to legume researchers. This article reviews the knowledge of this machinery in rhizobial systems.

TYPE III SECRETION GENES IN RHIZOBIAL STRAINS

A long recognized example of mutualistic relationship between bacteria and plants is the symbiosis between *Rhizobium* and legume, and it is mediated by TTSS. A complete rhizobial TTSS was first identified after sequencing the symbiotic plasmid of NGR234 (Freiberg *et al.,* 1997). Other TTSSs were found by sequencing of the genome of *Mesorhizobium loti* strain MAF303099 (Kaneko *et al.,* 2000) and the symbiotic island (a 410kb region containing nearly all of the known nodulation genes) of *Bradyrhizobium japonicum* USDA110 (Gottfert *et al.,* 2001), *nol* locus (cultivar specificity locus) *of Sinorhizobium fredii* USDA257 (Meinhardt *et al*1993) and HH103 (de Lyra *et al.,* 2000), and symbiotic plasmid p42d of *Rhizobium etli* strain CFN42 (Victor *et al.,* 2003). TTSS homologs have been identified in *B.elkanii* also.

Interestingly, a gene homologous with an elicitor of the hypersensitive response in plants, *hrpW,* is exclusively present in p42d *of R.etli.* This gene might form an operon with *pcrD,* which encodes a calcium-binding membrane protein that is also part of the type-III secretion system. No type III secreting systems have been identified in *Sinorhizobium meliloti, M.loti* ICMP3153, or in *Rhizobium leguminosarum.*

Rhizobial TTSS clusters contain 10 open reading frames (ORFs) that are homologous to *ysc* and *hrc* genes (Huek 1998; Bogdanove *et al.,* 1996). Ysc and Hrc proteins from *Yersinia* species and plant bacterial pathogens, respectively are absolutely required for the secretion of virulence factors from these strains. To differentiate symbiotic type III secretion system from pathogenic ones, the *hrc* homologues were proposed to be called *rhc* for *Rhizobium* conserved (Viprey *et al.,* 1998). The letter designating the gene identity was maintained. Thus, RhcN is homologous to YscN of *Yersinia* species and HrcN of plant pathogens.

The Hrc proteins enabling effector transport across the bacterial envelop are broadly conserved in the TTSSs of plant and animal pathogens and in the flagellar biogenesis system. In contrast, extracellular components of the TTSS machinery are more specialized and variable because of their interactions with host cell barriers. However, these extracellular proteins appear to perform similar functions in all phytopathogens and can be divided into three classes: Hrp pilus subunits, harpins, and translocator factors (Buttner and Bonas 2002). All phytopathogens appear to produce a Hrp pilus, which serves as a conduit for effector delivery as recently reviewed (He SY and Jin 2003). *Sinorhizobium fredii* USDA257 also produces surface appendages in a NodD1 and flavonoid dependent manner. These needle-like surface appendages resemble the Hrp pili of *P. syringae* strain DC3000. Five extracellular proteins secreted by *S. fredii* USDA257 is mediated by TTSS. However, direct evidence for the possibility of any one of them being secreted through the pili needs more evidence (Krishnan *et al.,* 1995; Krishnan, 2002; Annapurna and Krishnan 2003). Recent study by Krishnan et al (2003) have shown that extracellular proteins called Nops (nodulation outer proteins) involved in soybean cultivar-specific nodulation are associated with pilus-like surface appendages and exported by a type III protein secretion system in *S. fredii* USDA257. We still do not have evidence for any of the other rhizobia with TTSS sequences showing protein translocation through pili. Unlike the plant pathogenic bacteria where at least 4 *Oavr* genes and proteins have been identified which are responsible for host-pathogen specificity, very few TTSS secreted proteins have been identified in symbiotic bacteria. For example, the TTSS in NGR234 secretes NopX and Y4xL (Viprey *et al* 1998). Similarly not much is known about the translocation of TTSS effector proteins into host cells by accessory proteins called translocators. These have been extensively characterized in animal pathogens, not so well in plant pathogens, and almost nothing is known about them in rhizobia.

In rhizobia the genes encoding type III secretion machines are clustered within regions of 35-47 kilobases. Three groups of genes appear to be present and conserved in order in all the strains for which complete sequence is available. The three groups are *rhcC*1-*rhcU,* y4yQ-y4yS and y4xI-y4xK. Almost 98% homology is seen between strains NGR234 and *S.fredii* USDA257 TTSS sequences. Nucleotide sequence homology between NGR234 and *M. loti* MAF303099 and *B. japonicum* USDA110 is 75% and 68% respectively. Apart from these three gene clusters, some more ORFs have been observed in few strains only. The TTSS cluster of *S. fredii* USFDA257 contains 27 ORFs out of which 10 code for structural components of TTSS. However, we do not know whether these ORFs have any role in the *legume-Rhizobium* symbiosis. The presence of a type III secretion system does not obviously correlate with the host range of the symbiotic bacterium. *S. fredii* and *Bradyrhizobium* strains are all soybean symbionts while NGR234 induces the formation of non nitrogen fixing nodules on soybean.

FUNCTIONING OF TTSS

The TTSS genes of pathogenic bacteria are expressed in response to environmental factors that usually correspond to the conditions encountered during infection of a host (Hueck 1998). Extracellular proteins have been implicated in regulating legume nodulation. For example NodO of *R. leguminosarum* bv. *vicae* has been shown to play a role in extending the host range of this strain (Economou *et al.,* 1994; Sutton *et al.,* 1994). A calcium binding protein, calsymin of *R. etli* has been implicated in bacteroid development and nitrogen fixation (Xi *et al.,* 2000). Recent study has shown that some of the extracellular proteins secreted by TTSS are involved in regulating the nodulation efficiency in a host-dependent manner (Viprey *et al.,* 1998). Their study showed TTSS mutants of NGR234 (NGR rhcN and NGR y4xI) exhibit increased nodulation on *Pachyrhizus tuberosus* and decreased nodulation on *Thephrosia vogelii.* In NGR 234, transcription of TTSS-related genes requires the presence of flavonoids and two bacterial regulators, NodDl and y4xI. NodDl binds to consensus promoter regions called *nod-boxes* and, after exposure to flavonoids activates the transcription of the downstream genes (Marie *et al.,* 2001). Only one *nod-box* has been identified in the TTSS cluster upstream of y4xI. Y 4xI encodes a putative transcriptional activator belonging to the two-component regulatory family. In NGR234 its transcription requires NodDl and flavonoids (Viprey *et al.,* 1998; Perret *et al.,* 1999) similar to *S. fredii* USDA257 (Krishnan *et al.,* 1995;

Krishnan 2002). Plant flavonoids that activate the transcription of the *nod* genes of USDA257 also stimulate the production of novel filamentous appendages known as pili (Krishnan *et al.,* 2003).

The general model working in all rhizobia having TTSS is something like this: in the presence of flavonoids, NodD1 activates y4xI transcription, which in turn (directly or directly), activates the TTSS genes. In all the fully sequenced rhizobial TTSS industers, a potential *nod-box* consensus sequence is present upstream of y4xI; thus, a similar mechanism could be functioning in these strains. However, TTSS genes do not appear to be expressed through out the symbiosis. In USDA257 strain of *S. fredii,* one of the TTSS protein NopX was detected in the infection threads and was not present in the mature nodule of soybean (Krishnan 2002). Expression analysis indicates that most of the TTSS genes are either not transcribed or transcribed at low levels in nodules of *Vigna unguiculata* (Perret *et al.,* 1999). Undisputed functionality of TTSSs have been demonstrated in three rhizobial systems: NGR234, *S.fredii* HH103 and *S.fredii* USDA257 (Table 1). It was clearly established that this cluster of genes influence the nodulation efficiency in these strains. Nodulation efficiency of the wild-type strains was compared to TTSS mutants. Interestingly, abolition of protein secretion was found to affect nodule formation and functioning in different ways, ranging from no effect to a reduction or an increase in nodule number (Marie *et al* 2001). Mutations in *rhc*N, *rhc*J, *rhc*C and *tts*I altered the ability of USDA257 to form nodules on *Glycine max* and *Macroptilium atropurpureum* (Krishnan *et al* 2003). TTSS secreted proteins of symbiotic bacteria may elicit different response on their host legumes. Thus, on a particular host,

TABLE 1 *Symbiotic phenotypes of strains mutated in TTSS gene*

Rhizobial strain	No effect	Positive effect	Negative effect
NGR234	*G.max* cv. McCall(D) *G.max* cv. Peking (D) *L.leucocephala* (I) *Lotus japonicus*	*Flemingia congesta* (D) *T.vogelii* (I) *V.unguiculata* (D)	*C.juncea* (i) $\{Fix \rightarrow Fix^{+}\}$ *Pachyrhizus tuberosus* (D)
HH103 *S.fredii*	*Cajanus cajan* (D) *Crotolaria juncia* (I) *V.unguiculata* (D)	G *max.* cv. Williams (D) (R)	*Erythrina variegata* (D) $\{Fix^{-} \rightarrow Fix^{+}\}$
USFOA257 *S.fredii*	*G.max* cv Peking		*Erythrina species* (D) $\{Fix^{-} \rightarrow Fix^{+}\}$ *G.max* cv. McCall (D) $\{Fix^{-} \rightarrow Fix^{+}\}$

D: determinate nodule; I: indeterminate nodule; R: reduction in competetiveness; $\{Fix^{-} \rightarrow Fix^{+}\}$: conversion of pseudonodules to nitrogen fixing nodules.

the TTSS-secreted proteins may trigger plant defense responses that in turn may exert a negative influence on nodulation. In contrast TTSS secreted proteins on another host may negate or avoid the plant defense response, and may have a positive effect on nodulation.

The *hrp* systems (for hypersensitive reaction and, pathogenicity) of *P. syringae* and *X. campestris* have been implicated in the suppression of basal defenses long before these systems were shown to deliver effector proteins. As of yet, TTSS effectors that function as defense suppressors have not been described for any bacterial phytopathogens other than *P. syringae*. However, NolP from *Rhizobium* sp. NGR234 was shown recently to suppress induction of PR genes in tobacco, suggesting that the role of TTSS in symbiotic rhizobia may be suppression (Bartsev *et* al., 2004).

Conclusions

TTSS genes have been identified in a number of rhizobia and their role in symbiosis is being demonstrated. Protein secretion by this apparatus seem to affect nodulation in varying degrees in different host legumes. Possibly different flavonoids which modulate the expression levels of the secreted proteins produced by different host may be one of the reasons. Alternatively, the secreted protein may be recognized in varying degrees by the host. Unlike plant pathogenic bacteria there is no evidence so far, that *Rhizobium* secreted Nops are injected into the host cytoplasm. Although we do not fully understand the functioning of Nops, they seem to be targeted to the right location and appear at appropriate time to have some role in mediating symbiotic interactions with the host legume.

References

Annapurna, K. and Krishnan H.B. 2003. Molecular aspects of soybean cultivar specific nodulation by *Sinorhizobium fredii* USDA257. *Ind. J. Exptl. Biol.* 41: 1114-1123.

Bartsev, A.V., Deakin, W.J., Boukli, N.M., McAlvin, C.B., Stacey, G et al. 2004. NopL, an effector protein of *Rhizobium* sp. NGR234, thwarts activation of plant defense reactions. *Plant Physiol.* 134: 871-879.

Bogdanove, A.J., Beer, S.V., Bonas, U., Boucgher, C.A., Collmer, A., Collin, D.L., Cornelis, G.R., Huang, H.C., Hutcheson, S.W., Panopoulos, N.J et al .1996. Unified nomenclature for broadly conserved *hrp* genes of phytopathogenic bacteria. *Mol. Microbial.* 20: 681-683.

Broughton, W.J., J Jobbouri, S and Perret, X. 2000. Keys to symbiotic harmony. *J.Bacteriol.* 182: 5641-5652.

Buttner, D and Bonas, U. 2002. Port of entry—the type III secretion translocation. *Trends Microbiol.* 10: 186-192.

Cornelis, G.R. 2000. Type III secretion system: a bacterial device for close combat with cells of their eukaryotic host. *Phil. Trans. R. Soc. Lon.* **355**: 681-893.

Dale, C., Young, S.A., Haydon, D.T and Welburn, S.C. 2001. The insect endosymbiont *Sodalis glossinidius* utilizes a type III secretion system for cell invasion. *Proc. Natl. Acad. Sci.* USA. **98:** 1883-1888.

Economou. A., Davis, A.E., Johnson, A.W.B and Downie, J.A. 1994. The *Rhizobium leguminosarum* bv. *viciae nod*O gene can enable a *nod* mutant of *R.leguminosarum* bv. *trifolii* to nodulate vetch. *Microbiol.* **140**: 2341-2347.

Freiberg, C., Fellay, R., Bairoch, A., Broughton, W. J., Rosenthal, A and Perret, X. 1997. Molecular basis of symbiosis between *Rhizobium* and legumes. *Nature.* **387**: 394-401.

Galan, J.E and Collmer, A. 1999. Type III secretion machines: bacterial devices for protein delivery into host cells. *Science.* **284**: 1322-1328.

Gottfert, M., Rothlisberger, S., Kundig, C., Beck, C., Marty, R and Hennecke, H. 2001. Potential symbiosis specific genes uncovered by sequencing a 410-kilobqase DNA region of the *Bradyrhizobium japonicum* chromosome. *J.Bacteriol.* **183**: 1405-1412.

Haller, J.C., Carlson, S., Pederson, K.J. and Pierson, D.E. 2000. A chromosomally encodes type III secretion pathway in *Yersinia enterocolitica* is important in virulence. *Mol. Microbiol.* **36**: 1436-1446.

He SY 1998. Type III protein secretion systems in plant and animal pathogenic bacteria. *Ann. Rev. Phytopathol.* **36**: 363-392.

He SY and Jin, Q. 2003. The Hrp pilus: learning from flagella. *Curr. Opin. Microbiol.* **6**: 15-99.

Hueck, C.J. 1998. Type III protein secretion systems in bacterial pathogens of animals and plants. *Microbiol. Mol. Biol. Rev.* **62**: 379-433.

Kaneko, T., Nakamura, Y., Sato, S., Asamizu, E., Kato, T., Sasmoto, S., Watanabe, A., Idesawa, K., Ishikawa, A., Kawashima, K. et al 2000. Commplete genome structure of the nitrogen-fixing symbiotic bacterium *Mesorhizobium loti. DNA Res.* **7**: 331-338.

Krishnan, H.B., Kuo, C.L and Pueppke, S.G. 1995. Elaboration of flavonoid induced proteins by the nitrogen-fixing soybean symbiont *Rhizobium fredii* is regulated by both *nod*D_1 and *nod*D_2, and is dependent on the cultivar-specificity locus, *nolXWBTUV. Microbiol.* **141**: 2245-2251.

Krishnan, H.B.. 2002. NolX of *Sinorhizobium fredii* USDA257, a type III-secreted protein involved in host range determination, is localized in the infection threads of cowpea *(Vigna unguiculata* [L] Walp) and soybean *(Glycine max* [L] Merr.) nodules. *J. Bacteriol.* **184**: 831-839.

Krishnan, H. B., Lorio, J., Kim, W.S., Jiang, G., Kim, L., Deboer, M and Pueppke, S.G. 2003. Extracellular proteins involved in soybean cultivar-specific nodulation are associated with pilus-like surface appendages and exported by a Type III protein secretion system in *Sinorhizobium fredii* USDA257. *Mol. Plant Microb. Inter.* **16**: 617-625.

Long, S.R. 2001. Genes and signals in the *Rhizobium*-legume symbiosis. *Plant Physiol.* **125**: 69-72.

De Lyra, M.C.C.P., Ollero, F.J., Madinabeitia, N., Espuny, M.R., Bellogin, R.A., Cubo, M.T and Ruiz-Sainz, J.E. 2000. Characterization of a *nol*T mutant of *Sinorhizobium fredii* HHI03: its role in type III secretion protein. In *Fourth Europeon Nitrogen Fixation Conference* Abstract Book. Seville. Viceconsejeria.: 192.

Marie, C., Broughton, W.J and Deakin, W.J. 2001. *Rhizobium* type III secretion systems: legume charmers or alarmers? *Curro Opin. Plant. Biol.* **4**: 336-342.

Meinhardt, L. W., Krishnan, H.B., Balatti, P.A and Pueppke, S.G. 1993. Molecular cloning and characterization of a sym plasmid locus that regulates cultivar specific nodulation of soybean by *Rhizobium fredii* USDA257. *Mol. Microbiol.* **9**: 17-29.

Perret, X., Freiberg, C., Rosenthal, A., Broughton, W.J and Fellay, R. 1999. High resolution transcriptional analysis of the symbiotic plasmid of *Rhizobium* sp. NGR234. *Mol. Microbiol.* **32**: 415-425.

Perret, X., Staehelin, C and Broughton, W.J. 2000. Molecular basis of symbiotic promiscuity. *Microbiol. Mol. Biol. Rev.* **64**; 180-201.

Shea, J.E., Hensel, M., Gleeson, C and Holden, D.W. 1996. Identification of a virulence locus encoding a second type III secretion system in *Salmonella typhimurium. Proc. Natl. Acad. Sci.* USA. **93**: 2593-2597.

Sutton, J. M., Lea, E.J and Downie, J.A. 1994. The nodulation signaling protein NodO from *Rhizobium leguminosarum* bv. *viciae* forms ion channels in membranes. *Proc. Natl. Acad. Sci.* USA. **91:** 9990-9994.

Victor Gonzales et al 2003. The mosaic structure of the symbiotic plasmid of *R. etli* CFN42 and its relation to other symbiotic genome compartments. *Genome Biol.* **4**: 36.

Viprey, V., Del Greco, A., Golinowski, W., Broughton, W.J. and Perret, X. 1998. Symbiotic implications of type III protein secretion machinery in *Rhizobium. Mol. Microbiol.* **28**: 1381-1389.

Xi, C., Schoeters, E., Vanderleyden, J and Michiels, J. 2000. Symbiosis-specific expression of *Rhizobium etli cas*A encoding a secreted calmodulin-related protein: *Proc. Natl. Acad. Sci.* USA. **97**: 1114-1119.

Frontiers in Plant Sciences (Pages 355-372)
Eds. K.G. Mukerji, K.V.B.R. Tilak, S.M. Reddy, L.V. Gangwane, P. Prakash, I.K. Kunwar

Molecular Methods for Characterization of Yeasts Associated with Foods and Beverages

Kakoli Biswas and G.S. Prasad*

ABSTRACT

The application of yeasts is essential in the production of fermented foods and beverages. Unfortunately, yeasts are also invading microorganisms, contaminating yeasts, frequently referred to as wild yeasts, may grow during processing of foods. Yeasts are traditionally identified by growing on various selective solid media, morphological and physiological tests. No single method or medium allows the characterization of all yeasts, which means that complementary methods are required. Molecular techniques provide several alternative methods, which are time saving and more accurate. Hence, they have become increasingly important tools for characterization of microbes. Several molecular techniques are being used for characterization of yeast strains. Among such methods, Randomly Amplified Polymorphic DNA (RAPD), microsatellite finger-printing, Restriction fragment length polymorphism (RFLP) of rRNA genes, mitochondrial DNA RFLP, electrophoretic karyotyping etc. have gained widespread importance in characterization of yeast strains. However, sequencing of appropriate regions of ribosomal RNA gene cluster or other genes is essential for accurate identification of yeasts.

Key words: Characterization, yeasts, food, molecular techniques, PCR, RFLP, RAPD, rDNA, sequencing.

Author's address: Microbial Type Culture Collection and Gene Bank (MTCC) Institute of Microbial Technology (IMTECH) Council of Scientific and Industrial Research (CSIR) Sector-39A, Chandigarh – 160 036, India

*Author for correspondence, E-mail: prasad@imtech.res.in

INTRODUCTION

Man has exploited yeasts for centuries, fermentation of grape juice and the use of fermented foods were first recorded in Sumeria and Babylon in 6th century B.C., and since then man has successfully used yeasts to modify less palatable food sources. Yeasts are very important in the production of bread, beer, wine and other alcoholic drinks to wide range of other products including ethanol for fuel, yeast extracts, pigment, and other specific substances for food and feeds, as well as biochemicals for the pharmaceutical industry. Some yeasts are also used as expression hosts for heterologous production of enzymes. However, yeasts are also known to cause diseases in human and animals and also play the role as spoilage organisms in foods and beverages. Strains belonging to genera *Brettanomyces, Candida, Cryptococcus, Debaryomyces, Dekkera, Endomyces, Filobasidium, Hanseniaspora, Kloeckera, Kluyveromyces, Pichia, Rhodotorula, Schizosaccharomyces, Torulaspora, Torulopsis, Zygosaccharomyces* are frequent contaminants in beverage industry. (Deak and Beuchat,1996; Jespersen and Jakobsen,1996; van der Aa Kuhle and Jespersen,1998). This dual role of yeasts calls for a detailed insight of a proper characterization and systematic studies especially in several food and beverage processing industries.

Yeasts are unicellular fungi and can be members of ascomycotina, basidiomycotina or deuteromycotina. Like filamentous fungi, ascomycetous and basidiomycetous yeasts frequently reproduce in nature or in the laboratory as their anamorphic states, and hence posses anamorphic names in the deuteromycotina. Because of this it is common to see the dual names of these yeasts in literature. It is difficult to establish anamorph-teleomorph connections in the growth media used in the laboratory, because only a few isolates show both stages in their life cycle (Kreger van Rij, 1984). The differentiation of taxa is usually achieved by comparison of morphological traits and physiological features. Traditionally morphological and physiological characterization has provided a wealth of systematic information. Over the years these methods have been used to establish series of yeast classification system (Barnett et al., 1990; Kreger van Rij, 1984). However, these methods are laborious and time consuming and in some cases provide doubtful identification (Scheda and Yarrow 1968; Kurtzman 1994). Therefore alternative approaches have been developed to accurately identify and differentiate yeast species. Rapid and simple microbiological characterization systems in the sources and routes of contamination in the food fermentation process and in the food production chain may improve the microbiological quality of foods. Since the food industry is highly interested in the route and source of contamination, there is a growing demand for typing techniques which enable differentiation at the strain level. In the past few years, the development and use of molecular techniques have provided new dimensions to our knowledge on characterization and identification of yeast strains from different traditional foods, beverages, and wine industries. Such techniques are useful to check the presence of undesired yeasts and fungi contamination throughout the complete chain from the raw material to the final product. The development of these techniques has provided a new approach to establish several genetic differences between the species and had acted as an important tool in the food industry. The present review deals with characterization and identification of yeast strains from foods and beverages using some of the commonly used molecular techniques, as it is not possible to cover all the techniques, we have confined ourselves to the most commonly used techniques only.

Molecular Characterization of Yeasts

The use of molecular methods to estimate genetic relatedness among the yeasts had its beginnings in comparisons of nuclear DNA complementarities. Strains defined as conspecific from measurements of DNA relatedness were shown to differ in glucose fermentation, nitrate assimilation, and formation of pseudohyphae and true hyphae (Kurtzman and Phaff, 1987). These characters had been considered phylogenetically important and were used to define species and genera. However, comparisons of nuclear DNA relatedness resolve only the genetic distance of sibling species and thus provide no information about broader relationships (Kurtzman 1987). Hence, several other molecular methods have been attempted for the purpose of characterization and identification of yeasts. Among these, DNA fingerprinting methods like Randomly Amplified Polymorphic DNA (RAPD), Restriction Fragment Length Polymorphism (RFLP) of different regions of ribosomal RNA gene cluster, RFLP of mtochondrial DNA, Pulsed Filed Gel Electrophoresis (PFGE), and sequencing of the different rDNA are the most commonly used methods.

Randomly Amplified Polymorphic DNA (RAPD)

Randomly Amplified Polymorphic DNA (RAPD) involves random primers of approximately 10 bases, amplicons throughout the genome are targeted and amplified (Williams et al., 1990). Amplified products are separated on an agarose gel, stained with ethidium bromide, visualized under UV light and strains are identified from matching band patterns. RAPD has evolved as the most popular method for DNA fingerprinting and has been widely used to characterize yeasts. RAPD assay allows discrimination at the species level, but small differences in patterns of different strains belonging to the same species occur, indicating that it also enables discrimination at the subspecies level. In their early work Baleiras-Couto et al., (1994,1995) used RAPD technique for differentiation of spoilage yeasts and food-borne yeasts. RAPD assay allowed discrimination between all species of *S. cerevisiae* and *Zygosaccharomyces* genus. This group also assessed genomic variability among *S. cerevisiae* strains using RAPD analysis (Baleiras-Couto et al., 1996a). Intraspecific variations among the strains of yeasts, mainly from the wineries could be easily detected and are reproducible by RAPD-PCR. Quesada and Cenis (1995) suggested that to get a good differentiation of strains of yeasts, it was necessary to use accumulated information of several RAPD reactions. They found in their study using RAPD-PCR, 12 *S. cerevisiae* strains, 4 of 5 *Rhodotorula mucilaginosa* strains could be differentiated among six genera with a very low level of genetic polymorphism. Prillinger et al., (1999) could correctly identify only 39 out of 85 yeast isolates from brine and different cheeses from Europe, using the traditional phenotypic approach, which were confirmed by the RAPD-PCR data. RAPD was found to be useful for identification of yeasts present in dairy products, forty eight dairy associated yeasts were clearly assigned to the species level when compared with the distinctive band profiles of the type strains of different yeast species (Andrighetto et al., 2000).

RAPD has revealed the intraspecific variations in commercial winery yeast strains in Brazilian industries, the yeast strains could be divided into 3 groups in the cluster analysis (Echeverrigaray et al., 2000). Xufre et al., (2000) reported intraspecific variations among the enological strains of *S. cerevisiae* and *S. bayanus*. In a study on characterization of yeasts from Hungarian dairy products, Vasdinyei and Deak (2003) demonstrated that strains of *Geotrichum candidum* and *Debaromyces hansenii* could be further discriminated. In a recent study Foschino et al., (2004) used RAPD fingerprinting for identification of yeasts from sourdoughs of Italian sweet baked products, their study revealed a high degree of intraspesicific homogeneity in strains of *Candida humilis*. The same method, when applied to strains from different bakeries belonging to the *S. cerevisiae* group, highlighted high similarity between them and a certain distance from the reference strains, suggesting a genetic difference correlating with the adaptation to the habitat of recovery. This variation can be useful tool to recognize *S. cerevisiae* originating from sourdough. In another recent study on identification and characterization of contaminating yeasts in brewery, Barszczewski and Robak (2004) could differentiate between brewing yeast and wild yeast isolates by their RAPD patterns.

All these studies suggest that RAPD-PCR is an efficient tool to identify the different yeast species and to study the intraspecific variation up to the strain level from food, food products and wineries. Unfortunately, RAPD patterns are not always reproducible among or even within laboratories. In the development of a RAPD DNA fingerprinting system for a particular species, a number of oligonucleotide primers must be tested, and those that provide the best variability among independent isolates are selected.

Simple Sequence Repeat Finger-printing

A related approach is the amplification of fragments by PCR with oligonucleotides specific for simple repetitive DNA sequences (microsatellite-fingerprinting). Microsatellites are tandem repetitions of 10-100 motifs that, functions as recombinogenic hot spots, causing length polymorphisms and and represent mutative targets for molecular identification of different individuals (Jeffreys et al., 1985; Bishop et al., 2000). This technique was developed on the basis of previous knowledge achieved with the DNA fingerprinting analysis (Leickfield et al., 1993). PCR fingerprinting with the microsatellite oligonucleotides primers $(GTG)_5$, $(GACA)_4$, the phage M13 DNA and the M13 sequence GAGGTGGCGGTTCT has been used for yeast identification (Lavallee et al.,1994; Leickfeldt et al., 1993).

Identification of spoilage yeasts isolated from a food-production chain of mayonnaise and salad dressings, and from spoiled wine and beer was carried out using microsatellite fingerprinting (Baleiras Couto et al., 1996a,b). Primers $(GAC)_5$ and $(GTG)_5$ allowed discrimination between species of *Zygosaccharomyces* genus into *Z. bailli* and *Z. bisporus*. In addition, this method also allowed the discrimination of different types within the *Z. bailli*, which displayed PCR-fingerprinting types that appeared to be identical to some of those generated by strains isolated from one specific production line. This suggests that microsatellite fingerprinting is useful in tracing back the origin of spoilage outbreaks, and it can be applied in microbiological quality assurance monitoring systems in industrial environments. Similarly, 16 strains within the species *Saccharomyces cerevisiae* could be accertained to six different groups by PCR-fingerprinting with the primer $(GTG)_5$, compared with only 2 groups by $(GAC)_5$ (Baleiras Couto et al., 1996b). Microsatellite finger-printing was found to be the most effective method for genetic characterization of *S. cerevisiae* strains isolated during spontaneous fermentation of musts and wine, due to its high level of polymorphism, the speed and the ease of interpretation of genetic profiles obtained (Perez et al., 2001a, Gonzalez et al., 2001)). In another study, the same group (Perez et al., 2001b) evaluated the genetic variability generated by microsatellite markers in 51 *S. cerevisiae* isolates. The discriminating power produced by combining the information obtained by six microsatellite primers were very high proving this technique to be a powerful and agile tool for analyzing the genome of *S. cerevisiae* population. Bujodso et al., (2001) found microsatellite-PCR to be a reliable tool for differentiation of different strains of *Hanseniaspora uvarum* isolated in finger lake wineries. In a recent study on yeast isolates from traditional fermented foods of Himachal Pradesh, India, our group could differentiate 34 *S. cerevisiae* strains into 8 distinct groups using $(GTG)_5$ primer for microsatellite fingerprinting (Biswas et al., 2004). Minisatellite like sequences in SED1 gene were used for characterization of *S. cerevisiae* strains (Maringeli et al., 2004), they also demonstrated that SED1, AGA1, DAN4 and HSP150 genes in a wild population of *S. cerevisiae* are highly polymorphic in length and represent a sink of unexplored genetic variability.

Restriction Fragment Length Polymorphism (RFLP)

Another molecular technique, which has been widely used as a typing method is RFLP of the PCR amplified products of various regions of ribosomal RNA gene cluster. PCR-RFLP involves the amplification of the region of interest and then subjecting the PCR products to restriction analysis using suitable restriction endonucleases and visualizing restriction pattern under UV after running them on agarose gel. Among some of the earlier studies Huffman et al., (1992) had utilized RFLPs of the amplified ITS region to distinguish between *S. cerevisiae, S. bayanus* and *S. pastorianus* three phenotypically indistinguishable (sibling) species, but they could not separate *S. bayanus* and *S. pastorianus*. Guillamon et al., (1998) identified a total of 33 wine yeast species and strains using 3 restriction endonucleases (*HinfI*, *CfoI* and *HaeIII*) and also showed that the sexual and asexual state yielded the same pattern, confirming the previous report of anamorph-teleomorph connections. Baleiras Couto et al., (1996a) evaluated the efficiency of ITS and NTS-RFLP using TaqI and MseI of 15 individual strains of *S. cerevisiae* isolated from wine and beer. It proved to be suitable for generating intra-species specific patterns though only 3 different types of patterns were recognized upon digestion with TaqI. ITS-RFLP was used to identify a total of 132 yeast species belonging to 25 different genera, including teleomorphic and anamorphic ascomycetous and basidiomycetous yeasts. In many cases, the size of the PCR products and the restriction patterns obtained with endonucleases *CfoI*, *HaeIII* and *HinfI* yielded a unique profile for each species. (Esteve-Zarzoso, 1999). However, Xufre et al., (2000) could not find variation among 6 strains of *S. cerevisiae* isolated from wine by ITS-RFLP using five different restriction enzymes. Jespersen et al., (2000) identified yeast contaminants from brewery and about 13% of the total isolates which could not be identified to species level based on their phenotypic characteristics, could be identified as *S. cerevisiae* and some as hybrids of *S. cerevisiae* and *S. bayanus*. Similar results were reported by Hansen and Jakobsen (2001), who found ITS-RFLP pattern of all strains identical and similar to the reference strain of *S. cerevisiae*. ITS-RFLP of yeasts isolated from West African sorghum beer indicated that these isolates were related to *S. cerevisiae* regardless of their phenotypic characteristics (Naumova et al., 2003).

There are several reports where this technique has been successfully used to characterize non-*Saccharomyces* yeasts isolated from various food products. To establish enzymatic activities of interest in yeasts used in wine making, Fernandez et al., (2000) typed 69 isolates exhibiting enzymatic activity, and 11 randomly selected isolates without activity by ITS-RFLP with one restriction enzyme. They found 13 different molecular profiles and isolates of each profile could be co-related closely to the phenotypic characteristics and enabled proper identification of those species. The species identified were *Pichia membranifaciens, Kluveromyces thermotolerans, Metschnikowia pulcherrima, Brettanomyces clausenii, Pichia anomola* and *Candida stellata.* ITS-RFLP with DdeI enzyme could clearly differentiate *Hanseniaspora uvarum* and *H.osmophila* strains after making a comparison with their type strains (Bujdoso et al., 2001). Esteve-Zarzoso et al., (2001) could differentiate the flor yeast (wine under a velum) and non-flor yeast using ITS-RFLP technique. However, in a study on isolates of *Debaryomyces hansenii* from surface ripened cheese, Petersen et al., (2001) found ITS-RFLP ineffective for strain typing of *D. hansenii* as out of 9 restriction enzyme used only TaqI demonstrated diversity within the strains. This enzyme divided the 20 strains in three groups. *Pichia anomola* which is frequently the cause of spoilage of yogurts, could be effectively identified by ITS-RFLP using four restriction enzymes (*HaeIII, RsaI, HinfI* and *Dpn II*) when compared to the type strain of *P.anomola* (Caggia et al., 2001). Seven non-*Saccharomyces* species and two *Saccharomyces* species were identified among the yeast strains isolated from orange fruit and juice in a spontaneous fermentation by ITS-RFLP with three restriction enzymes (*CfoI, HaeIII, HinfI*) (Heraz-Vasquez et al., 2003). ITS-RFLP allowed the distinction of yeast isolates as contaminants from brewery, where 19 different restriction pattern were obtained for 27 strains tested, using three restriction enzymes (*ScrFI, HaeIII* and *MspI*) and hence could be directly used in the brewing microbiology (Barzczewski and Robak 2004). In another recent study, Clemente-Jimenez et al., (2004), characterized the oenological properties of wine yeasts isolated during spontaneous fermentation of six varieties of grape must, and identified members of the genera *Hanseniaspora, Issatchenkia, Metschnikowa, Pichia,* and *Saccharomyces* by ITS-RFLP using three restriction enzymes (CfoI, HinfI, HaeIII). In our study on the isolates from traditional fermented foods and beverages of Himachal Pradesh, we could very easily differentiate the 45 yeast isolates by ITS-RFLP into 8 different patterns using 4 restriction enzymes (CfoI, HaeII, HinfI, MspI) (Biswas et al., 2004). From most of the available information in the literature on ITS-RFLP it is clear that this method is suitable for species as well as strain differentiation of both *Saccharomyces* and non-*Saccharomyces* species. Dlauchy et al., (1999) developed a method to simplify the identification of yeast strains based the restriction fragment patterns of PCR-amplified 18S and ITS region of ribosomal RNA gene cluster. A total 169 yeast strains representing 128 species associated mainly with food, wine, beer, and soft drinks were examined. The pattern of PCR amplicons digested with four different four-base-cutting restriction enzymes proved to be sufficient for strain identification.

Mitochondrial DNA RFLP

Mitochondrial DNA restriction fragment length polymorphism (mt DNA RFLP) appears as one of the most suitable methods to differentiate between strains. Analysis of restriction patterns of mtDNA for industrial strain differentiation was first carried out on brewery yeasts (Aigle et al., 1984; Lee and Knudsen 1985) and then on several natural and commercial enological strains (Hallet et al., 1988). A new mitochondrial restriction analysis method was developed (Querol et al., 1992), that simplified the characterization of *S. cerevisiae* wine yeast strains. This method consists of the standard miniprep isolation of total yeast DNA and the use of G+C rich restriction endonucleases, which recognize a high number of sites in the yeast nuclear DNA, but few sites in the mitochondrial DNA. This method is simple and yields results within three working days. This technique has successfully been used by other workers to characterize strains of yeast species (Martinez et al., 1995; Romano et al., 1996; Gu illamon et al., 1997). In a study of several strains of four sibling species of *Saccharomyces* (*S.bayanus, S.cerevisiae, S.pastorianus* and *S. paradoxus*) were characterized using mt-DNA-RFLP (Guillamon et al.,1998). From the patterns obtained with four cutter endonucleases (*AluI, DdeI, HinfI* and *RsaI*) it was possible to differentiate each species. *S. cerevisiae* and *S. paradoxus* presented a greater number of large fragments than *S. bayanus* and *S. pastorianus* with all the enzymes. Fernández-Espinar et al., (2001) examined the authenticity of 45 commercial *Saccharomyces* wine yeast strains by mtDNA restriction analysis with *HinfI*. This restriction enzyme revealed a high level of restriction fragment length polymorphism giving a total of 25 different restriction patterns.

Lopez et al., (2001) modified the method of Querol et al., (1992) and developed a more rapid method based on mtDNA RFLP which reduced the identification time to a great extent. The DNA obtained was directly digested by the endonucleases displaying clear restriction patterns useful for *S. cerevisiae* yeast strain differentiation as well as the strains belong to other food borne yeast species. In a comparative study on strain typing of *Debaromyces hansenii* isolates from dairy products, Petersen et al., (2001), evaluated mtDNA RFLP using HaeIII and HpaII. The 20 dairy isolates showed lower genetic variability than the strains obtained from CBS culture collection. This method showed that more than one strain was involved in surface ripening in some dairies indicating that the method could be useful for subspecies typing and investigation of the microbial succession among strains of *D. hansenii* during the ripening process. In our study on isolates of *S. cerevisiae* from traditional food and beverages of Himachal Pradesh India, mtDNA RFLP using the endonuclease *HinfI*, 35 strains could be divided into 6 groups (Biswas et al., 2004).

Denaturing Gradient Gel Electrophoresis (DGGE)

DGGE has been successfully used as a powerful tool to describe microbial diversity in their natural environment. In this method, DNA fragments directly isolated from the natural environment can represent the genetic diversity of a complex microbial community. Required genes are then amplified from the community DNA by PCR. DGGE can separate DNA fragments of identical length but with different nucleotide sequences. Separation of fragments is based on the decreased electrophoretic mobility of partially melted double-stranded DNA molecules in polyacrylamide gels containing a linear gradient of DNA denaturants. DNA fragments with different sequences may have different melting behavior, and therefore, stop migrating at different positions in a gradient gel, the resultant bands isolated and sequenced to permit strain identification. ITS-DGGE was used successfully to distinguish strains of *S.cerevisiae, S.paradoxus* and *S.bayanus/ S.pastorianus*. The performance of denaturing gradient gel electrophoresis (DGGE) for analysing yeasts associated with wine grapes was compared with cultural isolation on malt extract agar (MEA), DGGE method detected a greater diversity of species than agar plating (Prakitchaiwattana et al., 2004). Yeast biodiversity in raw milk was analyzed by traditional methods and by DGGE of the D1–D2 domain of the 26S rRNA gene (Cocolin et al., 2002). Twelve different species of yeast were identified by traditional plating methods. DGGE method showed the presence of additional biodiversity in the milk samples. This approach is a powerful and reliable tool to monitor yeast ecology in milk and milk products without the need of traditional isolation. The succession of yeast biota in four sourdoughs were studied using fungi-specific PCR-DGGE which revealed the presence of *Candida humilis, Debaromyces hansenii, S. cerevisiae,* and *S. uvarum* (Merothn et al., 2003).

Electrophoretic Karyotyping

Karyotyping is the process of describing the chromosome(s) of an organism in terms of number, size and shape. Clasically, karyotyping involves a spread of condensed meosis I chromosomes followed by their visualization under light or electron microscopy. With the invention of pulsed-field gel electrophoresis (PFGE) by Carle and Olson (1985), chromosome-sized fragments of the yeast genome were readily separated in a gel. PFGE techniques have been experimented on numerous taxa (Johnston and Mortimer, 1986; De Jonge et al., 1986). Other variations of this technique, Orthogonal-Field-Alternation Gel Electrophoresis (OFAGE), Contour-clamped Homogeneous Electric Field gel electrophoresis (CHEF), Field Inversion Gele Electrophoresis are also being used (Johnston et al., 1988; Sor and Fukuhara, 1989). Transverse Alternating Field Electrophoresis (TAFE) has been used successfully to differentiate several yeasts strains by their chromosomal DNA patterns. Vezinhet et al., (1990) compared the results of TAFE and mt-DNA-RFLP profiles of 22 enological strains of *S. cerevisiae* and the contribution of these two methods to the strain identification and industrial control, based on TAFE gel banding patterns the 22 strains could be divided into 2 groups. PFGE performed on 38 isolates of *Saccharomyces* brewing contaminants and on 15 lager brewing yeasts (Jespersen et al., 2000). Their work revealed that polymorphism existed among the *Saccharomyces* brewing contaminants, based on the chromosome profiles the majority of these strains could be grouped as *S. cerevisiae* or *S. bayanus/S. pastorianus*. Further this technique was able to differentiate between almost all brewing contaminants and to separate them from specific lager brewing yeast. Similarly Hansen and Jakobsen (2001) examined the PFGE profiles of yeasts isolated from dairy products and observed that *Saccharomyces* sp. isolated from blue veined cheese exhibited similar PFGE chromosome profiles. Fernandez-Espinar et al., (2001) examined the authenticity of commercial wine yeast strains using mt RFLP and PFGE and reported that combination of these two techniques gave 13 different patterns for the 28 strains.

Sequence based identification

Most of the sequence based characterization methods for yeast strains have used the DNA sequences from the ribosomal gene cluster. The genes of this cluster are usually present in large numbers as tandem repeats along the chromosomes but they evolve as a single unit. The rDNA cluster comprises of three structural regions coding for the 5.8S, small subunit ribosomal (SSU or 18S) RNA and large subunit (LSU or 28S) ribosomal RNA genes. These three genes are separated from each other by internal transcribed spacers (ITS). The cluster of three genes along the chromosome is separated by the intergenic spacer region (IGS). The three regions, structural genes, transcribed spacers and the non-transcribed spacers evolve at different rates, and their sequences can be used to discriminate between taxa at different levels. White et al., (1990) designed a series of primers for amplification of different regions of ribosomal RNA gene cluster, they are being used extensively for amplification and sequencing ribosomal RNA genes by several workers.

The small subunit (SSU) rRNA gene has been used for the inference of phylogeny of ascomycetous yeasts (Cai et al., 1996; James et al., 1994). Cappa and Cocconcelli (2001) identified fungi and yeast from dairy products by means of SSU rRNA sequence analysis. However, most of this data exist one strain of single species making the assessment of the variability of closely related strains difficult. In general, the SSU gene is more conserved than the LSU gene and its utility for identification purposes appears rather limited as the determination of almost complete gene sequences of about 1800 nucleotides is often required to detect sufficient variation.

Sequencing of ITS region was used for identification of yeast strains isolated from orange fruit and juice in a spontaneous fermentation. The strains were identified as *Candida tropicalis, Clavispora lusitaniae, Hanseniaspora uvarum, Pichia anomola, P. fermentas, Rhodotorula mucilaginosa, S. cerevisiae, S. unisporus* and *Trichosporon asahii* (Heras-Vazquez et al., 2003). Sequencing of the ITS region permitted a consistent classification of sourdough yeasts for Italian sweet baked products (Foschino et al., 2004). In a study on molecular characterization of six varieties of grape must Clemente-Jimenez et al., (2004) could differentiate 1000 yeast isolates, 165 for each grape variety into six groups using ITS-RFLP as *Candida stellata, Issatchenkia terricola, Metschinkowia pulcherrima, Pichia fermentans, Saccharomyces cerevisiae* and *Hanseniospora uvarum*. The seventh profile could be confirmed by sequencing of ITS region as *Issatchenkia orientalis*. ITS sequencing was used to confirm the identity of 11 different non-saccharomyces yeast strains as *Issatchenkia orientalis, Pichia anomola, Pichia fermentans, Pichia membranifaciens, Saccharomycopsis fibuliger* and *Torulaspora delbruckii* (Biswas et al., 2004). Hence the ITS sequences turned out to be a reliable tool for the identification of strains at the species level. The discriminative potential of the ITS region to investigate species boundaries was successfully used by some other workers (Fernadez-Espinar,2000; Egli and Henickling, 2001; Noumova et al., 2003). However, as of now the ITS sequence database is not complete and may not be useful for routine identification of all yeast strains.

Kurtzman and Robnett (1998) proposed that sequencing or the D1/D2 domain of sufficient for identification all the ascomycetous yeast species. The D1/D2 domain is of about 600 bp, and shows 0–1% divergence in conspecific strains. Comparisons of two and more strains from 103 ascomycetous yeast species by Kurtzman and Robnett (1998) showed intra-species variation of up to two nucleotides in the D1/D2 region of most strains. The D1/D2 variable domain of the large subunit gene (LSU) has been sequenced for all the yeasts (Kurtzman and Robnett, 1998, Fell et al. 2000) and has become prime target for DNA sequence based identification of yeasts.

Molecular techniques are useful for characterization of yeasts associated with foods and beverages, no single method is sufficient for identifying species and different strains of same species. However the application of molecular techniques provide opportunities to re-evaluate the characterization of different taxa and shed light on the phylogenetic relationship where morphological studies fail, and provide new insights into the origin of yeasts as well as identify new genera. Methods such as PCR-RFLP of 18S, ITS and DGGE are useful in differentiation and identification of species. However, these methods may not be suitable for differentiating strains of same species. RAPD, microsatellite finger-printing, mitochondrila DNA-RFLP and electrophoretic karyotyping are useful for identification of species and some times strains of same species. Sequencing of appropriate genes is essential for accurate identification of yeast species. Sequencing of D1/D2 variable domain of LSU rRNA gene is the widely accepted method for identification of yeast species.

References

Aigle, M., Erbs, D. and Moll, M. 1984. Some molecular structures in the genome of larger brewing yeast. *American Society of Brewing Chemistry.* 42: 1-7.

Andrighetto, C., Psomas, E., Tzanetakis, N., Suzzi, G. and Lombardi, A. 2000. Randomly amplified polymorphic DNA (RAPD) PCR for the identification of yeast isolated from dairy products. *Letters in Applied Microbiology*. 30: 5-9.

Baleiras-Couto, M.M., Eijsma, B., Hofstra, H., Huis in't Veld, J.H.J. and van der Vossen, J.M.B.M., 1996a. Evaluation of molecular typing techniques to assign genetic diversity among strains of *Saccharomyces cerevisiae. Applied and Environmental Microbiology.* 62: 41-46

Baleiras-Couto, M.M., Hartog, B.J., Huis in't Veld, J.H.J., Hofstra, H. and van der Vossen, J.M.B.M. 1996b. Identification of spoilage yeast in a food-production chain by microsetelite polymerase chain reaction fingerprinting. *Food Microbiology.* 13: 59-67.

Baleiras-Couto, M.M., van der Vossen, J.M.B.M., Hofstra, H. and Huis in't Veld, J.H.J. 1994. RAPD analysis: a rapid technique for differentiation of spoilage yeasts. *International Journal of Food Microbiology.* 24: 249-260.

Baleiras-Couto, M.M., Vogels, J.T.W.E., Hofstra, H., Huis in't Veld, J.H.J. and van der Vossen, J.M.B.M. 1995. Random amplified polymorphic DNA and restriction enzyme analysis of PCR amplified rDNA in taxonomy: two identification techniques for food borne yeasts. *Journal of applied Bacteriology.* 79: 525-535.

Barnett, J.A., Payne, R. W. and Yarrow, D. 1990. *Yeasts: Characteristics and Identification,* 2nd ed. Cambridge University Press, Cambridge.

Barszczewski, W. and Robak, M. 2004. Differentiation of contaminating yeasts in brewery by PCR-based techniques. *Food Microbiology.* 21: 227-231.

Bishop A.J.R., Louis, E.J. and Borts, R.H. 2000. Minisatellite variants generated in yeast meiosis involve DNA removal during gene conversion. *Genetics*. 156: 7-20

Biswas, K., Prasad, G.S., Mayilraj, S., Thakur, N., Bhalla, T.C., Kanwar, S.S. and Chakrabarti, T. 2004. Polyphasic characterization of yeast strains originating from traditional fermented foods of Himalayan region of India. (Unpublished results)

Bujdoso, G., Egli, C.M. and Henick-Kling, T. 2001. Characterization of *Hanseniaspora* (*Kloeckera*) strains isolated in Fingerlakes Wineries using physiological and molecular techniques. *Food Technology and Biotechnology.* 39: 83-91.

Caggia, C., Restuccia, C., Pulverenti, A. and Giudici, P. 2001. Identification of *Picchia anomola* isolated from yogurt by RFLP of ITS region. *International Journal of Food Microbiology.* 71: 71-73.

Cai,J., Roberts, I.N. and Collins, M.D. 1996. Phylogenetic relationships among members of the ascomycetous yeast genera *Brettanomyces, Debaryomyces, Dekkera,* and *Kluyveromyces* deduced by small-subunit rRNA gene sequences. *International Journal of Systematic Bacteriology.* 46: 542-549.

Cappa, F. and Cocconcelli, P.S. 2001. 69: 157-160. Identification of fungi from dairy products by means of 18S rRNA analysis. *International Journal of Microbiology.* 69:157-160.

Carle, G.F. and Olson, M.V. 1985. An electrophoretic karyotype for yeast. *Proceedings of National Academy of Sciences, USA.* 82: 3756-3760.

Clemente-Jimenez, J.M., Mingorance-Cazrola, L., Martinez-Rodriguez. S., Heraz-Vasquez, F.J.L. and Rodriguez-Vico, F. 2004. Molecular characterization and oenological properties of wine yeasts isolated during spontaneous fermentation of six varieties of grape must. *Food Microbiology.* 21: 149-155.

Cocolin, L., Aggioa, D., Manzanoa, M., Cantonib, C., and Comi, G. 2002. An application of PCR-DGGE analysis to profile the yeast populations in raw milk. *International Dairy Journal* 12: 407-411

Deak, T. and Beuchat, L.R. 1996. *Hand Book of Food Spoilage Yeasts.* Boca Ratson, Florida, CRC Press, Florida, USA.

De Jonge, P., De Jonge, F.C.M., Meijers, R., Steensma H.Y. and Scheffers, W.A. 1986. Orthogonal-field-alteration gel electrophoresis banding patterns of DNA from yeasts. *Yeast.* 22: 193-204.

Dlauchy, D., Tornai-Lehoczki, J. and Peter, G. 1999. Restriction enzyme analysis of PCR amplified rDNA as a taxonomic tool in yeast identification. *Systematic and Applied Microbiology*. 22: 445-453.

Echeverrigaray, S., Paese-Toresan, S. and Carrau, J.L. 2000. RAPD marker polymorphism among commercial winery yeast strains. *World Journal of Microbiology and Biotechnology.* 16: 143-146.

Egli, C.M. and Henick-Kling, T. 2001. Identification of *Brettanomyces/Dekkera* species based on polymorphism in the rRNA internal transcribed spacer region. *American Journal of Enology and Viticulture.* 52: 241-247.

Esteve-Zarzoso, B., Belloch, C., Uruburu, F. and Querol, A. 1999. Identification of yeasts by RFLP analysis of 5.8S rRNA gene and the two ribosomal internal transcribed spacers. *International Journal of Systematic Bacteriology.* 49: 329-337.

Esteve-Zarzoso, B., Peris-Toran, M.J., Garcia-Maiquez, E., Uruburu, F. and Querol, A. 2001. Yeast population dynamics during the fermentation and ageing of the sherry wines. *Applied and Environmental Microbiology.* 67: 2056-2061.

Fell, J.W., Boekhout, T., Fonseca, A., Scorzetti. G. and Statzell-Tallman, A. 2000. Biodiversity and systematics of basidiomycetous yeasts as determined by large-subunit rDNA D1/D2 domain sequence analysis. International Journal of Systematic and Evolutionary Microbiology 50: 1351-1371.

Fernandez, M., Ubeda, J,F. and Briones, A.I. 2000. Typing of non-*Saccharomyces* yeasts with enzymatic activities of interest in winemaking. *International Journal of Food Microbiology.* 59: 29-36.

Fernandez-Espinar, M.T., Esteve-Zarzoso, B., Querol, A. and Barrio, E. 2000. RFLP analysis of the ribosomal internal transcribed spacers and 5.8S rRNA gene region of the genus *Saccharomyces*: a fast method for species identification and the differentiation of flor yeasts. *Antonie van Leeuwenhoek.* 78: 87-97.

Fernandez-Espinar, M.T., Lopez, V., Ramon, D., Bartra, E. and Querol, A. 2001 Study of the authenticity of commercial wine yeast strains by molecular techniques. *International Journal of Food Microbiology.* 70: 1-10.

Foschino R., Gallina, S., Andrighetto, C., Rossetti, L. and Galli, A. 2004. Comparison of cultural methods for identification and molecular investigation of yeasts from sourdoughs for Italian sweet baked products. *FEMS Yeast Research.* 4: 609-618

Gonzalez, T.A., Jubany, S., Carrau, F.M. and Gaggero, C. 2001. Differentiation of industrial wine yeast strains using microsatellite markers. *Letters in Applied Microbiology.* 33: 71-75.

Guillamon, J.M., Sabate, J., Barrio, E., Cano, J. and Querol, A. 1998 rapid identification of wine yeast species based on RFLP analysis of the ribosomal internal transcribed spacer (ITS) region. *Archives of Microbiology.* 169: 387-392.

Guillamon, J.M., Sanchez. And Huerta, T. 1997. Rapid characterization of wild and collection strains of the genus *Zygosaccharomyces* according to mitochondrial DNA patterns. *FEMS Microbiology Letters.* 147: 267-272.

Hallet, J.N., Craneguy, B., Zucca, J. and Poulard. 1988. Characterization de differentes souches industrielles de leuvres oenologiques par les profile de restriction de leur AND mitochondrial. *Progress in Agriculture and Viticulture.* 105: 328-333.

Hansen, T.K. and Jakobsen, M. 2001. Taxonomical and technological characteristics of *Saccharomyces* spp. Associated with blue veined cheese. *International Journal of Food Microbiology.* 69: 59-68.

Heraz-Vasquez, F.J.L., Mingorance-Cazrola, L., Clemente-Jimenez, J.M. and Rodriguez-Vico, F. 2003. Identification of yeast species from orange fruit and juice by RFLP and sequence analysis of the 5.8S rRNA gene and the two internal transcribed spacers. *FEMS Yeast Research.* 3: 3-9.

Huffman, J.L., Molina, F.I. and Jong, S.C.1992. Authentication of the ATCC strains in the *Saccharomyces cerevisiae* complex by PCR fingerprinting. *Experimental Mycology*. 16: 316-319.

James, S.A., Collins, M.D. and Roberts, I.N. 1994. Genetic interrelationship by small-subunit rRNA gene sequences. *Yeast*. 10: 871-881.

Jeffreys, A.J., Wilson, V. and Thein, S.L. 1985. Individual specific fingerprints of human DNA. *Nature*. 316: 76-79.

Jespersen,L. and Jakobsen, M. 1996.Specific spoilage organisms in breweries and laboratory media for their detection *International Journal of Food Microbiology* 33:139-155.

Jespersen, L., van der Aa Kuhle, A. and Petersen, K. M. 2000. Phenotypic and genetic diversity of *Saccharomyces* contaminants isolated from lager breweries and their phylogenetic relationship with brewing yeasts. *International Journal of Food Microbiology*. 60: 43-53.

Johnston, J.R. and Mortimer R.K. 1986. Electrophoretic karyotyping of laboratory and commercial strains of *Saccharomyces* and other yeasts. *International Journal of Systematic Bacteriology*. 36: 569-572.

Johnston, J.R., Contopoulou, C.R. and Mortimer R.K. 1988. Karyotyping of strains of several genera by field inversion gel electrophoresis. *Yeast*. 4: 191-198.

Kreger-van Rij, N.J.W. 1984. The Yeasts: a taxonomic study, 3rd ed. Elsevier Science Publishers B.V., Amsterdam, Netherlands.

Kurtzman, C.P. 1987. Prediction of biological relatedness among yeasts from comparisons of nuclear DNA complementarity. *Studies in Mycology*. 30: 459-468.

Kurtzman, C.P. 1994. Molecular taxonomy of Yeasts. *Yeast*. 10: 1727-1740.

Kurtzman, C.P., and Phaff, H.J. 1987. Molecular taxonomy. In: *The Yeasts* vol. 1. *Biology of yeasts* (Eds. Rose, A.H. and Harrison, J.S.) pp. 63-69. Academic Press, London.

Kurtzman, C.P. Robnett, C.J.1998. Identification and phylogeny of ascomycetous yeasts from analysis of nuclear large subunit (26S) ribosomal DNA partial sequences. Antonie Van Leeuwenhoek.73: 331-371.

Lavallee, F., Salva, Y., Lamy, S., Thomas, D.Y., Degre, R. and Dulau, L. 1994. PCR and DNA fingerprinting used as quality control in the production of wine yeast strains. *American Journal of Enology and Viticulture*. 45: 86-91.

Lee, S.Y. and Knudsen F.B. 1985. Differentiation of brewery yeast strains by restriction endonuclease analysis of their mitochondrial DNA. *Journal of Institute of Brewery*. 91: 169-173.

Leickfeldt, E., Meyer, W. and Borner, T. 1993. Rapid identification and differentiation of yeasts by DNA and PCR fingerprinting. *Journal of Basic Microbiology*. 33: 413-426.

Lopez, V., Querol, A., Ramon, D. and Fernandez-Espinar, M.T. 2001. A simplified procedure to analyse mitochondrial DNA from industrial yeasts. *International Journal of Food Microbiology*. 68: 75-81.

Marinangeli, P., Angelozzi, D., Ciani, M., Clementi, F. and Mannazzu, I. 2004. Minisatellites in *Saccharomyces cerevisiae* genes encoding cell wall proteins: anew way towards wine strain characterization. *FEMS Yeast Reserarch*. 4: 427-435.

Martinez, P., Codon, A.C., Perez, L., Benitez, T., 1995. Physiological and molecular characterization of flour yeast: polymorphism of flour yeast populations. *Yeast.* 11. 1399-1411.

Merothn, C.B., Hammes, W.P. and Hertel, C. 2003. Identification and population dynamics of yeasts in sourdough fermentation process by PCR-Denaturing Gradient Gel Electrophoresis. *Applied and Environmental Microbiology.* 69: 7453-7461.

Naumova, E.S., Korshunova, I.V., Jespersen, L. and Naumov, G.I. 2003. Molecular genetic identification of *Saccharomyces sensu stircto* strains from African soghum beer. *FEMS Yeast Research.* 3: 177-184.

Perez, M.A., Gallego, F.J. and Hidalgo, P. 2001a. Evaluation of molecular techniques for the genetic characterization of *Saccharomyces cerevisiae* strains. *FEMS Microbiology Letters.* 205: 375-378.

Perez, M.A., Gallego, F.J., Martinez, I. and Hidalgo, P. 2001b. Detection, distribution and selection of microsatellites (SSRs) in the genome of the yeast *Saccharomyces cerevisiae* as molecular markers. *Letters in Applied Microbiology.* 33: 461-466

Petersen, K.M., Moller, P.L. and Jespersen, L. 2001. DNA typing methods for differentiation of *Debaromyces hansenii* strains and other yeasts related to surface ripened cheeses. *International Journal of Food Microbiology.* 69: 11-24.

Prakitchaiwattana, C.J., Fleet, G.H., and Heard, G.M. 2004. Application and evaluation of denaturing gradient gel electrophoresis to analyse the yeast ecology of wine grapes. *FEMS Yeast Research* 4: 865-877

Prillinger, H., Molnar, O., Eliskases-Lechner, F. and Lopandic, K. 1999. Phenotypic and genotypic identification of yeasts from cheese. *Antonie van Leeuwenhoek.* 75: 267-283.

Querol, A., Barrio, E. and Ramon, D. 1992. A comparative study of different methods of yeast strain characterization. *Systematic and Applied Microbiology.* 15: 439-446.

Quesada, M. and Cenis, J.L. 1995. Use of random amplified polymorphic DNA (RAPD-PCR) in characterization of wine yeasts. *American Journal of Enology and Viticulture.* 46: 204-208.

Romano, A., Casaregola, S., Torre, P. and Guilladin, C. 1996. Use of RAPD and mitochondrial DNA RFLP for typing of *Candida zelanoides* and *Debaromyces hansenii* yeast strains isolated from cheese. *Systematic and Applied Microbiology* 19: 255-264.

Scheda, R. and Yarrow, D. 1968. Variations in the fermentative pattern of some *Saccharomyces* species. *Archiv fur Mikrobiologie.* 55: 209-225.

Sor, F. and Fukuhara, H. 1989. Analysis of chromosomal DNA patterns of the genus *Kluveromyces. Yeast.* 5: 1-10.

van der Aa Kuhle, A. and Jespersen, L.1998 Detection and identification of wild yeasts in lager breweries. *International Journal of Food Microbiology* 43: 205-213

Vasdinyei, R. and Deak, T. 2003. Characterization of yeast isolates originating from Hungarian dairy products using traditional and identification techniques. *International Journal of Food Microbiology.* 86: 123-130.

Vezinhet, F., Blondin, B. and Hallet, J.N. 1990. Chromosomal DNA patterns and mitochondrial DNA polymorphism as tools for identification of enological strains of *Saccharomyces cerevisiae*. *Applied Microbiology and Biotechnology.* 32: 568-571.

White, T. J., Bruns, T., Lee, E., Taylor, J. 1990. Amplification and direct sequencing of fungal ribosomal RNA genes for phylogenetics. In *PCR Protocols: A Guide to Methods and Applications*. (Eds. M.A. Innis., D.H. Gelfand., J.J. Snisky., and T.J. White). pp. 315-322. Academic Press, San Deigo,California, USA.

Williams, J.G.K., Kubelik, A.R., Livak, K.J., Rafalski, J.A. and Tingey, S.V. 1990. DNA- polymorphisms amplified by arbitrary primers are useful as genetic markers. *Nucleic Acids Research.* 18: 6531-6535.

Xufre, A., Simoes, F., Gfrio, F., Clemente, A. and Amaral-Collco, A.M. 2000. Use of RAPD analysis for differentiation among six enological *Saccharomyces* spp. strains. *Food Technology and Biotechnology* 38: 53-58.

Frontiers in Plant Sciences (Pages 373-380)
Eds. K.G. Mukerji, K.V.B.R. Tilak, S.M. Reddy, L.V. Gangwane, P. Prakash, I.K. Kunwar

Hypercellulolytic Mutants of *Humicola grisea* var. *thermoidea* from Sunderban Mangrove Forest

Kavita Johri and A.K. Jaitly

ABSTRACT

Mutants of *Humicola grisea* var. *thermoidea* were produced by the use of a mutagen 5-Bromo uracil. Out of 16 different auxotrophic mutants five were selected having increase in exoglucanase, endoglucanase and β-glucosidase activity. The maximum Increase in exoglucanase and β-glucosidase activity was obtained in HGBUM-5 (1.97 U/mg Protein, 26% and 0.48 U/mg Protein, 40%) while the maximum increase in endoglucanase activity was observed in HGBUM-3 (0.95 U/mg protein. 26%). Selected mutants differ in their amino-acid requirement and phenotypic characters.

Key words: Mangroves, Thermophiles, Mutants, Hypercellulolytic, Cellulases.

INTRODUCTION

Mangroves are complex ecosystem having characteristics of both terrestrial and marine life. The value of Mangrove is based largely on the production of organic matter which has resulted from the fall of plant debris and leaf litter. The presence of vast quantity of wood debris leaf litter and other organic matter provide suitable substrates for the growth and activity of microbes.

Thermophilic micro-organisms have been proved more important than mesophiles in biodegradation of agricultural, forest products and in the production of thermostable and industrially important enzymes (Gooes

Author's address: Microbiology Laboratory
Department of Plant Science, M.J.P. Rohilkhand University, Bareilly 243005 (U.P.), India

et al. 2000; Mouchacca- Jean 2000). During the past several years attention has been paid on utilization of biorefuse and renewable sources for the production of economically important products, i.e., glucose, alcohal, acetone and single Cell protein (Gutierrez and Tengerdy 1998; Tahoun and Ibrahim 1999; Golias *et al.* 2000; Kannangara *et al.* 2000; Oh-Kyeong *et al.* 2000; Van-wyk and Leogale 2001; Haros-Monica *et al.* 2002) The production of above products from biorefuse opened a new area in the field of food, energy and chemicals to conserve current world energy sources. (Kanotra and Mathur 1993; Lee and Saddler 1995; Van-wyk *et al.* 2000). Cellulose and lignocellulose substrates, which are most abundant in plant residues, form substantial resources for the production of economically important more utilizable products (Van-wyk 1997; Bollok and Reczey 2000; Krishna and Chowdary 2000; Yu-Peigiang *et al.* 2003).

In our laboratory, we have isolated thermophilic fungi from Sunderban mangrove forest and screened them for cellulase profile. These organisms were found to grow at a wide range of pH (acidic to alkaline) and at elevated temperature. An attempt has been made to produce hypercellulolytic mutants of one of the isolate of mangrove forest with regard to cellulolytic activity and the results have been presented in this paper.

MATERIALS AND METHODS

Micro-organism

The thermophilic fungus *Humicola grisea* var. *thermoidea* isolated from Sunderban mangrove forest was used for the study. Culture was maintained on Yeast powder soluble starch medium (YPSS). Culture was stored in microbiology laboratory, Department of plant science, M.J.P. Rohilkhand university, Bareilly (U.P.).

Enzyme Preparation

Crude cellulase from *Humicola grisea* var. *thermoidea* was prepared by growing the fungus in Glucose ammonium dihydrogen phosphate medium (Bhat and Maheshwari 1987) supplemented with 1% alkali treated sugarcane baggase in citrate phosphate buffer pH 6.0 for different days, i.e., 3rd, 5th and 7th day, at their optimum temperature (50°C). 25 ml of medium was poured in 100 ml Erlenmeyer flask, autoclaved and inoculated with 4 mm disc of freshly grown test fungus. Flasks were taken out at 3rd, 5th and 7th day and centrifuged at 6000 rpm for 10 minutes. The supernatant was used as enzyme.

Enzyme Activity

Cellulase activity was determined by evaluating exoglucanase, endoglucanase and β-glucosidase activity according to IUPAC (Ghosh 1987). Enzyme activity was expressed in terms of specific activity (Cellulase units per mg of protein as estimated by Lowery's method (1951)).

Mutation

Mutants were prepared by treating the test fungus with 5-Bromo uracil. Spore suspension of the test fungus was treated with 5-Bromo uracil solution for different time intervals. Spores were harvested at different time and washed with sterile distilled water. Spores were plated on solid yeast powder soluble starch medium and untreated spores were taken as control. Agar plates were incubated at 45°C, appearing colonies were isolated and selected as mutants. The selected mutants were showed the phenotypic changes as compared to wild type. Mutant thus obtained were tested for their specific requirement of the amino acid by growing them on replica plates with complete medium and on medium eliminated with particular amino acid.

Results

Data for the cellulase activity of the mutants of *Humicola grisea* var. *thermoidea* obtained by the treatment with 5-Bromo uracil are presented in Table 1. A total of 16 mutants were isolated of which only five were selected named as HGBUM-1, HGBUM-2, HGBUM-3, HGBUM-4 and HGBUM-5.

The mutants showed different level of increase in exoglucanase, endoglucanase and β glucosidase activity. Wild type showed the maximum exoglucanase activity on day 7th while the mutant HGBUM-2, HGBM-4 and HGBUM-5 showed the maximum activity on day 5th. The mutants HGBUM-1 and HGBUM-3 showed maximum activity on 7th day (like wild type). The percentage increase in exoglucanase activity in HGBUM-1 to HGBUM-5 mutants over wild type observed was 37, 36, 45, 43 and 46% respectively (Table-2). It was noticed that all the mutants showed comparatively less increase in endoglucanase activity than exoglucanase and β-glucosidase activity. Maximum increase in endoglucanase activity was observed for mutant HGBUM-3 (26%) where as maximum increase in β-glucosidase activity was observed for mutant HGBUM-5 (40%) on day 5th.

The selected mutants differ in their requirement of amino acid for their growth. The mutant HGBUM-1 and HGBUM-3 are marked for methionine, HGBUM-2 for valine. HGBUM-4 for aspartic acid and HGUBM-5 for

TABLE 1 *Cellulase Profile of* Humicola grisea var. thermoidea *wild type and its mutants*

Isolates	*Mutagen*	*Days*	*Specific activity (u/mg) protein*		
			Exoglucanase	Endoglucanase	β-Glucosidase
Wild Type	5 Bromo uracil	3	0.90	0.59	0.22
		5	0.98	0.65	0.26
		7	1.06	0.70	0.29
HGBUM-1		3	1.09	0.44	0.27
		5	1.41	0.70	0.29
		7	1.70	0.77	0.33
HGBUM-2		3	1.15	0.45	0.27
		5	1.65	0.78	0.40
		7	1.41	0.76	0.47
HGBUM-3		3	0.94	0.57	0.25
		5	1.84	0.95	0.37
		7	1.92	0.90	0.44
HGBUM-4		3	0.97	0.46	0.21
		5	1.86	0.58	0.27
		7	1.74	0.58	0.21
HGBUM-5		3	1.57	0.59	0.33
		5	1.97	0.73	0.48
		7	1.50	0.80	0.40

Enzyme production on Bhat and Maheshwari (1987) medium (ph 7.0) at 45°C

TABLE 2 *Percent increase in Cellulase activity of mutants of* Humicola grisea var. thermoidea

Mutants	*Exoglucanase*	*Endoglucanase*	*β-Glucosidase*
	% increase	*% increase*	*% increase*
HGBUM-1	37	9	12
HGBUM-2	36	10	38
HGBUM-3	45	26	34
HGBUM-4	43	5	-7
HGBUM-5	46	13	40

cystein. Their colony diameter, pattern of growth and colony characteristics were different from that of the wild type (Table-3). Generally the mutant showed lesser rate of growth than wild type.

Discussion

During the present study, the mutagen used was 5-Bromo uracii which acts as a base analogue and thus caused the structural changes into the DNA

TABLE 3 *Hypercellulolytic mutants of* Humicola grisea var. thermoidea *and its characteristics*

S.No.	*Isolates*	*Genotype*	*Colony characteristics*
1	Wild	Prototype	Colony in two zones - outer is greyish pink (Puffy) and inner is grey, spores brown
2	HGBUM-1	Methonine	Colony in two zones - outer is whitish pink and inner is grey, spores light brown.
3.	HGBUM-2	Valine	Colony in three zones - outer is whitish pink, middle and inner are greyish pink, spores light brown.
4.	HGBUM-3	Methionine	Colony in two zones - outer is white and middle is brownish pink, spores brown
5	HGBUM-4	Aspartic acid	Colony in two zones - outer is pinkish white, puffy and inner is pink, spore light brown
6.	HGBUM-5	Cysteine	Colony in two zones - outer is whitish grey and inner is brownish pink, spores reddish brown.

Growth on Yeast powder soluble starch medium at temperature 45°C and pH 7-0.

by replacing thiamine and also occasionally having faulty pairing with guanine. As a result many mutants were obtained, of which five mutants have been selected as an increase producer of cellulase complex than wild type (Table-2). The mutants obtained were characterized as auxotrophic mutants which showed the marked difference in their morphology from the wild type but these distinctions have not been found any correlation with the enzyme activity of mutants. The similar results were also obtained by many workers (Sharma and Goel 1989; Nozima *et al.* 1995; Kvachadze and Yashvili 1997; Melo *et al.* 1997; Petrova *et al.* 2000; Basaram *et al.* 2000; Pocas *et al.* 2000; Biosset-Claire *et al.* 2001). Mutants showed an increase up to 46% for exoglucanase, 26% for endoglucanase and 40% for β-glucosidase. Thus these mutants shall play an important role in waste utilization technology.

Acknowledgement

The author KJ is thankful to the "Head" Department of Plant Science Rohilkhand university campus, for providing necessary facilities during course of study.

References

Basaran Pervin, Basaran Nese, Hang Yong-D (2000) Isolation and characterization of *Pichia stipitis* mutants with enhanced xylanase activity. *World J of Microbiol Bioech* 16(6): 545-550.

Bhat KM and Maheshwari R (1987) *Sporotrichum thermophile* growth, cellulose degradation and cellulase activity. *Appl Enviorn Microb* 53: 2175-2182.

Boisset Claire, Petrequin Cecile, Chanzy Henri, Henrissat, Bernard, Schulein M (2001) Optimized mixtures of recombinant *Humicola insolens* cellulases for the biodegradation of crystalline cellulose. *Biotech, Bioeng* 72(3): 339-345.

Bollok M and Reczey K (2000) Cellulase enzyme production by various fungal strains on different carban sources. Acta-Alimentaria 29(2): 155-168.

Ghosh TK (1987) Measurement of cellulase activity. *Pure Appl Chem* 59: 257-268.

Golias H, Dumsaday GJ, Stanley-Grant A, Pamment-Neville B(2000) Characteristics of cellulases preparation affecting the simultaneous saccharification and fermentation of cellulose to ethanol. *Biotech Letters* 22(7): 617-621.

Gooes I, Gomes J, Gomes DJ, Steiner W(2000) Simultaneous production of high activities of thermostable endoglucanase and beta-glucosidase by the wild thermophilic fungus *Thermoascus aurantiacus. Appl microb. Biotech.* 53: 461-468.

Gutierrez-correa M and Tengerady RP (1997) Xylanase Production by fungal mixed culture solid substrate fermentation on sugarcane baggase. Biotech letters 19: 665-667

Haros-Monica, Rosell-Cristina-M, Benedito-carmon (2002) Effect of different carbohydrases on fresh bread texture and bread staling. *European-Food-Research Technology.* 215: 425-430

Kanotra S and Mathur RS (1993) Biodegradation of Paddy straw with cellulolytic fungi and its application on wheat crop. *Bioresource Tech* 47: 185-188.

Kannangara T, Utkhede RS, Paul JW, Punja ZK (2000) Effect of mesophilic and thermophilic composts on suppression of *Fusarium* root and stem rot of greenhouse Cucumber. *Cand J Microb* 46: 1021-1028.

Krishna S Hari and Chowdary GV (2000) Optimization of simultaneous saccharification and fermentation for the production of ethanol from lignocellulosic biomass. *J Agricultural Food Chemistry* 48: 1971-1976.

Kvachadze LL and Yashvili TS (1997) Selection of a *Chaetomium thermophile* strain producing thermostable cellulases. *Mikrobiologiya* 66: 644-649.

Lee D, Yu AHC and Saddler JN (1995) Evaluation of cellulose recycling strategies for the hydrolysis of lignocellulosic substrates. *Biotech Bioeng* 45: 328-336

Lowery OH, Rosenbregh NJ, Farr AL and Randall RJ (1951) Protein measurement with the Folin phenol reagent. *J Biol Chem* 193: 265-275.

Melo IS, Faull JL, Graeme-Cook KA (1997) Relationship between in vitro celilulse production of UV-induced mutants of *Trichoderma harizanum* and their bean rhizosphere competence. Myco research 101(11): 1389-1392.

Mouchancca-Jean (2000) Thermophilic fungi and applied research: A synopsis of name changes and synonymics. world J of Microbiology and Biotechnology 16(8-9): 881-888.

Nojima Y, Yagi T, Miyakawa T, Matsuzaki H, Hatano T and Fukul S (1995) Extra cellular formation of triglycerides from glucose by a mutant strain of *Trichosporon. J Ferment Bioeng* 80: 88-90

Oh-Kyeong K, Kim SW, Jeong YS, Hong SI(a) (2000) Bioconversion of cellulose into ethanol by nonisothermal simultaneous saccharification and fermentation. *Appl Biochem Biotech* 89: 15-30.

Petrova SD, Ilieva SZ, Bakalova NG, Atev AP, Bhat MK and Kolev DM (a) (2000) Production and characterization of extracellular alpha-amylases from the thermophilic fungus *Thermomyces lanuginosus* (wild and mutant strains) *Biotech Letters* 22: 1619-1624.

Pocas-Fonscca-Marcio-J, Silva-Pereira Ildinete, Rocha-Bruno-B, Azevedo Maristella-de-o (2000) Substrate dependent differential" expression of *Humicola grisea* var. *thermoidea* cellobiohydrolase genes. *Cand J Microb* 46: 749-752.

Sharma VK and Goel R (1987) High cellulase producing mutants of *Sporotrichum thermophile. J Gen Appl Microb* 35: 163-166.

Tahoun MK and Ibrahim AA (1999) Conversion of cellulosic substrate into fermentable sugar by recombinant fungi strains. Zeitschrift-fuer-Lebensmittel-unter-suchung and Forschung 208(1): 65-68.

Van-wyk JPH (1997) Cellulase adsorption, desorption and cellose saccharification during enzymatic hydrolysis of cellulose materials. *Biotech letters* 19, 8: 775-778.

Van-wyk JPH, Mogale MA, Seseng TA (2000) Saccharification of used paper with different cellulases. *Biotech Letters* 22: 491494.

Van-wyk JPH and Leogale PR (2001) Saccharification of waste water mixtures with cellulases from *Penicillium funiculosum. Biotech Letters* 23: 1849-1852

Yu-Peigiang, Mckinnon-John J, Maenz-David D, Olkowski-Andrzej-A, Racz-Vern J, Christensen David A (2003) Enzymic release of reducing sugars from oat hulls by cellulase as influenced by *Aspergillus* ferulic acid osterase and *Trichoderma* xylanase. *J Agricultural Food Chemistry* 1; 51: 218-223.

Frontiers in Plant Sciences (Pages 381-394)
Eds. K.G. Mukerji, K.V.B.R. Tilak, S.M. Reddy, L.V. Gangwane, P. Prakash, I.K. Kunwar

Calcium Mobilization in *Neurospora crassa*

Naveena Lavanya Latha, J. and Maruthi Mohan, P.*

ABSTRACT

Cell wall bound calcium constitutes a significant fraction (25%) in *Neurospora crassa*. The role of surface bound calcium was studied with a view to understand its equilibrium with intracellular Ca^{2+} pools. Cell wall Ca^{2+} levels were found to increase with calcium concentrations of growth medium, reaching saturation by 0.75 mM, after which it concentrates nearly 50% of the total Ca^{2+}. Growth of *N.crassa* in Ca^{2+}-free medium resulted in decrease of cell wall calcium with a corresponding increase in growth and intracellular levels of Ca^{2+}. Calcium removal from surface of mycelia and incubation in a Ca^{2+}-free medium resulted in mobilization of 75% of intracellular Ca^{2+} pool to cell wall fraction. Verapamil, nifedipine, neomycin and wortmanin inhibited Ca^{2+} mobilization, while azide inhibited 40% of the same. Concanamycin A, a specific inhibitor of vacuolar H^+-*ATPase* inhibited Ca^{2+} mobilization by more than 95%. Vacuolar mutant of *N.crassa vma-5* defective in vacuolar Ca^{2+} transport failed to mobilize Ca^{2+} on to surface fraction, while the same was unaffected in a calcineurin mutant (*Cnb-1*) of *N.crassa*. The role of wall bound Ca^{2+} in toxic metal ion interactions and its homeostasis is discussed.

Key words: Calcium mobilization, homeostasis, vacuoles, cell surface, *Neurospora crassa*

INTRODUCTION

Ca^{2+} is an essential ion required for growth, metabolism and differentiation of yeast and fungi (Gadd, 1992). It acts as a second messenger for

*Author's address: Chairman, Board of studies, Department of Biochemistry, Osmania University, Hyderabad-500 007, India.

transducing intracellular signals responsible for a plethora of cellular functions. Fungi maintain low intracellular levels of free Ca^{2+} ions by a variety of transporters located on plasma membrane and membranes of intracellular organelles. Fungal vacuoles are the major intracellular storage source for Ca^{2+} ions. Ca^{2+} transport across plasma membrane in fungi is suggested to occur through gated channels sensitive to membrane potential (Stroobant and Scarborough. 1979). Efflux of Ca^{2+} in *N.crassa* was suggested to be by Ca^{2+}/H^{+}-antiporter located in plasma membrane. Ca^{2+} transport across vacuolar membranes in *N.crassa* was shown to be by Ca^{2+}/H^{+} antiporter driven by proton-motive force generated by H^{+} ATPase (Miller *et al.,* 1990). Though the release of Ca^{2+} ions from vacuoles was suggested to be by IP3 sensitive ligand-gated channels in *N.crassa* (Cornelius *et al.,* 1989), the recent genome sequence data could not identify such receptor genes (Galagon *et al.,* 2003). Mutants of *N.crassa* with defects in Ca^{2+} transport were characterized with defects in sub-units of vacuolar H^{+}*ATPase* (Bowman and Bowman, 2000) resulting in morphological defects in growth.

The role of Ca^{2+} in maintenance of apical dominance and polarized hyphal growth was demonstrated in *N.crassa*. Highest levels of Ca^{2+} were shown to be present at the growing hyphal tip and a decrease in medium Ca^{2+} concentration or loss of intracellular levels using ionophore A23186, resulted in abnormal branching and stunted growth (Schmidt and Harold, 1988). In '*frost*' and '*spray*' mutants of *N.crassa*, growth in presence of excess Ca^{2+} restored normal growth, while inducing low levels of Ca^{2+} with verapamil in wild type caused hyperbranching (Dicker and Turian, 1990). The recent genome sequence data of *N.crassa* indicated at least 25 proteins to be necessary for Ca^{2+} signaling (Galagon *et al.,* 2003). However the identification of the main components remains to be defined in this model fungus.

Apart from vacuoles significant quantities of extracellular Ca^{2+} was observed on surface fraction of *N.crassa*, but its role has not been investigated (Schmidt and Harold, 1988). Our present study is aimed at understanding the importance of surface-bound Ca^{2+} and its regulation in *N.crassa*.

MATERIALS AND METHODS

Fungal Strains

Neurospora crassa wild (#4200), calcineurin mutant of *N.crassa Cnb-1* (G 7416-732) were obtained from Fungal Genetics Stock Centeer, University of Kansas Medical Center, Kansas city, USA. Vacuolar mutant of *N.crassa* (*vma-5*) was obtained from Dr. B.J. Bowmann (Uuniversity of

California, Santa Cruz, U.S.A.). *Aspergillus niger, Aspergillus. flavus, Mucor racemosus, Rhizopus chinensis, Pencillium citriinum, Trichoderma viride* were obtained from Plant Pathology section of Botany Department, Osmania University, Hyderabad.

Media and Growth Conditions

Media composition and growth conditions for *Neurospora crassa* and metal ion estimation procedures were described in earlier studies (Maruthi Mohan and Sastry, 1983; Sajani and Maruthi Mohan, 1998). *N. crassa* was grown in 10 ml medium in 50 ml conical flasks for 72 h at 28 ± 1°C. The harvested mycelia were washed thrice with glass-distilled water and pressed free of excess moisture with filter paper. The mycelia were processed as per the protocol and then dried for over night in hot-air oven at 60-80°C. Growth of the mycelia was quantitated by measuring weight of mycelia.

Growth and Distribution of Ca^{2+} Between Surface-bound and Intracellular Fractions

N.crassa conidia (10^7) were inoculated in 50ml conical flasks containing 10 ml basal medium containing Ca^{2+} (0-2.5 mM) and incubated for 72 h at 28 ± 1°C. The mycelia were harvested, washed with deionized water and suspended in 10 ml of EGTA (10mM) for 5 min to remove surface-bound Ca^{2+}. The mycelia were then dried and Ca^{2+} content remaining (presumed mostly as intracellular) was estimated after acid digestion (Maruthi Mohan and Sastry, 1983) of both EGTA lechate and mycelia by Atomic absorption spectrophotometer (Perkin Elmer model 2380) or by Flame photometer.

In a separate experiment, conidial suspension of *N.crassa* was inoculated in 10 ml medium containing normal levels of Ca^{2+} and incubated for 72 h at 28 ± 1°C. The mycelial mats were harvested at different incubation periods, washed thrice with glass-distilled water. The surface bound Ca^{2+} was removed by floating the mycelia in 10 ml solution of EDTA or EGTA (10 mM) for 5 min, dried and then subjected to wet acid digestion for intracellular Ca^{2+} content.

Note: Most of the experiments were also repeated with $^{45}Ca^{2+}$ (specific activity =157mCi/g) purchased from BRIT, Mumbai.

Ca^{2+} Mobilization Assay and Effect of Channel Blockers and Inhibitors

The surface-bound Ca^{2+} from mycelial mats of *N.crassa* was leached with EGTA, and mycelia were incubated in Ca^{2+}-free basal medium for 5 min.

Ca^{2+} mobilized on to cell surface was checked by once again suspending the mycelia in EGTA solution. Again the cycle of incubation and leaching with EGTA was repeated for 4-5 cycles. Surface-bound Ca^{2+} released at each cycle and Ca^{2+} remaining with mycelia was estimated after acid digestion.

In other experiments EDTA, citrate and succinate were also used in place of EGTA. In control experiment mycelia were suspended in deionized distilled water alone. In experiments with channel blockers and inhibitors mycelial mats were suspended in EGTA (10 mM) for 2 min to remove surface Ca^{2+}, washed thoroughly with distilled water and resuspended in Ca^{2+}-free medium containing channel blocker/inhibitor for 30 min and then checked for Ca^{2+} mobilization on to surface after leaching with EGTA. nifidipine, verapamil, sodium azide, neomycin and concanamycin A are products of Sigma chemicals Co, St Louis, USA.

Ca^{2+} Mobilization During Growth

N.crassa mycelial mats grown for 2 days were harvested and reincubated in Ca^{2+}-free basal medium and allowed to continue growth. Mycelia were collected at various time points and Ca^{2+} distribution on surface and intracellular fractions was determined as described above by AAS.

REPRODUCIBILITY

All experiments unless indicated were repeated thrice in duplicates and average values ± standard deviation are shown.

RESULTS

Calcium Requirements for Growth and Distribution in *Neurospora crassa*

Growth of *N.crassa* was studied as a function of calcium concentration in growth medium. The data shown in Fig-1 indicates that optimal growth is obtained between 0.5 to 0.75 mM calcium and at higher concentrations there is a progressive decrease in growth by about 25% at the highest concentration (2.5 mM). Calcium distribution between wall-bound and intracellular fractions was determined in the mycelia. While intracellular calcium levels paralleled growth up to 0.75 mM of medium calcium, surface wall-bound calcium increased dramatically and accounted for nearly 50% of total calcium above 0.75 mM. The distribution of calcium between cell wall and intracellular fractions as a function of growth was studied up to

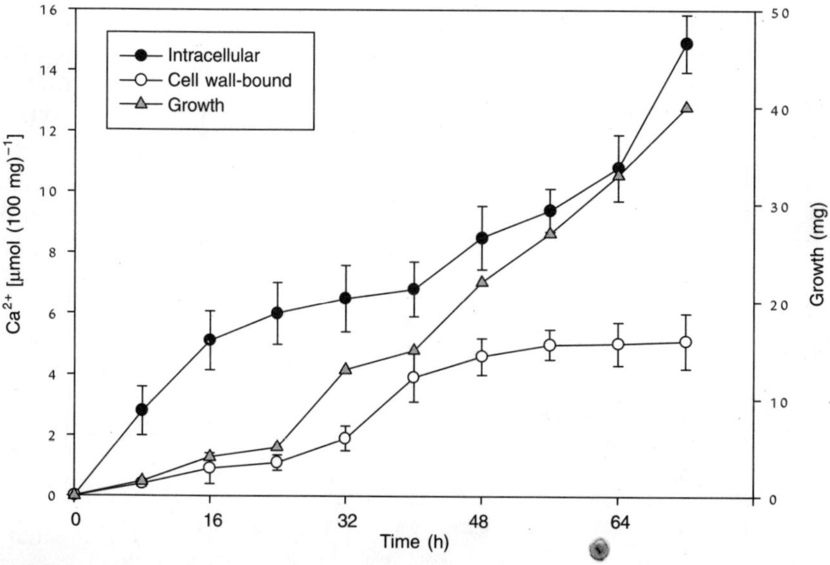

Fig. 1 Distribution of calcium during growth

Figure-1 Ca^{2+} distribution during growth in N.crassa

N.crassa conidiospores (10^7) were inoculated in to 10 ml basal medium and incubated under stationary conditions at 28 ± 1°C. Mycelia were harvested at indicated times and surface-bound Ca^{2+} was leached with 10 ml of EGTA (10 mM) and Ca^{2+} remaining (intracellular) after acid digestion of dried mycelia were estimated by AAS.

72 h by using optimal calcium concentration of 0.75 mM (300 ppm). The results presented in Fig-2 show increase in intracellular calcium levels from 0-5 μ moles (100 mg)$^{-1}$] at 16 h to about 15 μ moles (100 mg)$^{-1}$] at 72 h growth. In case of wall-bound fraction of calcium, significant quantities are observed only after 30 h growth. About 1 μ mole (100 mg)$^{-1}$] calcium is observed at 30 h, thereafter a progressive increase is seen which reaches up to 5 μ moles by 72 h growth of *N.crassa*. A point to be noted here is that at earlier time points of growth, calcium is predominantly intracellular and surface wall-bound calcium fraction becomes significant only at later time points of growth under the experimental conditions. Apart from calcium, Mg was the only other element present in significant quantities [2 μmoles (100 mg)$^{-1}$] on cell wall of mycelia and constitutes for 18% of the total mycelial magnesium content [12.4 ± 0.7 μ moles (100 mg)$^{-1}$]. Copper, manganese and molybdenum were found only in trace levels.

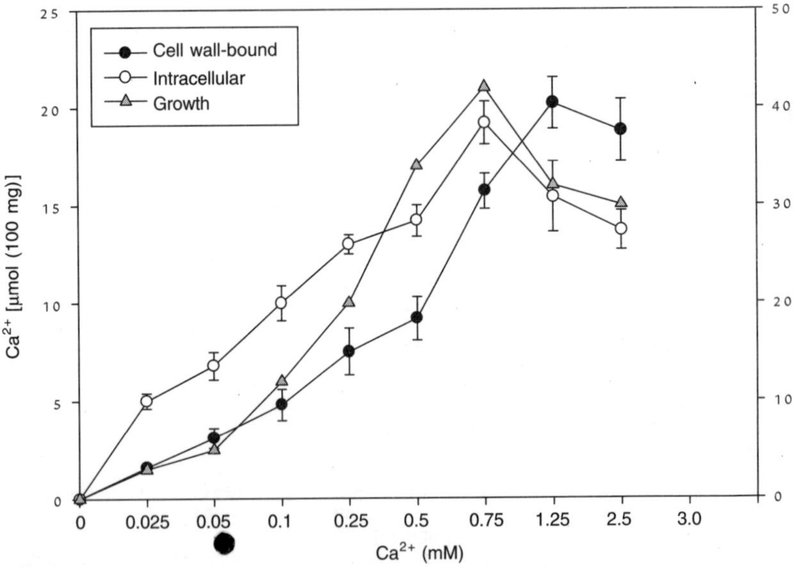

Fig. 2 Distribution of calcium in *N.crassa*

Figure-2 Effect of Ca^{2+} on growth and distribution in N.crassa
N.crassa conidiospores were inoculated in Ca^{2+}-containing medium (0-2.5 mM) and after 72 h of incubation mycelia were harvested and surface and intracellular levels of Ca^{2+} were estimated.

Cell Wall-bound Calcium is Required for Growth in Deficiency Conditions

In order to determine if wall bound calcium serves as storage source, *N.crassa* was grown for 48 h in normal medium and transferred to calcium-free medium, amd incubated further for 24 h. The data show that wall-bound calcium decreases with corresponding increase in growth, while intracellular calcium levels increased proportionally (Fig-3). Wall-bound calcium at 48 h was about 4.7 µ moles $(100\ mg)^{-1}$ which decreased to 2 µ moles $(100\ mg)^{-1}$ at 72 h. The corresponding increase in intracellular calcium was from 7 mmoles $(100\ mg)^{-1}$ at 48 h to 13 µ moles $(100\ mg)^{-1}$ at 72 h. Growth almost doubled during this 24 h period. This indicates that cell wall-bound calcium serves as storage source, under deficiency conditions.

Removal of Cell Wall-bound Calcium Results in Rapid Replacement from Intracellular Pool

Since a significant quantity of total calcium was observed on cell wall of mycelia, it was interest to see if the mycelia can reversibly bind calcium.

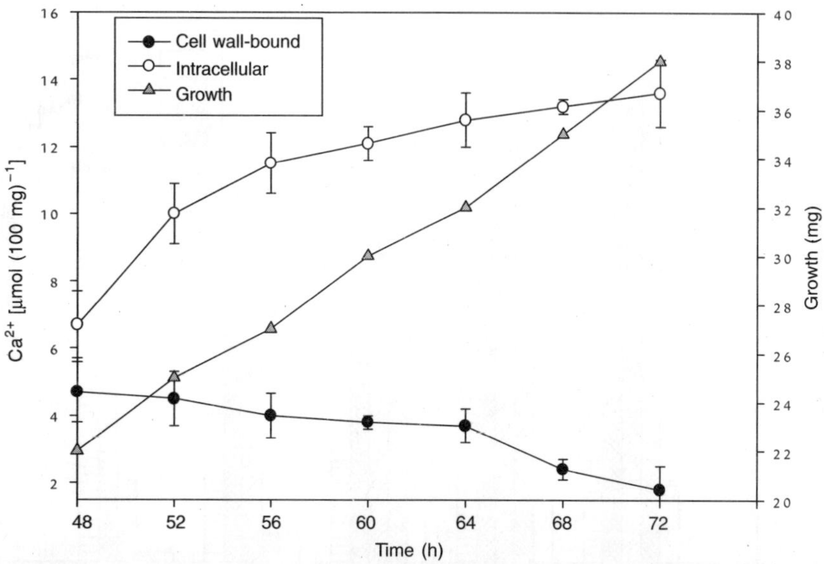

Fig. 3 Calcium distribution in Ca-depleted conditions

Figure-3 Ca^{2+} distribution on surafce and intracellular fractions during growth
After allowing growth for 48 h mycelial mats of *N.crassa* were harvested and re-incubated in Ca^{2+}-free basal medium and allowed to grow. At different incubation periods surface-bound and intracellular Ca^{2+} levels were estimated by AAS. A typical graph of two separate experiments is shown.

When cell wall-bound calcium was removed with EGTA, mycelial mats sequestered calcium immediately from medium. Calcium desorbed mycelia when suspended in calcium-free medium for 5 min regained more than 80% of initially bound quantities of calcium on cell wall fraction [about 4-5 µ moles $(100\ mg)^{-1}$]. When calcium was repeatedly desorbed with EGTA, calcium reappeared on cell wall fraction, though to a lesser quantity by 4th cycle (1 µ mole) (Fig-4). A total of 10-13 µmoles calcium $(100\ mg)^{-1}$, which accounts for about 75% of the total intracellular pool of calcium is mobilized to cell wall by 4th cycle and further mobilization was not significant. Calcium-depleted mycelia were able to accumulate calcium back to original levels and normal growth was restored, though with a distinct lag period. In order to see if calcium mobilization is a specific process EDTA was used in place of EGTA, which can chelate both calcium and magnesium ions. Our data showed that while calcium appeared rapidly on cell wall-fraction, magnesium was not observed. Further when succinate and citrate were used to remove cell wall-bound calcium, similar results were observed though quantitatively 20-50% lower amounts were leached at each cycle as compared to EDTA or EGTA.

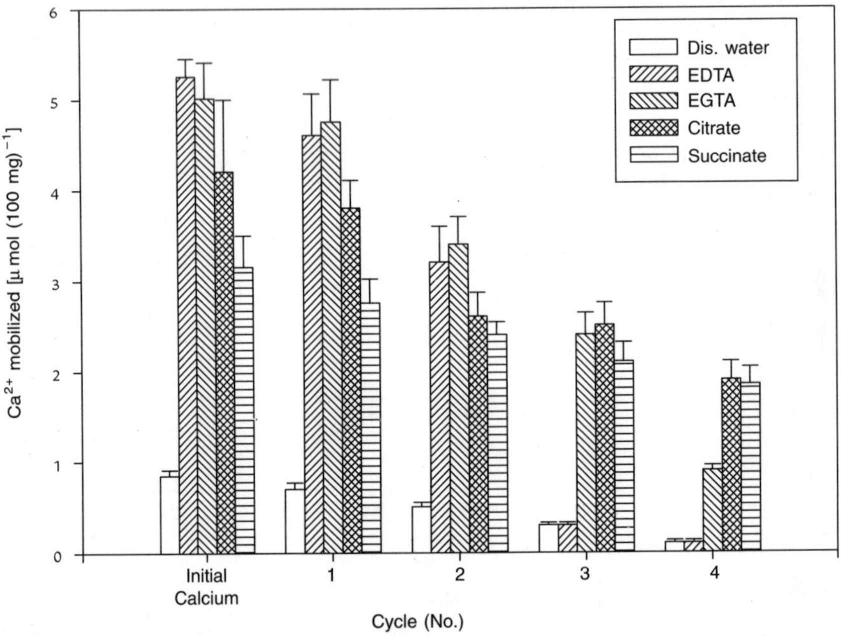

Fig. 4 Calcium mobilization on to cell wall

Figure-4 Ca^{2+} mobilization to surface fraction in N.crassa

Mycelial mats (3d) were treated with 10 ml solutions of indicated chelators (10 mM) to remove initial surface-bound Ca^{2+} and Mg^{2+} resuspended in Ca^{2+}/Mg^{2+}-free medium for 5 min. The Ca^{2+} and Mg^{2+} reappearing on surface fraction were once again removed with chelators (cycle-1) andabove cycle was repeated. Ca^{2+} and Mg^{2+} released were estimated by AAS and that remaining with mycelia at the end was also estimated following acid digestion. See, methods for details.

Effect of Inhibitors on Ca^{2+} Mobilization

Cytosolic free ionic calcium is maintained at very low levels (~100nM) in the cytoplasm and a sudden increase in these levels are dealt by excluding the same through calcium channels. Hence well-known channel blockers verapamil and nifedidpine were used to study their effect on calcium mobilization to cell wall fraction. The results indicate complete inhibition of calcium mobilization by 10 µM concentration of both the channel blockers used. On the other hand, metabolic inhibitor, sodium azide even at 100 µM concentration caused only 40% inhibition (table-1).

Vacuoles are the major storage site for calcium in *N.crassa* and the release of calcium from vacuoles has been shown to be dependent on IP3. Also a specific inhibitor of *N.crassa* vacuolar membrane V-type *ATPase,* concanamycin A involved in calcium transport was used to study the calcium mobilization. Concanamycin A inhibited calcium mobilization to wall fraction by more than 95% at 0.5 µM. Neomycin, which blocks IP3

TABLE 1 *Effect of incubating with inhibitors on calcium release in to medium*

Inhibitor	*Concentration (μM)*	*Calcium [μmol 10 ml⁻¹ 100 mg⁻¹](%)*
Control	-	4.8 ± 0.4 (100)
Sodium azide	100.0	2.9± 0.27 (60)
Verapamil	1.0	0.6 ± 0.06 (84)
Nifedipine	1.0	ND
Concanamycin A	0.5	0.26 ± 0 (5)
Neomycin	0.1	0.05± 0 (<1)
Wortmanin	20	0.5± 0.1(<10)
Calmidazolium	100	5.2± 0.5 (105)
Trifluoperazine	100	4.6± 0.3 (98)

Mycelial mats (3 days) were treated with 10 ml of EGTA (10mM) to remove calcium and resuspended in calcium-free medium containing specific inhibitors and calcium released into the EGTA solution was estimated by AAS.

turnover inhibited the calcium mobilization by 99% even at 0.1 μM concentration (table-1). Wortmanin, the specific IP3 inhibitor also blocked calcium mobilization by >90% at 20 μM. Inhibitors of calmodulin pathway calmidazolium (CaMZ) and trifluoperazine (TFP) had no effect on calcium mobilization on to cell wall (table-1).

Calcium Mobilization in Mutant Strains

In further experiments a mutant of *N.crassa* defective in vacuolar membrane *H^+-ATPase (vma-5)* was used in experiments to study calcium-mobilization to cell wall. The data obtained (table-2) clearly indicate that when cell wall

TABLE 2 *Calcium release in to EGTA solution in mutants of Neurospora*

Organism	*Initial calcium*	*Calcium rel;eased*
	[μmol (10 ml⁻¹100mg)⁻¹]	
N.crassa (wild)	5.07 ± 0.6	4.7 ± 0.4
vma-5	4.8 ± 0.45	ND
Cnb-1	5.15 ± 0.5	4.05 ± 0.3
N.crassa cor	6.2 ± 0.7	5.4 ± 0.5
N.crassa CSM	4.6 ± 0.4	2.1 ± 0.3
Aspergillus niger	8.0 ± 0.6	4.8 ± 0.4
A.flavus	4.3 ± 0.2	2.6 ± 0.2
M. racemosus	2.8 ± 0.3	1.7 ± 0.1
R.chinenses	4.2 ± 0.5	2.4 ± 0.4
P. citriinum	6.1 ± 0.4	3.1 ± 0.2

Cell wall-bound calcium of *vma-5*, *Cnb-1* and wild type *N.crassa* was leached with 10 ml of EGTA solution and mycelia resuspended in calcium-free medium for 5 min and calcium appearing on cell wall fraction was monitored as described in methods.

calcium was removed the same was not detectable in *vma-5* mutant of *N.crassa*. While in a calcineurin mutant (*Cnb-1*), calcium reappeared as in wild type *N.crassa*. It should be noted that the normal levels of cell wall-bound calcium was significantly less (~15%) in the two *N.crassa* mutants (*vma-5 and Cnb-1*) used in the study when compared to wild type.

Also, cobalt resistant (*cor*) and sensitive mutants (CSM) of *N.crassa* which have significant differences in cobalt uptake were used to evaluate calcium distribution. In *cor* mutant which has transport-block and binds more cobalt to wall-fraction, higher levels of cell wall-bound calcium was observed when compared to wild type. While in the cobalt-sensitive mutant (CSM) which is a hyperaccumulator and binds less cobalt to cell wall, lower levels of calcium were observed in wall fraction. In order to see if calcium mobilization is a common phenomenon in related fungi, several related species of fungi were tested by similar experimental protocols. Calcium mobilization on to cell wall fraction was observed in all the five fungal species examined (table-2).

DISCUSSION

Ca^{2+} was earlier shown to be required for growth of *Neurospora crassa* (Schmidt and Harold, 1988). However, differences were noted with respect to Ca^{2+} concentrations required for biomass production, and hyphal growth. While biomass production at 100 μM Ca^{2+} was normal, morphological abnormalities were observed in hyphal growth. It should be noted that the above studies were conducted at pH 7.0, which is not optimal for *N.crassa*. In the present study, total Ca^{2+} levels increased with growth almost 5-fold from 8 h to 72 h time period. However cell wall–bound Ca^{2+} levels were lower than 10% of total mycelial Ca^{2+} up to 24 h. This increased up to 25% at 72 h growth period. Further, growth of *N.crassa* was depended on Ca^{2+} concentration of the medium. Optimal growth up to 250 μm was observed above which it becomes a little toxic. Ca^{2+} distribution pattern indicated that at above optimal levels of Ca^{2+} in medium, cell wall-bound Ca^{2+} increased more than 2-fold and accounted for more than 50% of total Ca^{2+}. High concentration of Ca^{2+} in external medium is known to inhibit hyphal growth (Jackson and Heath, 1993), however, whether this is due to interactions with cell wall or a general toxic response is not known. The present result indicates that growth inhibition caused by excess Ca^{2+} could be due to increased cell-wall Ca^{2+} levels, which in turn could inhibit hyphal growth.

Ca^{2+} homeostasis is tightly regulated in all organisms and has been extensively investigated in yeast and fungi (Eilam *et al.*, 1985; Pitt and Ugalde, 1984). However, the distribution of Ca^{2+} between cell wall-bound and intracellular faction has not been investigated to our knowledge. Schmidt and Harold (1988) were the first to report the presence of Ca^{2+} on cell wall-faction in *N.crassa*. Our present study show for the first time that cell wall-bound Ca^{2+} is carefully monitored in *N.crassa* and maintained at steady levels in mature mycelia. Removal of cell wall–bound Ca^{2+} results in rapid replacement from intracellular storage pool. The process of calcium mobilization was found to be ceased within a time period of 5 min. Vacuoles are the major storage site for Ca^{2+} in *N.crassa* (Cornelius and Nakashima 1987). Concanamycin-A, a specific antibiotic inhibitor of *H^+-ATPase* of vacuolar membrane inhibits Ca^{2+}-mobilization to cell wall-faction by 95% at low concentrations (0.5 µm). Channel blockers, verapamil and nifedipine also inhibit Ca^{2+} mobilization completely in a dose-responsive manner. A general inhibitor of IP_3 turnover (neomycin) also blocks Ca^{2+} mobilization implicating it to be involved in the signaling pathway. Though IP3 was shown in vitro to release Ca^{2+} from vacuoles (Cornelius *et al, 1989)*, the receptor for the same could not be identified from the recent genome sequence data of *N.crassa* (Galagon et al, 2003). All the above experiments clearly point to the fact that majority of Ca^{2+} mobilized to cell wall fraction is from vacuolar source. This is further confirmed by the use of vacuolar mutant of *N.crassa* (*vma-5*) which is defective in one of the H^+-ATPase subunit of vacuolar membrane in *N.crassa* and is defective in Ca^{2+} transport (Bowman and Bowman, 2000). Sodium azide, a metabolic inhibitor had only 40% inhibitory effect on Ca^{2+} mobilization, while calmodulin inhibitors, trifluoperazine (TFP) and calmidazolium (CaMz) did not have any effect. Consequently in calcineurin mutant (*Cnb-1*), Ca^{2+} mobilization was normal like in wild type *N. crassa* (Kothe and Free 1998). Ca^{2+} was shown to be required for spore germination and an increase in CaM related activities were reported in *N.crassa* (Praveen Rao *et al.,* 1997). Though calmodulin-calcineurin are the major central pathways for modulating Ca^{2+} concentration effects, the present study does not implicate their role in Ca^{2+} mobilization to cell wall.

Most of the studies involving Ca^{2+} are related to apical growth, dominance and branching in filamentous fungi (Reissig and Kinney, 1983; Reiss *et al,* 1985; Brownlee and Wood, 1986). The steep Ca^{2+}-gradient with high concentrations of Ca^{2+} at tip were shown in *N.crassa* (Schmidt and Harold. 1988) and related fungi (Jackson and Heath, 1993). In all the above studies the knowledge about cell wall-bound Ca^{2+} was scanty or not available and hence intracellular homeostasis of Ca^{2+} was the major focus for discussion. In most of the studies vacuolar Ca^{2+} transport has been received considerable attention (Cornelius *et al.,* 1989; Bowman and Bowman, 2000; Cornelius, and Nakashima 1987). Vacuolar mutants with defect in the Ca^{2+} transport or *H^+-ATPase* ionic pump which maintains ionic gradient and couples Ca^{2+} transport were in general shown to suffer from morphological abnormalities. Using ionophores viz A12387 or inhibitors such as verapamil and nifedipine, abnormal branching has been induced in *N.crassa* and related fungi (Dicker and Turian, 1990; Tachikawa *et al,* 1987). On the other hand in 'frost' and 'spray' mutants of *N.crassa,* high levels of external Ca^{2+} was shown to allow normal growth of hyphae like in wild type. In these mutants Ca^{2+} was not observed in hyphal tips. High levels of external Ca^{2+} corrected morphological defect and restored high Ca^{2+} at growing tips (Dicker and Turian, 1990). In the above studies 1 mM verapamil was used when compared to the present study wherein 1 μM itself produced significant inhibitory effect on Ca^{2+} mobilization. It should be noted that in most studies shaker cultures which favour larger numbers of hyphal growth tips were used, while our study utilized stationary cultures which result in mycelial mats with fewer hyphal tips. We feel that this is closer to physiological conditions for growth of fungi, which infact grow commonly on semi-solid substrates.

Another logical function for Ca^{2+} on cell wall emanates from growth experiment of our present study. When *N.crassa* was allowed to continue growth in Ca-free medium, mobilization of cell wall-bound Ca^{2+} was observed, resulting in increased intracellular Ca^{2+} and growth. These data suggest for the presence of Ca^{2+} sensor which monitors both intracellular and cell wall-bond intracellular Ca^{2+}. Though Ca transport (efflux) across plasma membrane in *N.crassa* in suggested to be by a H^+/Ca^{2+} antiporter and influx through channels, no definitive information is available to our knowledge in *N.crassa* (Gadd G.M., 1993). The recent genome analysis showed about twenty putative transporters for calcium highlighting the importance in cell physiology of this organism (Borkovich *et al.*, 2004). Our present results clearly suggest the importance of cell wall-bound Ca^{2+} and its possible regulation and branching in *N.crassa.*

Cell walls in general have been implicated in conferring tolerance to toxic metal ions (Gadd G.M., 1993; Venkateswerlu and Stotzsky 1989) though clear experimental evidence is not available. The present data for the first time provides evidence for the cell wall in protection against toxic metal ions. It is tempting to suggest the Ca^{2+} on cell wall acts as a sensor for toxic metal ions (or other pertubants) in the environment and its displacement signals intracellular Ca^{2+} pools to be mobilized (IP3 signaling pathway?), which ultimately displaces toxic metal ions. This suggestion finds favour with the use of cobalt-resistant and sensitive mutants employed in the present study. While the resistant mutant has more cell wall-Ca^{2+} and binds more cobalt than wild type *N.crassa,* it is vice-versa for the sensitive mutant. A preliminary experiment showed that in *cor* mutant, cell wall-bound Ca^{2+} is replaced more rapidly and consequently displaces cell wall-bound cobalt ions, while it is opposite in case of the sensitive mutant of *N.crassa*. Another experiment supporting the above suggestion is that most of the wall-bound cobalt is released back into the medium. This release was shown to be due to rapid appearance of intracellular calcium on to cell wall fraction. Verapamil which blocks Ca mobilization, resulted in cobalt remaining to wall-bound fraction. Also vacuolar mutant of*N.crassa* were suggested to be hypersensitive to toxic metal ions (Bowman and Bowman, 2000), which further lends support our hypothesis. Our present study for the first time highlights the importance of cell wall-bound Ca^{2+} in growth and defense against toxic metal ion responses.

Acknowledgements

The authors (PMM & JNLL) would like to thank the University Grants Commission [UGC-SAP(DRS-1)] and Council Of Scientific & Industrial Research, New Delhi, for financial assistance.

References

Boprkovich, K.A. et al. 2004. Lessons from the genome sequence of *Neurospora crassa*: Tracing the path from genomic blue print to multicellular organism. *Microbiol. Moleoc boil. Rev.*, 68, 1-108

Bowman, E.J., Kendle, R. and Bowman, B.J. 2000. Disruption of *vma-1*, the gene encoding the catalytic subunit of the vacuole H^+-ATPase, causes severe morphological changes in *Neurospora crassa. J. Biol. Chem.* 275:167-176.

Brownlee, C. and Wood, J.W. 1986. A gradient of cytoplasmic free calcium in growing rhizoid cells of *Fucus serratus*. *Nature* 320: 624-26.

Cornelius, G., Gebour, G and Techel, D. 1989. Inositol triphosphate induces calcium release from *Neurospora crassa* vacuoles. *Biochem. Biophys. Res.Commun*. 162:852-856.

Cornelius, G. and H. Nakashima 1987. Vacuoles play a decesive role in Ca^{2+} homeostasis in *Neurospora crassa*. *J. Gen. Microbiol*. 133: 2341-2347.

Dicker, W. and Turian, G. 1990. Calcium deficiencies and apical hyper branching in wild type and frost and spray morphological mutants of *Neurospora crassa*. *J. Gen.Microbiol.* 136: 1413-1420.

Eilam, Y., Lavi, H and Grossowicz, N. 1985. Calcium homeostasis maintained by a vacuolar calcium transport system in the yeast *Saccharomyces cerevisiae*. *J. Gen.Microbiol*. 131: 623-629.

Gadd, G.M. 1992. Calcium signaling in fungi. In *'The growing fungus'* (ed. N.A.R Gow & G.M.Gadd) pp 183-210 Chapman & Hall: London.

Gadd G.M. 1993. Interactions of fungi with toxic metals. *New Phytology.* 124, 25-60.

Galagon et al, 2003. The genome sequence of filamentous fungus *Neurospora crassa*. Nature, 422, 859-68.

Jackson, S.L. and Heath, I.B. 1993. Roles of calcium ions in hyphal tip growth. *Microbiol. Rev.* 57:367-382.

Kothe, G.O. and Free, S.J. 1998. Calcineurin subunit is required for normal vegetative growth in *Neurospora crassa. Fungal Gen. Biol.* 23: 248-258

Maruthi Mohan, P. and K.S. Sastry 1983. Interrelationships in trace element metabolism in metal toxicities in nickel-resistant strains of *Neurospora crassa*. . *Biochem.J.* 212, 205-211.

Miller,A.J., Vogg, G and Sanders, D. 1990. Cytosolic calcium homeostasis in fungi: roles of plasma membrane transport in intracellular sequestration of calcium. *Proc. Nat. Acad. Sc.*, USA, 87, 9348-52.

Pitt and U.O. Ugalde. 1984. Ca^{2+} in fungi. *Plant Cell Environ*. 7: 467-475.

Praveen Rao, J., Reena, G and Subramanyam, C. 1997. Calmodulin-dependent protein phosphorylation durin conidial germination and growth of *Neuraspora crassa. Mycol. Res* 101: 1484-1488.

Reiss, H.D., Herth, W. and Nobling, R. 1985. Development of membrane and calcium-gradients during pollen germination of *Lilium longiflorum*. *Planta* 163: 84-90.

Reissig, J.L. and S.G. Kinney 1983. Ca^{2+} as a branching signal in *Neurospora crassa*. J. *Bacteriol*. 154:1397-1402.

Sajani, L.S. and Maruthi Mohan, P. 1998. Characterization of cobalt-resistant strain of *Neurospora crassa. Biometals,* 10, 175-183.

Schmidt, J. and F.M. Harold 1988. Dual roles for Ca^{2+} ions in apical growth of *Neurospora crassa. J. Gen. Microbiol*. 134: 2623-2631.

Stroobant, P. and G.A. Scarborough 1979. Active transport of Ca^{2+} in *Neurospora* plasma membrane vesicles. *Proc. Natl. Acad. Sci*. USA 767: 3102-3106.

Tachikawa, T., Miyala, T., Tsuchiya, E. and Fukui, S. 1987. A rapid and transient increase of cellular calcium in response to mating pheromone in *Saccharomyces cerevisiae. Agri. Biol. Chem.*, 51: 1209-1210.

Venkateswerlu, G. and Stotzsky, G. 1989. Binding of metals by cell walls of *Cunninghamella blakesleeana* grown in the presence grown in the presence of copper or cobalt. *Applied Microbiol. and Biotech*., 31, 619-625.

SECTION 4

MICROBIAL APPLICATION

Frontiers in Plant Sciences (Pages 397-422)
Eds. K.G. Mukerji, K.V.B.R. Tilak, S.M. Reddy, L.V. Gangwane, P. Prakash, I.K. Kunwar

Antibiotics: Past, Present, and Future—A Review

K.R. Aneja* and Parikshit Tyagi

ABSTRACT

The antibiotic (literally meaning "against life") era began in the middle of the last century, with introduction into widespread civilian use of first powerful agent, penicillin, discovered by Alexander Fleming in 1929 and followed by many other classes of such "Wonder drugs". Some of which that are in clinical practice include streptomycin, chloramphenicol, oxytetracycline, bacitracin, polymyxin, neomycin, erythromycin and carbomycin that were discovered between 1929 and 1960. During 1980 to 1989 medical science recognized the mechanism and risks of antibiotic resistance among the pathogenic microorganisms with the results fever and fever antibiotics maintained potency against rapidly mutated bacteria. It is now well accepted that the current arsenal of antibiotics is not going to be enough in the fight against pathogenic microorganisms in coming decades. Keeping in view this fact searches are to be made for the isolation of novel compounds to fight against deadly pathogenic microorganisms from the natural sources. In this review, the brief account of antibiotics discovered from 1929 to 2004 and the recent progress made in recovering the novel structures from the soil is summarized.

Key words: Natural antibiotics, Antimicrobial resistance, Soil screening, Fungi

For about five millennia, give or take a few centuries, human beings have treated their aches and pains with mind-boggling creativity. In 3500 BC, Sumerian physicians treated ailing patients with a beer soup heavily laced with ground snake skins and turtle shells. About 2500 years ago, ancient Greeks wrote about the use of moulds as healing agents. There are tales of

Author's address: Department of Microbiology, Kurukshetra University, Kurukshetra-136119, India
*Corresponding author's email: anejakr@rediffmail.com

the use of moulds by the ancient Chinese and Australian Aborigines to treat infections. 3000 yrs ago, Chinese applied moldy soybean curd to wounds to speed healing. Papyrus scrolls from the 16th century BC indicate that Egyptians recommended rubbing moldy bread on lesions. In the early days of microbiology, attempts were made to use extracts derived from fungal cultures to prevent surgical wound infection. Egyptian doctors dispense exotic concoctions, some of which are based on sound medical practice about 1500 BC. But it wasn't until the mid-19th century that Louis Pasteur observed that some microorganisms could destroy others, a phenomenon Pasteur believed could be useful in medicine (Hoel and Williams, 1997).

The word antibiotic has been in use much before the discovery of the first antibiotic. However, the word in its present restrictive and precise meaning was used and interpreted by Waksman. In 1939 Waksman defined **antibiotics** as *'chemical compounds derived from microorganisms, which have the capacity of inhibiting the growth of and even destroying other microorganisms in dilute solutions'*. Several other definitions have been forwarded from time to time. Pasteur and Joubert (1877) found that aerobic bacteria antagonized growth of *Bacillus anthracis.* Furthermore, they detected that experimental anthrax in susceptible animals could be repressed by simultaneous inoculation of various non-pathogenic bacteria. Since then many reports have appeared concerning the antibiotic activity of spore-forming bacilli (Primgsheim, 1920; Chambers and Weidman, 1928; Weiland, 1936). Gratia and Dath (1924) were among the first to begin a systematic search for antagonistic organisms from natural sources. They extracted a lytic agent from a mold and used it in successful treatment of staphylococcal skin infections.

The last days of the 19th and the first decades of the 20th centuries saw the birth of aspirin and barbiturates. In 1909, German bacteriologist, Paul Ehrlich, discovered the first chemical cure for disease, **salvarsan** an arsenic compound that cured early syphilis. Antibiotics indeed changed medicine revolutionarily, permitting control of many diseases that previously caused much morbidity and mortality even in otherwise healthy people-pneumonia, meningitis, and tuberculosis. In 1929, the most important and historical event of Fleming's discovery of penicillin took place ushering in the antibiotic era. Alexander Fleming a Scottish physician, who had devoted much of his career to finding methods for treating wound infections, recognized the antibacterial effect of a fungal metabolite that had grown as a contaminant on agar plate streaked with a bacterium*Staphylococcus aureus.* He noticed that the bacterial colonies around the fungus were transparent, because the cells were lysing. The metabolite was named penicillin because the fungal contaminant was identified as *Penicillium notatum*. Fleming found that penicillin was effective against many Gram-positive bacteria in laboratory conditions (Fleming, 1929). However, he could not purify this compound because of its instability, and it was not until the period of the Second World War (1939-1945) that two other British scientists, Florey and Chain, working in the USA, managed to produce the antibiotic on an industrial scale for widespread use. Its lack of toxicity was confirmed and its activity as a chemotherapeutic agent was demonstrated in mice and they were able to produce penicillin on an industrial scale for widespread use. All three scientists, Fleming, Florey and Chain shared the Nobel Prize in 1945 for this work, and rightly so- penicillin rapidly became "the wonder drug", which saved literally millions of life (Florey *et al.,* 1949). The phenomenal success of penicillin led to the search for other antibiotic-producing microorganisms, especially from soil environment. One of the early successes was the discovery of the streptomycin in 1943 by Salmen Waksman, from the soil actinomycetes, *Streptomyces griseus* (Waksman *et al.,* 1944). As a result of these vast screening programs, a considerable number of antibiotics from bacteria, actinomycetes and moulds have been made available namely bacitracin in 1945 (Johnson *et al.,* 1945), polymyxin and chloramphenicol in 1947 (Benedict and Longhykke, 1947; Ehrlich *et al.,* 1947), chlortetracycline in 1948 (Duggar, 1948), neomycin in 1949 (Waksman and Lechevalier, 1949), oxytetracycline in 1950 (Finlay *et al.,* 1950), synnematin in 1951(Gottshall *et al.,* 1951) and erythromycin and carbomycin in 1952 (McGuire, 1952; Tanner, 1952). Later several hundreds of compounds with antibiotic activity have been isolated from microorganisms over the years, but only a few of them are in clinical use (See Table 1).

Apart from the natural antibiotics, a number of synthetic antibiotics were also reported. Gerhard Domagk, a 38-year-old German pharmacologist, in 1934 discovered that dye used to tint cloth seemed to cure streptococcal infections in mice (Robinson, 1976). The quinolones, established with the introduction of nalidixic acid in 1962. A series of derivatives released in the 1970's only marginally improved on the properties of nalidixic acid. However, in the early 1980's it was discovered that a derivative of nalidixic acid containing a fluorine atom at position C-6, possessed excellent broad-spectrum activity *in vivo,* and unlike nalidixic acid showed a very low tendency to elicit resistance in microorganisms. Several fluoroquinolone derivatives namely enrofloxacin, norfloxacin, and ciprofloxacin were developed (King, 2000).

SCREENING OF ANTIBIOTICS FROM VARIOUS HABITATS

The majority of today's most effective classes of antimicrobials originated many decades ago as natural products isolated from soil-colonizing

TABLE 1 *Some major antibiotics and other antimicrobial substances discovered between 1929 and 2000*

Antibiotic	*Year of Discovery*	*Microorganism*
Penicillin	1929	*Penicillium notatum*
Sulfonamides (Prontosil)	1932	Synthetic
Tyrothricin	1939	*Bacillus brevis*
Gramicidin	1939	*B. brevis*
Fumagillin	1940	*Aspergillus fumigatus*
Streptomycin	1943	*Streptomyces griseus*
Bacitracin	1943	*Bacillus subtilis*
Cephalosporins	1945	*Cephalosporium acremonium*
Chloramphenicol	1947	*Streptomyces venezuelae*
Chlortetracycline	1947	*Streptomyces aureofaciens*
Polymyxin	1949	*B. polymyxa*
Neomycin	1949	*Streptomyces fradiae*
Nystatin	1951	*Streptomyces nousei*
Rimocidin	1951	*Streptomyces rimosus*
Erythromycin	1952	*Streptomyces erythreus*
Vancomycin	1956	*Streptomyces orientalis*
Kanamycin	1957	*Streptomyces kanamyceticus*
Methicillin	1960	Semi-synthetic (Peni. dervt.)
Ampicillin	1961	Semi-synthetic (Peni. dervt.)
Nalidixic Acid	1962	Synthetic (1st gen. Quinolone)
Gentamycin	1963	*Micromonospora* sp.
Clindamycin	1967	*Streptomyces lincolnensis*
Trimethoprim	1970	Synthetic
Cotrimoxazole	1970's	Synthetic
Enrofloxacin	1980's	Synthetic (2nd gen. Quinolone)
Norfloxacin	1980's	Synthetic (2nd gen. Quinolone)
Ciprofloxacin	1980's	Synthetic (2nd gen. Quinolone)
Fluoroquinolone	1990's	Synthetic (3rd gen. Quinolone)
Oxazolidinones	1999	Synthetic

microorganisms. Bacteria, actinomycetes and fungi are most preferably used as screening sources from various habitats. Fungi are well known as prolific producers of biologically active natural products. A significant number of these metabolites have found importance as pharmaceuticals or agrochemicals, or as useful lead compounds in the development of such agents. However, the fungi investigated to date represent only a small fraction (perhaps 5%) of estimated fungal biodiversity, as many fungi are niche-selective, and many ecological niches remain relatively unexplored as sources of fungi for screening (Gloer, 2004). For modern pharmaceutical research, biodiversity offer a vast source of chemical diversity and yield unusual and unexpected lead structures (Young, 1999). Natural products can derive not only lead structures but can often yield ready-made antibiotics, a number of such compounds are described here (See Table 2).

TABLE 2 *Brief account of bioactive metabolites isolated between 1999 and 2004 from various microorganisms*

Metabolite	*Year of Discovery*	*Microorganism*	*Activity*
Bis(2-ethylexyl) phthalate (DEMP), 2(4-hydroxyphenyl) -2-oxoacetaldehyde oxime (PHBA)	1994	*Penicillium olsonii*	Antibacterial and antifungal
Cladospirone bisepoxide	1994	*Cladosporium* sp.	Antibacterial and antifungal
Harizianic acid	1994	*Trichoderma harzianum*	Antibacterial
Sporothriolide	1994	*Sporothrix* sp.	Antimicrobial and herbicidal
Discosiolide	1994	*Discosia* sp.	Antimicrobial and herbicidal
4-epi-ethisolide	1994	*Pezicula livida*	Antimicrobial and herbicidal
Tricholin	1994	*Trichoderma viridae*	Antifungal
Darlucin A (1) and Darlucin B (2)	1995	*Sphaerellopsis filum*	Antibacterial, antifungal and weakly cytotoxic
Mycorrhizin A, Lachnumon, echloromycorrhizin A(12), papyracons A(13), B(14) and C(15)	1995	*Lachnum papyraceum*	Antimicrobial, cytotoxic, nematicidal and phytotoxic
Fusarielin A, B, C and D	1995	*Fusarium* sp.	Antifungal and cytotoxic
Favolon (1)	1995	*Favolaschia* sp.	Antifungal
NK 374200	1995	*Taralomyces* sp.	Antimicrobial, anti-mosquito larval and cytotoxic
Fusacandins A and B	1995	*Fusarium sambucinum*	Antifungal
Stachybocins A, B and C	1995	*Stachybotrys* sp.	Inhibitory to I-ET-1 binding of rat ET_A, human ET_A and human Et_n receptors
Halymecins A (1), B (2) and C (3)	1996	*Fusarium* sp.	Antimicroalgal
Halymecins D (4) and E (5)	1996	*Acremonium* sp.	Antimicroalgal
Culture broth	1996	*Sorangium cellulosum*	Antifungal
Cyclopentenone derivative	1996	*Trichoderma koningii*	Antibacterial and antifungal
Ascosteroside	1996	*Ascotricha amphitricha*	Antifungal
Papyracon D(6), 6-O-methylpapyracon B (9), 6-O-methylpapyracon C (10), lachnumfuran A (11) and lachnumlactone A (l2a)	1996	*L. papyraceum*	Weakly nematicidal and cytotoxic
Benesudon	1997	*Mollisia benesuada*	Antimicrobial, cytotoxic and phytotoxic

(Table Contd.)

(Table Contd.)

Tricyclic diterpene α-pyrones	1997	*Aspergillus* sp. NF 00659	Insecticidal
Aspergillimides	1997	*Aspergillus* strain IMI 337664	Antihelminthis
Aspochalasin F(1) and G(2)	1997	*Aspergillus* sp. FO-4282	Antimicrobial and weakly cytotoxic
Sch 56396 (1)	1997	*Tolypocladium* sp.	Anticancer
TMC-52 A-D and TMC-151 A-F	1999	*Gliocladium* sp.	Moderately cytotoxic
Prenylated p-terphenyl	1999	*Aspergillus candidus*	Antitumor
Tetranorditerpenoids, oidiodendrolites A(3), B(4), C (5) and oidiodendronic acid (7)	1999	*Oidiodendron truncatum*	Antifungal
Flagranones A (2), B (3) and, C (4)	1999	*Duddingtonia flagrans*	Antibacterial
Glisiprinins	2000	*Neobulgaria pura* HA A07-97	Fungicidal
1B-96212	2000	*Micromonospora* sp. L-25-ES25-008	Strongly cytotoxic
FR901469	2000	Unidentified fungus No.11243	Antifungal
Angucyclinone	2000	*Streptomyces antibioticus* Tu 6040	Antibacterial
Efrapeptins	2000	*Tolypocladium* sp.	Antifungal and antibacterial
BE-54238A and B	2000	*Streptomyces* sp. A54238	Cytotoxic
Tricyclic diterpenoid	2000	*Nocardia brasiliensis* IFM0406	Cytotoxic
Pluraflavin A ($C_4H_{54}N_2O_{14}$), pluraflavin B ($C_{43}H_{56}N_2O_{15}$) and pluraflavin E ($C_{36}H_{41}NO_{14}$)	2001	*Saccharothrix* sp. DSM 12931	Cytotoxic
CJ-17665 (I)	2001	*Aspergillus ocraceus* CL41582	Antibacterial and cytotoxic
L970843 and L970844	2001	Unidentified fungus HIL Y-903146	Antibacterial and antifungal
CJ-19784 (I)	2001	*Acanthostigmella* sp. CL12082	Antifungal
Cladospolide D	2001	*Cladosporium* sp.	Antibacterial and antifungal
Extract 1	2001	*Acremonium* sp. HKI 0230	Antibacterial
Tropolone named cordytropolone (1)	2001	*Cordyceps* sp. BCC 1681	Antimalarial
FR198248	2001	*Aspergillus terreus*	Anti-Influenza
TMC-69	2001	*Chrysosporium* sp. TC 1068	Antitumor

(Table Contd.)

(Table Contd.)

Clavariopsins	2001	*Clavariopsis aquatica* AJ117363	Antibacterial and antifungal
Epicorazine (1)	2001	*Basidiomycetes* sp.	Antibacterial and antifungal
Funicone	2002	*Talaromyces flavus*, FKI-0076	Anti-candida
Red pigment (quinone)	2002	*Fusarium solani*	Antifungal
2-oxo-succinimide polyketide	2002	*Vaginatispora aquatica* HK1821	Antibacterial and antifungal
Stachytlin and acetylstachytlin	2002	*Stachybotrys* sp. RF-7260	Antiviral
Simocyclinones D4 (1) and D8 (2)	2002	*Streptomyces antibioticus* Tu 6040	Antibacterial and antifungal
GEXI	2002	*Streptomyces* sp. GEX IA	Antibacterial and antifungal
Miyakamides A1 A2, B1 and B2	2002	*Aspergillus flavus*	Phytotoxic
YM-202204 (1)	2002	*Phoma* sp. Q60596	Antifungal
Altersetin	2002	*Alternaria* sp.	Antibacterial and antifungal
Hydroxysordarin and Neosordarin	2002	*Sordaria araneosa*	Antifungal
Atroviridins A-C	2002	*Trichoderma atroviride*	Antibacterial and antifungal
Hexacyclinol (1)	2002	*Panus rudis* HKI 0254	Antibacterial and antifungal
GTRI-BB	2002	*Micromonospora* sp. SA-246	Antibacterial, antifungal and cytotoxic
Sch 484129 (I) and Sch 484130 (2)	2002	*Basidiomycetes* sp.	Antifungal
Polycyclic xanthone IB-0020X	2003	*Actinomadura* sp.	Antimicrobial
Monordens C, D and E	2003	*Humicola* sp. FO-2942	Antifungal and cytotoxic
8-O-methylaverufin (1b) and 8-O-dimethylaverantin (2b)	2003	*Penicillium chyrsogenum*	Antibacterial and antifungal
Diketopiperazine (1)	2003	Unknown fungus	Antifungal
SCH643432 (1) and (2)	2003	*Paecilomyces varioti*	Antifungal
Acremonidins A-E (1-5)	2003	*Acremonium* sp.	Antibacterial
Enniatins H (3) and I (4)	2003	*Verticillium hemipterigenum* BCC 1449	Antiplasmodial and antimycobacterial
Nocathiacin I, II and III	2003	*Nocardia* sp. WW-2651	Antibacterial
Compounds (1~3)	2003	*Aspergillus ostianus*	Antibacterial
Bafilolide R176502	2003	*Micromonospora* sp.	Cytotoxic
IB-00208	2003	*Actinomadura* sp.	Antibacterial and antitumor
Beauvericins D, E and F	2004	*Beauveria* sp. FKI-1366	Anti-candida
Crocacin A	2004	*Chondromyces crocatus*	Antifungal and cytotoxic
Macrolides (pladienolides)	2004	*Streptomyces platellsis* Mer-11107	Antitumor

(Table Contd.)

(Table Contd.)

ICM030IA-H (1-8)	2004	*Aspergillus* sp. F1491	Anti-angiogenic
SF2809-1, SF2809-1I, SF2809-1II, SF2809-IY, SF2809-V and SF2809-VI	2004	*Dactylosporangium* sp.	Chymase inhibitors
Vertihemipterin A	2004	*Verticillium hemipterigenum* BCC 2370	Antiviral and cytotoxic
Pacciloxazinc	2004	*Paecilomyces* sp. BAUA3058	Insecticidal
Vanillobiocin, isovanillobiocin and declovanillobiocin	2004	*Streptomyces roseochromogenes*	Antibacterial
α, β unsaturated aldehydes	2004	*Cladosporium* sp.	Antibacterial
Retamycin	2004	*Streptomyces olindensis*	Antitumor

ANTIBIOTICS FROM GENUS *PENICILLIUM*

The first antibiotic to be used in medicine, penicillin derived from a fungus *Penicillium notatum*, is still one of the most useful and important antibiotic (Demain and Elander, 1999). In 1994 Amade, and coworkers have been isolated two metabolites as bis(2-ethylexyl) phthalate (DEMP) and 2(4-hydroxyphenyl)-2-oxoacetaldehyde oxime (PHBA) reported as a natural compound in the (E)-s-cis configuration from *Penicillium olsonii*. The strain was isolated from terminal leaves of *Suaeda fruticosa*. They showed inhibitory activity against gram-positive bacteria and phytopathogenic fungi (Amade *et al.*, 1994).

Maskey and coworkers in 2003 elucidated the structure and biological activity of 8-*O*-methylaverufin (1b) and 8-*O*-dimethylaverantin (2b) as a new antifungal agent from *P. chrysogenum*. The crude extract was inactive against *Streptomyces viridochromogenes*, *E. coli* and *Candida albicans*. 8-*O*-dimethylaverufin (1b) showed activity against the fungus *Mucor miechi*. While 8-*O*-dimethylaverantin (2b) lacks antibacterial activity but possess antifungal activity (Maskey *et al.*, 2003).

ANTIBIOTICS FROM GENUS *ASPERGILLUS*

During a search for new pesticide compounds from microbial metabolites, Suzuki *et al.* (1997) found that *Aspergillus* sp. NF 00659 produced new tricyclic diterpene α-pyrones antibiotics with potent insecticidal activities against the larvae of mosquito, *Culex pipiens molestus* and *Plutella xylostella*. They showed potent antitumor activities against human ovarian carcinoma A2780 and human colorectal adenocarcinoma SW-480 cells.

Fang and coworkers in 1997 during the course of screening program for new antimalarial antibiotics, isolated from strain FO-4282, a fungal soil isolate and was identified as *Aspergillus* sp. which produce two new components of the aspochalasins nature. Compound 2, showed moderate *in vitro* antibacterial and antifungal activity (Fang *et al.*, 1997).

Two members of a novel class of antihelmintics, the aspergillimides have been isolated from the *Aspergillus* strain IMI 337664 by Banks *et al.*, 1997.

Stead *et al.* (1999) reported a prenylated *p*-terphenyl metabolite from the fungus *A. candidus* was isolated from dung. The compound possesses cytotoxic activity against a range of tumor and other hyperproliferative cell lines.

Sugie and coworkers in 2001 reported a new antibiotic, CJ-17,665 (I) produced by *A. ocraceus* CL41582. The fungus was isolated from a soil sample collected in Venezuela. The antibiotic is reported to be effective against multi-drug resistant *Staphylococcus aureus* (MDRS), *Streptococcus pyogens* and *Enterococcus faecalis*. It showed cytotoxicity against HeLa cells with an IC_{50} of 1.1 μg/ml (Sugie *et al.*, 2001).

Nishihara *et al.* (2001) isolated a novel anti-influenza agent FR198248 from the cultural broth of *A. terreus* stain No. 13830. FRI 98248 was a new type of hydroxyl benzaldehyde compound, showed anti-influenza virus activity in Madin-Darby Canine kidney (MDCK) cells *in vitro*.

Shiomi *et al.* (2002) isolated new antibiotics, miyakamides A1, A2, B1 and B2 from the cultured broth of *A. flavus* Link *var. columnaris* FKI-0739. Miyakamides showed growth inhibitory activity against brine shrimp.

In the course of screening program for inhibitors of angiogenesis, novel substances designated as ICM030IA-H (1-8) were isolated from the culture broth of *Aspergillus* sp. F1491 by Kumagai *et al.* (2004). ICM030ls inhibited the growth of human umbilical vein endothelial cells (HUVECs) and showed significant anti-angiogenic activity in the angiogenesis model using rat aorta cultured in fibrin gel. lCM0301s showed very low cytotoxicity against various tumor cells.

ANTIBIOTIC FROM GENUS *TRICHODERMA*

Sawa *et al.* (1994) isolated a new antibiotic harizianic acid (1) from culture filtrate of a strain *Trichoderma harziamum* SP-307 from water sample collected at Hisoshima, Japan. The antibiotic shows antimicrobial activity against *Pasteurella piscicida* sp. 6395.

Lin and his team in 1994, isolated tricholin, a ribosome inactivating protein from culture broth of *T. viridae*, have been shown to exert fungicidal effects on *Rhizoctonia solani* (Lin *et al.*, 1994).

In 1996, Mukhopadhyay and his team isolated a fungus *T. koningii*, which produce a cyclopentenone derivative, active against *Candida albicans* from the piece of rotten wood in Mulund, Bombay. The culture filtrate exhibited bioactivity against *S. aureus* 209P, *E. coli* 9632, *C. albicans*, *A. niger*, *P. digitatum* 135, *F. culmorum* 100, *T. mentagrophytes* and *Cladosporium* sp (Mukhopadhyay., *et al.*, 1996).

Oh *et al.* (2002) isolated seven new peptaibols, atroviridins A-C from the culture broth of *T. atroviride*, strain F80317. These compounds showed antibacterial activity against Gram-positive bacteria and phytopathogenic fungi and exhibited significant cytotoxicity to human cancer cell lines *in vitro*.

ANTIBIOTICS FROM GENUS *FUSARIUM*

Kobayashi and coworkers in 1995 obtained a new antifungal antibiotic fusarielin A and three related compounds, fusarielins B, C and D from a culture of a *Fusarium* sp. K432. The culture broth was found to induce curling of mycelium of *P. oryzae*. Fusarielin A (1) showed a narrow antifungal spectrum and its cytotoxicity against HeLa S3 (Kobayashi *et al.*, 1995).

Jackson *et al.*, 1995 isolated fusacandins A and B novel antifungal antibiotics of papulacandin class from *F. sambucinum*. The fusacandin inhibit (1,3)-β-glucan synthesis.

Chen and coworkers in 1996 isolated novel antimicroalgal substances halymecins A (1), B (2) and C (3) from *Fusarium* sp. and halymecins D (4) and E (5) from an *Acremonium* sp. The structures of these halymecins were determined based on extensive 2D NMR spectroscopy studies as well as mass spectral data. Halymecin A showed antimicroalgal activity against *Skeletanema costatum* (Chen *et al.*, 1996).

Nazareth and Koshy, 2002 isolate a fungal species *F. solani* (Mart.) that produces an extracellular diffusible red pigment. It is shown to be a quinone in nature with antimycotic property and effective against human pathogenic fungi *C. albicans* and an onychomycotic fungus.

ANTIBIOTICS FROM GENUS *CLADOSPORIUM*

Petersen and Moerker, 1994 investigated culture extract of the fungus strain F-24707, initially classified as *Cladosporium* and isolated a unique active substance, Cladospirone bisepoxide. It showed selective activity against yeast, filamentous fungi and bacteria.

A new antibiotic named cladospolide D was isolated by Zhang *et al.* (2001) from *Cladosporium* sp., FT-0012, a fungus isolated from a sponge sample collected in Pohnpei island, federated state of Micronesia. The growth of *Mucor racemosus* and *Pyricularia oryzae* was inhibited by cladospolide D.

Besides the soil, marine habitats also proved to be a source of fungal bioactive natural products. Long-chain α, β unsaturated aldehydes with antibiotic activity were isolated from the culture of the fungus *Cladosporium* sp., obtained from an intertidal marine sediment (Gallo *et al.*, 2004).

ANTIBIOTICS FROM GENUS BASIDIOMYCETES

Dornberger *et al.* (2001) isolated an antibacterial metabolite named epicorazine (1) from the culture broth of basidiomycetes, *Stereum hirsutam* HKI 0195. Antibiotic (1) inhibited growth of Gram-positive bacteria including methicillin resistant *Staphylococces aureus* (MRSA) and vancomycin-resistant *Entercocous faecalis* (VRE) and *C. albicans.*

Chu and coworkers in 2003 isolated two novel antibiotics, Sch 484129 (I) and Sch 484130 (2) from fungal culture identified as a Basidiomycete. Both components were identified as glycolipids and displayed antifungal activity against *Saccharomyces* and *Aspergillus* strains (Chu *et al.*, 2003).

ANTIBIOTICS FROM GENUS *STREPTOMYCES*

Schimana and Fiedler, 2000 detected two novel angucyclinone antibiotics, simocyclinones D4 and D8 in the mycelium extract of *Streptomyces antibioticus* Tu 6040 by HPLC-diode-array and HPLC electronspray-mass-spectrometry screening. The compounds show antibiotic activities against Gram-positive bacteria and cytostatic effects on various tumor cell lines.

Tsukamoto *et al.* (2000) isolated new cytotoxic substances, designated BE-54238A and B from the culture broth of *Streptomyces* sp. A54238. BE54238A and B exhibited cytotoxic activity against murine and human tumor cell lines.

Sakai and coworkers in 2002 isolated six structurally related antitumor antibiotics named GEXI compounds from a culture broth of *Streptomyces* sp. GEX IA and was identified as a known herbicide, herboxidiene, structurally interested by the tetrahydropyran moiety. All GEX I compounds showed cytotoxicity with IC_{50} values of 0.0037-0.99 ìm against human tumor cell lines *in vitro* (Sakai *et al.*, 2002).

Retamycin is an anthracycline antibiotic complex with potent antitumour activity, similar to daunorubicin and doxorubicin (adriamycin), showing positive results in the treatment of human leukaemia's. It is produced by submerged cultivations of *S. olindensis*. As well as other anthracyclines used in cancer chemotherapy, retamycin intercalates into DNA and presumably inhibits DNA replication and transcription (Pamboukian and Facciotti, 2004)

In 2004 Freitag and his team isolated three new antibiotics, vanillobiocin, isovanillobiocin and declovanillobiocin, from the culture broth of a clo-*Q*-defective mutant of the clorobiocin producer *S. roseochromogenes,* which is blocked in the biosynthesis of the prenylated 4-hydroxybenzoic acid moiety or clorobiocin. All three compounds had lower antibiotic activity against *B. subtilis* than Clorobiocin (Freitag *et al.*, 2004).

Sakai and coworkers in 2004 isolated seven new macrolides having a 12-membered ring, termed as pladienolides were isolated from the fermentation broth of *S. platellsis* Mer-11107. They also demonstrated growth-inhibitory activity against U25l human glioma cells *in vitro* (Sakai *et al.*, 2004).

ANTIBIOTICS PRODUCED BY OTHER IMPORTANT GENERA

Krohan and his team in 1994 isolated sporothriolide, discosiolide and 4-epi-ethisolide three new furofurandiones from the *Sporothrix* sp., *Discosia* sp. and *Pezicula livida* respectively. The strains were selected for their antibacterial, antifungal and herbicidal activity. The antimicrobial activity was tested against *B. megterium, E. coli, Ustilogo violacea, Mycotypha microspora, Euroterium repens, Chlorella fusca, Lepidium sativum* and *Medicago sativa* showed herbicidal activity (Krohan *et al.*, 1994).

In 1995 Zapf and coworkers isolated two new xanthocillin type antibiotics Darlucin A (1) and Darlucin B (2) from fermentation broth of *Sphaerellopsis filum* (*Darluca filum*). Both compounds exhibit antibacterial, antifungal and weak cytotoxic activity (Zapf *et al.*, 1995).

Stadler and Anke (1995) isolated eight novel bioactive metabolites form submerged culture of ascomycetes, *Lachnum papyraceum (*Karst). All compounds exhibit antimicrobial, cytotoxic, nematicidal and phytotoxic activities.

Anke *et al.* (1995) described the fermentation, isolation and biological characterization of of favolon (1), a new antifungal sterol from culture of *Favolaschia* sp. 87129 found growing on the wood in a forest close to Kolobo, Ethiopia.

Morino and coworkers in 1995 during the course of physico-chemical screening, isolated a novel microbial secondary metabolite with an adenine moiety was isolated from the culture broth of fungus *Taralomyces* sp. The compound NK 374200 found to have anti-mosquito larval activity.

Cytotoxicity against HeLa cell was not observed at 160mg and no acute toxicity observed in mice at 160mg/kg (Morino *et al.*, 1995).

In 1995 Nakamura and his team isolated Stachybocins A, B and C novel endothelin (ET) receptor antagonists from the culture filtrate of *Stachybotrys* sp. They showed the inhibitory activity of I-ET-1 binding to rat ET_A, human ET_A and human Et_n receptors (Nakamura *et al.*, 1995).

An antifungal activity against *Mucor hiemalis* was detected in the culture broth of *Sorangium cellulosum* (Myxococcales) strain Socc90 by Gerth *et al.* (1996). The producing strain was isolated from a soil sample. The epothilons showed a high cytotoxicity for animal cells.

Gorman *et al.* (1996) isolated ascosteroside, a novel antifungal compound from the culture broth of *Ascotricha amphitricha.* It was active against yeasts such as *C. albicans* and *Saccharomyces cerevisiae* and against filamentous fungi but showed no activity against bacteria. It was non-toxic in mammalian cells at concentrations up to 150µM. In a mouse model, the compound afforded protection comparable to that of ketoconazole.

Shan and coworkers in 1996 isolated five new minor metabolites papyracon D(6), 6-*O*-methylpapyracon B (9), 6-*O*-methylpapyracon C (10), lachnumfuran A (11) and lachnumlactone A (l2a) together with the known chloromycorrhizinol A (4) from extract of the culture fluid of the ascomycete *Lachnum papyraceum.* The nematicidal antibiotic and cytotoxic activities of the compounds were weaker (Shan *et al.*, 1996).

Thines *et al.* (1997) reported a novel metabolite, benesudon from a fungus *Mollisia benesuada* possessing antimicrobial, cytotoxic and phytotoxic activities. It was reported to be effective against both Gram-positive and Gram-negative bacteria. The highest antifungal activity was observed against *Paecilomyces variotii* at 1 µg/ml. Cytotoxic effect was observed towards L 1210, RBL-1, BHK-21 and B16-F1 cells.

Chu and coworkers in 1997, during the investigation of natural product program in anticancer therapy a new compound Sch 56396 (1) was isolated from the fermentation broth of a fungal culture SCF-0729 from dead twigs of old oak tree. The isolated culture was identified as *Tolypocladium*. The cytotoxicity study of 1 showed IC_{50} value of 39 ìM in the MTT assay (Chu *et al.*, 1997).

Kohno and his team in 1999 were working on microbial metabolites, especially on the products of *Hypocreales*. They reported novel cysteine proteinase inhibitors, TMC-52 A-D in 1998 and novel cytotoxic antibiotics, TMC-151 A-F, produced by *Gliocladium* sp. in 1999. All compounds showed moderate cytotoxic activity to various cell lines (Kohno *et al.*, 1999).

In their search for antifungal substances against *Candida albicans,* Hosoe and his team in 1999 isolated a fungus from decaying *Picea glauca* wood that was identified as *Oidiodendron truncatum.* From its culture filtrate they isolated four new tetranorditerpenoids, oidiodendrolites A(3), B(4), C(5) and oidiodendronic acid (7). The antifungal activity was found against the pathogenic yeast *Candida albicans* (Hosoe *et al.*, 1999).

Anderson *et al.* (1999) reported that the nematode-trapping fungus *Duddingtonia flagrans* produces flagranones A (2), B (3) and, C (4) compounds that possess antibiotic activity against *B. subtilis and S. aureofaciens.*

A novel bioactive macrolide, 1B-96212 has been isolated from the fermentation broth of a marine actinomycete, L-25-ES25-008 by Chimeno *et al.,* 2000. The strain belongs to the genus *Micromonospora.* The macrolide showed a very strong cytotoxic activity against P-388 and lower but significant activity against *A-549,* 11T-29, and M EL-28 cell lines. The macrolide antibiotic only showed antibiotic activity against*Micrococcus luteus* out of six strains tested.

Fujie *et al.* (2000) isolated novel FR901469 antifungal antibiotic produced by an unidentified fungus No.11243. FR901469 inhibited the activity of 1,3-β-glucan synthase from *Candida albicans* with an IC_{50} value of 0.05 μg/ml, and displayed greater inhibitory activity than other 1,3-β-glucan synthase inhibitors such as, WF11899A, echinocandin B, aculeacin A, and papulacandin.

Bandani *et al.* (2000) isolated *Tolypocladium* species produce which efrapeptins a group of toxic peptides *in vivo*. Efrapeptins were toxic to a wide range of insects when injected into the haemocoel. Efrapeptins also exhibited limited antifungal and antibacterial activity.

In 2000, Komaki and coworkers reported a novel tricyclic diterpenoid antibiotic, brasilicardin A from a strain of *Nocardia brasiliensis* IFM0406. Brasilicardin A was produced by the fermentation of *Nocardia brasiliensis.* IC_{50} values against various cell lines (mouse leukemia P388/ADR, P388, L1210, mouse mastocytoma P815, Jurkat CCRF-CEM, HL 60, human leukemia MOLT-4 and HeLa human cervical carcinoma) were tested (Komaki *et al.*, 2000).

Vertesy *et al.* (2001) isolated new pluraflavin type antibiotics pluraflavin A ($C_4H_{54}N_2O_{14}$), pluraflavin B ($C_{43}H_{56}N_2O_{15}$) and pluraflavin E ($C_{36}H_{41}NO_{14}$) from cultures of the *Saccharothrix* species DSM 12931. The outstanding biological characteristic of pluraflavin A is its powerful, organ-dependent cytostatic action: the IC_{50} in the colon carcinoma proliferation assay is in the subnanomolar range

Vijayakumar and coworkers in 2001 isolated a fungal strain (culture number HIL Y-903146) from soil sample collected in Chandigarh, India that produced two metabolites, L970843 and L970844. Both were reported as possessing antifungal and antibacterial activity (Vijayakumar *et al.*, 2001).

Watanabe *et al.*, 2001 reported a new antifungal agent, CJ-19784 (I), obtained from the fermentation broth of *Acanthostigmella* sp. CL12082, New York Botanical garden, NY, U.S.A. The isolated compound inhibits the growth of pathogenic fungi, *C. albicans, Cryptococcus neoformans* and *Aspergillus fumigatus* with IC_{50} values of 0.11, 20 and 0.54 ìg/ml respectively.

Grafe *et al.* (2001) disclosed recently the strain of a fungus *Acremonium* sp. HKI 0230 as producer of a narrow spectrum antibacterial compound. The strain *Acremonium* sp. HKI 0230 from Pismo Beach, Califormia, USA. The compound showed moderate antimicrobial activity against gram-positive bacteria, *B. subtilis* ATCC 6633 (MIC 25μg/ml).

Seephonkai and coworkers in 2001 came across a new antimalarial tropolone named cordytropolone (1) in the culture broth of *Cordyceps* sp. BCC 1681, insect pathogenic fungi. *Cordyceps* sp. from Wildlife Sanctury, Chantaburi provence, Thailand. Cordytropolone exhibited *in vitro* antimalarial activity with IC_{50} value of 2.2 μg/ml. However, this compound also showed cytotoxicity against two cancer cell lines (KB and BC-1) and Vero cells with IC_{50} value of 17, 2.2 and 11 μg/ml, respectively (Seephonkai *et al.*, 2001).

Hirano *et al.*, 2001 isolated a new antibiotic TMC-69 from broth of a fungal strain *Chrysosporium* sp. TC 1068. TMC 69 exhibit moderate *in vitro* cytotoxic activity. TMC-69-6H, a derivative of TMC-69 prepared by hydrogenation, possessed more potent *in vitro* cytotoxicity than TMC-69, and exhibited *in vivo* antitumor activity against murine P388 leukemia and B16 melanoma cells.

In 2001 Kaida and coworkers isolated clavariopsins from fermentation broth of *Clavariopsis aquatica* AJ117363. Clavariopsins showed *in vitro* antifungal activity against not only *A. fumigatus* but also, to a lesser extent, *A. niger* and *Candida albicans*. The producing organism, *Clavariopsis aquatieca* AJ117363 was isolated from submerged decaying leaves collected from a mountain stream at Takao, Japan (Kaida *et al.*, 2001).

Omura and his team in 2002, isolated funicone related compounds from a soil fungus *Talaromyces flavus*, FKI-0076 that shows anti-candida activity of micronazole. The compounds are expected to show very weak antifungal activity and exhibit synergic antifungal activity in the presence of an azole (Omura *et al.*, 2002).

He *et al.* (2002) isolated a new 2-oxo-succinimide polyketide, designated as oxasetin (1) from the fermentation broth of a fungus *Vaginatispora aquatica* HK1821 from decaying piece of wood submerged in the Lam Tsuen River in Tai Po, Hong Kong. Oxasetin (1) exhibit moderate *in vitro* activity against Gram-positive bacteria, including methicillin-resistant *S. aureus* (MRSA) and vancomycin-resistant *Enterococcus faecalis* (VRE).

Two novel compounds, stachytlin and acetylstachytlin have been isolated by solid-state fermentation of *Stachybotrys* sp. RF-7260 by Minagawa *et al.*, 2002. The structures of both metabolites, determined by detailed NMR and X-ray crystallographic analysis and found to be novel with a pentacyclic moiety including cis-fused decalin. Stachyflin showed antiviral activity against influenza A virus (H1N1) *in vitro* with an IC_{50} value of 0.003 μM.

Nagai *et al.* (2002) isolated a new antifungal antibiotic, YM-202204 (1), from the culture broth of marine fungus *Phoma* sp. Q60596. The structure of (1) was determined by several spectroscopic experiments as a new lactone compound. This antibiotic exhibited potent antifungal activities against *C. albicans, C. neoformans* and *A. fumigatus,* and also inhibited glycosyl-phosphatidyl-inositol (GPI)-anchoring in yeast cells.

Hellwig *et al.* (2002) isolated a novel antibacterial antibiotic, altersetin from the culture broth of two endophytic fungi *Alternaria* species. Altersetin is chemically related to equisetin and showed potent MIC against several pathogenic Gram-positive bacteria.

In 2002, Yeo and coworkers isolated GTRI-BB, a new cytotoxic isochromanquinone produced by *Micromonospora* sp. SA-246. Cytotoxicity of the purified compound was tested against a panel of human tumor cell lines (PC-3 prostate, ACHN renal, A549 lung, SW620 colon, K562 leukemia, Du145 prostate and UACC 62 melanoma) and active fraction showed cytotoxicity against a number of them. Compound was active against *S. aureus, B. subtilis* and *Sarcina lutea* (Yeo, *et al.*, 2002).

In the course of screening program Arai and his team in 2003 isolated new monordens C, D and E, known monordens A (radicicol), B and 5-methylsclerone from *Humicola* sp. All monordens cause the cell cycle arrest at G I and G2/M phases in Jurkat cells. But among them, monordens A and E show antifungal activity only against *A. niger* (Arai *et al.*, 2003).

Hegde *et al.* (2003) isolated two novel antifungal metabolites namely SCH643432 (1) and (2) from fungus *Paecilomyces varioti* from the soil sample. Both the compounds were active against *C. albicans,* other candidans, dermatophytes and *Aspergillus* sp.

Acremonidins A-E (1-5) were produced by fermentation of *Acremonium* sp. by He *et al.*, 2003. The structures of these compounds, containing a bridging keto group were determined by spectroscopic analysis. Acremonidines A and B showed moderate activity against Gram-positive bacteria including methicillin resistant *Streptococci* and vancomycin-resistant *Enterococci.*

Two new enniatins H (3) and I (4), whose substituents on 2-hydroxycarboxylic acid moieties were different from those of known compounds isolated, together with known enniatins B (1) and B4 (2) from the insect pathogenic fungus *Verticillium hemipterigenum* BCC 1449 by

Nilanonta *et al.*, 2003. They exhibited antiplasmodial and antimycobacterial activities.

Li *at el.* (2003) identified thiazolyl peptide antibiotics, Nocathiacin I, II and III in the culture of *Nocardia* sp. WW-12651 (ATCC 202099). Nocathiacin inhibit potent *in vitro* activity against wide spectrum of Gram-positive bacteria, including multiple drug-resistant pathogens such as methicillin-resistant *S. aureus* (MRSA), multiple drug-resistant *E. faceium* (MREF) and fully penicillin-resistant *Streptococcus pneumoniae* (PRSP) and demonstrate excellent *in vivo* efficacy in a systemic *S. aureus* infection in mice model.

Laakso *et al.* (2003) during the course of screening program to identify new antiproliferative agents discovered new bafilolide metabolites R176502. The metabolite was isolated from liquid fermentation culture of novel *Micromonospora* sp. found in African River bottom sediments. R176502 exhibited potency for inhibition of tumor cell proliferation in nM range of concentrations.

A new compound IB-00208, has been isolated from the fermentation broth of an actinomycete, *Actinomadura* sp. from marine environment by Cascon *et al.*, 2003. IB-00208 showed potent cytotoxic activity against both human and murine cell lines. Besides, this compound has a good antibiotic activity against Gram-positive organisms but poor activity against Gram-negative bacteria.

Linezolid, an oxazolidinone antimicrobial, represents the first significant synthetic compound class introduced to the market in less than 25 years, since compounds of the quinolone class were optimized into today's very successful fluoroquinolones (Mills, 2003).

Three new beauvericins, designated beauvericins D, E and F were isolated along with known beauvericin A, from the culture of *Beauveria* sp. FKI-1366. by Fukuda *et al.* (2004). These compounds potentiate miconazole activity against wild *C. albicans.*

Kunze *et al.* (2004) in the course of screening of myxobacteria for new biologically active compounds, crude extracts of the species *Chondromyces crocatus* were noticed for their high antifungal and cytotoxic activity. They show only weak activity against yeasts, but are highly cytostatic for different cultured mammalian cell lines by interfering with the actin cytoskeleton.

Tani and coworkers in 2004 isolated six novel chymase inhibitors, SF2809-I, SF2809-II, SF2809-III, SF2809-IV, SF2809-V and SF2809- VI, from the fermentation broth of an actinomycete strain SF2809. The strain was identified as *Dactylosporangium* sp. by morphological, chemotaxonomical and phylogenetic studies. These six novel compounds inhibited recombinant human chymase in the range between IC_{50} of 0.0l4 and 7.3 μM (Tani *et al.*, 2004).

Vertihemipterin A, the ascochlorin glycoside, and its aglycone, 4',5'-dihydro-4'hydroxyascochlorin, were isolated from the insect pathogenic fungus *Verticillium hemipterigenum* BCC 2370 by Seephnokai *et al.* (2004). Antiviral (HSV-I) and cytotoxic activities of these ascochlorin analogs were evaluated.

Kanai *et al.* (2004) isolated a novel nematicidal antibiotic, paeciloxazine from the culture broth of a fungus *Paecilomyces* BAUA3058 strain. Paeciloxazine has moderate cidal activity against *Rhabditis pseudoelongata* and is weakly active against some insects.

For the past sixty years, the use of antibiotic drugs has turned bacterial infections into treatable conditions, rather than the life-threatening scourges they once were. Today some 5,000 antibiotics are known. Many antibiotics are becoming ineffective against the bacterial infections they once routinely cured. Health officials in the United States are concerned that bacteria are turning into "superbugs," resistant to more and more antibiotics. Children, the elderly, and those with weakened immune systems (including chemotherapy and transplant patients) are particularly at risk. Antibiotics can fail to cure an illness because the bacteria are intrinsically resistant toward the drugs or because they acquire resistance. Antimicrobial resistance is a global problem and antibiotic resistance genes move with travelers from one country to another, making antibiotic resistance an international problem (Smith, 1999). O'Brien *et al.* (1985) documented the intercontinental spread of antibiotic resistant gene on a plasmid. Globalization increases the vulnerability of countries to imported diseases travel faster and further than even before (Yach and Bettcher, 1998 and Fidler, 1997). During 1990, a resistant *Pneumonococcus* species first identified in Spain was rapidly found in Argetina, Brazil, Chile, China, Coloumbia, Malasia, Mexico, phillipines, Republic of Korea, south Africa, Thiland, USA and Uruguay (WHO, 2000). Thus, it is obvious that we should continue to exploit the diversity and quality provided by natural products to facilitate and optimize the drug-discovery process, fights with microbial resistance and improves the pipeline for new antibiotics (Nielsen, 2002).

Acknowledgement

The authors are grateful to Dr. A.K. Chawla, Vice-Chancellor, Kurukshetra University, Kurukshetra for deep encouragement. The junior author Mr. Parikshit Tyagi is also thankful to Kurukshetra University, Kurukshetra for providing University Research Scholarship and laboratory facilities.

References

Amade, P., Mallea, M. and Bouaicha, N. 1994. Isolation, structural identification and biological activity of two metabolites produced by *Penicillium olsonii*. *J Antibiotics*. 47: 201-207

Anderson, M.G., Rickards, R.W. and Lacey, E. 1999. Structure of Flagranones A, B and C, cyclohexenoide antibiotics from the nematode-trapping fungus *Duddingtonia flagrans*. *J Antibiotics*. 52: 1023-1028

Anke, T., Werle, A and Zapf, S. 1995. Savalon, a new antifungal triterpenoid from a/' *Favolaschia* sp. *J Antibiotics*. 48: 725-726

Arai, M., Yamamoto, K., Namatame, I., Tomoda, H. and Omura, S. 2003. New monordens produced by arnidepsine-producing fungus *Humicola* sp. FO-2942. *J Antibiotics*. 56: 526-532

Bandani, A.R., Khambay, B.P.S., Faull, J.L., Newton, R., Deadman, M. and Butt, T.M. 2000. Production of efrapeptins by *Tolypocladium* species and evaluation of their insecticidal and antimicrobial properties. *Mycol Res*. 5: 537-544

Banks, R.M., Blanchflower, S.E., Everett, J.R. Manger, B.R. and Reading, C. 1997. Novel anthelmintic metabolites from an *Aspergillus* species; the Aspergillimides. *J Antibiotics*. 50: 840-846

Benedict, R.G. and Langlykke, A.F. 1947. Antibiotic activity of *B. polymyxa*. *J Bact*. 54: 24

Cascon, L.M., Romero, F., Vazquez, F.E., Gravalos, D. and Puentes, J.L.F. 2003. IB-00208, a new cytotoxic polycyclic xanthone produced by marine derived *Actinomadura*. *J Antibiotics*. 56: 219-225

Chambers, S.O. and Weidman, F.D. 1928. A fungistatic strain of *Bacillus subtilis* isolated from normal toes. *Arch Derm Syph*. 18: 568

Chen, C., Imamura, N., Nishijima, N., Adachi, K., Sakai, M. and Sano, H. 1996. Halymecins, new antimicroalgal substances produced by fungi isolated from marine algae. *J Antibiotics*. 49: 998-1005

Chimeno, R.I.F., Canedo, L., Espliego, F., Gravalos, D., Calle, F.D.L., Puentes, J.L.F. and Romero, F. 2000. IB-96212, a novel cytotoxic macrolide produced by a marine *Micromonospora*. *J Antibiotics*. 53: 474-478

Chu, M., Mierzwa, R., Xu, L., He, L., Langsdorf, E., Beyazova, M., Terracciano, J., Pafel, M., Ciian, T. and Gullo, V. 2003. Isolation and characterization of two new antifungal antibiotics from a Basidiomycete. *J Antibiotics*. 56: 9-15

Chu, M., Mierzwa, T.R., Patel, M. and Puar, M.S. 1997. Sch 56396: a new *c-fos* proto-oncogene inhibitor produced by the fungus *Tolypocladium* sp. 1997. *J Antibiotics*. 50: 1061-1063

Demain, A.L. and Elander, R.P. 1999. Beta-lactam antibiotics: past, present and future. *Antonie van Leeuwenhoek*. 75: 5-19

Dornberger, K., Schlegel, B., Luhmann, U., Dashse, H.M. and Kleinwachter, P., 2001. Epicorazine, an antimicrobial metabolites from *Stereum hirutum* HKI 0195. *J Antibiotics*. 54: 521-525

Duggar, B.M. 1948. Chlortetracycline, a product on continuing searches for new antibiotics. *Ann NY Acad Sci*. 51: 177

Ehrlich, J., Bratz, Q.R., Smith, R.M., Joslyn, D.H. and Burkholder, P.R. 1947. Chloromycetin, a new antibiotic from soil actinomycete. *Science.* 106: 417

Eilbert, F., Anke, H. and Sterner, O. 2000. Neobulgarones A-F from cultures of *Neobulgaria pura,* new inhibitor of appressorium formation of *Magnaporthegrisea. J Antibiotics.* 53: 1123-1129

Fang, F., Ut, H., Shimoi, K., Masuma, R., Yamaguchi, Y., Zhang, C.G., Zhang, X.W., Tanaka, Y. and Omura, S. 1997. Two new components of the aspochalasins produced by *Aspergillus* sp. *J Antibiotics.* 50: 919-925

Fidler, D. 1997. The globalization of public health: emerging infectious diseases and international relation. *Indian Journal of Global Legal Studies.* 5: 11-51

Finlay, A.C., Hobby, G.L., Pan, S.Y. and Regna, P.P. 1950. Oxytetracycline, a new antibiotic. *Science.* 111: 85

Fleming, A. 1929. On the antibacterial action of a *Penicillium,* with special reference to their use in the isolation of *B. influenzae. J Exp Pathol.* 10: 226-236

Florey, H.W., Chain, E.B., Heatley, N.G., Jennings, M.A., Sanders, A.G., Abranam, E.P. and Florey, M.E. 1949. *Antibiotics.* Oxford University Press, London

Freitag, A., Galm, U., Li, S.M. and Heide, L. 2004. New Aminocoumarin, antibiotics from a cloQ-defective mutant of the clorobiocin producer *S. roseochromogenes* DS12.976. *J Antibiotics.* 57: 205-209

Fujie, A., Iwamoto, T., Muramatsu, H., Okudaira, T., Nitta, K., Nakanishi, T., Sakamoto, K., Hori, Y., Hino, M., Hashimoto, S. and Okuhara, M. 2000. FR901469, **a** novel antifungal antibiotic from an unidentified fungus no.11243. *J Antibiotics.* 53: 912-919

Fukuda, T., Arai, M., Yamaguchi, Y., Masuma, R., Tomoda, H. and Omura, S. 2004. New Beauvericins, potentiators of antifungal miconazole activity produced by *Beauveria* sp. FKI-1366. *J Antibiotics.* 57: 110-116

Gallo, M.L, Seldes, A.M. and Cabrera, G.M. 2004. Antibiotic long-chain á, â-unsaturated aldehydes from the culture of the marine fungus *Cladosporium* sp. *Biochemical Systematics and Ecology.* 32: 554-551

Gerth, K., Bedorf, N., Hofle, G., Irschik, H. and Relclienbach, H. 1996. Epothilons A and B: antifungal and cytotoxic compounds from *Sorangium cellulosum* (Myxococcales). *J Antibiotics.* 49: 560-563

Gloer, J.B. 2004. New bioactive natural products from fungi: an ecology-based approach. At web site: www.uiowa.edu/~chemdept/faculty/gloer/

Gorman, J.A., Chang, L.P., Clark, J., Gustavson, D.R., Lam, K.S., Member, S.W., Pirnir, D., Ricca, C., Fernandes, P.B. and O'Sullivan, J. 1996. Ascosteroside, a new agent from *Ascotricha amphitricha,* taxonomy, fermentation and biological activity. *J Antibiotics.* 49: 547-552

Gottshall, R.Y., Roberts, J.M., Portwood, L.M. and Jennings, J.C. 1951. Synnematin, an antibiotic produced by *Tilachlidium. Fioc Soc Exp Biol NY* . 76: 307-311

Grafe, U., Dorfelt, H, Schlergel, B. and Kleinwachter, P. 2001. Spriobenzofuran, a new bioactive metabolite from *Acremonium* sp. HK0230. *J Antibiotics.* 54: 526-527

Gratia, A. and Dath, S. 1924. Proprietes bacteriolytiques de certaines moisissures. *Compte Rendu Biol Paris.* 91:1442

He, H., Bigelis, R., Solum, E.H., Greenstein, M. and Carter, G.J. 2003. Acremonidines, new polyketide derived antibiotics produced by *Acremonium* sp., LL-cyan 416. *J Antibiotics.* 56: 923-930

He, H., Janso, J.E., Yang, H.Y., Singh, M.P., Berner, V.S., Greenstein, M. and Carter, G.T. 2002. Oxasetin, a new antibacterial polyketide produced by fungus *Veginatiospora aquatica* HK1821. *J Antibiotics.* 55: 821-825

Hegde, V.R., Silver, J., Patel, M., Gullo, V.P., Puar, M.S., Das, P.R. and Loebenberg, D. 2003. Novel fungal metabolites as cell wall active antifungal: fermentation, isolation, physico-chemical properties, structure and biological activity. *J Antibiotics.* 56: 437-447

Hellwig, V., Grothe, T., Bartschmid, A.M., Endermann, R., Geschke, F.U., Henkel, T. and Stadler, M. 2002. Alterstin, a new antibiotic from cultures of endophytic *Alternaria* sp. *J Antibiotics.* 55: 881-892

Hirano, K, Kohno, I, Tsunoda, S., Nishio, M., Kishi, N., Okuda, T., Kawano, K, Komatsubara, S. and Nakanishi, N. 2001. TMC - 69, a new antitumor antibiotic with ,,, CDC 25 A inhibitory activity produced by *Chrysosporium* sp. TCI 068. *J. Antibiotics.* 54: 421-427

Hoel, D. and Williams, D.N. 1997. Antibiotics: Past, present, and future. Unearthing nature's magic bullets. *Post Graduate Medicine.* 101: 20

Hosoe, T., Nozava, K., Lumley, T.C., Currah, R.S., Furukshima, k., Takizava, K., Miyaji, M. and Kawai, K. 1999. Tetranorditerpene lactone, potent antifungal antibiotics from human pathogenic yeast, from a unique species of *Oidiodendron. J Antibiotics.* 47: 1591-1597

Jackson, M., Frost, D.J., Karwowski, J.P., Humphrey, P.E., Dahod, S.K., Choi, W.S., Brandt, K., Malmberg, L.H., Rasmussen, R.R., Scherr, M.H., Flamm, R.K., Kadam, S. and McAlpine, J.B. 1995. Fusacandin A and B novel antifungal antibiotics of the papulacandin class from *Fusarium sambucinum. J Antibiotics.* 48: 608-613

Johnson, B.A., Anker, S.H. and Meleney, F.L. 1945. Bacitracin a new antibiotic produced by a member of *Bacillus subtilis* group. *Science.* 102: 376

Kaida, K., Fudoi, R., Kameyama, T., Tubaki, K., Suzuki, Y., Ojika, M. and Sakagami, Y. 2001. New cyclic depsipeptide antibiotic, Clavariopsin A and B, produced by an aquatic *Hyphomycetes, Clavariopsis aquatica. J. Antibiotics.* 54: 17-21

Kanai, Y., Fujimaki, T., Kochi, S., Konno, H., Kanazawa, S. and Tokumasu, S. 2004. Paeciloxazine, a novel nematicidal antibiotic from *Paecilomyces* sp. *J Antibiotics.* 57: 24-28

King, D.E., Malone, R. and Lilley, S.H. 2000. New classifications and update on the quinolone antibiotics. *American Academy of Family Physicians.* 59: 234-241

Kobayashi, H., Sunaga, R., Furihata, K., Morisaki, N. and Iwasaki, S. 1995. Isolation and structures of an antifungal antibiotic, fusarielin A, and related compounds produced by a *Fusarium* sp. 1995. *J Antibiotics.* 48: 42-52

Kohno, J., Asai,Y., Nishio, M., Sakurai, M., Kawano, K., Hiramaisi, H, Kameda, N. Kishi, N., kuda, T. and Komatsubara, S. 1999. TMC 171A, B, C and TMC-154. Novel polyketide antibiotics produced by *Gliocladium* sp. TC1282. *J. Antibiotics.* 52: 1114-1123

Komaki, H., Tanaka, Y., Yazama, K., Takagi, H., Ando, A., Nagata, Y. and Mikami, Y. 2000. Antitumor activity of brasilicardin A, a novel terpenoid antibiotic from *Nocardia brasiliensis. J Antibiotics.* 53: 75-77

Krohan, K., Ludewig, K., Aust, H.J., Draeger, S. and Schulz, B. 1994. Biologically active metabolite from fungi. *J Antibiotics.* 47: 113-118

Kumagai, H., Someno, T., Dobashlt, K., Isshiklt, K., Ishlzuka, M. and Ikeda, D. 2004. ICM030Is, New angiogenesis inhibitors from *Aspergillus* sp. F-I491. *J Antibiotics.* 57: 97-103

Kunze, B., Jansena, R., Hafle, G. and Reichenbach, H. 2004. Ajudazols, new inhibitors of the mitochondrial electron transport from *chondromyces crocatus,* production, antimicrobial activity and mechanism of action. *J Antibiotics.* 57: 151-155

Laakso, J.A., Mocck, U.M., Dun, J.V., Wouters, W. and Janicot, M. 2003. R176502, a new bafilolide metabolite with potent antiproliferative activity from a novel *Micromonospora* sp. *J Antibiotics.* 56: 909-916

Li, W., Leet, J.E., Ax, H.A., Grustavson, D.R., Brown, D.M., Turner, L., Brown, K., Yang, H., Tomc, J.F. and Lam, K.S. 2003. Nocathiacins, new thiazolyl peptide antibiotics from *Nocardia* sp. *J Antibiotics.* 56: 226-231

Lin, A., Lee, T.M. and Rern, J.C. 1994. Tricholin, a new antifungal agent from *Trichoderma viride* its action in biological control of *Rhizoctonia solani. J Antibiotics.* 47: 799-805

Maskey, R.P., Wollny, I.G. and Laatsch, H. 2003. Isolation, structure elucidation and biological activity of 8-*O*-methylaverufin (1b) and 8-*O*-dimethylaverantin (2b) as new antifungal agents from *Penicillium chrysogenum. J Antibiotics.* 56: 459-463

McGuire, J.M. 1952. Erythromycin, a new antibiotic. *Antibiotics and Chemotherapy* 2: 281

Mills, S.D. 2003. The role of genomics in antimicrobial discovery. *J Antimicrobial Chemotherapy.* 51: 749-752

Minagawa, K., Kouzuki, S., Yoshimoto, J., Kawamura, Y., Tani, H., Iwata, T., Terui, Y., Nakai, H., Yagi, S., Hattori, N., Fujiwara, T. and Kamigauchi, T. 2002. Stachyflin and acetylstachyflin, novel anti-influenza A virus substances produced by *Stachybotrys* sp. RF-7260. *J Antibiotics.* 55: 155-164

Morino, T., Nishimoto, M., Masuda, A., Fujita, S., Nishikiori, T. and Saito, S. 1995. NK374200, a novel insecticidal agent from *Taralomyces,* found by physico-chemical screening. *J Antibiotics.* 48: 1509-1510

Mukhopadhyay, T., Roy, K., Sawant, S.N., Deshmukh, S.K. and Ganguli, B.N. 1996. On an unstable antifungal metabolite from *T. koningii,* isolation and structure elucidation of a new cyclopentenone derivative. *J Antibiotics.* 49: 210-211

Nagai, K., Kamigiri, K., Matsumotob, H., Kawano, Y., Yamaoka, M., Shimoi, H., Watanabe, M. and Suzuki, K. 2002. YM-202204, a new antifungal antibiotic produced by marine fungus *Phoma* sp. *J Antibiotics.* 55: 1036-1041

Nakamura, M., Ito, Y., Ogawa, K., Michisuji, Y., Sato, S., Takada, M., Hayashi, M., Yaginuma, S. and Yamamoto, S. 1995. Stachbocins, novel endothelin receptor antagonists, produced by *Stachybotrys* sp. M6222. *J Antibiotics.* 48: 1389-1395

Nazareth, S. and Koshy, B. 2002. Antimycotic activity of a fungal pigment against pathogenic fungi. *Indian Journal of Microbiology.* 42: 241-243

Nielsen, J. 2002. Combinatorial synthesis of natural products. *Current Opinion in Chemical Biology.* 6: 297–305

Nilanonta, C., Isaka, M., Chanphen, R., Thongom, N., Tanticharoenb, M. and Thebtaranonthb, Y. 2003. Unusual enniatins produced by the insect pathogenic fungus *Verticillium hemipterigenum*: isolation and studies on precursor directed biosynthesis. *Tetrahedron.* 59: 1015-1020

Nishihara, Y., Tsujii, E., Yamagishi, Y., Sakamoto, K., Tsurumi, Y., Furukawa, S., Ohtsu, R., Kino, T., Hino, M., Yamashita, M. and Hashimoto, S. 2001. FR 198248, a new antiinfluenza agent isolated from *A. tereus* No. 13830. *J Antibiotics.* 54: 136-143

O'Brien, T.F., Del Pilar Pla, M. and Mayer, K.H. 1985. Intercontinental spread of a new antibiotic resistance gene on an epidemic plasmid. *Science.* 230: 87–88

Oh, S.U., Yun, B.S., Lee, S.J., Kim, J.H. and Yoo, I.D. 2002. Atroviridin A-C and neoatroviridin A-D, novel peptaibol antibiotics produced by *T. atroviride* F80317. *J Antibiotics.* 55: 557-564

Omura, S., Yamaguchi, Y., Masuma, R, Tabata, N., Wang, H, Okuda, T., Tomoda, H and Arai, M. 2002. Funicone related compounds potentiator of antifungal miconazole activity produced by *Talaromyces flavus* FKI-0076. *J Antibiotics.* 55: 172-180

Pamboukian, C.R.D. and Facciotti, M.C.R. 2004. Production of antitumor retamycin during continuous fermentatios of *S. olindensis. Process Biochemistry* (*Article in press*)

Pasteur, L. and Joubert, S.F. 1877. Chimie physiologique carbon et septocemie. *C R Soc De Biol.* 85: 101

Petersen, F. and Moerker, T. 1994. Production of Cladospirone Bisepoxide, anew fungal metabolite. *J Antibiotics.* 47: 1098-ll03

Primgsheim, E.G. 1920. Zur Verbilligeng and Verscharfung der Indolreaktion, Abt.1, 82: 318Zb.1 Bakt., 2Abt., 51: 72

Robinson, D. 1976. *The Miracle Finders: The Stories behind the most important Breakthroughs in Medicine.* David McKay, New York

Sakai, T., Smvteshimat, T., Matsufujlt, M., Kawamurat, N., Dooashlt and Mizuj, Y. 2004. Pladicnolides, new substances from culture of *Streptomyces platellsis* Mer-11107. *J Antibiotics.* 57: 173-179

Sakai, Y., Yoshida, T., Ochiai, K., Uosaki, Y., Saitoh, Y., Tanaka, F., Akiyama, T., Akinaga, S. and Mlzukami, T. 2002. GEXI Compounds, novel antitumor antibiotics related to herboxidiene, produced by fungus *Streptomyces* sp. *J Antibiotics.* 55: 855-867

Sawa, R., Mori, Y., Linuma, H., Naganawa, H., Hamada, M., Yosida, S., Furutani, H., Kajimura, Y., Fuwa, T. and Takeuchi, T. 1994. Harzianic Acid, A new antimicrobial antibiotic from a Fungus. *J. Antibiotics.* 47: 731-732

Schimana, J. and Fiedler, H.P. 2000. Simocyclinones, novel cytostatic angucyclinone antibiotics produced by *Streptomyces antibioticus* Tu 6040. *J Antibiotics.* 53: 779-787

Seephnokai, P., Isaka, M., Kittakoop, P., Boonnudomlap, U. and Thebtaranonth, Y. 2004. A novel ascochlorin glycoside from the insect pathogenic fungus *Verticillium hemipterigenum* BCC 2370. *J Antibiotics.* 57: 10-16

Seephonkai, P., Isaka, M., Kittakoop, P., Trakulnaleamsai, S., Rattanajak, R., Tanticharoen, M. and Thebtaranonth, Y. 2001. A new tropolone from the insect pathogenic fungus *Cordyceps* sp. BCC 1681. *J Antibiotics.* 54: 751-752

Shan, R., Stadler, M. and Sterner, O. 1996. New metabolites with nematicidal and antimicrobial activities from the ascomycete *Lachnum papyraceum* (Karst.) Karst. *J Antibiotics.* 49: 447-452

Shiomi, K., Hatae, K., Yamaguchi, Y., Masuma, R., Tomoda, H., Kobayashi, S. and Omura, S. 2002. New antibiotics Miyakamides produced by a fungus. *J Antibiotics.* 55: 952-961.

Smith, R. 1999. Antimicrobial resistance: the importance of developing long term policy. *Bulletin of World Health Organization.* 77: 862

Stadler, M. and Anke, H. 1995. Metabolites with nematicidal and antimicrobial activities from Ascomycete *Lachnum papyraceum* (Krast.) Krast. *J Antibiotics.* 48: 261-266

Stead, P., Affleck, K., Sidebottom, P.J., Taylor, N.L., Drake, C.S., Todd, M., Jowett, A. and Webb, G. 1999. Isolation and characterization of a prenylated p- terphenyl metabolite of *A. candidus* possessing potent and selective cytotoxic activity: studies on mechanism of action. *J Antibiotics.* 52: 89-95

Sugie, Y., Hirai, H., Inagaki, T., Ishiguro, M., Kim, Y.J., Kojima, Y., Sakakibara, T., Sakemi, S., Sugiura, A., Suzuki, Y., Brennan, L., Duignan, J., Huang, L.H., Sutcliffe, J. and Kojima, N. 2001. A new antibiotic CJ17, 665 from *Aspergillus ochraceus. J Antibiotics.* 54: 911-916

Suzuki, K., Kuwahara, A., Yushida, H., Fujita, S., Nishikiori, T. and Nakagava, T. 1997. NF00659A_1, A_2, A_3, B_1 and B_2, novel antitumor antibiotics produced by *Aspergillus* sp. NF00659. *J Antibiotics.* 50: 344-347

Tani, M., Gyobu, Y., Sasaki, T., Takenouchlb, O., Kawamurab, T., Kamimurab, T. and Haradab, T. 2004. SF2809 Compounds, novel chymase inhibitors from *Dactylosporangium* sp. *J Antibiotics.* 57: 83-88

Tanner, F.W. 1952. Some properties of carbomycin. *Antibiotics* and *Chemotherapy.* 2: 441

Thines, E, Arendholz, W.R. and Anke, H. 1997. Benesudon, a new antibiotic from culture of *Mollisia benesuada* (Tul.) Phill. *J Antibiotics.* 50: 13-17

Tsukamoto, M., Nakajima, S., Murooka, K., Hirayama, M., Hirano, K., Yoshida, S., Kojiri, K. and Suda, H. 2000. New cytotoxic agents, BE-54238A and B, produced by a *Streptomycete. J Antibiotics.* 53: 26-32

Vertesy, L., Barboneh, F.P., Cashmenh, E., Decker, H., Ehrlich, K., Jordan, B., Knauf, M., Schummer, D., Seweth, M.P. Wink, J. and Sehbert, G. 2001. Pluraflavins, potent antitumor antibiotics f.rom *Sacchqrothrix* sp. DSM 12931. *J Antibiotics.* 54: 718-729

Vijayakumar, E.K.S., Roy, K., Hiremath, C.P., Deshmukh, S.K. and Mukhopadhyay, T. 2001. L970843 and L970844, Two new antifungal metabolites from an unidentified - fungal species HIL Y-903146. *J Antibiotics.* 54: 973-976

Waksman, S.A. and Lechevalier, H.A. 1949. Neomycin, a new antibiotic active against streptomycin resistant bacteria including tuberculosis organism. *Science.* 109: 395

Waksman, S.A., Schatz, A. and Bugie, E. 1944. Bacteriostatic and bacteriocidal substances produced by soil actinomycetes. *J Bact.* 40: 609

Watanabe, S., Hirai, H., Kato, Y., Nishida, H., Saito, T., Yashikawa, N., Parkinson, T. and Kojima, Y. 2001. CJ-19, 784, a new antifungal agent from a fungus *Acanthostigmella* sp. *J Antibiotics.* 54: 1031-1035

Weiland, P. 1936. Bakterizide Wirkeng von Mesentaricus feltraten auf Diphtheriebazillen. *Zbl Bakt Abtz Orig.* 136: 451

WHO. 2000. Overcoming antimicrobial resistance. Geneva: World Health Organization. WHO/CDS/2000.

Yach, D. and Bettcher, D. 1998. The globalization of public health: threats and oppertunities. *American Journal of Public Health.* 88: 735-8.

Yeo, W.H., Yun, B.S., Kim, Y.S., Yu, S.H., Kim, H.M., Yoo, I.D. and Kim, Y.H. 2002. GTRI-BB, a new cytotoxic isochromanquinone produced by *Micromonospora* sp. SA-246. *J Antibiotics.* 55: 511-515

Young, R.N. 1999. Importance of biodiversity to the modern pharmaceutical industry. *Pure Appl Chem.* 71: 1655-1661

Zapf, S., Hobfeld, M., Anke, H., Velten, R and Staglich, W. 1995. Darlucins A and B, New isocyanide antibiotics trom *Sphaerellopsis filum (Darluca filum)*. *J Antibiotics.* 48: 36-41

Zhang, H., Tomoda, H., Tabata, H., Miura, H., Namikoshi, M., Yamaguchi, Y., Masuma, R. and Omura, S. 2001. Cladospolide D, a new 12 membered macrolide antibiotic producred by *Cladosporium* sp. FT-0012. *J Antibiotics.* 54: 635-641

Frontiers in Plant Sciences (Pages 423-446)
Eds. K.G. Mukerji, K.V.B.R. Tilak, S.M. Reddy, L.V. Gangwane, P. Prakash, I.K. Kunwar

Ergot Alkaloids: Industrial Perspectives

A. Mehta and P. Mehta

ABSTRACT

Ergot alkaloids are pharmacologically active compounds mainly produced by *Claviceps purpurea*. The ergot alkaloids are separated into four main structurally groups. Ergot alkaloid is derived from tryptophan, mevalonic acid and methionine. Cultural conditions suitable for metabolism of primary substances are not suitable for alkaloid metabolism. The key to successful production of ergot alkaloids is in obtaining the proper strain of the fungus. Gene transfer via protoplast fusion has opened up new possibilities in both the cellular and molecular aspects of *Claviceps* genetic engineering. The solid-substrate fermentation produces a more stable requiring less energy in smaller fermenters with easier downstream processing measures. Solid-surface fermentation has been seen as a viable option for industrial scale production of ergot alkaloids. The commercialization of ergot alkaloids and their production made possible through biotechnology requires a coordinated coupling of unit operation in order to develop efficient processes. The growth of the fungus and alkaloid formation in submerged batch fermentation can be described by mathematical model, which can be used in automatic process control and optimization of an industrial process. Pure ergot alkaloids and their semisynthetic derivatives are being used as drugs at low concentrations in medicine.

Key words: Ergot, *Claviceps purpurea*, Industry, Medicine

Author's address: Microbial Technology Laboratory, Department of Botany, Dr. H.S. Gour University, Sagar-470.003, India
E.mail: pradeepunisagar@hotmail.com

INTRODUCTION

Ergot alkaloids a group of pharmacologically active compounds produced by a number of fungi mainly *Claviceps* sp. is referred to as secondary metabolites. They are also found in other fungi and higher plants. Ergot (sclerotium of *Claviceps purpurea*) develops on plants of rye, cereals and in florets of grasses. In ancient time, the medieval midwives used to collect the fungus from naturally infected plants and used it in the induction of childbirth and in the control of postpartum bleeding. The role of ergot (*Secale cornutum*) has undergone important changes from a dreaded toxic parasite to an important source of biologically effective substances. *Claviceps purpurea* is the first fungus, apart from the yeast, which was biotechnologically exploited without its existence known. The modern ergot research dates back to the extraction of the first crystalline alkaloid (ergotinine) from sclerotia in 1875, to the major pharmacological discoveries of ergotoxine in 1906 and the discovery of the first clinically used compound, ergotamine in 1918. At the beginning of fifties the biosynthesis of ergot alkaloids received much attention from a subject of fruitful speculation to the region of experimental research. Toxic effects of ergot compounds have been recognized since the discovery that consumption of infected cereals infected by parasitic fungus, *Claviceps* could cause severe illness (ergotism). It is of particular interest that even before the ergot was identified as a cause of ergotism, it was evidently used in obstetrics to enhance muscle contraction and was also used for abortions. Many deaths were brought about by the ergot extract because during these early days of medicinal use, neither its physiological action nor the appropriate dosage was understood. At the end of 19th century the ergot was banned as a drug in Germany. Later, with increasing understanding of pharmacology, it was found that chemical alterations of various alkaloids improved their medical efficiency. Further, after a correlation between clinical indications and therapeutical dosage was developed, pure ergot alkaloids and their semisynthetic derivatives are being used as drugs at low concentrations extensively in medicine, at present.

The industrial production of ergot alkaloids has gained importance, as the supply of natural grown ergots is not sufficient. The microbial technology of ergot alkaloids has two approaches: (i) the acquisition of knowledge for producing microorganism, (ii) and the development of large-scale processes to exploit such knowledge. Initially, chemical properties of the alkaloids were studied while biologists studied the physiological aspects of the synthesis of alkaloids, i.e. for both agronomists who were trying to obtain grain with lowest possible alkaloid content and for pharmacologists, who on the other hand were interested in high alkaloid yields. The chemistry, biosynthesis, physiology, biochemistry, genetics, biotechnology, and therapeutical applications of ergot alkaloids have been extensively studied (Berde and Sturmer, 1978; Mukherjee and Menge, 2000; Rehacek and Mehta, 1991, 1993; Flieger *et al.*, 2003). The present review gives an overview of ergot alkaloids with industrial perspectives.

Sources

In nature, primarily the fungus *Claviceps* forms the ergot alkaloids. More than 50 different alkaloids have been isolated from this ascomycetous genus, which has about 50 species. All *Claviceps* species form ergot sclerotia on various plants. Atleast another 100 plants can be attacked by ergot to form the sphacelial but not the sclerotial stage of the fungus (Rehacek and Sajdl, 1990). Ergot alkaloids have also been found from organisms other than *Claviceps* (Flieger *et al.*, 1997; Kozlovsky 1999).

Fungi imperfectii: *Aspergillus clavatus, A.flavus, A.japonicus, A. tamarii, A. versicolor, A. oryzae, A. fumigatus, A.nidulans, Penicillium chermesinum, P. concavorugulosum,*
P. fumigatus, P.viridicatum, P. sizovae, P. rugulosum, P. rubrum , P. kapuscinskii, P.palitans, P.aurantiovirens, P.camembertii, P.clavigerum, P. crustosum, P. griseofulvin, P. roquefortii

Hypocreales: *Balansia claviceps, B. epichloe, B. obtecta, B. strangulans, Epichloe typhinna, Neotyphodium coenophialum, N. lolii, Hyphomyces aurantius, Sepedonium sp.*

Zygomycetes: *Rhizopus arrhizus, R. nigricans, Cunninghmella blakesleana, Mucor hiemalis.*

Basidiomycetes: *Corticum caeruleum, Lenzites trabea, Pellicularia filamentosa*

Connvolvulaceae: *Argyreia nervosa, Ipomoea argyrophylla, I. Hildebrandtii, I.ruberosa, I.violacea, I. rubrocaerulea, I. piurensis, Rivea corymbosa, Stictocardia tiliifolia*

Chemistry

The ergot alkaloids constitute the largest known group of nitrogenous fungal metabolites. The common part of chemical structure of ergot alkaloids is the tetracyclic ergoling ring system (**Fig. 1**). Ergoline, a common part of ergot alkaloids is formed from tryptophan (Taber and Vining, 1959), mevalonic acid (Groger *et al.*, 1961) and methionine (Baxter *et al.*, 1961).

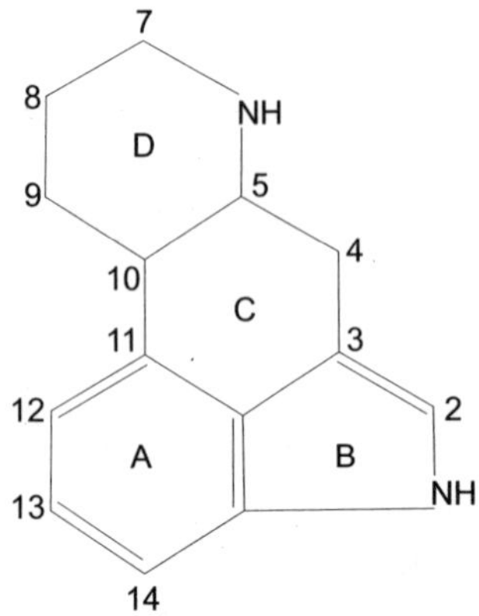

Fig. 1 Ergoline

Classification

The ergot alkaloids can be separated into four main structurally groups: (i) clavine alkaloids, (ii) lysergic acids, (iii) simple lysergic acid amides and (iv) peptide alkaloids (**Fig. 2**).

Clavines

The clavines are (**Fig. 3**) substituted 6,8-dimethylergolines and include a few members, namely the chanoclavines, with a 6,7-seco D ring. The degree of oxidation is a criterion for further differentiation in the group of clavine alkaloids. Some examples are: chanoclavine, norchanoclavine, paliclavine, paspaclavine, agroclavine, elymoclavine, etc.

Lysergic acid and its simple derivatives

The derivatives of lysergic acid are mostly amides, in which the amide part is a small peptide or a simple alkylamide. Non-peptide amides of lysergic acid found in ergot are ergometrine, lysergic acid 2-hydroxyethylamide, lysergic acid amide (ergine), and paspalic acid (**Fig. 4**). Some examples of lysergic acid derivatives are: ergometrine (ergonovine, ergobasine), lysergic acid 2- hydroxyethylamide, lysergic acid amide (ergine), lysergic acid

Ergopeptines

Ergotamine was the first alkaloid to be isolated from ergot crop and became commonly used compound for therapeutics. Ergotamine and other classical

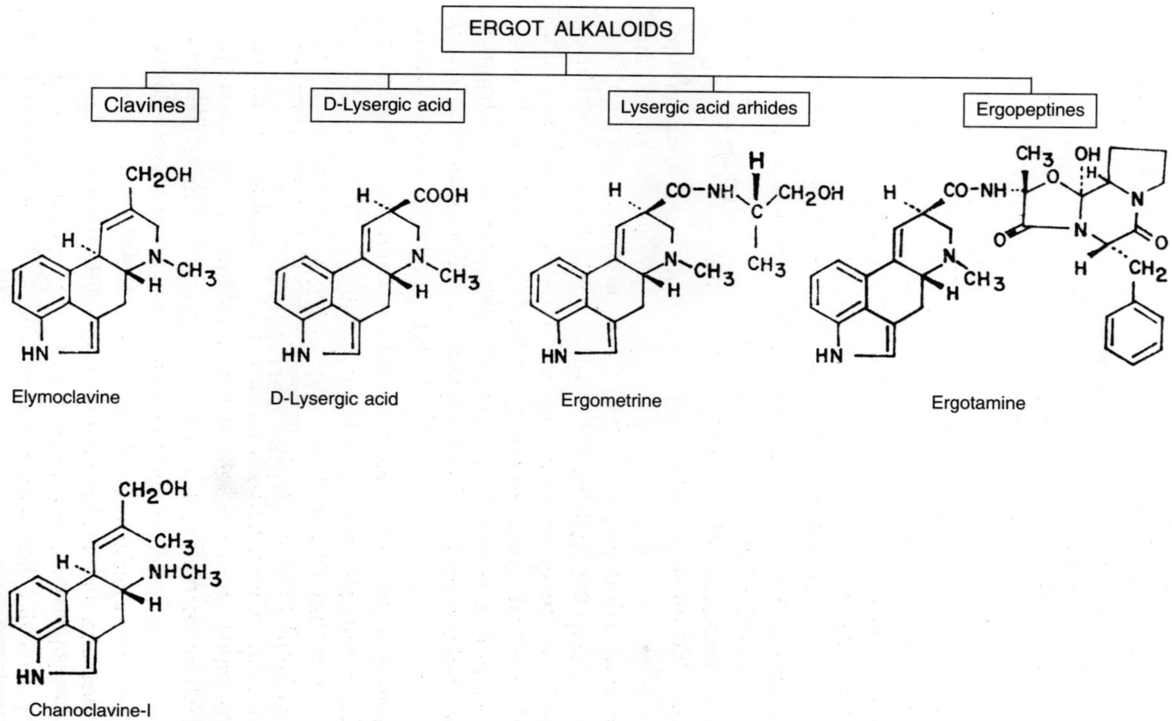

Fig. 2 Characterstic representative of the main groups of ergot alkaloids

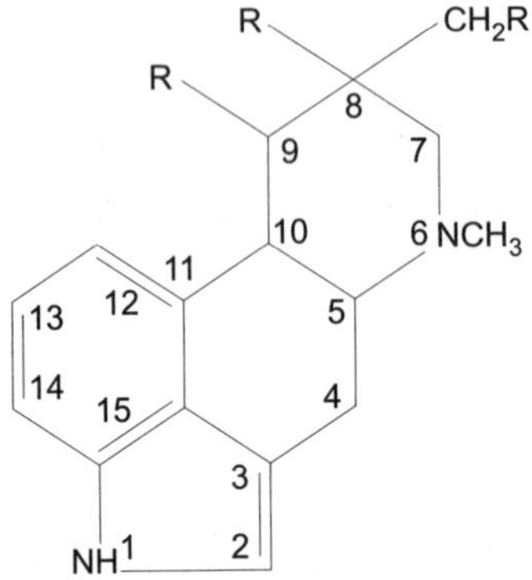

Fig. 3 Clavine

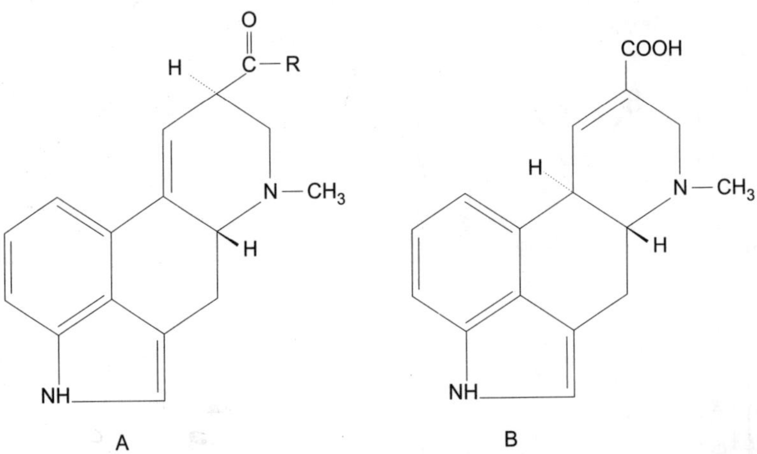

Name	*R*
Ergometrine (ergonovine, ergobasine)	CH_3 / -HN–CH / CH_2OH
Lysergic acid 2-hydroxyethylamide	CH_3 / -HN–CH-OH
Lysergic acid amide (ergine)	$-NH_2$
Lysergic acid	-OH

Fig. 4 Simple derivatives of lysergic acid (A) and paspalic acid (B)

peptide ergot alkaloids (ergopeptines) are composed of lysergic acid and an L- proline containing complex tripeptide moiety. Some examples are: ergotamine, ergosine, ergovaline, ergostine, ergoptine, ergonine, ergobutine, ergocrystine, ergocryptine, ergokryptine, ergocornine, and ergobutyrine.

Biogenesis

Ergot alkaloid is derived from tryptophan, mevalonic acid and methionine (Birch *et al.*, 1960; Groger *et al.*, 1960; Taylor and Ramsted, 1960) (**Fig. 5**). Tryptophan has also been found to be a factor in the induction and depression of enzymes catalyzing alkaloid formation (Krupinski *et al.*, 1976). In *Claviceps purpurea* the formation of conidia is inversely dependent on the synthesis of alkaloids for those saprophytic strains, which partially retain parasitic development, i.e. differentiation of the sphacelial phase to the conidial or the sclerotial phase or both (Mehta 1984). In contrast, in those strains for which the conidial phase, the phase of vegetative growth and the alkaloid phase are not clearly separated, the conidia formation occurs simultaneously with alkaloid synthesis. Physiological conditions necessary for conidiation and germination of conidia are different from conditions controlling alkaloid synthesis (Rehacek *et al.*, 1974). The morphological development of the mycelium in submerged *Claviceps* cultures show significant differences between high-yielding and degenerated cultures (Flieger *et al.*, 1982). The character of fat hyphae and sclerotia like cells of submerged mycelium is reminiscent of the plectenchymatic structure of production sclerotia of parasitic cultures. Such sclerotia like cells influence the alkaloid production in *C. fusiformis* (Dickerson *et al.*, 1970). Cultural conditions suitable for metabolism of primary substances are not suitable for alkaloid metabolism (Taber and Vining, 1963). To obtain over production of alkaloids, high citrate or Krebs cycle intermediate level in the medium is required (Pazoutova *et al.*, 1981). The inability of submerged *Claviceps* cultures to grow on hexoses in absence of Krebs cycle intermediates comes from the parasitic way of life where the level of citrate and malate in the host plant phloem sap is high. Another important fact is that clavine alkaloids and simple derivatives of lysergic acid are extracellular products with high solubility in the culture medium and feedbacks their own biosynthesis (Flieger *et al.*, 1988) while peptide ergot alkaloids (ergopeptines) are mostly intracellular and are accumulated in lipid droplets (oleosomes) with no influence on the metaboloism (Neumann *et al.*, 1979).

Clavine alkaloids lie at the beginning of the biogenetic order and classic ergot peptide alkaloids at the end. The oxidation order agroclavine – elymoclavine- lysergic acid is the main biogenetic pathway (**Fig. 5**). A variety of clavine alkaloids is formed due to action of various enzymes which

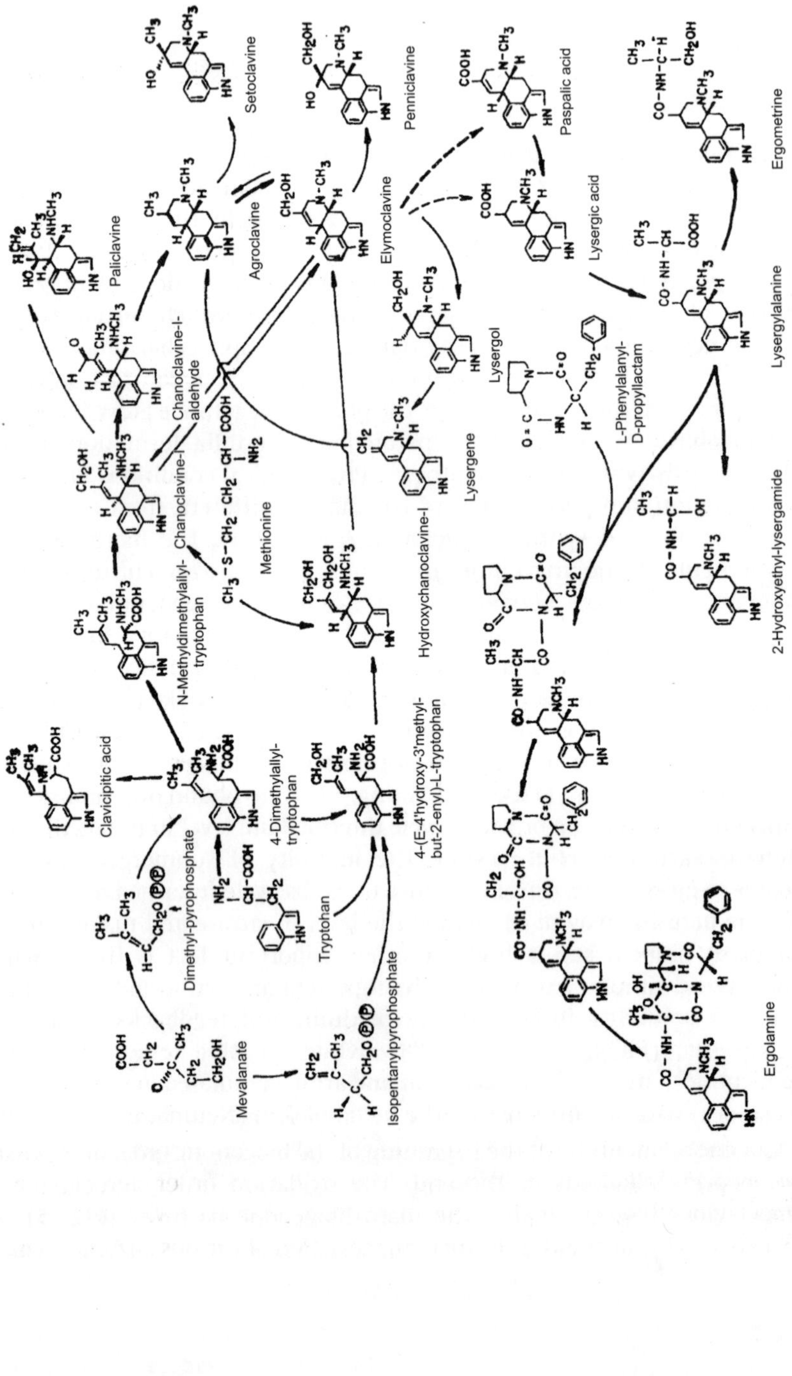

Fig. 5 Biogenesis of ergot alkaloids

direct elymoclavine and agroclavine from the main biosynthetic route. Such type of shunt activity is particularly evident in strains that lack the complete pathway and are unable to synthesize subsituted lysergic acid (Vining 1980). The rate of alkaloid synthesis and the activity of Krebs cycle were proved to be inversely proportional. The level of the Krebs cycle enzymes decreased in favour of glyoxylate enzymes. Citrate is an intermediate, which has an important role in metabolic regulation. The region of formation of ergot alkaloids is not generally the same as where alkaloids are found in increased concentrations. In *Ipomoea* the alkaloids are formed in leaves while they are found accumulated in seeds, which do not produce the alkaloids. The clavine alkaloids synthesis takes place in endoplasmic reticulum which later forms vacuoles containing lipoproteins (Vorisek and Rehacek, 1978). Ergot alkaloids are not waste physiologically inert products of cell metabolism (Rehacek 1974). Ergometrine added to a submerged culture of *Claviceps paspali* increases the alkaloid production by 120% (Rehacek 1974). Ergotamine and ergometrine reduce the intensity of conidia formation in a productive culture of *Claviceps purpurea* (Rehacek *et al.*, 1974). The effect of tryptophan on the enzyme system in the ergot alkaloid synthesis (Robbers *et al.*, 1972) suggests participation of the alkaloids in the feedback inhibition and repression of the enzyme-forming system.

INDUSTRIAL PERSPECTIVES

Strain

The key to successful production of ergot alkaloids is in obtaining the proper strain of the fungus. There are three main processes used for the preparation of *Claviceps* strains for saprophytic culture: (i) plating of plectenchymatic tissue from the surface sterilized sclerotia on an agar growth medium (Mantle 1969), (ii) plating of honeydew drops containing conidia formed at early stage of *Claviceps* infection (Pazoutova *et al.*, 2002), (iii) trapping of sexual ascospores ejected from fruiting bodies on germinated sclerotia. By this method monosporic culture can be obtained (Vasarhelyi *et al.*, 1980). In *Claviceps* sp. high alkaloid production is accompanied with thick walled round having full cytoplasm with long, thin vacuolated cells (Mehta 1984).

Strain Improvement

The classical methods (selection pressure, mutagenesis and recombination) used for the strain improvement are, to some extent, more complicated with *Claviceps* due to incomplete information on cell nucleus. Strains used for saprophytic cultivation might be heterokaryotic and homokaryotic (Didek-

Brumec *et al.*, 1991). Mutagenesis followed by subsequent selection of strain is an important technique in increasing the yield of alkaloids (Didek-Brumec *et al.*, 1987). An ergocristine producing *C. purpurea* strain showed 180-fold increase in alkaloid production after eight-step mutation-selection with different mutagens (Kobel and Sanglier, 1978). To increase the mutation frequency the protoplasts prepared from spores of selected strains were used (Zalai *et al.*, 1990). Hyphal fragments are rather unsuitable for mutagenesis due to the higher number of nuclei. Even protoplast formation from young mycelium and subsequent regeneration without any mutagenic treatment yielded strains with different properties (Schumann *et al.*, 1987). Protoplast fusion technique is beginning to find useful applications either in producing improved mutant strains by intraspecific crosses or in formation of novel spectrum of products by interspecific hybrids (Socic and Porekar, 1992).

Maintenance and Long-term Preservation

Cultures are the basic ingredient of microbiological work and at present there is much concern regarding the need to conserve genetic resources. Methods applied for long-term preservation have been recently reviewed (Hunter-Cervera and Belt, 1996). Two main methods are applied for preserving sporulating cultures (i) deep-freezing and (ii) maintaining of cultures on rye grains or agar slants placed in refrigerator. Non-sporulating strains due to higher sensitivity to conservation procedure are frequently preserved as cultures on the agar plates. The most universal technique applied for preservation is keeping of lyophilized cultures and cultures frozen in liquid nitrogen (Pertot *et al.*, 1990).

Saprophytic Cultivation of *Claviceps* sp

The first saprophytic cultivation of *Claviceps* on artificial nutritional media dates back to 1922 (Bonn 1922). The first attempt to industrial production of ergot alkaloids was isolation of clavine alkaloids from submerged cultures of different *Claviceps* sp. (Abe *et al.*,1956; Abe and Yamatodani, 1955). Later conditions for saprophytic production of simple derivatives of lysergic acid were developed using different strains of *C. paspali* (Arcamone *et al.*, 1961). Only 5 years later a new isolate of *C. purpurea* was found to produce ergotamine under submerged conditions (Amici *et al.*, 1966). Since that time all types of ergot alkaloids for direct use as therapeutic agents or precursors for the preparation of semisynthetic drugs could be obtained by fermentation.

Production in Parasitic Culture

Bekesy developed the first large technology for the parasitic production of ergot on rye in 1934 (Minghetti and Perellino, 1999). For a long time due to severe problems in early fermentation processes, the field cultivation was the major technology in Europe. Recently, the methods of cultivation, inoculation and harvesting have been improved (Nemeth 1999). Presently, the field production technology has reached 1,000 – 2,000 kg sclerotia/ha (Nemeth 1999). Since the alkaloid content in sclerotia of commercial strains averages 1% and more, an alkaloid yield of 8 – 10 kg per hactare can be achieved, even if losses during the purification process are taken into account (Udwardy 1980). The main technical problem in this technology is the generation and storage of high quality inocula, which is mostly achieved by using deep-frozen conidia suspensions. However, several disadvantages of this method have led to a significant reduction of parasitic alkaloid production: (i) production is too dependent on environmental parameters; (ii) the areas in which the climate is adequate for such production are limited, (iii) protection of the valuable production strains is difficult, and (iv) the purification procedures are expensive. Therefore the present proportion of alkaloids produced in parasitic culture worldwide is low, although exact figures are not available (Tudzynski *et al.*, 2001).

Genetic Engineering

Many field isolates of *C. purpurea* produce significant amounts of ergot alkaloids only in the parasitic phase and not in axenic culture. Thus the screening of different ergot strains from field surveys or various culture collections over the years has yielded interesting strains producing different specific alkaloids (or desired mixtures) at moderate rates in plants, but they were only suited (if at all) for parasitic production and not for submerse production of alkaloids. Therefore, ergot industrial strain improvement programs were launched, aimed at an increased capacity of strains to produce high amounts of a specific alkaloid. This was mainly achieved, as in other strain improvement programs of secondary-metabolite-producing fungi, by a stepwise mutagenesis selection process. Molecular genetic techniques have recently been successfully applied to *Claviceps* sp. (Pazoutova and Parbery, 1999; Tudzynski *et al.*, 1999) and have considerably widened the options for genetic research. They have been used for (i) detailed characterization of strains for evolutionay studies, (ii) for a molecular analysis of the biosynthetic pathway, and (iii) for detailed regulatory studies.These molecular approaches will allow the design of recombinant strains with improved production properties (e.g. by enhancing expression of genes of interest or by creating defined detection mutants) or that produce new compounds, which are not accessible by chemical synthesis.Ergot alkaloids are synthesized by the enzymic transformation of small precursor molecules and hence they are not direct gene products. In biogenesis of *Claviceps* the relevant structural and regulatory genes have not been mapped, therefore, isolation and cloning of the entire pathway of structural genes for alkaloids would be difficult. Gene transfer via protoplast fusion has opened up new possibilities in both the cellular and molecular aspects of *Claviceps* genetic engineering. There are three potentially profitable avenues which may help to bridge the gap between academic and industrial genetics of *Claviceps,* (i) the construction of a plasmid that replicates autonomously in *Claviceps* and carries easily selectable markers, (ii) the development of transformation systems with drug resistance markers for use in species lacking available auxotrophic mutants and (iii) the study of genes controlling the regulation of entire pathway. The recent indentification of a cluster of genes (Tudzynski *et al.*, 1999) involved in ergot alkaloid biosynthesis in *C. purpurea* and the availability of molecular genetic techniques allow the development of strategies for rational drug design of ergoline-related drugs by enzymic engineering and by biocombinatorial approaches (Tudzynski *et al.*, 2001). Cultural conditions affect the fermentation broth directly as well indirectly (**Fig. 6**).

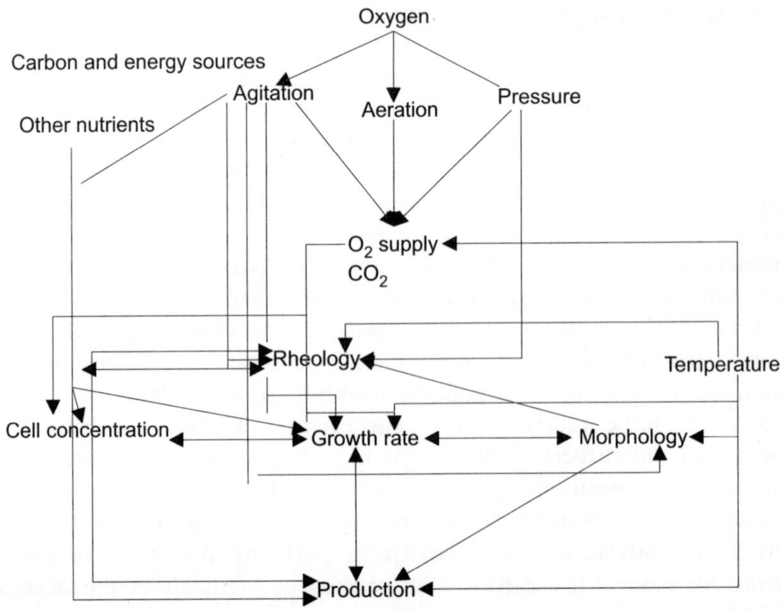

Fig. 6 Effect of cultivation on the culture growth rate and metabolits production

Fermentation

All technologies developed for industrial scale have the same aims, i.e. the maximum production of desired ergot alkaloids, the minimum amount of accompanying contaminants (other alkaloids, toxins, etc.), the shortest possible time of fermentation, the minimized cost of medium, energy, equipments, and labor. For the production of ergot alkaloids, different fermentation technologies have been employed, i.e. (i) stationary cultivation on liquid or solid medium (Trejo-Hernandez *et al.*, 1993) (ii) submerged cultivations (Kobel and Sanglier, 1986) also adapted to semicontinuous or continuous processes (Knopp and Rehm, 1984). In some cases the immobilized microorganisms were used for the production of ergot alkaloids under condition of submerged fermentation (Komel *et al.*, 1985; Kren 1991).

Stationary Surface Cultivation

The development of processes for the production of ergot alkaloids under conditions of stationary cultivation on liquid media started in early sixties (Adams, 1962). The stationary surface cultivation on agar slants was commonly used for preparation of starting cultures. When transformed to industrial scale this technology showed some limitations mostly due to difficulties with control of aseptic conditions of large surfaces. Vlcek and Kybal (1974) developed technology for stationary cultivation of *C. purpurea* in plastic bags partially filled with inoculated liquid medium. This procedure was used for production of ergotoxins and later adapted to production of asexual spores of *C. purpurea* for field production.

Submerged Cultivation

Several workers have got success in isolating strains of *Claviceps purpurea* and their cultivation under submerged conditions to produce ergot peptide alkaloids (ergopeptines), simple derivatives of lysergic acid and clavine alkaloids (Vorisek and Rehacek, 1976). Submerged fermentation in laboratory scale, i.e. shaker cultivation, is a primary step in getting knowledge of production microorganism physiology, biosynthesis, sporulation, stability, and influence of medium composition on production of alkaloids. In industrial scale, submerged fermentation in shaker culture is mostly used for preparation of sufficient amount of inoculum for further cultivation step. During the inoculum preparation an optimal state of the culture for biosynthesis of ergot alkaloids in the production step can be established (Socic *et al.*,1986). Usually, the whole industrial process consists of four steps i.e. (i) shaker culture, (ii) pre-inoculating fermentation,

(iii) seed fermentation, and (iv) production fermentation followed by downstream processing (Malinka 1999). Industrial production of alkaloid mixture (6 g/L) by submerged fermentation of different strains of *Claviceps* sp. (Pazoutova and Tudzynski 1999) was reported in a number of patents (Kren et al., 1988; Trinn *et al.*, 1990). In contrast to high production and well elaborated procedure of submerged cultivation the industrial production of clavine alkaloids does not receive much attention due to limited amount of procedures leading to preparation of desirable final products. Production of lysergic acid and its simple derivatives is of high technological and industrial interest due to relatively simple procedure of chemical modifications to semisynthetic ergot preparations (**Table-1**). The basic studies on biosynthesis, physiology and production of simple derivatives of lysergic acid were done on relatively limited number of *Claviceps* strains (Bumbova *et al.*, 1991). Flieger *et al.*, (1989 a, b) using *C. paspali* CCM 8061 developed a technology of aggressive bioconversion of clavine alkaloids to simple derivatives of lysergic acid with total production of nearly 6 g/L in batch cultivation and about 3 g/L in industrial fermentor. Different strains of *C. purpurea,* as the only producers of ergopeptines, have been described for their submerged production. Ergotamine producing strains and their cultivation are the best-studied processes due to direct therapeutical use of ergotamine.

TABLE 1 *Industrial submerged production (s) of ergot alkaloids (Rehacek 1983)*

Alkaloid	*Company*			
	A	B	C	D
Ergocornine	s	s	s	s
Ergocrystine	s	s	–	s
Ergokryptine	s	s	s	s
Ergotamine	–	s	s	–
Ergometrine	–	–	s	–
Lysergic acid	–	s	–	–
Paspalic acid	s	–	–	s

A Biochemie GmbH, Kundl/Tirol, Austria
B Fermitalia S.p.A., Milano, Italy
C Gedeon Richter, Budapest, Hungary
D Sandoz-Wander AG, Basal, Switzerland

Fermentor Design

Many laboratories have used different fermentors from time to time. In scaling-up a 'conventional' stirred tank fermentor must provide for atleast and the lowest possible cost and should satisfy all requirements (good heat, mass transfer, perfect sterilization, aseptic procedures, reliable foam control, good bulk mixing, simle, rapid and thorough cleaning methods, reliable and appropriate monitoring control systems, proven construction methods, appropriate materials of construction and multiple product application).

SOLID SUBSTRATE FERMENTATION

Solid substrate fermentation describes the microbiological transformation of biological materials in their natural state, in contrast with liquid of submerged fermentation, which is carried out in dilute solutions or slurries. Recently, Robbinson *et al.*, (2001) have proposed a solid-state (substrate) fermentation (SSF) for the production of enzymes and secondary metabolites to meet the global demand. Production of ergot alkaloids by SSF was found 3.9 times greater than that produced by *C. fusiformis* in SmF (Trejo-Hernandez *et al.*, 1993). Products of secondary metabolites by submerged liquid fermentations (SmF) have been found to be less efficient than SSF. The use of antifoam chemicals and the sheer stresses caused by stirring prevented high yields in SmF. Better air circulation can also be achieved in SSF increasing ergot alkaloid yields further (Balakrishnan and Pandey, 1996). The solid substrate fermentation has been shown to produce a more stable requiring less energy in smaller fermenters with easier downstream processing measures. Also, by removing the cost and trouble associated with anti-foaming chemicals and by maximising yield production, SSF may be seen as a viable option for industrial scale production of ergot alkaloids.

Scale-up

There are 3 basic ways to get ergot alkaloids: (i) by isolation from the sclerotia of *Claviceps purpurea* collected from the field (95% of the world production), (ii) by extraction of saprophytic cultures of other ascomycetes especially of the *Claviceps* genus and (iii) by partial/total synthesis. Sclerotia of *Claviceps purpurea* collected from rye may contain 0.8% peptide alkaloids on dry weight basis which is an equivalent of about 500 mg/L alkaloid level in a fermentation broth of 6% mycelial dry weight (Udwardy 1980). Since the naturally grown ergot is not sufficient, also as its production is labour intensive and wheather dependent, the industrial production of ergot alkaloids has gained importance. In some cases, the saprophytic production of ergot alkaloids is more economical than parasitic cultivation on rye plants. In fermentors the process is more presily controlled, not dependent on specific season and the production and isolation of alkaloids can take place on the same site and possible to apply some manipulations, which are not feasible in field conditions and above all clavines and simple lysergic acid derivatives are easier to manufacture than ergopeptines. In bioreactors the alkaloid yield may exceed 20% of the mycelial dry mass while natural sclerotia yield is normally less than 1% or even less than 0.1% in some cases. For optimization and scale-up process certain facts should be taken into account (Kobel and Sanglier, 1978) (i) the production strains especially *Claviceps* should be continually tested to maintain the optimal quality of selected production strain and thus maximally eliminate the biological effects given by transfer of cultures, ageing and other external influences, (ii) long cultivation period in absence of antibiotics require very high standard of sterility in operation and equipment, (iii) balanced aeration and stirring are required due to the sensitivity of the *Claviceps* cultures to the stress and high oxygen tension, (iv) alkaloid production varies with the composition of media especially with use of antifoam agents which could cause considerable loss in alkaloid yield, (v) as bulk of the peptide alkaloids are mainly confined in mycelial cells, therefore large mycelial mass with high alkaloid content should be obtained. Recently the application of oxygen vectors to *Claviceps purpurea* cultivation has been reported (Menge *et al.*, 2001). The classical problem of the large-scale fermentations is an optimal supply of oxygen to growing *Claviceps* sp. High oxygen demand in the exponential growth phase can be met by addition of different hydrocarbons (Gilmanov *et al.*, 1996) or perfluorocarbons (Menge *et al.*, 2001) to shake flask and/or stirred reactors. Perfluorocarbons were also applied successfully in cultivations of other microorganisms (Lowe *et al.*, 1998).The saprophytic *Claviceps* strains are prerequisite for modern production of ergot alkaloids. Some saprophytic strains of *Claviceps* for industrial production of ergot alkaloids are worth mentioning:

Clavines: *C. fusiformis* B 35, *C. fusiformis* w1, *C. fusiformis* F 27, *C. purpurea* Pepty 695/ch, *C. purpurea* 129, *C. purpurea* CP 7 and *C. paspali* SD 58

Simple lysegic acid derivatives: *C. paspali* NRRL 3027, *C. paspali* Li 342/SE 60

Ergopeptines:

(i) ergotamine: *C. purpurea* ATCC 15383, *C. purpurea* FI 275, *C. purpurea* OKI 56/1970

(ii) ergosine: *C. purpurea* MUT 168

(iii) ergocristine: *C. purpurea* ATCC 20 103 (IMI 131669, FIS 40), *C. purpurea* Ech 420

(iv) ergocornine and ergosine: *C. purpurea* FI 43/14

(v) ergotamine and ergotryptine: *C. purpurea* FI 32/17, *C. purpurea* IC/39/20

(vi) ergocornine and ergokryptine: *C. purpurea* Exy 20, *C. pur*purea OKI 00022, *C.purpurea* DM 838 (IMET PA 13), *C. purpurea* Pepty 695

(vii) dihydroergosine: *Sphacelia sorghi*

Processing and Cost Engineering

The commercialization of ergot alkaloids and their production made possible through biotechnology requires a coordinated coupling of unit operations in order to develop efficient processes. There are many factors involved in scaling down the solids and liquid handling processes. After fermentation to sepatate cell and broth membrane filters are widely in use in recent time in addition to conventional processes (centrifugation and filteration). New developments in membranes as well as improved process configuration continue to make this an attractive tool in biochemical process synthesis. Electrodialysis is important operation for purification and concentration of low molecular weight salts. Improvements in process configuration and membrane design make this technique attractive in a variety of applications.

MODERN TECHNOLOGY

Modern research into ergot alkaloids can draw on a very considerable number and variety of truely new means and methodologies, which are in continuous rapid evolution. A full range of features which involves (i) computerization of documentation (ii) automation of various experimental activities (iii) mathematical modeling (iv) calculation of atomic, molecular and crystal structures and their relevant physical, chemical, biological and therapeutical properties and (v) statistical design of experimentation and the interpretation thereof with translation into mathematical models.

MATHEMATICAL MODELING

The growth of the fungus and alkaloid formation in submerged batch fermentation can be described by mathematical model, which can be used in automatic process control and optimization of an industrial process. Grm *et al.*, (1980) while working with *C.purpurea* proposed a growth model based on morphological features of a cell population during the fermentation process. Pazoutova *et al.*, (1981), however, proposed another model accentuating the antagonistic effects of phosphate on growth and alkaloid production. Another possibility for process optimization lies in feeding the culture and thus prolonging the idiophase of the fermentation process (Milicic *et al.*, 1987). This process which seems to be one of the perspectives in modern alkaloid production (Socic and Porekar, 1992). The activation-inhibition kinetics of clavine alkaloids production was evaluated for two *Claviceps fusiformis* strains (Flieger *et al.*, 1988) and it was found that feedback inhibition could be eliminated by combination of fermentation and separation units in a closed loop. Increased efficiency (more than 100 %) of the fermentation process was also found when inducers of cytochrome P-450 were used (Rylko *et al.*, 1988).

Medicinal Use

The ergot alkaloids have many potential therapeutic uses some of which have been realized and others have not survived the critical test of time. Ergot alkaloids are of prime interest to pharmacologists because of their multiple actions on utrine and vascular muscle, inhibitors of prolactin secretion, as a tool for studying the mechanism of sympathetic nervous system, to produce stimulation of dopaminergic receptors and in the treatmnt of parkinsonism. The ergot and its semi-synthetic derivatives have been widely used in many diseases (acromegaly, amenorrhea-galactorrhea, suppression of postpartum lactation, treatment of breast cancer and possibly cancer of prostrate, angina pectoris, enterics (Demain 1999) and in infectious diseases like glaucoma and herpes zoster). A new group of alkaloids has been reported which interferes with DNA synthesis in mammalian cells. Ergot alkaloids can be chemically manipulated to form derivatives that possess altered pharmacologic effects. Preparations containing ergot alkaloids or their semi-synthetic derivatives are also applied for the treatment of migrain (ergotamine), high blood pressure (dihydrocornine, dihydrokryptine) and obstetrics (ergometrine). Ingestion of isoergine and lysergic acid diethylamide (LSD) are known to cause marked hallucinations (DeGroot *et al.*, 1998). Ergot alkaloids are not simple pharmacological or therapeutical entity as they also have wide range of biological effects (Berde 1984). The clavine alkaloids cause apathy, drowsiness and sedation in many mammalian systems and have antimicrobial activity against a series of human pathogenic bacteria and *Candida albicans* (Eich *et al.*, 1984), hypotensive and oxytoxic properties and are able to inhibit prolactin secretion (Edwardson 1968). Several other ergot derivatives influence the secretion of several hormones and act at the central nervous system. Ergot has 3 main types of actions (i) central (ii) peripheral and (iii) neurohormonal. LSD is pharmacologically inactive but its various derivatives display pharmacological actions which include (i) sympatholytic effect-inhibition of the hormones adrenalin, noradrenalin and serotonin and therby a dilation of veins and decrease in blood pressure and (ii) effect on uterus via contraction of smooth muscle. Nicergolin has been found to be of therapeutic value in senile dementia. Ergometrine and its semi-synthetic derivatives like methylergometrine have been widely used in postpartum to prevent hemorrhage. Since early time ergot alkaloids have shown their effects on the reproduction processes of animals, therefore ergot derivatives (ergotoxine and ergocornine) have been tried as antifertility agents in rat and can also terminate early pregnancy in both rat and mice (Sommer and Buchanan, 1955). Lisuride was also introduced into therapy as an antimigraine drug and has a high dopaminergic activity. Bromokryptine with dopaminergic effects on DA receptors are used in the therapy of Parkinson's disease and for curing acromegaly. Pergolide, a synthetic ergoline has been found to have a positive effect on parkinsonism (Liberman *et al.*, 1979). The ergot alkaloids can be chemically maniputated to form derivatives that possess altered pharmacologic effects. Moreover, after obtaining a correlation between clinical indications and therapeutical dosage, pure ergot alkaloids and their semi-synthetic derivatives can be used in medicine extensively in future.

References

Abe, M. and Yamatodani, S. 1955. A new water-soluble ergot alkaloid, molliclavine. *Bull. Agric. Chem. Soc.* (Japan) 19:161-162.

Abe, M., Yamatodani, S., Yamano, T. and Kusumoto, M. 1956. Isolation of two new water-soluble alkaloids pyroclavine and costaclavine. *Bull Agric Chem Soc.* (Japan). 20:59-60.

Adams, R.A.1962. Process for the production of ergot alkaloids. US pat 3117917.

Amici, A.M., Minghetti, A., Scotti, T., Spalla, C. and Tognoli, L. 1966. Production of ergotamine by a strain of *Claviceps purpurea* (Fr.) Tul. *Experientia,* 22: 415-418.

Arcamone, F., Chain, E.B., Ferreti, A., Minghetti, A., Pennela, P., Tonolo, A and Vero, I. 1961. Production of a new lysergic acid derivative in submerged cultures by a strain of *Claviceps paspali* Stevens & Hall. *Proc. Royal. Soc*. B 155:26-54.

Baumert, A., Erge, D., Groeger, D., Maier, W., Schmauder, H-P., Schumann, B., Breuel, K. and Hoehne, I. 1979. Verfahren zur herstellung von Mutterkornalkaloiden. DD Pat 234171.

Baxter, R.M., Kandel, S. and Okany, A. 1961. Studies on the mode of incorporation of mevalonic acid into ergot alkaloids. *Tetrahedron Lett.* 596-600

Balakrishnan, K. and Pandey, A. 1996. Production of biologically active secondary metabolites in solid-state fermentation. *J. Sci. Ind. Res.* 55: 365-372.

Berde, B. 1984. The future of ergot research. *J. Roy. Soc. Med*. 77: 5.

Berde, B. and Sturmer, E. 1978. Introduction to the pharmacology of ergot alkaloids and related compounds. In: *Ergot Alkaloids and Related Compounds* , Vol 49 (Eds. Berde, B. and Schild, O). pp. 1-28. Springer Verlag Berlin.

Bonn, W.W. 1922. A preliminary study of *Claviceps purpurea* in culture. *Amer. Jour. Bot.* 9:339-353.

Brich, A.J., McLoughlin, B.J. and Smith, H. 1960. The biosynthesis of the ergot alkaloids. *Tetrahedron Lett*. 7: 1-3.

Bumbova-Linhartova, R., Flieger, M., Sedmera, P. and Zima, J. 1991. New aspects of Submerged fermentation of *Claviceps paspali*. *Appl. Microbiol. Biotech*. 34:703-706.

De Groot, A., Dongen, P.W.J.V., Vree, T.B., Hekster, Y.A. and Roomalen, J.V. 1998. Ergot alkaloids: Current status and review of clinical pharmacology and therapeutic use compared with other oxytocics in obsterics and gynaecology. *Drugs*. 56:523-535.

Demain, A.L. 1999. Pharmaceutically active secondary metabolites of microorganisms. *Appl. Microbiol. Biotech*. 52 : 455-463.

Dickerson, A.G., Mantle, P.G. and Szezyrbak, C.K. 1970. Autolysis of extracellular glucans produced *in vitro* by a strain of *Claviceps fusiformis*. *J. Gen. Microbiol*. 60:403-415.

Didek-Brumec, M, Gaberc-Porekar, V., Alacevic, M., Druskovic, B. and Socic, H. 1991. Characterization of sectored colonies of a high-yielding *Claviceps purpurea* strains. *J. Basic. Microbiol*. 31:27-35.

Edwardson, J.A. 1968. The effect of agroclavine an alkaloid on pregnancy and lactation in the rat. *Brit. J. Pharma*. 33: 215.

Eich, E., Eichberg, D. and Muller, W.E.G. 1984. Clavines: New antibiotics with cytostatic activity. *Biochem. Physiol*. 33: 523-526.

Fleiger, M., Sedmera, P., Vokoun, J., Ricicova, A. and Rehacek, Z. 1982. Separation of four isomers of lysergic acid hydroxyethylamide by liquid chromatography and their spectroscopic identification. *J. Chromatogr*. 236: 453-459.

Flieger, M., Votruba, J., Kren, V., Pazoutova, S., Rylko, V., Sajdl, P. and Rehacek, Z. 1988. Physiological control and process kinetics of clavine production by *Claviceps purpurea* 129 and 129/35. *Appl. Microbiol. Biotechnol*. 29:181-185.

Flieger, M., Linhartova, R., Rehacek, Z., Saidl, P., Stuchlik, J., Malinka, Z., Harazim, P., Cvak, L. and Bremek, J. 1989 a. Industrial strain of microorganism *Claviceps paspali* Stevens et Hall. CCM 8061 CZ Pat 276430 (In Czech).

Flieger, M., Linhartova, R., Rehacek, Z., Saidl, P., Stuchlik, J., Malinka, Z., Harazim, P., Cvak, L. and Bremek, J.1989 b. Process for production of lysergic acid simple derivatives. CZ pat 281 196 (In Czech).

Flieger, M., Wurst, M. and Shelby, R. 1997. Ergot alkaloids – sources, structures and analytical methods. *Folia Microbiol*. 42: 3-30.

Flieger, M., Mehta, P. and A. Mehta. 2003. Biotechnological potential of ergot alkaloids. In: *Handbook of Fungal Biotechnology* , 2nd Edition (Eds. Arora,D.K., Bridge, P.D. and Bhatnagar, D.). pp. 21: 91-99. Marcel and Dekkar, New York, USA.

Gilmanov, V., Flieger, M. and Dymshits, V. 1996. Influence of hydrocarbons on some *Claviceps* cultures. *J. Appl. Bacteriol*. 81:678-680.

Grm, B., Mele, M. and Kremser, M. 1980. Model of growth and ergot alkaloid production by *Claviceps purpurea*. *Biotechnol. Bioengg.* 22: 255-270.

Groger, D., Mothes, K., Simon, H., Floss, H.G. and Weygand, F.1960. Uber den einbau von mevalonsaure in das ergolinsystem der clavine-alkaloide. *Z. Naturforsch*. 15: 141-143.

Groger, D., Tylerm V.E.Jr. and Dusenberry, J.E. 1961. Investigations of the alkaloid of paspalum ergot. *Lloydia*. 24: 97-102.

Hunter-Cervera, J.C. and Belt, A. 1996. *Maintening Cultures for Biotechnology and Industry*. Academic Press, New York.

Kobel, H. and Sanglier, J.J. 1978. Formation of ergotoxine alkaloids by fermentation and attempts to control their biosynthesis. In: *Antibiotics and other secondary metabolites* (Eds. Hutter, R., Leisinger, T., Nuesch, J. and Wehrli, W.).pp. 233-242. Academic Press, New York

Komel, R., Rozman, D., Puc, A. and Socic, H. 1985. Effect of immobilization on the stability of *Claviceps purpurea* protoplasts. *Appl. Microbiol. Biotechnol.* 23:106-109.

Kopp, B. and Rehm, H.J. 1984. Semicontinuous cultivation of immobilized *Claviceps purpurea. Appl. Microbial. Biotechnol.* 19:141-145.

Kozlovsky, A.G. 1999. Producers of ergot alkaloids out of *Claviceps* genus. In: *Medicinal and aromatic plant-industrial profiles. Ergot, the genus Claviceps.* (Eds. Kren, V. and Cvak, L.). pp. 479-499. Harwood Acedemic, Chur.

Kren, V. 1991. Bioconversion of ergot alkaloids. In: *Advances in Biochemical Engineering/Biotechnology*. (Ed. Fiechter, A.). pp.123-144. Springer Verlag, Germany.

Kren, V., Sajdl, P., Rehacek, Z., Malinka, Z. and Bremek, J.1988. Industrial strain of *Claviceps purpurea* (Fr.) Tul 88-EP/1988 for elymoclavine preparation. CZ Pat 272562 (In Czech).

Krupinski, V.M., Robbers J.E. and Floss, H.G. 1976. Physiological studies of ergot. Induction of alkaloid synthesis by tryptophan at the enzymatic level. *J. Bacteriol.* 125: 158-165.

Lowe, K.C., Davey, M.R and Power, J.B. 1998. Perfluorochemicals: their applications and benefit to cell culture. *Trends Biotechnol*. 16:272-277.

Liberman, A.N., Neibowitz, M., Neophytides, A., Kuperschmit, M., Mehl, S., Kleinberg, D., Serby, M. and Goldstein, M. 1979. Pergolide and lisuride for Parkinson's disease. *Lance*t. 2: 1129-1130.

Malinka, Z. 1999. Saprophytic cultivation of *Claviceps*. In: *Medicinal and aromatic plant-industrial profiles. Ergot, the genus Claviceps*. *Ergot the genus Claviceps.* (Eds. Kren, V. and Cvak, L.). pp. 321-371. Harwood Academic Publishers.

Mantle, P.G. 1969. Development of alkaloid production in vitro by a strain of *Claviceps purpurea* from *Spartia towsendii*. *Trans. B.r Mycol. So.c* 52:381-392.

Mehta, P. 1984. Submerged fermentation of clavine alkaloids. *Ph.D. Thesis*. Czech Academy of Sciences, Praha, Czech Republic, pp. 202.

Menge, M., Mukherjee J. and Scheper, T. 2001. Application of oxygen vectors to *Claviceps purpurea* cultivation. *Appl. Microbiol. Biotechnol*. 55: 411-416.

Milicic, S., Kremser, M., Gaberc-Porekar, V., Didek-Brumec, M. and Socic, H. 1987. Correlation between growth velocity and biosynthesis of ergot alkaloids in *Claviceps purpurea batch* fermentation. *Appl. Microbiol. Biotechnol*. 27:117-120.

Minghetti, A., Crespi-Perelino, N. 1999. The history of ergot. In: *Medicinal and aromatic plant-industrial profiles. Ergot, the genus Claviceps.(Eds. Kren, V. and Cvak, L.)* pp. 1-24. Harwood Acedamic Publishers. Chur.

Mukherjee, J. and Menge, M. 2000. Progress and prospects of ergot alkaloid research. *Adv. Biochem. Engg. Biotech* . 68: 1-20.

Nemeth, E. 1999. Parasitic production of ergot alkaloids. In: *Medicinal and aromatic plant-industrial profiles. Ergot, the genus Claviceps* . (Eds. Kren, V., Cvak, L). pp. 303-319. Harwood Academic. Chur.

Neumann, D., Loseckem, W., Maierm, W. and Groger, D. 1979. Localization of alkaloids in sclerotia and suspension cultures of *Claviceps purpurea* (Fr.) Tul. *Biochem. Physiol. Pflzen.* 174: 504-508.

Pazoutova, S., Votrubam J. and Rehacek, Z. 1981. A mathematical model of growth and alkaloid production in submerged culture of *C.purpurea*. *Biotech Bioengg.* 23: 2837-2849.

Pazoutova, S. and Parbery, D. 1999. The taxonomy and phylogeny of *Claviceps.* In: *Medicinal and aromatic plant-industrial profiles. Ergot, the genus Claviceps.* (Eds. Kren, V. and Cvak, L.). pp. 55-77. Harwood Acedemic, Chur.

Pazoutova, S. and Tudzynskim P, 1999. *Claviceps* sp. PRL 1980 (ATCC 26245), 59 and Pepty 695/ch-I: Their True Story. *Mycol Res.* 103/8:1044-1048.

Pazoutova, S., Johnsonm N. and Rajasab, A.H. 2002. *Heteropogon triticeus*, a new host of *Claviceps sorghi* in India. *J. Phytopathol.* 150:1-4.

Pertot, E., Gaberc-Porekar, V. and Socic, H. 1990. Isolation and characterization of an alkaloid-blocked mutant of *Claviceps paspali. J. Basic. Microbiol.* 30:51-56.

Rehacek, Z. 1974. Ergot alkaloids and some problems of the physiology of there formation. *Zbl. Bakt*. II. 129: 20-49.

Rehacek, Z., Kozova J., Sajdl, P. and Vorisek, J. 1974. Physiology of conidial formation in submerged cultures of *C. purpurea* (Fr.) Tul. producing alkaloids. *Can. J Microbiol.* 20: 1323-1329.

Rehacek, Z. 1983. New trends in ergot biosynthesis. *Process Biochemistry.* 22-33.

Rehacek, Z. and Sajdl, P. 1990. *Ergot alkaloids: Chemistry, Biological Effects, Biotechnology* . Czech Academy of Sciences, Academia, Praha, Czech Republic, pp 376.

Rehacek, Z. and Mehta, P. 1991. *Claviceps* strains for saprophytic production of ergot alkaloids. *The Genet. Engg. Biotechnol.* London 11:6: 19-21.

Rehacek, Z. and Mehta, P. 1993. Biological effects of ergot alkaloids. In: *Fungal Ecology and Biotechnology* (Eds. Rai, B., Arora, D.K., Dubey, N.K. and Sharma P.D). pp. 275-287. Rastogi Publications, Meerut, India.

Robbers, J.E., Robertson, L.W., Horneman, K.M., Jindra, A. and Floss, H.G. 1972. Physiological studies on ergot. Further studies on the induction of alkaloid synthesis by tryptophan and its inhibition by phosphate. *J Bacteriol.* 112: 791-796.

Robinson, T., Singh, D. and Nigam, P. 2001. Solid-state fermentation: A promising microbial technology for secondary metabolite production. *Appl Microbiol Biotech*. 55: 284-89.

Rosazza, J.P., Kelleher W.J and Schwarting, A.E. 1967. Production of lysergic acid derivatives in submerged culture. IV. Inorganic nutrition studies with *Claviceps paspali*. *Appl Microbiol.* 15:1270-1283.

Schumann, B., Maier, W. and Groeger, D. 1987. Characterization of some *Claviceps* strains derived from regenerated protoplasts. *Z. Naturforsch.* 42:381-386.

Socic, H., Gaberc-Porekar, V., Pertot, E., Puc, A. and Milicic, S. 1986. Developmental studies of *Claviceps paspali* seed cultures for the submerged production of lysergic acid derivatives. *J. Basic. Microbiol.* 26:533-539.

Socic, H. and Gaberc-Porekar, V. 1992. Biosynthesis and physiology of ergot alkaloids. In: *Handbook of Applied Mycology, Fungal Biotechnology* , Vol 4 (Eds Arora, D.K., Elander, R.P. and Mukerje, K.G).pp. 475-515. Marcel Dekkar Inc NewYork.

Sommer, A.F. and Buchanan, A.R. 1955. Effects of ergot alkaloids on pregnancy and lactation in the albino rat. *Amer. J. Physiol.* 18: 296-300.

Taber, W.A. and Vining, L.C. 1959. Tryptophan as a precursor of ergot alkaloids. *Chem. Ind.* London 1218-1219.

Taber, W.A. and Vining, L.C. 1963. Physiology of alkaloid production by *Claviceps purpurea* (Fr.) Tul. Correlation with changes in mycelial polyol, carbohydrate, lipid and phosphorous containing compounds. *Can. J. Microbiol*. 9: 1-14.

Taylor, E.M. and Ramsted, E. 1960. Biosynthesis of lysergic acid in ergot. *Nature.* 188: 494-495.

Trejo-Hernandez, M..R., Lonsane, B.K., Raimbault, M. and Roussos, S. 1993. Spectra of ergot alkaloids produced by *Claviceps purpurea* 1029 C in solid-state fermentation system. Influence of the composition of liquid medium used for impregnating sugar-cane pith bagasse. *Process Biochem.* 28:23-27.

Trinn, M., Manczinger, L., Polestyukne, N.A., Kordik, G., Pecsne, R.A., Zalai, K., Beszedics, G., Ferenczy, L., Nagy, L., Robicsek, K. and Szegedi, M. 1990. Process for the elymoclavine production and preparation of new strain. HU Pat 209325 (In Hungarian).

Tudzynski, P., Holter, K., Correia, T., Arntz, C., Grammel, N. and Keller, U. 1999. Evidence for an ergot alkaloid gene cluster in *Claviceps purpurea*. *Mole. Genet* 261: 133-141.

Tudzynski, P., Correia, T. and Keller, U. 2001. Biotechnology and genetics of ergot alkaloids. *Appl. Microbiol. Biotech* . 57: 593-605.

Udwardy, N.E. 1980. Consideration of the development of an ergot alkaloid fermentation process. *Process Biochem.* 15: 5-8.

Vasarhelyi, G., Tetenyi, P., Zambo, I., Kinicky, M., Balassa-Barkanyi, I., Kiss, J., Recs-Razso, A. and Zsoka-Somkuti, E. 1980. Strain of *Claviceps purpurea*, virulent on rye for ergot alkaloids, especially ergocristine production and process for preparation of it. DE Pat. 3006989 (In German).

Vining, L.C. 1980. Conversion of alkaloid and nitrogenous xenobiotics. In: *Economic Microbiology* (Ed. Rose, A.H). pp. 523-573. Academic Press, NewYork.

Vlcek, V. and Kybal, J. 1974, Equipment for cultivation of microorganisms. CS Patent 172552 (In Czech).

Vorisek, J. and Rehacek, Z. 1976. Origin of sclerotia like cells in submerged *Claviceps purpurea* producing clavine alkaloids. *Arch. Microbial*. 107:321-327.

Vorisek, J. and Rehacek, Z. 1978. Fine structure localization of alkaloid synthesis in endoplasmic reticulum of submerged *C. purpurea*. *Arch. Microbiol.* 117: 297-302.

Zalai, K., Kordik, G., Manczinger, I., Pecsne, R.A., Beszedics, G., Polestyukne, N.A., Ferenczy, I., Olasz, K., Szegedi, M. and Trinn, M. 1990. Process for production of ergolen compounds, mainly ergometrine, and process of colour selection. HU Pat 209324 (In Hungarian).

Frontiers in Plant Sciences (Pages 447-464)
Eds. K.G. Mukerji, K.V.B.R. Tilak, S.M. Reddy, L.V. Gangwane, P. Prakash, I.K. Kunwar

Prospect of MMIR (Mixed Microbial Inoculum for Reclamation) for Bioremediation and Ecorestoration of Chromite Mine Waste Area

A.K. Misra, H.N. Thatoi, M.M. Pattnaik, B. Dutta & G.S. Padhi

ABSTRACT

Intensive chromite extraction through opencast mining system in Orissa is leading to large-scale deforestation and serious environmental pollution problems due to dumping of minerals and overburdens, which needs urgent attention for stabilization. Restoration of mine waste areas is extremely difficult due to existence of hostile soil condition. The present work deals with an attempt for stabilization and restoration of chromite mine waste areas which included strategies like screening of suitable tree, pasture and grass species for development of three tier sustainable vegetation and enrichment of soil quality through inoculation of suitable beneficial soil microbes like nitrogen fixers (*Rhizobium*/*Azotobacter*) and phosphorus mobilisers like AM fungi etc. after careful selection. The microbial inocula applied in a consortia (Mixed microbial inoculum for reclamation or MMIR) taking coir pith decomposed by a lignocellulolytic fungus as a source of organic carbon.

Analysis of soil quality and natural regeneration capacity of young and old dumps of Sukinda chromite mines indicates decrease in soil pH and electrical conductivity (E. C.) but there is steady increase in water holding capacity and nutrient contents like organic carbon, available nitrogen and available phosphorus. Field experiments for plant survival were carried out in chromite

Author's address: Laboratory of Microbiology, Post-graduate Department of Botany, Utkal University, Bhubaneswar – 751004, Orissa

mine waste soil taking tree species (10 tree legumes and 8 non-legumes). Species like *Pongamia pinata, Albizzia lebbeck, Acacia nilotica* among legumes and non-legumes like *Syzizium cumini, Azadiracta indica, Grevelia petidifolia* appears more promising showing better tolerance capacity, with less disease syndromes while other plants showed drastic reduction in growth with associated syndromes like chlorosis, leaf burning etc. Pasture species like *Tephrosia candida, Saccharum munja, Cymbopogon flexiosus* also showed good survival capacity. Further nursery experiments on application of MMIR were undertaken with *Pongamia pinnata* and *Tephrosia candida* was seen to improve plant growth and survivability with improved drought tolerance capacity along with significant improvement in soil nutrient status. *Zea mays* was used as biocarrier for field mycorrhization and showed good growth response under MMIR application. Interestingly *Zea mays* under MMIR application was seen to immobilize heavy metals in its roots and helped in bioremediation of chromite mine waste soil.

Key words: Chromite mine waste soil, ecorestoration, *Rhizobium*, AM fungi, legumes, MMIR, bioremediation.

INTRODUCTION

Orissa is one of the major mineral producing states of India, with major minerals such as iron, chromium, manganese, coal etc. Most of the mining sites are located in the thickly forested, Eastern Ghats regions of the state. Sukinda valley, in the district of Jajpur, Orissa, is accounting for 90% of total chromite deposits of the country. With a view to augment economic growth, extensive and indiscriminate mining activities are being carried out in this region mostly operated through opencast system. This has resulted in serious environmental problems through large-scale deforestation, removal of fertile topsoil and indiscriminate dumping of overburdens in the nearby areas, leading to vast areas of wasteland formation. The surface mining control and reclamation act of 1977 requires restoration of mine wasteland to pre-mining condition, and productivity in so far as possible. With advance technologies land can be saved or soil can be restored approximately to its previous configuration. However, restoration of productivity capacity of the land and its previously existing vegetation has not been possible because of existence of hostile environment condition (Cordell *et. al.*, 1987). The mining overburden wastes are generally sparsely vegetated with few selected plant communities that barely survive under hostile environmental situations (Seighardt, 1990). The poor regeneration capacity is attributed to poor soil conditions (extremely low nutrient content, imbalanced pH along with low soil microbial populations related to nutrient recycling of the soil) in chromite mine soil has been studied by Misra *et. al.* (1990). The problems of chromite mine is further accentuated due to leaching of toxic heavy metal ions contaminating adjacent forest and crop fields which warrant immediate attention for stabilization of soil and restoration of ecosystem through revegetation. Seedlings of the population of the plant species that can tolerate heavy metals can only survive in these dumps and through their biomass production leads to further improvements of mine soil (Seighardt, 1987). However, enhanced growth is possible through biotechnological application of microbes related to nitrogen, phosphorus and carbon cycle in appropriate plant-microbe combination. Deficiencies of soil nutrients (N, P and organic C) also play major role in mine area soil restoration, which needs to be replenished (Misra *et. al.*, 1994).

Strategy Followed

Earlier studies on the chromite mine waste area reclamation by Misra *et. al.* (1994) have indicated good prospect of application of beneficial microbes viz. *Rhizobium* and AM fungi in improving growth of plants in chromite mine waste soil under nursery condition. In the present work field and nursery experiments were carried out to take up ecorestoration of chromite mine waste areas of adopting the strategy of developing three-tier vegetation through introduction of selected species of leguminous and non-leguminous tree, pasture and grass capable of surviving under the heavy metal toxic situation of mine overburden. Since plants show extremely poor growth response, they were further inoculated with compatible, heavy metal tolerant, suitable strains of nitrogen fixers (*Rhizobium/Azotobacter*) along with selected AM fungi with a view for improvement of growth and nitrogen fixation with enhanced nutrient uptake. As the mine overburden soils are also extremely deficient in organic carbon content, experiments were also carried out to develop a suitable carrier for the microbes for field application using decomposed coir pith (by a suitable lignocellulolytic fungi) as a source of carbon as well. *Zea mays*, an upland growing cereal crop of tropics that grows well in wide variety of soil condition is known to have affinity for AM fungi association. In the present experiment *Zea mays* was used as a carrier for field mycorrhization of mine waste areas.

MATERIALS AND METHODS

Study Site

Field studies were carried out in TISCO chromite mine overburden dumps at Kaliapani situated in the Sukinda valley located within latitudes 20° 53′ and 21° 05′ and longitudes 85° 40′ and 85° 53′ in Jajpur district of Orissa from October 1998 to December 2000. It is a narrow westerly slopping valley bounded to the north, east and south by steep hills. The valley contains the largest chromiferous field mass in Orissa occurring sporadically over an area of 420 sq. km. Being flanked by Daitari hill reserve forest in the west and

Mahagiri hill reserve forest on the east. The elevation of the field varies from 166 to 208 meter above mean sea level. There is one perennial water source is a thin stream, Damsala nala. The distance of these two places from the state capital Bhubaneswar is about 140 km. by road.

Assessment of Soil Quality of Young and Old Dumps

Soil samples were collected from young and old chromite mine overburden dumps and its nearby areas and brought to the laboratory of Microbiology, Post Graduate Department of Botany, Utkal University to study the physico-chemical and microbial properties of the soil. The physico-chemical properties viz. pH, electrical conductivity, water holding capacity, organic carbon, nitrogen and phosphorus content of the soil was estimated following standard method of Jackson (1973). Further, the study of microbial population viz. general bacteria, total fungi, actinomycetes and microbes related to nitrogen cycles (*Rhizobium, Nitrosomonas, Nitrobacter,* ammonifiers, denitrifiers) of the soil sample were determined following standard plate technique of Alexander (1962) using different media viz. nutrient agar medium (bacteria), czapeck dox (fungi), yeast mannitol agar medium or YMA (*Rhizobium*), Winogradsky's medium (*Nitrosomonas*), Omeliansky medium (*Nitrobacter*), urea-phosphate-chloride agar (ammonifiers), asparagin-nitrate-citrate-agar (denitrifying bacteria) and soil extract agar medium (actinomycetes).

Field Survivability Experiment to Study Responses of Tree and Pasture Species

A field experiment was undertaken in chromite mine over-burden dump area of Sukinda operated by TISCO during August 1998. A large area of size (1acre) was selected for our plantation work. The area was leveled before furrowing and these furrows were meant for sowing seeds and seedlings. The entire land was demarcated and different plots were marked for different tree and pasture species. The seedlings of different tree species were planted in different plots. Seeds of pasture legumes were sown in plots of size (6′ × 6′) in furrows with 1′ wide. Growth in terms and visual toxic syndromes due to heavy metal toxic ions were recorded from time to time. Readings were taken after 6 months.

Microbial Inoculum Preparation

Rhizobium isolation and maintenance

Rhizobial strains were isolated from nodules of different leguminous host species belonging to wild and cultivated legume grown in different ecological regions of Orissa following method as described by Vincent (1970). Pure culture of these strains has been prepared from single colony isolates on YMA slopes and broths. 10 ml. of suspension from the broth culture containing 10^9 cells ml^{-1} suspension were treated with germinated seeds for 3 hours in sterile petridishes before sowing.

Mycorrhizae mass inoculum preparation

Four different strains of Mycorrhizae such as *Glomus mossae, Glomus fasciculatum, Gigaspora gilmoreae* and *Gigaspora margarita* were collected from Microbiology Division of Indian Agricultural Research Institute, I. A. R. I. as starter inoculum. They were multiplied in our laboratory under pot culture condition using *Zea mays, Pennisetum* sp. as hosts. The soil for pot culture was prepared by mixing soil and sand in ratio 1:1 (w/w) and were sterilized along with pot at 121°C and 15 lb. Pressure. Mycorrhiza inoculum consisted of soil, spores and root fragments of the host. Approximately 1 g. of the inoculum was applied per polypot 2-3 cm. below the surface of soil in a furrow and covered with a layer of soil. The soil was amended with 1% rock phosphate to supplement phosphorus supply.

Development of Mixed Microbial Inoculum for Reclamation (MMIR)

Coir pith (CP) is a lignocellulolytic waste material produced by the coir industries is a rich source of organic carbon and is available in plenty in Orissa. Coir pith which was used as a substrate of carrier for microbial inoculum brought from Sakhigopal of Puri district near Bhubaneswar, was decomposed in cement tank (3m × 1.5m × 1m) was mixed with lignocellulolytic fungi grown in bottles and was used as inoculum for inducing decomposition. Decomposition was carried out under anaerobic condition for 3 months.

Bioremediation effect of MMIR application was undertaken with *Zea mays* studying uptake of heavy metals like chromium, iron and manganese using atomic absorption spectrophotometer (AAS) GBC 932 AA make.

RESULTS AND DISCUSSION

Soil Quality of Young and Old Chromite Mine Overburden Dumps

Soil quality in terms of physico-chemical property and microbial activity of new and old dumps up to 10 years is given in table – 1 (a, b & c). Physico-chemical properties like pH, electrical conductivity, water holding capacity,

TABLE 1a Physico-chemical properties of soils of young and old dumps of Sukinda chromite mines over-burden wastes.

Soil Samples	*pH*	E.C. (*mili.mho.cm*$^{-2}$)	*W.H.C* (%)	*Org. C* (%)	*Avail. P* (*Kg.ha*$^{-1}$)	*Total N* (%)
Fallow land	7.3	0.480	30.194	1.786	42.00	6.5
New Dump	6.8	0.126	37.381	0.031	24.50	0.0
Regenerated Area Near Mine Dump	6.5	0.144	43.951	2.909	22.40	3.0
Mine Dump with 5 yrs. Vegetation	6.3	0.08	41.91	0.29	5.45	4.5
Dump with 10 yrs. Vegetation	5.9	0.087	58.720	3.636	26.8	7.5

TABLE 1b *Microbial population (General bacteria, total fungi and actinomycetes) of soils of young and old dumps of Sukinda chromite mines over-burden wastes.*

Soil Samples	*Number of Colonies X* 10^5 g^{-1} *soil*		
	General Bacteria	*Total Fungi*	*Actinomycetes*
Fallow land	79.33±8.09	14.00±5.60	0.30±0.30
New Dump	44.33±5.04	10.67±4.67	0.00±0.00
Regenerated Area Near Mine Dump	21.33±2.33	5.67±2.01	0.30±0.30
Mine Dump with 5 yr. Vegetation	12.33±0.88	5.00±1.15	0.00±0.00
Mine Dump with 10 yr. Vegetation	87.33±16.18	5.33±2.40	0.30±0.30

TABLE 1c *Microbial population (related to nitrogen cycle) of soils of young and old dumps of Sukinda chromite mines over-burden wastes.*

Soil Samples	*Number of Colonies X* 10^5 g^{-1} *soil*				
	Rhizobium	*Nitrosomonas*	*Nitrobacter*	*Ammonifiers*	*Denitrifiers*
Fallow land	2.00±0.30	1.00±1.00	0.60±0.30	.30±0.80	0.30±0.30
New Dump	1.67±1.20	0.60±0.60	0.60±0.30	4.00±1.15	3.67±1.50
Regenerated Area Near Mine Dump	0.30±0.30	3.67±1.60	0.60±0.30	23.00±4.16	32.00±3.10
Mine Dump with 5 yr. Vegetation	1.00±0.00	2.30±0.50	2.00±0.00	8.00±6.00	12.30±3.30
Mine Dump with 10 yr. Vegetation	3.33±1.20	3.00±0.00	2.33±0.33	11.30±5.00	31.00±6.30

organic carbon, phosphorus and nitrogen etc. have been measured up to 10 years old dump. It was observed that old overburden dumps showed sharp decrease in pH and E. C. with age, whereas there was steady increase in water holding capacity with age where plantation has been done. In the regenerated area and old dumps were also showed significant increase in nutrients like organic carbon, available phosphorus and total nitrogen (Table – 1a). Increase in soil acidity of old chromite dumps may be due to availability of H^+ as a result of increased activity of microorganisms.

Analysis of microbial activity was also done related to the general bacteria, total fungi and actinomycetes. In new dump the population were extremely low and the population increased up to the level of normal fallow land with about 10 years (Table – 1b). Analysis of microbial population of bacterial related to nitrogen cycle viz. *Rhizobium, Nitrosomonas, Nitrobacter,* ammonifiers, denitrifiers were also assessed for young, old and neighboring area. New dump showed extremely low population of most these bacteria. However, with vegetation there was steady increase in all types of microbes almost equal to fallow land. In general it is observed that the population of all types of microorganisms related to nitrogen cycle (e.g. *Nitrosomonas, Nitrobacter,* ammonifier and denitrifier) increased with age of the dumps except *Rhizobium.* The population of total fungi also shows a decreasing trend. Among the bacteria of nitrogen cycle it is interesting to know that population of *Nitrosomonas* and *Nitrobacter* was seen to increase along with the ammonifiers and denitrifiers. But the increase in population of ammonifiers and denitrifiers is much higher than the nitrifiers indicating higher ammonification than nitrification (Table – 1c). This indicates soil compaction and creation of anaerobic condition. Similar higher ammonification rate has also been observed in open cast coal mine sites of UK by Williamson and Johnson (1998). The reason for decrease in *Rhizobium* population may be due to absence of leguminous plants.

Field Response of Tree, Pasture Species Grown in Chromite Mine Waste Soil

Basing on the analysis of soil as well as micro flora further nursery experiments were conducted to screen out the most suitable plant species (tree and pastures) to be grown in mine waste soil. In total 18 tree species (10 legumes and 8 non-legumes) and 4 pasture legumes (Table- 2 a, b) were grown in chromite mine waste soil (CMWS) for their growth and survival response. General observation showed retarded shoot length and plant biomass in almost all types of tested plant species under mine soil. *Dalbergia sissoo* and *Leucaena leucocephala* were most affected due to mine waste soil and showed retardation along with burning syndromes. There is continuous

TABLE 2a *Field response of tree species grown in chromite mine waste soil after 6 months of plantation*

Species	Shoot length (cm)	Disease syndromes
Dalbergia sissoo	71.07 ± 9.52	Top burning and chlorosis
Albizzia lebbeck	67.92 ± 8.24	Same as above Leaves more yellowish in colour
Pongamia pinnata	81.00 ± 5.25	Luxuriant growth like normal well adopted Brown and burning spots in old leaves.
Leucaena leucocephala	43.70 ± 7.12	Prominent chlorosis in leaves, Plants become bushy, top burning, surviving well but no growth branches develop at the base like bush.
Acacia nilotica	43.15 ± 5.40	Surviving well, Chlorosis present
Acacia catechu	15.00 ± 0.00	Surviving well, slow growing
Pterocarpus marsupium	26.6 ± 0.00	Extremely poor growth
Albizzia procera	34.5 ± 4.31	Surviving well, but very slow growth
Tamarindus indica	17.16 ± 1.33	Not surviving
Azadiracta indica	59.12 ± 4.60	Well-survived symptoms as above present.
Syzigium cumini	63.07 ± 0.00	Chlorosis though not prominent yet exists. Burning spots in leaves, Good growth
Artocarpus heterophylla	37.00 ± 0.00	Not surviving
Emblica officinalis	56.00 ± 0.00	Surviving well, leaf browning
Glericidium maculata	35.50 ± 0.00	Poor surviving
Simarauba glauca	10.00 ± 0.00	Well surviving
Greviella pteridifolia	72.00 ± 16.08	Surviving well, chlorosis syndrome
Dendrocalamus strictus	33.80 ± 0.00	Creeping type, Bushy
Melia azadiracta	26.00 ± 0.00	Surviving

TABLE 2b *Field response of pasture legumes grown in chromite mine waste soil after 6 months of plantation*

Species	Shoot length (cm)	Plant dry weight (g.)	Disease syndromes
Tephrosia candida	8.06 ± 0.26	0.728 ± 0.13	Dwarf
Stylosanthes hamata	4.24 ± 0.24	0.442 ± 0.09	Chlorosis and dwarf
Mimosa pudica	3.74 ± 0.32	0.394 ± 0.09	Severe chlorosis, stunted growth
Mimosa invisa	3.10 ± 0.24	0.446 ± 0.09	Severe chlorsis, stunted growth

degeneration of shoot from the top. Plants usually exhibit chlorotic symptoms with burning spots on the leaves. While such chlorotic syndrome is most prominent in case of *Albizzia lebbeck, L. leucocephala, D. sissoo, Azadiracta indica*. Other species showed disease symptoms to a lesser extent. Among all tree species *Pongamia pinnata* showed the best shoot growth in pure chromite mine overburden soil and was chosen as the most suitable plant for mine area reclamation process (Table – 2a). Pasture legumes generally showed stunted growth in pure chromite mine soil and except *Tephrosia candida.* Most of the pasture species showed severe chlorosis symptoms. Among all the pasture legumes *T. candida* was the best suited to the chromite mine soil as it showed the maximum growth in terms of shoot length and plant dry weight (Table – 2b). Among all the plant species *P. pinnata* and *T. candida* were found to be the most promising in terms of growth response and plant biomass production under nursery condition. However, in terms of nodulation response, *P. pinnata, Pterocarpus marsupium, Acacia nilotica, L. leucocephala* were better performing than other species. Eight non-legume species viz. *Griviella pteridifolia, Syzizium cumini, Semarouba glauca, Azadirachta indica, Emblica officinalis, Annona squamosa* grown in the field showed comparatively poor growth response in chromite mine overburden soil compared to garden soil. Pasture legumes viz. *T. candida, Stylosanthes hamata, Mimosa pudica* and *Mimosa invisa* in chromite mine soil under nursery conditions showed a drastic reduction in all the growth parameters viz. shoot and root lengths, fresh and dry weights in comparison to garden soil. The pasture legumes also showed the disease symptoms like chlorosis dwarfism and poor shoot and root growth.

In reclamation programme legumes are considered important as pioneer vegetation because of their nitrogen fixing ability and thriving capacity in hostile environmental conditions provided they are inoculated with suitable microorganisms like *Rhizobium* and AM fungi (Misra, 1994). The symbiotic association of legume-*Rhizobium* has been reported to fix 24-584 N ha^{-1} annually in different legumes (Sindhu and Dadarwal, 2000).

Plant Growth Responses Under MMIR Application

The microbes selected for ecorestoration study included nitrogen fixers (*Rhizobium* and *Azotobacter*), phosphate solubiliser/mobilisers such as AM fungi. While *Rhizobium* enhances the nitrogen fixing capacity of the legume plant and also contributes to the soil fertility, AM fungi improve phosphate uptake in plants. Arbuscular mycorhizal fungi (AM fungi) are soil microorganisms, those are associated with majority of higher plants through mutual symbiosis providing a direct link between soil and plant roots. AM fungi can contribute to plant growth, particularly in disturbed or heavy metal contaminated sites by increasing plant access to relatively immobile minerals such as P (Vivas *et. al.*, 2003; Yao *et. al.*, 2003), improving soil texture by binding soil particles into stable aggregates that resist wind and water erosion (Rilling and Steinberg, 2002; Steinberg and Rillig, 2003) and by binding heavy metals into roots that restricts their translocation into shoot tissue (Dehn and Schuepp, 1989; Kaldorf *et.* al,1999). This helps in better establishment and growth. Besides phosphorus uptake (Mosse, 1977), mycorrhizal fungi are known to increase drought tolerance of host plants (Osonubi *et. al.*, 1991) and uptake of other minerals viz. Ca, Mg and K (Cooper and Tinker, 1978). From our previous experiments it was observed that mycorrhiza could enhance growth, nodulation and biomass of the legumes grown in mine waste soils (Misra, 1994). Leguminous plants respond well, particularly to AM fungi infection and help in enhancement of atmospheric nitrogen fixation through improved P uptake (Bagyaraj *et. al.*, 1979). From our AM fungi screening experiment it was found that in chromite mine soil, *G. mosseae* was the most suitable AM fungi for *P. pinnata, T. candida* and *Zea mays*. Therefore, *G. mosseae* was used, as AM inoculant in our further experiment. *Zea mays* has seen to have greater affinity to AM fungi inoculation under chromite mine soil condition. Hence, *Z. mays* was used as a biocarrier for field mycorrhization experiments.

Mine soil is extremely deficient in organic nutrients particularly organic carbon, nitrogen and phosphorus and need to be supplemented from external sources. Addition of decomposed CP, a cheap source of organic nutrient produced from coconut husk is used for this purpose. Together with selected microorganisms it was developed as a mixed microbial inoculum for reclamation (MMIR) for our plantation work. Apart from organic nutrient supplementation required for speedy growth of plants it also helps in survival of the introduced microbes. Some of the properties of CP viz. availability as fine particle, greater water holding capacity, resistant to pathogenic infections makes it suitable to be used as a carrier of biofertilisers. Being a lignocellulolytic material, nutrients are not easily available to the plant provided it is decomposed by suitable lignocellulolytic microorganism. Studies indicated that *Pleurotus* sp. accelerates the decomposition rate and releases the nitrogen and phosphorus present in the CP. Application of *Pleurotus* sp. stimulated the particular nitrogen transforming bacteria viz. *Nitrosomonas, Nitrobacter* and ammonifiers. When the decomposed CP was inoculated with suitable *Rhizobium, Azotobacter* and AM fungi then it further accelerates the nutrient cycle and helps in improvement of growth and nodulation capacity of the introduced plants. Taking this concept into consideration MMIR was developed taking suitable cellulose decomposer, nitrogen fixer, phosphate mobiliser etc.

Further experiments were also conducted for evaluation of optimum dose of CP application under nursery condition taking *P. pinnata, T. candida* and *Z. mays* using suitable strains of *Rhizobium/ Azotobacter*, AM fungi with different concentrations of CP as MMIR (Table – 3,4 &5). Under chromite mine soil decomposed CP was applied in the dose of 10%, 20% and 30%. Effect of MMIR application was studied for growth, nodulation, as well as soil available nutrient content (organic C, N and P). Growth and nodulation responses were studied in terms of shoot length, plant dry weight, nodule number and nodule dry weight of the test plants. It was observed that under lower dose of decomposed CP (20%) there was optimum alleviation of stress

TABLE 3 *Effect of MMIR application on growth and nodulation of Pongamia pinnata grown in chromite mine waste soil with different concentrations of decomposed coir pith.*

Parameters	*Treatment*	*Chromite mine waste soil (CMWS)*	CMWS+10% *Coir pith*	CMWS+20% *Coir pith*	CMWS+30% *Coir pith*
Shoot length (cm)	Control	8.00±0.34	9.35±0.49	11.73±1.03	10.85±0.72
	MMIR treated (*Rhizo.*+AMF.)	9.50±0.75	12.95±0.89	13.50±0.61	11.73±0.17
Plant dry weight (g.)	Control	0.75±0.08	0.99±0.21	0.71±0.01	0.93±0.07
	MMIR treated (*Rhizo.*+AMF.)	0.90±0.08	0.86±0.02	1.52±0.21	1.34±0.09
Nodule number	Control	3.60±1.20	27.30±4.49	8.60±4.30	10.00±1.70
	MMIR treated (*Rhizo.*+AMF.)	8.60±1.76	11.60±2.60	18.60±6.30	11.60±0.88
Nodule dry weight (mg)	Control	24.40±1.40	40.20±4.90	35.20±4.80	19.50±2.15
	MMIR treated (*Rhizo.*+AMF.)	30.20±3.80	30.30±3.10	41.30±5.10	45.90±4.50

effects and toxic syndromes in *P. pinnata*. But higher concentration of CP did not show further stimulating effect indicating an upper dose limit at 20% decomposed CP (Table - 3). Pasture legumes did not show such limitation and there was a steady increase in growth parameters with increase in coir pith concentration (Table - 4).

Growth response of *Zea mays* under decomposed CP and MMIR indicated that there was significant increase in shoot length and plant dry weight under both decomposed CP and MMIR application condition but under higher concentration of CP (beyond 20%) efficiency of MMIR application was not very promising. CP is a rich source of organic carbon. Our study indicated decomposed CP releases substantial amount of nitrogen, phosphorus and available organic carbon. This may be the reason the plants under higher concentrations of CP, showed better plant growth (shoot length, plant dry weight). As the plant may be directly using the available nutrition. However application of MMIR with higher concentration of CP (30%) was seen to be functioning synergistically in improving root growth (Table - 5). From the table – 5 it was seen that there was significant increase in root length of *Zea mays* under MMIR treatment being maximum at 30% CP application.

TABLE 4 *Effect of MMIR application on growth and nodulation of Tephrosia candida (Roxb.) grown in chromite mine waste soil with different concentrations of decomposed coir pith*

Parameters	*Treatment*	*Chromite mine waste soil (CMWS)*	CMWS+10 % Coir pith	CMWS+20 % Coir pith	CMWS+30 % Coir pith
Shoot length (cm)	Control	10.17±0.17	17.16±0.73	27.76±	28.66±
	MMIR treated (*Rhizo.*+AMF.)	21.50±0.96	31.76±0.72	33.66±0.67	38.33±1.20
Plant dry weight (g.)	Control	0.15±0.005	0.50±0.038	0.76±0.01	1.40±0.03
	MMIR treated (*Rhizo.*+AMF.)	0.52±0.011	0.94±0.008	1.03±0.009	1.67±0.31
Nodule number	Control	0.00±0.00	3.67±0.33	2.33±0.66	3.50±0.64
	MMIR treated (*Rhizo.*+AMF.)	2.00±1.00	5.00±1.22	4.00±1.73	9.00±1.00
Nodule dry weight (g.)	Control	0.01±0.00	0.02±0.005	0.018±0.00	0.01±0.0005
	MMIR treated (*Rhizo.*+AMF.)	0.003±0.0008	0.016±0.006	0.006±0.0001	0.02±0.0005

Rhizo. - *Rhizobium* AMF - AM fungi

Analysis of the heavy metal uptake (chromium, iron and manganese) of *Z. mays* grown in chromite mine waste soil under MMIR treatment has been given in Table – 6. Heavy metal uptake in *Z. mays* was more under untreated chromite mine waste soil. When plants were grown under MMIR treatment, there was significant decrease in chromium uptake but there was increase in uptake of iron and manganese.

However, mostly the heavy metal ions are seen to accumulate in the root region immobilizing movement of heavy metal to shoot region, thereby alleviation of the heavy metal toxicity syndromes of the plant. Similar to the present observation alleviation of Al and Zn toxicity has also been observed by several workers (Heggo *et. al.*, 1990; Weissenhorn *et. al.*, 1995; Dueck *et. al.*, 1986; Tao, 1997). It has been suggested that heavy metal tolerant AM fungi could protect the plant from harmful effect of excessive heavy metal (Sylvia and Williams, 1992; Leyval *et. al.*, 1997). Several biological and physical mechanisms have been proposed to explain the metal tolerance of host plants which includes (a) immobilization of metals in the fungal

TABLE 5 *Effect of MMIR application on growth and nodulation of Zea mays grown in chromite mine waste soil with different concentrations of decomposed coir pith*

Parameters	*Treatment*	*Chromite mine waste soil (CMWS)*	CMWS+10% Coir pith	CMWS+20% Coir pith	CMWS+30% Coir pith
Shoot length (cm)	Control	23.40±2.29	40.8±5.99	38.4±4.07	37.60±4.02
	MMIR treated (*Rhizo.*+AMF.)	31.00±2.12	38.4±5.79	33.38±3.15	31.60±2.77
Root length (cm)	Control	48.4±5.08	47.67±1.86	47.00±5.13	57.60±4.03
	MMIR treated (*Rhizo.*+AMF.)	47.2±6.48	50.80±3.86	52.20±3.25	59.00±6.99
Plant dry weight (g.)	Control	0.88±0.05	2.07±0.66	3.07±0.57	4.11±1.15
	MMIR treated (*Rhizo.*+AMF.)	1.23±0.12	2.39±0.43	2.39±0.56	1.87±0.33
S.D.W/R.D.W	Control	0.48±0.11	0.96±0.24	1.19±0.24	1.19±0.07
	MMIR treated (*Rhizo.*+AMF.)	0.66±0.95	1.60±0.39	1.73±0.36	1.42±0.23

biomass (Zhu *et. al.*, 2001; Li and Christie, 2000) (b) reduced transfer as indicated by enhanced root shoot ratio (Joner *et. al.*, 2000; Tullio *et. al.*, 2003) (c) intracellular precipitation of cation by phosphate (Turnau *et. al.*, 1993). Recently observations have also indicated removal of hexavalent chromium from industrial effluents by CP application, which serves as an adsorbent (Singh and Srivastava, 2002). In the present work we have used both AM fungi and CP substrate. Therefore, the striking alleviation of heavy metal toxic syndromes from *Z. mays* and reduction of heavy metal ion concentrations from the shoot due to both adsorption by CP and partitioning.

Soil quality of chromite mine waste areas in terms of organic carbon, total nitrogen and available phosphorus was further studied in *T. candida* plant after MMIR treatment (Fig. 1). It was found that at 10% CP application both control and MMIR treatment showed maximum nutrient value which was steadily decreased with increase in CP concentration. Except organic carbon, nitrogen and phosphorus showed significant improvement over control. The trend of changes in the nutrient value due to MMIR application is seen to be different in all three types of nutrient content. There was general decrease in organic carbon content under MMIR

TABLE 6 *Comparative analysis of different heavy metals in Zea mays grown in chromite mine soil amended with decomposed coir pith under microbial inoculation.*

Types of heavy metals	*Treatment*	*Garden soil*	*Chromite mine waste soil (CMWS)*	*CMWS +20% Coir pith*	*CMWS +30% Coir pith*	
					Shoot	*Root*
Chromium (ppm)	Control	0.00	13.49	0.597	0.546	6.192
	MMIR treated (*Rhizo.*+AMF.)	0.00	0.382	3.300	0.663	11.545
Iron (ppm)	Control	29.35	29.59	30.95	19.54	30.16
	MMIR treated (*Rhizo.*+AMF.)	0.00	29.57	26.87	22.69	30.20
Manganese (ppm)	Control	0.34	0.41	0.85	0.11	0.86
	MMIR treated (*Rhizo.*+AMF.)	0.00	0.35	0.18	0.14	0.37

Rhizo. - *Rhizobium* AMF - AM fungi
S.D.W./R.D.W. - Ratio of shoot to root dry weight

application condition in comparison to pure CP treatment. However, there was significant increase in the total N and available P under MMIR treatment, which were optimum at 10% CP amendment. The CP used was decomposed by a lignocellulolytic fungi with a view to release the nutrients. The maximum impact on stimulation of plant growth was found to be at 10%. Beyond that concentration the steady decline may be due to the excess utilization of nutrients by microorganisms, which showed a high population (data unpublished). In the present treatment nitrogen fixers (*Rhizobium/ Azotobacter*) as well as phosphate mobilisers were used. From the increase in nitrogen and phosphorus content under MMIR treatment. Obviously due to the enhanced efficiency of the inocula under CP treatment which showed optimum activity of 10%. The retardation in efficiency of *Rhizobium* and AM fungi under high nutritional availability is known (Smith and Gianinazzi-Pearson, 1988).

Field Response of Tree and Pasture Species Under MMIR Application and Slope Stabilisation

Taking all the above concepts into consideration a revegetation trial in chromite mine over burden dumps of Sukinda during 2001-2003 was done by taking 402 seedlings belonging to 18 tree species and 6 pasture species

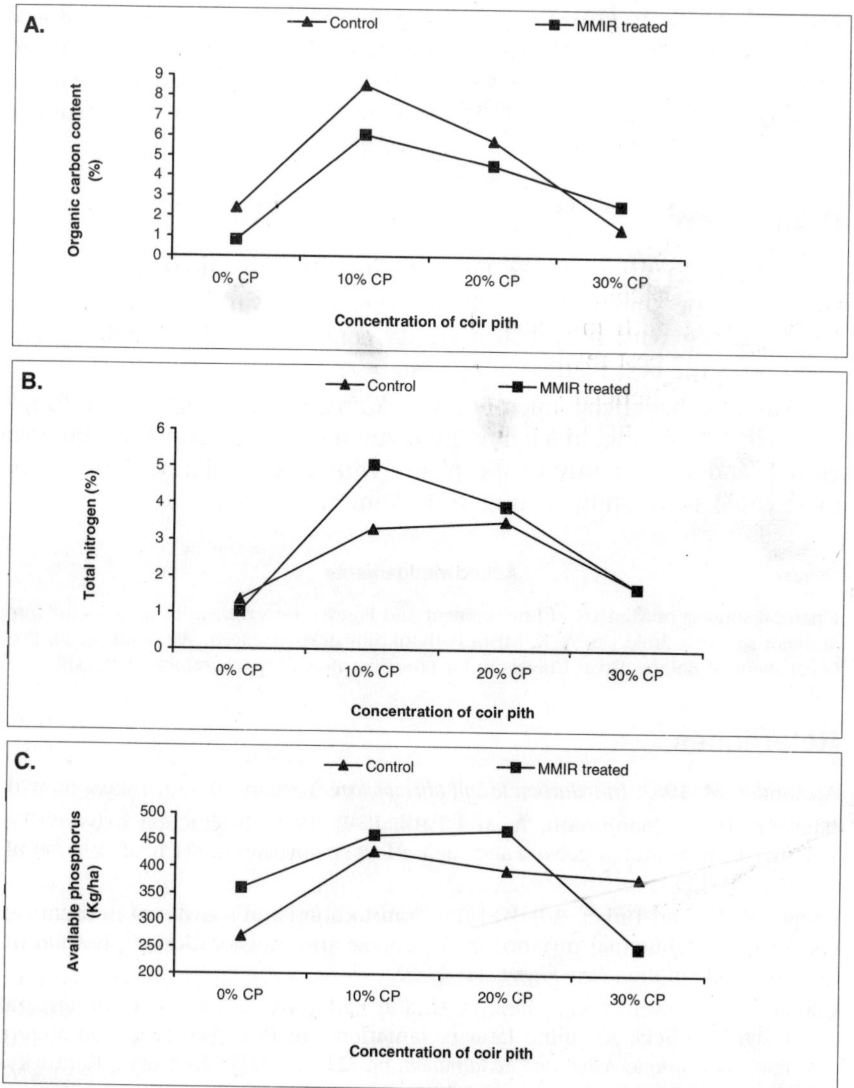

A. Organic carbon content B. Total nitrogen C. Available phosphorus

Fig. 1 Nutrient status of Tephrosia candida (Roxb.) grown in chromite mine waste soil under MMIR application (with different concentration of decomposed coir pith) at 90 DAS

in about 3 acres of land b. Application of In the field MMIR was applied at the rate of 100 g. per plant that is in the ratio of chromite mine soil: Coir pith: Sand::3:1:1. Tree and pasture plants with MMIR application under field conditions of chromite mine waste soil showed drastic improvement of plant growth, nodulation and rhizospheric microbial activity. The plants also did not show any disease symptoms as observed under pure chromite mine waste soil (unpublished data). In the over burden slopes are very steep and extremely prone to erosion, so plantation would not be successful unless slopes are stabilised. So the species like Agave, *Vetivera zizanoides, Eulaliopsis binata, Saccharum munja, Eulaliopsis binata, Prosopis juliflora* and *Acacia nilotica* (thickly sown) in the slopes were done, which produced good establishment results (unpublished data).

Conclusion

From the observations it can be concluded that an integrated approach covering slope stabilisation, soil moisture conservation, enhancing soil fertility along with microbial diversity combined with suitable species selection is the best method of ecosystem restoration

Suitable beneficial microbes viz. *Rhizobium/Azotobacter*, AM fungi, lignocellulolytic fungi in a lignocellulolytic substrate (carrier) can improve growth and survivability of the plants with bioremediation activity and these could successfully be used to reclaim mine waste soil.

Acknowledgements

Financial support of Ministry of Environment and Forests, Government of India in the form of major research project to A. K. Misra is thankfully acknowledged. We thank Head, P. G. Department of Botany, Utkal University for providing necessary laboratory facilities.

References

Alexander, M. 1962. *Introduction to Soil Microbiology*. Toppan Co. Ltd. Tokyo. pp-470.

Bagyaraj, D. A., Manjunath, A. and Patil, R. B. 1979. Interaction between VA mycorrhizae and *Rhizobium* and their effect on soybean in the field. *New Phytol.* 82: 141-145.

Cooper, K. M. and Tinker, P. B. H. 1978. Translocation and transfer of nutrients in vesicular arbuscular mycrorrhiza II. Uptake and translocation of phosphorus, zinc and sulphur. *New Phytol.* 81: 43-52.

Cordell, E. C., Owen, J. H., Mark, D. H. and Farley, M. E. 1987. Ectomycorrhizal fungi-beneficial for mine land reclamation. In: *Proc. Nat. Symp. On Mining Hydrology, Sedimentology and Reclamation.* pp 321-326. Univ. Kentucky, Kentucky.

Dehn, B. and Schuepp, H. 1989. Influence of VA mycorrhizae on the uptake and distribution of heavy metals in plants. *Agric. Ecosyst. Environ.* 29: 79-83.

Dueck, T. A., Visser, P. and Ernst, W. H. O. 1986. Vesicular-arbuscular mycorrhizae decrease zinc toxicity to grasses growing in zinc-polluted soil. *Soil Biol. Biochem.* 18: 331-333.

Heggo, A., Angle, J. S. and Chaney, R. L., 1990. Effects of vesicular arbuscular mycorrhizal fungi on heavy-metal uptake by soybeans. *Soil Biol. Biochem.* 22 : 865-869.

Jackson, M. L. 1973. Soil chemical analysis. Prentice Hall of India Limited, New Delhi.

Joner, E. J., Leyval, C. and Briones, R. 2000. Metal binding capacity of arbuscular mycorrhizal mycelium. *Biol. Fertil. Soils.* 226: 227-234.

Kaldorf, M., Kuhn, A. J., Schroder, W. H., Hildebrandt, U. and Bothe, H. 1999. Selective element deposits in maize colonized by a heavy metal tolerance conferring arbuscular mycorrhizal fungus. *J. Plant Physiol.* 154: 718-728.

Leyval, C., Turnau, K. and Haselwandter, K. 1997. Effect of heavy metal pollution on mycorrhizal colonization and function: physiological, ecological and applied aspects. *Mycorrhiza*. 7: 139-153.

Li, X. L. and Christie, P. 2000. Changes in soil solution Zn and pH and uptake of Zn by arbuscular myucorrhizal red clover in Zn-contaminated soil. *Chemosphere*. 42: 201-207.

Misra, A. K., Patnaik, R. K., Thatoi, H. N. and Padhi, G. S. 1990. Assessment of beneficial soil microflora of iron and chromite overburden mine area of Oriss-As an index of study of impact of heavy metal toxicity of soil fertility. In: *Environmental Impact of Industrial and Mining Activity*. (ed. L. N. Patnaik). pp. 263-269. New World Environment Series, Ashis Publishing house, New Delhi.

Misra, A. K. 1994. Study on growth and N_2 fixation ability of some leguminous plant species for reclamation of mine spoilt areas of Eastern Ghats of Orissa. Final Technical Report submitted to Ministry of Environment and Forests, Government of India.

Mosse, B. 1977. The role of mycorrhiza in legume nutrition on marginal soils. In: exploiting the legume-*Rhizobium* symbiosis in tropical agriculture. (eds. Vincent, J. M. Whitney, A.S. and Bose, J.). pp. 275-292. *College of Tropical Agriculture Misc. Publ.* 145, Univ. Hawaii.

Osonubi, O., Mulongoy, K., Wotlye, O. O., Atayese, M. O., and Ukai, D. U. O. 1991. Effect of ecomycorrhizal and vesicular arbuscular mycorrhial fungi on drought tolerance of four leguminous woody seedlings. *Plant and Soil.* 136: 131-143.

Rilling, M. C. and Steinberg, P. D. 2002. Glomalin production by an arbuscular mycorrhizal fungus: a mechanism of habitat modification. *Soil Biol. Biochem.* 34: 1371-1374.

Seighardt, H. 1987. Schwermetalland Nahrelementgehalte von Pflanzen Bodenproben Schwermetall haltiger halden in Raunbleiberg in Karnten (Osterreich) I. Krautige Pflanzen. Z. Pflanzenernacher. Bodenkd. 150:129-134.

Seighardt, H. 1990. Heavy metal uptake and distribution in *Silene vulgaris* and *Minuartia verna* growing on mining dump material containing lead and zinc. *Plant and Soil.* 123: 107-111.

Sindhu S.S.and Dadarwal K.R. 2000. Competition for nodulation among rhizobia in legume-*Rhizobium* symbiosis. *Indian Journal of Microbiology*. 4: 211-246.

Singh, D.K. and Srivastava, B. 2002. Removal of hexavalent chromium from industrial effluents using non-conventional adsorbents. In: *Proceeding of the national seminar on management of industrial effluents and wastes.* (eds.) Y. V. Swamy, S. N. Das, K. M. Parida and Vibhuti N. Misra. pp. 45-51. Allied Publishers Pvt. Limited.

Smith, S. E. and Gianinazzi-Pearson, V. 1988. Physiological interaction between symbionts in Vesicular arbuscular plants. *Ann. Rev. Plant Physiol. Mol. Bio.* 39: 221-244.

Steinberg, P. D. and Rilling, M. C. 2003. Differential decomposition of arbuscular mycorhizal fungal hyphae and glomalin. *Soil Biol. Biochem.* 35: 191-194.

Sylvia, D. M. and Williams, S. E. 1992. Vesicular-arbuscular mycorrhizae and environmental stresses. In *Mycorrhizae in Sustainable Agriculture* (eds. Bethlenfalvay, G. J. and Linderman, R. G.). pp. 101-124. ASA No. 54, Madison, USA.

Tao, H. Q. 1997. Effect of arbuscular mycorrhiza on resistance of red clover to heavy metal Zn and Cd pollution. MS thesis, China Agricultural University, Beijing.

Tullio, M., Pierandrei, F., Salerno, A. and Rea, E. 2003. Tolerance to cadmium of vesicular arbuscular mycorrhizae spores isolated from a cadmium-polluted and unpolluted soil. *Biol.Fertil. Soils.* 37: 211-214.

Turnau, K., Kottke, I. and Oberwinkler, F. 1993. Element localization in mycorrhizal roots of *Pteridium aquilinum* L. Kuhn collected from experimental plots treated with cadmium dust. *New Phytol.* 123: 313-324.

Vincent, J. M. 1970. A manual of practical study of the root nodule bacteria. IBH Handbook No. 15. pp 164. Blackwell Scientific Publication, Oxford and Edinburg.

Vivas, A., Marulanda, A., Gomez, M., Barea, J. M. and Azcon, R., 2003. Physiological characteristics (SDH and ALP activities) of arbuscular mycorrhizal colonization as affected by *Bacillus thuringiensis* inoculation under two phophorus levels. *Soil Biol. Biochem.* 35: 987-996.

Weissenhorn, I., Leyval, C. and Berthelin, J. 1995. Bioavailability of heavy metals and abundance of arbuscular mycorrhiza in a soil polluted by atmospheric deposition from a smelter. *Biol. Fertil. Soil,* 19: 22-28.

Weissenhorn, I., Leyval, C., Belgy, G., and Berthelin, J. 1995.Arbuscular mycorrhizal contribution to heavy metal uptake by maize (*Zea mays* L.) in pot culture with contaminated soil, *Mycorrhiza*. 5: 245-251.

Williamson, J. C. and Johnson, D. B. 1998. Key microbial interactions affecting restoration success of opencast coal mine sites in the UK: A Review. In: *Problems of wasteland development and role of microbes.* (eds.A. K. Misra). pp. 35-46. AMIFEM publication

Yao, Q., Li, X., Weidang, A. and Christie, P., 2003. Bi-directional transfer of phosphorus between red clover and perennial ryegrass via arbuscular mycorrhizal hyphal links. *Eur. J. Soil Biol.* 39: 47-54.

Zhu, Y. G., Christie, P. and Laidlaw, A. S. 2001. Uptake of Zn by arbuscular mycorrhizal white clover from Zn-contaminated soil. *Chemosphere* 42: 193-199.

Frontiers in Plant Sciences (Pages 465-480)
Eds. K.G. Mukerji, K.V.B.R. Tilak, S.M. Reddy, L.V. Gangwane, P. Prakash, I.K. Kunwar

Why Performance of PSMs was not Guaranteed?

M.G. Patil, A.B. Chaudhari, N.V. Phirke, S.K. Talegaonkar and R.M. Kothari*

ABSTRACT

Phosphate solubilizing microbes (PSMs) constitute an integral part of phosphorus cycle, catalyzing the inter-conversion of insoluble and organic phosphates into soluble form for absorption, growth and productivity by the plant system. However, it was commonly observed that PSMs catalyzed efficient phosphate solubilization under *in vitro* conditions, but could not necessarily match the same performance in natural soils. This prompted us to look into possible causes of their inefficiency *in situ*. It was found that (i) matrix for their maintenance during storage, (ii) nature of packaging material, (iii) conditions (humidity and temperature) of storage, (iv) geo-climatic (hot/dry/humid/cool) conditions of application, (v) shelf-life of PSMs after commercial production, (vi) type (acidic/neutral/alkaline/saline) of soil, (vii) absence/lack of organic carbon in soil, (viii) antagonism by the local microbial population, (ix) PSM-specificity (bacterial vs fungal), (x) substrate-specificity, (xi) nature of acids secreted by PSMs and (xii) wrong agronomic practices (toxicity as a result of broadcasting of chemical fertilizers, spray of weedicides/pesticides etc.) were responsible for poor performance of PSMs.

Key words: PSMs, soil conditioner, agro-climatic-specificity, substrate-specificity, wrong agronomic practices.

INTRODUCTION

Although Bajpai and Sundara Rao (1971) demonstrated solubilization of inorganic phosphates in soil and its subsequent uptake by plants, it was

Author's address: School of Life Sciences, North Maharashtra University, Jalgaon – 425 001 (MS)
Jain Irrigation Systems Ltd., Food Park, Jain Valley, Jalgaon – 425 001 (MS)
E-mail: foodpark_jains@sancharnet.in

commonly observed that efficiency of PSMs exhibited in laboratory was in most cases not matched in soils. This observation is replete in literature under various experimental conditions, in different soil types, for different crops and using different PSMs, in the hands of many researchers (Patil *et al.*, 2001). While conducting the largest known ecological survey of Maharashtra State, spread over 239 taluquas of 25 districts, we too came across factors which were apparently responsible for the poor performance of efficient PSMs, throwing some light as to why PSM performance was not guaranteed (Patil, 2002). These factors are focused in this article.

MATERIALS AND METHODS

AR grade chemicals and microbial media were procured from M/s. Hi-Media, Mumbai. Triple distilled water was used for analysis.

Press mud cake (PMC) was obtained from Madhukar Co-op. Sugar Mill, Faizpur. Alternate matrices for PSMs were charcoal powder, saw dust and black cotton soil procured locally. Soil conditioner was prepared from banana farm waste by solid state fermentation under warm, humid and ambient conditions as per Phirke *et al.* (2001).

All matrices were analyzed by standard methods for moisture/ash content, reducing sugars, CHNP etc. as per Tandon (1993).

High density polyethylene (HDPE), thin cotton and thick cotton bags for packaging PSMs on matrices during storage were procured locally.

Soil types were analyzed as per Martin and Focht (1989)

Aspergillus awamorii (NCIM–1188) was procured from National Collection of Industrial Microbes (NCIM), NCL, Pune to serve as a standard fungal PSM.

Pikovskaya - agar medium (1948) was used for enumerating and maintaining PSMs. The petri-plates were incubated at 28 ± 2°C up to two weeks for isolating PSMs on the basis of zone of solubilization (clearance) of insoluble phosphate around them.

Microscopic examination for enumeration of total viable count (TVC) and broad identification of PSMs (fungal vs bacteria) was carried out as per Bergey's Manual of Systematic Bacteriology (1994).

Parameters Monitored for Performance of PSMs

Performance of PSMs was monitored as a function of the following parameters:

(i) Matrices for PSMs' maintenance during storage.

(ii) Nature of packaging material.

(iii) Conditions (humidity and temperature) of storage.
(iv) Geo-climatic (hot/dry/humid/cool) conditions of application.
(v) Shelf-life of PSMs after commercial production.
(vi) Type (acidic/neutral/alkaline/saline) of soil.
(vii) Absence/lack of organic carbon in soil.
(viii) Antagonism by the local microbial population.
(ix) PSM-specificity.
(x) Substrate-specificity
(xi) Nature of acids secreted by PSMs.
(xii) Wrong agronomic practices

Typical Experimental Set-up

In 100 ml Pikovskaya medium in a 500 ml capacity Ehrlenmeyer's flask, 5% (v/v) spore suspension of PSMs was incubated at 28 ± 2°C on a rotary shaker (125 rpm) for 7 days. Tri-calcium phosphate (TCP) was the substrate (insoluble phosphate) as it is a major constituent of soils of Maharashtra. At an interval of 24 h, aliquot of broth was aseptically withdrawn from the flasks, filtered through Whatman filter paper No. 42, filtrate analyzed for pH and soluble phosphate, estimated as per Anusuya and Jayarajan (1998).

Performance of PSMs as a function of the above parameters was assessed using the above experimental set-up.

RESULTS AND DISCUSSION

Only one parameter was considered at a given time so that a cause and effect relationship could be established. Once optimal results were obtained with a given parameter, the same was considered as optimal and incorporated into a procedure for the optimization of the next parameter so that an integrated strategy of PSM efficacy/maintenance/performance could be evolved.

Matrices for PSMs' Maintenance During Storage

One kg of each matrix (PMC, charcoal, saw dust, black cotton soil and soil conditioner) was individually mixed thoroughly with the PSM broth (100 ml cultivation medium having approximately 10^9 TVC/ml) and packed in HDPE bags and stored at 27°C away from heat and light. Their samples were removed at monthly interval and total viable count per g was determined as a function of storage duration. The results are summarized in Table 1.

TABLE 1 *Viability of PSM as a function of matrix*

Matrix	*TVC per gm of matrix* after storage for months*					
	1	*2*	*3*	*4*	*5*	*6*
PMC+20% potato	27.0	33.0	30.0	27.0	23.0	17.0
Charcoal	27.0	17.0	10.0	4.0	1.0	1.0
Saw dust	27.0	15.0	12.0	8.0	3.0	2.0
Black cotton soil	27.0	20.0	14.0	7.0	3.0	2.0
Soil conditioner	27.0	26.0	23.0	21.0	20.0	17.0

*expressed as 10^9

From Table 1, it is clear that PMC + potato provides the most viable matrix, placing soil conditioner at second rank, leaving behind charcoal, saw dust and black cotton soil. Total viable count increased in the second month, peaked in the third month and started gradually declining thereafter. This may probably be due to availability of reducing sugars from PMC and potato. That potato served as an adjuvant for promoting the growth of PSMs is not surprising by virtue of its reserve carbohydrate content and its potential to supply sustenance (nutrition and moisture) for microbes over a long period. That soil conditioner also served equally well is not surprising in view of its multi-faceted properties (Sharma *et al.,* 1999).

Remaining three matrices, which are used in several commercial bio-fertilizer preparations, have little organic carbon/nutrition/moisture to serve as an incentive for the viability of PSMs, leave alone their growth under storage conditions. In fact, it was observed that lignin in the raw saw dust showed lethal effect, giving sharp decline in TVC (Ramamurthy *et al.*, 1996). Hence, for studying the effect of remaining factors on PSM viability, only PMC with potato has been examined.

Physically, PMC is a soft, spongy, fluffy and amorphous blackish brown by-product of sugar mills, generated perennially. Its proximate analysis by Unni (1987) is summarized in Table 2.

It is obvious from Table 2 that PMC contains significant amount of carbon, nitrogen, wax and steroids. Besides, it is a rich source of micro-nutrients (Unni, 1987). These ingredients not only serve as invaluable nutrition for soil microbes, but also as a source of plant nutrients/plant growth regulators/plant protecting chemicals in a sustained manner, rendering it to be the most valuable matrix for PSM preservation (Manohar Rao, 1997).

Another reason for more TVC in PMC for longer period could be its porous texture, which ensures release of CO_2 and dissipation of heat from

TABLE 2 *Typical analytical profile of PMC*

Constituent	*Percent*
Total solids	100.0
Volatile solids	70.0
Ash (non-volatile solids)	30.0
Cellulose	25.0
Hemicellulose	27.0
Solubles	38.0
Total carbohydrates	52.0
Organic carbon	24-27
Wax (benzene extract)	9.0
Steroids	7.2
Proteins	7.0
C/N ratio	30:1
Lignin	4.0
Reducing sugars	0.4

microbial metabolism, sparing PSMs from CO_2 toxicity and heat stress, respectively. Thus, availability of (i) food on sustained basis, (ii) air for respiration due to its fluffy nature, (iii) space for the release of CO_2 due to it porosity and (iv) avenue for dissipation of metabolic heat in PMC appears responsible for high viability of PSMs as a function of time. These attributes lack in commercial matrices rendering short shelf-life to PSMs.

Nature of Packaging Material

Lower percent survival of bio-fertilizers for a limited period (2-3 months), observed in many commercial packages led to a speculation that packaging in polyethylene bags may be injurious to them by restricting an entry of O_2 for respiration, exit of CO_2 and easy dissipation of metabolic heat generated by them. Therefore, to remove these suspected constraints, thin and thick cotton bags were explored for packaging, as in the case of high yielding seeds.

Using PMC+20% potato as a matrix, TVC monitoring as a function of packaging material showed that thin and thick cotton fabric bags, irrespective of porosity, were inferior to HDPE bags, as they preserved PSMs poorly, giving less TVC upon storage. Most probable reason for the failure of cotton bags in preservation of TVC of PSM was desiccation of matrix as a function of storage period, not observed in polyethylene bags.

HDPE bags are used for packing of bio-fertilizers due to their (i) ease of handling during packing, storage, transport and use, (ii) autoclavability, (iii) permeability to the entry of fresh air and exit of CO_2/metabolic heat and (iv) tolerance to γ-radiation (Burton, 1979). Apparently, the postulate that HDPE bags lacked these attributes was misplaced.

Conditions (Humidity and Temperature) of Storage

Using PMC + 20% potato as a matrix and storage at 27°C, it was found that moisture (humidity) below 30% expedited reduction in TVC, probably placing reverse osmotic pressure on the TVC, by reason of which slow desiccation led to gradual decrease in their population by the end of four months. On the contrary, experimental lot with 35% humidity exhibited superior preservation. These results are summarized in Table 3.

TABLE 3 *Viability of PSM as a function of moisture in storage matrix*

Humidity %	*TVC per gm of matrix * after storage for months*			
	1	2	3	4
20-25	26.5	23.5	20.9	16.0
26-30	26.4	27.6	25.1	18.2
31-35	26.4	32.8	30.0	25.0

*expressed as 10^9

Using PMC + 20% potato as a matrix and storage of PSMs at different temperatures, it was observed that 22°C provided optimum viability during storage as against viability either at 37°C or under ambient conditions, where warehouse temperature fluctuated (30°C at night and 42-45°C during day time). The take home lesson from these observations is that PMC fortified PSM packages should either be sold within four months or should be stored at 22°C for viability for a longer period. These results are summarized in Table 4.

Optimum temperature is vital for the growth of PSMs and enzymatic activity. However, desiccation as a result of storage at 37°C or higher ambient temperature expedited death of microbes. Thus, a cool chain, besides PMC, was necessary for maintaining high PSM count to be viable in field conditions.

TABLE 4 *Viability of PSM as a function of temperature of storage*

Temperature of storage °C	*TVC per gm of matrix * after storage for months*			
	1	2	3	4
22	27.0	29.5	30.3	28.5
37	27.0	28.0	25.4	20.0
Ambient	27.0	26.0	21.9	17.6

*Expressed as 10^9

Geo-climatic (Hot/Dry/Humid/Cool) Conditions

Presence or absence of moisture in soil as well as environment affected microbial activity (Patil *et al.*, 2001). From the decreased population of PSMs in several districts of Maharashtra. Patil (2002) found that PSMs from dry and hot climatic conditions do not survive in humid and cool soils and the reverse was equally true. It was also observed that while compact soils (viz. black cotton soil after drying) inhibited the efficiency of PSMs, presumably due to hindrance in colony formation and mobility, loose soils promoted it.

Dry weather induced desiccation of soil, resulting in to its compact texture, while irrigation reversed its texture, promoting the microbial activity. Similarly, lack of organic manure (OM) imparted compactness to soil, while its fluffy nature imparted porous texture. Besides porosity, moisture holding capacity of OM spared PSMs from heat-stress, thereby catalyzing more solubilization of phosphate. The take home lesson was cool and moist environ in soil was obligatory for their survival and perpetual functioning.

Shelf-life of PSMs After Commercial Production

Two fresh commercial preparations of PSMs were procured locally for comparing their TVC with that of the standard PSM (maintained on PMC + 20% potato) at 27°C. These results are summarized in Table 5.

TABLE 5 *Viability of PSM as a function of shelf-life*

PSM preparation	*TVC per gm of matrix * after storage for months*			
	1	2	3	4
Commercial –1	18.5	14.9	11.0	7.9
Commercial – 2	19.7	15.1	11.9	7.0
Standard PSM	26.5	31.8	29.7	26.3

*Expressed as 10^9

A steep decline in TVC noted in both commercial preparations, vis-a-vis stable TVC profile in standard PSM, indicated that essential parameters such as nutrition and moisture are probably lacking in commercial preparations (especially in matrix) due to which their rate of decay was faster as compared to that of standard.

Type (Acidic/Neutral/Alkaline/Saline) of Soil

In view of the above studies, soils appeared to play significant role. To settle this point, alkaline, acidic, saline and neutral soils were taken in 9 : 1 (w/w) ratio (soil : SC), sterilized and inoculated with fungal/bacterial PSM. The soil was purposefully maintained at 27°C and 40% moisture for 72 h, during which period count of both PSMs was taken by serial dilution of soil suspension. The results are summarized in Table 6.

TABLE 6 *Profiles of fungal/bacterial PSMs as a function of soil type*

Soil type	*TVC per gm of matrix **					
	Zero h		*48 h*		*72 h*	
	F	B	F	B	F	B
Alkaline	6.0	90	3.0	10.0	0.1	0.2
Saline	7.0	100	3.0	12.0	0.08	0.15
Acidic	10.0	100	19.0	15.0	0.2	0.3
Neutral	170	27.0	12.0	200.00	0.4	250.0

*Expressed as 10^9; F = Fungal PSM; B = Bacterial PSM

It is clear from Table 6 that within 72 h in (i) alkaline soil, fungal PSM registered a 60 fold decrease, while bacterial PSM, a 450 fold decrease, (ii) saline soil, fungal PSM registered a 100 fold decrease, while bacterial PSM, a 1000 fold decrease, (iii) acidic soil, fungal PSM, a 50 fold decrease, while bacterial PSM, a 500 fold decrease and (iv) neutral soil, fungal PSM, a 300 fold decrease, while no effect on bacterial PSM. From these observations, it is clear that in spite of soil being conditioned to be conducive for the growth of PSMs (by fortification with SC, maintenance of 40% moisture and ambient temperature), there was a precipitous decrease in their population in all soils (except in neutral soil, only for bacterial PSM).

These observations have a major implication as to why alien bio-fertilizers have failed or did not make any perceptible impact on productivity in alien soils. Since alien soils could not provide a matrix with which the PSMs could adapt, in all probability it appears that regional isolates alone will play their legitimate role in solubilizing phosphate, however weak they may appear in laboratory conditions. Thus, isolation and application of geo-climatically adapted PSMs appears obligatory.

This appears logical, since soil pH and/or salinity affects cell division, growth and population density of micro-flora, which in turn affects their solubilization potential. While bacteria solubilize phosphate maximally near the neutral pH, fungi achieve the same objective between pH 4.0 – 5.0 (Surange *et al.*, 1997).

Secondly, phosphate from fertilizers is progressively (often as much as 90%) immobilized due to its reactivity with soil (Varade, 1994). While phosphorus content in average soils is about 0.05% (w/w) (Barber, 1984) and may vary largely due to differences in soil type and weathering (Stevenson, 1986), calcareous soils irrigated with saline water showed large scale immobilization of phosphate as calcium phosphate, rendering only 0.1% solubility and availability to plants at neutral to alkaline pH (Chhalla *et al.*, 1995).

Absence/Lack of Organic Carbon in Soil

Organic carbon (SC) not only provides nutrition to the microbial population in the rhizosphere, it also safeguards their population from desiccation by virtue of its moisture holding and insulation providing capacity against adverse thermal effect during harsh summer (Ramamurthy *et al.*, 1996).

Therefore, it is inconceivable that in the absence of vital factors (nutrition, moisture and cool environ), PSM could survive. Since we have forgotten this age-old truth established by traditional farming, soil microbial activity is short-circuited, which is apparently the main reason for failure in solubilizing insoluble phosphate and consequent poor agri-yield. This has been traced to molecular level by Datta and Banik (1994), who found that soils rich in organic matter (SOM) flourished PSMs, generated organic/inorganic acids and solubilized more phosphate.

The above realities are independently substantiated by Dadarwal (1997), who noted that in an organic matter poor soils, microbial activity was low, except in rhizosphere of growing plants by root-associated fungi (mycorrhizae) and actinomycetes, which symbiotically helped in the solubilization and absorption of phosphate.

Antagonism by Local Microbial Population

An observation by Kundu and Gaur (1980) that exogenous inoculants always attained rich population in sterile vis-à-vis non-sterile soil indicated competitive role of indigenous micro-flora. Subsequently, Amor*et al.* (1985)

postulated that a rich indigenous population of microbes, presumably competes with an exogenous inoculant for nutrients and space.

On the contrary, there are also instances, where an inoculant proved efficient in non-sterile soil compared to sterile one. It appeared as if rhizosphere offered a nutrient-rich environment for the growth of PSMs than the bulk soil (Banik and Dey, 1982). It is true, therefore, that either laboratory experiments do not accurately assess PSMs' competitiveness in the rhizosphere (due to failure in mimicking micro-environment of rhizosphere), or soils inhabiting large indigenous PSM population may be masking the effect of an exogenous inoculant (Thomas *et al.*, 1985).

Phirke *et al.* (2001) traced inefficient performance of exogenous PSMs to tough competition between local isolates and exogenously added PSMs for survival and colonization and suggested the application of soil conditioner (SC) as a pre-requisite for their successful colonization and positive contribution. A potent inoculant, therefore, must be capable of overcoming this competition for successfully colonizing the roots in non-sterile soil, subject to availability of nutrients and moisture provided by SC.

PSM Specificity

Normally, TCP is solubilized with an equal ease by fungi and bacteria. However, fungi solubilized RP more easily than bacteria. Similarly, Singh *et al.*, (1980) found *A. awamorii* more efficient solubilizer of RP compared to *Bacillus/Pseudomonas.* Versatility of fungi over bacteria is traceable to three factors : (i) bacteria are efficient only in the presence of pentose and hexose in the culture medium, whereas fungi are equally efficient in the presence of pentose, hexose as well as disaccharides, (ii) fungi are easy to colonize and rapid to adopt in the surrounding soil and (iii) fungi are active in a broad pH range, whereas bacteria are active only above neutral pH (Thomas *et al.* 1998). In both organisms, solubilization enhanced as a function of (a) aeration, (b) temperature between 25-30°C, (c) chemical composition of RP and (d) its particle size; high grade RPs with fine particle size dissolved readily than low grade RPs with coarse particle size.

Use of RP as a substrate in neutral to alkaline soil led to increase in crop yield, presumably due to increased efficiency of PSMs specialized in the solubilization of RP (Tilak, 1998). Thus, nature and particle size of the substrate and local pH in the soil rendered or retarded phosphate solubilization.

Substrate-specificity

Phosphate interaction with N, K, Fe, Zn, Cu, B, Mg, Mn and Mo determined its availability to plants. Similarly, anions like SO_4^{-2}, MoO_4^{-2} and BO_3^{-3}

compete with PO_4^{-3} for adsorption/absorption sites. All these interactions could be additive or adverse as reviewed by Tandon (1992).

Amount of TCP solubilized by efficient PSMs in Pikovskaya medium was always higher compared to that in NBRI-P medium (Patil, 2002). This difference gives rise to illusion for anticipation of more solubilization, which may not happen *in situ* conditions.

Bardiya and Gaur (1974) found that tri-calcium phosphate (TCP) was easily solubilized than RP by different PSMs. As TCP is a major phosphate entity in soil, TCP - solubilizing PSMs are found in large number (Dharmwal *et al.*, 1989). Other PSMs solubilize rock phosphate (RP), Iron phosphate (Fe-P), aluminium phosphate (Al-P) and organic phosphate (OP) by secretion of either acid(s) or corresponding extra-cellular phosphatases. Secretion of acids allows pH drop to solubilize phosphates.

Khera and Balloli (1994) also observed that depending on soil type, a given crop preferred red elemental phosphate or Al-P or Fe-P or Ca-P. While this work has laid down basic foundation of phosphate preferences by crops, its correlation with micro-flora remains to be attended.

Nature of Acids Secreted by PSMs

When soil pH decreases from 9.0 to 7.0, $Ca_3\ (PO_4)_2$ in the soil is solubilized by chelating Ca^{2+} ions. This process depends on the number/position of hydroxyl group(s) of an organic acid secreted by PSMs. Calcium is chelated (i) to a small extent with a mono-basic acid (viz. lactic acid), (ii) moderately with tri-basic acid (viz. citric acid), (iii) strongly with dibasic acids (viz. malic acid) and (iv) by a-substitution than b-substitution on hydroxyl group of an aliphatic acid.

Secondly, carbon dioxide, which is final product of organic matter decomposition by the rhizospheric microbes, forms carbonic acid, thereby solubilizes $Ca_3\ (PO_4)_2$ to $CaHPO_4$ or $Ca\ (H_2PO_4)_2$.

Thirdly, a large number of aerobic autotrophs use salts for metabolism, releasing sulphuric and nitric acid, which in turn augmented the release of soluble phosphate in the rhizosphere.

Wrong Agronomic Practices

Application of chemical fertilizers by broadcasting or in an indiscriminate manner created local toxicity, reducing microbial population, even by an established plant tonic, urea. Therefore, use of chemical fertilizers had to be in extremely dilute solution, preferably on a daily basis so that while plants received nutrition, microbial activity sustained or enhanced. The same was more acutely applicable to the spray of pesticides. It is our experience that soil enriched by organic manure (fortified with neem seed cake)/vermi-compost not only solubilized more phosphate, but was also least susceptible to pest attack. This strategy not only permitted full nutrition to plant system, it did not short-circuit the microbial processes.

Cause–Effect and Solution

One of the main reasons as to why PSMs application by farmers has not been picked up is a drastic reduction (from 1×10^9 to 1×10^5) in the number of their live cells upon ambient storage for 3-6 months. Charcoal powder or saw dust hardly constituted an ideal matrix for their stabilization due to their inherent composition, lacking food and moisture for the survival of PSMs. Meager number (1×10^5) of live PSM cells, which get thousands of fold diluted by soil, can not overcome the constraint of numbers against outnumbered local micro-flora and local environment, resulting into apparently no positive effect on productivity. To be effective, for their stabilization during the storage, a matrix is needed, which provides sustained release of moisture and nutrients to sustain original number of live cells for minimum 6 months during storage.

For this purpose, Talegaonkar (2000) has reviewed factors governing the choice of matrices, their essential characteristics and advantages/ disadvantages in using them in sterile/non-sterile state. To overcome the constraint in viability of commercial preparations, a simple, economical, guaranteedly effective and easy to practice protocol for farmers has been devised. It involves 16 h incubation of one pack of commercial bio-fertilizer preparation at ambient temperature in plastic container having one liter water fortified with 15% jaggery, 1% common salt and 5% finely pulverized groundnut cake. This leads to minimum 100 – 1000 fold increase in TVC, increasing the chances of PSM viability. Farmers should be aware that only a fresh commercial preparation is viable on the farm and its viability is drastically compromised as a function of its age and conditions during storage.

Conclusions

This study has led to identification of PSMs from acidic, alkaline, neutral and saline soils of Maharashtra State exposed to hot, dry and humid environments. Out of 1181 farmers, whose soils were mapped for PSMs, only 13 were familiar with cumulative effect of soil conditioner and bio-fertilizers (PSMs). This has highlighted the necessity of educating the farmers in order to strengthen agricultural base with modern practices to eradicate stagnation in productivity.

Failure in using soil-specific PSMs is a main cause for the inefficiency of PSMs. Therefore, PSMs adapted to local conditions need to be used rather than bringing alien culture, which is not adapted to local geo-climatic conditions. Circumstantial evidence showed that those soils which have less organic carbon/less moisture/more salinity/higher temperature showed less/weak population of PSMs.

While presence of sizeable TVC per g soil was noted, corresponding PSMs were small in number, indicating that TVC per g of soil has no relevance to number of PSMs. Majority of PSMs have extremely weak phosphate solubilizing activity, while minority is characterized by strong solubilizing activity, keeping rest of the PSMs in between, giving rise to a library of PSMs to throw light on significance of their differential performance.

References

Amor, A. M. J., Frontera, G. M. and Janeiro, A. R. 1985. Microorganisms solubilizing insoluble phosphate mineral compounds in the rhizosphere of rape plant. *Agri. Microbiol.*, 17 : 97-101.

Anusuya, D. and Jayarajan, R. 1998. Solubilization of phosphate by *Trichoderma viride. Current Sci.*, 75 : 464-466.

Bajpai, P. D. and Sundara Rao, W. V. B. 1971. Phosphate solubilizing bacteria : soil inoculation with phosphorus solubilizing bacteria. *J. Soil Plant Nutr.,* 17 : 46-63.

Banik, S. and Dey, B. K. 1982. Available phosphate content of an alluvial soil as influenced by inoculation of isolated phosphate solubilizing microorganisms. *Plant and Soil*, 69 : 353-364.

Barber, S. A. 1984. *Soil Nutrient Bio-availability*. John Wiley & Sons, New York, USA.

Bardiya, M. C. and Gaur, A. C. 1974. Isolation and screening of microorganisms dissolving low grade rock phosphate. *Folia Microbiol.*, 19 : 386-389.

Burton, J. C. 1979. Rhizobium Species. In : *Microbial Technology.* (Ed. Peppler, H. J.) 1 : 29-58. Academic Press, Los Angelis, California, USA.

Chhalla, O., Vadivelu, S. and Sehgal, J. 1995. Soils of Maharashtra for optimizing land use : executive summary. NBBS Publ.,54b: 37-48, Nagpur, India.

Dadarwal, K. R. 1997. Soil micro-organisms for sustainable crop production. *Phosphate solubilization and mobilization through micro-organisms.* pp. 293-304, Sci. Publ., Jodhpur, India.

Datta, M. and Banik, S. 1994. Effect of poultry manure and phosphate dissolving bacteria on rice in acidic soil. *Ind. J. Agric. Sci.*, 64 : 791-793.

Dharmwal, N. S., Singh, R. B. and Rai, R. 1989. Isolation of phosphate solubilizers from different sources. *Curr. Sci.,* 58 : 570-571.

Khera, M. S. and Balloli, S. S. 1994. Phosphorus fractions in soils and their importance in nutrition of crops. In : *Phosphorus Research in India,* (Ed. Dev, G.), pp. 39-42, Proc. of Group Discussion, Ind. Agri. Res. Inst., New Delhi, India.

Kundu, B. S. and Gaur, A. C. 1980. Establishment of nitrogen fixing and phosphate solubilizing bacteria in the rhizosphere and their effect on the nutrient uptake of wheat crop. *Plant and Soil,* 57 : 223-230.

Manohar Rao, P. J. 1997. Filter cake and its uses. In : Industrial utilization of sugarcane and its products. (Ed. Manohar Rao, P. J.), ISPCK Publ., New Delhi, India.

Martin, J. P. and Focht, D. D. 1989. Biological properties of soils. In : *Soils for management of orgnaic wastes and waste waters* . (Eds. Elliott, L. F. and Stevenson, F. J.). pp. 114-169. American Soc. of Agron., Madison, Wisconsin, USA.

Patil, M. G., Chaudhari, A. B., Sayyad, R. Z. and Chincholkar, S. B. 2001. Phosphate solubilizing microbes : A potential bio-inoculant for efficient use of phosphate fertilizer. In : *Bioinoculant for sustainable Agriculture and Forestry* (Eds. Reddy, S. M., Reddy, S. R., Singarachary, M. A. and Girisham, S.). pp. 107-118, Scientific Publishers (India), Jodhpur, India.

Patil, M. G. 2002. Studies on efficient phosphate solubilizing microbes in different agro-climatic zones of Maharashtra State. *Ph.D. Dissertation,* North Maharashtra University, Jalgaon, India.

Phirke, N. V., Patil, R. P., Chincholkar, S. B. and Kothari, R. M. 2001. Recycling of banana pseudostem waste for economical production of quality banana. *Resources, Conservation & Recycling,* 31 : 347-353.

Pikovskaya, R. I. 1948. Mobilization of phosphorus in soil in connection with vital activity of some microbial sps. *Mikrobiologiya,* 17 : 362-370.

Ramamurthy, V., Sharma, R. K., Yadav, K. R., Kaur, J., Dev Vrat and Kothari, R. M. 1996. *Volvariella*-treated eucalyptus saw dust stimulates wheat and onion growth. *Biodegradation*, 7 : 121-127.

Sharma, R. K., Ramamurthy, V. and Kothari, R. M. 1999. Soil conditioner – A vital bio-resource for sustainable agriculture and forestry. In : *Bio-resource Biotechnology* (Eds. Cocking, E. C. and Kanniyan, S.), pp. 82-108, Assoc. Publ. Co., New Delhi, India.

Singh, C. P., Mishra, M. M. and Yadav, K. S. 1980. Solubilization of insoluble phosphate by fungi. *Annals Microbiol. (Paris)*, 131B : 289-296.

Stevenson, F. J. 1986. *Cycles of soil carbon, nitrogen, phosphorus, sulphur and micronutrients*. John Wiley & Sons, New York, USA.

Surange, S., Wollum, A. G., Kumar, N. and Nautiyal, C. S. 1997. Characterization of *Rhizobium* from root nodules of leguminous trees growing in alkaline soils. *Can. J. Microbiol.*, 43: 891-894.

Talegaonkar, S. K. 2000. Application of pressmud for plant growth promotion. *Ph.D. Dissertation*, North Maharashtra University, Jalgaon, India.

Tandon, H. L. S. (Ed). 1992. Management of nutrient interactions in agriculture. In *Fertilizers, organic manures, recyclable wastes and bio-fertilizers.* 3: 148-160, FDCO, New Delhi, India.

Tandon, H. L. S. (Ed). 1993. *Methods of analysis of soils, plants, waters, and fertilizers* . FDCO, New Delhi.

Thomas, G. V., Shantaram, M. V. and Saraswathy, N. 1985. Occurrence and activity of phosphate solubilizing fungi from coconut plantation soils. *Plant and Soil,* 87 : 357-364.

Tilak, K. V. B. R. 1998. *Bacterial fertilizers :* Technical Bulletin, Indian Council of Agricultural Research, New Delhi.

Unni, B. G. 1987. Production of bio-gas from sugarcane press mud. *Energy Management,* 11 : 23-28.

Varade, S. B. 1994. Management of vertisols as a natural resource. In : *Natural resources management for sustainable agriculture and environment.* (Ed. Deb, D. L.), pp. 159-166, Angkor Publ. (P) Ltd., New Delhi, India.

Frontiers in Plant Sciences (Pages 481-496)
Eds. K.G. Mukerji, K.V.B.R. Tilak, S.M. Reddy, L.V. Gangwane, P. Prakash, I.K. Kunwar

Mushroom Nutriceuticals: An Emerging Health Care Aid

[1]Deepak K. Rahi, [2]R.C. Rajak & [3]A.K. Pandey

ABSTRACT

For many years, mankind has benefited from green plants as a source of food drugs and herbal remedies. Fungi especially the mushrooms on the other hand, have not been considered in any significant way. During the past fifty years, several advancement in medicine came from lower fungi such as molds and yeasts while the mushrooms instead of having the great diversity and health giving benefits could not find place in mainstream. However, this is changing very rapidly which is apparent from the prominent emergence of mushrooms as a major source of nutriceuticals (refined/partially defined food extractives) or health food. The mushrooms are now recognized as a source of nutriceuticals in nutrient balancing, strengthening human immune system, enhancing natural body resistance and lowering proneness to disease. The present article discusses general concepts, active principles involved, potential types, significance and various research and developmental activities of mushroom nutriceuticals in detail.

Key words: Mushrooms/Nutraceuticals/Nutriceuticals/Dietary supplements/ Polysaccharides

INTRODUCTION

It is understood that human beings have constantly been in search of new substances that can improve biological functions and make people fitter and healthier. Recently the society has turned towards plants, herbs, and foods

Author's address: [1]Department of Microbiology, Panjab University, Chandigarh-160 014
[2]Institute of Basic Sciences, Bundelkhand University, Jhansi-248 128
[3]Department of Biological Science, R.D. University, Jabalpur-482 001

as the sources of these enhancers. These products have been called variously as vitamins, dietary supplements, functional foods, phytochemicals, nutraceuticals, nutriceuticals etc. (Brower, 1998 & Zeisel, 1999; Chang & Miles, 1996). Mushrooms, in this regard are now beginning to receive much deserved attention for their very real health giving qualities. It is their flavor and texture for which they have long been devoured by the mankind. Scientific appreciation of their nutritional and medicinal attributes is a recent phenomenon. Mushrooms have from nutrition point of view a distinct place in human diet which otherwise consists of items either of plant or animal origin. Mushrooms are perhaps the only fungi deliberately and knowingly consumed by human beings and they complement and supplement the human diet with various ingredients not encountered or deficient in food items of plants and animal origin. Besides, unique chemical composition makes them suitable for specific groups suffering with certain physiological disorders or ailments. This makes the mushrooms an ultimate health food. Consumers are deeply concerned about how their health care is managed, administered and priced. They are frustrated with the expensive, high tech disease-treatment approach predominant in modern medicine and are seeking complementary or alternative beneficial products from cheaper, easily availalable source for effective treatment. Nutraceuticals particularly from mushrooms (known as mushroom nutriceuticals) can be appropriate and appealing. The term 'nutraceutical' was coined in 1979 by Stephen Deflice, founder and chairman of the foundation for Innovation in Medicine located in Cranford, New Jersy. It was defined 'as a food or part of food, that provides medical or health benefits, including the prevention and treatment of diseases (Biesalski, 2001). Nutracuticals may range from isolated nutrients, herbal products, dietary supplements, nutraceuticals and diets to genetically engineered 'designer' foods and processed products such as cereals, soups and beverages. Doubtlessly, many of these prroducts possess pertinent physiological functions and valueable biological activities (Andlaur & Furst, 2002). The Dietary Supplement Health and Education Act of 1994, expanded the definition of nutraceuticals to include vitamins, minerals, herbs, botanicals, amino acids and any dietary substances used by humans to fortify the diet by increasing total dietary intake (Whiteman, 2001; Stauffer, 1999). The term 'mushroom nutriceutical' has been coined by Chang and Buswell (1996), to describe those compounds that have considerable potential as dietary supplements, used for the enhancement of health and prevention of various human diseases. The edible mushrooms with remarkable nutriceutical properties include the species of *Lentinus (Lentinula), Auricularia, Hericium, Grifola, Flamulina, Pleurotus, Lactarius, Pisolithus, tinctorius, Trermella Russula, Agaricus, Cordyceps* etc. A mushroom nutriceutical is a refined/partially defined extractives from either mycelium or fruiting bodies, which is consumed in capsule or tablet form as a dietary supplement (not as a regular food) with potential therapeutic applications. Dietary supplements are ingredients extracted from foods, herbs and plants that are taken without further modification for their presumed health enhancing benefits .The term 'dietary supplements' was formally defined by the US administration in 1994 as 'a product intended to supplement the diet to enhance health'. For the last two decades, the derivatives from medicinal mushroom, for example, *Ganoderma lucidum, Coriolus versicolor,* and *Lentinula edodes* etc. have received great attention for being products of improving biological functions, thus, making people fitter and healthier. In some cases, these dietary supplements were used in the prevention and treatment of various human diseases. Furthermore, these products have side effects that frequently accompany the synthetic drugs traditionally used for these purposes. For example, cancer patients may suffer from severe pain even after the treatments of radiotherapy and chemotherapy. However, constantly intake of good quality *Ganoderma* nutriceuticals will likely reduce the requirement of synthetic drugs. At the same time due to the enhancement of the immune systems, it can help people to reduce the possibility of being affected by diseases. The market value of mushroom dietary supplement products worldwide is about 5 – 6 billion US $ per year (Wasser *et al,* 2000). The market value of *Ganoderma*-mushroom based dietary supplements alone in 1995 was estimated to be 1,628.4 million US $ (Chang & Buswell, 1999). The important certified mushroom nutriceutical products are given in table 1 and 2.

BACKGROUND

Mushrooms have long been valued as highly tasty and nutritional foods by many societies throughout the world (Chang & Miles, 1989). Early civilization by trial and error built up a practical knowledge of those suitable to eat and those to be avoided, for example, poisonous or even psychotrophic. However, in the orient several thousand years ago, there was recognition that many edible and certain non edible mushrooms could have valuable health benefits (Bensky & Gamble, 1993; Hobbs, 1995).The historical usage of these mushrooms are known to be as hot water extracts, concentrates, liquors or powders for health tonics, tinctures, teas, soups and herbal formulae etc. It was believed that the regular intake of these concentrates enhanced the immune responses of the human body, thereby increasing resistance to disease and in some cases causing regression of the disease state (Song *et al,* 1991).

These mushroom dietary supplements were used extensively in traditional Chinese medicines in various combinations, often with other

TABLE 1 Commercial Mushroom Nutriceutical Products (Extracts)

Mushroom species	Product Name	Price (in $)	Uses
Agaricus brasiliensis	XAB 100	19.95	Energizing tonic
Blend of *Cordyceps sinensis* & *Ganoderma lucidum*	Cordy Chi*	34.95	Energizing tonic
Cordyceps sinensis	Cordy-Gen	27.00	Nutritional supplements
Cordyceps sinensis	XCS100 & XCS 200	19.95	Immunomodulation
Ganoderma lucidum	Rei-Gen	22.00	Nutrtional supplements
Grifola frondosa	Grifo-Gen	25.00	Nutritional supplements
Grifola frondosa	XGF100 & XGF200	19.45 & 34.95	Strengthening health & resisting diseases
Ganoderma lucidum	XGL100 & XGL200	19.45 & 34.95	Health tonic
Hericium crinaceus	XHE100&XHE200	19.45 & 34.95	Beneficial to heart, kidneys, lungs, liver & spleen
Inonotus obliquus	XIO100	19.45	Nutritional supplements
Lentinus edodes	Shi-Gen (SHX)	22.00	Nutritional supplements
Lentinula edodes	XLE100	19.95	Strengthening health & resisting diseases
Pleurotus ostreatus	XPO100	19.95	Causing muscles & joints to release
Tremella fuciformis	Baimu-Gen (BAX)	22.00	Strengthening health & resisting diseases
Trametes versicolor	XTV100	19.95	Anticancer agent

Extract: 30-40 drops/day (0.75 ml–1.00 m/day) or as directed by health professionals
Capsule: 3-5 capsules/day (1800 mg-3000 mg/day) or as directed by health professionals
*: preparation is also available in capsule form

herbal products, to treat many medical conditions. A limited number of highly purified polysaccharide compounds derived from certain medicinal mushrooms are now being used, particularly in Japan, as pharmaceutical grade products. The ancient of Indians, Iranians, Scythians and Chinese used certain mushrooms in their ritualistic performances. A ritual drink having an intoxicating property, obtained from a hallucinogenic mushroom, has been known to the ancients. According to the eminent Soviet scholar, Bongard-Levin,1980 there is an evidence of familiarity with 'Bhanga'(a drink made up of mushroom) of the ancestors of the finnourgic (people of East and North Europe and Western Siberia) peoples and the religious conceptions of Indian-Iranians. Based on information derived from Vedic Hymns (a collection of over 1,000 hymns), the practice and application of 'Soma', (the juice of *Amanita muscaria*) was known in two different forms,

TABLE 2 *Commercial Mushroom Nutriceutical Products (Capsules)*

Mushroom species	*Product Name*	*Price(in $)*	*Uses*
Agaricus blazei	Agaricus blazei	29.95	Immunomodulation
Blend of: *Agaricus brasiliensis+cordyceps sinensis+Ganoderma lucidum+Grifola frondosa+Hericium erinaceus+Polyporus umbellatus+Trametes versicolor*	FP's Stamets7™ (S7C60)	29.95	Energizing tonic
Blend of: *Ganoderma lucidum+Lentinusedodes +Grifola frondosa+ Cordyceps sinensis+ Tremella fuciformis*	Myco-Forte™ (MYX102)	24.00	Health tonic
Blend of: *Cordyceps sinensis+Ganoderma lucidum+Lentinus edodes*	Cordyceps Plus (TMX102)	24.00	Health tonic
Blend of :*Cordyceps sinensis+Ganoderma lucidum*	CCC60/CCC120	24.95/39.95	Health stimulating properties
Cordyceps sinensis	CSC60/CSC120	24.95/39.95	Promotion of cardiovascular, sexual & immune system, energizing tonic
Grifola frondosa	GFC60/GFC120	24.95/39.95	Energizing tonic
Ganoderma lucidum	GLC60/GLC120	24.95/39.95	Health stimulating properties
Lentinus edodes	Enzolen™	39.95	Immunomodulation
Lentinus edodes	*Reishi Plus (REZ)	32.00	Immunomodulation

Extract: 30-40 drops/day (0.75 ml–1.00 m/day) or as directed by health professionals
Capsule: 3-5 capsules/day (1800 mg-3000 mg/day) or as directed by health professionals
*: preparation is also available in extract form

first, it can be eaten at raw mushroom stage or used in its squeezed-out juice (tawny yellow colour) form. Also there is the practice of taking it mixed with water alone or with water and milk or curds or honey; second, it can be taken in the urine of the *Amanita* eating person who used the first form.

Active Principles Involved in Mushroom Nutriceuticals

The curative and health giving properties of fungal origin have a long history. The invention of the so called 'wonder drug', penicillin was a landmark in this field. Since then many fungi are known for their pharmacological values. Mushrooms were treated as a special kind of organisms which were known to possess all the essential curative ingredients in adequate quantities. The prominent health giving beneficial ingredients of mushroom nutriceuticals are mostly polysaccharides which are different in their composition, mode of action, degrees of ramification and polymerization etc. yet they all are immuno stimulants with no side effects and all of them functioning in strengthening health and increase immunity. Moreover, their popular and widespread use in traditional health practices and preparations is attributed to their general tonic and medicinal properties. The last two decades bear witness to an upsurge of academic and dietary interest in mushrooms as a source of bioactive compounds. Latter day application of modern analytical techniques has in several cases revealed a variety of mushroom derived compounds such as polysaccharides, triterpinoides, lectins, steroids, adenosine, PSK, PSP, PSPC, proteins and protein bound polysaccharids, eritadine butyric acid etc. which exhibit anti-cancer, anti-viral, immunoprotentiating, hypocholesterolaemic, hepato protective and genoprotective activities (Wasser & Wies, 1999; Shi *et al,* 2002; Table 3).

NUTRICEUTICALS VERSUS PHARMACEUTICALS AND FUNCTIONAL FOODS

Many people are confusing nutriceuticals with functional foods which are consumed as part of the normal diet and may have been modified/enriched in some way to provide health giving benefits. Since the latter exhibit relatively low potency, there is little if any scope for over dose/toxicity. As consequence, the degree of control by regulatory agencies is very low and such products are readily available from general retail outlets. A nutriceutical, on the other hand, as stated earlier is a refined/partially defined food extractives which is consumed in the form of capsules or tablets as a dietary supplement (not a food), and which has potential therapeutic applications. A regular intake may enhance the immune response of the human body thereby, increasing resistance to disease and in some cases, cause regression of a disease state. Since, most nutriceuticals are extracted from edible mushroom species, there is again little scope for toxicity and overdose. Although some degree of control by the regulatory agencies is both necessary and desirable, accessibility would be comparable with over-the-counter medicines. Nor are nutriceuticals to be confused with pharmaceuticals which are almost invariably a defined chemical preparation the specifications of which are listed in a pharmacopoeia. Pharmaceuticals possess medicinal properties and are mainly used therapeutically for the treatment of specific medical conditions. Since many

TABLE 3 *Active principle involved in nutriceutically important mushroom species*

Active principle	*Pharmacodynamics*	*Mushroom species*	*Source*
Polysaccharides/ Proteins	Immunomodulatory Anticancer, Hepatoprotective, Hypoglycemic, Antihistamine release, Anti- angiogeinic, Antiviral	*Ganoderma lucidum, Lentinus edodes*	Mizuno,1992; Kim *et al*, 1993; Sohn, 1995; Hikino *et al*, 1985; Chang *et al*, 1989
Triterpenes	Cytotoxic tumor, Anti-HIV, Hypolipidemic, Hypotensive, Antiplatelet aggregation, Cardioactive, immunomodulatory	*Ganoderma lucidum*	Toth *et al*, 1983; Luu, 1995; Lie *et al*, 1988; Komodo *et al*, 1989, Morigawa et al, 1986, Wang *et al*, 1986; Hattori, 1995; Luu, 1992; Lindequist, 1995
Ling Zhi-8 (LZ-8;MW 12430 da)/110 amino acids	Anti hypersensitivity, anti-autoimmune diabetes, Anti hepatitis B, immunomodulatory	*Ganoderma lucidum*	Kino *et al*, 1989
Adenosine & derivatives (5'-deoxy-5' methylsulfinyl adenosine)	Antiplatelet aggregation	*Ganoderma lucidum*	Shimizu, 1985; Kawagishi, 1995
Organic germanium (Carboxyl ethyl-germanium sesquinoside) GE-132	Antitumor	*Ganoderma lucidum*	Sate & Waguchi, 1979; Quian & Zhang, 1993
Hirustic acid	Antibacterial	*Many species*	Pai *et al*,1990;Chen & Miles, 1996
E-b methoxy acrylate	Antibiotic effect	*Odemansiella radiata*	Pai *et al*,1990;Chen & Miles, 1996
Volvatoxin/ Flammutoxin	Cardiac tonic	*Volvariella spp.*	Pai *et al*,1990;Chen & Miles, 1996
Eritadenine	Decrease cholesterol	*Collybia velutipes*	Pai *et al*,1990;Chen & Miles, 1996
Peptide, Glycogen, Ganodran, Glucan	Decrease blood sugar	*Ganoderma lucidum, Grifola frondosa*	Pai *et al*,1990; Chen & Miles, 1996
Marasmic acid	Analgesic/Sedative effect	*Marasmius androsaceus*	Pai *et al*,1990;Chen & Miles, 1996

pharmaceuticals have a high potency, there is often a considerable toxicity potential and, for this and other reasons (for example, purity, sterility) their production and usage is subject to close control by the regulatory agencies. There is a wide range of availability, from over the counter medicines to strictly controlled dangerous drugs. Some mushroom derived compounds for example, Lentinan (a purified polysaccharide from *L. edodes* used in the treatment of cancers and infectious diseases) and Schizophyllan (a purified polysaccharide extracted from *Schizophyllum commune* and used to treat cervical cancer) fall into the pharmaceutical category (Chihara,1993; Pai *et al*, 1990). As more becomes known about the active components of mushroom nutriceuticals, The purity and quality control standards improve these compounds and may qualify as pharmaceuticals. The main features of nutraceuticals, nutriceuticals and pharmaceuticals are presented in table 4.

Relevant Concerns

Quality control: There is no doubt that mushroom based products can serve as superior dietary supplements but how exactly these products work is still a matter of conjecture, in spite, various laboratory and human trials have proved repeatedly their potentiality. However, people unfamiliar with the field may ask that if this mushroom has such beneficial effects, why don't then chemists isolate the active component so that it can be marketed as a drug? The answer is 'yes' or 'no'if it is yes, the main focus is on single active component of the mushroom, and on certain disease, for example, cancer or stroke. The products are prescription drugs. If the answer is no, the main focus is on a group of compounds and their quality. The products are dietary supplement which includes a number of bioactive components instead of a single active component. They all contribute to the beneficial effect of the mushroom. For an example, the protection against cancer afforded by *Ganoderma* products could be attributed to several compounds present in *Ganoderma*, (glucans, lectins, terpenoids, steroids, nucleic acids and immuno modulatory proteins such as Ling Zhi-8). These compounds and possibly other compounds are yet to be identified in *Ganoderma*, contributing concertdly to the anticancer, antiviral, antitumor, antibacterial and immunomodulating properties of *Ganoderma*. It could be tentatively concluded that mushroom nutriceuticals are of multifunctional values with concerted effects. However, the problems that mushroom based dietary supplement are so diverse, and there are currently no standard protocols for guranteeing their product quality, critical testing, dosage, mode of administration and interactions with other drugs (Chang & Buswell, 1999). Besides, there is no necessary information regarding contraindications, precautions and adverse reactions of these mushroom nutriceuticals.

TABLE 4 *Features of distinction among Nutraceutical, Nutriceutical and Pharmaceutical*

	Composition	*Usage/ activity*	*Mode of delivery*	*Degree of control by regulatory agencies*	*Acessibility*
Nutraceutical	Modified/ enriched food material	Maintenance of healthy diet low potency	Consumed as part of normal diet. Little if any scope for overdose/ toxicity	Low	Readily available from general retail outlets
Nutriceutical	Refined/ partially defined food extractives	Dietary supplement, medicinal properties, Use primarily prophylactic but without potential therapeutic values	Capsule/ tablet, low toxicity, little scope for overdose	Medium	Comparable with over the counter medicines
Pharmaceutical	A most invariably a defined chemical preparation with medicinal properties, specification listed in a pharmacopoeia	Specific medical treatment, mostly therapeutic usage	Oral administration/ inhalation/ topical/ injection/ high potency, toxic potential High production, purity and usage subject to high level of control	High production, purity and usage subject to high level of control	Ranges from over-the-counter medicines to strictly controlled dangerous drugs

Source: Chang & Buswell, 1996

Contraindications: Myco-beta-glucans and myco-polysaccharides are contraindicated in those who are hypersensitive to mushrooms, to mushroom extracts and to any components in a mushroom containing supplement, a mushroom polysaccharide-containing supplement or a mushroom beta-glucan-containing supplements.

Precautions: Pregnant womens and nursing mothers should avoid supplementation with mushroom extracts and should avoid mushroom polysaccharide and beta-glucan-containing supplements. Those with cancer, immune deficiencies and other medical problems should use only mushroom supplementation, mushroom polysaccharide or mushroom beta-glucan supplements under medical supervision.

Adverse reactions: The most commonly reported adverse reactions from the use of the various beta-glucan-containing fungal products are gasterointestinal, including nausea and epigastric distress. Eosinophilia has been reported in subjects taking four grams daily of shiitake mushroom powder. Contact dermatitis has also been reported in some cases, from the handling of shiitake mushrooms.

Thus, the lack of quality control is a major area of concern for mushroom nutriceuticals. As mentioned above, these can have serious consequences. Adulterations and numerous other types of impurity of nutriceuticals conceivably remain undetected simply because there is an almost total absence of specific quality control. Absence of quality control not only increases the risk to the consumer, but also results in a total lack of impetus to conduct adequate research that demonstrates the potential benefits of these nutriceuticals or ensures their safety (Ernst, 2001; Chang, 2000).

Exploration and utilization of mushroom diversity: The research and development on mushrooms during the last few decades gave documentary evidence that mushrooms can have a great potential to offer the health-care system for human beings in 21st century. In many area where modern medicines may find it difficult to provide remedies, their tasks would be alleviated with the helpful assistance by mushroom nutriceuticals. Therefore the search for beneficial nutriceutical products from mushroom origin should be continued and undescribed species can be viewed as massive potential resource for production of a number of new products useful to man. Genetic diversity of various indigenous mushrooms is continuously under threat of extinction as result of growth exploitation, environment-unfriendly harvesting techniques, loss of growth habitat and unmonitored trade of macrofloral wealth (Rajak *et al,* 2002). Therefore, there is a need to investigate and explore the biodiversity of new, undescribed, lesser known, wild mushrooms and to screen out them for their nutriceutical potential. This should be done in a well defined time frame otherwise it is feared that the rich diversity of mushrooms may be lost without our having known them and their potential. As far as the India is concerned, the climatic conditions prevalent in various biogeographic regions made it a vast emporium of these macrofloral wealth (Rahi *et al,* 2002a, b).

Other concerns: Maximum work on mushroom have been done only on their cultivation aspect. Though it possess several other properties as well, but almost remained secondary. There are many reasons for this. The lack of governmental and industrial support for research, though, is relatively inexpensive, requiring neither huge research establishments nor massive complex equipment. The lack of interest of academician in mushroom science, without which the development of nutriceuticals from mushroom is difficult. The greatest needs are for creation of national and regional laboratories or centers of fundamental and applied research on mushrooms and for creating awareness among local people regarding consumption and utilization of mushroom based nutriceuticals. Besides, the problem of keeping the methodology and protocols of nutriceutical research secret is one of the important obstacle in exploring mushrooms for new nutriceutical capabilities. Maximum researchers working on this aspect do not trust their competitors and colleagues. They should share their experience and knowledge in order to enlarge and promote the research on mushroom nutriceuticals for human welfare.

REGULATION AND MONITORING

There is a serious need for legal control which are essential both to increase and maintain consumer confidence, and to meet current and future standards set by the regulatory authorities (Chang & Buswell, 1999). This is a reasonable requirement in protecting public health. In order to regulate the manufacturing and marketing of nutraceuticals including mushroom nutriceuticals, USA has passed a watershed legislation in 1994 known as the Dietary Supplement Health and Education Act under which there is a provision for keeping eyes on manufacturing agencies/establishments that whether they are functioning as per standard manufacturing procedures of the Act or not. (Bass & Young, 1996).

The similar law for regulation and monitoring of manufacturing, importing and marketing of health foods/dietary supplements and other nutraceuticals including mushroom nutriceuticals in India is still in its infantry stages of development. However, the country's central drug control department has put certain rules and regulations in place for dietary supplements, but it will take a long time for different states to cooperate and implement and some states even have rejected the model/structure when their own rules and regulations conflicted. Besides, a new autonomous association, to fulfill some of the requirements of these issues has also been constituted in India. At present, the Indian Health and Dietary Supplement Association is conducting the duties of regulation and monitoring of pharmaceuticals, nutraceuticals and herbal products and also monitor the direct selling and other services of concerning industries and companies. They are planning to affiliate with International Alliance of Dietary Supplement Associations to improve the qualities of their products.

CONCLUSION

It is now increasingly recognized that correct diet controls modulates many functions of the human body and consequently, participates in the maintenance of good health or homeostasis, necessary to reduce the risk of many diseases. Arising from the awareness of the relationship between diet and disease there has evolved the concept of dietary supplement or nutriceutical (refined or partially defined food extractives). Therefore, human beings have constantly searched for new substances that can steadily and efficiently improve biological functions of human body. The mushrooms in this regard can play a crucial role by virtue of its profound nutritional and medicinal atributes. Mushrooms have long been valued as tasty, nutritional foods, by different societies world wide. The relationship between mushrooms and man can be traced far back into antiquity. The nutriceuticalproperties in mushroom based preparation is due to the presence of various biologically active component like polysaccharides and protein bound polysaccharide. However, mushroom nutriceuticals should not be confused with pharmaceuticals ('real' medicines) that are almost invariably a defined chemical preparation and therefore cannot be claimed to cure diseases, however, the medicines with such properties can be used for treatment and there is evidence of beneficial treatment of some diseases by consumption of nutriceuticals or extracted biologically active compounds from mushrooms.Several nutriceutical products from mushrooms have been produced, tested and marketed and are in use as dietary supplement on a daily basis for strengthening and perfecting many different physiological systems of human body. They can be used in the form of extracts, capsules, syrups, teas, soups and herbal formulations. Such relatively crude dietary supplement have an immense market in the orient and increasing world wide. There is no doubt that in future mushroom nutriceuticals will play an important role in development of dietary supplements and it will of course bring health benefits quicker than would normally be the case through eating conventionally healthy food alone. But this required the proper regulation and monitoring of certain quality control parameters including purity, safety, efficacy, dosage and possible side effects. Therefore, nutriceutical uses of mushrooms demand further intensive research for its wider application. Even though a variety of mushrooms are well known for their edibility and nutritive values and a number of them are under cultivation, the possibility of their maximum exploitation for food and human health is yet to be explored.

References

Andlauer, W. and Furst, P. 2002. Nutraceuticals: a piece of history, present status and outlook. *Food Research International.* 35: 171-176.

Bass, I.S. and Young, A.L. 1996. Dietary supplements Health and Education Act. Washington DC: The Food and Drug Law Institute.

Bensky, D. and Gamble, A. 1993. Chinese Materia Medica 2nd Ed. East Land Press, Seattle.

Biesalski, H.K. 2001. Nutraceuticals: link between nutrition and Medicine (eds. Kramer, K., Hoppe, P.P., Packer, L.) Nutraceuticals in health and disease prevention. New York: Marcel Dekker Inc: p 1-26.

Bongard-Levin, G.M. 1980. The origin of Aryans-Scythia to India, Arnold-Heinemann Publishers (India) Pvt. Ltd, New Delhi (translated in English by H.C. Gupta).

Brower, V. 1998. Nutraceuticals: Poised for a healthy slice of the healthcare market? *Nature Biotechnology* 16: 728-731.

Chang, S.T. and Miles, P.G. 1989. Edible Mushrooms and Their Cultivation. CRC Press, Boca Raton.

Chang, S.T. and Buswell, J.A. 1996. Mushroom nutraceuticals. *World Journal of Microbiology and Biotechnology* 12: 473-476.

Chang, S.T. and Buswell, J.A. 1999. *Ganoderma lucidum* (Curt Fr.) P. Karst (Aphyllophoromycetideae) D.A. mushrooming medicinal mushroom. *International J. Medicinal Mushrooms* 1: 139-146.

Chang, J. 2000. Medicinal herbs: drugs or dietary supplements? *Biochem Pharmaceut.* 59: 211-219.

Chen, A.W. and Miles, P.G. 1996. Biomedical research and the application of mushroom nutriceuticals from *Ganoderma lucidum* (ed. Royse, D.J.) Mushroom Biology and Mushroom Products, Penn State Univ. 153-159.

Chihara, G. 1993. Medical aspects of Lentinan isolated from *Lentinus edodes* (Berk) Sing. In Mushroom biology and Mushroom Products (eds. Chang, S.T., Buswell, J.A. and Chiu, S.W.) Chinese University Press, Hong Kong pp 261-266.

Ernst, E 2001. Functional foods, Nutraceuticals, Designer foods: innocent fad or counter productive marketing ploy. *Eur. J. Clin. Pharmacol.* 57: 353-355.

Hattori, M. 1995. A New monitoring system of culured myocardial cell motion-effect of crude drug including the fruiting bodies of *Ganoderma lucidum* on the spontaneous beating of myocardial cells in culture. Recent advances in *Ganoderma lucidum* research. (eds. B.K. Kim, I.H. Kim and Y.S.). The Pharmaceutical Society of Korea, Seol, Korea, pp. 11-22.

Hikkino, H., Konno, C., Mirin, Y. and Hayashi, T. 1985. Antidiabetic drugs, isolation and hypoglycemic activity of ganoderma A and ganderma C, glycans of *Ganoderma lucidum* fruitbodies, Planta Medica 51: 339-340.

Hobbs, C. 1995. Medicinal mushrooms: an exploration of tradition healing and culture. Botanica Press, Santa Cruz, C.A. 172.

Jong, S.C., Brimingham, J.M., and Pai, S.H. 1991. Immunomodulatory substances of fungal orogin. Journal of Immunology and Immunopharmacology. 3: 115-122.

Kawagashi, H. 1995. Isolation of 5′-deoxy-5′methyl sulfinylaldenosive as platelet aggregation inhibitor from Ganoderma lucidum. In: Recent Advances in *Ganoderma lucidum* Research (Kim, B.K. et. al., eds) Seoul Korea.

Kim, B.K., Kim, H.W. and Choi, E.C. 1993. Anti-HIV activity of *Ganoderma lucidum* *J. Biol. Chem.* 264: 472-478.

Kino, K.Y., Yamaoka, K., Watanabe, J., Shimizu, K. and Tsunoo, H. 1989. Isolation and charactarization of new immunomodulatory proteins Ling Zhi-8 (LZ-8) from *Ganoderma lucidum*. *J. Biol. Chem.*. 264: 479-485.

Komodo, Y., Nakamura, H., Ishiham, S., Uchida, M., Kohda, H. and Yamasaki, K. 1985. Structure of New terpenoid constituents of *Ganoderma lucidum* (Fr.)Karst. (Polyporaceae) *Chem. Pharm. Bull.* 33: 4829-4835.

Lindequist, U. 1995. Structure and biological activity of triterpenes, polysaccharides and other constituents of *Ganoderma lucidum*. In: Recent Advances in Ganoderma lucidum Research (eds. B.K. Kim, I.H. Kim and Y.S.). Pharmacological Society of Korea, Seol Korea pp. 61-92.

Liu, K.C., Phounsavan, S.F., Huang, R.L., Liao, C., Hsu, S.Y. and Wang, K.G. 1988. Pharmacological and liver functional studies on the mycelium of Gnoderma lucidum. *Chinese Pharmaceut.* J. 40: 21-29.

Luu, B. 1992. Chemistry and Biology of oxysterols and oxytriterpenes. 4th International Symposium on *Ganoderma lucidum* Seoul Korea.

Luu, B. 1995. Does *Ganoderma lucidum* contain anti HIV agents: Occurrence and antiviral activities of oxysterols and oxylanosterols. In: Recent Advances in *Ganoderma lucidum* Research (eds. B.K. Kim, I.H. Kim and Y.S.). Pharmacological Society of Korea, Seol Korea pp. 23-28.

Mizuno, T. 1992. Comparative studies on host mediated antitimour polysaccharides isolated from fruiting body and mycelium of the *Ganoderma* species fungi, *Ganoderma lucidum, G. tsuage,* and *G. applanatum*. 4th International Symposium on *Ganoderma lucidum*. Seoul, Korea 21-29.

Pai, S.H., Jong, S.C. and Low, D.W. 1990. Usages of mushrooms *Bioindustry* 1: 126-131.

Quian, F. and Zhiang, S. 1993. Studies on medicinal mushrooms and germanium, G.E. 132. First International Conference on Mushroom Biology and Mushroom Products. Hong Kong. Abst. P.144.

Rahi, D.K., Shukla, K.K., Pandey, A.K. and Rajak, R.C., 2002. Nutritional potential of *Termitomyces heimii*: an important wild edible tribal mushroom species of M.P. *J. Basic Appl. Mycol.* 1(1): 36-38.

Rahi, D.K., Shukla, K.K., Rajak, R.C. and Pandey, A.K. 2002. Nutritional profile of wild edible mushroom genera of central India: *Russula & Lactarius Indian J. Mushrooms* (1&2): 21-23.

Rajak, R.C., Rahi, D.K., Shukla, K.K. and Pandey, A.K., 2002. Diversity and systematics of Agaricales of Central India. In: Frontiers of Fungal Diversity in India. (Eds Rao, G.P., Manoharachary, C., Bhat, D.J., Rajak, R.C., & Lakhanpal, T.N.) Intrnational Book Distributing Company, Lucknow 297-312.

Sate, H & Iwaguchi, T. 1979. Antitumour activity of new novel organogermanium compound GC-132. *Cancer Chemotherapy* 6: 79-83.

Shi, Y. L., James, A.E., Benzie, I.F.F. and Buswell, J.A. 2002. Mushroom derived preparations in the prevention of H2O2 induced oxidative damage to cellular DNA. Teratogenesis, Carcinogenesis and Mutagenesis. 22: 103-111.

Shimuzu, a. Yano, T., Saito, Y., and Inada, Y. 1985. Isolation of an inhibitor of platelet aggregation from a fungus Ganoderma lucidum *Chem. Pharm. Bull.* 33: 3012-3015.

Sohn, D.H. 1995. Antifibriotic effects of polysaccharides from *Ganoderma lucidum* in experimental hepatic cirrhosis model. In: Recent Advances in *Ganoderma lucidum* Research (eds. B.K. Kim, I.H. Kim and Y.S.). Pharmacological Society of Korea, Seol Korea pp. 29-41.

Stauffer, J. E. 1999. Nutraceuticals. *Cereal Foods World.* 44: 115-117.

Toth, J.O., Luu, B. and Ourisson, G. 1983. Ganoderic acid T-Z, cytotoxic triterpenes from *Ganoderma lucidum* (Polyporaceae). *Tetrahedron Lett.* 24: 1081-1084.

Wang, C. N., Chen, J.C., Shiao, M.S. and Wang, C.T. 1991. The inhibition of human platelet function by Ganodermic acid S. *Biochem.* J. 277: 189-197.

Wasser, S.P. and Weis, A.L. 1999. Medicinal properties of substances occurring in higher Basidiomycetes mushrooms: current perspectives. *International Journal of Medicinal Mushroom* 1: 31-62.

Wasser, S.P., Nevo, E., Sokolov, D., Reshetnikov, S. and Timor Tismenetsky, M. 2000. Dietary supplements from medicinal mushrooms: diversity of types and variety of regulations. *International Journal of Medicinal Mushrooms* 2: 1-19.

Whitman, M. 2001. Understanding the perceived need for complementary and alternative nutraceuticals: Life styl issues. *Clin. J. Oncol. Nurs.* 5: 190-194.

Zeisel, S. 1999. Regulation of 'Nutraceuticals' *Science* 285: 1853-1866.

Frontiers in Plant Sciences (Pages 497-514)
Eds. K.G. Mukerji, K.V.B.R. Tilak, S.M. Reddy, L.V. Gangwane, P. Prakash, I.K. Kunwar

Role and Control of Microorganisms Involved in Dental Caries and Periodontal Diseases—A Review

K.R. Aneja*, Pranay Jain and Raman Aneja

ABSTRACT

Periodontal diseases and Dental caries are probably the most widespread gum and root infections in humans, affecting virtually all adults and children throughout the world. While the Periodontal diseases are marked by overwhelmed inflammatory response of the gingival tissue to the plaque accumulation in the form of Periodontitis and Gingivitis, the Dental caries, on the other hand, is the dissolution of the teeth enamel due to the plaque. Although bacteria are responsible for the onset and progression of the disease, other factors have been linked to increased susceptibility and include age, poor oral hygiene, tobacco smoking, stress, systemic diseases, decreased immunocompetence and genetic traits. Microbial diagnosis is significant in the treatment planning of these infections. *Streptococcus mutans* and anaerobic lactobacilli are common in dental caries, *Actinomyces* and *Peptostreptococcus* spp. in gingivitis, *Actinobacillus actinomycetemcomitans* and *Porphyromonas gingivalis* in periodontitis. A multi-factorial antimicrobial approach is necessary for the successful management of the destructive periodontal disease and for that of dental caries. Commonly used drugs in periodontal diseases and dental caries include, Metronidazole, Clindamycin, Amoxicillin and Ciprofloxacin. Soil provides a vast array of microbial isolates capable of producing secondary metabolites that can be used biotechnologically for the synthesis of novel antibiotics that would be effective in controlling the infectious microorganisms in the near future.

Key words: Periodontitis, Gingivitis, Dental caries, Antibiotics, Soil Microflora

Author's address: Department of Microbiology, Kurukshetra University, Kurukshetra- 136 119, India
*Corresponding Author's Email: anejakr@rediffmail.com

INTRODUCTION

The mouth supports a varied and dense population of microbial flora. Common bacterial diseases of the oral cavity are gingivitis, periodontitis, periodontosis, dental caries, abscesses associated with teeth, pharyngitis and tonsillitis. Of these, the two common diseases namely periodontitis and dental caries have affected mankind since the earliest of times, and studies from ancient India and China have indicated that periodontitis has been observed for many centuries (Boghani, 2000).

Periodontal disease is the general description applied to the inflammatory response of the gingiva and surrounding connective tissue to the bacterial plaque accumulations on the teeth (van Winkelhoff, 2003; Boghani, 2000). Based on the damage caused to the gums and teeth, periodontal disease is of three types: (i) The first type of the chronic periodontal disease referred to as **Marginal gingivitis** or **Chronic Marginal** *Gingivitis* is a mild condition involving a painless accumulation of blood (hyperemia) of the gingival margin without gross pus formation; (ii) The second type is referred to as **Ulcerative Gingivitis**, also known as **acute gingivitis**, **Vincent's infection**, and **Pyorrhea simplex**, is an acute inflammatory disease involving destruction of exposed gingival tissue; and (ii) The third type, **Periodontoclasia**, also known as **Periodontitis** and **Pyorrhea profunda**, is characterized by an extension of the inflammatory response into the deeper periodontal structures with a protracted destruction of the periodontal membrane (van Winkelhoff, 2003). The classification of periodontitis, as delineated by the American Academy of periodontology (Neville *et al.*, 2002) is given in Table 1. Tooth decay, also termed as dental caries is an infectious microbial disease that results in localized dissolution and destruction of the calcified tissues of the teeth (Ross and Holbrook, 1994).

According to Neville *et al.* (2002), Gingivitis is an inflammation limited to the soft tissues that surround the teeth. The primary types of gingivitis are: plaque related gingivitis, necrotizing ulcerative gingivitis, medication influenced gingivitis, allergic gingivitis, specific infection related gingivitis and dermatosis related gingivitis.

Gingivitis and periodontitis are chronic bacterial infections. As in other infections, the host bacterial interactions determine the nature and extent of the resulting disease. Pathogenic microorganisms may cause disease indirectly by stimulating a host response or directly by producing toxins and invading the tissues. Overwhelming evidences show that organisms present in microbial plaque constitute the primary and possibly, the only extrinsic etiologic agent participating in the causation of gingivitis and periodontitis (Loe *et al.* , 1965). At present, the prevalence of dental caries in India is approximately 60-65%, while the prevalence of periodontal disease is approximately 65-100 % (Saini *et al.*, 2003).

TABLE 1 *Classification of periodontitis**

Name of the disease	*Subcategories*
A. Chronic periodontitis	• Localized chronic periodontitis • Generalized chronic periodontitis
B. Aggressive periodontitis	• Localized aggressive periodontitis • Generalized aggressive periodontitis
C. Periodontitis as a manifestation of systemic diseases	• Associated with hematologic disorders • Associated with genetic disorders
D. Necrotizing periodontal diseases	• Necrotizing ulcerative gingivitis • Necrotizing ulcerative periodontitis
E. Abscesses of the periodontium	• Gingival abscess • Periodontal abscess • Pericoronal abscess
F. Periodontitis associated with endodontic lesions.	———

*Modified from Neville *et al.* (2002).

Microbiological information of the subgingival plaque is being given emphasis in diagnosis and treatment planning in patients with severe periodontitis (van Winkelhoff, 2003). This approach is based on a number of facts as mentioned: Periodontitis is a bacterial disease and knowledge of the infectious agents can, in principle, improve the diagnosis and treatment of the disease and may contribute to the prognosis. Based on the microbiology of the subgingival plaque in the patient, periodontitis is considered as a collection of disease rather than one disease entity. Not all patients show a favourable treatment response upon mechanical periodontal treatment. Rationale use of antimicrobial agents is based on the knowledge of the causative agents of the disease.

A multifactorial antimicrobial approach is necessary for the successful management of destructive periodontal disease. Effective antimicrobial periodontal therapy aims to overwhelm periodontal pathogens with aggressive initial therapy and prevent previously suppressed pathogens from rising up a new through daily oral hygiene measures and frequent professional cleaning (Slots and Jorgensen, 2000).

This review deals with the various periodontal and dental caries microbiota, the mechanism involved in these microbial infections, their treatment and control by various antimicrobial agents and the significance of soil microflora in producing antibiotics which will be useful in treating these infections and also in overcoming the resistance developed amongst the pathogens to the known antibiotics.

PERIODONTAL MICROBIOTA

Destructive periodontal disease is an infectious process, and hence, may not be managed by anti- infective intervention. Causes may include nonspecific bacterial plaque, as in gingivitis and mild forms of chronic (adult) periodontitis, or specific bacterial infections as in aggressive (early onset) periodontitis. Various microbial pathogens involved with periodontal disease and dental caries include aerobic and anaerobic bacteria, spirochetes, molds and yeasts and protozoa (Table 2) and these have been extensively discussed by Saini *et al*. (2003), van Winkelhoff (2003), Schuster (1999), Socransky (1977) and Slots (1997).

Recent techniques that have been applied to detect periodontal pathogens are immunoassays (ELISA, immuno-flourescence), enzyme

TABLE 2 *Microorganisms associated with Periodontal disease and Dental caries*

Microorganisms	*Representattive species*
A. Bacteria Aerobic Gram positive cocci	Viridans Group Streptococci • *Streptococcus milleri* group • *Streptococcus mutans* group
Aerobic Gram positive bacilli	*Rothia dentocariosa*
Aerobic Gram negative coccobacilli	*Actinobacillus* spp. *Campylobacter* spp. *Capnocytophaga* spp. *Eikenella* spp.
Anaerobic Gram positive rods	*Actinomyces* spp. *Peptostreptococcus* spp. *Eubacterium* spp.
Anaerobic Gram negative rods	*Porphyromonas* spp. *Fusobacterium* spp. *Prevotella* spp. *Bacteroides* spp.
Anaerobic Gram negative cocci	*Veillonella* spp.
Spirochetes	*Treponema* spp.
B. Molds & Yeasts	*Candida albicans* C. dubliniensis C. glabrata, C. parapsilosis *Saccharomyces cerevisiae*
C. Protozoa	*Entamoeba gingivalis* *Trichomonas tenax*

tests, DNA probes (whole genomic probes), anaerobic culture and polymerase chain reaction (PCR). In daily practice, the aerobic culture, staining, biochemical tests, the anaerobic culture technique and the PCR technique are used most frequently (van Winkelhoff, 2003).

Most subgingival pathogens are strict anaerobic Gram negative bacteria, but Gram positive strict anaerobic bacteria have also been implied. Of the various pathogens revealed, *Streptococcus mutans* and anaerobic lactobacilli are common in dental caries; *Actinomyces* and *Peptostreptococcus* spp. in gingivitis; *Actinobacillus actinomycetemcomitans* and *Porphyromonas gingivalis* in periodontitis (Saini *et al.*, 2003; van Winkelhoff, 2003). The number of different bacterial species that have been isolated from dental plaque is well over four hundred. Yet, the pathogens that have been associated with progression of periodontitis is limited to less than twenty (van Winkelfhoff, 2003). Table 3 reveals the association between various periodontal pathogens and progression of periodontal disease at various stages.

TABLE 3 *Association between putative periodontal pathogens and progression of periodontal disease.**

Very Strong	*Strong*	*Moderate*	*Early stage of investigation*
Actinobacillus actinomycetemcomitans *Porphyromonas gingivalis* Spirochetes of acute necrotising gingivitis	*Bacteroides forsythus* *Prevotella intermedia* *Eubacterium nucleatum* *Treponema denticola*	*Streptococcus intermedius* *Peptostreptococcus micros* *Fusobacterium nucleatum* *Campylobacter rectus*	*Selenomonas* spp. *Enteric rods* *Pseudomonas* spp. *Staphylococcus* spp. Cytomegalovirus Epstein Barr virus type 1

**Modified from Slots, 1997; van Winkelhoff, 2003*

Systematic studies on the anaerobic bacterial flora of orodental infections reported from India are few. Saini *et al.* (2003) undertook the study to isolate and identify aerobic as well as anaerobic microbes from patients of adult periodontitis, gingivitis and deep seated dental caries and findings were compared against the normal healthy individuals. They found that aerobic/anaerobic ratio was high in normal flora as compared to dental caries, gingivitis and periodontitis. Ninety seven per cent of orodental infections were polymicrobial and three or more microbes were found in 84 % cases of study groups as compared to 28% in controls. The ratio of aerobes, facultative anaerobes to anaerobes decreased as the lesion progressed from normal gingiva to dental caries to gingivitis to periodontitis as shown in Table 4. Maximum number of anaerobes were isolated from the cases of periodontitis.

TABLE 4 *Aerobes, Facultative Anaerobes and Anaerobes isolated from gingivitis, periodontitis and normal flora.**

Isolates	*Normal Flora*	*Gingivitis*	*Periodontitis*
Streptococcus mutans	+	+	+
Non mutans streptococci	+	+	–
S. sanguis	+	+	+
S. mitior	+	+	+
Others			
Staphylococcus aureus	–	+	+
Coagulase negative staphylococci	+	+	+
Lactobacilli (Aerobic)	–	–	–
Actinobacillus actinomycetemcomitans	+	+	+
Klebseilla	+	+	+
Escherichia coli	–	–	+
Pseudomonas	+	–	+
Non fermenters	+	+	+
Yeast Cells	+	–	–
Gram positive bacilli			
Lactobacilli	–	+	+
Actinomyces	+	+	+
A. naeslundii	+	+	+
A. viscosus	+	+	+
A. israeli	+	+	+
Propionibacterium	–	+	+
Bifidobacterium	+	+	+
Gram negative bacilli			
Prevotella melaninogenica	–	+	+
Prevotella intermedia	–	+	+
Porphyromonas gingivalis	–	+	+
Fusobacterium	+	+	+
Gram negative cocci			
Veillonella	+	+	+
Gram positive cocci			
Peptostreptococcus spp.	+	+	+

**Modified from Saini et al. (2003)*

The majority of subgingival bacteria are probably commensal bacteria and are considered as endogenous pathogens. This implicate that the prevalence of these bacterial species in the oral cavity is similar both in periodontal health and in disease. These species may accumulate in the subgingival area, and over time, may comprise significant part of the subgingival plaque. This may not be true for *Actinobacillus actinomycetemcomitans* and *Porphyromonas gingivalis*. These pathogens have characteristics of exogenous pathogens. This presumption is based on the finding that they have a low occurrence in periodontally healthy subjects (Saini *et al.*, 2003; Griffen *et al.*, 1998; van Winkelhoff, 2003).

Fungal species have gained considerable importance as opportunistic pathogens of immunocompromised individuals, particularly those infected with human immunodeficiency virus (HIV) (Hazen, 1995; Phelan *et al.*, 1997). The cultivable microflora from HIV positive patients with periodontal diseases have been found to be comprised of periodontopathic species similar to those found in HIV negative subjects with periodontitis, mainly Gram negative bacteria (Barr, 1995; Brady *et al.*, 1996). The presence of yeasts was a striking feature of the subgingival plaque samples from some AIDS patients in one study. Among the subgingival samples, *Candida albicans, C. dubliniensis, C. glabrata, C. parapsilosis* and *Saccharomyces cerevisiae* are the yeast recovered (Jabra-Rizk *et al.*, 2001).

Ghayoumi and coworkers (2002) have implicated *Dialister pneumosintes* as a candidate periodontal pathogen. They found the association of subgingival *D. pneumosintes* with demographic variables (age and gender) and the presence of subgingival periodontopathic *Bacteroides forsythus* and *Porphyromonas gingivalis*. Subgingival *D. pneumosintes* occurred with significantly higher prevalence in older individuals and was closely associated with subgingival *B. forsythus*. According to Ghayoumi *et al.* (2002), *D. pneumosintes* may play an important role in the microbial complex responsible for destructive periodontal disease.

THE PERIODONTAL DISEASE AND DENTAL CARIES: MECHANISM INVOLVED

The most important new finding concerning periodontal disease is the realization that these clinical entities are really specific infections. These infections are unusual in that massive or even obvious bacterial invasion of the tissues is rarely encountered. Rather bacteria in the plaque touching the tissue elaborate various compounds such as H_2S, NH_3, amines, endotoxins, enzymes (such as collagenase) and antigens, all of which penetrate the gingiva and elicit an inflammatory response. This inflammatory response, although overwhelmingly protective, appears to be responsible for net loss of periodontal supporting tissue, and leads to the periodontal pocket formation, loosening of the teeth, and eventual tooth loss (van Winkelhoff, 2003).

Although bacteria are essential for the onset and progression of the disease, they are probably not enough. This concept is based on the fact that periodontal pathogens can occur in a subject without clinical signs of the disease. In order for the disease to develop, a susceptible host is required. The nature of susceptibility for destructive periodontal disease has long been an enigma. Several factors have been linked to increased susceptibility and these include age, poor oral hygiene, tobacco smoking, stress, systemic diseases, decreased immunocompetence (neutropenia) and genetic traits (gene polymorphisms). None of these factors by themselves are probably capable to produce the disease, but it seems likely that a set of risk factors is necessary to initiate the disease process. For this reason, periodontitis is considered as a multifactorial disorder (van Winkelhoff, 2003).

Establishing a cause and effect relationship between a single microorganism in oral flora and caries has been very difficult. Most of the investigators believe that development of caries of enamel is preceded by the formation of microbial plaque in the tooth (Madigan *et al.*, 2000). As dental plaque accumulates and acid products are formed, dental caries (tooth decay) is the usual result. Thus, tooth decay is an infectious disease caused by microorganisms. The role of the oral flora in tooth decay has now been well established through studies on germ free animals. The smooth surfaces of the teeth that are exposed to frequent cleaning by the tongue, cheek, saliva or toothbrush or to the abrasive action of food mastication are relatively resistant to dental caries. The tooth surfaces in crevices, where food particles can be retained are the sites where tooth decay predominates. Thus, the shape of the teeth is important (Madigan *et al.*, 2000).

Madigan and coworkers (2000) have found a correlation between the dental caries and sugars present in the mouth of a person. According to them, diets high in sugars are especially cariogenic because lactic acid bacteria ferment the sugars to lactic acid, which causes the decalcification of the enamel of the tooth. Once breakdown of the hard tissue has begun, proteolysis of the matrix of the tooth enamel occurs through the action of proteolytic enzymes released by bacteria. Microorganisms further penetrates into the decomposing matrix, but the later stages of the process may be exceedingly slow and are often highly complex. The structure of the calcified tissue also plays an important role in the extent of dental caries (Madigan *et al.*, 2000).

van Winkelhoff (2003) classified periodontitis on the basis of the occurrence of microorganisms in the host. Periodontitis associated with endogenous (commensal) pathogens be considered an *opportunistic infection* whereas periodontitis associated with exogenous pathogens may be considered a **true infection**. This view has clinical implications and allows a differentiated treatment strategy; endogenous pathogens may only need suppression, but exogenous pathogens should be eliminated in susceptible subjects.

TREATMENT AND CONTROL

A view to the past indicates that in ancient times, India was probably the most advanced country in dental health services in the world. In the old Ayurvedic literature of India, details of gum diseases and forms of treatments have been described. The practice of oral hygiene was included in daily rituals (D' Silva, 1999). Ancient scriptures such as Vedas, Puranas etc. proposed that the natural dentition could be preserved by appropriate periodontal treatment. Today, Ayurvedic treatment is still very popular amongst rural and uneducated people in India. For the treatment of chronic periodontitis, massaging of the gums with oils and ointments prepared from herbs is still advised. It has been reported that some of the Ayurvedic preparations have various ingredients which have analgesic, anti-inflammatory properties as well as being able to facilitate cleansing and mouth freshening (D' Silva, 1999).

Most periodontal treatment in present times is based on the modern allopathic concept of defining etiology of the diseases, analyzing signs and symptoms and executing treatment based on this concept. The research carried out in India is based on allopathic concept with special emphasis on Indian environment and Indian habits (Boghani, 1991).

Periodontal disease is probably the most widespread inflammatory disorder in humans, affecting virtually all adults and children throughout the world. Severity of periodontal disease ranges from mild gingival inflammation to advanced periodontal attachment loss, sometimes resulting in loss of teeth. Historically, a predominantly mechanical approach to treatment has been employed, and outcomes have often been disappointing (Slots and Jorgensen, 2000). Use of antimicrobial agents in periodontal treatment has often been haphazard and empirical. However, there is evidence that the addition of appropriate antimicrobial therapy to traditional periodontal treatment may substantially enhance clinical outcomes and reduce the need for costly surgical procedures.

Current antimicrobial periodontal therapy employs mechanical debridement performed with and without surgery, antibiotics and antiseptics (Slots and Jorgensen, 2000). A proficient cost effective antimicrobial periodontal treatment may offer populations who are currently unable to receive adequate care due to lack of financial resources, a practical and valuable means of maintaining a functional periodontum and dentition for a prolonged period (Slots, 1997).

New products and treatment modalities for the management of periodontal disease continue to offer a larger number of choices, many of which involve antimicrobials. Specific pathogenic bacteria play a central role in the etiology and pathogenesis of destructive periodontal disease. Under suitable conditions, periodontal pathogens colonize the subgingival environment and are incorporated into a tenacious biofilm. Successful prevention and treatment of periodontitis is contingent upon effective control of the periodontopathic bacteria. This is accomplished by professional treatment of diseased periodontal sites and patient performed plaque control. Appropriate antimicrobial agents that can be administered systemically (antibiotics) or via local delivery may enhance eradication or marked suppression of the subgingival pathogens (Slots and Jorgensen, 2000; Mombelli *et al.*, 2002).

The choice of antimicrobial therapy is empirical and based on the experience. Another approach is microbiological analysis of the subgingival microflora which establishes the presence and levels of periodontal pathogens. Consequently, the selected antimicrobial therapy can be based on known susceptibility profiles of the target microorganisms. Using this protocol, therapy failure with an empirically selected antibiotic can be avoided (van Winkelhoff, 2003).

IN VITRO SUSCEPTIBILITY OF PERIODONTAL AND DENTAL CARIES MICROORGANISMS TO ANTIMICROBIAL AGENTS

Extensive studies have been carried out involving putative periodontal and dental caries pathogens to a vast array of antimicrobial agents. Wilson and coworkers in 1990 investigated the effects of Minocycline on subgingival plaque samples from patients with chronic periodontitis. The results implied that adequate reductions in the numbers of viable subgingival plaque bacteria are unlikely to occur after exposure to Minocycline at concentrations attainable in gingival crevicular fluid after systemic administration.

Hoshino and coworkers (1996) undertook a study to clarify the anti-bacterial effect of a mixture of Ciprofloxacin, Metronidazole and Minocycline, with and without the addition of Rifampicin, on bacteria taken from infected dentine of root canal walls. The efficacy was also determined against bacteria of carious dentine and infected pulps which may be the precursory bacteria of infected root dentine. The respective drug alone (10, 25, 50 and 75 μg/ml) substantially decreased the bacteria recovery, but could not kill all the bacteria. Bacteria taken from carious dentine and infected pulps were also sensitive to the drug combination. These results indicated that the bactericidal efficacy of the drug combination is sufficiently potent to eradicate bacteria from the infected dentine of root canals.

The study on the sensitiveness of slow growing anaerobic bacteria to antibiotics is delicate when the technical motives are considered that make it difficult to transpose the standard methods frequently used in microbiological laboratories. The main methods used to determine susceptibility to antibiotics are-Disk Diffusion test and antibiotics containing microdilution plates. Kamagate *et al.* (2001) carried out a study to compare the effectiveness of each of these methods on severe anaerobes bacteria isolated in subgingival flora of patients suffering from developing periodontitis (rapidly progressive periodontitis, refractory periodontitis, active stage of adult chronic periodontitis). The observed bacteria were *Porphyromonas gingivalis, Prevotella intermedia, Fusobacterium nucleatum, Campylobacter rectus,* and *Peptostreptococcus micros.* The drugs used were Ampicillin, Tetracycline, Erythromycin and Metronidazole. The comparison of the MIC of each of these methods permitted to show a strict correlation in the results observed with these three methods, if only the growth of the severe anaerobes on agar medium did not exceed 72 hours.

Effects of three different types of short term applications (1-3 times during a week) of Chlorhexidine on the susceptibility of 863 clinical isolates of *Streptococcus mutans* and 53 isolates of *Streptococcus sobrinus* from 58 subjects were studied by Jarvinen and coworkers (1995). Chlorhexidine resistant isolates were not found either before or after the treatment. The MIC's to Chlorhexidine of all isolates of *S. mutans* were lesser or equal to 1 μg/ml and of *S. sobrinus* were also susceptible to Ampicillin, Penicillin, Cefuroxime and Tetracycline.

Buchman *et al.* (2003) tested a hypothesis that the bacterial susceptibilities in aggressive periodontitis change upon administration of systemic antibiotics as adjunct to periodontal therapy. The periodontal pathogens investigated prior to and one year after periodontal therapy were tested sensitive to the antimicrobial agents. In aggressive periodontitis, changes in bacterial susceptibility upon the administration of systemic antibiotics were associated with the limited number of isolates tested following therapy.

Milazzo and coworkers (2003) compared the *in vitro* activity of Faropenem, an oral penem, with those of Penicillin, Cefotixin, Clindamycin, Erythromycin and Metronidazole against anaerobic pathogenic isolates involved in systemic infections. Faropenem show structural similarities with both Penicillins and Cephalosporins, and are characterized by a broad antibacterial spectrum, a potent Penicillin binding protein affinity and good â-lactamase stability. The organisms tested comprised *Porphyromonas gingivalis, Prevotella* spp., *P. melaninogenica, P. intermedia, Actinomyces* sp. *Fusobacterium nucleatum, Peptostreptococcus* spp. and *Bacteroides forsythus*. The antimicrobial properties of Faropenem were investigated by studying MIC's, MBC's, time-kill kinetics and post antibiotic effect. Faropenem was highly active against all the anaerobes tested and was bactericidal against both â-lactamase positive and negative anaerobes, with a maximum bactericidal effect at 10 X MIC at between 12 and 24 hrs. Faropenem may be useful for treating infections caused by periodontal bacteria or oral flora. Table 5 depicts the spectrum of various antimicrobial agents on selected dental pathogens.

TABLE 5 *Spectrum of activity of antimicrobial agents against selected oral microorganisms**

	Viridans Group Streptococci	*Anaerobic* Gram negative bacilli	*Actinobacillus* Spp.	*Capnocytophaga* spp.	*Eikenella coorodans*	*Actinomyces* spp.
Penicillin	+/–	+/–	+/–	+	+	+
Amoxicillin	+/–	+/–	+/–	+	+	+
Amoxicillin Clavulanate	+/–	+/–	+/–	+	+	+
Oral Cephalosporins*	–	–	+/–	–	–	+
Erythromycin	+/–	–	–	–	–	+
Azithromycin/ Clarithromycin	+/–	–	–	–	–	+
Ciprofloxacin/ Levofloxacin	+/–	–	+	+	+	–
Clindamycin	+	+/–	–	+	–	+
Metronidazole	–	+	–	–	–	–
Doxycycline/ Tetracycline	+/–	–	+	+	+	+

*Includes agents such as Cephalexin, Cefitixime, Cefluroxime axetil, Cefprozil.
*Modified from van Winkelhoff, 2003; Slots, 1996

SYSTEMIC ANTIBIOTICS

Empirical antibiotic therapy may be used for periodontal diseases with known microbial etiologies, such as acute necrotizing ulcerative gingivitis, that is caused by anaerobic organisms that can be eradicated by systemic

Amoxicillin-Metronidazole combination therapy. However, even the most careful clinical examination cannot delineate the likely microbial pathogens in most types of periodontitis nor determine their susceptibility to various antibiotics. Microbiological analysis of subgingival plaque provides valuable information that will facilitate accurate diagnosis and optimal treatment planning. Slots (1996) suggested that therapeutic antibiotic therapy in periodontitis may be deferred until the results of culture, nucleic acid based diagnostic tests and antimicrobial susceptibility determination are available.

Systemic antibiotics should be used singly or in combination. Single agent antibiotic therapy in current periodontitis includes, Metronidazole, Clindamycin and Ciprofloxacin. Common combination therapy in periodontitis includes Metronidazole and Amoxicillin, and Metronidazole and Ciprofloxacin (van Winkelhoff, 2003).

LOCALLY DELIVERED ANTIBIOTICS

Controlled release devices that contain Tetracycline-HCL, Doxycycline, Minocycline, Metronidazole for direct pocket placement are commercially available in various countries. In controlled release drug delivery, the antimicrobial agent is released during an extensive period of time by zero order drug release kinetics (Rams and Slots, 1996). Locally applied tetracycline fibres have been shown to have some clinical benefits up to six months following therapy when used in conjunction with scaling and root planning (Newman *et al.*, 1994).

LOCALLY DELIVERED ANTISEPTICS

Mechanical root debridement to remove dental calculus is important in periodontal therapy, but is frequently inadequate in curing severe periodontal infections (Slots, 1996). To augment mechanical debridement, topical antiseptics may be used to kill periodontal pathogens during initial therapy and to suppress their repopulation during maintenance therapy. To have therapeutic value, pharmaceuticals must be delivered at an effective concentration and for an adequate length of time (Slots, 1996).

Subgingival antimicrobials can be administered either by non controlled delivery for fast acting pharmaceuticals or by controlled drug delivery devices for slow acting agents. Povidone-iodine is able to kill microorganisms rapidly enough to overcome the drug-diluting effect of gingival crevicular fluid and is effective when delivered by subgingival irrigation (Nakagawa and Saito, 1990). Sodium hypochlorite is a strong oxidizing agent that possesses numerous attractive properties for antiseptic use , include rapid bactericidal action, ease of use and very low cost. Irrigation with dilute (0.5%) sodium hypochlorite solution has been shown to cause significantly greater and longer lasting reduction in plaque and gingivitis than irrigation with water (Lang and Brecx, 1986). In contrast, Chlorhexidine digluconate mouthrinses at concentration of 0.1% to 0.2%, which have a long history of use in controlling supragingival plaque and gingivitis (Lang and Brecx, 1986), may require controlled- release delivery to exert effective killing of subgingival microorganisms (Caufield *et al.*, 1987). Also, chlorhexidine tends to be more bactericidal for gram positive organisms populating supragingival plaque than for gram negative species inhabiting periodontal pockets (Slots, 1997).

SOIL AS A SOURCE OF ANTIBIOTICS

The soil remains the best economical source that harbours the microorganisms which produce antibiotics and soil all around the world are continuously screened for the isolation of new antibiotic producing microorganisms. Antibiotics are microbial products or their derivatives that kill susceptible microorganisms and inhibit their growth even when used at low concentrations, while harming the host as little as possible (Aneja, 2003).

Penicillin, the first antibiotic to be used therapeutically was discovered from a mold *Penicillium notatum* by the Scottish physician, Alexander Fleming in 1929 and a new medical era was born. With the discovery of a means of producing large quantities of penicillin, and the successful treatment of once fatal infectious diseases, a concerted effort was begun to search for more antibiotics. The search was expanded to include other fungi, as well as algae, animals, bacteria and plants. Although none of the antibiotics discovered since penicillin have had the same notoriety, many are just as important and many are more commonly used than penicillin, especially the over- the – counter antibiotics used as topical applications. Most have been bacterial metabolites, or more specifically, Actinomycetes, which are mycelia producing bacteria that were once thought to be fungi. Unlike those antibiotics isolated from fungi, some of those from Actinomycetes did not require intravenous injections nor were they taken orally. Some examples of such antibiotics include bacitracin and neosporin (Singh, 1998).

Most of the antibacterial agents act on bacterial wall synthesis or protein synthesis. Peptidoglycan is one of the major wall targets because it is only found in bcteria. Some other compounds target bacterial protein synthesis, because bacterial ribosomes (termed 70 S ribosomes) are different from the ribosomes of humans (80S) and other eukaryotic organisms. Similarly the one antifungal agent (griseofulvin) binds specifically to the tubulin proteins that make up the microtubules of fungal cells, these tubulins are different from tubulins of humans (Singh, 1998).

As the putative periodontal and dental caries pathogens are developing resistance against the commercially known antibiotics, the need of hour is to explore soil microflora for discovering new bacterial, fungal and actinomycetes isolates in order to synthesize novel antibiotics that can be used in the treatment and control of not only the periodontitis and dental caries, but also of other dreadful diseases for which there has been no optimal antibiotic yet discovered.

THE FUTURE PROSPECTS

The last 25 years have brought unprecedented advances to our understanding of periodontal disease. In 1970, periodontitis was believed to effect most individuals over the age of 35 years, to progress steadily in an individual once initiated until teeth were lost, to be the primary cause of tooth loss in adults, to be caused by the bacterial mass accumulating on the tooth surface and subgingivally, and to involve the host in some fashion or another (Williams *et al.*, 1996). However, impressive research advances made in the recent years in the epidemiology of periodontal disease, the specific bacterial etiology of periodontal disease, and the immunoinflammatory mediators of periodontal tissue destruction have greatly altered our view of periodontal disease.

Impressive research into new ways to diagnose the periodontal disease is well underway. Investigators are seeking new ways to diagnose an individual's degree of risk for periodontal disease initiation susceptibility to disease progression, level of disease 'activity' and the likely response to treatment and recurrance of active disease. New diagnostic tests should greatly advance our ability to more accurately and specifically diagnose periodontal disease. The future also looks promising for new treatment strategies to slow or arrest periodontal disease progression. The bacterial specificity of periodontal disease etiology revealed since 1970 has logically led to the use of antibiotics in periodontitis treatment. In the late 1980's, the concept of locally delivering antibiotics to the periodontal pockets was introduced, and subsequent clinical trials have indicated that it is possible to reduce pocket depth and inflammation with tetracycline locally delivered to the periodontal pockets (Williams, 1996).

Likely, we have barely scratched the surface in studying the efficacy of locally delivery antimicrobial agents to alter the progression of periodontal disease. As new agents are developed and better delivery systems to the periodontal pockets are developed, the future should see a variety of antimicrobial agents available, which can slow periodontal disease progression. The future also holds promise for slowing periodontal disease progression by blocking inflammatory pathways important in periodontal tissue destruction. Clinical trials of Flubiprofen, Naproxen and Ketoprofen indicate that it is possible to slow periodontal disease progression with non steroidal anti-inflammatory drug which inhibit one destructive pathway. In addition, data from animal models indicate that chemically modified tetracycline as an inhibitor of collagenase can slow disease progression in animals (Williams, 1996). Again, we have likely only just begun to explore the wide range of molecular mediators of tissue destruction, which may be targeted for blocking and thereby slow or arrest periodontal disease progression.

Acknowledgement

Pranay Jain, working as a research scholar in the Department of Microbiology, Kurukshetra University, Kurukshetra is grateful to the honourable Vice Chancellor, Dr.. A.K. Chawla for providing the opportunity to carry out research work.

References

Aneja, K.R. 2003. *Experiments in Microbiology, Plant Pathology and Biotechnology.* 4th ed. New Age International Publishers, New Delhi, 608 pp.

Barr, C.E. 1995. Periodontal problems related to HIV-1 infection. *Adv. Dental Res.* **9**: 147-151.

Boghani, C.P 1991. Periodontal education. *Indian Soc. Periodontol.* **15**: 1.

Boghani, C.P. 2000. Progress of periodontal research and practice in India. In: *Progress of periodontal research and practice in Asian Pacific countries*. (Eds.) Bartold, P.M. and Ishikawa, L. Asian Pacific Society of Periodontology Bulletin.

Brady, L.J., Walker, C., Oxford, G.E. Stewart, C., Magnusson, I. and McArthur, W. 1996. Oral diseases, mycology and periodontal microbiology of HIV-1 infected women. *Oral Microbiol. Immunol.* **6**: 371-380.

Buchman, R., Muller, R.F., Van Dyke T.E. and Lange, D.E. 2003. Change of antibiotic susceptibility following periodontal therapy. A pilot study in aggressive periodontal disease. *J. Clin. Periodontol.* **30**: 222-229.

Caufield, P.W., Allen, D.N. and Childers, N.K. 1987. *In vitro* susceptibilities of suspected periodontopathic anaerobes as determined by membrane transfer assay. *Antimicrob. Agents Chemother.* **31**: 1989-1993.

D' Silva 1999. The effect of neem gel on plaque and gingivitis. *J. Indian Soc. Periodontol.* **2**: 63.

Ghayoumi, N., Chen, C., and Slots, J. 2002 . *Dialister pneumosintes*, a new putative periodontal pathogen. *J. Periodontol. Res.* **37**: 75-78.

Gilfillan, G.D., Sullivan, D.J., Haynes, K., Parkinson, T., Coleman, D.C. and Gow, N.A.R 1998. *Candida dubliniensis*: phylogeny and putative virulence factors. *Microbiology* **144**: 829-838.

Griffen, A.L., Becker, M.R., Lyons, S.R., Moeschberger, M.L. and Leys, E.J. 1998. Prevalance of *Porphyromonas gingivalis* and periodontal health status. *J. Clin. Microbiol.*. **36**: 3239-3242.

Haffajee, A.D. and Socransky. S.S. 1994. Microbial etiological agents of destructive periodontal disease. *Periodontol. 2000* **5**: 78-111.

Hazen, K. 1995. New and emerging yeast pathogens. *Clin. Microbiol. Rev*. **8**: 462-478.

Hoshino, E., Kurihara-Ando, N., Sato, I., and Iwaku, M. 1996. *In vitro* antibacterial susceptibility of bacteria taken from infected root dentine to a mixture of Ciprofloxacin, Metronidazole and Minocycline. *Int. Endod. J.* **29**: 125-130.

Jabra-Rizk, M.A., Baqui, A.A.M.A., Kelley, J.I., Falkler, Jr., W.A., Merz, W.G. and Meiller, T.F. 1999. Identification of *Candida dubliniensis* in a prospective study of patients in the United States. *J. Clin. Microbiol.* **37**: 321-326.

Jabra-Rizk, M.A., Ferreira, S.M.S. Sabet, M., Falkler, W.A., Merz, W.G. and Meiller, T.F. 2001. Recovery of *Candida dubliniensis* and other Yeasts from Human Immunodeficiency Virus associated periodontal lesions. *J. Clin. Microbiol*. **39**: 4520-4522.

Jarvinen, H., Pienihakkinen, K., Huovinen, P. and Tenovuo, J. 1995. Susceptibility of *Streptococcus mutans* and *Streptococcus sobrinus* to antimicrobial agents after short term oral chlorhexidine treatment. *Eur. J. Oral Sci.* **103**: 32-35.

Kamagate, A., Kone, D., Coulibaly, N.T. and Brou, E. 2001. A comparative study of various evaluation methods of the antibiotic sensitivity of strict anaerobic bacteria of the subgingival flora. *Odontostomatol Trop.* **24**: 9-12.

Lang, N.P. and Brecx, M.C. 1986. Chlorhexidine digluconate- an agent for chemical plaque control and prevention of gingival inflammation. *J. Periodontal Res.* **21**: 74-89.

Loe, H., Thellade, E. and Jensen, S.B. 1965. Experimental gingivitis in man. *J. Periodontol.* **36**: 177-187.

Madigan, M.T., Martinko, J.M. and Parker, J. 2000. *Brock Biology of Microorganisms.* 9th ed. Prentice Hall Inc. USA, 991 pp

Michael, G., Jorgensen, D.D.S. and Slots, J. 2002. The ins and outs of periodontal antimicrobial therapy. *J. California Dental Assoc.* At Website: www. nih.nlm.gov.in

Milazzo, I., Blandino, G., Nicolitti, G. and Speciale, A. 2003. Faropenem, a new oral penem: antibacterial activity against selected anaerobic and fastidious periodontal isolates. *J. Antimicrob. Chemother.* **51**: 721-725.

Mombelli, A., Schmid, B., Rutar, A. and Lang, N.P. 2002. Local antibiotic therapy guided by microbiological diagnosis. *Clin. Periodontol.* **29**: 743-749.

Nakagawa, T and Saito, A. 1990. Bacterial effects on subgingival bacteria of irrigation with a povidone-iodine solution (neojodin). *Bull. Tokyo Dent. Coll.* **31**: 199-203.

Neville, B.W., Damm, D.D., Allen, C.M. and Bouquot, J.E. 2002. *Oral and Maxillofacial Pathology*. 2nd ed. Elsevier. 844pp.

Newman, M.G., Komman, K.S. and Doherty, F.M. 1994. A 6-Month multi-center evaluation of adjunctive tetracycline fiber therapy used in conjunction with scaling and root planning in maintenance patients: clinical results. *J. Periodontol*. **65**: 685-691.

Phelan, J.A., Begg, M.D., Lamster, I.B., Gorman, J., Mitchell-Lewis, D., Bucklan, R.D. and el-Sadr, W..M. 1997. Oral candidiasis in HIV infection; predictive value and comparison of findings in injecting drug users and homosexual men. *J. Oral Pathol. Med.* **26**: 237-243.

Rams, T.E. and Slots, J. 1996. Local delivery of antimicobial agents in the periodontal pocket. *Periodontal 2000* **10**: 139-159.

Ross, P.W. and W.P. Holbrook 1994. Dental Plaque. In: *Clinical & Oral Microbiology*. Blackwell Scientific Publishers, Glasgow.

Saini, S., Aparna, Khanna, P., Mahajan, A., Arora, D.R., Saini, O.P. and Gupta, Naveen 2003. Antibiotic susceptibility of bacterial isolates in gingivitis and periodontitis. *J. Indian Dent. Assoc*. **74**: 216-222.

Schuster, G.S. 1999. Oral flora and pathogenic organisms. *Infectious Disease Clinics of North America.* **13**: 757-773.

Singh, B.D. 1998. *Biotechnology*. 1st ed. Kalyani Publishers, New Delhi. 574pp.

Slots, J. 1996. Systemic antibiotics in periodontics. *J. Periodontol.* **67**: 831-838.

Slots, J. 1997. Subgingival microflora and periodontal disease. *J. Clin. Periodontol*. **6**: 351-382.

Slots, J. and Jorgensen, M.G. 2000. Efficient antimicrobial treatment in periodontal maintenance care. *J.Am. Dent. Assoc.* **131**: 1293-1304.

Socransky, S.S. 1977. Microbiology of periodontal disease- present status and future considerations. *J. Periodontol.* **48**: 497-504.

van Winkelhoff, A.J. 2003. Microbiology in Diagnosis and treatment planning in periodontics. *Int. J. Dent. Hygiene* **1**: 131-137.

William, R.C., Beck, J.D. and Offenbacher, S.N. 1996. The impact of new technologies to diagnose and treat periodontal disease. A look to the future. *J. Clin. Periodontol.* **23**: 299-305.

Wilson, M., O'Connor, B. and Newman, H.N. 1990. Effect of Minocycline on subgingival plaque bacyeria. *J. Appl. Bacteriol.* **69**: 228-234.

Frontiers in Plant Sciences (Pages 515-532)
Eds. K.G. Mukerji, K.V.B.R. Tilak, S.M. Reddy, L.V. Gangwane, P. Prakash, I.K. Kunwar

Yeast–An Ideal Eukaryotic System for Modern Biology–An Overview

S. Ram Reddy and S.M. Reddy

ABSTRACT

E. coli has been the favourite organism for molecular studies particularly when the aim of the experiment is to study the basic features of molecular biology such as gene structure and function. However, eukaryotic gene functions cannot be investigated in a prokaryotic system for a variety of reasons thus necessitating the absolute requirement of eukaryotic system. Of many eukaryotic systems, especially *S. cerevisiae* proves to be an excellent system because it offers multitude of facilities for a spectrum of molecular investigations. It also satisfactorily fulfills the requirements and demands of biotechnologists. A range of plasmid vectors with varying characters, some of which can function as the shuttle vectors, have been developed which provide a wide range of options for the molecular biologists. Yeast is proving to be an ideal system to understand protein-protein and protein-ligand interactions. Similarly, it has become possible to trace out a gene for particular cellular function and conversely, connect a gene to a particular metabolic function. *Saccharomyces* continues to serve as an excellent eukaryotic model system and experimental tool in the future until an alternate and better system is discovered.

Keywords: *Saccharomyces cerevisiae,* plasmid vectors, recombinant proteins, promoters, two-hybrid system.

Prokaryotes like *E. coli* are excellent systems for cloning foreign genes including the eukaryotic genes and for producing heterologous (recombinant proteins) from cloned eukaryotic complimentary DNAs

Author's address: Department of Microbiology, Kakatiya University, Warangal-506 009, A.P.

(cDNAs). However, prokaryotic systems present certain constraints. For example, to determine the effects of experimentally induced changes on the function and expression of eukaryotic genes, the rearranged sequences must be taken out of the bacteria in which they were cloned and re-introduced into a eukaryotic organism. There are many functions unique to eukaryotic cells and they are to be studied and understood in eukaryotic systems only, e.g., localization of ATP-generating systems to mitochondria, association of DNA with histones, mitosis and meiosis and obligate differentiation of cells and thus the genetic control of such functions must be assessed in eukaryotic environment. Another limitation is that prokaryotes cannot be used for the expression of eukaryotic genes for the simple reason that eukaryotic proteins undergo a variety of post-translational modifications like proper folding, glycosylation, phosphorylation, formation of disulfide bridges etc. (Rai and Padh, 2001). Eukaryotic proteins that have been synthesized in bacteria either are unstable or lack biological activity. In addition, despite careful purification procedures, bacterial compounds that are toxic or that cause a rise in body temperature in humans and animals (pyrogens) may contaminate the final product. A human protein intended for medical use is generally required to be identical to the natural protein in its biochemical, biophysical and functional properties. To overcome these drawbacks, investigators have developed eukaryotic systems for cloning and expression of foreign genes. These include, yeast, insect cells, mammalian cells, slime molds (*Dictyostelium discoideum*) (Kitt and Possee, 1993; Kaufman, 1990 and Dittrich *et al.*, 1994). Nevertheless, it should be borne in mind that no single eukaryotic host cell system is ideal for gene cloning or expression. It is, therefore, necessary to examine different eukaryotic systems to find out the one that serves the targeted function.

Yeast–An Ideal Model

For a variety of reasons, the common yeast, *Saccharomyces cerevisiae* and more recently other yeasts like *Pichia pastoris* have been used extensively as an environment for molecular studies and expression host for production of heterologous proteins. They exhibit many ideal characteristics : yeast cells are much easier to grow and manipulate than plant and animal cells. They grow much rapidly (doubling time 90 min). Fortunately, the cellular biochemistry and regulation of yeasts are very like those of higher eukaryotes. For instance, the signal transduction and transcription regulation by mammalian steroid receptors can be mimicked in strains of *S. cerevisiae* expressing receptor sequences (Metzger *et al.*, 1988 and Schina and Yamomoto, 1988). There are many yeast homologues of human genes, e.g., those involved in cell division. Thus yeasts can be very good surrogate host for studying the structure and function of eukaryotic gene products. Added advantages are availability of complete genomic sequences of extranuclear stable high copy plasmids and ability to synthesize target proteins. Yeast is the favoured alternative host for expression of foreign proteins for research, industrial and medical use. As a food organism, it is highly acceptable for production of pharmaceutical proteins. It has been listed by the U.S. FDA as a generally recognized as safe (GRAS) organism. Several strong promoters have been isolated from this yeast and characterized, and a naturally occurring plasmid called the 2 mm plasmid, can be used in endogenous yeast expression vector systems. *S. cerevisiae* is capable of carrying out many post translational modifications. Further, it normally secretes so few proteins that, when it is engineered for extracellular release of a heterologous protein, the product can be readily purified (Glick and Pasternak, 2001).

At the moment this lower eukaryote proves to be a good compromise between bacteria on one side and mammalian cell lines on the other.

Genome

S. cerevisiae has a haploid chromosome number of 16. The entire *Saccharomyces* genome strain S 288C was sequenced and available on the *Saccharomyces* Genome Database (SGD) at http://genome-www.stanford.edu /saccharomyces. The site has a variety of tools for sequence analysis that are particularly useful for the *Saccharomyces* research, including gene and restriction maps of the chromosome. The site is inter-connected with genome databases for other genetic model organisms and sites for protein analysis.

Saccharomyces genome contains more than 13 million base pairs (13 Mbp) and more than 6000 open reading frames (ORFs). Each ORF is named to indicate the chromosome number (A for chromosome I to P for chromosome XVI), whether the gene is found on the right (R) or left (L) arm of the chromosome (i.e., to right or left of the centromere) and the ORF number. All ORFs are numbered on each chromosome arm starting at the centromere and going in the direction of the telomere regardless of which strand is the coding strand. Finally, the direction of the transcription is indicated by a W (for Watson, the upperstrand) or C (for Crick, the lower strand) depending on which strand is the coding strand. Thus ORF YBR 288C is found on the right arm of chromosome II. It is the 288 the ORF from the centromere and lower strand is the coding strand; that is, it is transcribed from right to left which for the right arm means towards the centromere.

Plasmid Vectors

Plasmid vectors, capable of self-replication, are required if the heterologous DNA introduced into the yeast is to be maintained in an extrachromosomal state. Four types of plasmid vectors have been developed viz., yeast episomal plasmids (YEPs), yeast replicating plasmids (Yrps), yeast centromere plasmids (Ycps) and yeast artificial chromosomes (YACs). Interestingly, all of them have certain common features : First, they all contain unique target sites for a number of restriction endonucleases, secondly, they can all replicate in *E. coli*, often at high copy number and finally they all employ markers that can be selected readily in yeast and which will often complement the corresponding mutations in *E. coli*. The four most commonly used markers are *His 3, Leu 2, Trp 1* and *Ura 3*. However, the two yeast selectable markers, *Ura 3* and *Lys 2*, have the advantage of offering both positive and negative selection. Positive selection is for complementation of auxotrophy, whereas, negative selection is for ability to grow on medium containing a compound that inhibits the growth of cells expressing the wild-type function.

2 μm plasmid

The development of cloning vectors for yeast has been stimulated greately by the discovery of a plasmid that is present in most strains of *S. cerevisiae*. This plasmid is infact an excellent basis for a cloning vector (Broach, 1982). It is 6 kb in size, which is ideal for a vector and exists in the yeast cell at a copy number between 70 and 200. Replication makes use of plasmid origin, several enzymes provided by the host cell and the proteins coded by the *REPI* and *REP2* genes carried by the plasmid.

Yeast Episomal Plasmids (YEps)

Vector derived from the 2 μm plasmid are called yeast episomal plasmids (YEPs). These were first constructed by Beggs (1978). Some YEPs contain the entire 2 μm plasmid others include just 2 μm origin of replication. An example of latter type is YEP 13, a shuttle vector. This plasmid is 6.3 kb in size, has a copy number of 50-100 per haploid cell and YEPs may remain independent of main chromosome or get integrated and in the later case recombination events may result in it being excised again.

Yeast Integrative Plasmids (YIps)

YIPs are basically bacterial plasmids carrying a yeast gene. An example is YIP5, which is pBR 322 with an inserted *URA 3* gene. A YIP cannot replicate as a plasmid as it does not contain any parts of the 2 μm plasmid and instead depends for its survival on integration into yeast chromosomal DNA.

Yeast Replicating Plasmids (YRps)

YRPs were initially constructed by Struhl *et al.* (1979). These are able to multiply as independent plasmids because they carry a chromosomal DNA sequence, known as autonomously replicating sequence, *ars* that includes an origin of replication (Huang and Kowalski, 1993). Replication origins are known to be located very close to several yeast genes, including one or two which can be used as selectable markers.

Yeast Centromere Plasmids (YCps)

Using a YRP vector, Clarke and Carbon (1980) isolated a number of hybrid plasmids containing DNA segments from around the centromere linked *leu 2, cdc 10,* and *pgk* loci on chromosome III of yeast. Similar to plasmids carrying *ars*, most of the recombinants were unstable in yeast. However, one of them was maintained stably through mitosis and meiosis. Structurally, plasmid-borne centromere sequences have the same distinctive chromatin structure that occurs in the centromere region of yeast chromosomes (Bloom and Carbon, 1982).

Yeast Artificial Chromosomes (YAC)

Szostak and Blackburn (1982) developed the first vector which could be maintained as a linear molecule, thereby mimicking a chromosome, by cloning yeast telomeres into YRp. Such vectors are known as yeast artificial chromosomes (YACs). The advantage with YACs is that their stability increases as the size of the insert increases. Thus they can be used as vehicles for genes that are too long to be cloned as a single fragment in an *E. coli* vector. YACs opened the way to the studies of the functions and modes of expression of genes that had previously been intractable to analysis by rDNA techniques YACs are equally important in the production of gene libraries.

Retrovirus Like Vectors

The genome of *S. cerevisiae* contains 30-40 copies of a 5.9 kb mobile genetic element called Ty (Fulton *et al.*, 1987). This transposable element shares many structural and functional features with retroviruses and the copia element of *Drosophila*. Ty consists of a central region containing two long open reading frames (ORFs) flanked by two identical terminal 334 bp repeats called delta. Each delta element contains a promoter as well as sequences recognized by the transposing enzyme. New copies of the transposon arise by a replicative process involving reverse transcription that can tranpose to many sites in the host DNA (Boeke *et al.,* 1988 and Jacobs *et al.,* 1988).

The properties and advantages and disadvantages with the above described vectors are summarized in Table 1.

Selection of the Vector

There are a wide range of facilities offered by the yeasts for cloning the genes. The first of these relates to the potential use of yeast as a cloning host for the overproduction of proteins of commercial value. Yeasts offer a number of advantages as discussed earlier. A second reason for cloning genes in yeast is the ability to clone a large piece of DNA. YACs offer a convenient way to clone large DNA fragment targeted for sequencing (Burk *et al.,* 1987). For many biologists the primary purpose of the cloning is to understand what particular genes do *in vivo*. Thus, most of the applications of yeast vectors have been in the surrogate genetics of the yeast. The availability of different kinds of vectors with different properties enables the yeast geneticists to perform manipulations in yeast, like those long available to *E. coli* geneticists with their sex factors and transducing phages. Thus cloned genes can be used in conventional genetic analysis (Orr-Weaver *et al.,* 1981; Struhl, 1983 and Stearns *et al.,* 1990).

Inspite of wide range of facilities offered by yeasts, no single strain and/ or vector is suitable for all purposes. One has to select an appropriate combination to suit to his requirements. Three factors come into play when deciding which type of yeast vector is most suitable for particular cloning experiment. The first of these is transformation frequency, a measure of the number of transformants that can be obtained per microgram of plasmid DNA. A high transformation frequency is necessary if a large number of recombinants are needed, or if the starting DNA is in short supply. YEps have the highest transformation frequency. YRps are also quite productive, but a YIp yields less than 1000 transformants per μg. The copy number figures are also important if the objective is to obtain protein from the cloned gene, as the more copies there are of the gene the greater the expected yield of the protein product. YEps and YRps have the highest copy numbers : 20-50 and 5-100 respectively. In contrast, a YIp is usually present at just one copy per cell. Finally, stability of the recombinants. YIps produce very stable recombinants, on the other hand YRp and Yep recombinants are extremely unstable. Thus, a YIp is the vector of choice if the needs of the

TABLE 1 *Properties of the different yeast vectors (from Primrose et al., 2001)*

Vector	*Transformation frequency*	*Copy no./cell*	*Loss in non-selective medium*	*Disadvantages*	*Advantages*
YIp	10^2 transformants per μg DNA	1	Much less than 1% per generation	1. Low transformation freqnecy 2. Can only be recovered from yeast by cutting chromosomal DNA with restriction endonuclease which does not cleave original vector containing cloned gene	1. Of all vectors, this kind give most stable maintenance of cloned genes. 2. An integrated YIp plasmid behaves as an oridinary genetic marker, e.g., a diploid heterozygous for an integrated plasmid segregates the plasmid in a Mendelian fashion. 3. Most useful for surrogate genetics of yeast.
YEp	10^3-10^5 transformants per μg DNA	25-200	1% per generation	Novel recombinants generated *in vivo* by recombination with endogenous 2 μm plasmid	1. Readily recovered from yeast. 2. High copy number. 3. High transformation frequency. 4. Very useful for complementation studies.
YRp	10^4 transformants per μg DNA	1-20	Much greater than 1% per generation but can get chromosomal integration	Instability of transformants	1. Readily recovered from yeast. 2. High copy number. Note that the copy number is usually less than that of YEp vectors but this ay be useful if cloning gene whose product is deleterious to the cell if produced in excess. 3. High transformation frequency. 4. Very useful for complementation studies. 5. Can integrate into the chromosome.
YCp	10^4 transformants per μg DNA	1-2	Less than 1%	Low copy number makes recovery from yeast more difficult than that with YEp or YRp vectors	1. Low copy number is useful if product of cloned gene is deleterious to cell. 2. High transformation frequency. 3. Very useful for complementation studies. 4. At meiosis generally shows Mendelian segregation.
YAC		1-2	Depends on length: the longer the YAC the more stable it is	Difficult to map by standard techniques	1. High-capacity cloning system permitting DNA molecules greater than 40 kb to be cloned. 2. Can amplify large DNA molecules in a simple genetic background.
Ty	Depends on vector used to introduce Try into cell	-20	Stable, since integrated into chromosome	Needs to be introduced into cell in another vector.	Can get amplification following chromosomal integration.

experiment dictate that the recombinant yeast cells must retain the cloned gene for longer periods in culture.

Transformation in Yeasts

Yeasts and other fungi are not naturally transformable and artificial means have to be used for introducing foreign DNA. Glick and Pasternak (2001) identifies three basic techniques that are commonly used to transform yeasts. Exogenous DNA can be added to yeast cells that (i) have had their cell walls removed either chemically or enzymatically (protoplasts), (ii) have been treated with lithium acetate, or (iii) are subjected to electroporation. DNA can also be introduced into yeast by conjugation. Heinemann and Sprague (1989) and Siroski *et al.* (1990) found that enterobacterial plasmids, such as R 751 (Inc PB) and F (Inc F), could facilitate plasmid transfer from *E. coli* to *S. cerevisiae* and *Schizosaccharomyces pombe*. A part of the T_i plasmid of *Agrobacterium tumefaciens* can also be conjugately transferred to protoplasts of *S. cerevisiae* (Bundock *et al.*, 1995).

In transformation process, unlike in other organisms where non-homologous integrations are more common, homologous recombinations are common in yeast (Kudla and Nicolas, 1992). This is probably due to the integration of the plasmid in novel locations of the genomes i.e., illegilimate recombination (Hinnen *et al.,* 1978). Kudla and Nicolas (1992) have constructed a novel vector which allows integration of a cloned DNA sequence at different sites in the genome. This feature is provided by the inclusion in the vector of a repeated yeast *sigma* sequnce present in approximately 20-30 copies per genome and spread over most or all of the 16 chromosomes. If the T-DNA from T_i plasmid is transferred to yeast, it too will insert in different parts of the genome by illegitimate recombination (Bundock and Hooykaas, 1996).

Expression of Cloned Genes

As has already been pointed out, for a variety of reasons, *S. cerevisiae* has been used for the production of recombinant proteins by inserting the gene of interest in a suitable vector (Table-2). There are three classes of *S. cerevisiae* expression vectors : (i) episomal vectors, (ii) integrating vectors and (iii) yeast artificial chromosomes (YACs). Of these, episomal vectors have been used extensively for the production of either intra or extra cellular heterologous proteins. The promoter is the most important component of an expression vector. This is because the promoter controls the very first stage of gene expression (Brown, 2001) and determines the rate at which mRNA is synthesized. The amount of recombinant protein obtained, therefore, depends to a great extent on the nature of the promoter carried by the expression vector. In *S. cerevisiae* cloned genes are often placed under the control of GAL promoter (Fig. 1), which is normally upstream of the gene coding for galactose epimerase, an enzyme involved in the metabolism of galactose. The GAL promoter is induced by galactose, providing a straight forward system for regulating expression of a cloned foreign gene. The other useful promoters are PHO_5 and CUPI (Gellissen and Hollenberg, 1997). Promoters for other yeasts are shown in Table 3. Most yeast expression vectors also carry a termination sequence from *S. cerevisiae* gene because animal termination signals do not work effectively in yeast.

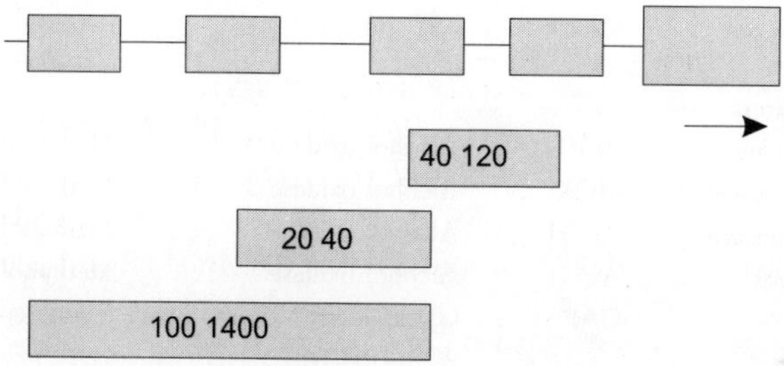

Fig. 1 Structure of typical promoters

TABLE 2 *Recombinant proteins produced by S. cerevisiae expression system*

Vaccines
Hepatitis B virus surface antigen
Malaria circumsporozoite protein
HIV-1 envelope protein
Diagnostics
Hepatitis C virus protein
HIV-1 antigens
Human therapeutic agents
Epidermal growth factor
Insulin
Insulin like growth factor
Platelet derived growth factor
Proinsulin
Fibroblast growth factor
Granulocyte-macrophage colony stimulating factor
α_1 antitrypsin
Blood coagulation factor XIIIa

Although, *S. cerevisiae* retains the loyalty of many molecular biologists, there are other microbial eukaryotes, that might be equally if not more effective in recombinant protein synthesis. In particular, *Pichia pastoris*, a

TABLE 3 *Common fungal promoters used for manipulation of gene expression (from Primrose et al., 2001)*

Species	*Promoter*	*Gene*	*Regulation*
General			
S. cerevisiae	PGK	Phosphoglycerate kinase	Glucose-induced
	GAL1	Galactokinase	Galactose-induced
	PHOS	Acid phosphatase	Phosphate-repressed
	ADH2	Alcohol dehydrogenase II	Glucose-repressed
	CUP1	Copper metallothionein	Copper-induced
	MFα1	Mating factor α1	Constitutive
Candida albicans	MET3	ATP sulphur lyase	Repressed by methionine and cysteine
Methanol utilizers			
Candida boidnii	AOD1	Alcohol oxidase	Methanol-induced
Hansenula polymorpha	MOX	Alcohol oxidase	Methanol-induced
Pichia methanolica	AUG1	Alcohol oxidase	Methanol-induced
Pichia pastoris	AOX1	Alcohol oxidase	Methanol-induced
	GAP	Glyceraldehyde-3-phosphate dehydrogenase	Strong constitutive
	FLD1	Formaldehyde dehydrogenase	Methanol- or methyl amine induced
	PEX8	Peroxin	Methanol-induced
	YPT1	Secretory GTPase	Medium constitutive
Lactose utilizer			
Kluyveromyces lactis	LAC4	β-Galactosidase	Lactose-induced
	PGK	Phosphoglycerate kinase	Strong constitutive
	ADH4	Alcohol dehydrogenase	Ethanol-induced
Starch utilizer			
Schwanniomyces occidentalis	AMY1	α-Amylase	Maltose- or starch-induced
	GAM1	Glucoamylase	Maltose- or starch-induced
Xylose utilizer			
Pichia stipitis	XYL1		Xylose-induced
Alkane utilizer			
Yarrowia lipolytica	XPR2	Extracellular protease	Petone-induced
	TEF	Translation elongation factor	Strong constitutive
	RPS7	Ribosomal protein S7	Strong constitutive

second species of yeasts, is able to synthesize large amount of recombinant protein (upto 30% of the total cell protein) and its glycosylation abilities are very similar to those of animal cells. This methylotrophic yeast has proved to be extremely popular as hosts for the over expression of heterologous proteins. The reasons for this are : (i) the alcohol oxidase (AOX1) promoter is one of the strongest and most regulatable promoters known. (ii) it is possible to stably integrate expression plasmids at specific sites in the genome in either single or multiple copies. (iii) the strains can be cultivated to very high density. To date over 300 foreign proteins have been produced in *P. pastoris* (Cereghino and Cregg, 1999 and 2000).

Specialist Vectors

Some specialist yeast vectors have been designed to incorporate the useful features. For example, an f1 origin to permit sequencing of inserts production of the cloned gene product as a purification fusion. These vectors include YES of *S. cerevisiae* and pPICZ of *Pichia*. Two aspects of these vectors need explanation i.e., secretion and surface display.

The nascent proteins synthesized after necessary processing may accumulate in the cytoplasm of the host or may be secreted into the growth medium. The term direct expression is often used to describe vector systems producing proteins that accumulate in the cytoplasm. A number of research groups have developed such direct expression vector systems for yeast and they all have the same basic features. The production of the human enzyme superoxide dismutase illustrates the process of heterologous gene expression in *S. cerevisiae.* Polypeptides defined for secretion have a hydrophobic amino-terminal extension, which is responsible for translocation to the ER (Blobel and Dobber Stein, 1975). Further, glycosylation of the polypeptide is a prerequisite for secretion. Secretion system must be used for recombinant proteins that require either O-linked or N-linked sugars for biological activity. To facilitate protein secretion in yeasts, the pre-pro-X-factor, or leader peptide (signal peptide), coding sequences of the yeast mating type factor $\alpha 1$ gene is cloned in front (just upstream) of the cDNA that encodes the desired protein. This gene fusion creates a protein that can be efficiently secreted by yeast (Loison *et al.,* 1988). Additional strategies have been devised to enhance the secretion of recombinant protein by *S. cerevisiae* strains. For example, it was of interest to determine if the over production of a naturally occurring secretory system enzyme such as protein disulfide isomerase, which promotes the proper folding of proteins during secretion, would increase the release of a recombinant protein.

Yeast Surface Display

S. cerevisiae can be used to elucidate and dissect the function of a protein in a manner similar to phage display systems. Both of these can be used to detect protein-ligand interactions and to select proteins with altered binding capacity (Shusta *et al.,* 1999). Yeast surface display makes use of the cell surface receptor α-agglutinin (Aga) which is a glycoprotein consisting of two subunits viz., Aga1p and Aga2p. In order to achieve surface display, the appropriate gene is inserted at the C-terminus of a vector borne Aga2p gene under the control of the GAL promoter. The construct is then transformed into a yeast strain carrying a chromosomal copy of the Aga I p gene, also under the control of the GAL I promoter. If the cloned gene has been inserted in the correct translational reading frame, its gene product will be synthesized as a fusion with Aga2p subunit. The fusion product will associate with the Aga1p subunit within the secretary pathway and be exported to the cell surface (Boder and Wittrup, 1997).

Detecting Protein—Protein Interaction

In the past, protein-protein interactions have been carried out by *in vitro* studies based on their interactions under non-physiological conditions. This kind of experiments have resulted in non-specific interactions and thus had led to misinterpretation of results. To avoid such artefactual results, Stanley Fields *et al.* (1994) at SUNY, Stony Brook have developed yeast two hybrid system which could test the interaction of two cloned genes through their protein products under physiological conditions in *S. cerevisiae.* Chen *et al.* (1991) have made use of the properties of the GAL 4 protein to develop a method for detecting protein-protein interactions. The GAL 4 protein has separate domains for the binding to UAS DNA and for transcriptional activation. Plasmids were constructed which encode two hybrid proteins. One carries the DNA binding domain-gene X chimera (i.e., bait plasmid) and another carries transcriptional activation domain-gene Ychimera in "prey" (i.e., prey plasmid). When these two (bait and prey) plasmids were introduced into an appropriate strain of budding yeast, *S. cerevisiae*, hybrid proteins were synthesized, DNA binding domain-X protein hybrid will recruit transcriptional activation domain-Y protein hybrid for transcriptional activation of a downstream reporter gene, if protein X and Y interact. If there was no interaction between X and Y proteins, the downstream gene will not be activated. This method is known as the two-hyrbid system, which has become a major tool in the study of protein-protein interactions and protein ligand interactions. Colas and Brent (1998) and Uetz and Hughes (2000) have provided the excellent reviews on the subject.

A major operational problem with the two hybrid system have been the high frequency of false positives i.e., cells in which the reporter gene is active even though the bait and prey do not interact in nature. Nevertheless, solutions to these problems were worked out (James *et al.*, 1996 and Xu *et al.*, 1997). To overcome other limitations different improved versions of the two hybrid system were developed recently post translational modifications

(Osborne *et al.*, 1996), protein peptide interactions (Yang *et al.*, 1995; Colas *et al.*, 1996; Geyer *et al.*, 1999; Norman *et al.*, 1999; Staglijar *et al.*, 1996), three component systems (Lee *et al.*, 1995; Chiu *et al.*, 1994; Tomashek *et al.*, 1996), bait and hook (Sen Gupta *et al.*, 1999; Licitra and Liu, 1996) and reverse two-hybrid system (Vidal *et al.*, 1996a,b; Huang and Schreiber, 1997).

Variations of the two-hybrid analysis method have also been developed to explore DNA-protein and RNA-protein interactions. One hybrid analysis is used to identify proteins that interact with a specific DNA sequence. This sequence is inserted into the promoter of reporter gene. For example, in place of the *Lex A* operator site of the *lac Z* reporter. Any hybrid proteins capable of binding to the DNA sequence of interest will activate the reporter gene expression (Michels, 2002). Three hybrid analysis was developed to study the RNA-protein interaction (Kraemer *et al.*, 2000). The third hybrid construct in this method is an RNA hybrid molecule. The bait hybrid protein binds one portion of this hybrid RNA. The other portion of the hybrid RNA is the sequence of interest. Fusion proteins of the 'prey' library are screened for their ability to bind to this RNA sequence of interest that is tethered to the reporter gene promoter via the bait hybrid system.

Identifying Genes for Cellular Functions

Genome-sequencing project enabled us to sequence the genes coding for almost all biochemical activities of *S. cerevisiae*. Logically, the next task is to connect specific biochemical activities with particular gene. *S. cerevisiae* genome consists of about 6000 genes. Of these, some 3600 could be assigned a function either on the basis of previous studies that had been carried out with yeast or because of the yeast gene homology to a gene that had been studied in another organism. This left 2400 genes as orphan genes. Martzen *et al.* (1999) have developed a rapid and sensitive method for determine the function of orphan genes by using microtitre method. The startng point was a yeast vector containing the glutathione S-transferase (GST) gene under the control of the CUPI promoter. A large number of different DNA fragments bearing yeast ORFs were cloned in this vector to generate GST-ORF fusions. In all, 6144 constructs were made and these were transformed back into *S. cerevisiae*.

Assigning Gene Functions

Determining the function of a particular gene product usually involves determining the expression profile of the gene, the subcellular location of the protein, and the phenotype of a null strain lacking the protein. Conditional alleles of the genes are often created as an additional tool. Ross-Macolonald *et al.* (1997) have developed a multifunctional, transposon-based system that simultaneously generates constructs for all of the above analyses and is suitable for mutagenesis of any given *S. cerevisiae* gene. The transposons used by them carry a reporter gene (β-galactosidase or green fluorescent protein) lacking a promoter. In-frame fusions between a yeast coding region and the transposon can be detected by β-galactosidase activity or fluorescence. The transposon insertion creates a truncation of the gene, thereby generating a null phenotype. The method is also adaptable to genome-wide analyses, using a high-throughput screening methodology (Ross-Macdonald *et al.*, 1999). Winzeler *et al.* (1999) employed a polymerase chain reaction methodology for the large scale mutational analysis of the yeast genome by deleting the genes individually. For each gene deletion cassette was constructed that contained a kanamycin resistance gene, two 'molecular bar codes' and yeast sequences homologous to the upstream and downstream flanking sequences. This construct was amplified by PCR and transformed into diploid yeast strain, where the deletion cassette replaces one of the two chromosomal gene sequences. The diploid strains were then allowed to sporulate and those haploid segregants that are viable are recovered.

The applications of genome-wide mutagenesis techniques to the understanding of the interactions of different cell components have been excellently reviewed by Delneri *et al.* (2001); Vidan and Snyder (2001).

Conclusions

Despite the absence of multicelled development, yeast, are not very different from other eukaryotes at the genetic level. In many respects it is superior to other eukaryotic model system like *Coenorhabditis elegans*, *Drosophila melanogaster*, mammalian cells and *Arabidopsis thalliana*.

Two yeasts, *S. cerevisiae* and *Pichia pastoris,* are proving to be promising eukaryotic systems for molecular studies. Many of the problems encountered with prokaryotic systems have been solved with the yeast system. Though the yeasts are naturally non-transformable a number of techniques have been developed to transfer the foreign genes. The greatest advantage offered by yeasts, but not available with other eukaryotic systems, is the availability of a large number of extrachromosomal plasmid vector to suit to variety of requirements. Through YAC it has now become possible to clone a large segments of DNA for cloning purpose. Human genome project (HGP) primarily employes the yeast YAC for the cloning and development of gene libraries of cloned genes.

S. cerevisiae also provides some excellent promoters for gene expression studies. GAL, an inducible promoter and AOXI a strong and regulatable promoter are assets of the system offered to molecular biologists. Proteomics, by definition large scale characterization of entire protein complement of a cell or a tissue or an organ makes use of yeast two hybrid system. The initial problems faced through systems are now solved to a large extent through improved versions of this system. Similarly, post genomics or functional genomics which is the analysis of a genome sequence to locate the genes, control sequences and other interesting features followed by various experiments to determine the functions of any unknown genes that are discovered owes a lot to the yeast system. *Saccharomyces* is an excellent model for the more complex systems and one that is far more amenable to genetic analysis, atleast in the near future. It is important to keep in mind that gene functions identified in*Saccharomyces* define the components to build a eukaryotic cell. Gene functions involved in the assembly of cells into multicelled organism will be missing from the *Saccharomyces* protein set. Nevertheless, if one is interested in basic eukaryotic cell functions, *Saccharomyces* is a very valuable tool. By and large, yeast especially *S. cerevisiae* has become an indispensable system to work by molecular biologists.

References

Beggs J.D. (1978). Transformation of yeast by a replicating hybrid plasmid. *Nature* 275: 104-109.

Blobel G. & Dobberstein B. (1975). Transfer of proteins across membranes. I. Presence of proteolytically processed and unprocessed nascent immunoglobulin light chains on membrane bound ribosomes of murine myeloma. *J. Cell Biol.*, 67: 835-851.

Bloom K.S. & Carbon J. (1982). Yeast centromere DNA is a unique and highly ordered structure in chromosomes and small circular minichromosomes. *Cell*, 29: 305-317.

Boder E.T. & Wittrup K.D. (1997). Yeast surface display for screening combinatorial polypeptide libraries. *Nature Biotechnol.* 15: 553-558.

Boeke J.D., Xu H & Fink G.R. (1988). A general method for the chromosomal amplification of genes in yeast. *Science*, 239: 280-282.

Broach J.R. (1982). The yeat 2 mm circle. *Cell*, 28: 203-204.

Brown T.A. (2001). Gene-cloning and DNA analysis, 4th ed., Blackwell Science.

Bundock P. & Hooykaas P.J.J. (1996). Integration of *Agrobacterium tumifaciens* T-DNA in the *Saccharomyces cerevisiae* genome by illegitimate recombination. *Proc. Nat. Acad. Sci.*, USA 93: 15272-15275.

Bundock P., Den Dulk-Ras A., Beijersbergen A. & Hooykaas P.J.J. (1995). Trans-kingdom T-DNA transfer from *Agrobacterium tumefaciens* to *Saccharomyces cerevisiae*. EMBOJ, 14: 3206-3214.

Burke D.T., Carle G.F. & Olson M.V. (1987). Cloning of large segments of exogenous DNA into yeast by means of artificial chromosome vectors. *Science* 236: 806-813.

Cereghin J.L. & Cregg J.M. (1999). Applications of yeast in biotechnology : protein production and genetic analysis. *Curr. Opin. Biotechnol.*, 10: 422-427.

Cereghino J.L. & Cregg J.M. (2000). Heterologous protein expression in the methylotrophic yeast *Pichia pastoris*. *FEMS Microbiol. Rev.,* 24: 45-66.

Chen K. & Arnold F.H. (1991). Enzyme engineering for non-aqueous solvents : random mutagenesis to enhance activity of subtilisin E in polar organic media. *Biotechnology* 9: 1073-1077.

Chiu M.I., Katz H. & Berlin V. (1994). RAPT1, a mammalian homolog of yeast Tor, interacts with the FKBP 12/rapamycin complex. *Proc. Nat. Acad. Sci. USA*, 91: 12574-12578.

Clarke L. & Carbon J. (1980). Isolation of a yeast centromere and construction of functional small circular chromosomes. *Nature,* 287: 504-509.

Colas P. & Brent R. (1998). The impact of the two hybrid and related methods on Biotechnology. *Trends Biotechnol.* 16: 355-363.

Colas P., Cohen B., Jessen T. et al. (1996). Genetic selection of peptide aptamers that recognize and inhibit cyclindependent-kinase 2. *Nature,* 380: 548-550.

Delneri D., Brancia F.L. & Oliver S.G. (2001). Towards a truly integrative biology through the functional genomics of yeast. *Curr. Opin. Biotechnol.* 12: 87-91.

Dittrich, W., Williams, K.L. and Slade, M.B. (1994) Biotechnlogy (NY), 12: 614-618.

Fields S. & Sternglanz R. (1994). The two-hybrid system : an assay for protein-protein interaction. *Trends Genet.,* 10: 286-292.

Fulton A.M., Adams S.E., Mellor J., Kingsman S.M. & Kingsman A.J. (1987). The organisatin and expression of the yeast transposon, *Ty. Microbiol. Sci.* 4: 180-185.

Gellissen G. & Hollenberg C.P. (1997). Applications of yeast in gene expression studies : a comparison of *Sacchoromyces cerevisiae, Hansenula polynorpha* and *Kluveromyces lactis* – A review. *Gene,* 190: 87-97.

Geyer C.R., Colman Lerner A. & Brent R. (1999). Mutagenesis by peptide aptomers identifies genetic network members and pathway connections. *Proc. Nat. Acad. Sci. USA*, 96: 8567-8572.

Glick B.R., Pasternak J.J. (1998). Molecular Biotechnology ASM Press, Washington, D.C.

Heinemann J.A. & Sprague G.F. (1989). Bacterial conjugative plasmids mobilize DNA transfer between bacteria and yeast. *Nature* 340: 205-209.

Hinnen A., Hicks J.B. & Fink G.R. (1978). Transformation of yeast. *Proc. Nat. Acad. Sci. USA,* 75: 1929-1933.

Huang J. & Schreiber S.L. (1997). A yeast genetic system for selecting small molecule inhibitors of protein-protein interactions in nanodroplets. *Proc. Nat. Acad. Sci. USA*, 94: 13396-13401.

Huang R.Y. & Kowalski D. (1993). A DNA unwinding element and an ARS consensus comprimise a replication origin within a yeast chromosome. *EMBOJ*, 12: 4521-4531.

Jacobs E., Dewerchin M. & Boeke J.D. (1988). Retrovirus like vectors for *Saccharomyces cereisiae* : integration of foreign genes controlled by efficient promoters into yeast chromosomal DNA. *Gene*, 67: 259-269.

James P., Halladay J. & Craig E.A. (1996). Genomic libraries and a host strain designed for highly efficient two-hybrid selection in yeast. *Genetics*, 144: 1425-1436.

Kaufman R.J. (1990a). Strategies for obtained high level expression in mammalian cells. *Technique*, 2: 221-236.

Kaufman R.J. (1990b). Vectors used for expression in mammalian cells. *Methods Enzymol,* 185: 487-511.

Kitts P.A. & Possee R.D. (1993). A method for producing recombinant baculovirus expression vectors at high frequency. *Biotechniques*, 14: 810-817.

Kraemer B., Zhang B., Sen Gupta S., Fields S. and Wichens M. (2000). Using the yeast three hybrid system to detect and analyze RNA-protein interactions. *Meth. Enzymol.* 328: 297-321.

Kudla B. & Nicolas A. (1992). A multisite integrative cassette for the yast *Saccharomyces cerevisiae*. *Gene,* 119: 49-56.

Lee J.W., Choi H.S., Gyuris J., Brent R. & Moore D.D. (1995). Two classes of proteins dependent on either the presence or absence of thyroid hormone for interaction with the thyroid hormone receptor. *Mol. Endocrinol,* 9: 243-254.

Licitra E.J. & Liu J.O. (1996). A three-hybrid system for detecting small ligand protein receptor interactions. *Proc. Nat. Acad. Sci. USA*, 93: 12817-12821.

Loison G., Findeti, A., Bernard S., Nguyen-Juilleret M., Marquet M., Riehl Bellon N., Carvallo D., Guella-Santos L., Brown S.W., Courtney M., Roitsch C. and Lemaine Y. (1988). Expression and secretion in *S. cerevisiae* of biologically active leach hirudin. *Biotechnology*, 6: 72-77.

Martzen M.R., McCraith S.M., Spinelli S.L. et al. (1999). A biochemical genomics approach for identifying genes by the activity of their products. *Science*, 286: 1153-1155.

Metzger D., White J.H. & Chambon P. (1988). The human estrogen receptor functions in yeast. *Nature*, 334: 31-36.

Michels C.A. (2002). Genetic techniques for biological research, John Wiley & Sons Ltd.

Norman T.C., Smith D.L., Sorges P.K. et al. (1999). Genetic selection of peptide inhibitors of biological pathways. *Science,* 285: 591-595.

Orr-Weaver T.L., Szostak J.W. & Rothstein R.L. (1981). Yeast transformation : A model system for the study of recombination. *Proc. Nat. Acad. Sci. USA,* 78: 6354-6358.

Osborne M.A., Zenner G., Lubinus M. et al. (1996). The inositol 5'-phosphotase 5H1P binds to immunoreceptor signaling motifs and responds to high affinity IgE receptor aggregation. *J. Biol. Chem.*, 271: 29271-29278.

Rai M. and Padh H. (2001). Expression systems for production of heterologous proteins. *Curr. Sci.,* 80: 1121-1128.

Ross-macdonald P. & 17 others (1999). Large-scale analysis of the yeast genome by transposon tagging and gene disruption. *Nature*, 402: 413-418.

Ross-Macdonald P., Sheehan A., Roeder G.S. & Snyder M. (1997). A multipurpose transporon system for analyzing protein production, localization and further in *Saccharomyces cerevisiae*. *Proc. Nat. Acad. Sci. USA*, 94: 190-195.

Schena M. & Yamamoto K.R. (1988). Mammalian glucocorticoid receptor derivatives enhance transcription in yeast. *Science,* 241: 965-967.

SenGupta D.J., Wickens M. & Fields S. (1999). Identification of RNAs that bind to a specific protein using the yeast three hybrid system. *RNA*, 5: 596-601.

Shusta E.V., Van Antwerp J. & Wittrup K.D. (1999). Biosynthetic polypeptide libraries. *Curr. Opin. Biotechnol.,* 10: 117-122.

Sikorski R.S., Michuad W., Levin H.L. Boeke J.D. & Hieter P. (1990). Trans-kingdom promiscuity. *Nature*, 345: 581-582.

Staglijar I., Borquin J.P. & Schaffner W. (1996). Use of the two-hybrid system and random sonicated DNA to identify the interaction domain of a protein. *Biotechnfiques* 21: 430-432.

Stearns T., Ma H. & Botstein D. (1990). Manipulating the yeast genome using plasmid vectors. *Methods Enzymol.,* 185: 280-297.

Struhl K. (1983). The new yeast genetics. *Nature*, 305: 391-397.

Struhl K., Stinchcomb D.T., Schorer S. & Davis R.W. (1979). High frequency transformation of yeast : autonomous replication of hybrid DNA molecules. *Proc. Nat. Acad. Sci. USA*, 76: 1035-1039.

Szostak J.W. & Blackburn E.H. (1982). Cloning yeast telomeres on linear plasmid vector. *Cell*, 29: 245-255.

Tomashek J.J., Sonnenburg J.L., Artimovich J.M. & Klionsky D.J. (1996). Resolution of subunit interactions and cytoplasmic subcomplexes of the yast vacuolar proton translocating ATPase. *J. Biol. Chem.* 271: 10397-10404.

Uetz P. & Hughes R.E. (2000). Systematic and large scale two-hybrids screens. *Curr. Opin. Microbiol.,* 3: 303-8.

Vidal M., Brachmann R.K., Fatteaey A., Harlow E. & Boeke J.D. (1996a). Reverse two-hyrbid and one-hybrid systems to detect dissociation of protein-protein and DNA protein interaction. *Proc. Nat. Acad. Sci. USA,* 93: 10315-10320.

Vidal M., Braun P., Chen E., Boeke J.D. & Harlow E. (1996b). Genetic characterization of a mammalian protein-protein interaction domain by using a yeast reverse two-hyrbid system. *Proc. Nat. Acad. Sci. USA*, 93: 10321-10326.

Vidans & Snyder M. (2001). Large-scale mutagenesis : Yeast genetics in the genome era. *Curr. Opin. Biotechnol.,* 12: 28-34.

Winzeler E.A. & 25 others (1999). Functional characterization of the *S. cerevisiae* genome by gene deletion and parallel analysis. *Science,* 285: 901-906.

Xu, C.W., Mendelsohn A. & Brent R. (1997). Cells that register logical relationships among proteins. *Proc. Nat. Acad. Sci. USA*, 94: 12473-12478.

Yang M., Wu Z. & Fields S. (1995). Protein-peptide interactions analysed with the yeast two-hybrid system. *Nucl. Acids Res.,* 23: 1152-1156.

SECTION 5

BIOLOGICAL CONTROL

Frontiers in Plant Sciences (Pages 535-548)
Eds. K.G. Mukerji, K.V.B.R. Tilak, S.M. Reddy, L.V. Gangwane, P. Prakash, I.K. Kunwar

Application of Microorganism in the Biological Control of Plant Pathogens

A.B. Ade, L.V. Gangawane and V.C. Khilare

ABSTRACT

Now a days biological control of plant pathogens is advocated. The *Rhizobium*, a symbiotic N_2 fixer can be used as biological control. The transgenic antagonistic *Rhizobium* has been developed in this laboratory although the literature indicates the application of many microorganisms for the integrated management of plant pathogens.

Key words: *Pseudomonas fluorescens*, transgenic antagonistic *Rhizobium*, *Trichoderma* spp., Mycorrhiza, *Bacillus* spp.

Garrett (1965) was first to use the term biological control of a disease through some biological agencies. Before this Sanford and Broadfoot (1931) provided the experimental evidence that wheat seedling disease caused by *Ophiobolus graminis* could be completely suppressed by the antagonistic action of co-inoculated species of fungi and bacteria. Recently many reports are coming on biological control of plant pathogens as pesticidal compounds are found to be hazardous to the environment. Weindling (1932) was the first to show parasitism by *Trichoderma viride* on several soil fungi. Since then several reports are coming on biological control of plant pathogens by various organisms. Apart from *Triochoderma* spp., *Aspergillus terreus* (Reddy and Reddy, 1994), *Gliocladium virens* (Rajeshari *et al.*, 1998; Prasad and Rajeshwaran, 1999), many bacteria have been used to manage

---

Author's address: Soil Microbiology and Pesticides Laboratory, Department of Botany, Dr. Babasaheb Ambedkar Marathwada University, Aurangabad 431004, MS, India

different soil-borne pathogens. During the last decade more than 32 diseases have been controlled by different species of *Trichoderma* (Table 1).

Since the mycorrhizal importance was investigated by Harley (1959) many researchers came out with the phenomenon of phosphate solubilization and now a days VAM is constantly advocated as biofertilizer (Janaki Rani and Manoharachary, 1994; Kunwar and Manoharachary, 1998). Ansuya (1995) studied the influence of inoculation of VAM (*Glomus fasciculatum*) and the pathogen (*Fusarium oxysporum* f. sp. *ciceri*) either singly or jointly on the growth of *Cicer* plants. Plant growth, dry matter production and population of mycorrhizal sporocarps were significantly increased when VAM was inoculated prior to the introduction of pathogen. Use of only pathogen on the other hand resulted in increased wilt incidence and death of plants.

The work of Millard and Taylor (1927) and Sanford (1926) on the control of potato scab by green manuring indicated the beginning of antagonisms among the microbes in the soil. They also postulated that soil microbes can be manipulated by various soil treatments and control of soil-borne diseases may be achieved (Gangawane and Deshpande, 1976, 1977). Some workers also isolated antagonistic microorganisms from the rhizosphere. Nandi and Nandi (1972) indicated that *Aspergillus wentii* from the rhizosphere of groundnut was antagonistic to *Sclerotium rolfsii* causing stem blight of groundnut. Mall (1972) isolated 4 species of fungi and 21 species of streptomyces antagonistic to *Fusarium oxysporum* in the rhizosphere of coriander. Singh (1983) determined the presence of 3 isolates of *Trichoderma* and two isolates of *Aspergillus* causing lysis of *Colletotrichum falcatum* from the rhizosphere of sugarcane. Population dynamics of biocontrol agents and pathogens in soil and rhizosphere of various crops have very well been discussed by Paulitz (2000). Similarly Chakraborty *et al.* (2003) isolated large number of fungi and bacteria antagonistic to *Fomes pionaoensis* and *Poria hypobruned* from the rhizosphere of tea. The interaction between 29 isolates of *Rhizobium* and *in vitro* growth of 3 strains of *Phytophthora* spp. was investigated to determine the potential of these bacteria on biological control against root rot of subterranean clover. 13 out of 29 rhizobial isolates caused significant reduction in the hyphal growth of *Phytophthora* isolates in Australia (Simpfendorfer *et al.*, 1999). Integrated management of web blight (*Rhizoctonia solani*) urd and mung bean was studied by Dubey and Singh (2003) applying certain fungi (*Gliocladium virens, Trichoderma viride*) extracts of *Pongamia glabra,* fungicide carboxin along with *Rhizobium* as foliar spray and found that disease intensity was reduced up to 93.7%.

Now-a-days use of *Trichoderma* sp. for the management of various crop diseases has been increased. Many formulations by different manufacturers are also coming in the market. Table 1 show that nearly 32 crop diseases

TABLE 1 *Biocontrol of crop disease by Trichoderma spp*

Crop and Disease	Trichoderma species	Author and Year
1. Damping off of sugarbeet (*Pythium aphanidermatum*)	*T. harzianum*	Sawant and Mukhopadhyaya (1990)
2. Root rot of cotton (*Rhizoctonia bataticola*)	*T. harzianum*	Ratnalikar and Gaikwad (1991)
3. Paddy blast (*Pyricularia oryzae*)	*T. viride*	Rasal (1991)
4. Wilt of egg plant (*Sclerotium rolfsii*)	*T. harzianum*	Datar (1991)
5. Muskmelon wilt (*Fusarium solani*)	*T. viride*	Asalmol and Sen (1991)
6. Red rot of sugarcane (*Colletotrichum falcatum*)	*T. harzianum*	Singh (1991)
7. Pink stalk of maize (*Acremonium strictum*)	*T. harzianum*	Dwivedi and Sangam Lal (1991)
8. Slow wilt of black pepper (*Fusarium oxysporum*)	*T. viride*	Mane and Raut (1991)
9. Root rot pathogens of different crops (*Bipolaris sorokiniana, Fusarium javanicum*)	Eight spp. of *Trichoderma*	Kumar *et al.* (1991)
10. Damping off of tobacco (*Pythium aphanidermatum*)	*Trichoderma* spp.	Raju (1992)
11. Collar and root rot of mothbean (*Rhizoctonia bataticola*)	*T. harzianum*, *T. viride*	Chaudhary and Mirashi (1992)
12. Root rot of wheat (*Fusarium avenaceum*)	*T. viride*	Kumar *et al.* (1993)
13. Stem canker of mulberry (*Botryodiplodia theobromae*)	*T. viride, T. koningini, T. pseudokonigii*	Gupta *et al.* (1995)
14. Dry root rot of caster (*Macrophomina phaseolina*)	*T. harzianum*	Humayun *et al.* (2003)
15. Wilt of pigeonpea (*Fusarium udum*)	*T. harzianum* T. MTR-35	Bhatnagar (1995)
16. Stem blight of groundnut (*Sclerotium rolfsii*)	*T. harzianum*	Desai (1975)
17. Root rot of chickpea (*Macrophomina phaseolina*)	*Trichoderma* sp.	Selvarajan and Jeyarajan (1996)
18. Gray mold of chickpea (*Botrytis* sp.)	*Trichoderma* sp.	Mukherjee *et al.* (1997)
19. Charcoal rot of cowpea (*Macrophomina phaseolina*)	*Trichoderma viride*, *T. harzianum*	Ushamalini *et al.* (1997)

(Table 1 Contd.)

(Table 1 Contd.)

Crop and Disease	Trichoderma species	Author and Year
20. Mungbean root rot (*Macrophomina phaseolina*)	*T. viride*	Raghuchander *et al.* (1997)
21. Jasmine wilt (*Sclerotium rolfsii*)	*T. harzianum*	Alice et al. (1998)
22. Stem rot of chickpea (*Sclerotinia sclerotiorum*)	*T. harzianum*	Sharma *et al.* (1999)
23. Root rot of sunflower (*Rhizoctonia bataticola*)	*T. harzianum*	Aggarwal *et al.* (1999)
24. Wilt of pigeonpea (*Fusarium udum*)	*T. harzianum*	Pandey and Upadhyay (1999)
25. Root rot and collar rot of sunflower (*Sclerotium rolfsii*)	*T. harzianum*, *T. koningi*	Prasad *et al.* (1999)
26. Collar rot of groundnut (*Aspergillus niger*)	*T. viride*	Raju and Krishnamurthy (2000)
27. Rhizome rot of ginger (*Pythium myriotylum*)	*T. viride*	Mathur *et al.* (2000)
28. Stem rot of ginger (*Sclerotium rolfsii*)	*T. harzianum*	Biswas and Sen (2000)
29. White root rot of apple (*Dermatophora necatrix*)	*T. harzianum*	Sharma and Sharma (2001)
30. Damping off of forest nurseries (*Pythium aphanidermatum*)	*T. virens, T. hamatum, T.longibrachiatus*	Arya and Kaushik (2001)
31. Wilt of chillies (*Fusarium solani*)	*T. harzianum*	Kelaiya *et al.* (2003)

have been controlled by different species of *Trichoderma* from 1990 to 2003. Similarly many reports are available on the use of different bacteria. These bacteria include various species of *Bacillus* sp., *Enterobacter* sp., *Pantoea agglonemeus*, *Cedeca derisiae* and *Azotobacter* (Table 2). Interestingly use of *Pseudomonas solanacearum, P. auruginosa, P. aureofaciens* and *P. fluorescens* for the biological management of different pathogens is increasing day by day. Table 3 indicates that 34 crop diseases could be managed with *Pseudomonas* spp. The pathogens are mainly soil-borne.

In this laboratory attempts have been made to develop transgenic antagonistic Rhizobium against dry root rot and aflaroot of groundnut. The antagonistic gene from *Pseudomonas fluorescens* has been cloned in the *Rhizobium* nodulating groundnut. The plasmid profile and curing indicated that the gene is present on the plasmid of *Pseudomonas* which is transferred to *Rhizobium* and visualized in the plasmid profile.

TABLE 2 *Earlier reports on biocontrol of plant pathogens by bacteria other than Pseudomonas spp.*

Disease of pathogen	*Bacterium*	*Author and country*
1. Tomato wilt (*Fusarium oxysporum*)	*Azotobacter* sp. *Bacillus* sp.	Kapoor and Kar (1989), India
2. Bacterial wilt of banana (*Pseudomonas solanacearum*)	*Bacillus* sp.	Anuratha and Gnanamanickam (1990), India
3. Dry root rot of chickpea (*Rhizoctonia bataticola*)	*Bacillus subtilis*	Saxena and Moly Saxena (1995), India
4. Seedling wilt of tomato (*Fusarium oxysporum*)	*Bacillus subtilis*	Padmodaya and Reddy (1998), India
5. Pepper damping off (*Phytophthora capsici*)	*Bacillus subtilis*	Sharifi Tehrani and Omati (1999), Iran
6. Root rot of soybean (*Rhizoctonia solani*)	*Bacillus megaterium*	Zheng and Sinclair (2000), USA
7. Sheath flight of rice (*Rhizoctonia solani*)	*Bacillus megaterium*	Pengnoo *et al.* (2000), Thailand
8. Wilt of pigeon pea (*Rhizoctonia solani*)	*Bacillus megaterium*	Bapat Sangita and Shah (2000), USA
9. Bacterial wilt/Brown rot of potato (*Ralstonia solanacearum*)	*Bacillus* spp.	Dhanbir Singh and Rana (2001), India
10. Root rot of angelica trees (*Phytophthora cactorum*)	*Enterobacter cloacae* *Serratia ficaria*	Okamoto Hiroshi *et al.* (2000), Japan.
11. Leaf spot of mungbean (*Xanthomonas axonopodis* pv. *Vignae radiatae*)	*Bacillus* spp.	Borah *et al.* (2000), India.
12. Tomato diseases (aerial plants) (*Pseudomonas syringae* pv. *Tomato*)	*Pantoed agglonerous* *Cedecea davisiae*	Romeiro *et al.* (2000), Brazil
13. Dry root rot of chickpea (*Rhizoctonia bataticola*)	*Bacillus subtilis*	Sindham *et al.* (2002), India
14. Bacterial blight of rice (*Xanthomonas oryzae* pv *oryzae*)	*Bacillus subtilis*	Manmeet Manav and Thind (2002), India
15. Basal rot of balsam (*Sclerotium rolfsii*)	*Bacillus* sp.	Girija and Umamaheshwaran (2003), India
16. Damping off of true potato (*Pythium aphanidermatum*)	*Bacillus subtilis*	Lakra (2003), India

TABLE 3 *Earlier reports of management of crop diseases by Pseudomonas spp.*

	Disease of pathogen	*Pseudomonas spp.*	*Author and country*
1.	Bacterial wilt of tobacco (*Ralstonia solanacearum*)	*Pseudomonas solanacearum*	Chen *et al.* (1982), China
2.	Bacterial wilt of tobacco (*Ralstonia solanacearum*)	*Pseudomonas solanacearum*	Luo & Wang (1983), China
3.	Bacterial wilt of tobacco (*Ralstonia solanacearum*)	*Pseudomonas solanacearum*	Chen and Echondi (1984), China
4.	Bacterial wilt of tobacco (*Ralstonia solanacearum*)	*Pseudomonas solanacearum*	Yi (1991), Korea
5.	Root rot and damping off of cotton (*Rhizoctonia solani*)	*Pseudomonas fluorescens*	Laha *et al.* (1992) India
6.	Bacterial wilt of potato (*Pseudomonas solanacearum*)	*Pseudomonas fluorescens*	Shekhawat *et al.* (1993), India
7.	Rots and damping off of crops (*Rhizoctonia* and *Pythium*)	*Pseudomonas fluorescens*	Saxena *et al.* (1995), India
8.	Sclerotial wilt of cotton (*Sclerotium rolfsii*)	*Pseudomonas fluorescens*	Laha *et al.* (1996), India
9.	Root rot of cotton (*Macrophomina phaseolina, Rhizoctonia solani, Fusarium solani*)	*Pseudomonas fluorescens*	Saxena *et al.* (1995), India
10.	Leaf spot on caster bean (*Alternaria ricini*)	*Pseudomonas fluorescens*	Da Silva *et al.* (1998), Brazil
11.	Tomato bacterial wilt (*Ralstonia solanacearum*)	*Pseudomonas fluorescens*	Abdulla *et al.* (1999), Saudi Arabia
12.	Root rot of mungbean and mash bean (*Ralstonia solanacearum*)	*Pseudomonas fluorescens*	Siddiqui *et al.* (1999) Pakistan
13.	Fusarium crown and root rot of tomato (*Fusarium oxysporum*)	*Pseudomonas* spp.	Sharif Tehrani (1999), Iran
14.	Pepper damping off (*Phytophthora capsici*)	*Pseudomonas* spp.	Sharif Tehrani (1999), Iran
15.	Fusarium wilt (*Fusarium* spp.)	*Pseudomonas aeruginosa*	Perez Naranjo (1999), Columbia
16.	Root rot disease of crop plants (*Macrophomina phaseolina, Fusarium solani, Rhizoctonia solani*)	*Pseudomonas aeruginosa*	Izhar *et al.* (1999), Pakistan
17.	Bacterial blight of cotton (*Xanthomonas axonopodis* pv *malvacearum*)	*Pseudomonas fluorescens*	Mandal *et al.* (1999), India

(Table 3 Contd.)

(Table 3 Contd.)

Disease of pathogen	Pseudomonas spp.	Author and country
18. Root rot of cotton (*Rhizoctonia* spp.)	*Pseudomonas fluorescens*	Anith *et al.* (1999), India
19. Dollar spot as creeping bentgrass pathogens	*Pseudomonas aureofaciens*	Powel *et al.* (2000), USA
20. Root rot of sesamum (*Macrophomina phaseolina*)	*Pseudomonas fluorescens*	Karunanithi *et al.* (2000), India
21. Safflower root rot (*Rhizoctonia bataticola*)	*Pseudomonas fluorescens*	Prashanthi (2000), India
22. Collar rot of teak seedlings (*Rhizoctonia solani*)	*Pseudomonas fluorescens*	Ramesh (2000), India
23. Damping off disease of tomato (*Pythium aphanidermatum*)	*Pseudomonas fluorescens*	Manoranjitham *et al.* (2000), India
24. Sclerotium rot of sunflower (*Sclerotium rolfsii*)	*Pseudomonas* spp.	Rangeshwaran & Prasad (2000), India
25. Root rot of *Asparagus* (*Phytophthora megasperma*)	*Pseudomonas aureofaciens*	Godfrey *et al.* (2000), New Zealand
26. Root rot of crop plants (*Pythium ultimum, Fusarium solani, F. oxysporum*)	*Pseudomonas ameofaciens*	Paulitz *et al.* (2000), Canada
27. Collar rot of teak (*Rhizoctonia solani*)	*Pseudomonas fluorescens*	Ramesh (2000), India
28. Root rot of cucumber (*Pythium aphanidermatum*)	*Pseudomonas putida*	Ongena *et al.* (2000)
29. Damping off disease of tomato (*Pythium aphanidermatum*)	*Pseudomonas fluorescens*	Manoranjitham *et al.* (2000), India
30. Fruit rot of chilli (*Colletotrichum capsici*)	*Pseudomonas fluorescens*	Ramamoorthy and Samiyappan (2001), India
31. Seedling blight of sunflower (*Alternaria helianthi*)	*Pseudomonas fluorescens*	Prasad and Kulshreshtha (2002), India
32. Bacterial blight of rice (*Xanthomonas oryzae* pv *oryzae*)	*Pseudomonas fluorescens*	Manmeet Manav and Thind (2002), India
33. Dry root rot of chickpea (*Rhizoctonia bataticola*)	*Pseudomonas fluorescens*	Sindham *et al.* (2002), India
34. Root rot of groundnut (*Macrophomina phaseolina*)	*Pseudomonas fluorescens*	Charitadevi *et al.* (2003), India

References

Abdulla, M.Y., Al-Mihanna, A.A., Al Rokibah, A.A. and Ibrahim, G.H. 1999. Tomato bacterial wilt in Saudi Arabia and the use of antagonistic bacteria for its control. *Annals Agric. Sci*. (Cairo) **44**: 511-521.

Aggarwal Ashok, Kumar J., Sharma Dipti and Mehrotra, R.S. 1999. Biological control of root rot of sunflower using *Trichoderma harzianum* and a mycorrhizal fungi. *J. Indian Bot. Soc*., **78**: 87-90.

Alice, D., Ramamoorthy, V., Muthusamy, M. and Seetharaman, K. 1998. Biocontrol of Jasmine wilt incited by *Sclerotiorum rolfsii* Sacc. *Indian J. Plant Prot*. **26(1)**: 64-67.

Anith, K.N., Tilak, K.V.B.R. and Manmohandas, T.P. 1999. Analysis of mutation affecting antifungal property of a fluorescent *Pseudomonas* sp. during cotton *Rhizoctonia* interaction. *Indian Phytopath*. **52(4)**: 366-369.

Ansuya, D. 1995. V.A. Mycorrhiza – A biological tool to control chickpea (*Cicer arietinum* L.) *Fusarium* wilt. *Indian J. Mycol. Pl. Pathol*. **25(1&2)**: 131.

Anuratha, C.S. and Gnanamanickam, S.S. 1990. Biological control of bacterial wilt caused by *Pseudomonas solanacearum* in India with antagonistic bacteria. *Plant Soil* **124**: 109-116.

Arya Sanjay and Kaushik, **J.C.** 2001. Efficacy of *Trichoderma species* and *Gliocladium virens* as biocontrol agents against damping off of forest nurseries. *Pl. Dis. Res*. **16(1)**: 46-51.

Asalmol, M.N. and Sen, B. 1991. Mycoparasitism of *Aspergillus niger* and *Trichoderma viride* against *Fusarium solani* causing muskmelon wilt. *Indian Phytopath.* Supp. Issue, **44**: 17.

Bapat Sangita and Shah, A.K. 2000. Biological control of fusarial wilt of pigeon pea by *Bacillus brevis. Can. J. Microbiol*. **46(20)**: 125-132.

Bhatnagar Himani 1995. Use of commercial biofungicide against wilt of pigeon pea *Cajanus cajan* (L.) Millsp. *Indian J. Plant Prot.* **23**: 224-225.

Biswas Kajal Kumar and Sen Chitreswar 2000. Management of stem rot of groundnut caused by *Sclerotium rolfsii* through *Trichoderma harzianum. Indian Phytopath*. **53(3)**: 290-295.

Borah, P.K., Jindal, J.K. and Verma, J.P. 2000. Biological management of bacterial leaf spot of mungbean caused by *Xanthomonas axonopodis* pv. *vignae radiatae*. *Indian Phytopath*. **53(4)**: 384-394.

Chakraborty, U., Raychaudhary, P., Tandon, C. and Chakraborty, B.N. 2003. Screening of tea rhizosphere microorganisms antagonistic to root rot pathogens. *National Symposium on plant pathogens diversity in relation to plant health*. Jan. 16-18, p. 86.

Charitha Devi, M., Reddy, M.N. and Eswara Reddy, N.P. 2003. Biocontrol efficacy of *Pseudomonas* spp. against *Macrophomina phaseolina* the incitant of root rot of groundnut *Arachis hypogea* L. In: *Proc. of National Symposium on plant pathogen diversity in relation to plant health*. Jan. 16-18, p. 49.

Chaudhary, G.S. and Mirashi, R.P. 1992. Biological suppression of *Rhizoctonia bataticola* – collar and root rot pathogen of mothbean. *Indian Phytopath*. 43-44 Supp. Issue (II).

Chen, W.Y. and Echandi 1984. Effect of a virulent bacterium producing strains of *Pseudomonas solanacearum* on the control of bacterial wilt of tobacco. *Plant Pathol.* **33**: 245-253.

Chen, W.Y., Echandi, E. and Spurs, H.W. 1982. Protection of tobacco plants from bacterial wilt with a virulent bacterium promoting strains of *Pseudomonas solanacearum*. In: *Proceedings of the 5th International Conference on plant pathogenic bacteria,* cali, Columbia, pp. 482-492.

Datar, V.V. 1991. Biological control of wilt of egg plant caused by *Sclerotium rolfsii. Indian Phytopath*. Supp. Issue **44:** 16.

Desai Suseelendra 1975. Biological control of *Sclerotium rolfsii* using *Trichoderma* spp. evaluation of carrier systems. *Indian J. Mycol. Pl. Pathol*. **25(1&2)**: 132.

Dubey and Singh 2003. Integrated management of web blight of urd and mung bean. *National Symposium on plant pathogens diversity in relation to plant health*. January 16-18, 2003: 18.

Dwivedi, R.R. and Sangam Lal 1991. Antagonism of *Acremonium strictum, Cephalosporiumn maydis* and *Fusarium moniliformae* by *Trichoderma harzianum. Indian Phytopath*. Supp. Issue of Vol. **45**, LXIX.

Gangawane, L.V. and Deshpande, K.B. 1976. Agronomic treatment and changes in rhizosphere mycoflora of groundnut-I. Effect of organic amendments. *Kavaka* **4**: 21-24.

Gangawane, L.V. and Deshpande, K.B. 1977. Agronomic treatment and changes in rhizosphere mycoflora of groundnut-II. *Acta Agronomica Acad. Sci. Hungary*, **26**: 68-78.

Garrett, S.D. 1965. Rhizomorph behaviour in *Armilaaria mellea* (Vahl.) Quel. II. Logistics of infection. *Ann. Botany (London)* (N.S.) **20**: 193-209.

Girija, V.K. and Umamaheswaran, K. 2003. Basal rot of balsam and its management through bioagents. *Pl. Dis. Res.* **18(1)**: 52-55.

Godfrey Scott, A.C., Mark, W., Silby Peter G. Fallen and Khris Mohanty, H. 2000. Biological control of *Phytophthora megasperma* var. *sojae,* causal agent of *Phytophthora* rot of *Asparagus* by *Pseudomonas aureofaciens* PA 147. 2: A Preliminary field trial. *New Zealand Journal of Crop & Horticultural Science* **28(2)**: 97-103.

Gupta, Govindaiah, V.P., Bajpai, A.K. and Datta, R.K. 1995. Antagonistic potential of *Trichoderma* and *Gliocladium* sp. to *Botryodiplodia theobromae* infecting mulberry. *Indian J. Mycol. Pl. Pathol*. **25(1&2)**: 127.

Harley, J.L. 1959. In *Biology of Mycorrhiza*. Leonard Hill Books, London.

Humayun, P., Raja Ram Reddy, D. and Chandrashekhara Rao, K. 2003. Biological control of caster by dry root rot caused by *Macrophomina phaseolina* (Tarsi) Goid. *National Symposium on plant pathogens diversity in relation to plant health*, January 16-18: 122.

Izhar Ishrat Imran Ali, Siddiqui Syed Ehteshamal Haque, and Abdul Ghaffar 1999. Multiplication of plant growth promoting rhizobacteria for use in the control of root rot disease of crop plants. *Pakistan J. Botany* **31(2)**: 397-405.

Janaki Rani and Manoharachary, C. 1994. Occurrence and distribution of VAM fungi associated with safflower. *Indian Phytopath*. **47(3)**: 263-265.

Kapoor, I.J. and Kar, B. 1989. Antagonism of *Azotobacter* and *Bacillus* to *Fusarium oxysporum* f. sp. *lycopersici*. *Indian Phytopath*. **42(3)**: 400-404.

Karunanithi, K., Muthuswamy, M. and Seetharaman, K. 2000. Pyrolnitrin production by *Pseudomonas fluorescens* effective against *Macrophomina phaseolina*. *Crop Research (Hissar)* **19(2)**: 368-370.

Kelaiya, D.S., Jadhav, H.R. and Parikh, A.M. 2003. Biological control of *Fusarium solani* the incitant of wilt disease of chilli (*Capsicum annum* L.). *National Symposium on plant pathogens diversity in relation to plant health*. January 16-18, 2003: 38.

Kumar Pramod Chandra, C., Ganibova, L.V. and Velikanov, L.L. 1991. *Trichoderma* spp. as potential biocontrol agents. *Indian Phytopath*. Supp. Issue **45**: 111.

Kunwar, I.K. and Manoharachary, C. 1998. Colonization by Arbascular mycorrhizal and mycophyllous fungi in roots and scale leaves of garlic. *J. Indian Bot. Soc*. **77**: 201-203.

Laha, G.S., Singh, R.P. and Verma, J.P. 1992. Biocontrol of *Rhizoctonia solani* in cotton by fluorescent Pseudomonads. *Indian Phytopath*. **45(4)**: 412-415.

Laha, G.S., Verma, J.P. and Singh, R.P. 1996. Effectiveness of fluorescent Pseudomonads in the management of sclerotial wilt of cotton. *Indian Phytopath*. **49(1)**: 3-8.

Lakra, B.S. 2003. Management of damping off of the potato seedling by using bioagents and a fungicide. *Pl. Dis. Res*. **18(1)**: 73-74.

Luo, K. and Wang, Z. 1983. Study of bacterial wilt *Ralstania solanacearum* controlled by antagonistic and a virulent *Pseudomonas solanacearum*. *Acta Phytopathologica Sinica* **13**: 51-56.

Mall, O.P. 1972. Antagonistic microorganisms in the root regions of coriander. *Proc. 59th Ind. Sci. Cong*. Part III, Ab. 307.

Mandal, K.K., Singh, R.P. and Verma, J.P. 1999. Beneficial effects of indigenous cotton rhizobacteria on seed germinability, growth promotion and suppression of bacterial blight disease. *Indian Phyotpath*. **52(3)**: 228-235.

Mane, S.B. and Raut, S.P. 1991. Biocontrol of soil borne pathogenic fungi of black pepper with two *Trichoderma* species. *Indian Phytopath*. Supp. Issue of Vol. 45, C XIII.

Manmeet Manav and Thind, B.S. 2002. Management of bacterial blight of rice with bioagents. *Pl. Dis. Res*. **17(1)**: 21-28.

Manoranjitham, S.K., Prakasam, V., Rajappan, K. and Amutha, G. 2002. Effect of two antagonists on damping off disease of tomato. *Indian Phytopath*. **53(4)**: 441-443.

Mathur Ram Kusum, Lodha, B.C. and Webster, J. 2000. Evaluation of resistant biocontrol agents as seed treatments against ginger rhizome rot. *Indian Phytopath*. **53(4)**: 450-454.

Millard, W.A. and Taylor, C.B. 1927. Antagonism of microorganisms as the controlling factor in the inhibition of scab by green manuring. *Ann. Appl. Biol*. **14**: 202-215.

Mukherjee, P.K., Haware, M.P. and Raghu, K. 1997. Induction and evaluation of benomyl tolerant mutants of *Trichoderma viride* for biological control of *Botrytis* grey mold chickpea. *Indian Phytopath.* **50(4)**: 485-489.

Nandi, S.K. and Nandi, B. 1972. Role of antimetabolites of some rhizosphere organisms on seedling growth of groundnut. *Proc. 59th Ind. Sci. Cong*. Part III: Ab. 308.

Okamoto Hiroshi, Masmoru Sato, Yoshio Miyata, Masani Yoshikawa and Makoto Isaka 2000. Biocontrol of *Phytophthora* root rot of angelica trees by *Enterobacter cloacae* and *Sersatia ficaria* strains. *J. Gen. Pl. Patho*. **66(1)**: 86-94.

Ongena, M., Dayf, F., Jacques, P., Thonart, P., Benhamou, N., Paulitz, T.C. and Belanger, R.R. 2000. Systemic induction of phytoalexins in cucumber in response to treatments with fluorescent pseudomonads. *Plant Pathology (Oxford)* **49(4)**: 523-530.

Padmodaya, B. and Reddy, H.R. 1998. Screening of antagonists against *Fusarium oxysporum* f. sp. *lycopersici* causing seedling disease and wilt in tomato. *J. Mycol. Pl. Pathol*. **28(3)**:

Pandey, K.K. and Upadhyay, J.P. 1999. Comparative study of chemical, biological and integrated approach for management of *Fusarium* wilt of pigeon pea. *J. Mycol. Pl. Pathol*. **29(2)**: 214-216.

Paulitz Timothy, Brian Nowork, Thompson, Pascle Gamord, Evelyn Tsang and Joyce Loper 2000. A novel antifungal furanone from *Pseudomonas aureofaciens,* a biocontrol agent of fungal plant pathogens. *Journal of Chemical Ecology* **26(6)**: 1515-1524.

Paulitz, T.C. 2000. Population dynamics of biocontrol agents and pathogens in soils and rhizospheres. *Euro. J. Pl. Patho*. **106(5)**: 401-413.

Pengnoo Ashara, Chirasak Kusangniriyawang, Ladda Nilratna and Mana Kanjanamaneesathian 2000. Green house and field trials of the bacterial antagonists in pellet formations to suppress sheath blight of rice caused by *Rhizoctonia solani. Biocontrol (Dordrecht)* **45(2)**: 245-256.

Perez Naranjo, Junan C. and Jairo Leguizaman, C. 1999. The effect of the fertilizers *Pseudomonas aeruginosa* and benomyl in the field of *Asparagus* affected by *Fusarium* spp. at field level. *Fitopatologia Colombiana* **22(1-2)**: 82-87.

Powel, J.F., Vargas, J.M., Nair, M.G., Detweiler, A.R. and Chandra, A. 2000. Management of dollar spot on creeping bentgrass with metabolites of *Pseudomonas aureofaciens. Plant Disease* **84(1)**: 19-24.

Prasad, R.D. and Kulshreshtha 2002. Efficiency of some bacterial antagonists against seedling blight of sunflower caused by *Alternaria helianthi. Pl. Dis. Res.* **17(1)**: 65-66.

Prasad, R.D. and Rajeshwaran, R. 1999. Granular formulation of *Trichoderma* and *Gliocladium* spp. in biocontrol of *Rhizoctonia solani* of chickpea. *J. Mycol. Pl. Pathol*. **29(20)**: 222-226.

Prashanthi, S.K., Kulkarni Shrikant and Anahosur, A.H. 2000. Management of safflower root rot caused by *Rhizoctonia bataticola* by antagonistic microorganism. *Pl. Dis. Res.* **15(20)**: 146-150.

Raghuchander, T., Rajappan, K. and Samiappan, R. 1997. Evaluating methods of application of biocontrol agent in the control of mungbean root rot. *Indian Phytopath*. **50(2)**: 229-234.

Rajeshwari, B., Chandrashekhara Rao, K. and Pramod Kumar, C. 1998. Antagonism of *Trichoderma* and *Gliocladium* spp. against *M. phaseolina,* the incitant of dry root rot of mungbean. *Indian J. Plant Prot*. **26(1)**: 9-11.

Raju Ram Bhadra, M. and Krishnamurthy, K.V.M. 2000. Efficacy of *Trichoderma* spp. in the management of collar rot of groundnut caused by *Aspergillus niger* van Teighem. *Indian J. Plant Prot.* **28(20)**: 177-199.

Raju, C.A. 1992. *In vitro* studies on the antagonistic activity of *Trichoderma* spp. against *Pythium aphanidermatum*. *Indian Phytopath*. Supp. Issue of Vol. 45 LIX.

Ramamoorthy, V. and Samiyappan, R. 2001. Induction of defence related genes in *Pseudomonas fluorescens* – treated chilli plants in response to infection by *Colletotrichum capsici. J. Mycol. Pl. Pathol*. **31(2)**: 146-155.

Ramesh, K.R. 2000. Inhibition of *Rhizoctonia solani* caused agent for collar rot of teak seedlings by fungicides and biocontrol agents *in vitro* condition. *Indian Forester* **126(3)**: 284-288.

Rangeshwaran, R. and Prasad, R.D. 2000. Biological control of *Sclerotium* rot of sunflower. *Indian Phytopath*. **53(4)**: 444-449.

Rasal, R.H. 1991. Biocontrol of paddy blast by *Trichoderma viride. Indian Phytopath*. Supplementary issue of Vol. 44 XV.

Ratnalikar, K.K. and Gaikwad, S.J. 1991. Biological management of root rot of cotton caused by *Rhizoctonia bataticola. Indian Phytopath*. Supplementary Issue of Vol. 44 XV.

Reddy and Reddy 1994. Influence of some toxigenic fungi on production of *Trichoderma* by *Trichoderma viride*. *J. Indian Bot. Soc*. **73**: 209-211.

Romeiro, Reginaldo da S., Dora, M.S. Neves, Murilo G. de Carvalho and Revato Carver Filho 2000. Selection of bacterial residents in tomato canopy for the biocontrol of aerial part diseases of the crop. *Suma Phytopathologica* **26(2)**: 220-224.

Sanford, G.B. 1926. Some factors affecting the pathogenicity of *Actinomyces scabis. Phytopathology* 16: 525-547.

Sanford, G.B. and Broadfoot, W.C. 1931. Studies of the effect of soil inhabiting microorganisms on the virulence of *Ophiobolus graminis. Sci. Agri*. **11**: 512-518.

Sawant Indu, S. and Mukhopadhyay, A.N. 1990. Integration of metalaxyl with *Trichoderma harzianum* for the control of *Pythium* damping off in sugar beet. *Indian Phytopath*. **43(4)**: 535-541.

Saxena Deepak, Patel, K.R., Rao, K.K. and Mehta, M.H. 1995. Role of *fluorescent Pseudomonas* in controlling plant diseases. *Indian J. Mycol. Pl. Pathol*. **25(1&2)**: 131.

Saxena, D.R. and Moly Saxena 1995. Evaluation of antagonists against dry root rot pathogen of chickpea. *Indian J. Mycol. Pl. Pathol*. **25(1&2)**: 130.

Selvarajan, R. and Jeyarajan, R. 1996. Inhibition of chickpea root rot of pathogens, *Fusarium solani* and *Macrophomina phaseolina* by antagonists. *Indian J. Mycol. Pl. Pathol*. **26(3)**: 248-251.

Sharifi Tehrani, A., Zala, M., Moenne-Locoz, Y. and Defago, G. 1999. Biological control of Fusarium crown and root rot of tomato by fluorescent pseudomonad. *Mededelingur facultiet Landbowkundige en Topegepaste Biologische Wetenschappan Universitet Gent* **64(3b)**: 425-429.

Sharifi Tehrani, A. and Omati, F. 1999. Biocontrol of *Phytophthora capsici*, the causal agent of pepper damping off by antagonistic bacteria. *Mededelingur facultiet Landbowkundige en Topegepaste Biologische Wetenschappan Universitet Gent*. **64(3b)**: 419-423.

Sharma Mamta and Sharma, S.K. 2001. Biocontrol of *Dermatophora necatrix* causing white root rot of apple. *Pl. Dis. Res.* **16(1)**: 40-45.

Sharma, B.K., Verma, B.R. and Sharma B.K. 1999. Biocontrol of *Sclerotinia sclerotiorum* causing stem rot of chickpea. *Indian Phytopath*. **52(1)**: 44-46.

Shekhawat, G.S., Rajhpalsingh, V. and Chakrabarti, S.K. 1993. Possibility of biological management of potato bacterial wilt with microbial antagonists and latent viruses. *Indian Potato Association* **20**: 219-222.

Siddiqui, Imran, A.S., Ehteshamal Haque and Ghaffar, A. 1999. *Pseudomonas aeruginosa* and fungal; antagonists in the control of root knot rot disease complex on mungbean and mashbean. *Pakistan Journal of Nematology* **17(2)**: 155-167.

Silva Francisco de A.G., Cecilia do N. Peixoto, Sayanara M.P. de Assis Rosa, de L. Mariano L.R. and Isairas Padovan, P. 1998. Potential of fluorescent *Pseudomonas* spp. for biological control of *Alternaria ricini* on caster bean. *Brazi. Arch. Biol. Tech.*, **41(1)**: 91-102.

Simpfendorfer, S., Harden, T.J. and Murray, G.M. 1999. The *in vitro* inhibition of *Phytophthora cladestina* by some rhizobia and the possible role of *Rhizobium trifolii* in biological control of *Phytophthora* root rot of subterranean clover. *Australian J. Agric. Research* **50(8)**: 1469-1473.

Sindhan, G.S., Indra Hooda and Karvarra, S.S. 2002. Biological control of drug root rot of chickpea caused by *Rhizoctonia bataticola. Pl. Dis. Res*. **17(1)**: 68-71.

Singh Narendra 1983. Antagonistic activity of sugarcane rhizosphere soils against red rot pathogen. *Indian Bot. Reptr*. **2(2)**: 142-144.

Singh Narendra 1983. Biological control of red rot disease of sugarcane. *Indian Phytopath*. Supp. Issue. **45**: 65.

Ushamalini, C., Rajappan, K. and Kausalya Gangadharan 1997. Management of charcoal rot of cowpea using biocontrol agents and plant products. *Indian Phytopath*. **50(4)**: 504-507.

Weindling, R. 1932. *Trichoderma lignorum* as a parasite of other soil fungi. *Phytopathology* **22**: 837-845.

Yi, Y.K. 1991. Protection of tobacco plants from bacterial wilt with a virulent bacteria in producing strands of *Pseudomonas solvanacearum*. J. Korean Society Tobacco Science, **13**: 42-47.

Zheng, X.Y. and Sinclair, J.B. 2000. The effects of traits of *Bacillus megaterium* on seed and root colonization and their correlation with the suppression of *Rhizoctonia* root rot of soybean. *Biocontrol (Dordrecht)* **45(2)**: 223-243.

Frontiers in Plant Sciences (Pages 549-564)
Eds. K.G. Mukerji, K.V.B.R. Tilak, S.M. Reddy, L.V. Gangwane, P. Prakash, I.K. Kunwar

Biological Control of Postharvest Fungal Diseases of Fruits and Vegetables: Current Trends and Future Directions

A.R. Podile and Ch. Neeraja

ABSTRACT

Increased consumer awareness that diet and health are linked has resulted in a greater consumption of fruits and vegetables. They are being harvested, stored and transported during which the commodities are prone to microbial attack. The development of pathogen resistance to many key fungicides, negative public perception regarding the safety of fungicides has drastically reduced the use of fungicides. Biological control has emerged as one of the most promising alternative to chemicals. Several alternative approaches that include antagonistic microorganisms, induced resistance and natural fungal compounds are known for developing new and potentially safer technologies for postharvest disease control. Development of antagonistic microorganisms has been the most studied and substantial progress has been made in this area. Postharvest application of antagonists can be attributed to the uniqueness and relative simplicity of the postharvest system. The antagonistic activity was attributed to the production of antibiotics and enzymes like chitinase and glucanase. Antibiotic producing antagonists or antibiotics alone applied to postharvest fruits and vegetables leave a trace amount that on consumption result in adverse effects or human pathogens can develop resistance. The understanding of biochemistry of chitinolytic enzymes will make them more useful in a variety

Author's address: Department of Plant Sciences, School of Life Sciences, University of Hyderabad, Hyderabad-500046, India
E-mail: arpsl@uohyd.ernet.in; podilerao@yahoo.com

of processes in the near future. This review presents the current status of biological control of postharvest fungal diseases of fruits and vegetables and provides future directions.

Key words: Postharvest fungal diseases, fungicides, biological control, chitinases

WHT BIOLOGICAL CONTROL ?

In a hungry and competitive world where population increases day by day, reducing postharvest losses has become a major agricultural goal. Increased consumer awareness that diet and health are linked has resulted in a greater consumption of fruits. Fresh fruits and vegetables always draw the attention of an average consumer who look for quality food and often discard spoiled ones that result ultimately to an unaffordable loss. Fruits and vegetables are being harvested, transported to the market and stored, until sold they respire, undergo enzymatic breakdown and become softened followed by microbial attack. For highly perishable commodities, such as tomatoes, squash, peaches, as much as 30% of the harvested crop may be lost due to postharvest diseases before it reaches the consumer. A partial reduction in postharvest losses can significantly reduce the overall cost of production. Efforts to control factors like mechanical damage during harvesting, grading and packing can reduce the loss to a considerable extent. Postharvest decay adversely affects fruit industries; the reasons being that there are very few effective fungicide controls, people do not prefer fungicide treated commodities and the development of resistance by the causal pathogens (Mukerji and Garg, 1988a, b; Mukerji *et al.* 1999; Upadhyay *et al.*, 1997, 2000, 2001).

Developments in biological control of postharvest diseases are still in infancy, possibly because of little attention paid in this area. In this article, an effort is made to summarize the work done on biological control of postharvest diseases and the possibilities on the improvement of existing practices.

USE OF FUNGICIDES HAS SERIOUS LIMITATIONS FOR CONTROL OF POSTHARVEST DISEASES

In 1960s and 1970s, modern fungicides were extensively used to increase the shelf life of postharvest fruits and vegetables, the postharvest loss was 25% in United States (Janisiewicz and Korsten, 2002). Postharvest losses are estimated to range from 10-30% per year despite the use of modern storage facilities and techniques (Harvey, 1978). Infection by fungi and bacteria may occur during the growing season, at harvest time, during handling, storage, transport and marketing, or even after purchase by the consumer (Dennis, 1983). Safety and environmental concerns have forced the withdrawal of some fungicides from the market. In search of an alternative to synthetic fungicides it was proved that biological control would serve the purpose more efficiently to control postharvest diseases (Wilson *et al.*, 1991).

Fungicides are the primary means of controlling postharvest diseases and they have recently come under special scrutiny as posing potential oncogenic risks when applied to processed foods. Since postharvest fungicides are applied close to the time that fruits and vegetables are consumed, they present residual problems. Children are at higher risk than adults as their tissues are more susceptible to the carcinogenic nature of some commonly used fungicides. Even when acceptable fungicides were used to control postharvest diseases, pathogens developed resistance to them. Some of the postharvest chemicals like thiabendazole, dichloran and immazalil were used commonly. Resistance to thiabendazole and imazalil was widespread and their use was declined subsequently. Intensive and continuous use of fungicides for control of blue and green mold on citrus has led to resistance by the causal pathogens of these diseases (Holmes and Eckert, 1999). The fungicides that are available do not control all postharvest pathogens. *Mucor piriformis*, a major postharvest pathogen of apples and winter pears in the Pacific North is not controlled by any registered fungicide (Spotts and Dobson, 1989). This negative perception has promoted Governmental policies restricting use of fungicides. There is an urgent need to develop an alternative method that can efficiently control postharvest diseases of fruits and vegetables.

Research on the biological control has received its impetus in the year 1984 when *Bacillus subtilis* (B-3) was able to control brown rot of peaches. This organism was patented and was tested on semi-commercial packinghouse. One of the very successful stories of using biological control in postharvest diseases of fruits and vegetables were, the commercial products **Biosave 100, 110** and **Aspire** registered in United States by the US Environmental protection agency (EPA). In South Africa, commercial products registered were named as **Avogreen** and **Yeildplus**. Efforts are on to develop general and efficient methods of formulations, scaling-up, pilot-testing and involving genetic engineering inorder to come up with a solution to control postharvest diseases.

BIOLOGICAL CONTROL IS AN UPCOMING TECHNOLOGY

Biocontrol employs natural enemies of pathogens to eradicate or to control their population. It can be brought about either by introduction or by augmentation of one or more species of controlling organism or by a change in environmental conditions designed to favor the multiplication and activity of such organisms or by a combination of both procedures (Rivera *et al.*, 1974). Several promising approaches that include antagonistic microorganisms, induced resistance, and natural fungal compounds are known for developing new and potentially safer technologies for postharvest disease control. Specific causes of postharvest losses of fruits and vegetables may be classed as parasitic, nonparasitic, or physical (Cappellini and Ceponis, 1984). Among the proposed alternatives, development of antagonist microorganisms has been the most studied and substantial progress has been made in this area (Wisniewski and Wilson, 1992).

POSTHARVEST DISEASE SYSTEMS ARE HIGHLY SUITABLE FOR BIOCONTROL

The recent developments in the postharvest application of antagonists can be attributed to the uniqueness and relative simplicity of the postharvest system. Environmental conditions such as temperature and relative humidity can be managed to favor antagonist survival. Moreover, antagonists encounter minimal competition from other microorganisms due to the controlled environmental conditions (Janisiewicz and Korsten, 2002). Biocontrol of postharvest diseases tends to be more effective than under field conditions. A high concentration of the antagonists could be economically feasible rather than in field where it may not be cost effective. The period between harvesting and consumption includes a crucial stage where the fruits after harvest are stored, at relatively low temperatures. Extensive investigation is needed to find out the efficient antagonist that can be metabolically active even at lower temperature. The main strategy used to prevent postharvest decay is the application of antagonist to prevent pathogens from infecting fruit wounds after harvest, but the field application of biocontrol agents may also reduce the occurrence of postharvest decay to a significant extent.

On the other hand microbial metabolites and enzymes can also be exploited for the control of postharvest decay. But keeping in mind the safety concern of the human health, microbial metabolites including antibiotics cannot be preferred, and the fungal cell wall degrading enzymes could be focussed for postharvest disease control. Postharvest system will be advantageous as it could be modified according to the activity and stability of the concerned enzyme used.

ANTAGONISTIC MICROORGANISMS

Natural antagonists can be promoted and managed that already exist on fruit and vegetable surfaces or antagonists are artificially introduced against postharvest pathogens. Fruit surface is an excellent source of naturally occurring antagonists against postharvest decay. Many ecologically fit bacterial and yeast antagonists effective against postharvest decay of apple were isolated (Janisiewicz, 1991). When concentrated washings from the surface of the citrus fruit were plated out on agar medium only bacteria and yeast appeared, when these washings were diluted rot fungi appeared indicating that former is suppressing the growth of the later (Chalutz and Wilson, 1990). When fruits and vegetables are washed they are more susceptible to decay than those that are unwashed.

Artificially introduced antagonists will be productive area to explore, reasons being environmental conditions are maintained uncontrolled, the area for biological control agents are much more limited, application of elaborate biocontrol procedures are more cost effective than in field. Consumer losses due to postharvest diseases were recently identified (Wattal *et al.*, 2003). Biotechnological techniques are becoming increasingly valuable to elucidate the mechanisms of action of fungi and yeasts and provide genetic characterization and molecular markers to monitor the spread of these agents in controlling fungal diseases on vegetable crops (Punja and Utkhede, 2003).

Several antagonistic microorganisms were identified and are shown to be effective against a number of postharvest pathogens on a variety of harvested commodities. Presently, two antagonistic microorganisms, a yeast *Candida oleophila* and a bacterium *Pseudomonas syringae* are commercially available under the trade names **Aspire^{TM}** and **Biosave^{TM}**, respectively. Antagonistic microorganisms include bacteria, fungi and yeast.

Bacteria

Strains of *Bacillus pumilus* and *Pseudomonas fluorescens* that exhibit successful control of *Botrytis cinerea* have been identified (Swalding and Jeffries, 1998). Yam tubers inoculated with a spore suspension of *Bacillus subtilis* in potato dextrose broth showed a drastic reduction in the range and number of mycoflora (Okigbo, 2002). The antagonism was so effective that the normal tuber surface mycoflora was greatly reduced throughout the storage period of 5 months by a simple initial application of the antagonist. *Bacillus subtilis* was found to act against postharvest diseases of peaches and nectarines, apricots and plums (Pusey and Wilson, 1984). Bacteria were recovered and screened as biocontrol agents on cultivars red delicious apples primarily for the control of blue mold caused by *Botrytis cinerea* (Sholberg *et al.*, 1995).

Bacillus subtilis protects papaya against *R. nigricans*, causal agent of fruit rot (Dileep Kumar *et al.*, 1988). Biocontrol of postharvest brown rot and alternaria rot by *B. subtilis* and *Enterobacter aerogenes* were reported (Utkhede and Sholberg, 1986). *Alternaria citri* was controlled on citrus fruit by dipping wounded fruit in suspension of *B. subtilis* (Singh and Devarall, 1984).

Fungi

It was reported that *Botrytis* rot of strawberry was controlled with *Trichoderma* (Tronsmo and Dennis, 1977). Locally isolated antagonistic *T. viride* were observed to serve as a biocontrol agent in the management of postharvest pathogens of mango (Bhuvaneswari and Subba Rao, 2001). Using *Trichoderma viride* studies were made on *Botrytis cinerea* of postharvest rots in yams (*Dioscorea* spp.) (Okigbo and Ikediugwu, 2000).

Yeast

Yeasts such as *Pichia guilliermondis* (Wisniewski *et al.*, 1991) and *Cryptococcus laurentii*, that occur naturally on apple leaves, buds and fruits (Roberts, 1990) were applied for the control of postharvest decay on fruit. It was reported that yeast like *Cryptococcus infirmo-miniatus* and *Candida sake* successfully control brown rot and blue mold on sweet cherry (Spotts *et al.*, 1998) and three diseases of apple (Vinas *et al.*, 1998) respectively. The biocontrol activity of *Candida saitoana* was enhanced by the addition of glycolchitosan for the control of postharvest decay of apple and citrus fruit (El. Ghaouth *et al.*, 2000). Efficacy of *Candida saitoana* got increased when 0.2% (w/v) of 2-deoxy glucose was added for the control of some postharvest diseases of apple and early season citrus fruits (El. Ghaouth *et al.*, 2001; El. Ghaouth *et al.*, 2000). The application of a cell suspension of the BIO126 strain of *Metschnikowia pulcherrima* proved to be highly effective in the control of blue and gray mould, two of the most severe postharvest diseases on apple fruit (Spadaro *et al.*, 2004). It was even found that yeasts *Saccharomyces* spp., *Cryptococcus* spp., *Candida* spp., and *Rhodotorula* spp. could control postharvest diseases caused due to *Penicillium expansum* (Dan He *et al.*, 2003). $CaCl_2$ inhibited growth of the pathogen *R. stolonifer*, but did not affect the colony forming units of yeast *Candida guilliermondis* and *Pichia membranefaciens* (Tian *et al.*, 2002). Improvement of *Candida sake* biocontrol activity against postharvest decay by addition of ammonium molybdate was observed (Nunes *et al.*, 2002). *Pseudomonas syringae* and pink yeast *Sporobolomyces* when combined were found to have increased activity against blue mold caused by *P. expansum* than when individually applied (Janisiewicz and Bors, 1995). Yeast was also used as biocontrol agent for postharvest diseases on peaches and nectarines (Karabulut and Baykal, 2003). Mechanisms of action and efficacy of four isolates of the yeast *Metschnikowia pulcherrima* active against postharvest pathogens on apples (Spadaro *et al.*, 2002).

Abiotic and biotic factors that influence biocontrol activity can be determined with relative ease and can be manipulated to antagonistic advantage. Rapid colonization of fruit wounds by the antagonist is important for decay control; biocontrol can be improved by enhancing colonization (Mercier and Wilson, 1994; Janisiewicz *et al.*, 1992). Biocontrol of postharvest diseases can be improved by manipulating the environment, use of mixed cultures of antagonists, physiological and genetic manipulation of antagonists, modifications of formulations and integration with other methods.

MECHANISMS OF ACTION FOR BIOLOGICAL CONTROL AGENTS

Understanding the mode of action of antagonists is important for two reasons: (a) it will allow the development of more reliable procedures for the effective application of known antagonists; and (b) it should provide a rationale for selecting more effective antagonists. A basic knowledge on mode of action may reveal certain facts including identification of enzymes or metabolites that actually impart biocontrol activity to the antagonistic micro organism under investigation, and can direct us to study the biochemical aspects involved in the process to a great extent.

Several mechanisms of biocontrol have been suggested to be operating against postharvest rots on fruits including antibiosis, competition for space and nutrients, parasitism or direct interaction with the pathogen and induction of resistance in the host tissue (Droby and Chalutz, 1994). Often, more than one mechanism was implicated, but in no case has a sole mechanism been found responsible for biological control.

Antibiosis

Some of the most active biocontrol agents are bacteria producing antibiotics whose action partially determines their effectiveness. *Bacillus subtilis* produces gramicidin S (Edwards and Seddon, 2001), *Pseudomonas cepacia* synthesizes pyrrolnitrin, which used alone can control *Botrytis cinerea* and *Penicillium expansum* attacks on pome fruit (Janisiewicz *et al.*, 1991). Antibiosis could be due to antibiotics or hydrolytic enzymes produced by the biocontrol agent.

Several studies have demonstrated that many of the antibiotics produced by bacterial biocontrol agents have a broad-spectrum activity. Pyrrolnitrin has shown activity against *R. solani, B. cinerea, Verticillium dahliae* and *Sclerotinia sclerotiorum* (Ligon *et al.*, 2000). It was also reported to be active against several Gram-positive bacteria in particular *Streptomyces* sps. (El. Banna and Winkelmann, 1998). Zwittermycin A appeared to interact synergistically with Bt toxin produced by *B. thuringiensis* enhancing its insecticidal activity (Broderick *et al.*, 2000). The culture filtrate or purified antibiotics provided similar levels of control as achieved by the wild type of strain alone (Nakayama *et al.*, 1999). When the antibiotic production was inactivated, the ability of antagonistic bacteria was substantially reduced in controlling pathogen. Mutagenesis has been successfully used to demonstrate that antibiotic produced by *Pseudomonas spp*. (Thomashow and Weller, 1988; Hokeberg *et al.*, 1998) and *Bacillus cereus* (Silo-suh *et al.*, 1994) plays an important role in the biological control of plant diseases. Enhancement of antibiotic production was achieved by introduction or modification of antibiotic biosynthetic or regulatory genes. Multiple copies of the regulatory *gacA* gene introduced into *P. fluorescens* strain BL915 resulted in 2.5- fold increase in pyrrolnitrin production (Ligon *et al.*, 2000). Similar increase in pyrrolnitrin production in strain BL915 were obtained by PCR- based modification of the initiation codon of the chromosomal *gacA* gene or by replacing the native promoter of this regulatory gene with the strong *ptac* promoter from *E. coli*. In comparison with the wild type strain, the genetically modified strains showed an increase in biocontrol activity against *Rhizoctonia solani* (Ligon *et al.*, 2000).

The antagonistic activity was attributed to the production of enzymes like chitinase and glucanase. Chitin is abundantly present in fungal cell wall hence chitinases can be exploited to control fungal pathogens. Chitinase E was sprayed from a yam alone or together with β-1,3-glucanase directly onto the surface of a powdery mildew infecting strawberry berries and leaves, resulted in the complete degradation of the infection (Karasuda *et al.*, 2003). Postharvest diseases of fruits and vegetables can be controlled by using chitinases that are produced naturally by selected species of fungi and bacteria. A highly chitinolytic strain of *Serratia marcescens* was found to suppress the growth of *Botrytis* species *in vitro* (Someya, 2000). The role of extracellular chitinase in the biocontrol activity of *Trichoderma virens* was examined using genetically modified strains (Jong-Min Baek, 1999). The chitinase system of *Bacillus circulans* WL-12 has been studied extensively and has provided comprehensive findings on biochemical properties, structure-function relationships, the identification of essential amino acid residues for catalytic activity, and the mechanism by which multiple chitinases are generated (Alam, 1995). In a green house setting, *S.*

marcescens has been shown to control *B. cinerea, R. solani and Fusarium oxysporum. S. marcescens* controlled growth of *Sclerotinia minor*, the causal agent of basal drop diseases, in lettuce grown in green house. The *chiA* and *chiB* genes from *S. marcescens* have been transformed into other bacterial species like *P. fluorescens* and *E. coli* in an attempt to improve their ability to control fungal plant pathogens (Koby *et al.*, 1994; Sundheim *et* al., 1988) or to create new biocontrol agents (Shapira *et al.*, 1989; Oppenheim and Chet, 1992).

Competition

It exists between organisms that require the same resource for growth and the survival. Use of the resource by one organism reduces its availability for the other organism. Competition for space or nutrients usually takes place between closely related species. It was reported that the mechanism by which *Pseudomonas syringae* (Janisiewicz, 1987), *Acremonium breve* (Janisiewicz, 1988), and *Debaryomyces hansenii* (Wisniewski *et al.*, 1988) show their antagonistic nature on blue mold and gray mold respectively was due to nutritional competition (Chalutz *et al.*, 1988).

Parasitism

Antagonist and pathogen can interact also through direct parasitism. This is well documented when one fungus acts on another (hyperparasitism or mycoparisitism) and is affected by environmental factors, including nutrient availability. The hyphae of parasitic fungi penetrate their victims, sometimes with the aid of cell wall degrading enzymes. A strong adhesion of *P. guillermondii* antagonist cells to *B. cinerea* was observed *in vitro*, due to a lectin link(Wisniewski *et al.*, 1991; Jijakli and Lepoivre, 1998). Moreover *P. guilliermondii* shows a high activity of β-1, 3-glucanase enzymes that could result in the degradation of the fungal pathogen cell walls.

Induced Resistance

Some biocontrol agents can interact with the host tissue, particularly wounds, increasing the cicatrisation processes (Droby and Chalutz, 1994). This observation has suggested that yeast cell application induced resistance processes in the fruit skin. Some *Candida* strains are able to cause chemical and osmotic changes in apple tissues, favouring antagonist settlement (Mclaughlin *et al.*, 1990). In addition to controlling decays, *A. pullulans* can cause a transient increase in β-1, 3-glucanase, chitinase, and peroxidase activities on apple fruit (Ippolito *et al.*, 2000). The ability of

Candida saitoana to induce systemic resistance in apple fruit against *Botrytis cinerea* was investigated (El. Ghaouth *et al.*, 2003).

FUTURE PERSPECTIVES: CHITINASES CAN CHANGE THE SCENARIO

Natural compounds and microorganisms are abundantly present which can replace synthetic fungicides. A remarkable progress has been observed where biological control was implemented but still when compared to fungicidal application, biocontrol of postharvest diseases is in its infancy. Before biological control agents are formulated and applied as commercial products, it is important to explore strategies to enhance their activity. The conventional understanding of antagonist-pathogen interaction has to be modified and biotechnological aspects can be employed to design a product that can be powerful and economically feasible. It is through such an approach that full potential of the biocontrol agent can be realized. New avenues of research have to be proposed for advancing this needed technology. Biological control and fungicidal treatments have their own limitations, merits and demerits, so at this point it is unfair to equate both of them.

The introduction of antibiotic biosynthetic genes in heterologous, non-producing strains and subsequent evaluation of their ability to control plant diseases can also be exploited in postharvest biocontrol. In spite of understanding in detail about the antibiotic ability of some biological control bacteria, still this approach may not be exploited to control postharvest diseases of fruits and vegetables efficiently. Since biocontrol agents are applied close to the time that fruits and vegetables are consumed, human health concerns need to be focussed at this standpoint. Antibiotic producing antagonists or antibiotics alone when applied on to postharvest fruits and vegetables may leave a trace amount that on consumption may result in adverse effects. Moreover, pathogens against which antibiotics are targeted may develop resistance or can develop a mechanism with which they can degrade the biocontrol product thus used.

The main concern, related to the use of antibiotics in food products, is the development of human pathogens resistant to these compounds and the possible development of resistance in fruit pathogens. Even if antibiotic producers appear to be able to control wound infections established before antagonist application, at the moment, there are no such biocontrol agents registered for use on fruits (Spadaro and Gullino, 2004). Finally it can be concluded that use of antibiotic producing antagonists or antibiotics alone cannot meet the requirement efficiently hence considerable attention is needed to develop non-antibiotic producing antagonists, or to search enzymes, other compounds that can answer the problem.

An extensive understanding of enzymes such as chitinases may result in the discovery of a product that can be efficient enough and can solve the current problem under investigation. Chitinolysis plays an important role in biological control of plant diseases has been substantiated with increased disease control by chitin-supplemented application of chitinolytic biocontrol agents, greater field efficiency of chitinase preparations in disease control and enhanced biocontrol potential of genetically engineered organisms for chitinase overproduction. Biocontrol efficacy on non-chitinolytic bacteria was enhanced by expression of the chitinases from other organisms. Chitinases from different microbial sources have been purified, characterized and had highly significant antifungal activity. Instead of going for the whole cell in case of antagonistic microorganisms, it would be appropriate to identify the product responsible for the biocontrol activity. The success in using chitinases for different applications depends on the supply of highly active preparations at a reasonable cost. Most of the suppliers use either natural microbial biodiversity or genetically engineered chitinase overproducing microbial strains to obtain efficient preparations. The understanding of biochemistry of chitinolytic enzymes will make them more useful in a variety of processes in the near future (Patil *et al.*, 2000).

Most of the chitinases are acidic in nature but there are reports stating that some bacterial chitinases are alkaline. Hence manipulating the postharvest environment according to the need of the chitinase activity can prove better and could be economically feasible. A basic understanding of resistance and defense mechanisms in harvested fruits and vegetables should reveal a multitude of new approaches for control of postharvest diseases.

Significant progress has been made towards biological and integrated control of postharvest diseases on fruits. Postharvest conditions provide an ideal niche for biocontrol agents and their products since they are less subjected to sudden weather changes, and are often equipped with sophisticated climate control system (Spadaro and Gullino, 2004).

Considering the above successful reports, testing biological control agents for postharvest decay, jumping from the lab to postharvest environment is not as abrupt as that into the field. We can perhaps expect a more rapid application of laboratory results for postharvest biocontrol than that experienced with their field applications.

Finally, promising research on the use of antagonistic microorganisms and its products to control postharvest diseases of fruits and vegetables will be able to develop biological control which will be effective and safer to the human community and environment. In the near future, use of biological control will ensure us the delicacy of fruits and vegetables with no fungicide in it. Hence chitinases can be exploited to control postharvest diseases of fruits and vegetables thereby increasing the shelflife of the perishables and a safer healthy environment for the coming generation. At this moment, biological control should be viewed seriously as an important component of postharvest disease management if a permanent reduction of pesticide use is our goal.

Acknowledgements

The authors thank the AP-Netherlands Biotechnology Programme for sponsoring research work on biological control of fungal pathogens to the senior author.

References

Alam, M.M., Nikaidou, N., Tanaka, H.N. and Watanabe, T. 1995. Cloning and sequencing of *chiC* gene of *Bacillus circulans* WL-12 and relationship of its product to some other Chitinases and Chitinases like proteins. *J. Fermentation Bioengi*. **80**: 454-461.

Bhuvaneswari, V. and Subba Rao, M. 2001. Evaluation of *Trichoderma viride* antagonistic to post-harvest pathogens on mango. *Indian Phytopathol*. **54**: 493-494.

Broderick, N.A., Goodman, R.M., Raffa, K.F. and Handelsman, J. 2000. Synergy between zwittermycin A and *Bacillus thuringiensis* subsp. *kurstaki* against gypsy moth (Lepidoptera: Lymantriidae). *Environmental Entomology*. **29**: 101-107.

Chalutz, E., Droby, S. and Wilson, C.L. 1988. Mechanism of action of postharvest biological control agents. 5th Int. Congr. Plant Pathology. 422pp.

Chalutz, E. and Wilson, C.L. 1990. Postharvest biocontrol of green and blue mold and sour rot of citrus fruit by *Debaryomyces hansenii*. *Pl. Dis*. 74: 134-137.

Cappellini, R.A. and Ceponis, M.J. 1984. Postharvest losses in fresh fruits and vegetables. In: H.E. Moline (ed) *Postharvest Pathology of Fruits and Vegetables: Postharvest Losses in Perishable Crops,* Univ. Calif., Agric. Exp. Sta., Bull. No. 1914 (pub NE-87), pp. 24-30.

Dan He, Zheng, X.D., Yin, Y.M, Sun, P., and Zhang, H-Yin. 2003. Yeast application for controlling apple postharvest diseases associated with *Penicillium expansum*. *Botanical Bull. of Acad. Sinica*. **44**: 211-216.

Dennis, C. 1983. *Postharvest Pathology of Fruits and Vegetables*. Acad. Press. London.

Dileep Kumar, B.S., Podile, A.R. and Dube, H.C. 1988. Antagonistic activity of *Bacillus subtilis* towards *Rhizopus nigricans*. *J. Biological Control*. **2**: 42-44.

Droby, S. and Chalutz, E. 1994. Mode of action of biocontrol agents of postharvest diseases. In: Wilson, C.L., Wisniewski, M.E. (Eds.), *Biological control of postharvest diseases. Theory and practice*. CRC Press, Boca Raton, USA, pp. 63-75.

Edwards, S.G. and Seddon, B. 2001. Mode of antagonism of *Brevibacillus brevis* against *Botrytis cinerea in vitro*. *J. of Appl. Microbiol*. **91**. 652-659.

El-Banna, N. and Winklemann, G. 1998. Pyrrolnitrin from *Burkholderia cepacia*: antibiotic activity against fungi and novel activities against *Streptomycetes*. *J. Appl. Microbiol.* **85**: 69-78.

El-Ghaouth, A., Wilson, C.L. and Wisniewski, M.E. 2003. Control of postharvest decay of apple fruit with *Candida saitoana* and induction of defense responses. *Phytopathology*. **93**: 344-348.

El-Ghaouth, A., Smilanick, J.L, Brown, G.E., Ippolito, A. and Wilson, C.L. 2001. Control of decay of apple and citrus fruits in semi-commercial tests with *Candida saitoana* and 2 deoxy-D-glucose. *Biological Control*. **20**: 96-101.

El-Ghaouth, A., Smilanick, J.L., Wisniewski, M.E. and Wilson, C.L. 2000. Improved control of apple and citrus fruit decay with a combination of *Candida saitoana* and 2 deoxy-D-glucose. *Pl. Dis.* **84**: 249-253.

El-Ghaouth, A., Smilanick, J.L. and Wilson, C.L. 2000. Enhancement of the performance of *Candida saitoana* by the addition of glycholchitosan for the control of postharvest decay of apple and citrus fruits. *Postharvest Biol. Technol.* **19**: 103-110.

Harvey, J.M. 1978. Reduction of losses in fresh fruits and vegetables. *Annual Review of Phytopathol.* **16**: 321-341.

Hokeberg, M., Wright, S.A.I., Svensson, M., Lundgren, L.N. and Gerhardson, B. 1998. Mutants of *Pseudomonas chlororaphis* defective in the production of an antifungal metabolite express reduced biocontrol activity. Abstract Proceedings ICPP98, Edinburgh, Scotland.

Holmes, G.J. and **Eckert, J.W.** 1999. Sensitivity of *Penicillium digitatum* and *Penicillium italicum* to postharvest citrus fungicides in California. *Phytopathol.* **89**: 716-721.

Ippolito, A., El-Ghaouth, A., Wilson, C.L. and Wisniewski, M. 2000. Control of postharvest decay of apple fruit by *Aureobasidium pullulans* and induction of defense responses. *Postharvest Biol. Tech.* **19**: 265-272.

Janisiewicz, W.J. 1987. Postharvest biocontrol of blue mold on apple. *Phytopathol.* **77**: 481-485.

Janisiewicz, W.J. 1988. Biological control of postharvest diseases of apples with antagonist mixtures. *Phytopathol.* **78**: 194-198.

Janisiewicz, W.J. 1991. Control of postharvest diseases of fruits with biocontrol agents. In *The Biological control of Plant Diseases*, ed. J Bay-Petersen, pp. 56-68. Taipei, Taiwan: Food Fertil. Technol. Cent. Asian Pac. Reg.

Janisiewicz, W.J., Yourman, L., Roitman, J. and Mahoney, N. 1991. Postharvest control of blue mold and gray mold of apples and pears by dip treatments with pyrrolnitrin, a metabolite of *Pseudomonas cepacia*. *Pl. Dis.* **75**: 490-494.

Janisiewicz, W.J., Usall, J. and Bors, B. 1992. Nutritional enhancement of biocontrol of blue mold on apples. *Phytopathol.* **82**: 1364-70.

Janisiewicz, W.J. and Bors, B. 1995. Development of a microbial community of bacterial and yeast antagonists to control wound-invading postharvest pathogens of fruits. *Appl. Environmental Microbiol.* **61**: 3261-3267.

Janisiewicz, W.J. and Korsten, L. 2002. Biological control of postharvest diseases of fruits. *Ann. Rev. Phytopathol.* **40**: 411-441.

Jijakli, M.H. and Lepoivre, P. 1998. Characterization of an exo-β-1,3- glucanase produced by *Pichia anomala* strain K, antagonist of *Botrytis cinerea* on apples. *Phytopathol.* **88**: 335-343.

Jong-Min Baek, Howell, C.R., and Kenerley, C.M. 1999. The role of an extracellular chitinase from *Trichoderma virens* Gv29-8 in the biocontrol of *Rhizoctonia solani*. *Current Genetics.* **35**: 41-50.

Karasuda, S., Tanaka, S., Khaihara, H., Yamamoto, Y. and Koga, D. 2003. Plant chitinase as a possible biocontrol agent used instead of chemical fungicides. *Biosci. Biotechnol. Biochem.* **67**: 221-224.

Karabulut, O.A. and Baykal, N. 2003. Biological control of postharvest diseases of peaches and nectarines by yeasts. *Journal of Phytopathol.* **151**: 130-134.

Koby, S., Schickler, H., Chet, I. and Oppenheim, A.B. 1994. The chitinase encoding Tn7-based *chiA* gene endows *Pseudomonas fluorescens* with the capacity to control plant pathogens in soil. *Gene.* **147**: 81-83.

Ligon, J.M., Hill, D.S, Hammer, P.E., Torkewitz, N.R., Hofmann, D., Kempf, H.J. and Van Pee, K.H. 2000. Natural products with antifungal activity from *Pseudomonas* biocontrol bacteria. *Pest Management Sci.* **56**: 688-695.

Mclaughlin, R.J., Wisniewski, M.E., Wilson, C.L. and Chalutz, E. 1990. Effects of inoculum concentration and salt solutions on biological control of postharvest diseases of apple with *Candida* sp. *Phytopathol.* **80**: 456-461.

Mercier, J. and Wilson, C.L. 1994. Colonization of apple wounds by naturally occurring microflora and introduced *Candida oleophila* and their effect on infection by *Botrytis cinerea* during storage. *Biological control.* **4**: 138-44.

Mukerji, K.G. and Garg K.L. (eds.) 1988a, Biocontrol of Plant Diseases. Vol 1. CRC Press, Florida, p. 211.

Mukerji, K.G. and Garg K.L. (eds.) 1988b, Biocontrol of Plant Diseases. Vol 2. CRC Press, Florida, p. 198.

Mukerji, K.G., Chamola, B.P. and Upadhyay, R.K. (eds.) 1997. Biotechnological Approaches in Biocontrol of Plant Pathogens, Kluwer Academic/Plemem Publishers, New York, Dordrecht, London. p. 255.

Nakayama, T., Homma, Y., Hashidoko, Y., Mizutani, J. and Tahara, S. 1999. Possible role of Xanthobaccins produced by *Stenotrophomonas* sp strain SB-K88 in suppression of sugarbeet damping off disease. *Appl. Environmental Microbiol.* **55**: 4334-4339.

Nunes, C., Usall, J., Teixido, N. and Vinas, I. 2002. Improvement of *Candida sake* biocontrol activity against postharvest decay by the addition of ammonium molybdate. *J. Appl. Microbiol.* **92**: 927-935.

Okigbo, R.N. and Ikediugwu, F.E.O. 2000. Studies on biological control of postharvest rots in yams (*Dioscorea* species) using *Trichoderma viride*. *J. Phytopathol.* **148**: 351-355.

Okigbo, R.N. 2002. Mycoflora of tuber surface of white yam (*Dioscorea rotundata* poir) and postharvest control control of pathogens with *Bacillus subtilis*. *Mycopathologia.* **156**: 81-85.

Oppenheim, A.B. and Chet, I. 1992. Cloned chitinases in fungal plant-pathogen control strategies. *Trends in Biotechnol.* **10**: 392-394.

Patil, S.R., Ghormade, V. and Deshpande, M.V. 2000. Chitinolytic enzymes: an exploration. *Enzyme and Microbial Technol.* **26**: 473-483.

Punja, Z.K. and Utkhede, R.S. 2003. Using fungi and yeasts to manage vegetable crop diseases. *Trends in Biotechnol.* **21**: 400-407.

Pusey, P.L. and Wilson, C.L. 1984. Postharvest biological control of stone fruit brown rot by *Bacillus subtilis. Pl. Dis.* **68**: 753-756.

Rivera, J.R., Gonalez, M.A. and Cuevas-Ruz, J. 1974. Sprout inhibition in yam by gamma radiation. *J. Agric. Univ. Puerto Rico.* **58**: 330-337.

Roberts, R.G. 1990. Biological control of grey mold of apple by *Cryptococus laurentii. Phytopathology.* **80**: 526-530.

Shapira, R., Ordentlich, A., Chet, I. and Oppenheim, A.B. 1989. Control of plant diseases by chitinase expressed from cloned cDNA in *Escherichia coli. Phytopathology*. **79**: 1246

Spadaro, D., Vola, R. and Piano, S. 2002. Mechanisms of action and efficacy of four isolates of the yeast *Metschnikowia pulcherrima* active against postharvest pathogens on apples. *Postharvest Biol. Technol.* **24**: 123-134.

Spadaro, D. and Gullino, M.L. 2004. State of the art and future prospects of the biological control of postharvest fruit diseases. *Internat. J. Food Microbiol.* **91**: 185-194.

Spadaro, D., Garibaldi, A. and Gullino, M.L. 2004. Control of *Penicillium expansum* and *Botrytis cinerea* on apple combining a biocontrol agent with hot water dipping and acibenzolar-S-methyl, baking soda, or ethanol application. *Postharvest Biol. and Technol.* **33**: 141-151.

Spotts, R.A., Cervantes, L.A., Facteau, T.J. and Chand-goyal, T. 1998. Control of brown rots and blue mold of sweet cherry with pre-harvest iprodione, postharvest *Cryptococcus infirmo-miniatus*, and modified packaging. *Pl. Dis*. **82**: 1158-1160.

Silo-Suh, L.A., Lethbridge, B.J., Raffel, S.I., He, H.Y., Clardy, J. and Handelsman, J. 1994. Biological–activities of two fungistatic antibiotics produced by *Bacillus cereus* UW85. *Appl. Environmental Microbiol.* **60**: 2023-2030.

Someya, N., Kataoka, N., Komagata, T., Hirayae, K., Hibi, T. and Akutsu, K. 2000. Biological control of cyclamen soil borne diseases by *Serratia marcescens. Pl. Dis.* **84**: 334.

Sholberg, P.L., Marchi, A. and Bechard, J. 1995. Biocontrol of postharvest diseases of apple using *Bacillus* species isolated from stored apples. *Can. J. Microbiol.* **41**: 247-252.

Singh, V. and Deverall, B.J. 1984. *Bacillus subtilis* as a control agent against fungal pathogens of citrus fruit. *Tr. Br. Mycol. Soc.* **83**: 487-490.

Spotts, R.A. and Dobson, R.L. 1989. Effect of the experimental fungicide RH886 on *Mucor piriformis. Pesticide Sci*. **25**: 391-399.

Sundheim, L., Poplawsky, A.R., and Ellingboe, A.H. 1988. Molecular cloning of two chitinase genes from *Serratia marcescens* and their expression in *Pseudomonas* species. *Physiological and Mol. Pl. Pathol.* **33**: 483.

Swalding, I.R. and Jeffries, P.J. 1998. Antagonistic properties of two biocontrol agents of grey mould diseases. *Biocontrol Sci. Techol.* **8**: 439-448.

Tian, S.P., Fan, Q., Xu, Y. and Jiang, A.L. 2002. Effect of calcium on biocontrol activity of yeast antagonists against postharvest fungal pathogen *Rhizopus stolonifer. Pl. Pathol.* **51**: 352-358.

Thomashow, L.S. and Weller, D.M. 1988. Role of phenazine antibiotic from *Pseudomonas fluorescens* in biological control of *Gaeumannomyces graminis* var. *tritici. J. Bacteriology.* **170**: 3499-3508.

Tronsmo, A. and Dennis, C. 1977. The use of *Trichoderma* species to control strawberry fruit rots. *Netherland Journal of Pl. Pathol.* **83**: 449-455.

Upadhyay, R.K., Mukerji, K.G. and Rajak, R.L. (eds.) 1997. IPM System in Agriculture, Vol. 2. Biocontrol in Emerging Biotechnology, Aditya Books Pvt. Ltd., New Delhi, p. 620.

Upadhyay R.K., Mukerji, K.G. and Chamola, B.P. (eds.) 2000. Biocontrol Potential and Its Exploritation in Sustainable Agriculture, Volume I, Kluwer Academic Publishers, New York, Dordrecht, London, p. 294.

Upadhyay, R.K., Mukerji, K.G. and Chamola, B.P. (eds.) 2001. Biocontrol Potential and Its Exploritation in Sustainable Agriculture. Volume II. Kluwer Academic/ Plemum Publishers, New York, Dordrecht, London. p. 421.

Utkhede, R.S. and Sholberg, P.L. 1986. *In vitro* inhibition of plant pathogens by *Bacillus subtilis* and *Enterobacter aerogenes* and *in vivo* control of two postharvest cherry diseases. *Can. J. Microbiol.* **32**: 963-967.

Vinas, I., Usall, J., Texido, N. and Sanchis, V. 1998. Biological control of major postharvest pathogens in apple with *Candida sake. Internat. J. Food Microbiol.* **40**: 9-16.

Wilson, C.L., Wisniewski, M.E., Biles, C.L., MacLaughlin, R., Chalutz, E. and Droby, E. 1991. Biological control of post-harvest diseases of fruits and vegetables: alternatives to synthetic fungicides. *Crop Protection.* **10**: 172-177.

Wisniewski, M., Wilson, C., Chalutz, E. and Hershberger, W. 1988. Biological control of postharvest diseases of fruits: inhibition of *Botrytis* rot on apples by antagonistic yeast. *Proc. Electron Microscopy Soc. of America.* **46**: 290-291.

Wisniewski, M., Biles, C., Droby, S., McLaughlin, R., Wilson, C.L. and Chalutz, E. 1991. Mode of action of the postharvest biocontrol yeast *Pichia guilliermondii.* 1. charecterization of attachment to *Botrytis cinerea. Physiological and Mol. Pl. Pathol.* **39**: 245-258

Wisniewski, M.E. and Wilson, C.L. 1992. Biological control of postharvest diseases of fruits and vegetables; recent advances. *Horticultural Sci.* **27**: 94-98.

Wattal, S.K., Razdan, V.K. and Puttoo, B.L. 2003. Post-harvest diseases of cherry and consumer losses. *Indian Phytopathol.* **56**: 194-195.

SECTION 6

MYCORRHIZA

Frontiers in Plant Sciences (Pages 567-582)
Eds. K.G. Mukerji, K.V.B.R. Tilak, S.M. Reddy, L.V. Gangwane, P. Prakash, I.K. Kunwar

Plant–Fungus Interaction During Development of Arbuscular Mycorrhiza

Rupam Kapoor

ABSTRACT

Compared with the interactions between plants and other microorganisms, AM symbiosis have extremely low specificity and high compatibility between plant fungal symbionts, which may reflect both their mutualistic biotrophic nature and the very long period of coevolution. The development of AM colonization involves several well – defined stages. The interaction between plant and fungus starts before they come into physical contact. Hyphal growth and branching are induced by roots of host plants, followed by the formation of appressoria which are thought to be key event in recognition leading to the hyphal penetration of the roots. By contrast, in non-host plants, the signals inducing hyphal growth and branching, and recognition at the root surface, are apparently switched off. Following the penetration of host root epidermal cells, hyphae grows inter–or intracellularly in the root cortex and intracellular hyphae forms arbuscules. Exchange of P and carbohydrate between plant cells and fungal hyphae occurs across these plant–fungus interfaces.

Key words: Root, fungus, interaction, arbuscular mycorrhiza, gene expression.

INTRODUCTION

Significant steps have been made over recent years towards unravelling mechanisms governing arbuscular mycorrhizal symbiosis, particularly in

Author address: Department of Botany, University of Delhi, Delhi-110007

relation to molecular host responses to the mycosymbionts and the plant genes involved (Harrison 1998). Develpment and functioning of arbuscular mycorrhiza is governed by morpohogenetic programmes in both partners and evidence for an interplay of signalling/recognition events can be found at different stages during plant fungal interactions.

SIGNALLING BETWEEN AM FUNGI AND PLANTS

Role of Root Exudates

Germination of AM fungal spores and the initial growth of hyphal germ tubes can occur in the absence of the plant. Experimental evidence indicates that the quality and source of the exudates play an important role in triggering germination. For example, the exudates from non-host plants such as *Brassica* spp. or *Lupinus* spp. do not stimulate germination (Giovannetti et al. 1993). In this case, lack of recognition of the non-host is due to inhibitory compounds or due to lack of a signal (Schreiner and Koide, 1992, 1993; Vierhe ilig et al., 1996).

Over the years, evidence has accumulated that roots emit a volatile signal that stimulates the directional growth of the AM fungus towards them (Annapurna et al., 1996). One of the prime candidate for this volatile signal is CO_2 which can stimulate extensive hyphal growth of some AM fungi (Becard and Piche 1989; Balaji et al. 1995; Chabot et al. 1992). It has been suggested that CO_2 is an essential carbon source for hyphal growth, and it may be involved in the catabolism of lipids in the growing hypha. Thus, the possibility of a dual role for CO_2 in AM mycorhizal biology as both a trigger for germination and a carbon source cannot be ruled out (Gadkar et al. 2001). Some of the activities are probably due to flavanoid and phenolic compounds that stimulate the growth of some AM fungal species while inhibiting others (Douds et al. 1996; Poulin et al. 1993; Tsai and Phillips 1991). The particular molecule(s) responsible for eliciting hyphal branching is unknown but, based on its estimated size, it could be a phenolic or flavonoid derivative (Giovannetti et al. 1996). Since the flavonoid compounds are active at very low concentrtions, it is assumed they do not have a nutritional effect but rather that they act as signals to stimulate or inhibit growth. Based on the structures of these molecules, it is suggested that the A and C rings of the isoflavonoid and the hydroxyl group at position A-7 are important features for recognitions by the receptor (Poulin et al. 1997). Although flavanoid derivatives can influence the initial stages of the fungal life cycle, experiments with flavanoid-deficient mutants of maize indicate that they are not essential for the development of the symbiosis (Becard et al. 1995).

Signal compounds as yet unknown, could also exist; these may be analogous to the oligosaccharins that act as signal molecules in both plant–pathogen and legume–*Rhizobium* spp. interactions. Mycorrhizal fungi have been demonstrated to have a weak cellulase and endopolygalacturonase activities, and both of these enzymes have the ability to catalyze the release of oligosaccharides or oligosaccharins from the cell wall (Fry et al. 1993). The latter could trigger the colonization and spread of the fungus by a cascade of events that are autoregulated and controlled by the host (Salzer and Boller 2000).

Genes Expressed During early Interaction

Some 50 different plant genes have been analysed for expression in arbuscular mycorrhiza. Studies targeting known genes have focussed mostly on those encoding defence response molecules or nodulins in legume–*Rhizobium* interactions. The localized and/or low activation of such genes suggests that some signalling pathways in arbuscular mycorrhiza and other plant-microbe interactions may be shared (Albrecht et al. 1998; Gianinazzi-Pearson, 1996; Jayadev and Dubey, 2005).

Hypotheses to explain attenuated plant defence responses to symbiotic fungi include elicitor degradation, cell signalling through altered sugar fluxes and modified hormone balance (Blee and Anderson 2000; Lambias 2000). Cytokinins or chittoligosacchrides, released from glomalean walls, have been proposed as components of signal transduction pathway mediating nodulin gene induction in arbuscular mycorrhizal interactions (van Rhijn et al. 1997). Differential regulations are known for plant genes encoding membrane proteins which may be inolved in transport processes within the symbiosis (Gianinazzi-Pearson et al. 2000; Harrison 1998; Krajinski et al. 1998) and a number of other plant genes with putative functions in the symbiosis have been identified by non-targeted molecular analyses (Lapopin et al. 1999; Murphy et al. 1997 Lammers et al. 2005).

Partial transcriptome analysis by differential RNA display revealed modifications in gene expression early during pea–*Glomus mosseae* interactions which represented essentially up-regulated genes. It indicated frequent similarities in up-regulated genes in *G. mosseae* inoculated mutant and non-mutant pea genotypes. One of these has been identified as plant *Clp* serine protease (Roussel et al. 2001). In all organisms, proteolysis assures vital functions in the regulation of many metabolic processes and in cell responses to environmental stimuli. As regulatory enzymes, protease can control metabolic fluxes by regulating the level of key rate-limiting enzymes. As housekeeping enzymes, they maintain cell homeostasis and keep optimal metabolic activities by removing non-functional proteins. The latter functions becomes essential during environmental stresses where the occurrence of damaged protein increases (Porankiewicz et al. 1999).

RESPONSE OF AM FUNGUS TO THE PRESENCE OF A HOST

Hyphae elongate 20 times more slowly in the absence of host roots than in their presence (Becard and Piche 1989). AM fungi respond to host exudates with extensive hyphal growth and branching (Giovannetti et al. 1993). Despite the high mycelial growth in the presence of the roots, hyphae do not always appear to exhibit "directional growth" toward the roots until they are very close to the host (Mosse and Hepper 1975). Preliminary evidence suggests that a factor(s) that elicits branching of *Glomus mosseae* is a compound of < 500D (Giovannetti et al. 1996). Such a factor(s) is exuded from the roots of many host species but not at the same level in all cases. Nagahashi and Douds (2000) showed that, in response to a soluble host factor derived from the roots, the branching pattern of *Gigaspora gigantia* changed from dense to scattered. All mycotrophic plants produce a soluble factor(s), which induced hyphal branching in *Gigaspora margarita* (Buee et al. 2000). This branching factor was found to be strikingly absent from non-mycotrophic plant e.g. *Brassica* spp.

The root exudate can not only stimulate the branching capacity of the fungi, but at the same time, inhibitor molecules may also be involved in regulating the symbiotic event (David-Schwartz et al. 2001). Whether branching inhibitors are exuded from the host or are produced in the rhizosphere following exudation of other substances remains to be demonstrated but the fact that this phenomenon can be related to the genetic make up of the host opens new possibilities for the discovery of new complementary AM fungus – host interactions. Whether the same chemical acts at each check point or whether there are different signals remains to be determined.

As the main hypha approaches a root, it puts out a characteristic fan-shaped complex of lateral branches (Giovannetti et al. 1993). These fan-like structures were also observed by Karandashov et al. (2000) under *in vitro* conditions. Prevention of actual contact with the roots by means of the membrane sandwich technique did not, however, prevent the development of these fan–like structures. It can be concluded that in the presence of the host, specific morphogenesis of the fungus take place, process that the non-host plant is unable to elicit.

APPRESSORIUM FORMATION

The topographical or biochemical signals on the root surface may be necessary for appressorium formation (Gadkar et al. 2001). Although components of root exudates are capable of stimulating hyphal growth and branching, they are unable to elicit the formation of appressoria, which were observed only on intact plant roots (Giovannetti et al. 1993). Nagahashi and Douds (1997) showed that appressoria formed on isolated epidermal cell walls derived from carrot roots, but not on isolated cortical or vascular cell walls. Apart from attachment to isolated cell walls, formation of penetration hyphae was not observed, indicating the absence of a physiochemical signal intrensic to the living epidermal cells. *Gigaspora margarita* could form appressoria *in vitro* on purified epidermal cell walls isolated from carrot roots (host) but not on walls isolated from sugar beet, a non-host.

No plant signal has so far been implicated in appressorium formation, but there is evidence that a shoot factor(s) may be involved in the inhibition of appressorium formation in *Lupinus* spp. (Gianinazzi-Pearson and Gianinazzi 1992). The experiment confirms that the branching signal is either loosely bound to the wall or exuded from the roots, since the purified wall fragments did not elicit the extenisve hyphal branching observed in intact roots. Carbohydrate molecule act as signals in a number of other plant/fungal interactions and are likely candidates for the induction of appressorium in the AM symbiosis (Longman and Callow 1987).

PLANT RECOGNIZES AM FUNGI

In plant-pathogen interaction, plant respond to fungal attack by eliciting various mechanisms, some of which are well characterized; such as the plant defense response. Such resistance responses i.e. incompatible interactions, occur when plant recognize elicitor compounds in the presence of an invading pathogenic fungus. Although there are indications of elicitor involvement in the early stage of mycorrhizal formation, the elicited defense response is generally less vigorous than that observed in plant-pathogen interactions (Salzer and Boller 2000) and is often completely suppressed (David et al. 1998). In contrast to plants interaction with pathogens, plant-mycorrhizal associations are exceptional in being compatible. Infection by AM fungi appears to initiate some plant defense responses in the host tissue, but these do not seem to reach levels that would prevent colonization. Furthermore, expression of defense genes is localized to arbuscule – containing parenchyma cells, and the elicitation of the defense reaction in other cells of the roots is not overcome (Gianinazzi-Pearson 1996).

Weak, transient and in most cases, localized and uncoordinated host defense responses are elicited (Dumas–Gaudot et al., 2000). In addition, Shaul et al. (2000) discussed the possible involvement of a suppression mechanism in the *Glomus interadices* – tobacco interaction. Finally, Vierheilig et al. (1995) demonstrated that transgenic tobacco plants that over-express some of the PR gene do not restain AM fungal colonization. These inconsistencies in the expression of PR genes and of protein activities in relation to the AM fungal symbiosis have revealed that PR defense reactions do not necessarily respond to the AM fungal invasion via a typical defense–related pathway (Dumas–Gaudot et al. 2000). It might be concluded that signal reception or recognition of AM fungi by a plant does not elicit a typical plant defense response or, alternatively, that such a response is rapidly suppressed by parallel mechanism(s) induced by the AM fungus.

Harrison (1998) suggested that recognition of compatibility between the plant and the fungus terminated the elicitation of the plant defense response. The brief mobilization of defense responses may result in the production of suppressors by the mycorrhizal fungi which prevent recognition of elicitors (Lambias and Mohdy 1993). To date, no suppressor has been identified, and the alternative hypotheses have recently been presented by Salzer and Boller (2000) and Shaul et al. (2000). The existence of plants that exhibit defense response and plants that do not, as well as plants that suppress their defense responses during mycorrhizal formation, suggests the involvement of co-evolutionary processes in the development of this symbiosis.

PENETRATION OF THE ROOT

Appressorium formation is followed by the development of a penetration hypha and penetration of the root. This can occur in different ways : in some species, the hypha enters by forcing its way between two epidermal cells, whereas in other cases, the hypha penetrates an epidermal or root hair cell wall and grows through the cell (Bonfante-Fasolo 1984). The exact mechanisms involved in penetration are unknown; however, by analogy to a number of the biotrophic pathogens (Mendgen and Deising 1993), it has been suggested that specific, localized production of cell wall–degrading enzymes, in combination with mechanical force, may facilitate entry of the hyphae without inducing defense responses (Bonfante and Perotto 1995).

AM fungi produce exo- and endoglucanases, cellulases, xyloglucanases, and pectolytic enzymes including polygalacturonase (Garcia–Garrido et al. 1992; 1996; Garcia-Romer et al. 1991; Rejon-Palomares et al. 1996) all of which would expedite their passage through a cell wall.

A wide range of plant mutants on which the AM fungi can form appressoria but cannot develop further are direct proof that the plant controls this developmental step in the association. These referred to as *myc*$^{-1}$ are most frequently found among *nod*$^{-}$ fix^{-} mutants. Mutants blocked at this stage of this symbiosis have been described in *Pisum sativum* (Duc et al. 1989, Gianinazzi-Pearson et al. 1991), *Medicago sativa* (Bradbury et al. 1991; 1993) *Vicia faba* (Duc et al. 1989), *Phaseolus vulgaris* (Shirtliffe and Versey 1996), *Medicago truncatula* (Sagan et al. 1995), *Lotus japonicus* (Wegel et al. 1998), and *Lycopersicon esculentum* (Barker et al. 1998). The phenotypes of these mutant association are fairly similar at the morphological level and fall into two broadly defined groups. In association with the *P. sativum, L. esculentum* and *Medicago* mutants, the fungus forms appressoria that are frequently large and deformed and that become septate when the fungus fails to enter the root (Gianinazzi-Pearson et al. 1991).

In one of the *Medicago sativa* mutant lines, the number of appressoria formed on the mutant increases, a possible consequence of the failure to penetrate (Bradbury et al. 1991); however, increased numbers of appressoria were not reported for the *P. sativum* mutants (Gianinazzi-Pearson et al. 1991). The non-penetrating phenotype (myc^{-1}) belong to five complementation groups, which indicates that entry into the root is under complex genetic control. The traits segregate as single recessive loci and the conditions is determined by the roots (Gianinazzi-Pearson et al. 1991). Grafting wild-type scions onto mutant stocks did not rescue the mutants (Vierheilig and Piche 1996). Hairy roots prepared from these genotypes also maintain their non-mycorrhizal status (Balaji et al. 1994). Cytochemical analysis of one of the *P. sativum* and one of the *M. sativia* mutant interactions indicated tht cell wall depositions including callose and phenolics, were present in the walls of cells adjacent to the appressoria (Peterson and Bradbury 1995). Such depositions were not seen in wild-type interaction, which suggests that a defense has been elicited in these mutants. Based on these data, it is possible that a suppressor of defense response has been mutated such that the plants now views the fungus as a pathogen.

In the lotus mutants, the hyphae occasionally manage to overcome the block in the epidermis and growth from the deformed hyphae continues, producing normal internal mycorrhizal structures. Since there are indistinguishable from wild-type, it has been suggested that the mutated genes are not required for the later phases of growth (Wegel et al. 1998, Senoo et al. 2000).

Another phenotype, designated as myc^{-2} has been identified among Nod^{+} fix^{-} mutants in *Pisum sativum* (Gianinazzi-Pearson et al. 1991), *Medicago truncatula* (Sagan et al. 1995) and *M. sativa* (Bradbury et al. 1993). In this phenotype, fungi can penetrate the roots and colonize cortical cells, but arbuscular development is reduced to a few stumpy branches.

These different phenotypes of mycorrhiza mutants were thought to be entirely controlled by the altered genes in the plants (Gianinazzi-Pearson et al. 1991; Wegel et al. 1998). A mycorrhiza-defective mutant has been identified in a non-legume plant species, *Lycopersicon esculentum* (Barker et al. 1998). Differences in the development of *G. mosseae, G. intraradices* and *Gigospora margarita* were observed (Barker et al. 1998), which suggest that the interactions with *rmc* might vary with fungal genotypes.

The colonization of *rmc* by most of the fungal species tested was aborted either on the root surface, within the epidermal cells or outer cortical cells, suggests that a gene is involved in the "acceptance" or "susceptibility" of the root to colonization by AM fungi has been mutated in *rmc* and other genotypes resulling in 'exclusion' of or 'resistance' to a broad range of species of AM fungi. The site of action of this gene is presumably the epidermis and/or outer cortical cells of the root (Banfante-Fasolo et al. 2000).

The mechanism of the "resistance" is not clear. All mutants so far isolated are recessive, which does not immediately appear consistent with increased transcript accumulation of defense-related genes (Rutiz-Lozano et al. 1999) or with the increased deposition of phenolics and β (1-3) glucans observed in the interaction between *G. mosseae* and *P. sativum* myc^{-1} (Gollotte et al. 1993). The mutant might lack a receptor that normally exerts control on defence pathways and enables tissue penetration by the mycorrhizal fungi (Gollotte et al. 1993).

Gao et al. 2001 studied interaction between a mycorrhiza-defective tomato mutant, *rmc* and eight different species of arbuscular mycorrhizal (AM) fungi. The development of different fungal species with *rmc* was impaired at different steps. The details of structures formed by the species of AM fungi with *rmc* were different, resulting in different colonization phenotypes. The variations highlight the diversity of the fungal interactions with a single plant genotype and suggest that the colonization of different cell types of by AM fungi may not necessarily be under control only of specific plant genes, as proposed by Wegel et al. 1998. In the few cases, where the same fungi have been used to challenge different mutants they do not necessarily respond in the same way. For example, *G. intraradices* is completely excluded by *rmc*, but penetrates the epidermal cells of *Lotus japonicus* mutants (Wegel et al. 1998) and *G. versiforme* colonizes *rmc* extensively but cannot colonize *Medicago sativa* mutants (Bradbury et al. 1991). Despite having the same site of action and other potential similarities, the operative of the genes may be different, giving rise to apparent differences in specificity (Gao et al. 2001).

MORPHOLOGY OF ARBUSCULAR MYCORRHIZA

Following entry into root, internal development of the fungus is influenced by the plant, and a single species of fungus may show significantly different morphological growth patterns depending on the plant partner in the association (Jacquelinet – Jeanmougin and Gianinazzi-Pearson 1983).

Despite the extensive occurrence of AM both geographically and within plant taxa, it has been possible to define two morphological types, the *Arum*-type and *Paris*-type. These types are named after the plants in which they were first described, *Arum maculatum* and *Paris quadrifolia*. In *Arum*-type fungi form intercellular hyphae between cortical cells and intra-cellular arbuscules within them. *Paris*-type is characterized by extensive intracellular hyphal coils and arbuscular coil in the root cortex. In Paris-type the intercellular phase of colonization is virtually absent (Gallaud, 1905).

Studies of different plant and fungal combinations have suggested that AM morphological type is largely dependent on the plant species (Smith and Smith 1997). The role of plant in controlling the morphological type of AM was first demonstrated experimentally by Bannett (1958) who found that the same Glomalean isolate formed the *Arum* type in *Zea mays* and the *Paris* type in *Solanum tuberosum*. Brundrett and Kendrick (1988, 1990) suggested that the plants largely controls AM morphology for example via the presence or absence of extensive air-spaces in roots.

A few studies have revealed that morpohology of AM is not solely under plant control but is also influenced by fungal identity. Cavagnaro et al. 2001 investigated the influence of fungal identity on the morphology of AM. Three of the fungal species (*Glomus intraradices, G. mosseae, G. versiforme*) formed the *Arum*-type in association with wild-typoe tomato (*Lycopersicon esculentum*) whereas, *Gigaspora margarita, Glomus coronatum* and *Scutellospora calospora* formed the *Paris*-type.

Brundrett and Kendrick (1988, 1990) suggested that the formation of two types is related to the presence of continuous longitudinal air-spaces in the root cortex, so that hyphae grow along the intercellular air-spaces in *Arum*-type and interacellularly in their absence in *Paris*-type. Work of Cavagnaro et al. 2001 suggested, even if the presence and continuity of air-spaces is the defining morpohological character, fungal identity is also important. In a situation of an air-space with a diameter less than that of its own hyphae, the affect of fungal identity may depend on hyphal diameter or plasticity.

Where a fungus is unable to grow along an intercellular space, successful colonization will depend on the ability of the hyphae to penetrate the cell walls. Penetration of host roots is due to combination of mechanical and enzymatic mechanisms (Bonfante and Perotto, 1995). There may be differences in the enzymatic and mechanical characteristics of different fungal species. Within each of morpohological groups (*Arum* or *Paris*-type) variation is noted, for example in ratios of hyphal coils and arbusculate coils. If these structures have dissimilar roles, these ratios may reflect differences in functionality. There may also be differences in the molecular activity of the interfaces of the structures; however there have been no molecular studies of *Paris*-type AM.

Environmental factors such as soil phosphorus, temperature and light can influence the level of colonization and relative abundance of structure (Smith and Read, 1997), there are no evidence demonstrating that they change the morphological type of AM. However, flooding of soils can influence the diameter of air-spaces in roots.

DEVELOPMENT OF *ARUM*-TYPE MYCORRHIZA

In the *Arum*-type mycorrhiza, penetration of the root is initially followed by intercellular hyphal growth, although in some instances the fungus will penetrate the exodermis and form hyphal coils in the exodermal cells as it passes through (Bonfante-Fasolo 1984). Although an arbuscule develops within a cell, it remains essentially apoplastic as the plant plasma membrane extends to completely surround it. The fungal cell wall becomes progressively thinner as the arbuscule develops and consequently in these cells, there is an extensive intracellular interface in which the two symbionts are in extremely close contact, separated only by their membranes and a narrow plant-derived apoplast (Smith and Gianinazzi-Pearson 1988).

As the fungal hypha penetrates a cortical cell wall and begins to differentiate into an arbuscule, the invaded cell responds with fragmentation of the vacuole, migration of the nucleus to a control position within the cell and an increase in number of organelles (Balestrni et al. 1992; Bonfante and Perotto 1995). This response seems to be arbuscule specific and it does not occur in exodermal cells during the development of coils. The plasma membrane extends approximately fourfold to form peri-arbuscular membrane developing the arbuscule. Cytochemical analysis indicate that peri-arbuscular membrane has high levels of ATPase activity. Since H^+-ATPase activity gives rise to the proton gradient required for many secondary active transporter processes, the data support the suggestion that active transport of mutants may be occurring across this membrane. Immunocytochemical analysis have demonstrated the presence of a matrix composed of pectin, xyloglucans, non-esterified polygalacturonans, arabinogalactans, and hydroxy proline-rich glycoproteins (HRGP) in compartment formed between the peri-arbuscular membrane and the arbuscule (Balestrini et al. 1994, Bonfante and Perotto 1995; Gollotte et al. 1995).

The development of arbuscular interface is accompanied by many other alterations in the cortical cells. In *M. truncatula*, transcripts encoding enzymes of the flavanoid biosynthetic pathway, phenylalanine ammonia lyase (*PAL*) and Chalcone synthase (*CHS*) are induced. Flavanoids act as auxin transport inhibitors (Jacobs and Rubery 1988) and therefore alter the hormone balance in the roots. Hormone levels are known to change in mycorrhizal roots and are probably responsible for the induction of expression of some of the mycorrhiza induced gene (van Rhijn et al. 1997).

CONCLUSION

Over the past decade, the first insights into molecular mechanisms underlying development of the symbiosis have been achieved. Despite this progress, there is still a vast amount to learn, and the molecular approaches that have been initiated recently need to be interacted with other disciplines to address both development and functioning of the symbiosis.

References

Albrecht C., Geurts R., Lapeyrie F. and Bisseling T. 1998. Endomycorrhizae and rhizobial factors activate signal transduction pathways inducing *PsENOD5* and *PsENOD12* expression in which Sym8 is a common step. *Plant J.* 15: 605-15.

Annapurna, K., Tilak, K.V.B.R. and Mukerji, K.G. 1996. Arbuscular mycorrhizal symbiosis recognition and specificity. In: *Concepts in Mycorrhizal Research* (ed: Mukerji, K.G.), pp. 77-90. Kluwer Academic Publisher, Netherlands.

Balaji, B., Ba, A.M., LaRue, T.A., Tepfer, D. and Piche, Y. 1994. *Pisum sativum* mutants insensitive to nodulation are also insensitive to invasion in vitro by the mycorrhizal fungus, *Gigaspora margarita*. *Plant Sci.* 102: 195-203.

Balaji, B., Pulin, M.J., Vierheilig, H. and Piche, Y. 1995. Responses of an arbuscular mycorrhizal fungus, *Gigaspora margarita*, to exudates and volatiles from the Ri T-DNA-transformed roots of nonmycorrhizal and mycorrhizal mutants of *Pisum sativum* L. sparkle. *Exp. Mycol.* 19: 275-83.

Balestrini, R., Berta, G. and Bonfante, P. 1992. The plant nucleus in mycorrhizal roots: positional and structural modifications. *Biol. Cell* 75: 235-43.

Balestrini, R., Romera, C., Puigdomenech, P. and Bonfante, P. 1994. Location of a cell-wall hydroxyproline-rich glycoprotein, cellulose and β-1,3-glucans in apical and differentiated regions of maize mycorrhizal roots. *Planta* 195: 201-9.

Barker, S.J., Stummer, B., Gao, L., Dispain, I., O'Connor, P.J. and Smith S.E. 1998. A mutant in *Lycopersicon esculentum* Mill. with highly reduced VA mycorrhizal colonization: isolation and preliminary characterization. *Plant J.* 15: 791-797.

Barrett, J.T. 1958. Synthesis of mycorrhiza with pure cultures of *Rhizophagus*. *Phytopathology* 48: 391.

Becard, G. and Piche, Y. 1989. Fungal growth stimulation by CO_2 and root exudates in vesicular-arbuscular mycorrhizal symbiosis. *Appl. Environ., Microbiol.* 55: 2320-2325.

Becard, G., Taylor, L.P., Douds, D.D. Jr., Pfeffer, P.E. and Doner, L.W. 1995. Flavanoids are not necessary plant signals in arbuscular mycorrhizal symbiosis. *Mol. Plant Microbe Interact.* 8: 252-58.

Blee, K.A. and Anderson, A.J. 1996. Defense-related transcript accumulation in *Phaseolus vulgaris* L. colonized by the arbuscular mycorrhizal fungus *Glomus intraradices. Plant Physiol.* 110: 675-88.

Bonfante, P. and Perotto, S. 1995. Strategies of arbuscular mycorrhizal fungi when infecting host plants. *New Phytol.* 130: 3-21.

Bonfante, P., Genre, A., Faccio, A., Martini, I., Schauser, I., Stougaard, J., Webb, J. and Parniske, M. 2000. The *Lotus japonicus LjSym4* gene is required for the successful symbiotic infection of root epidermal cells. *Mol. Plant-Microbe Interact.* 13: 1109-1120.

Bonfante-Fasolo, P. 1984. Anatomy and morphology of VA mycorrhizae. In: *VA Mycorrhizae*, (eds. C.L. Powell, D.J. Bagyaraj) pp. 5-33. CRC Press, Boca Raton, FL.

Bradbury, S.M., Peterson, R.L. and Bowley, S.R. 1991. Interaction between three alfalfa nodulation genotypes and two *Glomus* species. *New Phytol.* 119: 115-20.

Bradbury, S.M., Peterson, R.L. and Bowley, S.R. 1993. Further evidence for a corelation between nodulation genotypes in alfalfa (*Medicago sativa* L.) and mycorrhiza formation. *New Phytol.* 124: 665-673.

Brundrett, M. and Kendrick, B. 1988. The mycorrhizal status, root anatomy and phenology of plants in a sugar maple forest. *Can. J. Bot.* 66: 1153-1173.

Brundrett, M. and Kendrick, B. 1990. The roots and mycorrhizas of herbaceous woodland plants: 1. Quantitative aspects of morphology. *New Phytol.* 114: 457-468.

Buee, M., Rossignol, M., Jauneau, A., Ranjeva, R. and Becard, G. 2000. The pre-symbiotic growth of arbuscular mycorrhizal fungi is induced by a branching factor partially purified from plant root exudates. *Mol. Plant Microbe Interact.* 13: 693-698.

Cavagnaro, T.R., Gao, L-L., Smith, F.A. and Smith, S.E. 2001. Morphology of arbuscular mycorrhizas is influenced by fungal identity. *New Phytol.* 151: 469-475.

Chabot, S., Bel-Rhlid, R., Chenevert, R. and Piche, Y. 1992. Hyphal growth promotion *in vitro* of the VA mycorrhizal fungus, *Gigaspora margarita* Becker & Hall, by the activity of structurally specific flavonoid compounds under CO_2-enriched conditions. *New Phytol.* 122: 461-67.

David, R., Itzhaki, H., Ginzberg, I., Gafni, Y. and Kapulnik, Y. 1998. Suppression of tobacco basic chitinase gene exporession in response to colonization by the arbuscular mycorrhizal fungus *Glomus intraradices. Mol. Plant Microbe Interact* 11: 489-497.

David-Schwartz, R., Badani, H., Wininger, S. Levy, A.A., Galili, G. and Kapulnik Y. 2001. Identification of a novel genetically controlled step in mycorrhizal colonization: plant resistance to infection by fungal spores but not to extraradical hyphae. *Plant J.* 27: 561-569.

Douds, D.D. Jr., Nagahashi, G. and Abney, G.D. 1996. The differential effects of cell wall associated phenolics, cell walls and cytosolic phenolics of host and non-host roots on the growth of two species of AM fungi. *New Phytol.* 133: 289-294.

Duc, G., Trouvelot, A., Gianinazzi-Pearson, V. and Gianinazzi, S. 1989. First report of non-mycorrhizal mutants (Myc^-) obtained in pea (*Pisum sativum* L.) and Fababean (*Vicia faba* L.) *Plant Sci.* 60: 215-222.

Dumas-Gaudot, E., Gollotte, A., Cordier, C., Gianinazzi, S., Gianinazzi-Pearson, V. 2000. Modulation of host defense systems. In: *Arbuscular Mycorrhizas : Phsyiology and Function* (eds. Y. Kapulnik, Y.D.D. Douds Jr.) pp. 173-200, Kluwer Academic Press, Dordrecht. The Netherlands.

Fry, S.C., Aldington, S., Hetherington, P.R. and Aitken, J. 1993. Oligosacchrides as signals and substrates in the plant cell wall. *Plant Physiol.* 103: 1-5.

Gadkar, V., David-Schwartz, R., Kunik, T. and Kapulnik, Y. 2001. Arbuscular mychorrhizal fungal colonization. Factors involved in host recognition. *Plant Physiol.* 127: 1493-1499.

Gallaud I. 1905. Etudes sur les mycorrhizes endotrophs. *Revue Generale de Botanique.* 17: 66-83.

Gao, L-L., Delp, G. and Smith, S.E. 2001. Colonization patterns in a mycorrhiza-defective mutant tomato vary with different arbuscular-mycorrhizal fungi. *New Phytol.* 151: 477-491.

Garcia-Garrido, J.M., Garcia Romera, I. and Ocampo, J.A. 1992. Cellulose production by the vesicular-arbuscular mycorrhizal fungus *Glomus mosseae* (Nicol & Gerd.) Gerd. and Trappe. *New Phytol.* 121: 221-26.

Garcia-Garrido, J.M., Garcia-Romera, I. Parra-Garcia, M.D. and Ocampo, J.A. 1996. Purification of an arbuscular mycorrhizal endoglucanase from onion roots colonized by *Glomus mosseae. Soil. Biol. Biochem.* 28: 1443-1449.

Garcia-Romera, I., Garcia-Garrido, J.M., Martinez-Molina, E. and Ocampo J.A. 1991. Production of pectinolytic enzymes in lettuce root colonized by *Glomus mosseae. Soil Biol. Biochem.* 23: 597-601.

Gianinazzi-Pearson V. and Gianinazzi S. 1992. Influence of intergenic grafts between host and non-host legumes on the formation of vesicular-arbuscular mycorrhiza. *New Phytol.* 120: 505-508.

Gianinazzi-Pearson V., Gianinazzi S., Guillemin J.P., Trouvelot A., Duc G. 1991. Genetic and cellular analysis of resistance to vesicular arubucular (VA) mycorrhizal fungi in poea mutants. In : *Advances in molecular genetics of plant-microbe interactions*. (eds. Hennecke H., Verma D.P.S.), pp. 336-342. Kluwer Academic Publishers, I. Dordrecht, The Netherlands.

Gianinazzi-Pearson, V. 1996. Plant cell responses to arbuscular rmycorrhizal fungi: getting to the root of the symbiosis. *Plant Cell* 8: 1871-1883.

Gianinazzi-Pearson, V., Arnould, C., Oufattole, M. Arango, M. and Gianinazzi, S. 2000. Differential activation of H^+-ATPase genes by an arbuscular mycorrhizal fungus in root cells of transgenic tobacco. *Planta* 211. 609-613.

Giovannetti, M., Sbrana, C., Avio, L., Citernesi, A.S. and Logi C. 1993. Differential hyphal morpohogenesis in arbuscular mycorrhizal fungi during pre-infection stages. *New Phytol.* 125: 587-593.

Giovannetti, M., Sbrana, C., Citernesi, A.S. and Avio, L. 1996. Analysis of factors involved in fungal recognition respones to host-derived signals by arbuscular mycorrhizal fungi. *New Phytol.* 133: 65-71.

Giovannetti, M., Shrana, C. and Logi, C. 1994. Early processes involved in host recognition by arbuscular mycorrhizal fungi. *New Phytol.* 127: 703-709.

Gollotte, A., Gianinazi-Pearson, V., Giovannetti, M, Sbrana, C., Avio, I. and Gianinazzi S. 1993. Cellular localization and cytochemical probing of resistance reactions to arbuscular mycorrhizal fungi in a locus a myc mutant of *Pisum sativum* L. *Planta* 191: 112-122.

Harrison, M. 1998. Molecular and cellular aspects of the arbuscular mycorrhizal symbiosis. *Annu. Rev. Plant Physiol. Plant Mol. Biol.* 50: 361-389.

Jacobs, M. and Rubery, P.H. 1988. Naturally occurring auxin transport regulators. *Science*. 241: 346-349.

Jacquelinet-Jeanmougin, S. and Gianinazzi-Pearson, V. 1983. Endomycorrhizas in the Gentinaceae. I. The fungus associated with *Gentiana lutea* L. *New Phytol.* 95: 663-666.

Jayadev, M.S.K. and Dubey, A.K. 2005. Molecular mechanism in the Symbiotic Association between Plant and Arbuscular Mycorrhizae. In: *Basic Research and Applications of Mycorrhizae* (eds. Rodila, G.K. and Varma, A). pp. 57-65. I.K. International Pvt. Ltd., New Delhi.

Karandashov, V., Kuzovkina, I., Hawkins, H.J. and George E. 2000. Growth and sporulation of the arbuscular mycorrhizal fungus *glomus caledonium* in dual culture with transformed carrot roots. *Mycorrhiza* 10: 23-28.

Krajinski, F., Martin Laurent, F., Gianinazzi, S., Gianinazzi-Person, V. and Franken, P. 1998. Cloning and analysis of psam2, a gene from *Pisum sativum* L. regulated in symbiotic arbuscular mycorrhiza and pathogenic root-fungus interactions. *Physiol. Mol. Plant Pathol.* 52: 297-307.

Lambais, M.R. 2000. Regulation of plant defense-related genes in arbuscular mycorrhizae. In: *Current Advances in Mycorrhizae Research*. (eds. G.K. Podia and D.D. Douds), pp. 45-59. APS Press, St. Paul, Minnesota.

Lambias, M.R. and Mehdy, M.C. 1993. Suppression of endochitinase, 1,3 endoglucanase, and chalcone isomerase expression in bean vesicular-arbuscular mycorrhizal roots under different soil pohosphate conditions. *Mol. Plant Microbe Interact.* 6: 75-83.

Lammers, P.J., Abulaker, J., Govindrajulu, M., Jun, J., Krijgsman, O. and de Jong, M. 2005. Quantitative analysis of gene expression in the Arbuscular Mycorrhizal Symbiosis. In: *Basic Research and Applications of Mycorrhizae* (eds. Rodila, G.K. and Varma, A.) pp. 67-82. I.K. International Pvt. Ltd., New Delhi.

Lapopin, L., Gianinazzi-Pearson, V. and Franken, P. 1999. Comparative differential RNA display analysis of arbuscular mycorrhiza in *Pisum sativum* wild type and a mutant defective in late stage development. *Plant Mol. Biol.* 41: 669-667.

Longman, D. and Callow, J.A. 1987. Specific saccharide residues are involved in the recognition of plant root surfaces by zoospores of *Pythium aphanidermatum*. *Physiol. Mol. Plant Pathol.* 30: 139-150.

Mendgen, K. and Deising, H. 1993. Tansley Review No. 48, Infection structures of fungal plant pathogens – a cytological and physiological evaluation. *New Phytol.* 124: 193-213.

Mosse, B. and Hepper, C.M. 1975. Vesicular-arbuscular mycorrhizal infections in root organ cultures. *Physiol. Plant Pathol.* 5: 215-233.

Murphy, P.J., Langrdge, P. and Smith S.E. 1997. Cloning plant genes differnetially expressed during colonization of roots of *Hordeum vulgare* by the vesicular arbuscular mycorrhizal fungus *Glomus intraradices. New Phytol.* 135: 291-3091.

Nagahashi, G., and Douds, D.D. 1997. Appressorium formation by AM fungi on isolated cell walls of carrot roots. *New Phytol.* 36: 299-304.

Nagahashi, G. and Douds, D.D. 2000. Partial separation of the root exudates components and their effects upon the growth of germinated spores of AM fungi. *Mycol. Res.* 104: 14353-1464.

Peterson, R.L. and Bradbury, S.M. 1995. Use of plant mutants, intraspecific variants and non-hosts in studying mycorrhiza formation an function. In: *Mycorrhiza Structure, Function, Molecular Biology and Biotechnology* (eds. A. Varma, B. Hock) pp. 157-180, Springer-Verlag Berlin.

Porankiewicz, J., Wang, J. and Clarke, A.K. 1999. New insights into the ATP-dependent Clp protease : *Escherichia coli* and beyond. *Mol. Microbiol.* 32: 449-458.

Poulin, M.J., Bel-Rhlid, R., Piche, Y. and Chenevert, R. 1993. Flavanoids released by carrot (*Daucus carrota*) seedlings stimulate hyphal development of vesicular arbuscular mycorrhizal fungi in the presence of optimal CO_2 enrichment. *J. Chem. Ecol.* 19: 2317-2327.

Poulin, M.J., Simard, J., Catford, J.G., Labrie, F. and Piche, Y. 1997. Response of symbiotic endomycorrizal fungi to estrogens and antiestrogens. *Mol. Plant. Microbe. Interact.* 10: 481-487.

Rejon-Palomares, A., Garcia-Garside, J.M., Ocampo, J.A. and Garcia-Romera, T. 1996. Presence of xyloglucan-hydrolyzing glucanases (xyloglucanases) in arbuscular mycorrhizal symbiosis. *Symbiosis* 21: 249-261.

Roussel, H., van Tuinen, P., Franken, P., Gininazzi, S. and Gianinazzi-Pearson, V. 2001. Signalling between arbuscular mycorrhizal fungi and plants: identiication of a gene expressed during early interactions by differential RNA display analysis. *Plant and Soil* 232: 13-19.

Rutz-Lozano, J.M., Roussel, H., Gininazzi, S. and Gianinazzi-Pearson, V. 1999. Defense genes are differntially induced by a mycorrhizal fungus and *Rhizobium* sp. in wild-type and symbiosis-defective pea genotypes. *Molecul. Plant-Microbe Interact.* 12: 976-984.

Sagan, M., Morandi, D., Tarenghi, E. and Duc G. 1995. Selection of nodulation and mycorrhizal mutant in the model plant *Mdicago trancatula* (Gaertn.) after $\propto$-ray mutagenesis. *Plant Sci.* 111: 63-71.

Salzer, P. and Boller, T. 2000. Elicitor induced reactions in mycorrhizal and their suppression. In: *Current Advances in Mycorrhizal Research*, (eds. G.K. Podilla, D.D. Douds), pp.1-10, APS Press, St. Paul, MN.

Schreiner, R.P. and Koide, R.T. 1992. Antifungal compounds from the roots of mycotrophic and non-mycotrophic plant species. *New Phytol.* 123: 99-105.

Schreiner, R.P. and Koide, R.T. 1993. Mustards, mustard oils and mycorrhizas. New Phytol. 123: 106-113.

Senoo, K., Solaiman, M.Z., Kawaguchi, M., Imaizumi-Anraku, H., Akao, S., Tanaka, A. and Obata, H. 2000. Isolation of two different phenotypes of mycorrhizal mutants in the model legume plant *Lotus japonicus* after EMS – treatment. *Plant Cell Physiol.* 41: 726-732.

Shaul, O., David, R., Sinvani, G., Ginzberg, I., Ganon, D., Wininger, S., Ben-Dor, B., Badani, H., Ovdat, N. and Kapulnik, Y. 2000. Plant defense responses during arbuscular mycorrhizal symbiosis. In: *Current Advances in Mycorrhizae Research* (eds. G.K. Podial, D.D. Douds), pp. 61-68. APS Press, St. Paul, MN.

Shirtliffe, S.J. and Vessey, J.K. 1996. A nodulation (Nod $^{+}$/Fix^{-}) mutant of *Phaseolus vulgaris* L. has nodule-like structures lacking peripheral vascular bundles (Pbv^{-}) and is resistant to mycorrhizal infection (Myc^{-}). *Plant Science* 118: 209-220.

Smith, F.A. and Smith, S.E. 1997. Structural diversity in (vesicular)-arbuscular mycorrhizal fungi. *New Phytol.* 137: 373-388.

Smith, S.E. 1995. Discoveries, discussions and directions in mycorrhizal research. In: Varma A., Hock B., eds. *Mycorrhiza*. Berlin Heidelberg. Germany: Springer-Verlag, 3-24.

Smith, S.E. and Gianinazzi-Pearson, V. 1988. Physiological interactions between symbionts in vesicular-arbuscular mycorrhizal plants. *Annu. Rev. Plant Physiol. Plant Mol. Biol.* 39: 221-244.

Smith, S.E. and Read, D.J. 1997. *Mycorrhizal Symbiosis*. Ed.2. Academic Press, San Diego.

Staples, R.C. and Macko, V. 1980. Formation of infection structures as a recognition response in fungi. *Experimental Mycology*. 4: 2-16.

Tsai, S.M. and Phillips, D.A. 1991. Flavanoids released naturally from alfalfa promote development of symbiotic *glomus* spores *in vitro*. *Appl. Environ. Microbiol.* 57: 1485-1488.

van Rhijn, P., Fang, Y., Galili, S., Shaul, O., Atzmon, N., Wininger, S., Eshed, Y., Kapulnik, Y., Lum, M., Li, Y. and To, V. 1997. Signal transduction pathways in forming arbuscular-mycorrhizae and *Rhizobium*-induced nodules may be conserved based on the expression of early nodulin genes in alfalfa mycorrhiza. Proc. Natl. Acad. Sci. USA 94: 5467-5472.

Vierheilig, H. and Piche, Y. 1996. Grafts between peas forming the arbuscular mycorrhizal symbiosis (Myc^{+}) and pea mutants resistant to AM fungi (Myc^{-}) show the same colonization characteristics as ungrafted plants. *J. Plant Physiol.* 147: 762-764.

Vierheilig, H., Alt, M., Lange, J., Gut Rella, M., Wiemken, A. and Boller, T. 1995. Colonization of transgenic tobacco constitutively expressing pathogenesis-related proteins by the vesicular-arbuscular ycorrhizal fungus *Glomus mosseae*. *Appl. Environ. Microbiol* 61: 3031-3034.

Vierheilig, H., Tseli, B., Alt, M., Raikhel, N., Wiemken, A. and Boller, T. 1996. Resistance of *Urtica dioica* to mycorrhizal colonization: A possible involvement of *Urtica dioica* agglutinin. *Plant Soil*. 183: 131-136.

Wegel, E., Schauser, L., Sandal, N., Stougaard, J. and Parniske, M. 1998. Mycorrhiza mutants of *Lotus japonicus* define genetically independent steps during symbiotic infection. *Molecul. Plant Microbe Interact.* 11: 933-936.

Frontiers in Plant Sciences (Pages 583-596)
Eds. K.G. Mukerji, K.V.B.R. Tilak, S.M. Reddy, L.V. Gangwane, P. Prakash, I.K. Kunwar

Arbuscular Mycorrhizal (AM) Fungi for Heavy Metal Remediation

P. Rajendran[1], N. Lakshmanan[2] and P. Gunasekaran[*]

ABSTRACT

Heavy metals produced by human activities such as burning fossil fuel, mining, smelting of metaliferous ores, municipal wastes, pesticides, fertilizers and industrial activities can not be chemically degraded. They are toxic, teratogenic, carcinogenic and mutagenic to living systems. Traditional physical and chemical metal remediation strategies are economically not viable and eco-friendly due to sludge production which needs further treatment. Current on-site management, or excavation, and subsequent disposal to a landfill site cause several problems posing a health risk. Bioremediation, use of variety of microorganisms including fungi, algae, bacteria and plants are the modern technologies that could eliminate the problems of conventional treatment procedures. Among the microbes, Arbuscular mycorrhizal (AM) fungi, forms a symbiotic association between the plants and certain soil fungi and have the capacity of greater retention of toxic metals in roots and lesser translocation to shoots. AM association with plant roots grown in heavy metal contaminated areas can alleviate the toxicity by accumulation and can accelerate revegetation. This review highlights heavy metal toxicity and the potential of AM fungi in the restoration of heavy metal contaminated sites.

Key words: Arbuscular mycorrhizae (AM), Heavy metal, Bioremediation.

Author's Address: [1]Department of Zoology, [2]Department of Botany, Vivekananda College, Tiruvedakam west, Madurai 625 217, India.

[*]Department of Microbial Technology, Centre for Excellence in Genomic Sciences, School of Biological Sciences, Madurai Kamaraj University, Madurai 625 021.

[*]Corresponding author: Phone: 91-452-2458478 Fax: 91-452-2459181 E. mail: pguna@eth.net

INTRODUCTION

Population explosion with concomitant industrial revolution of recent times has resulted in the discharge of many toxic chemicals in the environment which mainly constitutes heavy metals. Contamination of the environment by toxic metals, metalloids, organochemicals and radionuclides is of economic and environmental significance (Gadd *et al.,* 2004). Toxic heavy metal contamination of soil poses a major environmental and human health concern which is still in need of an effective and affordable technological solution for remediation strategy (Krishnamurthi, 2000). Metals are not biodegradable once they have entered the environment and they have cumulative effects at low levels in drinking and ground water (Roane and Pepper, 2000). At every level of food chains, the metals increase in concentration and are passed to the next higher level, a phenomenon called biomagnification (Paknikar *et al.,* 2003) The presence of heavy metals in the environment at concentrations above critical values stipulated by national and international regulatory bodies is considered unacceptable (Atkinson 1998).

Conventional procedures that are available for the removal and management of heavy metals, are not only costly and inefficient, but also increase the contaminant exposure to cleanup crews (Torresedy*et al.,* 1997). Furthermore, these phsico-chemical technologies used for soil remediation render the land useless for plant growth, as they remove all biological activities of soil (Gaur and Adholeya, 2004). However, biological treatment systems have been the preferred over the physical or chemical methods because they are environmentally and economically attractive (Bhamidimarri, 1998). Bio-related process in the synthesis of metal nanoparticles using microbes such as bacteria, yeasts, algae, actinnomycetes and fungi is gaining momentum due to the eco-friendly nature of the organisms which reduce the use of toxic chemicals in conventional strategies (Muralisastry *et al.,* 2003).

Ramkumar (2000) reported that fungi are good at bioaccumulation of heavy metals. AM fungal isolates adapted to local soil conditions can stimulate plant growth better than non-indigenous isolates. Isolation of indigenous stress-adapted AM fungi can be a potential tool for inoculation of plants in disturbed ecosystems (Dodd and Thompson, 1994). Plants which appear spontaneously in heavy metal contaminated places are frequently devoid of mycorrhizal symbiosis and are mostly characterized by poorly developed root and shoot biomass when heavy metals are present (Pawlowska *et al.,* 1996). Introduction of metal tolerant AM fungal inoculum into heavy metal contaminated soil will be a sustainable on-site technique for metal remediation.

CURRENT STATUS OF HEAVY METAL POLLUTION

The group of heavy metals are about 65 and are defined in a number of criteria such as their cationic-hydroxide formation, specific gravity greater than 5 g/ml, complex formation, hard-soft acids and bases, and, more recently, association with eutrophication and environmental toxicity. The heavy metals are classified into two groups essential and nonessential. Essential metals required for metabolic processes are as follows V, Se, Fe, Co, Ni, Cu, Zn and Mo which are required in trace amounts. Non essential heavy metals are Cd, Hg, Sn, Pb, Ag, Al; these have no metabolic function and are toxic even at very low concentration (Badar *et al.,* 2004).

Wastewaters containing toxic heavy metals and other harmful products spoil the soil, plants and aquatic life through industrial processes, domestic sewage discharge, street dust, land run off and fossil fuel burning. The various heavy metal sources that affect target components of the biosphere are provided in the Fig 1. Because of their abundance and toxicity, some heavy metals have been placed on the U.S.Environmental Protection Agency's (USEPA) list of 129 priority pollutants (Keith and Telliard, 1979).

EFFECT OF HEAVY METALS ON LIVING ORGANISMS

The growing industrialization and modern agriculture practices spread worldwide have adversely affected the eco-system. These practices leave persistent toxic metals and organic pollutants in the surrounding, which tend to accumulate and deteriorate the environment (Ahmend *et al.,* 2004). Traces of heavy metals such as Hg, Cd, Pb, As, Co, Mn, Fe and Cr have been identified as deleterious to aquatic ecosystem and human health. Wastes containing high concentrations of toxic metals either separately or in combination are extremely toxic to all organisms (Sharma, 2001). Metal concentration has been linked to birth defects, cancer, skin lesions, retardation leading to disabilities, liver and kidney damage and a host of other maladies (Trivedi & Gundap Raj, 1992). Toxic effects of heavy metals are given in the Table 1.

i. Microbes

The water and soil environment receives a variety of xenobiotic compounds that are foreign to living systems. Microbes in soil and water act upon these compounds by using them directly as substrates for the energy and biomass production (Singh and Dwivedi, 2004). Industrial wastewaters have impact on microorganisms depending on the nature and composition of the constituents. This can be: i.) Inhibition of biodegradation, ii.)

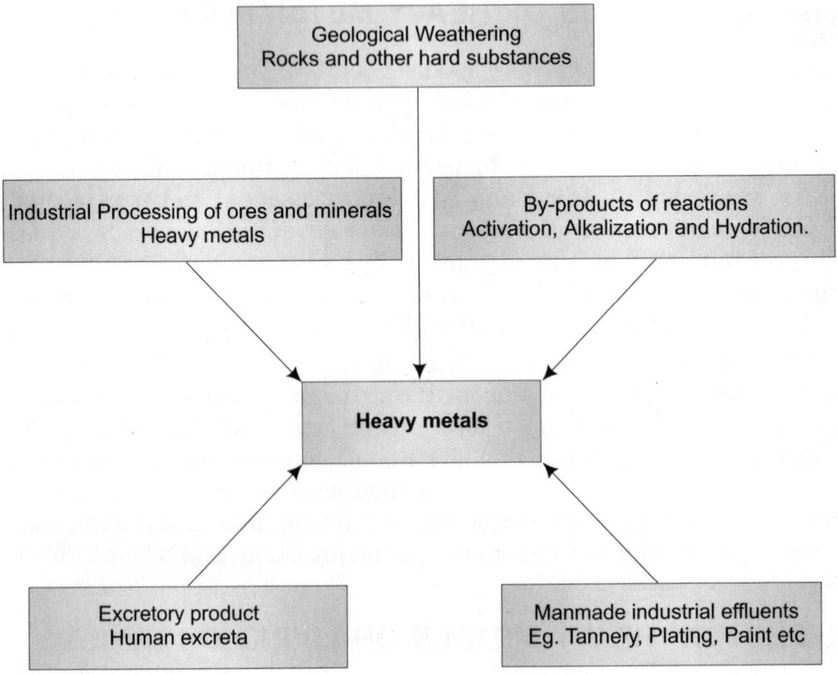

Fig. 1 Different sources for heavy metals.

Biodegradation, iii.) Incidental removal, iv.) Cometabolism and v.) Low concentration of substrate situation (Jogdand, 2004).

ii. Plants

Plants resist toxic effects of heavy metals by limiting cellular uptake (avoidance), or detoxify the metals. Possible mechanisms that govern heavy metal tolerance in plant cells are, binding of the metal with the cell wall, reduced metal transport across the cell membrane, active metal efflux, compartmentalization and chelation. The most-prevalent mechanism is chelation though the induction of metal-binding peptides and the formation of metal complexes. The members of the class III-MTs, the phytochelatins (PCs), are responsible for the remediation of complexes with heavy metals in plants (Salt *et al.,* 1995).

iii. Animals

Wastewaters discharged from industries change physical, chemical as well s the biological state of water and cause temporary or long term effects on

TABLE 1 *Major heavy metals, their sources and their effects*

Metal	*Major source*	*Use*	*Toxic effect*
Hg	Electrochemical industry	Coal, vinyl chlorides, electrical batteries.	Poisoning neurotoxic
Pb	Leaded gasoline, paints	Plastic, paint, pipe, batteries, gasoline, auto exhaust.	Neurotoxic.
As	Coal burning, ore processinng	Pesticides, coal, detergents.	Poisoning
Cd	Electroplatingwastes, All products containing Zn	Fertilizers, plastic, pigment.	Kidney damage, injury in cns and mental retardation.
Cr	Plating industry wastes	Tanning, paints, pigment, fungicide.	Nephritic, cancer and ulceration.
Zn	Mining operations, galvanizing units, dry cell batteries	Fertilizer.	Vomiting, renal damage and cramps.
Co	Industrial wastes	Vitamin B-12.	Diarrhea, low blood pressure and paralysis.
Se	Industrial wastes	Coal, sulfur.	Damage of liver, kidney, spleen and nervousness.
Ni	Electroplating industry, mining wastes	Electroplating.	Teratogenic, carcinogenic, genotoxic and mutagen.
Be	Coal burning, fungicides	Coal, rocket fuel.	Carcinogen, acute and chronic poison.

living organisms. Acute effects of heavy metals on fish includes gill damage and hypoxia, whereas chronic effects include alteration in enzyme activity, changes in energy substrates, osmoregulatory changes, and levels of induced metal binding protein (Harichandan, 2003). Direct acting pollutants of industrial wastewater create a lethal effect either through (a) their action on the epithelial surface of the gills, or (b) by absorption through the body affecting the internal structure and metabolism of fish. Oil wastes, acids and salts or heavy metals and phenols may produce their lethal effect by acting upon the mucous of the gills by clogging and coagulating them (Heukelekian, 1997).

iv. Human

Heavy metal contamination in soil and ground water from various anthropogenic sources such as industrial waste is of major concern because of its extreme toxicity and threat to human life (Burd *et al.*, 1998). Wastes generated by industrial activities have brought dangers to human beings and very little is known about the long term consequences of exposure to the chemicals. Tens of millions of tons of toxic or otherwise hazardous substances enter the environment every year. They cause cancer, delayed nervous damage, malformations in urban children, and mutagenic changes. Once the wastewater containing toxic effluents enter into the environment, they spread in a very complex way and may be converted into other substances which have different effect. "Minimata disease" episode due do the discharge of methyl mercury industrial wastewater that contains mercury in the sea of Japan was an evidence to prove contamination of the fish, which in turn caused neurological disorders to nearly two thousand people; about 400 of them have died (Fujiki, 1980).

HEAVY METAL REMEDIATION STRATEGIES

Metal ions are universal components of ecosystems and many organisms have evolved mechanisms to cope up with varying levels of these elements. Some metals are essential, as they are required for normal cellular growth. The heavy metals are mobilized and eventually accumulated through the food chain, leading to serious ecological and health problems. An effective and affordable technological solution is yet to be found for their removal from the environment (Salt *et al.*, 1995). High toxicity, environmental mobility and stability of heavy metals in the environment render their removal an absolute necessity (Raina Maier, 2000). High operation cost, sludge production, sophisticated instrumentation and high-energy requirement and metal precipitation are the constraints of the conventional effluent treatment processes. Mejare and Bulow (2001) reported that microorganisms and plants serve as efficient agents for removal of heavy metals.

AM AND ITS IMPORTANCE

Arbuscular mycorrhizal fungal associations are integral, functioning of plant roots and are widely recognized as enhancing plant growth on severely disturbed sites, including those contaminated with heavy metals (Gaur and Adholeya, 2004). The Arbuscular Mycorrhizae are classified as endomycorrhizae because the invading fungus is closely associated with the internal tissue of the host roots. The fungi of this category belong to Endogonales and they form characteristic fungal structures, the *arbuscules* and *vesicles* in the root cortex. The fungi form association in most of the Bryophytes, Pteridophytes, Gymnosperms, and Angiosperms include the species of *Endogone, Acaulospora, Gigaspora, Glomus* and *Sclerocystis* (Dwivedi and Dwivedi, 2004). Marx and Altman (1979) had reported AM can accelerate revegetation of severely degraded lands such as coal mines or waste sites containing high levels of heavy metals.

MYCOREMEDIATION OF HEAVY METALS

Galun *et al.* (1983) reported fungi are ubiquitous in distribution and defined the term *geomycology* as fungal capabilities of attacking and accumulating metals and minerals. Nordgren *et al.* (1985) suggested application of fungal biomass to decontaminate metal-polluted water or to recover economically valuable metals from natural or industrial water is an attractive proposal. Fungal bioremediation strategies are divided into three categories: i. the target compound is used as a carbon source, ii. The target compound is enzymatically attacked but is not used as a carbon source (cometabolism), and iii. The target compound is not metabolized at all but is taken up and concentrated within the organism (bioaccumulation). Although fungi participate in all three strategies, they are more proficient at cometabolism and bioaccumulation than at using xenobiotics as a sole carbon source (Ramkumar, 2000). Mycorrhizae reduce toxicity to plants by binding heavy metals. The efficiency of protection differs among fungal isolates and different heavy metals. Heavy metal binding occur in the mycelium, either in the hyphal cell walls or in the surrounding polysaccharide slime (Galli *et al.*, 1964). Metals are known to adsorb to fungal cell wall components, including cellulose, chitin and melanin. Binding within cells of ectomycorrhizal fungi are metallothionine like proteins, which have high cystine content, and perhaps polyphosphate granules. Wuertez and Mergeay (1997) reported, in general mycorrhizal fungi either increase the uptake of metals at low concentrations or reduce their uptake at toxic concentrations. A list of AM fungi that are useful in heavy metal remediation is provided in Table 2.

TABLE 2 *List of AM- fungi that are useful for heavy metal remediation.*

Heavy metal	*VAM-Fungi*
Cadmium	*Glomus mosseae, Glomus sp, Gigaspora sp & G. caledonium.*
Nickel	*Gigaspora* sp, *G. tenue & G. caledonium.*
Zinc	*Glomus* sp, *G. constrictum, G. ambisporum, G. fasiculatum, G. caledonium, G. versiforme* & *Scutellospora dipurpurescens.*
Lead	*G.intraradices* & *G. caledonium*
Copper	*G. caledonium*

*Cited from Gaur & Adholeya (2004)

AM FUNGI IN METAL REMEDIATION

AM fungi constitute an important functional component of the soil-plant system that is critical for sustainable productivity in degraded soils. The interaction between mycorrhizal colonization and accumulation of heavy metals and other toxic elements is an area of considerable interest. Most substrates of ectomycorrhizal plants are acidic, and in particular of many of the man made substrates such as mine-spoils which they colonize are conducive to the solubilization of potentially toxic metal ions. Amongst these Al and Fe are quantitatively the most significant in natural soil, while mine soils may be polluted by Ni, Pb, Zn and Cd, either separately or in combination. Colonization of clonal cuttings of *Betulla pubescens* by *Paxillus involutus* provides significant increase in resistance to Zn toxicity and the growth enhancement association with mycorrhizal formation is related to reduced transfer to the leaves (Denny and Wilkins, 1987).

A number of different mechanisms maybe involved, including tissue dilution of the toxic element due to interactions either P nutrition, sequestration of the toxic metal in the fungus, and development of tolerance by the fungus (Smith and Reed, 1997). El-Kherbawy *et al.* (1989) reported tissue dilution of a toxic element can occur as consequence of improved P nutrition and increased plant growth, even in situations where the uptake of heavy metal per plant is actually greatly increased. External mycelium of AM fungi provides a wider exploration of soil volumes by spreading beyond the root exploration zone, thus providing access to greater volume of heavy metals present in the rhizosphere (Malcova *et al.*,2003).

AM fungi isolated from heavy metal polluted soils are more tolerant to cadmium (Weissenhorn *et al.*, 1994). AM inoculated plants have the capacity of greater retention of toxic metals in roots and lesser translocation to shoots. In soils rich in Cd, both tissue concentration and total uptake to cadmium by plants are significantly lower in the mycorrhizal plants than in non-mycorrhizal plants (Leyvel *et al.*, 1995). Mycorrhizal plants are known to have higher levels of tolerance to toxic heavy metals, root pathogens, drought, high soil temperatures, salinity of soil, soil pH and transplantation shock. Occurrence of VAM fungi in metal polluted soils suggests possible adaptation of the VAM population to metals (Griffoen *et al.*, 1994). Jones and Hutchinson (1988) reported in Birch (*Betula papyrifera* Marsh.), growth, photosynthesis, transpiration and respiration was not affected due to nickel (85μM) in the presence of the fungi *Lactarius rufus* and *Scleroderma flavidum*. They have also suggested that the mycobiont binds nickel, reducing the amount available for translocation to the shoots. Tonin *et al.* (2001) suggests the accumulation of toxic metals such as Zn, Pb and Ni in the roots of metallophyte, such as *Viola calaminaria* in a non-toxic form is due to AM fungal association.

ADVANTAGES OF AM FUNGI IN HEAVY METAL REMEDIATION

VAM fungi can contribute to plant growth, particularly in disturbed or heavy metal contaminated sites, by increasing plant access to relatively immobile minerals such as phosphate (Vivas *et al.*, 2003). Marx and Altman (1979) reported that the AM fungi can accelerate the revegetation of severely degraded lands such as coal mines or waste sites containing high levels of heavy metals. Mycelial growth habit gives a competitive advantage over single cellular organisms such as bacteria and yeast, especially with respect to the colonization of insoluble substrates. Furthermore, fungi can rapidly ramify through substrates, digesting their way along by secreting a battery of extra-cellular degradative enzymes. Hyphal penetration provides a mechanical adjunct to the chemical breakdown affected by the secreted enzymes. The high surface to cell ratio characteristic of filaments maximizes both mechanical and enzymatic contact with the environment. The extracellular nature of the degradative enzymes enables the fungi to tolerate higher concentration of the toxic chemicals (Ramkumar, 2000).

Electron energy loss spectroscopical studies of Smith and Reed (1997) in mycorrhizal roots of *Pteridium aquilinum* exposed to heavy metals (Cd, Ti and Ba) showed that the metals were sequestered in the hyphae and not transferred to the plant. AM fungal metal tolerance includes adsorption onto plant or fungal cell walls present on and in plant tissues or chelation by such compounds as siderophores and metallothionens released by fungi or other rhizosphere microbes and sequestration by plant-derived compounds like phytochelatins or phytase (Joner and Leyval, 1997). The studies of del Val *et al.* (1999) showed AM fungal spores never disappeared completely in the soils that were applied with sewage-sludge containing higher concentrations of heavy metals such as Zn, Cd, Cu, Ni and Pb, suggesting certain adaptation of these indigenous fungi to the environmental stress.

FUTURE PROSPECTS

Volesky (2001) suggested that high cost effective and ecofriendly biosorption technology holds the key to heavy metal detoxification of the environment. Extended screening of microbiota will lead to isolation and identification of microbes with the desired traits. Aquatic fungi, anerobic fungi and mycorrhizal fungi in conjunction with pollutant-tolerant plants all provide opportunities for new research. Genetic engineering strategies can be utilized to introduce fungal genes for degradative enzymes into bacteria. However, competent fungi can be modified to grow in an extended range of environments. In the long run, the use of purified biopolymers from microbes as specific chemical binding agents for heavy metal remediation holds considerable promise. The application of both genetic and protein engineering techniques could conceivably lead to peptides or other biopolymers with enhanced metal specificity, stability and other useful properties. Identification of cheap and effective immobilizing matrices will help in developing efficient metal sequestering microbial systems for continuous treatment plants.

CONCLUSION

Microbes can significantly affect the distribution of heavy metals in the environment, since they have developed means to use them for their benefit. This clearly holds promise for effective, economical and eco-friendly metal bioremediation technology for industrial exploitation and pollution free environment. Indigenous AM fungal isolates existing naturally in heavy metal polluted soils are more tolerant than isolates from non-polluted soils, and are reported to efficiently colonize plant roots in heavy metal-stressed environments. Thus it is important to screen indigenous and heavy metal tolerant AM isolates to remediate contaminated soil. Phytoremediation combined with AM inoculation will be an appropriate novel strategy to combat heavy metal contamination. Screening of heavy metal tolerant AM fungi, genetic improvement and development of new methods to grow them in large numbers will be the future research thrust. Although several microbial metal bioremediation approaches are established to combat heavy metal pollution from anthropogenic and natural sources, none is yet in wide spread use. It is time to initiate more comprehensive interdisciplinary approach between mycologists and metallurgists to bring the process to land scale technology.

References

Ahmed, N., Badar, U. and Raihan, S. 2004. Resistance and accumulation of heavy metals by indigenous bacteria: Bioremediation. In: *Industrial and environmental biotechnology* (Eds: Ahmed, N., Qureshi, F.M. and Khan, O.Y). pp. 81-102. Horizon Scientific Press, Wymondham, UK.

Atkinson, B.W., Bux. and Kasan, H.C. 1998. Waste activated sludge remediation of metal plating effluents. *Water S A* 24 (4): 355-359.

Badar, U., Abbas, R. and Ahmed, N. 2004. Characterization of copper chromate resistant bacteria isolated from Karachi tanneries effluent. In: *Industrial and environmental biotechnology* (Eds. Ahmed, N., Qureshi., F.M. and Khan, O.Y.). pp. 43-54. Horizon Scientific Press, Wymondham, UK.

Bhamidimarri, R. 1998. Advances in biotechnology applications for remediation of toxic and inhibitory pollutants. In: *Advances in Biotechnology* (Eds. Ashok pandey.). pp. 395-403. Educational publishers and distributors, New Delhi.

Burd, G.I., Dixon, G. and Glick, B.R. 1998. A plant growth-promoting *Rhizobacterium* that decreases nickel toxicity in seedlings. *Appl Environ Microbiol.* 64: 3663-3668.

del Val, C., Barea, J.M. and Azcon-Augilar, C. 1999. Assessing the tolerance to heavy metals of Arbuscular mycorrhizal fungi isolated from sewage sludge-contaminated soil. *Appl. Soil. Ecol.* 11: 261-269.

Denny, H.J. and Wilkins, D.A. 1987. Zinc tolerance of mycorrhizal Betula. *New Phytologist* 99: 101-106.

Dodd, J.J. and Thompson, B.D. 1994. The screening and selection of inoculant Arbuscular mycorrhizal and ectomycorrhizal fungi. *Plant Soil.* 159: 149-158.

Dwivedi, S.K. and Dwivedi, P. 2004. Microbial interactions and biocides. In: *Environmental microbiology and biotechnology* (Eds. Singh, D.P. and Dwivedi, S.K.). pp. 204-217. New Age International (P) Limited Publishers, New Delhi.

El-Kherbawy, M., Angle, J.S., Heggo, A. and Chaney, R.L. 1989. Soil pH, rhizobia and vesicular Arbuscular mycorrhizae inoculation effects on growth and heavy metal uptake of alfalfa (*Medicago sativa* L.) *Biology and fertility of Soils* 8: 61-65.

Fujiki, N. 1980. The pollution of Minamata Bay by mercury and Minamata disease. In: *Contaminants and sediments* Vol. 2 (Edi. Baker, R.A.). pp. 493-500. Ann Arbor Science Publ., Ann Arbor, Michigan.

Gadd, G.M., Bridge, T.A.M., Gharieb, M..M., Sayer, J.A. and White, C. 2004. Microbial processes for solubilization or immobilization of metals and metalloids and their potential for environmental bioremediation. In: *Industrial and environmental biotechnology* (Eds. Ahmed, N, Qureshi., F.M. and Khan, O.Y.). pp. 55-80. Horizon Scientific Press, Wymondham UK.

Galun, M., Keller, P., Feldstein H., Galun, E., Sigel, S., and Siegel, B. 1983. Recovery of Uranium (VI) from solution ysing fungi. Release from uranium loaded Penicillium biomass. *Water Air Soil Pollut.* 20: 277-280.

Gaur, A. and Adholeya, A. 2004. Prospects of Arbuscular mycorrhizal fungi in phytoremediation of heavy metal contaminated soils. *Current Science.* 86: (4) 528-534.

Griffoen., W.A.J., Ietswart, J.H. and Ernst, W.H.O. 1994. Mycorrhizal infection of an *Agrostis capillaris* population on a copper contaminated soil. *Plant and Soil.* 158: 83-89.

Harichandan, N., Beura, B. and Panigrahi, A.K. 2003. Toxicological effects of chomium on a freshwater fish, *Tilapia mossambicus,* Peters and its ecological significance. *Indian J Environ and Ecoplan.* 7(1): 63-70.

Heukelekian, H. 1997. Stream pollution and self purification. In: *Industrial wastes-their disposal and treatment* (Eds. Rudolfs, W.). Allied scientific publishers, Bikanir, India.

Jogdand, S.N. 2004. *Environmental biotechnology.* Himalaya publishing House, Mumbai, 9 pp.

Joner, E.J. and Leyval. C. 1997. Uptake of ^{109}Cd by roots and hyphae of a *Glomus mossae/Trifolium subterraneum* mycorrhizal from soil amended with high and low concentrations of cadmium. *New Phytol.* 135: 353-360.

Jones, M.D. and Hutchinson, T.C. 1988. Nickel toxicity in mycorrhizal birch seedlings infected with *Lactarius rufus* or *Scleroderma flavidum*. *New Phytol.* 108: 451-549.

Keith, L.H. and Telliard, W.A. 1979. Priority pollutants: a perspective view. *Environ Sci Technol.* 13: 416-423.

Krishnamurthi, G.S.R. 2000. Speciation of heavy metals: An approach for remediation of contaminated soils. In: *Remediation engineering of contaminated soils* (Eds. Wise, D.L., Trantalo, D.J., Cichon, E.J., Inyang, H.I. and Stottmeister, U.). pp 693-709. Marcel Dekker Inc, New York.

Leyvel, WC., Belgy, G. and Berthelin, J. 1995. Arbuscular mycorrhizal contribution to heavy metal uptake in maize (*Zea mays* L.) in pot culture with contaminated soil. *Mycorrhiza.* 5: pp. 245-251.

Malcova, R.; Vosatka, M. and Gryndler, M. 2003. Effects of inoculation with *Glomus intraradices* on lead uptake by *Zeamys* L. and *Agrosis capillaries* L. *Appl Soil Ecol.* 23: 55-67.

Marx, D.H. and Altman, J.D. 1979. *Pisolithus tinctorius* ectomycorrhiza improve survival and growth of pine seedlings on acid coal spoil in Kentucky and Virginia. *Mycorrhizal Manual.* Springer, Berlin, pp.387-399.

Mejare. and Bulow, L. 2001. Metal–Binding proteins and peptides in bioremediation and phytoremediation of heavy metals. *Trends Biotech* 19: 67-73.

Muralisastry, Absar Ahmad, Islam Khan, M. and Rajiv Kumar. 2003. Biosynthesis of metal nanoparticles using fungi and actinnomycetes. *Curr. Science.* 85: (2)162-170.

Nordgren, A., Baath, E., and Soderstrom, B. 1985. Soil microfungi in an area polluted by heavy metals. *Can. J. Biol.* 63: 448-455.

Paknikar, K.M., Pethkar, A.V. and Puranik, P.R. 2003. Bioremediation of metalliferous wastes and products using inactivated microbial biomass. *IJB.* (2): 426-443.

Pawlowska, T.B., Blaszkowski, J. and Ruhling. 1996. The mycorrhizal status of plants colonizing a calamine spoil mound in southern Poland. *Mycorrhiza.* 6: 499-505.

Raina Maier. 2000. Lipid lather removes metals. *Environ Heal Per.* 108: 320-323.

Ramkumar. 2000. Biodegradation. In: *Environmental degradation.* pp. 1-5. Sarup and sons, New Delhi.

Roane, T.M. and Pepper, I.L. 2000. Microorganisms and metal pollution. In: *Environmental Microbiology* (Eds. Maier, R.M., Pepper, I.L. and Gerba, C.B). Academic press, London.

Salt, D.E., Blaylocy, M., Nanda, P.B.A., Kumar., Dushenkov, V., Ensley, B.D., Chet, I.and Raskin, I. 1995. Phytoremediation a novel strategy for removal of toxic metals from the environment using plants. *Biotech.*13:468-474.

Sharma, B.K. 2001. Industrial wastes and treatment processes. In: *Water pollution* (Edi: Sharma, B.K). pp. 155-203. Goel publishing House, Meerut.

Singh, D.P. and Dwivedi, S.K. 2004. Environmental perspectives of microbiology and biotechnology: an overview. In: *Environmental microbiology and biotechnology* (Eds:

Singh, D.P. and Dwivedi, S.K). 1-9. New age international (P) limited publishers, New Delhi.

Smith, S.E. and Reed, D.J. 1997. Mineral nutrition, heavy metal accumulation and water relations. In: *Mycorrhizal symbiosis,* Second edition, Academic press, Harcourt Brace& Company Publication, USA, 153 pp.

Tonin, C., Vandenkoornhuyse, P., Joner, E.J., Straczek, J. and Leyval, C. 2001. Assesment of Arbuscular mycorrhizal fungi diversity in the rhizosphere of *Viola calminaria* and effect of these fungi on heavy metal uptake by clover. *Mycorrhiza.* 10: 83-89.

Torresdey, J.L.G., Tricmann, K.J., Gonzalez, J.H. and Rodriguez.1997. Phytofiltration of hazardous metal ions by alfalfa: a study of calcium and magnesium interferences. *Journal of Hazardous Materials.* 56: 169-179.

Trivedi, R. P. and Gundap Raj. 1992. A study of industrial water pollution, In *Encyclopaedia of environmental sciences.* Akashdeep publishing house, New Delhi, pp.185.

Vivas, A., Marulanda, A., Gomez, M., Barea, J.M. Azcon. R. 2003. Pysiological characteristics (SDH and ALPactivities) of arbuscular mycorrhizal colonization as affected by *Bacillus thuringiensis* inoculation under two phosphorous levels. *Soil Biol. Biochem.* 35: 987- 996.

Volesky, B. 2001. Detoxification of metal-bearing effluents: biosorption for the next century. *Hydrometallurgy.* 59: 203-216.

Weissenhorn, I., Glashoff, A., Leyval, C. and Berthelin, J. 1994. Differential tolerance to Cd and Zn of arbuscular mycorrhizal (AM) fungal spores isolated from heavy metal polluted and unpolluted soils. *Plant and Soil.* 167: 189-196.

Wuertez, S. and Mergeay, M. 1997. The impact of heavy metals on soil microbial communities and their activities In: *Modern Soil Microbiology* (Eds. Elas, J.D.V., Trevors, J,T. and Wellington, E.M.H.). pp. 607-642. Marcel Dekker. Inc, New York.

Frontiers in Plant Sciences (Pages 597-616)
Eds. K.G. Mukerji, K.V.B.R. Tilak, S.M. Reddy, L.V. Gangwane, P. Prakash, I.K. Kunwar

Arbuscular Mycorrhizal Rhizosphere and its Impact on Plants Growth under Stress Conditions

Bhoopander Giri and K.G. Mukerji

ABSTRACT

The rhizosphere is a "cloud" of microbes, which literally surrounds plant roots, and this is vital for the plant's survival and growth. This area has the largest microbial activity of soil since it represents an important source of nutrients and a physical support for many microorganisms. The rhizosphere is a dynamic environment where microbial interactions take place constantly and may significantly affect plant growth and health. The microbial impact on plant growth is called the rhizosphere effect and is due to the microbial production of plant hormones, enzymes, or changes in the nutrient availability for the plant. The rhizosphere of mycorrhized plants is different from the rhizosphere of the same plant when non-mycorrhized. This is due to the prolongation of the root absorption ability in form of fungal external mycelium increases the nutrient depletion area surrounding the roots. The pattern of root exudation of mycorrhizal plants is altered and consequently physico-chemical soil properties, such as pH, moisture, nutrient content, organic matter or soil aggregation are normally modified around the mycorrhizal roots. AM fungi help the plant to cope with various kinds of stress such as soil pH, heavy metals, soil salinity, and water and drought stresses.

Key words: rhizosphere, mycorrhizosphere, salinity, heavy metal, drought stress

Author's address: Department of Botany, University of Delhi, Delhi 110007, India
E-mail: bhoopg@yahoo.com, kgmukerji@rediffmail.com

INTRODUCTION

Hiltner (1904) recognized the rhizosphere as the volume of soil in the immediate vicinity of the roots, which is predominantly affected by the activity of plants. The rhizosphere differs from the surrounding soil in most of the physico-chemical factors and a wide range of microorganisms colonizes this rhizosphere soil along with the rhizoplane (i.e., the root surface) (Phillips et al. 2003). The number of these microorganisms per gram of soil is much larger in the rhizosphere as compared with bulk soil. This increased microbial activity in vicinity of roots can be ascribed to root exudates, sloughed senescent root cells and mucigel, which have been described as rhizodeposition (Mukerji et al. 1997; Bansal et al. 2000).

In nature, most of the actively absorbing rootlets form symbiotic association with mycorrhizal fungi, which are ubiquitous soil inhabitants. The formation of symbiotic association with mycorrhizae significantly changes the physiology and/or morphology of roots and plants in general, leading to altered root exudation (Bansal and Mukerji 1994). The changes in root exudates affect the microbial communities around the roots, leading to the formation of the "mycorrhizosphere" (Mukerji et al. 1997). The mycorrhizosphere is the zone of soil where the physical, chemical and microbiological processes are influenced by plant roots and their associated mycorrhizal fungi. A major difference in the rhizosphere around the non-mycorrhizal roots and mycorrhizosphere effect is the presence of extramatrical hyphae of mycorrhizal fungi (Mukerji 2002). These extramatrical hyphae extend well beyond the roots into the bulk soil and are an important source of carbon to the soil organisms. The mycorrhizal hyphae increase the soil aggregation and in root association increase exudation, which favours the microbial growth around (Bansal and Mukerji 1996).

The mycorrhizosphere microbiota differ qualitatively as well as quantitatively from the rhizosphere of non-mycorrhizal plants. The soil microfauna influence the mycorrhiza formation as well as the host growth response (Fitter and Garbaye 1994). Many kinds of interactions occur between these microbial communities in the mycorrhizosphere and mycorrhizae. The interaction between the mycorrhizae and soil microorganisms may be mutualistic or competitive and they affect the establishment and functions of mycorrhizal symbionts as well as modify the interactions of plant with other symbionts or pathogens in soil.

THE RHIZOSPHERE

The rhizosphere is the region in which materials released from the root, and root metabolic activities such as respiration, affect microbes. Roots in the process of rhizodeposition release volatile, soluble, and particulate materials (Table 1). The rhizosphere microbes, after their growth on these materials and their cellular turnover, release nutrients, which can be utilized by plants. Plants and their rhizospheres are found in soils, in which the environment is primarily aerobic, and in many marine and freshwater environments in which oxygen is often limiting. The rhizosphere encompasses not only the region of nutrient uptake by the roots, but extends into the soil by action of root products and the trophic interactions that are affected by these products (Van der Putten et al. 2001). A growing root can be reached from the root tip to the crown, where different populations of soil biota have access to a continuous flow of organic substrates derived from the root. This infusion of organic substrates into the rhizosphere by plants explains why the biomass and activity of microbes and soil fauna are greater in the rhizosphere than the bulk soil (Parmelle et al. 1993; Bardgett et al. 1998). The root tip is the site of root growth and is characterized by rapidly dividing cells and secretions or exudates that lubricate the tip as it passes through the soil. The exudates and sloughed root cells provide carbon for bacteria and fungi, which in turn immobilize nitrogen and phosphorus. Farther up the root is the region of nutrient exchange, characterized by root hairs and lower rates of exudation, which stimulate additional microbial growth (Bringhurst et al. 2001).

TABLE 1 *Various kinds of spheres and material released in the soil.*

Terms	*Definition*
Rhizosphere	Region around the plant root where material released from the root modifies microbial populations and their activities
Endorhizosphere	Regions of the various cells layer of the root itself where microorganisms also colonize
Ectorhizosphere	An area surrounding the root and containing root hairs, plant and bacterial mucilage
Rhizoplane	Root surface that can be colonized by microorganisms
Mycorrhizosphere	The ectorhizosphere extend a substantial distance from the root with the development of mycorrhizal fungal associations. Material released from the fungus increase the microbial populations and their activities around the fungal hyphae
Spermosphere	The region around the germinating seed
Rhizodeposition	Release of materials from toots
Exudates	Compounds of low molecular weight that produced by plant cells into the root environment
Mucilages	Gelatinous organic materials released by the plant in the root cap region derived from the Golgi apparatus, polysaccharides hydrolysis, and epidermal materials

The food web that develops within the rhizosphere is complex, consisting of multiple assemblages of species that are supported by roots and their by-products. These assemblages dubbed as the root, bacterial, and fungal energy channels. Live roots form the base of the root energy channels. The root energy channel consists of root-feeding insects and nematodes, and microbes that engage in symbiotic relationships with plant roots (mycorrhizal fungi, rhizobia, *Frankia*). Detritus forms the bases of the bacterial and fungal channels. The bacterial energy channel is composed of bacteria, protozoa, rotifers, nematodes, and a few arthropods. The fungal energy channel largely consists of saprophytic fungi, arthropods, and nematodes.

Soil saprophytic bacteria that compose most of the microbial biomass in the rhizosphere are aquatic organisms and are more efficient in using the more labile root exudates than saprophytic fungi. In contrast, fungi are better adapted to utilize more resistant root cells and substrates than are bacteria. The bacterial energy channel represents a "fast cycle", while the fungal energy channel represents a "slow cycle".

A common suit of nematodes and arthropods predators links the root, bacterial, and fungal energy channels. The linkages between the energy channels tend to be weak at the trophic levels occupied by roots, bacteria and fungi, and strongest at the trophic levels occupied by predatory mites (Moore et al. 2003). The strength of the linkages between energy channels and the dominance of a given energy channel varies by the type of ecosystem, changes with disturbance, and affects nutrient turnover rates. The fungal energy channel tends to be more dominant in systems where the ratio of carbon to nitrogen (C:N) is high while the bacterial channel is more dominant in systems with narrow C:N ratios (Moore et al. 2003).

EVOLUTION OF THE RHIZOSPHERE

Plants surely encountered microorganisms in primordial soil as they moved from aquatic to terrestrial environments. Geochemical evidence for microorganisms exists from 2600 million years ago (MYA) and bacterial fossils dating to 1200 MYA are known (Horodyski and Knauth 1994; Watanabe et al. 2000). Although true roots with vascular tissue appeared perhaps 345 MYA, early terrestrial plants had a variety of underground structures, including stems and rhizoidal appendages (Raven and Edwards 2001), which were beset early on by bacteria and from at least 400 MYA by fungi (Taylor et al. 1995). Stomata were present approximately 410 MYA (Edwards et al. 1998), and thus water movement through the evolving soil food web towards early terrestrial plant tissues probably predated roots.

Phillips et al. (2003) pointed out that all plants in early terrestrial environment interacted with microorganisms. Those relationships in primordial soil predated vascular roots by 131-355 million years, depending on whether one documents the beginning of interactions by plant microfossils (Kenrick and Crane 1997) or by estimates based on protein data (Heckman et al. 2001). It is of thought that the complex rhizobial symbiosis with legumes evolved in a mere 75 million years after the Caesalpiniodeae group of legumes appeared, either estimate offers sufficient time for simpler mutualisms to develop (Phillips et al. 2003).

One cannot assess the extent to which primitive plants resisted microbial attacks, but the presence of their reasonably intact, fossilized remains shows that some protective mechanisms existed. Thus, it is reasonable to suggest that populations of epiphytic and endophytic microorganisms were an accepted fact of life for early land plants. The chemical residues of those microbial populations, as well as any signals released among the microorganisms must have been in close contact with early land plants. Under such conditions, a sifting of water-soluble microbial products for potentially important data on the water and mineral content of nearby environments probably occurred (Phillips et al. 2003).

MICROBIOLOGISTS VIEW OF THE ROOT ANATOMY

Vascular plants are widely distributed over the world. They are one of the most important links which humans have to nature. The vast majority of our food and fiber are directly derived from plants. Although it often is not evident, plant roots and their surrounding microbes (the rhizosphere) are important wherever plants are found: forests, grasslands, tundra, deserts, and wet areas such as marshes and mangrove swamps. The root of these plants is divided into three zones: 1) zone of cell division or meristematic activity 2) zone of cell elongation and 3) zone of cell maturation. Roots grow by the activity of apical meristems, which also form a root cap distally. The root cap is a dynamic, specialized organ that facilitates root penetration of soil, senses threats as well as bounty, and responds by transmitting signals that alter growth patterns. Root supports a unique modified microbial community in an environment termed the rhizosphere, the region influenced by the root and its activities. Microbes directly colonize root surfaces and also found under them, creating additional unique environments for microbes.

The term rhizosphere, which has been used for 100 years, is critical to understanding how plants interact with their environment. In essence, the microbes in the rhizosphere provide the critical link between plants, which require inorganic nutrients, and the environment, which contains the nutrients but often in organic and largely inaccessible forms.

Microbes also colonize the plant root surface (the rhizoplane). The colonization of the rhizoplane by microbes can involve specific attachment mechanisms. For *Agrobacterium thaliana,* which forms tumors in susceptible plants, this involves a two-step process of (i) loose binding to the cell surface and (ii) the synthesis of cellulose fibrils by the bacterium. This results in binding of the bacteria to the plant root surface. If mutant bacteria are used which do not have these attachment characteristics, they will not bind to the root surface.

Plants also have other microbes with which they develop unique relationships in the root environment, including the symbiotic nitrogen-fixing bacteria such as *Rhizobium*. These bacteria form nodules on susceptible legumes, and the fixation of nitrogen by filamentous actinomycetes of the genus *Frankia*, an association that occurs with a wide range of shrubs and woody plants. Another important group of microbes, which form direct associations with plants, includes the mycorrhizae or "fungus-roots", which occur in a wide variety of plants, considered to be one of the oldest plant-microbe associations. The nitrogen-fixing bacteria and the mycorrhizae form structures within the plant root, indicating these physiologically active relationships. The mycorrhizal hyphal network, supported by carbon derived from the plant, also releases organic carbon. Microbes grow around the mycorrhizal hyphae.

CHEMICAL SUBSTANCES IN THE RHIZOSPHERE

A general thought is that aerial parts (stem and leaves) contain greater biomass than root. This impression is misleading. For many plants the root:shoot ratio is positive-more of the plant mass is in the roots than in stems and leaves. The materials released by the plants include a wide variety of organic compounds (Table 2). The types of these substances are constantly changing due to a wide range of plant and environment-related factors. These factors can include temperature and moisture stress, fertilizer additions, herbage removal (both above- and below-ground), changes in sunlight, herbicide additions, plant age, and other changes in the plant's environment. The materials lost from plant roots can be 30-40% of the carbon fixed through photosynthesis.

The fine hairs are a critical part of the root system. These can be rapidly shed when environmental conditions become less suitable for plant growth. From cortical and epidermal cells of the root, mucilages, and soluble metabolic products (amino acids, sugars, organic acids, etc.), i.e. exudates, are also released. In addition, a variety of gaseous metabolites flow from the roots. The release of these different materials is described as the process of rhizodeposition. When the mucilages combine with microbes, soil colloids,

TABLE 2 *Compounds released by plant roots in the process of rhizodeposition*

Compound	*Exudate components*
Sugars	Glucose, fructose, sucrose, maltose, galactose, rhamnose, ribose, xylose, arabinose, raffinose, oligosacchride.
Amino compounds	Asparagine, α-alanine, glutamine, aspartic acid, leucine/isoleucine, serine, γ-aminobutyric acid, glycine, cystine/cysteine, methionine, phenylalanine, tyrosine, threonine, lysine, proline, tryptophane, β-alanine, arginine, homoserine, cystathionine.
Organic acids	Tartaric, oxalic, citric, malic, proponic, butyric, succinic, fumaric, glycolic, valeric, malonic.
Fatty acids and sterols	Palmitic, stearic, oleic, linoleic, linolenic acids, cholestrol, campesterol, stigmasterol, sitosterol.
Growth factors	Biotin, thiamine, niacin, pantothenate, choline, inositol, pyridoxine, ρ-aminobenzoic acid, *N*-methyl nicotinic acid.
Nucleotides, flavonines and enzymes	Flavonine, adenine, guanine, uridine/cytidine, phosphatase, invertase, amylase, protease, polygalacturonase.
Miscellaneous compounds	Auxins, scopoletin, fluorescent substances, hydrocyanic acid, glycosides, saponin (glucosides), organic phosphorus compounds, nematode-cyst or egg-hatching factors, nematode attractants, fungal mycelium-growth stimulants and inhibitors, zoospore attractants.

and soil organic matter. mucigels are formed which cover and protect the root tip.

MICROBIAL DIVERSITY IN THE RHIZOSPHERE

The rhizosphere has a "cloud" of microbes, which literally surrounds plant roots which has been vital for the plant's survival and growth. Plant roots create new environments for microbes; due to the increased levels of nutrients, microbial populations increase, often by 1000- to 10,000-fold, and a marked change in the composition of the microbial community occurs (Table 3), as indicated by the rhizosphere:soil (R:S) ratio for a soil. The number and types of microbes often increase along the root away from the tip of the plant root. The plant roots also respire (use oxygen), which changes the environment of the rhizosphere microbes.

TABLE 3 *Microbial diversity of major groups in the rhizospheric and non-rhizospheric soils*

Organism	*Microbes/g dry soil* *Rhizosphere soil*	*Non-rhizosphere soil*	*R:S ratio*
Bacteria	1200×10^6	53×10^6	23
Actinomycetes	46×10^6	7×10^6	7
Fungi	12×10^5	1×10^5	12
Algae	5×10^3	27×10^3	0.2

The microbial community, which develops in this changed rhizospheric environment will face additional challenges; many of the materials released from roots do not contain sufficient nitrogen, and some times phosphorus, to allow rapid microbial growth. This situation limits both the plant and the associated rhizosphere microbes.

The plant has an increasing demand for inorganic nutrients, which are often not available at a sufficient rate. The rhizosphere contains a wide variety of free-living and symbiotic nitrogen-fixing bacteria (Table 3), which make a major contribution to meet this demand, but at a high energetic cost for the plant. The filamentous fungi, including the free-living and mycorrhizal types, also play a unique role in making nutrients available for the plant, which cannot be provided, by most bacteria. The filamentous fungi in the rhizosphere have an extensive hyphal network. With this hyphal network, they can utilize carbon derived from the plant while obtaining their nitrogen and other limiting resources from outside of the immediate root zone.

Free-living nitrogen-fixing bacteria including the genera *Azotobactor, Azospirillum,* and *Azoarcus* are aboundent in the rhizosphere. In the presence of nitrogen-free or lower nitrogen-content substrates released from the root, these bacteria play an important role. They carry out associative nitrogen fixation and thus provide nutrient for the plant. The rhizosphere community not only has bacteria and fungi but also some protozoans and nematodes. These consumers feed on the nutrient-rich bacteria and fungi, leading to more rapid turnover of the microbes, which leads to an accelerated release of nutrients for plant use.

THE MYCORRHIZOSPHERE

The rhizosphere is defined as the narrow zone (1-2 mm) of soil around the plant roots, which is influenced by the presence and activity of the root. This area has the largest microbial activity of soil since it represents an important source of nutrients and a physical support for many microorganisms (Weller and Thomashow 1994; Varma et al. 1999). The constant release of exudate, cell debris, mucilage or lysates provides the nutrient requirements of most saprophytic microbes. The root itself is a perfect niche for some symbiotic microorganisms such as *Rhizobium* or mycorrhizal fungi, and for other microorganisms intimately associated with the roots such as the plant growth promoting rhizobacteria (PGPRs). The rhizosphere is a dynamic environment where microbial interactions take place constantly and may significantly affect plant growth and health. The microbial impact on plant growth is called the rhizosphere effect (R:S ratio) and is due to the microbial production of plant hormones, enzymes, or changes in the nutrient availability for the plant (Azcon-Aguilar and Barea 1992; Mukerji, 2002, Mukerji et al 1998).

The rhizosphere of mycorrhized plants is very different from the rhizosphere of the same plant when non-mycorrhized. First, the prolongation of the root absorption ability in form of fungal external mycelium increases the nutrient depletion area surrounding the roots. Secondly, the pattern of root exudation of mycorrhizal plants is altered and consequently physico-chemical soil properties, such as pH, moisture, nutrient content, organic matter or soil aggregation are normally modified around the mycorrhizal roots. The mycorrhizae act in modifying the nutrient availability for the rest of rhizospheric microorganisms, and also providing an additional ecological niche for these microorganisms. All these crucial changes due to mycorrhizal formation is due to mycorrhizosphere effect (Linderman 1988). The microbial components of this mycorrhizosphere have normally an increased activity, which, in turn, may affect plant growth and health, and also modify the behavior of its fungal counterpart. These complex interactions are crucial to understand the hidden world under the soil surface. Interactions among mycorrhizal fungi and other soil microorganisms are reciprocal i.e., mycorrhizal fungi affect the other microbes and other microbes in turn affect the mycorrhizal association. The interaction can be synergistic and antagonistic, summarized in the Tables 4 and 5 respectively.

TABLE 4 *Synergistic interactions between mycorrhizal fungi and other rhizosphere microorganisms (Modified from Bansal et al. 2000)*

Mycorrhizal fungi	Rhizosphere microorganisms
Glomus fasciculatum	**Frankia**
G. fasciculatum	*Streptomyces cinnamomeous*
Glomus mosseae	*Rhizobium leguminosarum*
G. versiforme	*R. loli*
Glomus fasciculatum	**Beijerinckia**
G. fasciculatum	*Azotobacter chroococcum*
G. fasciculatum	*Azospirillium brasilense*
G. versiforme	*Corynebacterium*
Glomus versiforme	*Pseudomonas sp.*
Endogone sp.	*Agrobacterium sp.*
	Pseudomonas sp.
G. macrocarpum	*Bacillus megaterium*
	Pseudomonas fluorescence
Glomus macrocarpum	*Cladosporium sp.*
	Gliocladium virens
G. intraradices	*Fusarium oxysporium f. sp. chrysanthemi*

TABLE 5 *Antagonistic interactions between mycorrhizal fungi and rhizosphere microorganisms (Modified after Bansal et al. 2000)*

Mycorrhizal fungi	Soil microorganisms
Glomus mosseae	*Phytophthora nicotianae* var. *parasitica*
G. *macrocarpum*	*Fusarium sp.*
G. *intraradices*	*Fusarium oxysporium f. sp. radicic-lycopersici*
G. *intraradices*	*Ordium lini*
G. *fasciculatum*	*Pythium ultimum*
G. *fasciculatum*	*Aphanomyces euteiches*
Glomus sp.	*Verticillium albo-atrium*
Glomus sp.	*Rhizoctonia solani*
G. *fasciculatum*	*Phytophthora perasitica*
G. *fasciculatum*	*P. fragariae*
G. *etunicatum*	*P. fragariae*
G. *fistulosum*	*Meloidogyne hapla*
Gigaspora margarita	*Pratylenchus vulnus*
G. *intraradices*	*Pratylenchus vulnus*
G. *etunicatum*	*Rhodopholus similis*
G. *manihotis*	M. *incognita*
G. *mosseae*	*Azotobacter chrocococcum*

WHAT ARE MYCORRHIZAL FUNGI?

Mycorrhizae provide an intimate link between the soil environment and the functional nutrient absorbing system of the plant. The modification of plant roots by symbiotic fungi into the distinct structures characteristic of mycorrhizae results in a unique and intriguing component of the rhizosphere. Since the first published description of a mycorrhizal association by Frank (1885), scientists continue to be challenged as to the role of mycorrhizae in the ecological and physiological context of plants. Although most research and observations on mycorrhizae have been concerned with nutrient uptake by mycorrhizae, especially immobile elements such as phosphorous, there is an increasing awareness of their potential importance in many diverse aspect of a plant's ability to grow and survive in natural and man-altered environment.

TYPES OF MYCORRHIZAL FUNGI

So far seven types of mycorrhizae have come into general use over the years on the basis of morphology and anatomy but also of either host plant taxonomy or fungal taxonomy (Srivastava et al. 1996; Smith and Read

1997; Raina et al 2000). These are: ectomycorrhiza, endomycorrhiza or arbuscular mucorrhiza, ericoid mycorrhiza, arbutoid mycorrhiza, monotropoid mycorrhiza, ect-endomycorrhiza and orchidaceous mycorrhiza.

FUNCTIONS OF MYCORRHIZAL FUNGI

In terrestrial ecosystems, AM fungi make various promises (Table 6) where the organic detritus-decomposer pathway accounts for the majority of energy flow and nutrient turnover. Microflora coupled with micro fauna in the soil are the major components of both biomass and activity affecting nutrient availability. Although soil bacteria and fungi generally immobilize mineral nutrients as carbon is consumed and thereby compete with plants for macronutrients, mycorrhizal fungi, because of their unique carbon strategies, can efficiently couple soil mineralization and nutrient uptake by the plant roots. In many cases, the mycorrhizal system actually "bridges" across the rhizosphere, and provides an organic link between the root and the bulk soil. In addition, AM fungi help the plant to cope with various kinds of stress such as soil pH, heavy metals, soil salinity, and water and

TABLE 6 *Promises made from AM fungi*

Promotes plant growth	Maintain plant and soil health
Bio-protection against root diseases (bacteria, fungi and nematodes)	Plant production with reduced fertilizers and pesticides
Nutrient acquisition	Plant size or biomass
Improved soil-root contact	Influence population dynamics of soil flora
Symbiosis alters host water relations	Revegetation of landscape or contaminated soils
Symbiosis alters root length, root architecture and root/shoot ratio	Biological hardening of tissue culture raised plants
Alters rate of water movement into, through and out of host plants	Effects on tissue hydration and leaf physiology
Postpones leaf dehydration	Alters leaf osmotic potential
Alters the number of photosynthetic units	Photosynthetic storage and export rates
Dissimilar symplastic solute pools	More effective scavenging of soil water
Effects on osmotic adjustment	Drought responses
Altered transpiration rates	Stomatal conductances to water vapor
Intrinsic leaf hydraulic or biochemical properties	Osmoprotect enzymes
Altered nodule number and their activity	Enhanced P acquisition
Altered total protein	Altered morphological and phenological effects
Altered leaf abscission	Altered leaf drop, necrosis and senescence
Altered leaf movements	Altered wilting of leaves
Altered recovery from wilting	Provide salt tolerance to plant

drought stresses. A brief account on the functioning of AM fungi under these stresses is described in this chapter.

i Impact of AM Fungi under Different Soil pH

Soil pH crucially affects development of mycorrhizal fungi by affecting solubility of several compounds. Most phosphate exists as insoluble complexes of Al and Fe at acidic pH and at alkaline pH phosphate exists as insoluble complexes of Ca and Mg while maximum solubility of phosphate occurs at neutral pH. However, inorganic forms are still largely insoluble. Many metals are insoluble under alkaline edaphic conditions but highly soluble at acidic conditions.

It is well demonstrated that mycorrhizal fungi vary tolerance of soil pH. Some grow only in low pH soils whereas others grow after modifying the soil pH with certain amount of lime (Giri et al. 2003a). A few mycorrhizae have tendency to grow at the pH from which they have been isolated (Giri and Mukerji 2003).

Soil pH is an important factor in studying the ecology of endomycorrhizal fungi. Low soil pH has profound effect on movement and uptake of P. Rhizosphere acidification affects fungal-soil-plant nutrients supply mechanisms (Giri et al. 2003a). Phosphorus diffusion rate increased with increase in acidity of soil. *Glomus* sp. extracted from a neutral soil showed best growth at 7 pH, while its growth was less at acidic and alkaline pH. Relative tolerance of *Glomus mosseae, G. fasciculatum,* and *G. macrocarpum* to graded pH levels (7.8-10.5) and their influence on P uptake in *Prosopis juliflora* were evaluated by Sidhu and Behl (1997). They found that increase in pH adversely affected growth, biomass and P concentration in seedlings. Chlamydospore formation by all three AM fungi decreased with increase in the rhizoshpere soil pH. However, application of AM resulted in a significant increase in seedlings root and shoot length, collar diameter and biomass production at high pH levels. Biomass production of mycorrhizal seedlings grown at pH 10.5 was equivalent to that of uninoculated plants at pH 7.8. In AM-colonized roots P concentration increased by 68% at pH 7.8 and 40% at pH 10.5. *Glomus fasciculatun* originally isolated form site having high pH (9.2) showed relatively high tolerance to pH ranging between 8.5-10.5 as compared to other species as exhibited by higher degree for chlamydospore formation and root colonization (Sidhu and Behl 1997).

Soil pH and available soil nutrients have a cumulative effect on the efficiency of AM fungi on plant growth. The increased P concentration led to high soil pH and reduced available Mn concentration in soil. This is probably due to precipitation of cation directly by excess of phosphate, which induced elevation in pH that increased growth of the plant.

ii Impact of AM Fungi under Heavy Metal Stress

Metal toxicity in soil may be induced by discharge of sewage or industrial pollutants into the soil, or due to presence of excessive quantities of metals in certain categories of soils. In soil they are present as free ions, soluble metal complexes, exchangeable metal ions, organically bound metals, precipitated or insoluble complexes such as oxides, carbonate and hydroxides or they may form part of the structure of silicate minerals (indigenous soil content). The toxicity of metals in the soil depends on their bioavailability, which may be defined as the availability of metals to be transferred from a soil compartment to living organisms. High metal concentration in soil is not only toxic to plant but also affects germination and growth of soil microorganisms. Amongst the myriad of soil microorganisms, mycorrhizal fungi are considered as integral functioning parts of plant roots and the fungi involved provide direct link between soil and plant roots. The influence of mycorrhizal fungi on plant nutrition is greater for elements with narrow diffusion zones around plant roots such as P (Smith and Read 1997; Giri et al. 1999)

There are only a few reports concerning interaction of arbuscular mycorrhiza with metals. Nonetheless, it is apparent that soils high in available metals do provide a habitat for specific AM fungi, which do provide some degree of protection to host plant from toxic metals by restricting uptake of metals to the plant or tolerating these themselves. *Medicago sativa* grown in mine spoils and waste sediments containing high amount of Zn and Cd showed significant mycorrhizal colonization although the number of AM fungal spores was lower than in an adjacent soil not altered by mining activity. *Agropyron trachycaulum* showed considerable mycorrhizal colonization in a subalpine coalmine spoil and on oil sand tailings amended with peat containing AM fungal propagules, but colonization was absent in the same species cultivated in the unamended oil sand, reveling them to be devoid of arbuscular mycorrhizal fungi. These results indicated that poor or absence of mycorrhizal propagules in some of the mine spoils, may have resulted in non-mycorrhizal colonization. However, mycorrhizal rather than non-mycorrhizal grasses could colonize polluted mining sites, suggesting that heavy metal tolerance was due to mycorrhizal association.

In *Festuca* and *Calamagrostis epigejos,* mycorrhizal colonization was observed when grown in coastal dunes contaminated by atmospheric deposition from a blast furnace (Duke et al. 1986). Similarly, extensive colonization of mycorrhizal fungi was observed in *Agrostis capillaris* in Zn and Cd contaminated site. Experiments conducted on three populations of *A. capillaris* using a sandy soil contaminated with smelter and limestone-derived clay with or without metals of natural origin, however, did not show significant difference in mycorrhizal root colonization between these populations.

In *Albizzia amara* high level of colonization by AM fungi was observed in agricultural soil and fly ash severely contaminated with heavy metals (Giri 2001). *Oxalis acetosella* grown in low pH soil polluted with Cd, Zn, and Pb showed even higher mycorrhizal colonization (Turnau et al. 1996). These results further substantiate the fact that mycorrhizal fungal colonization has potential to tolerate heavy metals particularly in case of AM fungi, which have originated from metal contaminated sites. In contrary, there are some reports demonstrating inhibition of mycorrhizal colonization in the presence of metals of different origin (Chao and Wang 1991; Weissenhorn et al. 1995; Joner et al. 2000).

Experiments conducted on maize in the metal contaminated soil showed that mycorrhizal colonization either increased plant biomass and decreased Cd, Cu, Zn and Mn concentrations in shoot- and root-tissues, or had no effect on growth and metal uptake depending on root density, plant growth conditions, and mycorrhizal inoculum (Weissenhorn et al. 1995). Loth and Hofner (1995) observed that mycorrhizal colonization increased uptake of Cu, Zn and Cd in *Avena sativa* roots from a highly contaminated soil but a reduced translocation to the aerial part. Weissenhorn and Leyval (1995) reported higher uptake of metal by mycorrhizal plants under high metal concentration. It was found that *G. fasciculatum* reduced negative effect of Zn on plant growth. However, they did not report the effect of mycorrhiza on the Zn concentration in shoot and root tissues.

The effect of AM fungi on plant metal uptake also depends on soil pH. With increasing soil pH, DTPA-extractable metals decrease, but at the same time AM fungi increased Cd, Zn and Mn uptake in the shoots of *Medicago sativa*. At a lower soil pH, mycorrhizal colonization decreased metal uptake. In both cases, mycorrhizal colonization enhanced alfalfa growth. It is unfortunate that most of the studies on heavy metal tolerance/uptake by mycorrhizal fungi have been performed in pots where it is not possible to separate the effect of the fungus and of the host plant on the mobilization and uptake of metals (Leyval et al. 1997). To differentiate fungus and host plant effects, plant containers with different compartments separating root and extraradical hyphae of *G. mosseae* from a sandy loam to clover was undertaken. In mycorrhizal clover, uptake of Cd increased by 55% in comparison to non-mycorrhizal plants. In the same plant, Burkert and Robson (1994) also reported transport of Zn in extraradical hyphae from a sandy soil. In the bean plants, higher amount of Cd, Zn and Cu was transferred by mycorrhizal extraradical hyphae.

The experiments carried out on the capacities of extramatrical hyphae to bind Cd and Zn have shown that AM fungal mycelium has a high metal sorption capacity and a CEC comparable to other microorganisms. Metal sorption by AM fungi was rapid and appeared mainly to be due to passive adsorption. It was also noticed that highest adsorption took place in a metal-tolerant *G. mosseae* isolate and intermediate for fungus isolated from a soil treated with metal contaminated waste. *G. mosseae* absorbed ten times more metals than commonly used biosorption organism *Rhizopus arrhizus* (Joner et al. 2000). These results confirm the AM involvement in plant protection against excess heavy metal uptake and more binding capacity of mycorrhizal fungi than others.

iii Impact of AM Fungi under Salt-stressed Conditions

Soil salinity is a problem of great concern. About one third of world's irrigated land and half of the land in arid, semi-arid and tropical regions is not in use due to salinity (Juniper and Abbott 1993; Giri and Chamola 1999; Giri et al. 2002; 2003a; Giri and Mukerji 2003; Giri et al. 2003b). Ten million hectares of irrigated agricultural land is abandoned annually. In India alone, 75 Mh land has lost their fertility because of the deposition of salts. Most of the areas of Uttar Pradesh (Indo-gangatic plane), Rajasthan and Haryana are adversely affected due to high salt concentration and they have lost their fertility (Giri et al. 2000). Thus, there is an urgent need for improving such degraded wastelands to combat the increasing pressure of alarmingly rising population on agriculture by low input technologies. Several microorganisms are known to have the ability of tolerating high salt concentrations. They can survive under a wide range of salinity. Among these microorganisms, mycorrhizal fungi are of growing concern (Table 7).

High salt concentration inhibits germination of AM fungal spores as well as growth of hyphae resulting in decreased growth and development of AM fungal density in soil (Juniper and Abbott 1993; Al-Karaki and Clark 1998; Al-Karaki et al. 2001). McMillen et al. (1998) found that increasing concentration of NaCl inhibits either the hyphal growth or the infectivity of hyphae and AM colonization of plant roots. This may be due to adverse effect of NaCl on the hyphal growth as well as altered supply of carbohydrates from the plant to the fungus.

In our laboratory the effect of AM fungi *Glomus fasciculatum* and *G. macrocarpum* was investigated on growth, development and nutritional responses of *Acacia auriculiformis*, under nursery and field conditions (Giri et al. 2003a). Plants were grown under a wide range of salinity, imposed by 3, 5 and 10 dS/m solutions of 1 N NaCl. Both AM fungi protected host plant against detrimental effect of salinity. The extent of AM response on growth as well as root colonization varied with the species, and salinity levels. Mycorrhiza–inoculated plants accumulated greater amount of P, and K while Na uptake lowered as salinity increased. The greater nutrient

TABLE 7 Plants and AM fungi that tolerate high salinity levels

Plants	*AM fungi*	EC	*References*
Bell pepper	*Glomus fasciculatum*	1-12	Hirrel and Gerdemann 1980
Bell pepper	*Gigaspora margarita*	1-12	Hirrel and Gerdemann 1980
Onion	*G. fasciculatum*	1-12	Hirrel and Gerdemann 1980
Onion	*Gigaspora margarita*	1-12	Hirrel and Gerdemann 1980
Indian ricegrass	*Entrophospora infrequences*	1.6-2.0	Stahl and Williams 1986
Indian ricegrass	*G. fasciculatum*	1.6-2.0	Stahl and Williams 1986
Indian ricegrass	*Glomus microcarpum*	1.6-2.0	Stahl and Williams 1986
Indian ricegrass	*Glomus mosseae*	1.6-2.0	Stahl and Williams 1986
Yellow sweetclover	*E. Infrequences*	1.6-2.0	Stahl and Williams 1986
Yellow sweetclover	*G. fasciculatum*	1.6-2.0	Stahl and Williams 1986
Yellow sweetclover	*G. microcarpum*	1.6-2.0	Stahl and Williams 1986
Yellow sweetclover	*G. mosseae*	1.6-2.0	Stahl and Williams 1986
Big sagebrush	*G. fasciculatum*	0.6	Stahl et al. 1988
Big sagebrush	*G. microcarpum*	0.6	Stahl et al. 1988
Big sagebrush	*E. Infrequences*	2.6-3.8	Stahl et al. 1988
Big sagebrush	*G. fasciculatum*	2.6-3.8	Stahl et al. 1988
Big sagebrush	*G. macrocarpum*	2.6-3.8	Stahl et al. 1988
Big sagebrush	*G. microcarpum*	2.6-3.8	Stahl et al. 1988
Big sagebrush	*G. mosseae*	2.6-3.8	Stahl et al. 1988
Acacia auriculiformis	*G. fasciculatum*	1-10	Giri et al. 2003b
Acacia auriculiformis	*G. macrocarpum*	1-10	Giri et al. 2003b
Sesbania aegyptiaca	*G. macrocarpum*	1-5	Giri and Mukerji 2004
Sesbania grandiflora	*G. macrocarpum*	1-5	Giri and Mukerji 2004

acquisition, change in root morphology and electrical conductivity of soil in response to AM colonization was observed during course of the study and this may be the possible mechanism to protect the plant from salt stress (Giri et al. 2003a). Inoculation of *Sesbania grandiflora* and *Sesbania aegyptiaca* with AM fungus *Glomus macrocarpum* had significant increase in growth and biomass production (Giri and Mukerji 2004). Under saline condition, *Sesbania* spp. had higher amount of Mg and reduced Na content in shoot-tissues; the increased Mg uptake and reduced sodium uptake helped in chlorophyll synthesis. AM fungus also increased establishment and survival of tree plants. Both the tree species were highly dependent on *G. macrocarpum* (Giri and Mukerji 1999, Giri et al. 1999)

The response of mycorrhizal and non-mycorrhizal *Olea europaea* under saline conditions with or without supplemental Ca resulted in less depolarization at cellular level (cell transmembrane electropotential) in roots of mycorrhizal than non-mycorrhizal plants. Supplemental Ca in the saline treatments had protective effect on membrane integrity canceling or reducing the differences in depolarization between mycorrhizal and non-mycorrhizal plants. Mycorrhizal roots accumulated greater quantities of Na, K, and Ca and exhibited a lower K:Na ratio, but in leaves, mycorrhizal plants had greater K:Na ratio than non-mycorrhizal plants (Rinaldelli and Mancuso 1996).

Mycorrhizal colonization brought about a noticeable improvement in salt-tolerance in olive plants, which was clearly demonstrated by trends in impedance parameters (Mancuso and Rinaldelli 1996). In saline conditions, the electrical impedance parameters in shoots and leaves of olive plants were studied to understand variations in extracellular resistance, intracellular resistance and the state of membrane in mycorrhizal and non-mycorrhizal plants. There was a reduction in extra- and intra cellular resistance for non-mycorrhizal plants with increased NaCl concentration (Mancuso and Rinaldelli 1996). Ezz and Nawar (1994) found that sour orange seedlings irrigated with saline water of 450 ppm salt, reduced total leaf chlorophyll, peroxidase activity, starch and total carbohydrate concentration in leaves and roots. Inoculation with *Glomus intraradices* increased total chlorophyll, polyphenol activity, leaf and root sugars, and carbohydrate concentrations but peroxidaes activity was not altered.

Acknowledgements

The senior author is thankful to Council of Scientific and Industrial Research, New Delhi for financial support.

References

Al-Karaki, G.N., and Clark, R.B. 1998. Growth, mineral acquisition and water use by mycorrhizal wheat grown under water stress. *J Plant. Nutr.* 21: 263-276.

Al-Karaki, G.N., Hammad, R., and Rusan, M. 2001. Response of two tomato cultivars differing in salt tolerance to inoculation with mycorrhizal fungi under salt stress. *Mycorrhiza* 11: 43-47.

Azcon-Aguilar, C., and Barea, J.M. 1992. Interaction between mycorrhizal fungi and other rhizosphere microorganisms. In: *Mycorrhizal functioning: an Integrative Plant-Fungal Process* (Ed. Allen, M.J.). pp. 163-198. Chapman and Hall, New York.

Bansal, M., Chamola, B.P., Sarwer, N., and Mukerji, K.G. 2000. Mycorrhizosphere: interaction between rhizosphere microflora and VAM fungi. In: *Mycorrhizal Biology* (Eds. Mukerji, K.G., Chamola, B.P., and Singh, J.). pp. 143-152. KA/PP, The Netherlands.

Bansal, M., and Mukerji, K.G. 1994. Positive correlation between root exudation and VAM induved changes in rhizosphere mycoflora. *Mycorrhiza* 5: 39-44.

Bansal, M., and Mukerji, K.G. 1996. Root exudates in rhizosphere biology. In: *Concepts in Applied Microbiology and Biotechnology* (Eds. Mukerji, K.G., and Singh, V.P.). pp 98-120. Aditya Book, New Delhi, India.

Bardgett, R.D., Wardle, D.A., and Yeates, G.W. 1998. Linking above- ground and below-ground food webs: how plant responses to foliar herbivory influences soil organisms. Soil Biol Biochem 30: 1867-1878.

Bringhurst, R.M., Cardon, Z.G., and Gage, D.J. 2001. Galactosides in the rhizosphere: utilization by *Sinorhizobium meliloti* and development of a biosensor. *Proc Nat Acad Sci* 98: 4540-4545.

Burkert, B. and Robson, A. 1994. ^{65}Zn uptake in subterranean clover (*Trifolium subterraneum* L.) by three vesicular arbuscular mycorrhizal fungi in a root-free sandy soil. *Soil Biol. Biochem* 26: 1117-1124.

Chao, C.C. and Wang, Y.P. 1991. Effects of heavy metals on vesicular arbuscular mycorrhizae and nitrogen fixation of soybean in major soil groups of Taiwan. *J Chin Agric Chem Soc* 29: 290-300.

Duke, E.R., Johnson, C.R., and Koch, R.E. 1986. Accumulation of phosphorus, dry matter and betaine during NaCl stress of split-root citrus seedlings colonized with vesicular-arbuscular mycorrhizal fungi on zero, one or two halves. *New Phytol.* 104: 583-590.

Edwards, D., Kerp, H., and Hass, H. 1998. Stomata in early land plants: an anatomical and ecophysiological approach. *J Exp. Bot.* 49: 255-278.

Ezz, T., Nawar, A. 1994. Salinity and mycorrhizal association in relation to carbohydrate status, leaf chlorophyll and activity of peroxidase and polyphenol oxidase enzymes in sour orange seedlings. *Alexandria J Agri Res* 39: 263-280.

Fitter, A.H. and Garbaye, J. 1994. Interaction between mycorrhizal fungi and other soil organisms. *Plant soil* 159: 123-132.

Frank, A.B. 1885. Uber die auf Wurzelymbioze beruhende emahrung gewisser baume durch unterirdische rilze. Ber. Dt. Bot. 3: 128-145.

Giri B (2001) Mycorrhization in afforestation of stressed habitats. Ph.D thesis, Department of Botany, University of Delhi, Delhi, India.

Giri, B., Kapoor, R., Lata, A., and Mukerji, K.G. 2003b. Preinoculation with arbuscular mycorrhizae helps *Acacia auriculiformis* in a degraded Indian wasteland soil. *Comm. Soil Sci. Plant Anal.* 35 (1&2): 193-204.

Giri, B., Kapoor, R., and Mukerji, K.G. 2002. VA Mycorrhizal/VAM technology in establishment of plants under salinity stress conditions. In: *Techniques in Mycorrhizal Studies* (Eds. Mukerji, K.G., Manoharachari, C. and Chamola, B.P.). Pp. 313-327. Kluwer Academic Publishers, The Netherlands.

Giri, B., Kapoor, R., and Mukerji, K.G. 2003a. Influence of arbuscular mycorrhizal fungi and salinity on growth, biomass, and mineral nutrition of *Acacia auriculiformis. Biol Fertil Soils* 38: 170-175.

Giri, B., Kaur, M., and Mukerji, K.G. 1999. Growth responses and Dependency of *Sesbania aegyptiaca* on vesicular arbuscular mycorrhiza in salt stressed soil. *Ann. Agric. Res.* 20: 109-112.

Giri, B., and Mukerji, K.G. 1999. Improved growth and productivity of *Sesbania grandiflora* Pers. under salinity stress through mycorrhizal technology. *J Phytol. Res.* 12: 35-38.

Giri, B., and Mukerji, K.G. 2004. Mycorrhizal inoculant alleviates salinity stress in *Sesbania aegyptiaca* Pres and *S. grandiflora* Pres under field conditions: evidence for improved magnesium decreased sodium uptake. *Mycorrhiza* 14: 307-312.

Giri, B., Kapoor, R., and Mukerji, K.G. 2002. 2000. *Sesbania aegyptiaca* Pers seedlings response to VA mycorrhization in two types of soil. *Phytomorphology* 50: 327-332.

Heckman, D.S., Geiser, D.M., Eidell, B.R., Stauffer, R.L., Kardos, N.L., and Hedges, S.B. 2001. Molecular evidence for the early colonization of land by fungi and plants. *Science* 293: 1129-1133.

Horodyski, R.J., and Knauth, L.P. 1994. Life on land in the Precambrian. *Science* 263: 494-498.

Joner, E., Roberto, B. and Leyval, C. 2000. Metal-binding capacity of arbuscular mycorrhizal mycelium. *Plant soil* 226: 227-234.

Juniper, S., and Abbott, L. 1993. Vesicular arbuscular mycorrhizas and soil salinity. *Mycorrhiza* 4: 45-57

Kenrick, P., and Crane, P.R. 1997. The origin and early evolution of plants on land. *Nature* 389: 33-39.

Leyval, C., Turnau, K. and Haselwandter, K. 1997. Effect of heavy metal pollution on mycorrhizal colonization and function: physiological, ecological and applied aspects. *Mycorrhiza* 7: 139-153.

Lindermann, R.G. 1988. Mycorrhizal interactions with the rhizosphere microflora. The mycorrhizosphere effect. *Phytophatol.* 78: 366-371.

Loth, F.G. and Hofner, W. 1995. Einfluss der VA-Mykorrhiza auf die Schwermetallaufnahmen von Hafer (*Avena sativa* L.) in Abhangigheit vom Kontaminationsgrad der Boden. *Z Pflanzenernaehr Bodenk.* 158: 339-345.

Mancuson, S., and Rinaldelli, E. 1996. Response of young mycorrhizal and non-mycorrhizal plants of olive tree (*Olea europaea* L.) to saline conditions. II. Dynamics of electrical impedance parameters of shoots and leaves. *Adv. Hort. Sci.* 10: 135-145.

Mc Millen, B.G., Juniper, S., and Abbott, L.K. 1998 Inhibition of hyphal growth of vesicular arbuscular mycorrhizal infection from fungus spores in soil containing sodium chloride. *Soil Biol Biochem* 30: 1639-1646.

Moore, J.C., McCann, K., Setala, H., and De Ruiter, P.C., 2003. Top-down is bottom–up: dose predation in the rhizosphere regulate aboveground dynamics? Ecology 84: 846-857.

Mukerji, K.G. 1999. Mycorrhiza in control of plant pathogens: molecular approaches. In: *Biotechnological Approaches in Biocontrol of Plant Pathogens* (Eds. Mukerji, K.G., Chamola, B.P., and Upadhyay, R.K.). pp. 135-156. KluwerAacademic/ Plenum Publishers, New York.

Mukerji, K.G. 2002. Rhizosphere biology. In: *Techniques in mycorrhizal studies* (Eds. Mukerji, K.G., Manoharachary, C. and Singh, J.). Pp. 87-102. Kluwer Academic Publishers, Dordrecht/Boston/ London.

Mukerji, K.G., Chamola, B.P., and Sharma, M. 1997. Mycorrhiza in control of plant pathogens. In: *Management of Threatening Plant Diseases of National Importance* (Eds. Agnihotri, V.P., Sarbhoy, A., and Singh, D.V.). pp. 297-314. MPH, New Delhi.

Parmelle, R.W., Ehrenfeld, J.G., and Tate, R.A. 1993. effects of pine roots on microorganisms, fauna and nitrogen availability in two soil horizons of a coniferous spodosol. *Biol. Fertil. Soils* 15: 113-119.

Phillips, D.A., Ferris, H., Cook, D.R., and Strong, D.R. 2003. Molecular control points in rhizosphere food webs. *Ecology* 84: 816-826.

Raina, S., Chamola, B.P. and Mukerji, K.G. 2000. Evolution of mycorrhiza. In: Mycorrhizal Biology (Eds Mukerji, K.G., Chamola, B.P. and singh, J.). pp. 1-27. Kluwer Academic/Plenum publishers,New York.

Raven, J.A., and Edwards, D. 2001. Roots: evolutionary origins and biogeochemical significance. J Exp. Bot. 52: 381-401.

Rinaldelli, E., and Mancuso, S. 1996. Response of young mycorrhizal and non-mycorrhizal plants of olive tree (*Olea europaea* L.) to saline conditions. I. Short-term electrophysiological and long-term vegetative salt effects.*Adv. Hort. Sci.* 10: 126-134.

Sidhu, O.P., and Behl, H.M. 1997. Response of three Glomus species on growth of *Prosopis juliflora* Swartz at high levels. *Symbiosis* 23: 23-24.

Smith, S.E., and Read, D.J. 1997. Mycorrhizal Symbiosis. Academic Press, San Diego, California, London, pp 605.

Srivastava, D., Kapoor, R., Srivastava, S.K., and Mukerji, K.G. 1996. Vesicular arbuscular mycorrhiza: an overview. In: *Concepts in Mycorrhizal Research* (Ed. Mukerji, K.G.). pp. 1-39. Kluwer Academic Publishers, The Netherlands.

Taylor, T.N., Remy, W., Hass, H., and Kerp, H. 1995. Fossil arbuscular mycorrhizae from the early Devonian. *Mycologia* 87: 560-573.

Turnau, K., Kottke, I. and Dexheimer, J. 1996. Toxic element filtering in *Rhizopogon roseolus/Pinus sylvestris* mycorrhizas collected from calamine dumps. *Mycol Res.* 100: 16-22.

Vander Putten, W.H., Vet L.E.M., Harvey, J.A., and Wackers, F.L. 2001. Linking above-ground and below-ground multitrophic interactions of plants, herbivores, pathogens, and their antagonists. *Trends Ecol. Evol.* 16: 547-554.

Watanaba, Y., Martini, J.E.J., and Ohmoto, H. 2000. Geochemical evidence for terrestrial ecosystem 2.6 billion years ago *Nature* 408: 574-578.

Weissenhorn, I. and Leyval, C. 1995. Root colonization of maize by a Cd-sensative and a Cd-tolerance *Glomus mosseae* and cadmium uptake in sand culture. Plant Soil 175: 233-238.

Weissenhorn, I., Leyval, C. and Berthelin, J. 1995. Bioavailability of heavy metal and abundance of arbuscular mycorrhiza in a soil polluted by atmospheric deposition from a smelter. *Biol. Fertil. Soils* 19: 22-28.

Weller, D.M., and Thomashow, L.S. 1994. Current challenges in introducing beneficial microorganisms unto the rhizosphere into the rhizosphere. In: *Molecular Ecology of Rhizosphere Microorganisms Biotechnology and the Release of GMOs* (Eds. Gara, F.O., Dowling, D.N., Boesten, B.). pp. 1-13. VCH Verlagsgesellschaft, Winheim.

Frontiers in Plant Sciences (Pages 617-642)
Eds. K.G. Mukerji, K.V.B.R. Tilak, S.M. Reddy, L.V. Gangwane, P. Prakash, I.K. Kunwar

Ectomycorrhizal Diversity and its Role in the Productivity of Forest Plants

Kamal Prasad[1], A.K. Pandey[2] and R.C. Rajak[+2]

INTRODUCTION

The fecilitation of plant nutrient uptake by fungi in symbiotic association with the roots of higher plant is emerging as one of the exciting frontiers for the enhancement of crop production. These micro organisms help in nutrient uptake by changing the amounts, concentrations and properties of minerals available to plants both in agriculture and forestry. The potential is not only for a substantial increase in conventional crop production but for the expansion of productivity of economic plants in areas that have an unfavourable climate and non-productive soils. These favourable shifts in mineral nutrition result in the growth, development and chemical composition of host plants. Mycorrhiza plays a dominant role, particularly in the uptake of less-mobile plant nutrients. These interactions are varied, widespread, common and ecologically significant for plant mineral nutrition.

The term "Mycorrhiza" is applied to associations of fungi with the absorbing organs of plants, strictly where the organ concerned is the root. The scope of the terms and the systems of nomenclature are controversial, partly due to diversity in nutritional interrelationships, in kinds of fungi and host plants involved and in morphological details of different mycorrhizas (Mukerji 1996, Mukerji et al. 2000, 2002).

[1]INCO DEV Res. Lab., Department of Botany, Maharshi Dayanand Saraswati University, Ajmer-305009 (Raj.) India.
[2]Department of Postgraduate Studies & Research in Biological Science, R.D. University, Jabalpur-482001 (M.P.) India.

Traditionally, the mycorrhizas are clarified as "Ectotrophic" and Endotrophic". In ectotrophic type of mycorrhiza, the fungal endophyte is predominantly exogenous, resulting in the formation of a fungal sheath around the root. The penetration of the hyphae is normally intercellular within the host plant. The fungal partners involved in ectomycorrhiza are mostly Basidiomycetes and few Ascomycetes, infecting many tropical, subtropical and temperate forest trees. Ectomycorrhizal research in India did not get under way until the early fifties when Chaudhuri (1945) first reported mycorrhizal association in *Abies spectablis, Cedrus deodara, Picea morinda, Pinus roxburghii* and *Taxus baccata*. However, the credit for firmly establishing mycorrhizal research goes to Bakshi who claimed ecytomycorrhizal association in *Abies pindrow, Cedrus deodara, Picea morinda* and *Pinus roxburghii* (Bakshi, 1957). As very few scientists were interested in tree research, Bakshi's findings did not impress many. There was complete ignorance about the significance of ectomycorrhizae on tree growth. Naturally, ectomycorrhizal research was relegated to a position of near obscurity.

About 95% of world's species of vescular plants are reported to be mycorrhizal (Trappe, 1977). Nevertheless, over 12,000 species have been scrutinized for mycorrhizal association including major economic plants.

Ectomycorrhizae are characterized by modification of roots and loss of absorbent hairs. The ectomycorrhizal fungus surrounds the roots and forms a mantle of mycelium. The external hyphae originating from the mantle explore the soil and help an absorption of nutrients and water, while the internal hyphae make a close contact with the roots. Four to five thousand members of higher fungi form ectomycorrhizae.

ECTOMYCORRHIZAL FUNGI IN SOIL

In ectotrophic mycorrhizal plants, roots are dimorphic, uninfected long roots and infected short roots exhibiting degrees of morphogenesis and an increase in absorbing surface. The short roots become enveloped by a fungal sheath (Mantle) which is well organised into structures characteristic of the fungal species. Hyphae from the sheath extend into the soil upto a few mm and many coalesce to form strands and can be traced to the fruit body of the fungus. Hyphae from the sheath extend into the roots and occupy the intercellular space in cortex as 'Harting net'. Nutrients from soil pass through the sheath into the cortical cells of root.

OCCURRENCE OFECTOMYCORRHIZAL

Ectomycorrhiza is a specialized roots organ, which is the result of a complex interaction between a plant and a compatible ectomycorrhizal fungus leading to a finely turned symbiosis (Harley and Smith, 1983). This type of mycorrhiza occurs in about 10% of the world flora, especially in trees belonging to the pinaceae (pine, larch, spruce and hemlock), fagaceae (oak, chestnut, and beech), betulceae (alder and birch), salicaceae (popular and willow), juglandaceae (hickory and pecan), myrtaceae (eucalpytus) and ericaeceae (arbutus). Some tree genera viz., *Almus, Eucalyptus, Casuarina, Cupressus. Juniperus, Tilia,* and *Ulmus* form both ectomycorrhizae and vesicular arbuscular mycorrhizae (VAM) depending on soil conditions and trees age (Harley and Smith, 1983; Harley, 1969; Hacskaylo, 1971; Marks and Kozolowski, 1973).

Around 5000 species of fungi all over the world are known to form ectomycorrhizae. These fungi predominantly belong to Basidiomycotina, and among them the mycorrhizal species belonging both to Hymenomycetes and Gastromycetes, in the genera *Boletus, Suillus, Russula, Cortinarius, Gomphidius, Hebeloma, Scleroderma* and *Pisolithus*. (Certain ascomyectous members in Eurotiales (*Coenococcurn geophilum*), Tuberales (Truffles), and Pezizales are also known to form ectomycorrhizae on trees (Trappe, 1962; Kendric and Berch, 1985; Gautam and Prasad, 2001). Ectomycorrhizal fungi, therefore, have a great diversity. Out of the genera listed above, the truffles occur below ground and are dispersed where forest animals eat these and then defeet in the fecal matter (Maser et al., 1978).

Spores or hyphae of the fungal symbionts when stimulated by root exudates grow vegetatively over the feeder root surface. Initially, the hyphae form a loose external weft at the point of contact with a root let and then penetrate outer root cells, soon forming a mantle tightly oppossed to root surface together with the harting net which surrounds outer cortical cells, but leave intact the plasmodesmata connecting one root cell to another. Ultimately, the labyrinthi ferous tissue accumulate mitochondria and golgi bodies indicating high rates of metabolic activity. Often the root cell reacts and the walls become invaginated from interaction with harting net (Duddridge and Read, 1984, Massicotte *et al.*, 1986; Melville *et al.*, 1988). These alterations of root cells called transfer cells, are thought to maximize the area for exchange of materials between symbionts (O'Dell *et al.*, 1992). Thereafter, the hyphae develop intercellularly round the root cortical cells and form the harting net which may completely replace middle lamellae between the cortical cells. Harting net is the main distinguishing feature of ectomycorrhizae (Harley and Smith, 1983).

The fruiting bodies are seasonal, but spores of ectomycorrhizal fungi can remain dormant for undetermined length of time. Consequently viable spores may always be available in the soil. Exudates from roots and several soil organisms can increase the spore germination of some ectomycorrhizal fungi (Fries, 1977, 1982).

Ectomycorrhizal colonization normally changes the feeder root morphology and colour. They may be unforked, bifurcate, nodular, multiforked or shaped otherwise. Their colour, which is usually determined by the colour of the mycelium of the fungal symbiont, may get black , red, yellow, brown, white or blends of these colours. Gross morphology of ectomycorrhizal roots is probably controlled by the host (Zak, 1973). Ho and Zak (1979) found a wide range of acid phosphatase activity among six species of ectomycorrhizal fungi. O'Dell *et al.*, (1992) pointed out that significance of these variation is not well established, but sometimes be inferred. Some evidence indicate that rhizomorph forming fungi improve host drought tolerance more than fungi lacking rhizomorphs (Dosskey*et al.*, 1990). Some ectomycorrhizal fungi produce substantial quantities of auxins and cytokinins, where as others do not (Ho. 1987a, 1987b, Ho and Trappe, 1987). Amounts and ratios of such growth regulators produced by the fungi can regulate pattern of root let branching and even shoot/root ratios of host seedlings, hence, differences among these fungi cause substantial differences in host morphology (Ho, 1987a). Antibiotic activity can also differ strikingly, with resultant differences in protection of root-lets from pathogens and microbial community development at mycorrhiza surface (Zak, 1969).

The functional importance of variations in structures such as mantle, branching pattern or intercellular hyphae is more difficult to interpret. Ectomycorrhizal colonization limited to the primary cortex does not spread beyond the endodermis or into meristem tissue of the feeder root (Mortin and Hilbert, 1991).

ECOLOGICAL STUDIES OF ECTOMYCORRHIZA

Sharma and Singh (1990) observed *Scleroderma verrucosum, S. areolatum, S. dictyosporum, Rhizopogon spp. Suillus spp., Astraeus hygrometricus, Amanita vaginala, Laccana caccato,* and *Lectarius sanguiflus* on *Pinus roxburghii* in tropical zone at an altitude of 700 m. At 1800 m. *Amanita spp., Lepista nuda, Lactariuz sangiflus, Suillus sibricus, Boletus edulis, Scleroderma, taxense, Rhizopogon rubescence* and *Thelephora terrestris* formed symbiosis with *Pinus roxburgii*.

The occurrence of *A. muscaria* fruit bodies on *Pinus patula* in Kodaikanal was influenced by rainfall and age of the plantation. *L. laccata* and Shola soils enhanced mycorrhizal formation in *P. patula* nursery seedlings (Last *et al.*, 1981).

CHARACTERIZATION AND IDENTIFICATION OF ECTOMYCORRIHZAL SYMBIONTS

Many attempts have been made to chracterize and identify the ectomycorrhizae since early twenties. Melin (1927) classified ectomycorrhizae of pines into four groups on the basis of gross morphology. Dominik (1956) incorporated many more categories, utilizing anatomical features. Both the methods were too general to give satisfactory results. Nevertheless they provided a direction to characterization and identification of ectomycorrhizae on roots of 14 species and varieties of Salix. Marx (1975) characterized seven mycorrhizae on *Pinus radiatia*. Rambelli (1966) dscribed and illustrated 10 mycorrhizae on roots of nursery seedlings of the same species. However, fungal symbionts were not identified in either of the studies. Chilvers (1968a) characterized eight mycorrhizae an *Eucalyptus* and also gave identities of two fungal symbionts.

Ceruti and Bussetti (1962) identified the fungal symbionts of five ectomycorrhizae of *Tilia* species. Fassi and Devecehi (1963) named four of the five mycorrhizae of *Pinus strobus* seedlings. Fontana and Centrella (1967) identified the mycorrhizal fungi of *Castanea, Corpinus, Fagus, Pinus* and *Quercus* species. Trappe (1965) dscribed a tuberculate mycorrhizae of Douglas fir.

Zak (1969) described two new ectrotrophic mycorrhizae of Douglas fir each of which was formed by a distinct strain of *Poria terrestris* (D.C. exfries) Sace. Zak (1971) classified and characterized ectomycorrhizae formed by *Rhizopogon vinicolor* with *Pseudotsuga* mengiesd roots and observed the tuberculate mycorrhizae of Douglas fir described earlier by Trappe (1965), to be actually formed by *R. vinicolor*.

Among the most important articles describing the characterization of more than one mycorrhizae by unknown fungi in nature are those of Chilvers (1968b) for Eucalyptus, Fontana (1962) for *Populus. Salix Carpinus bitulus* and *Fagus sivatica*. Mausi (1962) for *Quercus pauscdentata*, Marx (1975), for *Pinus radiata*, Mejstrik (1972) for *Albus rubra*, Pachlewskki (1968), for *Larix sudetice*. Pachlewaki (1968) for *Abies alba, Quercus* sp., *Picea excelsa* and *Carpinus betulus*, Rambelli (1967) for *Pinus radiata* and Viory (1981) for *Quercus petraca* and *Fagus silvalica*, Alexander (1981) and Acsai and Largent (1983) described 15 ectomycorrhizae for *Abies concolor* of which the mycobiont is known for six. They also described 13 mycorrhizae for *Pseudotsuga mengiesi*, of which three mycobionts were identified. Agerer (1986a) described comprehensively the methods applied for characterization and identification of mycorrhizae also summarized the available literature. Agerer (1986b) described and identified the mycorrhizae of spruce with *Lactarius deterrimnus, L. Picinus, Russula uchroleuca* and *R. xerampeliha* by tracing hyphae connections between the fruit bodies and the mycorrhizae. Trappe (1987) characterized, identified and compared the ectomycorrhizae formed by *Suillus tridentinus* and *Suillus aeruginascens* on *Larix decidua*. Agerer and Weiss (1989) characterized, identified and compared the naturally grown and synthesized ectomycorrhizae formed by *Thelephora terrestris* on *Picea abies*.

It is clear that various methods have been used by different scientists in the characterization and identification of mycorrhizae. Each method has contributed to the refinement of the techniques, though a lot remain to be done.

The ectophytes may be identified from the following character:

1. **Morphology of ectomycorrhizae**—Whether they are simplel bifurcate, coralloid, monopodial, simple racemose, pyramidal or punnate; the individual fork-whether elongate, short, sessile or club shaped; colour of short roots—whether white, creamishpale, pale-orange-brown, bright to yellow, chestnut brown or jet black; surface of mycorrhizae—whether smooth, with loose weft, surface having a white crust or surface densely tomentose.
2. **Mantle characters**—Such as mantle structure whether pseudoparenchymatous or prosenchymatous or palisade or loosely interwoven or lightly interwoven or duplex type; Mantle thickness - which depends on virulence of the associated symbiont and physical and chemical conditions of the soil; ornamentation of mantle: hyphae emerging from mantle, mantle with setose hyphae, bristles, tortuous or dendroid structures. Harting net-the depth of penetration of hyphae into the cortex may be a usable characters.
3. Establishing mycorrhizal relationship between the roots and sporophores of fungi.
4. **Taxonomical character** : Colour of pileus, attachment of gills, presence or absance of annulus, Volva, colour of spore deposit, size and shape of spores, their ornamentation; shape of cap-whether convex, conical, companulate, flat, funnel shaped, unbonate or umblicate; close and outer margin—whether straight, inrolled, incurvate, undulate, lebate, or striate; surface of cap-whether smooth or glabrous, hairy, downy, tomentose, velvety, fibrillose or scaly; pellicle-separable or non-separable; annulus-wheather fixed or movable, simple or compound.
5. Cultural characteristics
6. Pure culture synthesis to confirm specific host/fungus mycorrhizae association.
7. Chemical tests of tissues with melzer reagent and other chemicals such as potassium hydroxide and ammonium hydroxide (the two

chemicals used at 2–25% for dried material and 10% for fresh material), ferrous sulphate (10%), formalin (40%), phenol (2%), aniline (pure oil or 50% in water) which give distinct colour reactions in different tissues (Singer, 1975).

8. Flourescence of fungal tissue in long wave ultravoilet light.
9. **Microscopic characters:** Such as types of hyphae present, their diameter, hyphal form whether straight, undulating or dentritic, presence or absence of clamp connections and their frequency, types of branching, attached rhizomorphs - their anatomical features.

ECTOMYCORRHIZOSPHERE

Lakhanpal and Kumar (1984) for the first time reported eleven fungi and mycelium of some unidentified basidiomycetes from the mycorrhizosphere of *Picea smithiana* (wall.) Bioss. With predominance of members of Deuteromycotina. Sharma and Lakhanpal (1981) isolated 21 species of fungi from *Abies pindrow* Royle, and Lakhanpal and Kumar (1990) isolated 13 fungal species and one unidentified basidiomycete mycelium from *Pinus gerardiana* wall. In *A. pinarow* and *T. baccata,* he isolated one type of sterile basidiomycetous mycelium and in *A. spectabilis* and *P. roxburghii,* two types of such mycelia, which on comparison were found to be similar to the types isolated from mycorrhizal roots.

PHYSIOLOGICAL AND BIOCHEMICAL STUDIES OF ECTOMYCORRHIZA

The ectomycorrhizal fungi can be isolated from short roots, Spores and sporophores. Ectomycorrhizal fungi grow well in modified Melin-Norkran's medium and potato dextrose agar (PDA) medium (Bakshi, 1966). They are highly specialised in their nutritional requirements. They require soluble carbohydrates, vitamins and amino acids which they derive most, if not all, from the symbiotic niche in the primary tissues of roots.

The majority of mycorrhiza form estimated by Melin (1925) required sugars such as glucose as a source of carbon and some made fair growth on maltose, xylose and mannitol. They grew strongly on starch and inulin and no growth occured on cellulose.

It is by virtue of being specific in nutritional requirement, the ectomycorrhizal fungi colonize effectively only those substrates or soil locations where there exists either a significant free concentration of sugars or where sugars or simple organic compounds are being consistently released.

IN VITRO SYNTHESIS OF ECTOMYCORRHIZA

In vitro synthesis of ectomycorrhizae studies provide the most direct and scientifically rigorous means of determining the ability of a fungal isolate to form mycorrhizae. Melin (1921; 1922; 1923; 1936) successfully demonstrated for the first time that ectomycorrhizae could be produced in synthetic cultures by inoculating seedlings of *Picea abies*, *Pinus sylvestris* and *Lorix curopaea* with appropriate fungi. They used sand moistened with a nutrient solution as the substrate for synthesis cultures. But sand has a relatively small surface area, a low water holding capacity and needs to be chemically cleaned for removal of undesirable morganic and organic compounds before use in pure culture experiments.

Hacskaylo (1953) substituted sand with vermiculite in *in vitro* synthesis experiments as it is biolotically sterile, void of organic material, has no constituents unfavourable for growth and the buffering capacity is only slightly greater than that of an inert material like sand. Marx and Zak (1965) further improved the substrate by stabilizing the acidity with addition of finely ground sphagnum peat moss. Marx and Bryan (1970) synthesized ectomycorrhizae of *Thelophora terrestris* and *Pisolithus tinctorius* on different conifer hosts. They observed that mycorrhizae formed by *T. terrestris* were macroscopically and microscopically different from those of P. tinctorius, but mycorrhizae formed by different isolates of *T. terrestris* were indistinguishable from each other regardless of the host. Marx and Ross (1970) synthesized ectomycorrhizae on *Pinus taeda* by basidiospores of *T. terrestris*. Zak (1976) described pure culture synthesis of *Arotostaphylos uvaursi* mycorrhizae and *Laccaria laccata*, *Lactarius sanquifluus* and *P. tinctorius*. Zak (1976) also observed close resemblance between morphology of natural and synthesized ectomycorrhizae of Douglar fir, *Tsuga heterophylea* and *Pinus ponderosa* formed by the same fungi. Molina (1979) tested pure culturess of 28 ectomycorrhizal fungi for mycorrhiza formation with red alder. Fortin *et al.*, (1980) described synthesis of ectomycorrhiza on *Pinus strobus* seedlings within five days after inoculation with *Pisolithus tinctorius*. Duddridge *et al.*, (1980) synthesized mycorrhizal rhizomorphs with aseptically germinated seedlings of *Pinus sylvestris* and *Sutllus bovinus* myceelium and studied their role in water transport. Alexander (1981) described and compared ectomycorrhizae formed by *Lactorius rufus* and *Picea sitchensis* in the field and under aseptic condition. Nyland and Unestom (1982) observed the process of *in vitro* mycorrhiza formation in norway spruce using the fungus *Piloderma croceum*.

Fortin *et al.*, (1983) reviewed and evaluated different methods used for synthesizing ectomycorrhizae. Palm and Stewart (1984) reported mycorrhizal synthesis between *Pinus resinoja* and *Suillus americanus*. *S. brevipess*. *S. luteus*., and *S. newclbidipes* and between *Pinus strobus* and *S.*

americanus. S. brevipes, S. granulatus, S. pictus and *S. punctipes*. Warombrodt and Eschrich (1985) observed mycorrhizal synthesis between *Pinus sylvestris* and *S. variegatus* and compared the synthesized one with the naturally occurring one.

Duddridge (1986a,b) synthesized ectomycorrhiza between *Picea sitchensis, P. abies* and *Pinus sylvestris* and the fungus S. grevillei. Kannan and Natrajan (1987) reported pure culture synthesis of ectomycorrhizae between Pinus patula and *Scleroderma citrimum*. Agerer and Weiss (1989) achieved ectomycorrhizal synthesis between *Thelephora terrestris* on *Picea abies*. Theodorou and Reddell (1991) tested 11 species of ectomycorrhizal fungi from stands of *Pinus radiata* and *Eucalyptus spp.* for their abilities to initiate ectomycorrhizae formation. Kasuya *et al.*, (1992) observed mycorrhizal formation in six species and varieties of pine with six species of ectomycorrhizal fungi using an *in vitro* synthesis technique which allowed observation of mycorrhizal formation without disturbing the host plants.

It is clear that many host and fungus species have been tested and tried the world ever for establishing ectomycorrhizal synthesis and in many cases the fungi have been reisolated and their cultural characteristics have been compared with the naturally occurring ones, thus confirming and establishing their identities. But how many of these fungi will ultimately be used on a commercial scale, as *P. tinctorius*, remains to be seen. Nevertheless, when preliminary date are obtained as all these and many more from other parts of the world, some of these would certainly prove to be useful commercially. In most of these, the host range of different species still remain to be established.

NUTRIENT UPTAKE THROUGH ECTOMYCORRIZAE

Beneficial effects of ectomycorrhizal fungi on plant nutrition have been known (Bowen, 1973, Hatch, 1937, Melin, 1925). Hatch (1937) reported that mycorrhizal white pine weighted significantly more and contained more nitrogen, phosphorus and potassium (NPK) than did non-mycorrhizal plants. Hatch (1936) observed that in the same substrate plants with mycorrhizae observed 234%, more P, 75% more K and 86% more N than plants without mycorrhizae. The role of the ectomycorrhizal fungi in the liberation of nutrients from complex compounds in the forest soil has been very much discussed (Hatch, 1937, Melin, 1925, Bjorkman *et al.*, 1967). The inverse relationship between soil fertility and ectomycorrhizal infection has long been recognised (Hatch, 1937; Bjorkman, 1942) and ectomycorrhizal infection is often reduced by application of fertilizer (Richards and Wilson, 1963). Harley (1969; 1970) has discussed the results of a number of studies of macro nutrient uptake by ectomycorrhizal plants. In general, ectomycorrhizae have been found to increase uptake of NPK but few other elements have been investigated. Ectomycorrhizae increased the Ca content of *Pinus radiata* (Henderson and Stone 1970) Bowen (1973) has pointed out that the more uptake of trace elements by ectomycorrhizae trees. Bowen *et al.*, (1974) observed that excised mycorrhizal roots of *Pinus radiata* absorbed more zinc than non-mycorrhizal roots. Naturally occuring ectomycorrhizal fungal symbionts in association with *P. taeda* and *P. echninata* roots were as efficient as *Pisolithus tinctorius* in the uptake of B, Cu, Fe, Mo, Mn and Zn from sewage sluge applied to nursery beds (Berry and Marx, 1976).

In many observations involving mycorrhizae, the mycorrhizal association usually increased the growth of plants by enhancing the uptake of nutrients especially P (Harley and Smith, 1983; Kucey *et al.*, 1989; Bilgrami and Prasad, 1993; Prasad and Bilgrami, 1993; 1995, Prasad, 1995, 1997a, 1997b, 2002, Gautam *et al.*, 1998). Mycorrhizal fungi in association with plant root seen likely to increase P uptake by more thorough exploration of soil volume thereby making positionally unavailable "nutrients available" (Prasad; 1993). This is achieved by decreasing the distance for diffusion of phosphate ions and by increasing the surface area for absorption (Bowen, 1973; Prasad, 1993).

Ectomycorrhizae are capable of increasing P uptake, especially in low fertility soils in experiments using 32 P. It was found that the mantle of ectomycorrhizae accumulates and stores P (Harley and Brierlay, 1955). This accumulation remains consistant unless there is a dramatic change in the plants' P status. A Phosphorous deficiency in the plant may stimulate the release by the mantle to the plant some of its accumulated P. However, whether or not an ectotrophic mycorrhizae can store other nutriennts is questionable.

Majstrik (1975) studied the effect of mycorrhizal infection of *Pinus sylvestric* and *Picea abies* by two *Boletus* spp. on the accumulation of P. Following mycorrhizal development, increase of the dry weight of roots and shoots over controls was significant in all combinations. The P content in mycorrhizal seedlings was higher than in non-mycorrhizal controls. Ekwebelam (1979) investigated the effect of ectomycorrhizal fungi on the growth and nutrient uptake of caribeam pine seedlings. Nutrients content of K, Ca, Mg, Fe, B, Mn, Zn, Cu and Mo was also influenced by the ectotropic fungal symbiont (Mitchel *et al.*, 1984). Bolan (2001), Gautam and Prasad, (1997) and Prasad and Rajak (1991, 2000, 2001, 2002) reviewed the role of mycorrhizal fungi in the uptake of phosphorus by plants. Koide (1991) give a comprehensive review on nutrient supply, nutrient demand and plant response to mycorrhizal infection. Wallander and Nyland (1991) investigated the effects of excess N on carbohydrate concentration and mycorrhizal development of Pinus sylvestris seedlings. Roussean et al., (1992) studied the relationship between biomass of the mycorrhizal fungus *Pisolithus tinctorius* and P uptake in Loblolly pine seedlings.

INOCULUM AND INOCULATION TECHNIQUES OF ECTOMYCORRHIZA

Ectotrophic mycorrhizae may be initiated by several different kinds of inocula each having advantages and disadvantages in relation to the objective and economics of the inoculation programme.

Soil inocula taken from beneath the ectomycorrhizal host trees have been used extensively especially in developing countries (Mikola, 1970). Soil inoculum consists of soil already colonized by an ectomycorrhizal fungus a fungi. In bare root nurseries, upto 10% by volume of soil inoculum is incorporated to the soil before sowing. This type of inoculum has been used successfully in many cases. Gobalwin (1925) used two ounces of screened pine straw as inoculum for loblolly pine container seedling and found a significant increase in height and growth after three years. Parke *et al.*, (1983) reported enhance growth of *Douglas fir* container seedlings inoculated with litter and humus taken from beneath Douglas fir trees.

The other type of inoculum consists of spores or macerated fruiting bodies of some ectomycorrhizal mushrooms, puffballs, or truffles. Spore inoculum is prepared by blending freshly collected fruiting bodies with tap water at high speed for two to three minutes. Some beneficial microorganisms within and on the surface of mature fruiting bodies of various ectomycorrhizal fungi have been found. Spores can be applied to the seed before sowing (Marx and Bell, 1985, Marx *et al.*, 1984; Theodoroun, 1984; Theodorum and Bowen, 1973) or 6-12 weeks after sowing either with a standard watering can or through the irrigation system. Spores of different fungi have been successfully used to inoculate and stimulate growth of pines in Australia (Theodoroun and Bowen, 1970, 1973, Theodoroun, 1971) and South Africa (Donald, 1975). Marx (1976; 1980) and Ruchle (1980) have had similar success with inoculating *Pisolithus tinctorious* on to assorted pine spp. in the United States. However, one of the major disadvantages of spore inoculum of most ectomycorrhizal fungi is the lack of appropriate laboratory tests to determine spore viability. The other disadvantage is that sufficient sporophores of many fungi may not be available every year.

Another type of inoculum recommended for inoculuations in pure mycelial or vegetative inoculum of ectotrophic mycerrhizal fungi (Trappe, 1977; Marx, 1977a and Mikola, 1973). Molina and Palmer, (1982) studied isolation and maintenance of ectomycorrhizal pure cultures. Basically a pure culture of a particular fungus is obtained by isolating fungal material (Vegetative tissue explant) on to special media, that is than grown under aseptic conditions to produce inoculum.

Moser (1958a, b, c) was the first to make a serious attempt to produce vegetative inoculum of ectomycorrhizal fungi in Austria. For production of inoculum, mycelium of *Suillus plorans* was first grown in liquid culture than in the sterile peat moss. Takacs (1961, 1964, 1967) modified Moser's technique to produce inoculum for new pine nurseries established in formerly treeless area lacking native ectotrophic mycorrhizal fungi in Argentina.

Marx (1980) discussed in detail the early testing and development stages of producing viable inoculum of *Pisolithus tinctorius* for use in nursery and container growth seedling. Vermiculite based inoculum has been used successfully to form *P. tinctorius* ectomycorrhizae in fumingated soil on pine. Dramatic improvement in survival and growth of mycorrhizically tailored pine seedlings over naturally infected control seedling produced in containers or bare root nurseries have been reported from studies on acid coal spoils in Appalachia (Marx, 1977a, b; Marx and Artman, 1979; Walker *et al.*, 1980), Kaolin spoils in Georgia (Marx, 1977a), severly eroded site of the copper Basin in Tennessee (Berry and Marx, 1978) borrow pits in South Carolina (Ruchle 1980) and North Carolina (Goodwin, 1980) and prairie soil (Baer and Otta, 1981). Similar improvements in the performance of pine seedling have been reported on routine reforestration sites (Marx *et al.*, 1977; Mexal, 1980; Ruchle *et al.*, 1980 and Kais *et al.*, 1981).

FACTORS RESPONSIBLE FOR DEVELOPMENT OF ECTOMYCORRHIZA

The intensity of ectomycorrhizal infection is dependent on the intensity of sunlight under which seedlings are growing. During active photosynthesis, plant roots accumulate an excess of soluble carbohydrates which, together with root exudates are necessary for the fungal symbiont to infect roots. The infection is reduced with lowering of light intensity shading of nursery beds, if made, to protect seedlings from scorching sun, should be done for a limited time at midday.

Ectotrophic mycorrhizae are dependent on simple carbohydrate, utilize ammonium and organic nitrogen compounds, require thiamine, other vitamins and growth factors including root exhudates, a pH in the acid range (about pH-5.0) and a temperature range of 20°C to 25°C for optimum growth. These fungi are inhibited by phenolic produces in the hymns and are sensitive to competition from saprophytes.

Fertilisers added to nursery soil in high dosage have depressing effect on development of mycorrhiza. For adequate development of mycorrhiza, fertilizers should be added sparingly in the nursery.

Soil amendments including application of biocides to nursery soil initially retard mycorrhiza which, however, develops normally later on. Normal watering of nursery beds helps proper development of mycorrhiza.

ECTOMYCORRHIZAL INOCULUM, INOCULATION, GROWTH AND DEVELOPMENT OF FOREST SEEDLINGS.

The need of many species of forest trees for ectomycorrhizal association was initially observed when attempts to establish plantations of exotic pines routinely failed until the essential fungi were introduced (Briscoe, 1959; Clements, 1941; Gibson, 1963; Hatch 1936; Kessel, 1927; Van Suchtelen, 1962). The need of pine and oak seedlings for ectomycorrhizae has also been convincingly demonstrated in the afforestation of former treeless area, such as the grasslands of Russia and the Great Plains of the United States (Shemakhanova, 1962; Goss, 1960; Hatch, 1937; McComb, 1938; Rosendahl and Wilde, 1942; White, 1941).

The concept of improving field performance of tree seedlings by forming ectomycorrhizae on them in nurseries with specific fungi ecologically adapted to the plantings site was originally developed by Moser (1958a) in Austria. Using various modifications of Moser's technique and philosophy, Takacs (1967) in Argentina, Theodorou and Bowen (1970) in Australia, and Vozzo and Hacskaylo (1971) in the United States showed experimentally that field survival and growth of tree seedling with specific ectomycorrhizae exceeded the performance of seedling that lacked or had few natural ectomycorrhizae at planting. Moser (1958a, b) has shown that cembra seedling artificially inoculated with pure culture of *Suillus plorans* have a better survival and initiall growth after planting than the naturally inoculated seedlings. In Australia, artificial inoculation of *Pinus cembra* has now reached the stage of practical application. Few experiments have been conducted on the artificial inoculation of Scots pine (*Pinus sylvestris*) compared to natural inoculation with pure cultures of *Laccaria laccata* and *Hebeloma cyliudrosporum* improved growth in a bare root nursery (Le Tacon and Bouchard, 1986). Large scale inoculation experiments are being done in the United States with *Pisolithus tinctorius* on different pine spp. including *Pinus taeda, P. virginiana, P. ponderosa, P. strobus* and *P. resinosa* (Marx *et al.*, 1982 and Marx *et al.*, 1984).

Marx and Schenck (1983) reviewed the potential of mycorrhizal symbiosis in agricultural and forest productivity. Inoculation of containerized seedlings with pure cultures has been practised by many workers (Marx and Barnett, 1974; Dixon *et al.*, 1979; Pawuk *et al.*, 1980; Marx *et al.*, 1982; Maronek and Hendrix, 1979; Molina, 1980; 1982). Molina (1982) succeeded in inoculating containerized Douglas fir seedlings with four isolates of *Laccaria laccata*. Mason and Wilson (1985) have also reported improved growth of out planted Sitka spruce seedling is pre-inoculated with *L. laccata* or *Paxillus involutus*. Marx and Cordell (1989) described the use of specific ectomycorrhiza to improve artificial forestation practices. Effective mycorrhizal fungus inoculum along with the necessary equipment and technology for successful operational application in bare root and container nurseries is now available to nursery personnel. Marx *et al.*, (1991) produced *Pinus pinaster* seedlings with *Suillus granulatus* ectomycorrhizae and out planted them to produce edible fruit bodies of the fungus.

ECTOMYCORRHIZAL INTRODUCTION IN NURSERY

Introduction of ectomycorrhizal fungi is also done by planting mycorrhizal (mother) seedlings in transplant beds around which seedlings are planted. The infection posses from the mother seedlings to adjoining plants which form mycorrhiza. This method has the advantage over soil inoculum in instances where soil is to be carried to long distance at the risk of loss of viability of the inoculum, in addition to other drawbacks of using soil as already mentioned earlier. This is said to be the only method of inoculating *Pinus merkusii*, which is its natural home grows well in high rainfall and low attitudinal locations in the tropics. It may appear suitable for introduction in India in similar situation where *Pinus caribaea* is growing well. *P. merkusii* may even be more promising because of its higher mean annual increment (M.A.I) values and long fiber lengths in addition to its suitability as an industrial wood and timber over *P. caribaea*. Since it may not be possible to import mycorrhizal (Mother) seedlings of *P. merkusii*, efforts may be made to introduce mycorrhiza in the species using local inocula. In this way, the mycorrhizal specificing in the species will be known. In case mycorrhiza develops the species may be raised under trial plantations.

CONCLUSION

Apart from nutritional benefit to plants derived from ectomycorrhizal symbiosis, it is known to import disease resistance in plants. In *Pinus ectinate*, mycorrhizal roots resist infection due to *Phytophthora cinnomoni*, where as uninfected roots are highly infected. This is based on the hyphothesis that the rhizosphere and the sheath surface of mycorrhizal roots possess microflora different from this in non-mycorrhizal roots. This is an area of research needing study.

During prinking carriage and transportation of seedlings from the nursery to the planting site, non-mycorrhizal plants are more susceptible to desiccation than mycorrhizal plants. On the ground, uninfected roots present near the surface are more easily damaged by drought and frost as compared to mycorrhizal roots.

There is an increasing realisation about the usefulness of mycorrhizal association in plants among botanists, agriculturists, horticulturist and foresters to whom the science is of great practical and economic significance.

Refereences

Acsai, J. and D.L.Largent (1983). Fungi associated with *Arbutus menzienii, Anctostaphylos manzanita* and *Arctosuvaursi* in Central and Northern California, Mycologia **75**: 344-547.

Agerer, R. (1986a). Studies on ectomycorrhizae-II. Introducing remarks on characterization and identification. *Mycotaxon* **26**: 473-492.

Agerer, R. (1986b). Studies on ectomycorrhizae-III. Mycorrhizae formed by four fungi in genera *Lactarius* and *Russula* on spruce. Mycotaxon **27**: 1-59.

Agerer, R. and M. Weiss (1989). Studies on ectomycorrhizae-XX. Mycorrhiza formed by *T. terrestris* on Norway spruce. *Mycologia* **81**: 444-453.

Alexander, I. J. (1981). The *Picca sitchenses + Lactarius rufus* mycorrhizal association and its effects on seedling growth and development. *Trans. Brit. Mycol. Soc.* **76**: 417-423.

Atkinson, D. (1980). The distribution and effectiveness of roots of tree crops. *Hort Rev.* **2**: 424-430.

Baer, N.W. and J.D. Otta (1981). Out planging survival and growth of ponderosa pine seedlings inoculated with *P. tinctorius* in South Dakata. *For Sci* **27**: 277-280.

Bakshi, B.K. (1957): Occurrence of mycorrhizae on some Indian conifers. *Mycologia* **49**: 269-272.

Bakshi, B.K. (1966). Mycorrhizae in *Eucalyptus* in India. *Indian Forester* **94**: 19-20.

Bakshi, B.K., H.S.Thapar and B. Singh (1968). Mycorrhiza in blue pine spruce and deodar. *Proc. Nat Inst Sci* India **84**: 27-34.

Berry, C.P. and D.H.Marx (1976). Effect of *P. tinctorius* ectomycorrhizae on growth of Loblolly and Virginia pines in Tennessee copper Basin. *VSDA Forest Serv. Res. Nat. SE* **264**: 6.

Berry, C.P. and D.H. Marx (1978). *Pisolithus tinctorius* ectomycorrhiza on growth of Loblolly and Virginia pines in the Tennessee copper Basin. *VSDA Forest Serv Res Nat. SE Washington DC.* **264.**

Bilgrami, R.S. and K. Prasad (1993). Impact of VAM on level of sucrosc in *Saccharum officinurum* L. grown in amended soil. *Mycorrhiza News* **5(3)**: 6-7.

Bilan, M.V. (1971). Some aspects of tree root distribution. In: E. Hacskaylo *Mycorrhiza* (ed.). *VSDA For Ser. Misc. Publ.* 1189, Washington DC., 69-80.

Bjorkman, E. (1942). Uber die Bedingun-gen der Mykorrhiza abildung bet kilfer and Fichte. *Symb Bot Upsal* **6**: 2.

Bjorkman, E., G.Lundeberg and H.Nommik (1967). Distribution and balance of N^{15} labelled fertilizer nitrogen applied to young pine trees. *For, Sci.* **48**: 1-23.

Bolan, N.S. (1991). A critical review on the role of mycorrhizal fungi in the uptake of phosphorus by plants. *Plant and Soil* **134**: 189-207.

Bowen, G.D. (1973). Mineral nutrituion of ectomycorrhizae. In: *Ectomycorrhizae: Their Ecology and Physiology* (eds.) G.C. Marks and T.T. Kozlowski. Acedemic press, New York, PP. 151-205.

Bousquet, N., D. Mousin and L. Salsae (1986). Use of phytase by ectomycorrhizal fungi. In: *Physiology and Genetics* V. Giaminazzi, Pearson and S. Gianinazzi. *Mycorrhizae* (eds.) Proc. Ist *Euro Symp.* on Mycorrhizae dijan INRA, Parins, PP. 363-368.

Bowen, G.D., M.F. Skinner and D.I Bevege (1974). Zick uptake by mycorrhizal and uninfected roots of *Pinus radiate* and *Araucaria cunningghamii. Soil Boil Biochem* **6**: 141-144.

Briscoe, C.B. (1959). Early results of mycorrhizal inoculation of pine in puerto Rico. Caribbean For. **20**: 73-77.

Ceruti, A. and L. Bussetti (1962). *Suila simbtosi* mycorrhiza tratiglie *Boletus subtomentosus, Russula grises, Balsamia platysopora, Hysterangium clathorides. Alliona* **8**: 55-66.

Clements, I. B. (1941).The introduction of pines into Nyasaland. *Nyasald Agric* Q.J. **1**: 5-15.

Chaudhuri, H. (1945). *Mycorrhizae of forest trees.* Proceeding of the *32nd Indian Science Congress* **3**: 66.

Chilvers, G.A. (1968a). Some distinctive types of eucalyptus mycorrhiza. *Aust. J. Bot.* **26**: 49-70.

Chilvers, G.A. (1968b). Low power electron microscopy of the root cap region of eucalptus mycorrhizae. *New Phytol* **67**: 663-665.

Dixon, R.K., H.E. Garrett and G.S. Cox (1979). Containerized shortleaf pine seedlings show superior growth and ectomycorrhizal development with mist foliar fertilization. *South J. App. For.* **3**: 154-157.

Dominik, T.(1956). Vorschlage einer neues klassification der ectotrophen Mykorrhizen ouf morphologisch anatomishen Merkmalen begrundet. *Roczn. Nauk. Les,* **14**: 223-245.

Donald, D.G.M. (1975). Mycorrhizal inoculation for pines. *S. Afr. For. J.* **92**: 27-29.

Dosskey, M.G., R.G.Lindermann and L. Boersma (1990). Carbon sink stimulation of photosynthesis in Douglas fir seedlings by some ectomycorrhizas. *New Phytol* **115**: 269-274.

Duddridge, J.A. (1986a). The development and ultrastructure of ectomycorrhizas-III. Compatible and incompatible interactions between *Suillus grevillei* and 11 species of ectomycorrhizal hosts *in vitro* in the absence of exogenous carbohydrates. *New Phytol* **103**: 457-464.

Duddridge, J.A. (1986b). The development and ultrastructure of ectomycorrhizas-IV. Compatible and incompatible interactions between *S. grevillet* and a number of ectomycorrhizal hosts *in vitro* in the presence of exogenous carbohydrate. *New Phytol* **103**: 465-471.

Duddridge, J.A., A. Malibari and D.J. Read (1980). Structure and function of mycorrhizal rhizomorphs with special reference to their role in water transport. *Nature* **287**: 834-836.

Duddridge, J.A. and D.J. Read (1984). The development and ultrastructure of ectomycorrhizas-II. Ectomycorrhizae development of pine in vitro. *New Phytol* **96**: 575-582.

Ekwebelam, S.A. (1979). Effect of mycorrhizal fungi on the growth and nutrient uptake of caribbean pine seedlings, *Ind. For.* **75,** 750-757.

Fassi, B. and E. Devecehi (1963). Researches in ectotrophic mycorrhizae of Pinus strobus in nurseries-I. Description of some of the most common forms in piedmont. *Allionia* **8**: 133-152.

Fontana, A. and E. Centrella (1967). Ectomicorrize prodotte da fungi ipogei. *Allionia* **13**: 149-176.

Fontana, A. (1962). Researches on the mycorrhizae of the genus *Salix. Allionia* **8**: 67-85.

Fortin, J. A., Y. Piche and M. Lalonde (1980). Technique for observation of early morphological changes during ectomycorrhiza formation. *Can. J. Bot.*, **58**: 361-365.

Fortin, J. A., Y.Piche and C. Godbout (1983). Methods for synthesizing ectomycorrhizas and their effect on mycorrhizal development. *Plant and Soil* **11**: 275-284.

Fries, N. (1942). Uber die Sporenkeimung beieinigen Gasteromycetan und mycorrhizan Hymenonyceten. *Arch. Mikrobiol.* **12**: 266-284.

Fries, N. (1977). Germination of *Laccaria laccata* spores, *in vitro*. Mycologia **69**: 216-219.

Fries, N. (1982). Effects of plant roots and growing mycelia on basidiospore germination in mycorrhiza forming fungi. In: G.A. Lauren and J.F. Ammirati. *Arctic and Alpine Mycology* (ed.) University of Washington press, Seattle, 493-508.

Gautam, S.P. and K. Prasad (1997). V A Mycorrhiza-Importance and Biotechnological application. In : D.K. Maheshwari and R.C. Dubey (eds.). *Innovative Approaches in Microbiology.* Bishan Singh Mahendra Pal, Dehradun, India, PP 83-114.

Gautam, S.P., K. Prasad and A. Maitra (1998). Impact of *Glomus macrocarpum* (VAM) on protein content of *Dendrocalomus strictus* (Roxb.). Nees under field conditions. *Ind. J. App. Pure Biol.* **13(1)** 41-45.

Gibson, I.A.S. (1963). Eine Mitteilung uber die keifernmykorrhiza in den waldern kenias. In: W. Rawald and H. Byr. G. Fischer (eds.). *Mycorrhiza Int. Mycorrhiza Symp*. Jena. 49-51.

Gobalwin. F. (1925). Brfahrungen beider Anzucht von Mycorrhiza Impfinaterial. *Centralbi Ges Gerstw* **92**: 227-237.

Goodwin, O.C. (1980). Survival and growth of containerized loblolly virginia pine seedlings colonized with *P. tinctorius* on a borrow site in North Carolina. *W. For. Sery. Div. of For Res. Raleigh for Nole* **44**: P.5

Goss, R.W. (1960). Mycorrhizae of ponderosa pine in Bebraska grass land soils. *University of Nehraska, College of Agric. Res. Bull.* No. 192.

Hacskaylo, E. (1953). Pure culture synthesis of pine mycorrhizae in erra-lite. *Mycologia*, **45**: 971-975.

Hacskaylo, E. (1971). *Mycorrhizae. Proc.* Ist NACOM. US Government Printing office, Washington, D.C.

Hatch, A.B. (1936). The role of Mycorrhizae in afforestation. *J. For* **34**: 22-29.

Hatch, A.B. (1937). The physical basis of mycotrophy in the genus *Pinus. Black Rock For Bull.* 6.

Harley, J.L. (1969). The Biology of Mycorrhizae (2nd edn) Leonard Hill, London.

Harley, H.L. (1970). Mycorrhiza and nutrient uptake of forest trees: physiology of tree crops. Academic Press, London 163-178.

Harley, J.L. (1989). The significance of mycorrhiza. *Mycol.* Res. **92**: 129-139.

Harley, J.L. and J.K. Brierlay (1955). The uptake of phosphate by excised mycorrhizal roots of beech. VIII. Active transport of 32P from fungus to host during uptake, *New phytol* **54** : 296-301.

Harley, J.L. and S.E. Smith (1983). *Mycorrhizal Symbiosis*. Academic Press. London.

Henderson, G.A.and E.L. Stone (1970). Interactions of Phosphorous availability, mycorrhizae and soil fumingation on conifor seedlings. *Agron abst.*, 134.

Ho, I. (1987a). Comparison of eight *Pisolithus tinctorius* isolates for growth rate, enzyme activity and phytohormone production. *Can J. Res.* **17**: 31-35.

Ho, I. (1987b). Enzyme activity and phyto-hormone production of a mycorrhizal fungus, *Laccaria laccata. Can. J. For Res.* **17** : 855-858.

Ho, I. and J.M. Trappe (1987). Enzyme and growth substance of Rhizopogon species in relation to mycorrhizal hosts and infrageneric taxonomy. *Mycologia* **79**: 553-558.

Ho, I. and B. Zak (1979). Acid phosphatase activity of six ectomycorrhizal fungi. *Can. J. Bot.* **57**: 1203-1205.

Kais, A.G., G.A.Snow and D.H. Marx (1981). The effects of benomyl and *P. tinctorius* ectomycorrhizae on survival and growth of long leaf pine seedlings. *South J. Appl. For.* **5**: 189-195.

Kannan, K. and K. Natarajan (1987). Pure culture synthesis of *Pinus patula* ectomycorrhizae with *Scleroderma citrinum*. *Curr. Sci.* **56**: 1066-1068.

Kasuya, M.C.M., R.M.C. Muchovej, M.M. Bellei and A.C. Borges (1992). *In Vitro* ectomycorrhizal formation in six variaties of pine. *For Ecol. and Management* **47**: 127-134.

Kendric, B. and S. Berch (1985). Mycorrhizae: Application in Agriculture and Forestry. In: C.W. Robinson and J.A.Howell (eds.) Comprehensive Biotechnology. Pergamon Press NewYork 109-152.

Kessel, S.L. (1927). Soil organisation and the dependence of certain pine species on a biological soil factor. *Empire Forest J.* **6**: 70-74.

Koide, R.T. (1991). Nutrient supply, nutrient demand and plant response to mycorrhizal infection. *New Phytol* **117**: 365-386.

Kucey, R.M., N.H.H. Janzen and M.E. Leggett (1989). Microbially mediated increases in plant available phosphorus. *Adv Agron*. **42**: 199-229.

Lakhanpal, T.N. and S. Kumar (1984). Studies on mycorrhiza and mycorrhizosphere of Picea smithiana. *J. Tree. Sci.* **3**: 5-9.

Lakhanpal, T.N.and S. Kumar (1990). Fungal flora of mycorrhizoplane and mycorrhizosphere of chilgozapine. In: D.J.Bagyaraj and A. Manjunath, *Mycorrhizal Symbiasis and plant growth. Proc. Sec. Nat. Conf. on Mycorrhiza.* Reproduction/Graphic Typesetters, Bangalore, India PP 113-114.

Last, F.T.; P.A. Mason; J. Pelpham; K.A.S. Bhoja and M. Hussain (1981). Factors affecting the production fruit bodies of *Amanita musearia* in plantation of *Pinus putula*. *Acad. Soc. Plant* **90**: 91-98.

Le Tacon, F. and D. Bouchard (1986). Effects of different ectomycorrhizal fungi on growth of larch, Douglas fir, Scots pine and Norway spruce seedlings in fumigated nursery soil. *Acta Ecologia Applicata* **6**: 389-402.

Lyr, H. and G. Hoffman (1967). Growth rates and growth periodicity of tree roots. In: J.A. Romberger and P. Mikola (eds.). International Review of Forestry Research. Academic Press, New York. 181-236.

Majestrik, V.K. (1972). VAM of the species of *Milinietum coerulae.* New Phytol **71**: 883-890.

Majestrik, V.K. (1975). The effect of mycorrhizal infection of *P. sylvestric* and *Picca abies* by two *Boletus* species on the acc umulation of phosphorus. *New Phytol* **74**: 455-459.

Marks, G.C. and T.T. Kozolowski (1973). Ectomycorrhizae: Their ecology and physiology. Academic press, New York.

Martin, F.M. and J.L. Hilbert (1991). Morphological, biochemical and molecular changes during ectomycorrhiza development *Experientia* **47**: 321-331.

Maronek, D.M. and J.W. Hendrix (1979). Slow release fertilizer for optimising mycorrhizal production in pine seedlings by *Pisolithus tinctorius*. Abs. 4th North Am. Conf. on Mycorrhizae. Colorado University, Colorado.

Marx, D.H. (1975). Role of ectomycorrhizae in the protection of pine from root infection by *Phytophthora cinnamomi.* In: J.W. Bruehl (eds.). *Biology and Control of Soil borne plant pathogens. Am Phytopathol. Soc.*

Marx, D.H.(1976). Synthesis of ectomycorrhizae on loblolly pine seedlings with basidiospores of *Pisolithus tinctorius. For. Sci.* **22**: 13-20.

Marx, D.H. (1977a). The role of mycorrhizae in forest production. TAPPI Conf. paper. Annual Meeting at G.A.

Marx, D.H. (1977b). Tree host range and world distribution of the ectomycorrhizal fungus *Pisolithus tinctorius. Can. J. Microbiol* **23**: 217-223.

Marx, D.H. (1980). Role of mycorrhizae in forestation of surface mines. In: *Proc. Sym. Tree for Reclamation of interstate mining.* Lexington. Kentucky.

Marx D.H. and J.D. Artman (1979). *Pisolithus tinctorius* ectomycorrhizae improve survival and growth of pine seedlings on acid coal spoils in kentucky and Virrginia. *Reclam. Rev.* **2**: 23-31.

Marx, D.H. and W. Bell (1985). Formation of *Pisolithus* ectomycorrhizae on loblolly pine seedlings with spore pellet inoculum applied at different times. *VSDA Forest Serv. Res.* paper 249.

Marx, D.H. and W.C. Bryan (1970). Pure culture synthesis of ectomycorrhizae by *Thelephora terrestris* and *Pisolithus tinctorius* on different conifer hosts. *Can. J. Bot.* **48**: 639-643.

Marx, D.H. and W.C. Bryan (1975). Growth and ectomycorrhizal development of loblolly pine seedlings in fumigated soil infested with the fungal symbiont *Pisolithus tinctorius*. *For. Sci.* **22**: 245-254.

Marx, D.H. and J.P. Barnett (1974). Mycorrhiza on containerized forest tree seedlings. In: R.W. Times, W.I. Stein and W.E. Balmer. (edss.). *Proc North American containerized forest tree seedlings Symp.* Great Plain Agric. Cong. Publ. 85-92.

Marx, D.H. and C.E Cordell (1989). The use of specific ectomycorrhizas to improve artificial forestation practices. In: J.W. Whipps and R.D. Lumsden. (eds.). *Biotechnology of fungi for improving plant growth.* Cambridge University Press, Cambridge. 1-25.

Marx, D.H. and E.W. Ross (1970). Aseptic synthesis of ectomycorrhizae on P. taeda by basidiospores of *T. terrestris. Can. J Bot.* **48**: 197-249.

Marx, D.H. and N.C. Schenck (1983). Potential of mycorrhizal symbiosis in agricultural and forest productivity. In: T. Kommendahal and P.H. Williams. (edss.). *Challenging Problems in plant Health.* (eds) Am phytopathol. Soc. St. Paul. Min. 334-347.

Marx, D.H. and B. Zak (1965). Effect of pH on mycorrhizae formation of slash pine in aseptic culture. *For. Sci.* **11**: 66-75.

Marx, D.H.; W.C. Bryan and C.E. Cordell (1977). Survival and growth of pine seedlings with *Pisolithus* ectomycorrhizae after two years on reforestation sites in North Carolina and Florida. *For. Sci.* **23**: 367-375.

Marx, D.H.; J.L. Ruehle; D.S. Kenney; C.E. Cordell; J.W. Riffle; R. Molina; W.H. Pawuk; S. Nsvratil; R.W. Tinus and O.C. Goodwin (1982). Commercial vegetative inoculum of *Pisolithus tinctorius* and inoculation techniques for development of ectomycorrhizae on container grown tree seedlings. *For. Sci.* **28**: 272-400.

Marx, D.H.; C.E. Cordell; D.S. Kenney; J.G. Mexal; J.D. Artman; J.W. Riffle; R.J. Molina; (1984). Commercial vegetative inoculum of *Pisolithus tinctorius* and inoculation techniques for development of ectomycorrhizae on bare root tree seedlings. *For. Sci.* **25**: 1-101.

Marx, D.H.; J.L. Ruchle and C.E. Cordell (1991). Methods of studying nursery response of trees to specific ectomycorrhiza. In: J.R. Norris, D.J. Read and A.K. Varma. (eds.). *Methods in Microbiology.* (eds.) Academic Press. London. 383-441.

Maser, C.M., J.M. Trappe and R.A. Nursbaum (1978). Fungal-small mannual interrelationship with emphasis on Oregon coniferous forests. *Ecology* **59**: 799-809.

Mason, P. and J. Wilson (1985). The concept of succession in relation to the spread of sheathing mycorrhizal fungi on inoculated tree seedlings growing in unsterile soils. *Plant and Soil* **71**: 247-256.

Massicotte, H.B.; R.L. Peterson; C.A. Ackerley and Y. Piche (1986). Structure and ontogeny of *Alnus crispa, Alpova diplophloeus* ectomycorrhizae. *Can. J. Bot.* **64**: 177-192.

Mausi, K. (1962). A study of mycorrhiza of *Abies firma*, with special reference to its mycorrhizal fungus *Canthorellus floccousus*. *Mem Coll Sci Kyoto Imp University* B. **2**: 85-92.

McComb, A.L. (1938). The relation between mycorrhizae and the development and nutrient absorption of pine seedlings in a prairie nursery. *J. For.* **36** : 1148-1154.

Melin, E. (1921). Uber die Mycorrhizempilze von *Pinus suylvestris* and *Picea abies. Sv. Bot. Tidskr* **15**: 192-203.

Melin, E. (1922). Untersuchunger uber die Larox Mykorrhiza I. Synthesis der Mykorrhiza in Reinkultur. *Sv. Bot.Tidskr* **16**: 161-196.

Melin, E. (1923). Experimentelle Untersuchungen uber die konsitution and Oklogie der Mykorrhizen von *Pinus sylvestris* and *Picea abies. Mykol. Unters. Ber.* **2** : 72-331.

Melin, E. (1925). Untersuchungen uber die Bedeutung der Baummykoriza G. Fischer Jena.

Melin, E. (1927). Studier over barrtradsplatans utveckling rahumus-II. Mykorrhizas utbildining hos tallplantan iolika rahuimus former. *Meddeln St. Skogs Forsk Inst* **23**: 433-494

Melin, E. (1936). Methoden der experimentellen Untersuchung Mykotropher Pflangen Handb *Boil Arbeitsmeth* **11**: 1005-1008.

Melville, L.H.; H.B. Massicotte; C.A. Ackerlery and R.L. Peterson (1988). An ultrastructural study of modifications in *Drya sintegrifolia* and *Mebelomat cylindrosporum* during ectomycorrhiza formation. *Bot Gaz.* **149**: 408-418.

Mexal, J.G.(1980). Aspects of micorrhizal inoculation in relation to reforestation. *NZJ For. Sci.* **10**: 208-217.

Mikola, P. (1970). Mycorrhizal inoculation in afforestation. *Inst. Rev. For. Res.* **3**: 123-196.

Mikola, P.(1973). Application of mycorrhizal symbiosis in forestry practices. In: G.C. Marks and T T Kozlowski. (eds.). *Ectomycorrhizae: Their Ecology and Physiology.* Academic Press, New York. 387-411.

Mitchel, R.L.; G.S. Cot; R.K. Dixon; H.E. Garrett and I.L. Sander (1984). Inoculation of three *Quercus* species with eleven isolates of ectomycorrhizal fungi-II. Folior nutrient content and isolate effectiveness. *For. Sci.* **30**: 563-572.

Molina, R. (1979). Pure culture synthesis and host specificity of Red Alder mycorrhiza. *Can. J. Bot.* **57**: 1123-1125.

Molina, R. (1980). Ectomycorrhizal inoculation of containerized western conifer seedlings. *USDA Forest Serv. Res. Note. Sc.* 357.

Molina, R. (1982). Use of the ectomycorrhizal fungus *Laccaria laccata* in Forestry. Consistency between isolates in effective colonisation of containerized conifer seedlings. *Can. J. For. Res.* **12**: 469-473.

Molina, R. and J.G. Palmer (1982). Isolation, maintenance and pure culture monipulation of ectomycorrhizal fungi. In: N.C. Schenck. (eds.). *Methoods and Principles of Mycorrhizal Research. Am. Phytopath Soc. St. Paul Minn.* 115-129.

Moser, M. (1958a). Die Kunstlich Mykorrhiza Impfung und von forstpflanzen. *Forstwissenchaftliches cent.* **77**: 1-68.

Moser, M. (1958b): Die kunstliche Mykorrhizaimpfung von forstpflazen. *Die Torf streukultur von mykorrhizapilzen Forestissenschaftliches cent.* **77**: 257-320.

Moser, M. (1958c). Die kunstliche mykorrhizaimpfung von forstpflanzen. II. Die Torf streukultur von mykorrhiza pilzen. *Forestissenschaftliches cent.* **77:** 273-278

Mukerji, K.G. (ed.) 1996, *Cocepts in Mycorhizal Research.* Kluwer Academic Publishers, Netherlands. p. 374.

Mukerji, K.G., Chamola, B.P. and Singh, J. (ed.) 2000. *Mycorrhizal Biology.* Kluwer Academic/Plemin Publishers, New York, Dordsecht, London. p. 336.

Mukerji, K.G., Monoherachany, C. and Chamola, B.P. (eds.) 2002. *Techniques in Mycorbhizal Studies.* Kluwer Academic Publishers, Netherlands. p. 554.

Nylund, J.E. (1981). The formation of ectomycorrhiza in conifers: Structural and physiological studies with special reference to the microbiont *Piloderma croceum*. Ph. D. Thesis, University of Uppsala.

Nylund, J.E. and T. Unestam (1982). Structure and physiology of ectomycorrhizae I. The process of mycorrhiza formation in Norway spruce in vitro. *New Phytol.* **91**: 63-79.

O'Dell, T.E.; M.A. Castellano and J.M. Trappe (1992). Biology and application of ectomycorrhizal fungi. In: F.Blaine and Jr Metting. (eds.). *Soil Microbial Ecology—Application in Agriculture and Environmental Management.* Mercel Dekkar (eds.). Inc. NY.

Pachlewski, R. (1968). Untersuchunger uber Mykorrhiza-*Pilze der Fichte. Lactarius rufus* (Scop). Fr. und *Rhizopogon lutelus* Fr. et. Wordhunter naturlichen Bedingungen und in Reinkulturen Prace naturilichen Bedingungen und in Reinkulturen Prace. *Inst. Bad. Les.* **365**: 173-187.

Palm, M.E. and E.L. Stewart (1984). *In vitro* Synthesis of mycorrhizae between presumed specific and non-specific *Pinust Suillus* combinations. *Mycologia* **76**: 579-600.

Pawuk, H.W.; J.L. Ruehle and D.H. Marx (1980). Fungicide drenches effect ectomycorrhizal development of container grown *Pinus palustris* seedlings. *Can. J. For. Res.* **10**: 61-64.

Parke, I.I.; R.G. Linderman and C.H. Black (1983). The role of ectomycorrhizae in drought tolerance of Douglas fir seedlings. *New Phytol.* **95**: 83-95.

Prasad, K. (1995). Physico chemical characteristics of soil in relation to vesicular arbuscular mycorrhizal (*Glamus fasciculatum*) colonization in *Saccharum officinarum* L. *Journal of Phytological Research* **8(2)**: 201-205.

Prasad, K. (1997a). Effect of vesicular arbuscular mycorrhizal fungi, particularly *Glomus fasciculatum* and phosphates on chlorophyll content of *Saccharum officinarum L. Ind. J. Appl. Pure. Biol.* **12(2)** : 91-95.

Prasad, K. (1997b). Improvement of biomass yield of *Terminalia arjuna* L. through vasicular arbuscular mycorrhizal fungi (*Glomus fasciculatum*) under nursery and field condition. J. Phytol. Res. **10(1-2)**: 39-41).

Prasad, K. (1993). Studies on edological factor affecting VA mycorrhizal infection in sugarcane. Ph.D. thesis, BRA University, Muzafferpur.

Prasad, K. (1998). Application of vesicular arbuscular mycorrhizal (*Glomus fasciculatum*) and *Rhizobium* on biomass production of *Acacia nilotica* in saline and forest soil. In: A.M. Deshmukh (ed.) Biofertilizers and Biopesticides. Technoscience Publications, Jaipur, India. 27-33.

Prasad, K. and R.S. Bilgrami (1993). Effect of VAM on uptake of phosphate by sugarcane. *J. Ind. Bot. Cont.* **10**: 59-62.

Prasad, K. and R.S. Bilgrami (1995). Impact of VAM and Phosphates on biomass yield and sucrose content *of Saccharum officinarum* L. In: A. Adholgya and S. Singh (eds.) *Mycorrhizae: Biofertilizers for the future.* Tata Energy Research Institute. New Delhi, India. 465-470.

Prasad, K. and R.C. Rajak (1999). Recent advances in Mycorrhizal Taxonomy: Morphological and Molecular criteria. In: R.C. Rajak (eds.). *Microbial Biotechnology for sustainable Development and Productivity.* Vol. I Scientific Publication (India), Jodhpur, 62-72.

Prasad, K. and R.C. Rajak (2000). Biotechnological application of Mycorrhizae in reclamation of mined Dumps. In : M.K. Rai, Ajit Verma and R.C. Rajak (eds.). *Integrated Management of Plant resources.* Scientific Publishers (India) Jodhpur, India. 282-292.

Prasad, K. and R.C. Rajak (2001). Microbes and Wasteland Management: Challange ahead. In: A.K. Roy and S.K. Verma (ed.) Management of Wasteland to protect environment. Scientific Publishers (India) Jodhpur, India. 27-39.

Prasad, K. and R.C. Rajak (2002). Management of Plant pests through Biological entities: Possibilities and Prospects. In: P.C. Trivedi (ed.) Plant Pest Management, Aavishkar Publishers, Distributors, Jaipur, India. 75-99.

Rambelli, A. (1966). *Pinus radiata*. Nei suol rapporti contiambiente microbiodeterreno Pubbl. cont. sper. *Agr. Forest Rome* **11**: 271.

Rambelli, A. (1967). Atlas of some mycorrhizal forms observedon *Pinus radiata in Italy*. Pubbl. Centro Specrim Agric. Forest, Vol. IX, Suppl.

Richards, B.N. and G.L. Wilson (1963). Nutrient supply and mycorrhiza development in caribbean pine. *For. Sci.* **9**: 405-412.

Rogers, W.S. (1939). Root studies IX. The effect of light on growing apple roots. A trail with root observation boxes. *J. Pomol Horticul. Sci.* **17**: 131-140.

Rogers, W.S. and G.C. Head (1966). The roots of fruit plants. *J.R. Hort. Soc.* **41**: 199-205.

Rosendahl, R.O. and S.A. Wilde (1942). Occurence of ecto-trophic mycorrhizal fungi in soils of cutover areas and sand dunes. *Bull Ecol. Soc. Am.* **23**: 73-74.

Roussean, J.W.D.; C.P.P. Reid and R.J. English (1992). Relationship between biomass of the mycorrhizal fungus *P. tinctorious* and phosphorous uptake in loblolly pine seedlings. *Soil Biol. Biochem.* **24**: 183-184.

Ruehle, J.I. (1980). Growth of containerized lobloly pine with specific ectomycorrhizas after 2 years on an amended borrow pit. *Reclamation Review* **3**: 95-101.

Ruehle, J.L.; D.H. Marx and M. Abourouh (1980). Development of *P. tinctorius* and *Thelephoraterrestris* ectomycorrhizae on seedlings of coniferous trees important to morocco. *Ann. de. Re. Cherche. Forestiere. Nu. Maroc.* **21** : 281-296.

Sharma, J.R. and T.N. Lakhanpal (1981). Mycorrhiza form of species in the family Boletaceae. In : R.K.Khosla (ed.). *Improvement of Forests Biomass.* Society of Tree Scientists. Solen 455-457.

Sharma, B.M. and B.M. Singh (1990). Ectomycorrhizal fungi associated with different forest trees of Himachal Pradesh. In: B. L. Jalali and H Chand (eds.). *Current Trends in Mycorrhizal Research.* H.A.U. Press, Hissar, Haryana. 13-15.

Shemakhanova, N.M. (1962). Mycotrophy in woody plants. In: A. A. Imshenetski (ed.). Mycotrophy in plants Springfield US Deptt. of Commer. Washington DC. 329.

Singer, R. (1975). Agricales in Modern Taxonomy (3rd edn) *J. Cramer Vaduz.*

Takacs, A.E. (1961). Inoculation de species de pinoscon hongos formodores de micorrzas Silvicultura 15: 5-17.

Takacs A.E. (1964): Inoculation artificial de pinos de regiones subtropicales con hangocs fromadores de micorrhizas. *India supl. For.* **2**: 41-44.

Takacs, A.E. (1967). Production de cultivos puros de hongos micor rizogenos enel centro mcional de Investigeiciones Agropecuarias, Castelar. *India supl. For.* **4**: 83-87.

Theodorou, C. (1971). The phytase activity of the mycorrhizal fungus *Rhizopogon roseolous. Soil Biol. Biochem.* **3**: 89-90.

Theodorou, C. (1984). Inositol phosphate in needles of Pinus radiata and the phytase activity of mycorrhizal fungi. *Being. Proc. Congr Soil Sci.* **3**: 483-493.

Theodorou, C. and G.D. Bowen (1970). Mycorrhizal responses of radiata pine in experiments with different fungi. *Aust. For.* **34**: 183-191.

Theodorou, C. and G.D. Bowen (1973). Inoculation of seeds and soil with basidiospores of mycorrhizal fungi. *Soil Biol. Biochem.* **5**: 765-771.

Theodorou, C. and P. Reddell (1991). In vitro synthesis of ectomycorrhizas on Casuarinaceae with a range of mycorrhizal fungi *New phytol.* **118**: 279-288.

Trappe, J.M. (1962). Fungus associates of ectotrophic mycorrhizae. *Bot. Rev.* **28**: 538-606.

Trappe, J.M. (1965). Tuberculate Mycorrhizae of Douglas fir. *For. Sci.* **11**: 27-32.

Trappe, J.M. (1967). Pure culture synthesis of Douglas fir Mycorrhizae with species of *Hebeloma, Suillus, Rhizophogon and Astraus. For. Sci.* **13**: 121-130.

Trappe, J.M. (1977). Selection of fungi for ectomycorrhizal inoculation in nurseries. *Ann. Rev. Phytopath,* **15**: 203-222.

Trappe, J.M.(1987). Phylogenetic and ecologic aspects of mycotrophy in the angiosperms from an evolutionary standpoint. In: G.R. Safir (ed.). *Ecophysiology of VA Mycorrhizal Plants.* CRC Press, Boca Roton. 2-25.

Van Suchtelen, M.J. (1962). Mycorrhiza biz *Pinus spp* in de Tropen Med. Landbouwhogesch. Opzoekingsstn. *Staat. Gent.* **27**: 1104-1106.

Vazzo, J.H. and E. Hacskaylo (1971). Inoculation of Pinus caribaca with ectomycorrhizal fungi in Puerto Rico. *For. Sci.* **17**: 239-245.

Viory, H. (1981). Classification morphologique des ectomycorrhizas duchene et du-hetre dans lenordest de la France. *Eur J. For. Path.* **11**: 284-299.

Walker, R.F.; D.C. West and S.D. Melanghlin (1980). The development of ectomycorrhizae on contrainerized sweet birch and European alder seedlings for planting on low quality sites. *General technical report. VSAD Forest Service* **24**: 409-417.

Wallander, H. and J.E. Nyland (1991). Effect of excess on nitrogen on carbohydrate concentration and mycorrhizal development *Pinus sylvestris seedlings, New phytol.* **119**: 405-411.

Warmbrodt, R.D.and W.Eschrich (1985). Studies on the mycorrhizas of *Pinus sylvestris*, Produced *in vitro* with *S. varieqatus. New phytol.* **100**: 215-223.

White, D.P. (1941). Praire soil as a medium for tree growth. *Ecology* **22**: 308-407.

Zak, B. (1969). Characterization and classification of mycorrhizae of fir. *Pseudosuga menziesii, Poria terresteris* (blue and orange-staining strains). *Can. J. Bot.* **47**: 1833-1849.

Zak, B. (1971). Characterization and identification of Douglas fir mycorrhizae. In: E. Hacskaylo (ed.). *Mycorrhizae.* Washington DC. 38-53.

Zak. B. (1973). Classification of Ectomycorrhizae. In: G.C. Marks and T.T. Kozlowski (eds). *Ectomycorrhizae.* Academic press, New York. 43-78.

Zak. B. (1976). Pure culture synthesis of pacific modrone ectendomycorrhizae. *Mycologia* **68**: 362-369.

Frontiers in Plant Sciences (Pages 643-656)
Eds. K.G. Mukerji, K.V.B.R. Tilak, S.M. Reddy, L.V. Gangwane, P. Prakash, I.K. Kunwar

Effect of Different Isolates of *Laccaria fraterna* for Growth Improvement of *Eucalyptus grandis* Seedlings in the Nursery

Evelin Kramony and K. Natarajan

ABSTRACT

The effect of two isolates of the ectomycorrhizal fungus *Laccaria fraterna* on the seedling growth of *Eucalyptus grandis* was studied over a period of 10 months in the nursery. Several parameters were used to assess the seedling growth. The isolates of the fungus collected from Kodaikanal hills was found to increase the seedling growth better when compared to the Udagamandalam isolate.

Key words: *Laccaria fraterna*, Ectomycorrhiza, *Eucalyptus grandis*, seedling growth.

About 95% of the world's species of vascular plants are reported to be mycorrhizal. But ectomycorrhizal associations are formed only with approximately 10% of world flora (Marx and Cordell, 1990). Many ectomycorrhizal host species grow very poorly without mycorrhizal associations, even though the soil is highly fertile. In addition to their nutritional role, the fungi can deter root pathogens, protect seedlings from root toxins and promote seedling growth by production of growth regulators.

The first and most important step in any inoculation program in tree seedlings is the selection of ectomycorrhizal fungi. Many fungi are host

---

Author's address: Cas In Botany,University Of Madras,Guindy Campus, Chennai – 600 025.
E-Mail knraj42@yahoo.com

specific which is well documented in the literature (Cairney and Chambers, 1997). Variability among the isolates of the same species is observed which is an important criterion to be considered (Burgess *et al.*, 1995). *Eucalyptus* in association with mycorrhizal fungi has greater potentials to grow on sites of low nutrient stauts (Malajczuk *et al.*, 1990). *E. grandis* (Rose gum) is one of the premier forest species in Australia where it grows 43-55 m tall and 122-183 cm in diameter (Hodgson, 1976). Rose gum is one of the most important commercial eucalyptus, with more than one half million hectares planted in tropical and subtropical areas on 4 continents. *E. grandis* is best suited for medium elevation from 1,000 to 1,800 m above MSL with rainfall exceeding 1,500 mm (Schonau *et al.*, 1981).

Natarajan (1977) for the first time reported the fungus *Laccaria fraterna* from India (as *L. ohiensis*) being associated with *Eucalyptus globulus*. It is found mostly being associated with *Eucalyptus* spp. worldwide (Tommerup *et al.*, 1991). *L. fraterna* grows rapidly on several agar media and forms pyramidal ectomycorrhizas on eucalypt roots. Not much of work has been done on its association with *E. grandis* seedlings and in the present paper the results on the effect of two different isolates of *L. fraterna* on the growth and ectomycorrhizal formation in *E. grandis* seedlings in the nursery is being reported.

MATERIALS AND METHODS

Isolation of the Mycobiont

Fruit bodies of the ectomycorrhizal fungi were collected from under natural stands and brought to the laboratory (Natarajan *et al.*, 1988). The covered central tissue mass was picked with the help of sterile forceps and incoulated on to PDA or MMN medium under aseptic conditions. The emerging mycelial tips were then subcultured and maintained on PDA agar slants.

Inoculum Production: Vermiculite Carrier Method (Marx and Bryan, 1975)

Erlenmeyer flasks (1000 ml) containing 750 ml of vermiculite was supplemented with 375 ml of MMN medium. The flasks were autoclaved at 120°C for 20 minutes and inoculated with three 5 mm diameter discs. The whole unit was incubated in the dark at ambient temperature for 5 weeks. The contents of the flasks were collected in cheese cloth, leached thoroughly in sterilized cool distilled water. The leached inoculum was readily used for nursery experiment. Bulk density of the inoculum was approximately 395 g/L with 60% moisture. Bulk density (g/L) was determined by drying 100 g of inoculum at 85°C for 48 hours and the difference in weights between the fresh and oven dried inoculum was considered the moisture content. The vermiculite particles were stained with cotton blue and examined. Hyphae with clamp connections were assumed to be ectomycorrhizal fungi.

Nursery Experiment

The nursery experiment was carried out with both steam sterilized and unsterilized soil. The soil was sterilized by autoclaving at 15 1b pressure for 1 hour on 3 alternate days.

Experimental Design

The nursery experiment was laid at the glass house of University of Madras, Chennai. Steam sterilized and unsterilized soil. (2:1 ratio of Red Soil: Sand) were filled in nursery bags (2 Kg/bag).

Vermiculite inoculum of the ectomycorrhizal fungi were added to each bag according to each treatment and bags without inoculum served as control. The inoculated and uninoculated bags were arranged in randomized block design with 10 replicates.

Plant Material

Healthy and viably seeds of *Eucalyptus grandis* were obtained from Forest Seed Centre, Genetics Divisions, Coimbatore. The seeds were pre-germinated on a nursery bed of sterilized soil. One week-old seedlings were transferred to the bags. After 4 months of growth, 10 seedlings from each treatment were harvested and observed for various parameters like shoot height, root length, root collar diameter, shoot dry weight, root dry weight, total dry weight, total number of mycorrhizas, total number of nonmycorrhizas and percentage of mycorrhization. Seedlings were then harvested every month and observed for the growth parameters till 10th month.

After measuring the shoot height, root length and root collar diameter, the shoot and the root was separated for biomass determination. They were oven dried at 80°C for 48 hours and observed for dry weight.

Ectomycorrhizal Assessment

Seedlings were screened for the presence of ectomycorrhizas by manual counting under Wild M3 Stereomicroscope (Richards and Wilson, 1963).

Ectomycorrhizal Dependency

Ectomycorrhizal dependency was calculated by the method as followed by Plenchette *et al.,* (1983).

$$\text{EMD} = \frac{\text{Shoot dry wt. of mycorrhizal plant - Shoot dry wt. of nonmycorrhizal plant} \times 100}{\text{Shoot dry wt. of nonmycorrhizal plant}}$$

RESULTS

Shoot Height

Seedlings treated with mycorrhizal fungi had tremendous increase in the shoot height when compared to uninoculated control. Kodaikanal isolate of *L. fraterna* showed more improvement of shoot height of seedlings when compared to Udagamandalam isolate of *L. fraterna*. Gradual increase in the shoot height was observed with age of the seedlings in both sterilized and unsterilized soil (Table 1). The untreated control seedlings produced lowest shoot height in all the harvests.

TABLE 1 *Effect of* Laccaria fraterna *on the shoot height (in cm) of* Eucalyptus grandis *seedlings in the nursery*

ISOLATE	AGE OF THE SEEDLINGS (IN MONTHS)						
	4	5	6	7	8	9	10
	STERILIZED SOIL						
L. fraterna 1	35.45c	60.70c	76.68c	92.96c	116.0c	135.58c	192.9c
L. fraterna 2	20.55b	32.48b	42.22b	65.04b	80.25b	100.45b	142.0b
Control	6.98a	17.27a	21.94a	25.96a	28.68a	31.5a	34.52a
	UNSTERILIZED SOIL						
L. fraterna 1	35.01c	63.14c	79.9c	96.4c	117.8c	140.02c	202.2c
L. fraterna 2	21.92b	34.07b	44.62b	65.7b	77.90b	104.89b	154.05b
Control	7.05a	17.32a	21.92a	25.96a	28.96a	3.50a	35.04a

L. fraterna 1 – Kodaikanal isolate
L. fraterna 2 – Udagamandalam isolate
Means sharing a common letter in the same column within soil types are not significantly different at P= 0.05 level.

Root Length

The root length of the seedlings increased with age of the seedligs. The seedlings inoculated with mycorrhizal fungi produced longer roots than the untreated ones (Table 2). Statistically significant differences were observed among the treatments in both sterilized and unsterilized soil.

TABLE 2 *Effect of* Laccaria fraterna *on the root lenght (in cm) of* Eucalyptus grandis *seedlings in the nursery*

ISOLATE	AGE OF THE SEEDLINGS (IN MONTHS)						
	4	5	6	7	8	9	10
	STERILIZED SOIL						
L. fraterna 1	36.20c	59.50c	74.3c	91.4c	114.4c	135.0c	191.5c
L. fraterna 2	22.5b	31.95b	41.6b	64.68b	78.55b	96.1b	142.5b
Control	7.2a	17.7a	20.74a	25.86a	28.08a	31.56a	34.68a
	UNSTERILIZED SOIL						
L. fraterna 1	36.31c	63.05c	79.55c	95.0c	115.55c	137.6c	202.3c
L. fraterna 2	24.37b	33.2b	43.21b	64.58b	73.78b	101.0b	149.5b
Control	7.16a	16.72a	21.52a	25.28a	23.42a	31.6a	34.94a

L. fraterna 1 – Kodaikanal isolate
L. fraterna 2 – Udagamandalam isolate
Means sharing a common letter in the same column within soil types are not significantly different at P= 0.05 level.

Root Collar Diameter

Marked increase in the root collar diameter of *E. grandis* seedlings were observed in the treated plants when compared to control. In sterilized soil the seedlings inoculated with Kodaikanal isolate of *L. fraterna* revealed more root collar diameter than when treated with Udagamandalam isolate of *L.fraterna* (Table 3). Lowest root collar diameter was observed in the seedlings which were devoid of mycorrhizal fungi. Similar result was obtained in the seedlings grown in the unsterilized soil. Analysis of variance showed significant difference among isolates.

Shoot Dry Weight

Immense increase in the shoot dry weight was observed in seedlings treated with *L. fraterna*. Seedlings inoculated with Kodaikanal isolate produced pronounced shoot dry weight when compared to Udagamandalam isolate

TABLE 3 *Effect of* Laccaria fraterna *on the root collar diameter (in mm) of* Eucalyptus grandis *seedlings in the nursery*

ISOLATE	AGE OF THE SEEDLINGS (IN MONTHS)						
	4	5	6	7	8	9	10
				STERILIZED SOIL			
L. fraterna 1	3.35c	5.2c	6.62c	7.4c	8.29c	9.58c	14.35c
L. fraterna 2	1.78b	2.4b	3.01b	3.47b	4.05b	5.93b	6.77b
Control	0.76a	1.34a	1.76a	1.94a	2.20a	2.36a	2.60a
				UNSTERILIZED SOIL			
L. fraterna 1	3.26c	5.35c	6.67c	7.59c	8.38c	9.70c	15.3c
L. fraterna 2	1.8b	2.58b	3.10b	3.51b	3.87b	6.40b	6 .83b
Control	0.75a	1.40a	1.76a	1.94a	2.16a	2.36a	2.66a

L. fraterna 1 – Kodaikanal isolate
L. fraterna 2 – Udagamandalam isolate
Means sharing a common letter in the same column within soil types are not significantly different at P = 0.05 level.

TABLE 4 *Effect of* Laccaria fraterna *on the shoot dry weight (in g) of* Eucalyptus grandis *seedlings in the nursery*

ISOLATE	AGE OF THE SEEDLINGS (IN MONTHS)						
	4	5	6	7	8	9	10
				STERILIZED SOIL			
L. fraterna 1	1.93b	6.57c	10.62c	15.41c	20.26c	33.82c	61.75c
L. fraterna 2	1.48b	5.37b	7.64b	10.34b	14 .68b	17.30b	18.22b
Control	0.03a	0.15a	0.42a	1.28a	1.55a	1.82a	2.16a
				UNSTERILIZED SOIL			
L. fraterna 1	2.69c	7.16c	12.28c	16.13c	21.73c	40.41c	62.60c
L. fraterna 2	0.27b	5.86b	8.07b	11.53b	14.8b	18.39b	20.05b
Control	0.04a	0.15a	0.43a	1.3a	1.54a	1.85a	2.17a

L. fraterna 1 – Kodaikanal isolate
L. fraterna 2 – Udagamandalam isolate
Means sharing a common letter in the same column within soil types are not significantly different at P= 0.05 level.

of *L. fraterna* (Table 4). Both sterilized and unsterilized soil produced similar response. The seedlings which were not inoculated with mycorrhizal fungi gave the lowest shoot dry weight in all the harvests.

Root Dry Weight

Seedlings when treated with *L. fraterna* produced increased root dry weight than other treatments. Increase in root dry weight was observed with age of the seedlings. Kodaikanal isolate treated seedlings showed increased root biomass when compared to Udagamandalam isolate and untreated control (Table 5).

TABLE 5 *Effect of* Laccaria fraterna *on the root dry weight (in g) of* Eucalyptus grandis *seedlings in the nursery*

ISOLATE	AGE OF THE SEEDLINGS (IN MONTHS)						
	4	5	6	7	8	9	10
	STERILIZED SOIL						
L. fraterna 1	1.57b	5.58b	9.54b	13.50c	17.92c	33.35c	58.66c
L. fraterna 2	1.2b	3.4b	6.25b	6.14b	7.11b	14.71b	16.55b
Control	0.01a	0.08a	0.36a	1.13a	1.31a	1.59a	1.98a
	UNSTERILIZED SOIL						
L. fraterna 1	2.20b	5.12b	10.37c	13.09c	18.29c	37.29c	61.68c
L. fraterna 2	1.25b	6.03b	6.66b	9.27b	10.72b	14.92b	17.90b
Control	0.04a	0.08a	0.36a	1.15a	1.36a	1.67a	1.94a

L. fraterna 1 – Kodaikanal isolate
L. fraterna 2 – Udagamandalam isolate
Means sharing a common letter in the same column within soil types are not significantly different at P= 0.05 level.

Total Dry Weight

As observed in the shoot and root dry weight, total dry weight also increased with age of the seedlings. The treated seedlings produced more seedlings biomass when compared to control. Among the treated seedlings, Kodaikanal isolate was seen to be more effective than Udagamandalam isolate of *L. fraterna* (Table 6).

Shoot: Root Ratio

In the sterilized soil, unionculated seedlings showed more shoot: root ratio in the first harvest. But in the next three harvests, seedlings treated with Udagamandalam isolate showed more shoot: root ratio. No significant difference could be observed in the 8 and 10 months old seedlings among

TABLE 6 *Effect of* Laccaria fraterna *on the total dry weight (in g) of* Eucalyptus grandis *seedlings in the nursery*

ISOLATE	AGE OF THE SEEDLINGS (IN MONTHS)						
	4	5	6	7	8	9	10
	STERILIZED SOIL						
L. fraterna 1	3.51b	12.15c	20.16c	28.92c	38.18c	67.17c	120.41c
L. fraterna 2	2.68b	9.21b	13.89b	16.48b	21.79b	32.01b	34.77b
Control	0.05a	0.23a	0.78a	2.41a	2.86a	3.41a	4.15a
	UNSTERILIZED SOIL						
L. fraterna 1	4.89c	12.28c	22.66c	29.22c	40.02c	77.70c	124.28c
L. fraterna 2	1.52b	11.89b	14.73b	20.8b	25.52b	33.31b	37.95b
Control	0.05a	0.24a	0.79a	2.45a	2.90a	3.52a	4.11a

L. fraterna 1 – Kodaikanal isolate
L. fraterna 2 – Udagamandalam isolate
Means sharing a common letter in the same column within soil types are not significantly different at P= 0.05 level.

TABLE 7 *Effect of* Laccaria fraterna *on the shoot: root* Ratio *of* Eucalyptus grandis *seedlings in the nursery*

ISOLATE	AGE OF THE SEEDLINGS (IN MONTHS)						
	4	5	6	7	8	9	10
	STERILIZED SOIL						
L. fraterna 1	1.03a	1.16a	1.10a	1.13a	1.13a	1.07a	1.05a
L. fraterna 2	1.23a	1.48b	1.30b	1.29b	1.24a	1.30c	1.10a
Control	2.84b	1.87a	1.13a	1.14a	1.21a	1.14b	1.09a
	UNSTERILIZED SOIL						
L. fraterna 1	1.25a	1.14a	1.18a	1.22a	1.18ab	1.07a	1.01a
L. fraterna 2	1.22a	1.22a	1.39b	1.22a	1.28b	1.14b	1.14b
Control	1.51a	1.84a	2.07a	1.13a	1.15a	1.1ab	1.12b

L. fraterna 1 – Kodaikanal isolate
L. fraterna 2 – Udagamandalam isolate
Means sharing a common letter in the same column within soil types are not significantly different at P= 0.05 level.

all the treatments. Analysis of variance showed significant difference among isolates in the 9 months old seedlings (Table 7).

In the unsterilized soil, significant difference was not observed among treatments in the 4, 5 and 7 months old seedlings. But in the following harvest statistically significant difference was observed.

Total Number of Mycorrhizas

Inoculation of Kodaikanal isolate of *L. fraterna* increased the total number of mycorrhizas in both sterilized and unsterilized soil. Seedlings inoculated with Udagamandalam isolate of *L. fraterna* produced comparatively less number of mycorrhizas in all harvests (Table 8). The number of mycorrhizas increased with age of the seedlings in both the soil types.

TABLE 8 *Effect of* Laccaria fraterna *on the total number of mycorrhizas of* Eucalyptus *grandis seedlings in the nursery*

ISOLATE	AGE OF THE SEEDLINGS (IN MONTHS)						
	4	5	6	7	8	9	10
	STERILIZED SOIL						
L. fraterna 1	1498b	3468b	4414b	6335b	7668b	8475b	8987b
L. fraterna 2	794a	1139a	1565a	3880a	4801a	6944a	8411a
Control	-	-	-	-	-	-	-
	UNSTERILIZED SOIL						
L. fraterna 1	1536b	3854b	4700b	6354b	7849b	8638b	9608b
L. fraterna 2	883a	1231a	1607a	4017a	5004a	7248a	8309a
Control	-	-	-	-	-	-	-

L. fraterna 1 – Kodaikanal isolate
L. fraterna 2 – Udagamandalam isolate
Means sharing a common letter in the same column within soil types are not significantly different at P= 0.05 level.

Percentage of Mycorrhization

Percentage of colonization of *L. fraterna* on the roots of *E. grandis* showed statistically similar results in the 4 and 9 months old seedlings in the sterilized soil. The percentage of mycorrhization gradually increased up to 7 months and then decreased in the subsequent two months, with a rise in the mycorrhization at the 10th month (Table 9). Kodaikanal isolate of *L. fraterna* treated seedlings produced more mycorrhizal percentage than that of the seedlings treated with Udagamandalam isolate.

In the unsterilized soil, the percentage of mycorrhization was seen to increase and decrease in the consecutive harvests in the seedlings treated

TABLE 9 *Effect of* Laccaria fraterna *on the percentage of mycorrhization of* Eucalyptus grandis *seedlings in the nursery*

ISOLATE	AGE OF THE SEEDLINGS (IN MONTHS)						
	4	5	6	7	8	9	10
				STERILIZED SOIL			
L. fraterna 1	63.05a	68.6b	75.03b	78.99b	74.70b	70.10a	77.15b
L. fraterna 2	63.85a	63.00a	66.90a	69.44a	70.88a	69.30a	73.39a
Control	-	-	-	-	-	-	-
				UNSTERILIZED SOIL			
L. fraterna 1	57.22a	74.69b	73.56b	81.71b	70.88a	75.92b	74.47a
L. fraterna 2	67.66a	60.86a	59.79a	71.76a	75.44b	69.06a	78.34b
Control	-	-	-	-	-	-	-

L. fraterna 1 – Kodaikanal isolate
L. fraterna 2 – Udagamandalam isolate
Means sharing a common letter in the same column within soil types are not significantly different at P= 0.05 level.

with Kodaikanal isolate. But in the seedlings treated with Udagamandalam isolate there was a decrease in percentage of mycorrhization till the 6th month and a sudden increase in the next two harvests. The percentage of colonization again decreased in the 9 month old seedlings and thereby increasing in the next harvest. Analysis of variance showed statistically significant difference among isolates of *L. fraterna* in the rate of colonization.

DISCUSSION

Studies on ectomycorrhizal association in *Eucalyptus* spp., has shown that the association between *L. fraterna* and *E. grandis* is more common (Tommerup *et al.*, 1991). Abundant basidiomes have been found in disturbed sites such as rehabilitated and replanted cleared forest. It has been found in abundance in rehabilitated bauxite mines (Gardner and Malajczuk, 1988). *L. fraterna* is also been reported from Australia under *Acacia* and *Coffea* stands (Mueller and Vellinga, 1986). There have been reports of *L. fraterna* being associated with eucalypts and Mexican pines in Southern India (Irving *et al.*, 1985). Last and Watling (1998) reported *L. fraterna* for the first time in Britain from under *E.gunnii* plantation. It readily forms mycorrhizas with *E.camaldulensis*, *E. diversicolor*, *E. grandis*, *E. tereticornis* and *Allocasuarina littoralis*.

The lack of efficient form of inocula has been a limiting factor for the success of establishment of tree seedlings in the nursery and field conditions. Initially basidiospore inocula of *P. tinctorius* had been in practice by many workers to improve growth of tree seedlings and mycorrhizal development (Alvarez and Trappe, 1983; Marx *et al.*, 1989). However, spore inoculum has several disadvantages like spore germination efficiency, genetic diversity and the length of time to form mycorrhizas with the inoculated seedlings (Marx and Cordell, 1990). According to Trappe (1977) use of pure cultures of vegetative mycelium of ectomycorrhizal fungi is probably the best method to inoculate forest tree seedlings. In the present study pure culture of the fungus *L. fraterna* was used for inoculation studies.

In the present study the effect of two isolates of *L. fraterna* was monitored for their effect on growth response of *E. grandis* seedlings over a period of 10 months in the glass house. The experiment was carried out in sterilized and unsterilized soil. Ectomycorrhizal inoculation stimulated all the growth parameters when compared to uninoculated control seedlings. Not much difference could be observed between seedlings grown in sterilized and unsterilized soil.

Among the two isolates used, Kodiakanal isolate was found to be more productive then Udagamandalam isolate. The ability of selected fungal isolates to infect and influence growth of *E. grandis* seedlings assessed indicate considerable variation within the isolates. Similarly, variations among isolates of the same fungus was also observed by Burgess *et al.*, (1994). Variations were observed among different isolates of *Suillus* sp. to colonise pine roots by Jacobson and Miller (1992). This is because large genetic variation can be expected to occur when collected from different locations and tree hosts (Marx and Kenney, 1982).

L. fraterna inoculated seedlings were more dependent on the mycobiont throughout the course of study . Among the seedlings treated with the two isolates of *L. fraterna*, Kodaikanal isolate treated seedlings were more dependent than the Udagamandalama isolate treated seedlings in both sterilized and unsterilized soil. The present study clearly indicate the need to experiment with different isolates from different geographical regions before choosing a proper isolate for inoculation in a large scale in the nurseries.

Acknowledgements

Thanks are due to the Director, CAS in Botany for the facilities.

References

Alvarez, I. K. and Trappe, J. N. 1983. Effects of application rate and cold soaking pretreatment of *Pisolithus* spores on effectiveness as nursery inoculum of western conifers. *Can. J. For. Res.* 13: 533-537.

Burgess, T., Dell, B. and Malajczuk, N. 1994. Variation in mycorrhizal development and growth stimulation by 20 *Pisolithus* isolates inoculated on to *Eucalyptus grandis* W.Hill ex. Maiden. *New Phytol.* 127: 731-739.

Burgess, T., Malajczuk, N. and Dell, B. 1995. Variation in *Pisolithus* based on basidiome morphology, culture characteristics and analysis of polypeptides using ID SDS-PAGE. *Mycol. Res.* 99: 1-13.

Cairney, J.W.G. and Chambers, S.M. 1997. Interactions between *Pisolithus tinctorius* and its hosts: a review of current knowledge. *Mycorrhiza* 7: 117-131.

Gardner, J.H. and Malajczuk, N. 1988. Recolonisation of rehabilitated bauxite mine sites in Western Australia by mycorrhizal fungi. *For. Ecol.Manag.* 24: 27-42.

Irving, F., Crossley, A., Manson, P.A., Last, F.T., Wilson, J. and Natarajan, K. 1985. Characteristics of some species of *Laccaria*, a fungal genus of significance of forestry, temperate and tropical. *Proc. Ind. Acad. Sci.* 95: 321-331.

Jacobson, K.M. and Miller, O.K. 1992. Physiological variation between tree-associated populations of *Suillus granulatus* as determined by *in vitro* mycorrhizal synthesis experiments. *Can. J.Bot.* 70: 26-31.

Last, F.T. and Watling, R. 1998. First record of *Laccaria fraterna* in Britain. *Mycologist* 12: 62-63.

Malajczuk, N., Lapeyrie, F. and Garbaye, J. 1990. Infectivity of pine and eucalypt isolate of *Pisolithus tinctorius* on roots of *Eucalyptus urophylla in vitro*. I. Mycorrhiza formation in model systems. *New Phytol.* 114: 627-631.

Marx, D. H. and Bryan, W.C. 1975. Growth and ectomycorrhizal development of loblolly pine seedlings in fumigated soil infested with the fungal symbiont *Pisolithus tinctorius*. *For. Sci.* 21: 245-254.

Marx, D.H. and Cordell, C.E. 1990. Development of *Pisolithus tinctorius* ectomycorrhizae on loblolly pine seedlings from spores sprayed at different times and rates. *USDA. For.Serv.Res.Note*. S.E-376, IP.

Marx, D.H. and Kenney, D.S. 1982. Production of ectomycorrhizal fungus inoculum. In: Schenck, N.C. (ed.) Methods and principles of mycorrhizal research. *The American Phytopathological Society*, St.Paul, Minn. Pp. 131-146.

Marx, D.H., Cordell, C.E., Maul, S.B. and Ruehle, J.L. 1989. Ectomycorrhizal development on pine by *Pisolithus tinctorius* in bare-root and container seedling nurseries. II. Efficacy of various vegetative and spore inocula. *New For.*3: 57-66.

Natarajan, K. 1977. South Indian Agaricales III. *Kavaka* 5: 35-39.

Natarajan, K., Mohan, V. and Kaviyarasan, V. 1988. On some ectomycorrhizal fungi occurring in Southern India. *Kavaka* 16:1-7.

Plenchette, C., Fortin, J.A. and Furlan, V. 1983. Growth responses of several plant species to mycorrhizae in a soil of moderate P fertility. I. Mycorrhizal dependency under field conditions. *Pl. Soil* 70: 199-209.

Richards, B.N. and Wilson, G.L. 1963. Nutrient supply and mycorrhiza development in Carribean pine. *For. Sci*. 9: 405-412.

Schonau, A.P.G., Verloren von Themaat, R. and Boden, D.L. 1981. The importance of complete site preparation and fertilizing in the establishment of *Eucalyptus grandis*. *South African J. Forestry* 116:1-10.

Tommerup, C., Bougher, N.L. and Malajczuk, N. 1991. *Laccaria fraterna*, a common ectomycorrhizal fungus with mono-and bi-sporic basidia and multinucleate spores: comparison with the quadristerigmate, binucleate spore *L. laccata* and the hypogenous relative *Hydnangium carneum*. *Mycol. Res.* 95: 689-698.

Trappe, J.M. 1977. Selection of fungi for ectomycorrhizal inoculation in nurseries. *Ann. Rev. Phytopathol*. 15: 203-222.

Frontiers in Plant Sciences (Pages 657-700)
Eds. K.G. Mukerji, K.V.B.R. Tilak, S.M. Reddy, L.V. Gangwane, P. Prakash, I.K. Kunwar

Actinorhizal and Mycorrhizal Tripartite Symbioses in *Casuarina equisetifolia* for Forestry Practices

N. Raman, S. Elumalai* and S. Patharajan

ABSTRACT

Symbiotic microorganisms are present in great numbers on and near the feeder roots of trees and they play vital role in numerous physiological and biochemical processes. The microorganisms implicated in this symbiotic interaction are from 2 groups: bacteria and fungi. The bacteria group is implicated on nitrogen fixation (Pawlowski and Bisseling, 1996), while the fungus group is involved in the uptake of nutrients with low mobility (Diop, 1996; Gianinazzi-Pearson, 1996). Amoung the bacteria, which establish symbiotic association with dicotyledonous plants, nitrogen fixation is exclusively carried out by *Rhizobium* and *Frankia* in a specialized organ, the root nodule where atmospheric nitrogen is reduced to ammonium.

INTRODUCTION

The plants nodulated by *Frankia* species are known as actinorhizal plants (Torry and Tjepkema, 1979). To date, about 194 species and 25 genera of such plants have been identified. Unlike the *Rhizobium*-legume symbiosis, in which the host plants with a few exceptions belong to a single large family, actinorhizal plants are distributed among 8 families and 7 orders

Author's address: Centre for Advanced studies in Botany, University of Madras, Guindy campus, Chennai - 600 025. India.
E-mail: raman55n@hotmail.com
**Currently at Botany Department,Presidency College, Chennai - 600 005.*

(Baker and Schwintzer, 1990). These plants have a predilection to grow in marginally fertile soils and they often serve as pioneer species in successional plant community development. Representatives can be found in most climatic zones and they inhabit a variety ecosystems including arctic tundra (*Dryas* species), coastal dunes (*Casuarina*, *Hippophae*, *Myrica*, and *Elaeagnus* species), riparin (*Alnus* and *Myrica* species), glacial till (*Alnus* and *Dryas* species) forest (*Alnus*, *Casuarina*, *Coriaria* and *Shepheridia* species), Chapparal and xeric (*Casuarina*, *Purshia*, *Ceanothus*, *Cercocarpus*, *Comptonia* and *Cowania* species), and alpine (*Alnus* species). The input of fixed nitrogen by these plants is considerable, especially in older temperate region where indigenous legumes are absent or rare (Silver, 1971).

Since they often thrive on marginal soils, actinorhizal plants have current and potential applications in reclaiming and conditioning the soils, producing timber and pulp and acting as nurse, windbreak, ornamental and fuelwood plants (Diem and Dommergues, 1990), agroforestry, reclamation of marginal and disturbed lands in the humid and semi arid tropics (Midgley *et al.*, 1983). However, only one species of *Casuarina* namely *C. equisetifolia* is widely planted for these purposes (US National Research Council, 1984). *Casuarina* species occur naturally in peninsular India and they are found in environments ranging from tropical forests to arid woodlands and saline soil.

Casuarina species are unique trees possessing mycorrhizal and actinorhizal symbionts in their root system. The mycorrhiza of actinorhizal plants is essential in the field to obtain higher yields, especially when the plants are growing in phosphorus deficient soils. Atmospheric dinitrogen fixation rates in these plants are of the same order of magnitude as those observed in legumes. Thus, rates of over 100 kg N/ha/yr which is approximately the amount of nitrogen added as fertilizer in intensive forest exploitations have been reported for *C. equisetifolia* (Cervamtes and Rodriguez-Barrueco, 1992). Globally they have potential for integrating into schemes for addressing issues of pyrodenitrification and reforestation.

Out of 3.1 million hectares of coastal salt affected soils of India, a huge area is not under regular cultivation with agricultural crops. This area either remains barren due to high soil salinity or is covered mostly by various types of wild, useless, uneconomical plants. It suffers from various natural calamites including storm, flood etc., as a regular phenomena and in general, the area is low lying, mono-croped and affected with salinity problem. Area with human habitation and regular cultivation should be protected from the seawater ingress or inflow of tidal water with a circuit embarkment along with the peripheral boundary. Huge amount of expenditure and physical labour has to be incurred to maintain the embarkments every year. To overcome this, it is economical to give a vegetation cover in the coastal area. *C. equisetifolia* is a well known windbreaker along with its high tolerance to salinity and also social acceptability. The world's population increases and hunger persists. Productive lands decline continuously. Mycorrhizal and actinorhizal (bio-fertilizers) technology has great potential for combating the current and future global problems of supplying food, fodder and fuel.

ACTINORHIZAL *FRANKIA*

On the basis of morphology and cultural characterisitics, the actinomycete endophytes of non-legumes are placed in one family Frankiaceae with a single genus, *Frankia* (Becking, 1970; Becking, 1974; Normand *et al.*, 1996). A further classification to species level has been made and 10 species of *Frankia* could be recognized: *Frankia alni, F. elaeagni, F. ceanothi, F. discardiae, F. brunchorstii, F. casuarinae, F. coriariae, F. dryadis, F. purshiae* and *F. cercocarpi.* These *Frankia* species differed to some extent in their morphological characteristics (Becking, 1970; Becking, 1974).

Frankia has been described as a fungus (Brunchorst, 1886), a bacterium (Spratt, 1912; Koslova *et al.*, 1966). It is now well established that the endophyte is an actinomycete. Although filamentous, the organism is definitely prokaryotic since it lacks a nuclear membrane (Becking *et al.*, 1964). Besides this, a relation to fungi is excluded because of the narrow hyphae which are characteristic of actinomycetes, being 0.2-0.3 mm in diameter. In addition, the hyphae contain plasmalemmosomes associated with cross-wall formation. The morphology of the actinomycetous structures within the host cells has been extensively studied using transmission electron microscopy in *Alnus glutinosa* (Becking *et al.*, 1964). The endophytic hyphae are branched and septate but branching is not necessarily correlated with septum formation. Both light and electron microscopy (scanning) have indicated that the hyphae bear terminal swellings, which are called vesicles. These vesicular structures are spherical, measuring 3-5 mm in diameter, in *Alnus* and *Elaeagnus*, but club - shaped in *Casuarina*, *Myrica* and *Dryas* (Becking *et al.*, 1964). The vesicular structures are always situated at the tips of the hyphae and within the host cell near the periphery of the hyphal mass close to host cell wall. The vesicles have very elaborate internal structure with many membranes and incomplete cell walls. The vesicle probably serves as a site of nitrogen fixation as it possesses a strong tetrazolium-reducing capacity (Akkermans, 1971).

FRANKIAL NODULATION IN *CASUARINA EQUISETIFOLIA*

Casuarina grows in a diverse range of environments, from tropical forest to arid woodlands and coastal dunes (Barlow, 1983). Species of *Casuarina* are capable of forming symbiotic nitrogen fixing association with *Frankia* (Torrey, 1978) and may contribute significantly to the nitrogen economy of ecosystems (Dommergues, 1963; Silvester, 1977). However, field nodulation by *Casuarina* can be irregular (Lawrie, 1982). Hence reservation as to the ecological significance of N_2 fixation by many species have been expressed (Bond, 1976; Lawrie, 1982) but the absence of infective endophytes and environmental inhibition of nodulation have been suggested as a possible explanation (Lawrie, 1982; Dommergues *et al.*, 1984). Specificity amoung the *Casuarina* in root nodulation by *Frankia* is reported by Torrey and Racette (1989). Reddell *et al.* (1986) suggested that P levels may be critical in regulating nodulation of some *Casuarina* in the field. The absence of infective *Frankia* populations was responsible for the lack of nodulation in two soils under greenhouse conditions. Soil temperatures below 20°C (Reddell *et al.*, 1985), high levels of mineral N (Rodriguez-Barrueco, 1972), deficiencies of nutrients other than N and P (Robson and Loneragan, 1978) and low soil pH (Bond, 1957; Coyne, 1973) may also inhibit field nodulation. Low soil pH may inhibit root nodulation by *Frankia* in some *Allocasuarina* species on acid soils (Bond, 1957). Kurdali and Domenach (1991) reported that the rate of nodulation of the host is affected by nutrient supply and *Frankia* population, which decreased with soil depth. Gardner (1986) observed sporulation in the root nodules of *C. cunninghamiana* grown in greenhouse and inoculated with nodules suspensions. Flemming *et al.* (1988) found that *C. cunninghamiana* grows well when crushed nodule is used as inoculum source. Reddell and Bowen (1986) reported that host-endophyte specificity was expressed mainly at the generic level in a crushed nodule inoculation study involving 15 species in 3 genera of Casuarinaceae. Variation in soil depth, topographic position and host species in the capacity of soils are observed as possible reasons for nodulation in *Casuarina* and *Allocasuarina* seedlings (Dawson *et al.*, 1989). Van Dijk (1979) concluded that degree of nodulation depends on soil conditions and endophyte strains. Aerial nodules in *C. cunninghamiana* are observed by Prin *et al.* (1991).

External factors have the potential to bring about an imbalance in the symbiotic relationship. The role of pH, soil, soluble phosphorus and calcium levels, available nitrogen, conductivity of water, climate, light availability, etc. are significant in this regard (Quesada and Valiente, 1996; Quesada *et al.*, 1997; Olson *et al.*, 1998; Quesada, 1998). Righetti *et al.* (1986) studied the effect of soil moisture, P, K and microelements in the nodulation of actinorhizal plants. Growth and yield of *C. equisetifolia* plantations on the coastal sand dunes of Senegal are studied by Ndioye *et al.* (1993). Nitrogen fixing root nodule symbiosis in non-leguminous angiosperms *Alnus*,

Casuarina, Coriaria, Datisca and *Elaeagnus* species of Pakistan is reported by Chaudhary and Mirza (1988) and they found that some of the isolates are pigmented. Smolander *et al.* (1988) observed that survival of *Frankia* in soils with pH higher than 6.0 is better than at pH 4.5. Nodules formed by *Frankia* in roots of *Casuarina* are capable of nitrogen fixation at rates comparable to those found for effectively nodulated legumes (Torrey, 1978). Two field studies have shown the estimated fixation rates of 40-60 kg N/ha/yr in plantings of *C. equisetifolia* on coastal sand dunes in Senegal (Dommergues, 1963).

Lawrie (1982) examined field nodulation of 9 species of *Casuarina* and *Allocasuarina* in Victoria and found nodules on only 15% of 336 plants examined and at only 17% of the 77 sites. Baiting studies in a glasshouse confirmed field findings in 88% of the cases. Reddell *et al.* (1986) found that 4 of the 5 *Casuarina* species studied in natural populations are regularly nodulated in 88 sites examined, only 9 out of a total of 17 *Allocasuarina* species examined are found to bear nodules in field. Actinorhizal root nodules are coralloid structures consisting of multiple nodule lobes, each of which is a modified lateral root without root cap, with a superficial periderm, a central vascular bundle, and both infected and uninfected cortical cells (Pawlowski and Bisseling 1996; Franche *et al.*, 1998). *Frankia* infects the roots primarily by root-hair infection. Nodules formed have an internal anatomy similar to that of lateral roots with a cortical cylinder of vascular tissue, a cortical region in which the infected cells are found and a typical outer periderm layer (Ganesh *et al.*, 1994). In actinorhizal nodules, a pre-nodule consisting of infected and uninfected cells is, however, formed. The nodule premordium formation does not involve prenodule cells. Thus the prenodule apparently represents a primitive symbiotic organ (Laplaze *et al.*, 2000).

ULTRASTRUCTURE

In some actinorhizal nodules, there is no diffusion barrier between the soil atmosphere and the infected zone (Tjepkema, 1979, 1988; Wheeler *et al.*, 1979). In these nodules, intercellular air spaces extended into the zone of infected tissue and are connected to the soil atmosphere through lenticels or nodule roots. As a result, the PO_2 in the infected zone may be too close to atmosphere (Tjepkema, 1979, 1988). Zeng *et al.* (1989) reported that *Casuarina* may be unique among actinorhizal plants in its small number of intercellular air spaces in the zone of infected cells and the lack of continuity between these spaces. Root nodule initiation in *Gymnostoma* (Casuarinaceae) and *Sphepherdia* (Elaeagnaceae) induced by *Frankia* strain HEP GP I1 - is reported by Racette and Torrey (1989). Evidence that initial penetration occurs through root-hair infection, in association with root-hair deformation is first presented for *A. glutinosa* (Callaham *et al.*, 1979). Callaham and Torrey (1977) showed root hair infection in the nodulation of *Comptonia peregrina* and it is also confirmed for two species of *Myrica* and one species of *Casuarina* (Callaham *et al.*, 1979). The nodule is bounded by a cork periderm which contains lenticels through which gases can exchange with the atmosphere. In nodules subjected to conditions of restricted gas diffusion, e.g., growth in water culture, a mass of loosely packed, aerenchymatous cells proliferates from the lenticels, which may be important in aeration (Mc Vean, 1955).

ISOLATION OF *FRANKIA* FROM ROOT NODULES

Currently several isolation procedures are employed with variable success. The sucrose fractionation technique (Baker *et al.*, 1979), serial dilution or modified dilution technique (Lalonde and Calvert, 1979; Diem *et al.*, 1982), the selective incubation or microdissection techniques (Berry and Torrey, 1979; Quispel and Burgraaf, 1981), osmium tetroxide technique (Lalonde *et al.*, 1981) and filter exclusion technique (Benson, 1982) are some of the known techniques. More than one *Frankia* strain might coexist in same nodule (Reddle and Bowen, 1985). Isolation and characterization of *Frankia* from nodules of *Alnus, Casuarina, Coriaria, Datisca, Elaeagnus* and *Hippophae* have been reported by Chudhary and Mirza (1987). Identification and subgrouping of *Frankia* strains using sodium dodecyl sulphate-polyacrylamide gel electrophoresis have been shown by Gardes and Lalonde (1987). Torrey and Racette (1989) found that the pure culture isolates of *Frankia* made from *Casuarina* root nodules grow well on P medium (Burgraaf and Shipton, 1983) and BAP medium (Murry *et al.*, 1984). Moderate growth is shown on FMC medium (Benson, 1982), defined propionate medium (Baker and O' Keefe, 1984) and Qmod medium (Lalonde and Calvert, 1979).

MYCORRHIZAL ASSOCIATION IN *CASUARINA* SPECIES

Mycorrhizae are benificial symbiotic fungal associations with plant roots, which benefit plant growth by increasing the effective absorbing zone of the root through the hyphae, which explore the soil away from the root surface. Evidences clearly indicate the beneficial role of AM fungi to higher plants (Hirrel and Gerdemann, 1980; Mahadevan *et al.*, 1988; Khan, 1988; Sambandan *et al.*, 1991). Reports indicate the importance of mycorrhizal association to members of the Casuarinaceae (Rose, 1980; Gardner, 1986;

Reddell *et al.,* 1986). This family includes 2 genera, *Allocasuarina* and *Casuarina* and much of the recent debate has been focussed on the relative importance of the arbuscular mycorrhizae and or ectomycorrhizal association in members of each genus. *Allocasuarina* forms ectomycorrhizal association more commonly than *Casuarina* species (Bougher *et al.,* 1990 a; Thoen *et al.,* 1990; Theodorou and Reddel, 1991). *Eucalyptus* and *Casuarina* are generally grown in common forest nurseries and were inoculation with a single mycorrhizal fungal isolate for both plants in China (Gong-Mingqin, personal communication).

Bamber *et al.* (1980) have reported the ectotrophic mycorrhizal association of *Pisolithus tinctorius* with roots of *C. distyla*. Torrey (1983) made a survey of *Casuarina* species in south Florida, roots of *C. glauca* and *C. equisetifolia* in both wet and dry sites exhibited the presence of both ectotrophic and endotrophic mycorrhizae. Rose (1980) and Gardner (1986) suggested that the role of *Casuarina* as a pioneering plant may be assisted by the presence of mycorrhiza on the root system. This may also be helped by the presence of proteoid roots in members of the Casuarinaceae, tightly packed rows of rootlest probably structured to increase the ability of the plant to absorb nutrients (Diem and Dommergues, 1990). Rangarajan *et al.* (1988) reported that the results of dual inoculation of *Frankia* (crushed nodules) and *P. tinctorius* or *Rhizopogan luteolus* (spores) in *C. equsetifolia* revealed that nodulation and biomass production could be increased better than the individual inoculation. Growth and root structure of *Eucalyptus*, *Allocasuarina* and *Casuarina* seedlings inoculated with ectomycorrhizal fungi *P. tinctorius* and *Scleroderma cepa* were reported by Bougher *et al.* (1990 b). De la Cruz (1990) developed a technology of mycorrhizal tablets made from compressed mixtures of basidiospores of *P. tinctorius* and *S. cepa*. One mycorrhizal tablet was inoculated to each container with seedlings in the nursery during the seedling or picking operation. This was successful for *Pinus* sp. *Eucalyptus* sp. *Dipterocarpus* sp. and *C. equisetifolia.* Effects of AM fungi on growth of *C. glauca* seedlings were studied by Abou- El- Khair (1993).

Thoen *et al.* (1990) found that 3 isolates of *P. tinctorius* (two from Senegal, one from Australia) formed ectomycorrhizas with a mantle and Hartig net on *Eucalyptus grandis* but failed to form a Hartig net with *C. equisetifolia*. Theodorou and Reddell (1991) have synthesized ectomycorrhizae of Australian isolates of *P. tinctorius* and *S. cepa* and have observed that ectomycorrhizae either failed to form or formed infrequently with *C. equisetifolia* compared to *E. grandis* and *Allocasuarina littoralis.* Dell *et al.* (1994) have found that *S. cepa* and *P. tinctorius* showed broad host associations and have been found under *A. littoralis* and *C. equisetifolia* (Mallajczuk *et al.,* 1984; Theodorou and Reddell, 1991). The failure of the inoculated *C. equisetifolia* plants to form extensive ectomycorrhiza in soil, in open pots and in a glasshouse substantiates the observations by Reddell *et al.* (1986) that ectomycorrhizas are less common in *C. equisetifolia* than in *A. littoralis*. Thoen *et al.* (1990) grow inoculated plants in test tubes containing perlite-peat-Melin-Norkrans agar and reported that 3 isolates of *P. tinctorius* formed ectomycorrhizae on *C. equisetifolia* but these lack Hartig nets. The sheathed roots that formed with *C. equisetifolia* are typical of incompatible association. For example, Malajczuk *et al.* (1984) and Duddridge (1986) have observed hyphal colonization of root surfaces with incompatible partners such as *P. tinctorius*. Wong and Fortin (1990) suggested that hyphal envelopes may simply be result of growth on root exudates without requiring fungus-root attachment. However, the hyphal envelopes that formed on *C. equisetifolia* with *S. cepa* and *P. tinctorius* were compact. *C. equisetifolia* sheaths are mantles, specific to ectomycorrhizal development (Wong and Fortin, 1990). This suggests the array of signals (Mortin and Hilbert, 1991) between the roots of *C. equisetifolia* and the hyphae of *P. tinctorius* and *S. cepa* are incompatible for the sustained ectomycorrhizal development.

Harley (1970) stressed the importance to occurrence of dual symbiosis in primary colonizers like *C. equisetifolia*, which contained root nodules, infected with actinomycetes and roots with AM association. Gullan (1975) and Sward (1978) reported the occurrenence of ectomycorrhiza in *C. equisetifolia.* Khan (1974, 1978) observed the presence of endomycorrhizal fungal spores in soil harbouring *C. cunninghamiana* although he did not find AM infection. Rose (1980) found no ectomycorrhizae in *Casuarina* sp. but noted the presence of arbuscular mycorrhizal structures in the roots of *C. cunninghamiana* obtained from sand woodlands of Florida and Japan and in the roots of *C. equisetifolia* from marine and dunes of Florida. The spores of AM fungi, *Gigaspora gigantea* and *Gig. nigra* are found in the soil of *C. equisetifolia* and the spores of *Gig. coralloidea*, *Gig. gigantea* and *Glomus macrocarpum* are found in the soils of *C. cunninghamiana*. Raman *et al.* (1992) found AM association in 19 tree species out of 25 tree species in Mamandur forest of Tamil Nadu. Raman and Elumalai (1991) reported 8 species of AM fungi from rhizosphere soils of *Casuarina equisetifolia* growing in the Coramandal coastal soils. Sambandan *et al.* (1994) reported AM association of *Casuarina equisetifolia* grown in different soil types and isolated 10 species of *Glomus*, 2 of *Gigaspora* and 2 of *Sclerocystis* in the rhizosphere soils. Improved rate of growth in *Alnus glutinosa* (Hall and Maynard, 1979) and other actinorhizal tree species (Rose, 1980) resulting in a tetrapartite symbiosis (Gauither *et al.*, 1983; Gardner, 1986; Chatarpaul *et al.*, 1989) has been reported. AM hyphae are found in the nodular tissues and surrounded by ectomycorrhizal hyphae which penetrated the outer cortex of the *Alnus* nodule or the adjoining root. However such hyphae were hardly seen in the nodules containing the bacterioids.

FRANKIA INOCULATION TO AUGUMENT FERTILITY

Ferguson and Bond (1953) provided evidences for the fixation of nitrogen in the root nodules of *Alnus glutinosa* inoculated with nodule suspension. Nodulation occurred freely in a pH range of 5.4 to 8.0. Fessenden (1979), Hall and Maynard (1979) demonstrated the use of actinorhizal plants for forestry and amenity planting involving large scale production of seedlings or clonal plant material and their nodulation with selected strains of *Frankia*. Nodule homogenates were used to nodulate the plants produced for experimental studies or field plantations (Akkermans and Houwers, 1979). Large scale production of seedlings of nodulated *Alnus* sp. using pure cultures of *Frankia* was made by Fortin *et al*. (1981). Revegetation of 6000 hectares of land used for the hydro - electric project of the "La Grande River, James Bay, Northern Quebec is attempted with *Casuarina* spp. (Arsenault, 1983; 1984). Actinorhizal plants are popularly used as pioneer plants in the regenaration of wastelands. Actinorhizal trees are also valued for landscaping, providing shade and contributing to the beautification of parks and cities. (Diem and Dommergues. 1990). Apart from these practical aspects, the *Frankia*-actinorrhizal plant system provides enough food for thought to those interested in cell biology.

Vernaud *et al.* (1985) developed a new spot inoculation technique for the nodulation of *Alnus glutinosa* close to the point of inoculation. Large-scale inoculation of actinorhizal plants with *Frankia* is demonstrated by Perinet *et al.* (1985). The use of *Frankia* isolates is restricted because of their slow growth. Inadequate information is available on their growth requirements (Shipton and Burggraaf, 1983). Both these aspects limit the production of inoculum for nursery stock resulting in the use of crushed nodule suspensions on soil from existing plantations as inoculants (Torrey, 1982). Glasshouse trials of *Frankia* in promoting growth of seedlings of various species of *Casuarina* with different inoculation procedures reveal the effectiveness of *Frankia* (Rosbrook and Bowen, 1987; Megahed and El-Lakany, 1986; Sougoufara *et al.*, 1989; Stele *et al.*, 1989; Burleigh and Torrey, 1990).

MOLECULAR BIOLOGY OF ACTINORRHIZAL NODULES AND MYCORRHIZAE

The actinorrhizal plant symbiosis, protein specificity expressed at the time of nodules formation and functioning (Nap and Bisseling, 1990). These proteins called nodulins. According to the time of their appearance it is classified into 2 groups. The early nodulins and the late nodulins. The early nodulins are identified during the infection processes and nodule formation. The late nodulins which are expressed into the functioning nodules. Studies of the early nodulin genes during the nodule formation have provided some useful tools to evaluate the mode of action of bacteria nodulin factors. Scheres *et al.* (1990a) have studied about early nodulin gene *ENOD12* and it is induced in root epidermis by nodulin factors. Bauer *et al.*, 1997 have showed the *Medicago sativum* nodules and roots play a role in the differentiation process of reverse transcriptase polymerase reaction (RT-PCR) and transgenic techniques to suggest that *ENOD12A* is expressed adjacent to the meristematic zone. Another nodulin gene, ENOD5 (Scheres *et al.*, 1990b) is also expressed during infection process and nodule development. Garnas *et al.* (1996) and Mylone *et al.* (1995) have observed that *ENOD5*, *ENOD12*, *M+N8* and *Mt N12* encode prolin-rich proteins are extension components which participate on the infected thread wall structure.

The plant genes involved in symbiosis is fragmentary and information about frankial genes and their expression are meager. The morphological and physiological features of some of the steps involved in this process and try to extrapolate it to the composite endosymbiotic system. It seems that the actinorhizal plants undergo feedback regulation of symbiosis (Wall. 2000). Several genes expressed specifically or in enhanced levels in actinorhizal nodules, collectively referred to as actinorhizal nodulin genes, have been isolated from *Alnus glutinosa* and *Casuarina glauca* (Franche *et al.*, 1998; Guan *et al.*, 1998). Most of these genes have been classified as late actinorhizal nodulin genes, reflecting their involvement in final stages of nodule formation and functioning. Actinorhizal nodulin genes activated in early stages of nodule development have been identified in *A. glutinosa* (Ribeiro *et al.*, 1995; Pawlowski *et al.*, 1997). *Frankia*-Casuarinaceae symbiosis, a model system that is especially suited to the analysis of plant symbiotic gene expression since *C. glauca* and its close relative *Allocasuarina verticillata*, unlike *A. glutinosa* can be readily transformed and regenerated (Franche *et al.*, 1997; 1998).

A gene cg12 from *Casuarina glauca* has been cloned and characterized recently (Laplaze *et al.*, 2000). It was found to have a strong homology to subtilisin-like protease gene families of several plants, including the actinorhizal nodulin gene ag12 of *Alnus glutinosa*. This gene expressed early during nodule development and therefore, represents an early actinorhizal nodulin gene. Verghese and Misra (2000) have made attempts at identifying superior *Frankia* genotypes using molecular markers and have met with a limited success. Diouf *et al.* (2003) have reported three symbiosis

of rhizobial, actinorhizal and mycorrhizal associations appear that nodule and mycorrhizal formation are under the control of specific genes plant host, which could be specifically induced by bacterial or fungus products. Mycorrhiza induces a discreet modification in plant roots, actinorhizal and *Parasponia* nodule structure is very similar to roots and legume nodule is a new organ.

PHYSIOLOGY AND BIOCHEMISTRY OF *FRANKIA* AND ARBUSCULAR MYCORRHIZAL INOCULATED PLANTS

AM fungal associations with roots of actinorhizal plants enhances phosphorus uptake as well as increased nodulation and nitrogen fixation (Rose and Youngberg, 1981; Gauthier *et al.,* 1983). Diem and Gauthier (1982) have found that double inoculation of *C. equisetifolia* with *Glomus mosseae* and *Frankia* significantly improve the plant growth and nodulation. After six months of growth, mycorrhizal nodulated plants contain twice as much nitrogen as the seedlings that had been inoculated with *Frankia* only (Diem and Dommergues, 1990). Recent studies conducted by Duhoux *et al.* (2001) in *Gymnostoma* have shown that root nodules can also be colonized by arbuscular mycorrhizae. Plants and arbuscular mycorrhizal symbiotic associations, which stimulate hyphal growth of fungi by synergistic interaction between volatile compounds and exudates produced by roots (Becard and Piche, 1989).

Both endomycorrhizal fungi, *Glomus* sp. and symbiotic N_2 - fixing bacteria depend on the host for their carbon supply (Kucey and Paul, 1982), carbohydrates (Ocampo and Azcon, 1985), protein (Carling *et al.,* 1978) and lipids (Pacovsky and Fuller, 1988). Increased respiratory losses (Snellgrove *et al.,* 1982) from the AM symbiont could increase sink demand for photosynthesis and provide stimulus for higher photosynthetic rates. Koch and Johnson (1984) estimated that 6-10% of total photosynthates are directly allocated to AM related events, which would include metabolic processes of the symbiont. The photosynthetic rate was more in *Glycine-Glomus* symbiosis (Harris, *et al.,* 1985), and higher rate of CO_2 fixation per chloroplast in AM plants than the control. Allen *et al.* (1981) observed higher levels of chlorophyll in mycorrhizal plants indicating greater photosynthetic rates. Comparatively increased water relation and photosynthesis of mycorrhizal than non-mycorrhizal *Bouteloua gracilis* is recorded (Allen *et al.,* 1981). Increased CO_2 fixation rate in mycorrhizal plants was reported in many plants (Levy and Krikun, 1980; Allen *et al.,* 1981; Kucey and Paul, 1982; Snellgrove *et al.,* 1982; 1986; Bayne *et al.,* 1984). Koach *et al.* (1997) have reported garlic plants colonized by AM fungus showed high rates of photosynthesis than uncolonized plants. Radioactive labeling experiments showed the transport of photosynthate from the host to the fungus (Bevege *et al.*, 1975; Losel and Cooper, 1979). About 70-90% of the ^{14}C label present in both the roots and mycelium are in the form of soluble carbohydrates. Cluster roots formed in some of the actinorhizal plants may help P uptake from deficient or low P soils (Raman and Elumalai, 1992; Raman and Mahadevan, 1996). The occurrence of cluster root in field condition in *Casuarina* spp. has been recorded by Redell, (1986).

SUGAR

A direct correlation between the total sugars and *G. intraradix* infection is observed by Nemec and Guy (1982) and Same *et al.* (1983). Specific fungal sugars such as mannitol and trehalose are observed in ecto and ericoid mycorrhizae (Harley and Smith, 1983). Bevege *et al.* (1975) have proposed that a concentration gradient is maintained by the fungus converting the soluble sucrose from the host into soluble glycogen.

Ocampo and Azcon (1985) have found no direct relationship between sugar concentration in roots and AM infection in wheat and concluded that sugar in root extracts of these plants may not be a decisive factor in plant susceptibility to mycorrhizal infection. Dixon *et al.* (1988) reported decrease in exudation of reducing sugars and amino acids from roots of *Citrus aurantium* as soil and root P increased in mycorrhizal than non-mycorrhizal plants. Elevated levels of sucrose and pinitol in nodulated roots have been associated with increases in N_2 fixation (Streeter, 1980) while an influence of nitrogen supply on root carbohydrate composition has been documented (Bevege *et al.*, 1975; Ratnayke *et al.*, 1978; Patt *et al.*,1979; Barta, 1979; Minchin *et al.*, 1981; Snellgrove *et al.*, 1982;). Larger root mass typically accompanies mycorrhizal infections (Menge *et al.*, 1978). Nemec and Guy (1982) have reported that levels of reducing sugar relative to sucrose are greater after mycorrhizal infection in *C. aurantium* roots. Storage of carbohydrate occurs in the hyphae as glycogen (Bevage *et al.*, 1975). Accumulation of glycogen is high in hyphae containing large levels of lipid and in the arbuscular hyphae (Kinden and Brown, 1975; Bonfante - Fasolo, 1984). There is a variable correlation between mycorrhizal infection and total or reducing sugar content of root extracts (Azcon and Ocampo, 1981; Graham *et al.*, 1981; Nemac and Guy, 1982). In orchid seedlings, which are obligatory mycorrhizal symbionts, it has been shown to utilize exogenous trehalose supplied by the fungus in early growth (Smith, 1966, 1967).

Leaves of AM fungus-infected plants generally contain more sucrose, reducing sugars and starch than non-mycorrhizal control plants (Bevege *et al.*, 1975; Cox *et al.*, 1975; Snellgrove *et al.*, 1982; Dixon *et al.*, 1988). The carbohydrates status of roots of AM fungus infected plants is more variable. Roots of several AM fungus-infected plant species contain levels and patterns of soluble carbohydrates similar to those of non-mycorrhizal plants. (Bevage *et al.,* 1975; Hayman, 1975). Other studies have shown that more reducing sugars are present in roots of AM fungus-infected plants than in the control (Dixon *et al.,* 1988; Nemec and Guy, 1982), and increase in reducing sugars are often associated with reducing in starch in the root tissues (Nemec and Guy, 1982; Dixon *et al.,* 1988). Subramanian *et al.* (1997) observed that the root colonization with *G. intraradix* in two tropical maize cultivars had a beneficial effect on host plant drought tolerance by maintaining higher water potential, higher relative water content, higher P and sugar contents and high dry mass during drought and recovery periods.

Brown and Bethlenfalvay (1987) showed that higher rates of CO_2 exchange in AM fungus infected plants are associated with P nutrition. Mycorrhizal and non-mycorrhizal plants have equal rates of C-assimilation per unit of leaf area, but that mycorrhizal plants have higher assimilation rates expressed on a dry matter basis (Snellgrove *et al.,* 1982). At least 3 to 5% of the whole plant ^{14}C- labelled photosynthates is allocated to mycorrhizal-related events on one side of a split-root system (Koch and Johnson, 1984). Increased accumulations of free amino N in roots of mycorrhizal groundnuts and amino acid N in leaves of non-mycorrhizal *C. aurantium* have been reported (Nemec and Meredith, 1981; Krishna and Bagyaraj, 1983). The processes involved in the development of mycorrhizae on the root systems of woody plants should lead to increased rates of photosynthesis and translocation (Harley and Lewis, 1969). They have suggested that the fungal sink might indirectly increase the rate of photosynthesis, an idea consistent with the concept that photosynthesis is under some control by sinks demand. The mycorrhizae of conifers cause increased translocation of photosynthate to root and increased photosynthesis (Lister *et al.,* 1986). Studies with non-woody plants forming vesicular-arbuscular mycorrhizae have indicated stimulation of photosynthesis by mycorrhiza formation (Allen *et al.,* 1981).

Ho and Trappe (1973) recorded the accumulation of ^{14}C- labelled photosynthate in the lipid filled chlamydospores of *Gl. mosseae.* Autoradiographs of mycorrhizal roots and external mycelium, after translocation of ^{14}C-labelled photosynthate from the leaves indicated that lipid rich hyphae, vesicles and arbuscules contain a greater proportion of ^{14}C-labelled photosynthate in infected and uninfected roots of arbuscular mycorrhizal *Araucaria cunninghami.* High proportion of ^{14}C- lipid fraction is found in the fungal hyphae external to the root. Trehalose and mannitol, the characteristic carbohydratets of ectomycorrhizae have not been found in arbuscular mycorrhizal fungi (Bevege *et al*., 1975).

LIPID

Lipids are the major storage compounds in mycorrhizal roots. Infected roots contain more total lipids than uninfected roots (Cooper and Losel, 1978). Of the various lipid classes, neutral lipids especially triglycerides are dominant (Cooper and Losel, 1978; Nagy *et al.,* 1980). Mycorrhizal roots have a higher level of phospolipids than non-mycorrhizal roots (Cooper and Losel, 1978; Nagy *et al.,* 1980). Nagy *et al.* (1980) and Nordby *et al.* (1981) have observed the presence of unusual fatty acids in the roots of mycorrhizal *Citrus* sp. Pacovsky and Fuller (1988) showed a positive correlation between an isomer of hexadecanoic acid and vesicle number in the roots. Mycorrhizal plants also have higher sterol content.

However, Nagy *et al.* (1980) have found that roots of *C. aurantium* infected with *Gl. mosseae* has higher percentages of campesterol and free sterols and lower percentages of chloesterol and β-sitosterol than the non-mycorrhizal plants. Losel and Cooper (1979) demonstrated the presence of label in lipid components of mycorrhizal roots by supplying exogenous sources of labelled sucrose, acetate and glycerol to the plants. Synthesis of lipids by AM fungi provided a possible sink for photosynthate (Bevage *et al.,* 1975; Cox *et al.,* 1975; Cooper and Losel, 1978).

Oil droplets have been frequently recorded in the hyphae and spores of arbuscular mycorrhizal fungi (Mosse, 1959; Greenall 1963; Mosse and Bowen, 1968; Cooper, 1976). Cox and Sanders (1974) illustrate lipid droplets in the arbuscules, vesicles and intercellular hyphae of arbuscular mycorrhizal fungi using electron micrographs. Lewis (1975) points the need for information on lipid composition in arbuscular mycorrhizal, both of the oil droplets of hyphae and vesicles and of the extensive fungus and host membranes. The proliferation of host membrane observed in electron micrographs by Cox and Sanders (1974) particularly around arbuscules must partly mask the presence of smaller amount of fungal phospholipid. Accumulation of lipid in fungal structures is encountered in a wide range of host-parasite systems (Brennan and Losel, 1978). The unusual lipid-derived fatty acids found only in mycorrhizal roots are almost certainly synthesized by the AM fungus (Beibly and Kidby, 1980; Nagy *et al.,* 1980).

AMINO ACIDS

Leaf *et al.* (1959) have determined the free amino acids and related soluble nitrogen compounds in *Alnus glutinosa* root nodules with chromatographic methods. Citrulline is the predominant amino acid present, accompanied by smaller amounts of aspartic, glutamic and amino-butyric acids, arginine and other constituents. When molecular nitrogen labelled with 15_N is applied, the highest atom percent of 15_N was found in glutamic acid, followed by citrulline or aspartic acid. When citrulline is degraded to ammonia and ornithine, the ammonia liberated is richer in 15_N than even glutamic acid. Leaf *et al.* (1959) found asparagine as the predominant amino acid in *M. gale.* Glutamine and other amino acids are present in smaller amounts. The highest enrichment with 15_N is in the amide-N-of glutamine and the next highest is in the amide N - of asparagine and amino N- of glutamic acid.

Lower amount of amino acid content in leaves of mycorrhizal than non-mycorrhizal plants of *C. aurantium* is reported by Nemec and Meredith (1981). In contrast, 3 to 10 times more amino acid content and a much greater level of arginine and citrulline in mycorrhizal roots as compared to uninfected ones is reported by Young and Trappe (1972). Krishna and Bagyaraj (1983) studied the changes in free amino acids and protein fraction of groundnut due to *G. fasciculatium* inoculation and observe an increased accumulation of free amino acids due to mycorrhizal development though it fluctuated initially.

PROTEIN

Krishna and Bagyaraj (1983) have found an increased level of protein in *G. fasciculatum* inoculated *Arachis hypogia* roots. Roots of *G. mosseae* infected *Trifolium pratense* (Arines *et al.*, 1993) and *G. intraradix* inoculated *Nicotiana tabacum* and *Allium cepa* roots (Dumas *et al.*, 1994) showed enhanced protein content. Ramesh *et al.* (2000) have found *Gig. margarita* inoculated with *Pennisetum pedicellatum* plants exhibited higher level of protein. In general mycorrhizal roots contained significantly high level of protein. Protein levels in *P. pedicellatum* plants inoculated with three species of AM fungus show *Gig. margarita* inoculated plant found higher levels of proteins as compared with *G. mosseae* and *G. aggregatum* inoculated plants (Ramesh *et al.*, 2000). The protein profile of *Gig. margartia* colonized by *P. pedicellatum* roots showed difference in polypeptide (Ramesh, 1997).

ENZYME ACTIVITY

Phosphatases

Interspecific genetic variation of acid phosphatases activity in monocaryotic and dikaryotic populations of the ectomycorrhizal fungus *Hebeloma cylindrosporum* is reported by Meysselle *et al.* (1991). The ericoid mycorrhizal fungus *Hymenoscypus ericae* produces high level of acid phosphatases (Pearson and Read, 1975; Straker and Mitchell, 1986; Lemoine *et al.*, 1992). Dodd *et al.* (1987) finds increase acid phosphatase activity in mycorrhizal wheat and onion. Analysis of the acid phosphatases activity in six ectomycorrhizal fungi reveal that *Hebeloma crustuliniforme* showed highest activity followed by *Laccaria laccata, Amanita muscaria, Thelephora terrestis, Rhizopogon vincicolor* and *Piloderma bicolor* (Ho and Zak, 1979). Eight different isolates of *P. tinctorius* are low in acid and alkaline phosphatase activities (Ho, 1987). The acid and alkaline phosphatase activities in six different isolates of *L. laccata* are analysed by Ho and Zak. (1979), and they finds that the activity of acid phosphatase is higher than alkaline phosphates in all species. High acid phosphatase activity and tolerance to high concentrations of Cu and Ni are reported in *Laccaria laccata* (Periasamy and Raman, 1995) and *Amanita muscaria* (Raman *et al.*, 1998). Kroelhler *et al.* (1988) have demonstrated the activity of surface and extracellular soluble acid phosphatases produced by the ectomycorrhizal fungi *Cenococcum geophillum* and *Entoloma sericeum.* Raman *et al.* (2002) have reported that in *L. laccata* and *Suillus bovines* the acid phosphatase activity was increased with the increasing concentrations of Cr amended medium compared with control. Whereas the alkalline phosphatase activity of *L. laccata* was increased in 0.60 mM concentration of Cr than other treatments and control. But in *S. bovinus,* the alkaline phosphatase activity decreases in all the treatments when compared with control. Intracellular acid and alkaline phosphatase activity of *P. tinctorius* grown in Al and Cr amended medium enhanced on 15 d and declined on 30 and 45 d when compared with control (Srinivasan, 2000). Antibus *et al.* (1992) studied the effect of inorganic and organic phosphorus growth sources on enzyme activity and ^{32}P uptake in several ectomycorrhizal fungi i.e., *A. rubescens, E. sericeum, Hebeloma* sp., *Lactarius* sp., *Paxillus involutus, Scleroderma citrinum* and *C. geophillum* (Periaswamy and Raman, 1995).

Plant roots exude organic acids, e.g. malic and citric acids, and/or acid phosphatases under phosphorus deficiency (Hoffland *et al.*, 1989; Ohwaki and Hirata, 1992; Liu *et al.*, 1990; Marschener, 1998). This localized enhanced excretion of organic acids increases the effectiveness of exudates for the mobilization of nutrients such as phosphorus, zink, iron and manganese (Marschener, 1998). The population density and composition of symbiotic and non-infecting microorganisms in the rhizosphere can enhance root exudation and the concentration of organic acids, chelators and acid phosphatases released as ectoenzymes from roots, or from microorganisms including arbuscular mycorrhizal fungi as microbial metabolites (Meharg, 1994; Tarafdar and Marschner, 1994).

Straker and Mitchell (1986) characterised the acid phosphatases in endomycorrhizal fungi of the Ericaceae. MacDonald and Lewis (1978) studied the occurrence of some acid phosphatases and dehydrogenases in *G. mosseae*. Alkaline phosphatase was present in *G. mosseae* inoculated onion plant during growth phase (Gianinazzi and Gianninzzi-Pearson, 1979). Uptake and translocation of ^{32}P labelled phosphate to onion roots by endomycorrhizal fungi is demonstrated by Hattingh *et al.* (1973). Gianninazzi and Gianninzzi-Pearson (1979) have observed a significant acid phosphatase activity in little vacuolated, immature terminal arbuscule. Acid and alkaline phosphatase activity were increased in mycorrhizal plants as compared with non-mycorrhizal plants. However *Gig. margarita* inoculated plants displayed significantly high in acid and alkaline phosphatase activity as compared with *G. mosseae* and *G. aggregatum* (Ramesh *et al.*, 2000).

NITROGENASE

In the symbiosis of *A. incana* with *Frankia* sp., the nitrogenase activity measured on intact plants is easily disturbed on addition of ammonium to the root system (Huss-Danell *et al.*, 1982). The attempts to measure nitrogenase activity in root nodule homogenates of *Frankia* - symbioses show initially very low activities (Sloger and Silver, 1965). Later works (Akkermans *et al.*, 1977; Benson *et al.*, 1979; Van Straten *et al.*, 1977) show that nitrogenase activity can be measured if the whole procedure is done under strictly anaerobic condition. There is now convincing evidence that the nitrogen fixing activity is localized within the vesicles of the *Frankia* endophyte. However, a strong correlation exists between the presence of these structures and the ability of nodules to reduce acetylene to ethylene (Lechevalier *et al.*, 1982; Schwintzer *et al.*, 1982). There is evidence that actinorhizal nodules contain a pigment analogous to the leghaemoglobin which protects the nitrogenase in legume nodules. Torrey and Callaham (1982) have suggested that the lipid layers in the vesicle envelope could similarly create a barrier for gas exchange and thus protect the nitrogenase from the external oxygen. The vesicle has been identified as the site of nitrogenase in cultured *Frankia* (Tjepkema *et al.*,1980). *Frankia* can adapt to a wide range of oxygen levels and retain active nitrogenase layers laid down in the vesicle envelope (Parsons *et al.*, 1987). In accordance with observations on vesicles cluster in root nodules (Van Straten *et al.*, 1977), vesicles are the proposed site for nitrogen fixation (Tjepkema *et al.*, 1981; Noridge and Benson, 1986). The necessary protection of nitrogenase against oxygen is probably provided by the vesicles structures (Torrey and Callaham, 1982; Murry *et al.*, 1984).

HORMONES

The phytohormones play a pivotal role in various developmental processes in the plants. The mycorrhizal symbiosis now being recognized as also involved the phytohormones metabolism (Gianinazzi-pearson, 1996; Barker and Tagu, 2000). The increase in the vasculature, greater lignifications of the style, early flowering and other reproductive and anatomical changes observed in the arbuscular mycorrhizal plants may all be due to the altered hormonal balance (Krishna *et al.,* 1981). The developmental regulation of the mycorrhizae may be achieved through modification of endogenous phytohormone levels, or through the regulation of the phytohormone receptors in the fungus and/or the host (Bee and Anderson, 1998). The effect growth regulators as mediators of mycorrhizal sink strength may be considered, because their balance changes the symbiotic or parasitic association (Lewis, 1975). Auxin, gibberellin and cytokinin increased in AM infected plants (Allen*et al.*, 1980; Raman *et al.,* 1994). Endogenous modifications of phytohormone levels, associated with AM infection have also been described in specific host-fungus associations (Allen *et al.,* 1980). Dugassa *et al.* (1996) have been detected the hormonal balance in mycorrhizal plants at various symbiotic storage in which morphological differences between mycorrhizal and non-mycorrhizal plants.

INDOLE ACETIC ACID

Chromatography has been used by Madej and Haggblom (1985) to detect IAA in *P. tinctorius* and 3 species of *Suillus.* Gopinathan and Raman (1992) analyzed and identified auxin by GC-Mass in ectomycorrhizal fungi. Indole compounds secreted by *Frankia* are reported by Berry *et al.* (1989) and Hensen and Wheeler (1976). Phillips and Torrey (1970) have demonstrated the production of IAA, cytokinin and gibberellin like (GA) substances by *Rhizobium* in culture (Williams *et al.,* 1982). Orchid mycorrhizal fungi belonging to the endomycorrhizal group of fungi secreted indole acetic acid (IAA) in pure culture (Pais *et al.*,1986). Gopinathan and Raman (1992) have observed higher amounts of IAA production by *Pisolithus tinctorius* and *Laccaria laccata* whereas the least amounts of IAA production by *Amanita muscaria* and *Rhizopogon lutelolus* in *in vitro* culture medium.

Silver *et al.* (1966) have found low IAA concentrations in the root nodules of both *Myrica* and *Casuarina* and have stated that it is due to an active IAA oxidizing system in the root nodules. It is suggested that negative geotropic response of the roots associated with the nodules of *Myrica* and *Casuarina* can be explained by this activity producing a sub optimal IAA concentration. Dullart (1970 a,b) estimated spectofluorometrically, in the acid ether soluble fraction of methanol extracts of *A. glutinosa* root nodules, indole – 3 acetic – acid (IAA) in a concentration of 0.6 to 1.0 mg/kg fresh weight of nodules. Levels of IAA, cytokinin and gibberellin-like activity in actinomycetes and leguminous nodules are higher than in the roots (Newcomb *et al.*, 1977). There have been frequent speculations about the nodule auxins and cytokinins (Rodriguez – Barrueco, 1968; Becking, 1975). Berry *et al.* (1989) identified 5 indole compounds secreted by *Frankia* under defined cultural conditions by using gas chromatography and mass spectrometry techniques.

GIBBERLLIC ACID

Increase in GA level in shoots of AM plants over that of non-mycorrhizal plants is observed by Allen *et al.* (1982). *Gl. mosseae* is known to produce GA like substances (Barea and Azcon – Aguilar, 1982). Mycorrhizal inoculation is known to stimulate the ontogeny and delay the senescence of leaves mediated by altered hormonal level (Daft and Nicholson, 1969). Altered levels of gibberellin – like substances and abscisic acid in the host plant is reported by Allen *et al.* (1982). Dugassa *et al.* (1996) have been observed that Enhanced contents of gibberellin-like substances were found in mycorrhizal inoculated *Linum usitatissimum*. Raman and Elumalai, (1998) have reported the maximum amount of gibberellic acid in triple inoculation of *Frankia*, *G. fasciculatum* and *P. tinctorius* in both cladodes, roots and nodules of *C. equisetifolia*. Clapperton *et al.* (1985) have observed the GA_s levels in the roots of *Agropyron thrachycaulum* infected with a *Glomus* species have analysed by GC-MS found that infected roots had more GA_s than control roots. Shaul-Keinan *et al.* (2002) have reported gibberellins concentrations by GC-MS analysis in tobacco roots inoculated with mycorrhizae were significantly increased than non-mycorrhizal control.

Most significantly mycorrhizal colonization increased root concentrations of GA_1, which is the major biologically active gibberellin it is known to be involved in the plant growth regulation (Phinney, 1984; Ross *et al.*, 1993; Spray *et al.*, 1996) and increased GA_1 concentrations have been correlated with increased plant growth (Grunzweig *et al.*, 2000). An AM symbiosis could induced and increase in gibberellin production as been suggested to be involved in strengthening the carbohydrate sink of the fungi expression of sugar-related enzymes and the secretion pathway (Bee and Anderson, 1998). It is therefore possible that the higher gibberellin concentrations observed in our AM-treated plants indicate mycorrhizal activity related to the fungal carbohydrate requirements. Dugassa *et al.*

(1996) observed that the shoot of *Linum usitatissimum* growth was already improved by AM fungi therefore possible to increased gibberellin-like activity might have been a consequence of large plant size rather than a symbiosis-specific signal.

CYTOKININ

The observed low level of cytokinin in the non-mycorrhizal plants is due to the deficiency of minerals like P and K. According to the observations of Allen *et al.* (1980), the mycorrhizal infection increased cytokinin in both roots and shoots to a level higher than those found in the non-mycorrhizal plants. The enhancement of cytokinin production in mycorrhizal *C. aurantium* was associated with AM infection. *Gl. mosseae* is known to produce cytokinin *in vitro* during germination and growth (Barea and Azcon – Aguilar, 1982). The fungus may produce cytokinin in the roots. The higher cytokinin level may be due to the inhibition of cytokinin degradation by certain inhibitors produced by the fungus or the plant itself after being stimulated by the fungus (Allen *et al.*, 1980). The increased cytokinin levels through mycorrhizal association may facilitate further infection (Azcon *et al.*, 1978). The significantly greater cytokinin activity in the roots of AM fungal inoculated *C. jambhiri* seedlings is consistent with the production of zeatin and zeatin riboside by mycorrhizal fungi *in vitro* (Crafts and Miller, 1974; Barea and Azcon – Aguilar, 1982). Allen *et al.* (1980, 1982) have reported the increase of cytokinin and gibberllic acid contents in the plant tissues associated with the infection of AM fungi. Isolation and identification of endogenous cytokinins from mycorrhizal infected *Medicago sativa* is reported by Hashizume *et al.* (1985). Baas and Kuiper, 1989; Danneberg *et al.*, 1992 have reported the increased levels of cytokinin are found in mycorrhizaql infected *Plantago major* ssp. *Pleiosperma* and *zea mays*. Raman and Elumalai, (1996) have detected the significant amount of cytokinin in *Frankia* inoculated root nodules of *C. equisetifolia*. Shaul-Keinan *et al.*, 2002 have reported the tobacco plants under low P concentrations, the total cytokinins were found at significantly higher concentrations in mycorrhizal roots and leaves than in non-mycorrhizal controls. It is also important to note that the higher cytokinins levels in leaves of mycorrhizal plants could be attributed to translocation from the roots, where the synthesis of the hormone may be stimulated either by the plant or the AM fungus, rather than *de nova* synthesis in the leaves (Dixon *et al.*, 1988). Shaul-Keinan *et al.* (2002) have reported the cytokinin compounds (iPA-like compounds) of tobacco roots and leaves are accumulated high concentration in mycorrhizal inoculated plants than non-mycorrhizal control. Differences between mycorrhizal and non-mycorrhizal plants in total ZR-like compounds (cytokinin compound) have been reported (Ginzberg *et al.*, 1998; Druge and Schonbeck, 1992). Mycorrhizal and P nutrition have been shown to induce the accumulation of cytokinin molecules in roots and leaves (Baas and Kuiper, 1989; Torelli *et al.*, 2000).

Cytokinin undoubtedly play a role in the nodulation process in legumes and non-legumes. Cytokinin secretion by *Frankia* sp – HF PAr 13 in defined culture medium is studied by Berry *et al.* (1989). The hormonal involvement in ectomycorrhizal development is reviewed by Harley and Smith (1983). Zeatin and Zeatin ribonucleoside have been isolated from ectomycorrhizal fungi (Crafts and Miller, 1974). Increased cytokinin via mycorrhiza may trigger increase root infection levels (Azcon *et al.*,1978).

ELEMENTS ANALYSIS

Physiological adaptation such as selective uptake and exclusion of elements by roots can influence the accumulation of minerals in tissues. Garten (1978) maintained that nutrient levels in tissues, which reflect on differential uptake, translocation and storage reveal more about a species mineral element niche than soil nutrient levels. Mackkum *et al.* (1993) have found the elemental difference between *Myrica* and *Chamaedaphne*. Small (1972) has analyzed leaf blades of both species from two bogs in Ontario for Al, Mn, N and P. Langille and Maclean (1976) have analyzed leaves and twigs from two region of Nova Scotia for Ca, Co, Cu, Mn, Mo, P and Z. Macronutrients such as Mg and K, micronutrients such as B and Fe are nonessential but beneficial elements such as Na, Ni and Se serve important functions in plants (Marschner,1986) and are of interest when considering the overall nutrients status of different species. Information is also lacking on the allocation of elements between foliage and twigs in *Myrica* and *Chamaedaphne*. The quantity of foliar N resorbed in the fall can depends on site fertility and plant species, sites low in available N tend to have proportionately more N resorption from tree foliage in the fall transites than the site high in available N (Stachurski and Zimka, 1975). Minerals like N, P, K, Ca, S, Zn, Cu and Sr are absorbed from the soil by arbuscular mycorrhizal fungi and are translocated to the host plants (Mosse, 1957). Earlier studies (Baylis, 1967; Sanders and Tinker, 1971; Marschener and Dell, 1994; Fidelibus *et al.*, 2001) have revealed that mycorrhizal roots absorbed phosphate from low P status at a greater rate per unit length of root than non-mycorrhizal roots.

Since heavy metal uptake and tolerance depend on both plant and soil factors including soil microbes, we required information on interaction between plant root and their symbionts such as arbuscular mycorrhizal fungi and nitrogen fixing microbes. Mycorrhizae have also been reported in plants growing on heavy metal contaminated sites (Shetty *et al.*, 1995; Weissenhorn and Leyval, 1995; Pawlowski *et al.*, 1997; Chaudhry *et al.*, 1998, 1999) indicating these fungi have evolved a heavy metal tolerance. The mycorrhizal fungi are considered as an important tool in afforestation and rehabitation of degraded lands especially in mine spoils, eroded sites and polluted wastelands (Raman and Sambandan, 1998). Plants growing on the tannery effluent polluted soils were mycorrhizal and that their mycorrhizosphere contained AM fungal spores (Khan, 1999a; 1999b; Khan and Bari, 1997). Mycorrhizal infection may increase the uptake of Copper (Gildon and Tinker ,1983a,b), Zinc (Lambert *et al.*, 1979), Nickel (Killham and Firestone, 1983), Chloride and Sulphate (Buwalda *et al.*,1983). AM fungal isolates differ in their effect on heavy metal uptake by plants (Leyval *et al.*, 1997). Sambandan *et al.* (1992) reported AM association in 16 plant species out of 18 plant species in th heavy metal polluted soils of Tamil Nadu. Raman *et al.* (1993) investigated AM association in 14 plant species in magnesite mine spoils in Salem, Tamil Nadu. Thirteen AM fungal species were identified. *Gl. fasciculatum* and *Gig. gigantea* were found common in the magnesite mine spoil. Khan *et al.* (2000) have reported the role of plants, mycorrhizae and phytochelators in heavy metal contaminated land remediation. Similarly, Pawlowski *et al.* (2000) reported the effects of metal phytoextraction practices on the indigenous community of arbuscular mycorrhizal fungi at a metal contaminated landfill.

SALINITY EFFECT

Salinity is one of the major abiotic stress that reduces plant survival and development. Saline soils occupy approximately 7% of the earth's land surface and agronomic and forest crop production of the sites is relatively low (Barrett –Lenard *et al.*, 1986). Properties of saline soil which inhibit or reduce plant survival and development include unfavorable pH, imbalance of essential cations and anions and altered the soil structure and texture which reduce aeration and water holding capacity (Abrol and Sandhu, 1985; Bettenay, 1986).

Some *Casuarina* species are known to grow in saline coastal soils or near brackish waters, salt lakes and swamps (Doram and Hall, 1983). *C.equisetifolia* is one of the most salt- tolerant species (EL-Lakany and Luard, 1982). Differences in salt tolerance between species of *Casuarina* appear to depend on difference in ion uptake and distribution in individual organs of the plants (Clements *et al.*, 1983; Aswathapa and Bachelard, 1986). The highly salt tolerant species accumulate little Na^+ and Cl^- in their shoots and the concentrations of Na^+ and Cl^- decrease from old to young growing needles (Aswathapa and Bachelard, 1986). In actinorhizal symbionts, high salinity could affect nodule development, N_2 fixation and growth of the host plant. Reddell *et al.* (1986) have found *C. obsea* at sites with salinity levels up to 28 mg Cl^- (g soils)$^{-1}$ but in a glasshouse study, growth, nitrogen fixation and nodule weight of *C. obsea* declined when NaCl concentration is raised from 1.5 to 15 mg (g soil) $^{-1}$.

Ameliorating effect of *C. equisetifolia* on saline soils is demonstrated and screening for salinity and water logging tolerance in 5 *Casuarina* species is reported by Moezel *et al.* (1989). Nitrogen-fixation symbiosis in acid and saline zones and the application of it to Casuarinaceae was reported by Dommergues and Girgis (1993). The introduction of mycorrhizal fungi on some of these saline soil sites with unfavourable saline conditions may improve early plant survival and growth (Reddell *et al.*, 1986). Salinity on some symbiotic relationships (actinorhizal or legume *Rhizobium* symbiosis) has been assessed in a preliminary manner (Reddell *et al.*, 1986). Plants relying on symbiotic relationships for adequate mineral nutrition and water uptake may differ in salt tolerance, depending on the host and symbiont tolerance to soil salinity (Reddell *et al.*, 1986; Perry *et al.*, 1987; Marcar *et al.*, 1991).

Salt tolerance is important to the long term survival, reproduction and spread of mycorrhizal fungi in saline soils (Perry *et al.*, 1987). Certain plants that are known to be halophilic or tolerate high salinity levels are colonized by AMF (Sylvia and Williams, 1992). This was recently confirmed by Hildebrandt *et al.* (2001). Salt tolerance fungus show increased protein content. Saxena *et al.*, 1996 have reported the proteins have a role to play in salt tolerance. Torzilli, 1997 have reported the common trigger for the induction of heat shock protein seems to be denatured protein, which can be generated by both high temperatures and high salt concentration. The mechanisms of stress in fungal species, generally increased in most soluble amino acid (Smith *et al.*, 1990) but in others the increases are predominant in proline and or glutamate (Razak *et al.*, 1985). Gupta and Krishnamoorthy (1996) reported that the formation of arbuscular mycorrhizae is reduced by an increase in salt concentration but that plant growth is still improved by the symbiosis as compared with non-mycorrhizal plants growing under similar salt conditions.

In vitro halotolerance of *Frankia* isolates form *Allocasuarine, Casuarina,* and *Purshia* has reported (Shipton and Burggraaf, 1983) suggesting that host and endophyte has been co-selected for xerotolerance. Halotolerance may be important in the spread of noudlation within and between host plants and in the survival of free living *Frankia* in saline soils.

The symbiotic efficiency of partnership between the actinorhizal host and the *Frankia* strain seems to be defined broadly by the functional capacity of the host genotype, the *Frankia* genotype, and the environmental factors (Verghese and Misra, 2002). Though large gaps still exist in our understanding of the intricacies of this symbiotic process, a considerable progress has been made in recent years in several aspects of *Frankia* physiology, the physiology of the host and their molecular biology and genetics. Further advances in the subject may enable a better understanding of how this system functions. It would then be possible to envisage the possibility of a better utilization of the symbiosis to improve the production of actinorhizal species in forestry.

Acknowledgement

Thanks are due to Prof. N. Anand, Director, Centre for Advanced Studies in Botany, for laboratory facilities and encouragement.

References

Abou-El-Khair, K.S. 1993. Influence of cadmium on the growth of *Casuarina glauca, Eucalyptus camaldensis* and *Cuppressus sempervirens* seedlings grown in soil containing vesicular arbuscular mycorrhizal spores. *Alexandria J. Agri. Res.* **38:** 229-241.

Abrol, I.P. and Sandhu. S.S. 1985. Growth responses of *Eucalyptus tereticornis* and *Acacia nilotica* to selected methods of site preparation in a highly sodic soil. *Int. Tree crops J.* **3:** 171-183.

Akkermans, A.D.L. 1971. Nitrogen fixation and nodulation of *Alnus* and *Hippophae* under natural conditions. Ph.D., thesis, University of Leiden, The Netherlands.

Akkermans, A.D.L., Van Straten, J. and Rolofsen, W. 1977. Nitrogenase activity in nodule homogenates of *Alnus glutinosa*. In: *A comparison with the Rhizobium – Pea System*. In: *Recent developments in nitrogen fixation.* (Eds. Newton, W., Postgate, J.R. and Rodriguez – Barrueco, C.). pp. 591-603. Academic Press, London.

Akkermans, A.D.L. and Houwers, A. 1979. Symbiotic nitrogen fixers available for use in temperate forestry. In: *Symbiotic nitrogen fixation in the management of temperate forests.* (Eds. Gardon, J.C., Wheeler, C.T. and Perry, D.A.). pp. 23-36. Oregon State University, O.R. Coravallis.

Allen, M.F., Moore, T.S. and Christensen, M. 1980. Phytohormone change in *Bouteloua gracilis* infected by VAM fungi. I. Cytokinin increased in the host plant. *Can. J. Bot.* **58:** 371-374.

Allen, M.K., Smith, W.K. Moore T.S. and Christensen, M. 1981. Comparative water reactions and photosynthesis of mycorrhizal and non-mycorrhizal *Bouteloua gracilis* H.B.K.ex Stend. *New Phytol.* **88:** 683-693.

Allen, M.F., Moore, T.S. and Christensen, M. 1982. Phytohormone change in *Bouteloua gracilis* infected by VAM fungi. II. Altered levels of gibberellin - like substances and abscisic acid in the host plant. *Can. J. Bot.* **60:** 468-471.

Antibus, R.K., Sinsabaugh, R.L. and Linking, A.E. 1992. Phosphate activities and phosphorus uptake from inositol phosphate by ectomycorrhizal fungi. *Can. J. Bot.* **70:** 794-801.

Arines, J., Palma, J.M. and Vilarine, A. 1993. Comparison of protein patterns in non-mycorrhizal and vesicular-arbuscular mycorrhizal roots of red clover. *New Phytol.* **123:** 763-768.

Arsenault, P. 1983. Renaturalisation, rapport (1981). Societe d' Energie de la Baie james Direction de 1 Environment, Montreal, 96 pp.

Arsenault, P. 1984. Renaturalisation, rapport. (1982). Societe d' Energie de la Baie james Direction de l'environment, Tomes 1 et 2 montreal, 257 pp.

Aswathappa, N. and Bachelard, E.P. 1986. Ion regulation in the organs of *Casuarina* species differing in salt tolerance. *Aust. J. plant physiol.* **13:** 533-545.

Azcon, R. and Ocampo, J.A. 1981. Factors effecting the infection and mycorrhizal dependency of thirteen wheat cultivars. *New Phytol.* **87:** 667-685.

Azcon, R.L., Azcon, G., Aquilar, D.E. and Barea, J.M. 1978. Effects of plant hormones present in bacterial cultures on the formation and responses to VA mycorrhiza. *New Phytol.* **80:** 359-364.

Baas, R. and Kuiper, D. 1989. Effect of vesicular-arbuscular mycorrhizal infection and phosphate on *Plantago major* ssp. *pleiosperma* in relation to internal cytokinin concentrations. *Plant Physiol.* **76:** 211-215.

Backer, D.D. and Schwintzer, C.R. 1990. Introduction. In: *The biology of Frankia and actinorhizal plants.* (Eds. Schwintzer, C.T. and Tjepkema, J.D.). pp. 1-13. Academic Press, New York.

Baker, D., Torrey, J.G. and Kidd, G.H. 1979. Isolation by sucrose-density fractionation and cultivation *in vitro* of actinomycetes from Nitrogen fixing root nodules. *Nature (London).* **281:** 76-78.

Baker, D. and O'Keefe, D. 1984. A modified sucrose fractionation procedure for the isolation of *Frankia* from actinorhizal root nodules and soil samples. *Plant Soil* **78:** 23-28.

Bamber, R.K., Mulette, K. and MacKowski, C. 1980. Mycorrhizal studies. Research Report: *1977 - 1978, Forest commission of New South Wales,* pp. 70-72. Sydney, Australia.

Barea, J.M. and Azcon-Aguilar, C. 1982. Production of plant growth regulating substances of VAM fungus *Glomus mosseae*. *Appl. Environ. Microbiol.* **43:** 810-813.

Barlow, B.A. 1983. Casuarinas – A taxonomic and biogeographic review. In: *"Casuarina ecology, Management and Utilisation".* (Eds. Midgley, S.J., Turnbull, J.W. and Johnson R.D.). pp. 10-18. CSIRO-Melbourne.

Barrett Lennard, E.G., Malcom, C.V., Stren, W.R. and Wilkins, S.M. 1986. Forage and fuel production from salt- affected wasteland. In: *Proceedings of a Research for Development Seminar,* 19-27 May 1984, at Cunderdin, Western Australia, Elsevier, Amsterdam, 459 pp.

Barta, A.L. 1979. Effect of nitrogen supply on photosynthate accumulation, and acetylene reduction in bird foot trefoil. *Crop Sci.* **19:** 715-718.

Barker, S.J. and Tagu, D. 2000. The roles of auxins and cytokinins in mycorrhizal symbiosis. *J. Plant Growth Regulation* **19:** 144-154.

Bauer, P., Poirier, S., Ratet, P. and Kondorosi, A. 1997. MsENOD12 expression is linked to meristematic activity during development of indeterminate and determinate nodules and roots. *Mol. Plant Microbe Interact.* **10:** 39-49.

Baylis, G.T.S. 1967. Experiments on the ecological significance of phycomycetous mycorrhizas. *New Phytol.* **66:** 231-243.

Bayne, H.G., Brown, M.S. and Bethlenfalvay. G.J. 1984. Defoliation effects on mycorrhizal colonization, nitrogen fixation, photosynthesis in the *Glycine – Glomus - Rhizobium* symbiosis. *Physiol. Plant* **62:** 576-580.

Becard, G. and Piche, Y. 1989. Fungal growth stimulation by CO_2 and root exudates in vesicular arbuscular mycorrhizal symbiosis. *Appl. Environ. Microbiol.* **55:** 2320-2325.

Becking, J.H., De Boer, W.E. and Houwink, A.L. 1964. Electron microscopy of the endophyte of *Alnus glutinosa*. Antionie Van Leeuwenhoek. *J. Microbiol. Ser.* **30:** 343-376.

Becking, J.H. 1970. Frankiaceae fam nov. (Actinomycetales) with one new combination and six new species of the genus *Frankia.* Brunchorst, 1886, 174. *Int . J. Syst. Bateriol.* **20:** 201-220.

Becking, J.H. 1974. Family III. Frankiaceae. In: *"Bergey's manual of Determinative Bacteriology"* 8th ed. (Eds. Buchanan, R.E. and Gibbon, N.E.). pp. 701-706. Williams and Williams Co., Baltimore, U.S.A.

Becking, J.H. 1975. Root nodules in non-legumes, In: *Development and function of roots.* (Eds. Torrey, J.G. and Clarkson, D.T.). pp. 508-566. Academic press, London.

Bee, K.A. and Anderson, A.J. 1998. Regulation of arbuscule formation by carbon in the plant. *Plant* **16:** 523-530.

Beilby, J.P. and Kidby, D.K. 1980. Biochemistry of germinated and ungerminated spores of the vesicular arbuscular mycorrhizal fungus *Glomus caledonium*. Changes in neutral and polar lipids. *J. Lipid. Res.* **21:** 739-750.

Benson, D. 1982. Isolation of *Frankia* strains from alder actinorhizal root nodules. *Appl. Environ. Microbiol.* **44:** 461-465.

Benson, D.R., Arp, D.J. and Burris, R.H. 1979. Cell free nitrogenase and hydrogenase from actinorrhizal root nodules. *Science* **205:** 688-689.

Berry, A. and Torrey, J.G. 1979. Isolation and characterization *in vivo* and *in vitro* of an actinomycetous endophyte from *Alnus rubra* Bong. In: *Symbiotic nitrogen fixation in the management of temperate forests.* (Eds. Gordon, J.C. Wheeler, C.T. and Perry, D.A.). pp. 69-83. Forest Res. Lab, Oregon State Univ. Corvallis.

Berry, A.M., Kahn, R.K.S. and Booth, M.C. 1989. Identification of indole compounds secreted by *Frankia*. HF PAr 13 in defined culture medium. *Plant Soil* **118:** 205-209.

Bettenary, E. 1986. Salt affected soils in Australia. *Reclam. Reveg. Res.* **5:** 167-179.

Bevege, D.L., Bowen, G.D. and Skinner, M.F. 1975. Comparative carbohydrate physiology of ecto and endomycorrhizas. In: *Endomycorrhizas,* (Eds. Sanders, E.F. Mosse, B. and Tinker, P.B.). pp. 149-174. Academic Press, London.

Bond, G. 1957. The development and the significance of the root nodules of *Casuarina. Ann. Bot.* (London). **21:** 373-380.

Bond, G. 1976. The results of the IBP survey of root nodule formation in non-leguminous angiosperms. In: *Symbiotic nitrogen fixation in plants.* (Ed. Nutman, P.S.). pp. 443-476. Cambridge, London.

Bonfante, P. 1984. Anatomy and morphology. In: *VA mycorrhizae.* (Eds. Powell, C.L. and Bagyaraj, D.J.). pp 5-33. CRC press, Boca Raton. U.S.A.

Bougher, N.L., Dell, B. and Malajczuk, N. 1990 a. Growth and root structure of *Eucalyptus, Allocasuarina* and *Casuarina* seedlings inoculated with ectomycorrhizal fungi. In: *Proceedings of the 8th North American Conference on mycorrhizae. 5-8 September (1990).* (Eds. Allen, M.F. and Williams, S.E.). University of Wyoming, Jackson, Wyoming. 32 pp.

Bougher, N.L., Grove, T.S. and Malajczuk, N. 1990 b. Growth and phosphorus acquisition of Karri (*Eucalyptus diversicolor*) seedlings inoculated with ectomycorrhizal fungi in relation to phosphorus supply. *New Phytol.* **114:** 77-85.

Brennan, P.J. and Losel, D.M. 1978. Physiology of fungal lipids. Adv. Microb. Physiol. **17:** 47.

Brown, M.S. and Bethlenfalvay, G.T. 1987. *Glycine - Glomus - Rhizobium* symbiosis. IV. Interaction between the mycorrhizal and nitrogen - fixing endophytes. *Plant Cell Environ.* **10:** 607-612.

Brunchorst, J. 1886. Uber emige wurzelansch wellengen, bessouderes diejenigen von *Alnus* den Elaeagnaceae *Unters Bot. Inst. Tusign* **2:** 150-177.

Burggraaf, A.J.B and Shipton, W.A. 1983. Studies on the growth of *Frankia* isolates in relation to infectivity and nitrogen fixation (Acetylene reduction). *Can. J. Bot.* **61:** 2774-2782.

Burleigh, S. and Torrey, J.G. 1990. Effectiveness of different *Frankia* cell types as inocula for the actinorhizal plant *Casuarina*. *Appl. Environ. Microbiol.* **56:** 2565-2567.

Buwalda, J.G., Stribley, D.P. and Tinker, P.B. 1983. Increased uptake of anions by plants with VAM. *Plant Soil* **71:** 463-467.

Callaham, D. and Torrey, T.G. 1977. Pre - nodule formation and primary nodule development in roots of *Camptonia* (Myricaceae). *Can. J. Bot.* **55:** 2306-2318.

Callaham, D., Newcomb, W., Torrey, T.G. and Peterson, R.L. 1979. Root hair infection in actinomycete-induced root nodule initiation in *Casuarina myrica* and *Camptonia*. *Bot. Gaz.* (Chicago) **140** (suppl): S1-S9.

Carling, D.E., Richle, W.G., Brown, M.F. and Johnson, D.R. 1978. Effects of VAM fungus on nitrate reductase and nitrogenase activities in nodulating and non-nodulating soybeans. *Phytopathol.* **68:** 1590-1596.

Cervamtes, E. and Rodriguez-Barrueco, C. 1992. Relationships between the mycorrhizal and actinorhizal symbiosis in non-legumes. In: *Methods in Microbiology.* Vol. 24. (Eds. Norris, J.R., Read, D.J. and Verma, A.K.). pp. 417-432. Academic press, London.

Chatarpaul, L., Chakravarthy, P. and Subramanian, P. 1989. Studies in tetrapartite symbiosis. I. Role of ecto and endomycorrhizal fungi and *Frankia* on the growth performance of *Alnus incana*. *Plant Soil* **118:** 145-150.

Chaudhary, N.R. and Ram, T. 1961. Common methods of vegetative propagation as useful in forestry. *Proc. Tenth. Silvicult. Conf.* pp. 214-223. Dehra Dun. India.

Chaudhary, A.H. and Mirza, M.S. 1987. Isolation and characterization of *Frankia* from nodules of actinorhizal plants of Pakistan. *Physiol. Plant* **70:** 255-258.

Chaudhary, A.H. and Mirza, M.S. 1988. Nitrogen fixing root nodule symbiosis in non-leguminous angiosperms *Alnus, Casuarina, Coriaria, Datisca* and *Elaeagnus* species of Pakistan. Managing soil resources: *Proceedings of first national congress of soil science*. pp. 140-150. Lahore.

Chaudhry, T.M., Hayes, W.J., Khan, A.G. and Khoo, C.S. 1998. Phytoremediation-focussing on accumulator plants that remediate metal contaminated soils. *Aus J. Ecol. Toxicol.* **4:** 37-51.

Chaudhry, T.M., Hill, L., Khan, A.G. and Koek, C. 1999. Colonization of iron and zinc-contaminated dumped filter cake waste by microbes, plants and associated mycorrhizae. In: *Remediation and Management of Degraded Lands.* (Eds. Wong, M.H., Wong, J.W.C. and Barker, A.J.M.). pp. 275-283. CRC Press, Boca Raton, Florida.

Clapperton, M.J., Koshioka, M. and Pharis, R.P. 1985. The effect of infection by a vesicular-arbuscular mycorrhizal fungus on the gibberellin content of slender wheat grass. *Plant Physiol.* **77:** 9-82.

Clements, J., Campbell, L.C. and Nurjsjah, S. 1983. Germination and mineral ion concentrations of *Casuarina* species under saline conditions. *Aust. J. Bot.* **31:** 1-9.

Cooper, K.M. 1976. A field survey of mycorrhizas in New Zealand. *N.Z.J. Bot.* **14:** 169.

Cooper, K.M. and Losel, D.M. 1978. Lipid physiology of vesicular arbuscular mycorrhizae. I. Composition of lipids in root of onion, clover and ryegrass. *New Phytol.* **80:** 143-151.

Cox, G. and Sanders, F.E. 1974. Ultrastructure of the host fungus interface in vesicular arbuscular mycorrhizae. *New Phytol.* **73:** 901-912.

Cox, G., Sanders, F., Tinker, P.B. and Wild, J.A. 1975. Ultrastructural evidence relating to host-endophyte transfer in vesicular arbuscular mycorrhizae. In: *Endomycorrhizas*, (Eds. Sanders, F.E. Mosse, B. and Tinker, P.B.). pp. 297-312. Academic Press, New York.

Coyne, P.D. 1973. Some aspects of autecology of casuarinas with particular reference to nitrogen fixation. Ph.D., Thesis, Australian National University, Canberra.

Crafts, C.B. and Miller, C.O. 1974. Determination and identification of cytokinins produced by mycorrhizal fungi. *Plant Physiol.* **54:** 586-588.

Daft, M.J. and Nicolson, T.H. 1969. Effect of *Endogone* mycorrhiza on plant growth. II. Influence of inoculum concentration on growth and infection in tomato. *New Phytol.* **68:** 953-963.

Danneberg, G., Latus, C., Zimmer, W., Hundshagen, B., Schneider-Poetsch, H.J. and Bothe, H. 1992. Influence of vesicular arbuscular mycorrhizae on phytohormone balance in maize (*Zea mays* L.). *J. Plant Physiol.* **141:** 33-39.

Dawson, J.O., Kowalski, D.G. and Dart, P.J. 1989. Variation with soil depth, topographic position and host species in the capacity of soils from an Australian locale to nodulate *Casuarina* and *Allocasuarina* seedlings. *Plant Soil* **118:** 1-11.

De La Cruz, R.E. 1990. Current status of nursery and field application of ectomycorrhizas in the Philippines. In: *VIII – NACOM.* (Eds. Allen, M.F. and Williams, S.E.). Philippines. 75 p.

Dell, B., Malajczuk, N., Bougher, N.L. and Thomson, G. 1994. Development and function of *Pisolithus* and *Scleroderma* ectomycorrhizas formed *in vitro* with *Allocasuarina, Casuarina* and *Eucalyptus*. *Mycorrhiza* **5:** 129-138.

Diem, H.G., Gauthier, D. and Dommergues, Y. 1982. Isolation of *Frankia* from nodules of *Casuarina equisetifolia*. *Can. J. Microbiol.* **28:** 526-530.

Diem, H.G. and Dommergues, Y.R. 1990. Current and potential uses and management of Casuarinaceae in the tropics and subtropics. In: *The biology of Frankia and Actinorhizal plants*. (Eds. Schwintzer, C.R. and Tjepkema, J.D.). pp. 317-342. Academic press, New York.

Diem, H.G. and Guathier, D. 1982. Effect de I' interaction endomycorrhizienne (*Glornus mosscae*) sur la nodulation et la croissance de *Casuarina equisetibolia*. *Acad. Sci. Paris.* **294:** 215-217.

Diop, T.A. 1996. Les mycorrhizas a vesicles et arbuscules. *J. Fac. Sci.* **2:** 49-64.

Diouf, D., Diop, T.A. and Ndoye, I. 2003. Actinorhizal, mycorrhizal and rhizobial symbioses: how much do we know? *African Journal of Biotechnology* **2:** 1-7.

Dixon, R.K., Garratt, H.E. and Fox, G.S. 1988. Carbohydrate relationships of *Citrus jambhiri* inoculated with *Glomus fasciculatum*. *J. Am. Soc. Hort. Sci.* **113:** 239-242.

Dodd, J.C., Burton, C.C., Burnus, R.G. and Jeffries, P. 1987. Phosphatase activity associated with rhizosphere of plants infected with vesicular arbuscular mycorrhizal fungi. New Phytol. **7:** 163-172.

Dommergues, Y.R. 1963. Evaluation du taux de fixation de I' azote dans un sol dunaire reboise en filao (*Casuarina equisetifolia*). *Agrochimica* **7:** 335-340.

Dommergues, Y.R., Diem, H.G., Guathier, D.L., Dreybus, B.L. and Cornet, F. 1984. Nitrogen fixing trees in tropics. Potentialities and limitations. In: *"Advances in Nitrogen fixation Research"*. (Eds. Veeger, C., Newton, W.E., Martinus Nijhoff, M. and Junk, W.). pp. 7-13. The Hague.

Dommergues, Y.R. and Girgis, M.G. 1993. Nitrogen fixing symbiosis in acid and saline zones: Application to Casuarinaceae. Physiology of trees and shrubs in arid and semi-arid zones. Seminar Paris – Nancy (France). 1990, March 20 – April 6 John Libbey Eurotext. pp. 183-194.

Doram, J.C. and Hall, N. 1983. Notes on fifteen Australian *Casuarina* species. In: *Casuarina ecology, Management and Utilisation* (Eds. Midgeley, S.J., Turnbull, J.W. and Johnson, R.D.). pp. 19-52. Melbourne.

Druge, U. and Schonbeck, F. 1992. Effect of vesicular arbuscular mycorrhizal infection on transpiration, photosynthesis and growth of flax (*Linum usitatissimum*. L.) in relation to cytokinin levels. *J. Plant Physiol.* **141:** 40-48.

Duddridge, J.A. 1986. The development and ultrastructure of ectomycorrhizas. IV Compatible and incompatible interaction between *Suillus grevillei* (Klotzsch) Sing and a number of ectomycorrhizal hosts *in vitro* in the presence of exogenous carbohydrate. *New Phytol.* **103:** 457-464.

Dugassa, G.D., Van Alten, H. and Schonbeck, F. 1996. Effect of arbuscular mycorrhizae (AM) on health of *Linum usitatissimum* L. infected by fungal pathogens. *Plant Soil* **185:** 173-182.

Duhoux, E., Rinaudo, G., Diem, H.G., Auguy, F., Fernandez, D., Bogusz, D., Franche, C., Dommergues, Y. and Huguenin, B. 2001. Angiosperm *Gymnostoma* trees produce root nodules colonized by arbuscular mycorrhizal fungi related to *Glomus*. *New Phytol.* **149:** 115-125.

Dullart, J. 1970 a. The bioproduction of indole-3-acetic acid and related compounds in root nodules and roots of *Lupinus luteus* L. and by its rhizobial symbionts. *Acta. Bot. Nearl.* **19:** 573-615.

Dullart, J. 1970 b. The auxin content of root nodule and roots of *Alnus glutinosa*. *J. Exp. Bot.* **21:** 975-984.

Dumas, G.E., Guillaume, P., Tahiri, A.A., Gianinazzi-Pearson, V. and Gianinazzi, S. 1994. Changes in polypeptide patterns in tobacco roots by *Glomus* species. *Mycorrhiza* **4:** 215-221.

El-Lakany, M.H. and Luard, E.J. 1982. Comparative salt tolerance of selected *Casuarina* species. *Aust. For. Res.* **13:** 11-20.

Ferguson, T.P. and Bond, G. 1953. Observation on the formation and function of nodules of *Alnus glutinosa* (L.). *Ann. Bot.* **17:** 175-188.

Fessenden, R.J. 1979. Use of actinorhizal plants for land reclamation and amenity planting in the USA and Canada. In: *Symbiotic Nitrogen fixation in the Management of temperate Forests.* (Eds. Gordon, J.C., Wheeler, C.T. and Perry, D.A.). pp. 403-419. Oregon State University, Coravallis, OR.

Fidelibus, M.W., Martin, C.A. and Stuz, J.C. 2001. Geographic isolates of *Glomus* increased root growth and whole-plant transpiration of Citrus seedlings grown with high phosphorus. *Mycorrhiza* **10:** 231-236.

Flemming, A.J., Williams, E.R. and Turnbull, J.W. 1988. Growth and nodulation of provenances of *Casuarina equisetifolia* inoculated with a range of *Frankia* sources. *Aust. J. Bot.* **36:** 171-181.

Fortin, J.A., Lalonde, M. and Perinet, P. 1981. Inoculation of alder with an *in vitro* cultivated and efficient fixing microorganism at the industrial scale. *Proceedings of the third Bio-energy R and D.* Seminar, March, 24 - 25, 1981. pp. 42-46. CNRC, 19515, Ottawa, Canada.

Franche, C., Diouf, D., Le, Q.V., Bogusz, D., N'Diaya., A. Gherbi., H. Gobe., C. and Duhoux, E. 1997. Genetic transformation of the actinorhizal tree *Allocasuarina verticillata* by *Agrobacterium tumefaciens*. *Plant J.* **11:** 897-904.

Franche, C., Laplaze, L., Duhoux, E. and Bougsz, D. 1998. Actinorhizal symbiosis: Recent advances in plant molecular and genetic transformation studies. *Crit. Rev. Plant Sci.* **17:** 1-28.

Ganesh, G., Misra, A.K., Chapelon, C. and Normand, P. 1994. Morphological and molecular characterization of *Frankia* sp. isolates from nodules of *Alnus nepalensis* Don. *Arch. Microbiol.* **161:** 152-155.

Gardes, M. and Lalonde, M. 1987. Identification of subgrouping of *Frankia* strains using SDS gel electrophoresis. *Physiol. Plant.* **70:** 237-244.

Gardner, I.C. 1986. Mycorrhizal status of actinorhizal plants. MIRCEN. *J. Appl. Microbiol. Biotech.* **2:** 147-160.

Garnas, P., Niebel F de, C., Lescure, N. and Cullimore, J.V. 1996. Use of a subtractive hybridization approach to identify new *Medicago truncatula* genes induced during root nodule development. *Mol. Plant Microbe Interact.* **9:** 233-242.

Garten, C.T. 1978. Multivariate perspectives on the ecology of plant minerals element composition. *Am. Nat.* **112:** 533-544.

Gauthier, D., Diem, H.G. and Dommergues, Y. 1983. Preliminary results of research on *Frankia* and endomycorrhiza associated with *Casuarina equisetifolia.* In: *Casuarina ecology, Management and utilization.* (Eds. Midgley, S.J., Turnbull, J.W. and Johnson, R.D.). pp. 211-217. CSIRO, Melbourne, Australia.

Gianinazzi, S. and Gianinazzi-Pearson, V. 1979. Enzymatic studies on the metabolism of vesicular arbuscular mycorrhiza. III. Ultrastructural localization of acid and alkaline phosphatase in onion roots infected by *Glomus mosseae* Nicol. & Gerd. *New Phytol.* **82:** 127-132.

Gianinazzi-pearson, V. 1996. Plant cell responses to arbuscular mycorrhizal fungi: getting to the roots of the symbiosis. *Plant cell* **8:** 1871-1883.

Gildon, A. and Tinker, P.B. 1983 a. Interactions of vesicular arbuscular mycorrhizal infection and heavy metals in plants. I. The effect of heavy metals on the development of vesicular mycorrhizas. *New Phytol.* **95:** 247-261.

Gildon, A. and Tinker, P.B. 1983 b. Interactions of vesicular arbuscular mycorrhizal infection and heavy metals in plants. II. The effect of infections on uptake of copper. *New Phytol.* **95:** 262-268.

Ginzberg, I., David, R., Shaul, O., Elad, Y., Wininger, S., Ben-Dor, B., Badani, H., Fang, Y., van Rhijn, P., Li, Y., Hirsch, A.M. and Kapulnik, Y. 1998. *Glomus intraradices* colonization regulates gene expression in tobacco roots. *Symbiosis* **25:** 145-157.

Gopinathan, S. and Raman, N. 1992. Indole-3-acetic acid production in ectomycorrhizal fungi. *Ind. J. Exp. Biol.* **30:** 140-143.

Graham, J.H., Leonard, R.T. and Menge, J.A. 1981. Membrane-mediated decrease in root exudation responsible for phosphorus inhibition of vesicular arbuscular mycorrhiza formation. *Plant Physiol.* **68:** 548-552.

Greenall, J.M. 1963. The mycorrhizal endophytes of *Griselinia littoralis* (Cornaceae). *N.Z.J. Bot.* **1:** 389.

Grunzweig, J.M., Rabinowitch, H.D., Katan, J., Wodner, M. and Ben-Tal, Y. 2000. Involvement of endogenous gibberellins in the regulation of increased tomato shoot growth in solarized soil. *Plant Growth Regulation* **30:** 233-239.

Guan, C., Pawlowski, K. and Bisseling, T. 1998. Interaction between *Frankia* and actinorhizal plant. In: *Subcellular Biochemistry, Vol. 29, Plant - Microbe interactions.* (Eds. Biswas, B.B. and Das, H.K.) Plenum Press, New York. 165 pp.

Gullan, P.K. 1975. Vegetation of Crassbourne, Ph.D., Thesis, Monesh University, Victoria, Australia.

Gupta, R. and Krishnamoorthy, K.V. 1996. Response of mycorrhizal and non-mycorrhizal *Arachis hypogaea* to NaCl and acid stress. *Mycorrhiza* **6:** 145-149.

Hall, R.B. and Maynard, C.A. 1979. Consideration in the genetic improvement of alder. In: *Symbiotic N_2 fixation in the management of temperate forests.* (Eds. Gordon, J.C. Wheeler, C.T. and Perry, D.A.). pp. 322-344. Oregon State University, Forest Res. Lab. Corrallis.

Harley, J.L. and Lewis, D.H. 1969. The physiology of ectotrophic mycorrhizas. *Adv. Microb. Physiol.* **3:** 53.

Harley, J.L. 1970. The importance of microorganisms to colonizing plants. *Trans. Bot. Soc.* (Edinburgh). **41:** 65-70.

Harley, J.L. and Smith, S.E. 1983. *Mycorrhizal symbiosis.* Academic press, London and New York. 483 pp.

Harris, D., Pacovsky, R.S. and Paul, E.A. 1985. Carbon economy of soybean *Rhizobium – Glomus* associations. *New Phytol.* **101:** 427-440.

Hashizume, T., Kawakita, K. Senda, S. and Sugiyama, T. 1985. Isolation and identification of endogenous cytokinins from alfalfa (*Medicago sativa* L.) *Agric. Biol. Chem.* **49:** 3481-3484.

Hattingh, M.J., Gray, L.F. and Gerdemann, J.W. 1973. Uptake and translocation of ^{32}P labelled phosphate to onion roots by endomycorrhizal fungi. *Soil Science* **116:** 383-387.

Hayman, D.S. 1975. The occurrence of mycorrhizas in field crops as affected by soil fertility. In: *Endomycorrhizas,* (Eds. Sanders, F.E. Mosse, B. and Tinker, P.B.). pp. 495-509. Academic press, London.

Henson, I.E. and Wheeler, C.T. 1976. Hormones in plants bearing nitrogen fixing root nodules: Patial characterization of cytokinins from root nodules of *Alnus glutinosa,* L. Gaertn. *J. Exp. Bot.* **28:** 1076-1086.

Hildebrandt, U., Janetta, K., Ouziad, F., Renne, B., Nawrath, K. and Bothe, H. 2001. Arbuscular mycorrhizal colonization of halophytes in central European salt marshes. *Mycorrhiza* **10:** 175-183.

Hirrel, M.C. and Gerdemann, J.W. 1980. Improved growth of onion and bell pepper in saline soils by two vesicular arbuscular mycorrhizal fungi. *J. Soil Sci. Soc. Am.* **44:** 654-655.

Ho, I. and Trappe, J.M. 1973. Translocation of ^{14}C from *Festuca* plants to their endomycorrhizal fungi. *Nature* **244:** 30-31.

Ho, I. and Zak, B. 1979. Acid phosphatase activity of six ectomycorrhizal fungi. *Can. J. Bot.* **57:** 1203-1205.

Ho, I. 1987. Enzyme activity and phytohormone production of a mycorrhizal fungus, *Laccaria laccata*. *Can. J. For. Res.* **17:** 855-858.

Hoffland, E., Findenegg, G.R. and Nelemans, J.A. 1989. Solubilization of rock phosphate by rape: II. Local root exudation of organic acids as a response to P-starvation. *Plant Soil* **113:** 161-165.

Huss-Dannel, K., Roelofsen, W., Akkermans, A.D.L. and Meijer, P. 1982. Carbon metabolism of *Frankia* sp. in root nodules of *Alnus glutinosa* and *Hippophae rhamnoides*. *Physiol. Plant.* **54:** 461-466.

Khan, A.G. 1974. The occurrence of mycorrhiza in halophytes, hydrophytes and xerophytes and of *Endogone* spores in adjacent soils. *J. Gen. Microbiol.* **81:** 7-14.

Khan, A.G. 1978. Vesicular arbuscular mycorrhizae in plants colonizing black wastes from bituminous coal mining in the Illawarra region of New South Wales. *New Phytol.* **81:** 53-63.

Khan, A.G. 1988. Inoculum density of *Glomus mosseae* and growth of onion plants in unsterilized bituminous coal spoil. *Soil Biol. Biochem.* **20:** 749-753.

Khan, A.G. 1999 a. Occurrence of mycorrhizae and root nodules in plants growing on tannery effulent polluted soil. In: *International conference on the Biogeochemistry of trace elements (ICOBTE)*. pp. 174-175. Vienna, Austria.

Khan, A.G. 1999 b. Diversity of arbuscular mycorrhizal fungus populations in tannery effulent polluted soils. In: *Proc. of the 3rd International conference on Environmental chemistry and Geochemistry in the Tropics (GEOTROP)*, Baptist University, Hong Kong. 21 pp.

Khan, A.G. and Bari, A. 1997. Will deep rooted mycorrhizal popular trees do the dirty work in clean up of chromium contaminated soil polluted by the tannery industry in Pakistan by rooting-out the heavy metal. In: *Proc. Development and Environmental Impact Conference*. Ministry of Municipal and Rural Affairs, Riyadh, Kingdom of Saudi Arabia. pp. 199.

Khan, A.G., Kuek, C., Chaudhry, T.M., Khoo, C.S. and Hayes, W.J. 2000. Role of plants, mycorrhizae and phytochelators in heavy metal contaminated land remediation. *Chemosphere* **41:** 197-207.

Killham, K. and Firestone, M.K. 1983. VAM mediation of grass response to acidic and heavy metal depositions. *Plant Soil* **72:** 39-48.

Kinden, D.A. and Brown, M.F. 1975. Electron microscopy of vesicular arbuscular mycorrhizae of yellow popular. III. Host endophyte interactions during arbuscular development. *Can. J. Microbiol.* **21:** 1930-1939.

Koch, K.E. and Johnson, C.R. 1984. Photosynthate partitioning in split root citrus seedlings with mycorrhizal and non-mycorrhizal root systems. *Plant Physiol.* **75:** 26-30.

Koch, M., Tanami., Z., Bodani, H. and Wininger, S. 1997. Field application of vesicular arbuscular mycorrhizal fungi improved garlic yield in disinfected soil. *Mycorrhiza* **7:** 47-50.

Koslova, E.I., Badumyan, L.S. and Vendilo, M.V. 1966. Properties of bacteria from the nodules of sea buckthorn. *Mikrobiologiya* **35:** 699-706.

Krishna, K.R. and Bagyaraj, D.J. 1983. Changes in the free amino nitrogen and protein fractions of groundnut caused by inoculation with VA-mycorrhiza. *Ann. Bot.* **51:** 399-401.

Krishna, K.R., Suresh, H.M., Syamsunder, J. and Bagyaraj, D.J. 1981. Changes in the leaves of finger millet due to VA mycorrhizae infection. *New Phytol.* **87:** 717-722.

Krochler, C.J., Antibus, R.K. and Linkins, A.E. 1988. The effects of organic and inorganic phosphorus concentration on the acid phosphatase activity of ectomycorrhizal fungi. *Can. J. Bot.* **66:** 750-756.

Kucey, R.M.N. and Paul, E.A. 1982. Carbon flow, photosynthesis and N_2 fixation in mycorrhizal and nodulated faba beans (*Vicia faba* L.). *Soil Biol. Biochem.* **14:** 407-412.

Kurdali, F., Domemach, A.M. 1991. Evolution of nodulation capacity in *Alnus glutinosa* seedlings grown in soil sampled from various depths. *Acta Oecologica* **12:** 397-401.

Lalonde, M. and Calvert, H.E. 1979. Production of *Frankia* hyphae and spores as an infective inoculant for *Alnus* sp. In: *Symbiotic nitrogen fixation in the management of temperate forests.* (Eds. Gordon, J.C., Wheeler, C.T. and Perry, D.A.). pp. 95-110. Forest Research Lab. Oregon State University, Corvallis.

Lalonde, M., Calvert, H.E. and Pine, S. 1981. Isolation and use of *Frankia* strains in actinorhizal formation. In: *Current perspective in nitrogen fixation.* (Eds. Gibson, A.H. and Newton, W.E.). pp. 296 - 299. Australian Academy of Sciences, Canberra.

Lambert, D.H., Baker, D.E. and Cole, H. J. R. 1979. The role of mycorrhizae in the interactions of phosphorus with zinc, copper and other elements. *Soil Sci. Soc. Amer. J.* **43:** 970-980.

Langills, W.M. and Maclean, K.S. 1976. Some essential nutrient elements in forest plants as related to species, plant part, season and location. *Plant Soil* **45:** 17-26.

Laplaze, L., Rabeiro, A., Franche, C., Duhoux, E., Auguy, F., Bogusz, D. and Pawlowski, K. 2000. Characterization of a *Casuarina glauca* nodule - specific subtilisin - like protease gene, a homology of *Alnus glutinosa* ag 12. *Mol. Plant Microb. Interact.* **13:** 113-117.

Lawrie, A.C. 1982. Field nodulation in nine species of *Casuarina* in Victoria. *Aust. J. Bot.* **30:** 447-460.

Leaf, G., Gardner, I.C. and Bond, G. 1959. Observations on the composition and metabolism of nitrogen-fixing root nodules of *Myrica. Biochem. J.* **72:** 320-331.

Lechevalier, M., Baker, D. and Horriere, F. 1982. Physiology, chemistry, serology and infectivity of two *Frankia* isolates from *Alnus incana* (L.) Moench subsp. *rugosa* (Duroi) Clausen. Communication at the conference on *Frankia* Madison, U.S.A.

Lemoini, M.C., Gianninazzi-Pearson, V., Gianninazzi, S. and Straker, C.J. 1992. Occurrence and expression of acid phosphatase of *Hymenocyphus cricae* (Read) Korf and Kernan, in isolation or associated with plant roots. *Mycorrhiza* **1:** 137-146.

Levy, Y. and Krikun, J. 1980. Effects of vesicular arbuscular mycorrhiza on *Citrus jambhiri* water relations. *New Phytol.* **85:** 25-31.

Lewis, D.H. 1975. Comparitive aspects of carbon nutrition of mycorrhizas. In: *Endomycorrhizas*. (Eds. Sanders, E.F., Mosse, D. and Tinker, P.B.). pp. 119-148. Academic Press, New York.

Leyval, C., Turnau, K. and Haselwandter, K. 1997. Effect of heavy metal pollution on mycorrhizal colonization and function: Physiological, ecological and applied aspects. *Mycorrhiza* **7:** 139-153.

Lister, J.R., Slankis, V. Krotkov, G. and Nelson, C.D. 1986. The growth and physiology of *Pinus strobus* L. seedlings as affected by various nutritional levels of nitrogen and phosphorus. *Ann. Bot.* (NS). **32:** 33-43.

Liu, Z., Shi, W. and Fan, X. 1990. The rhiozosphere effects of phosphorus and iron in soil. *Trans. 14th Int. Cong. Soil Sci. Kyoto, Japan* **2:** 147-152.

Losel, D.M. and Cooper, K.M. 1979. Incorporation of ^{14}C labelled substrates by uninfected and VA mycorrhizal roots of onion. *New Phytol.* **83:** 415-426.

Mac Donald, R.M. and Lewis, M. 1978. The occurrence of some acid phosphatases and dehydrogenases in the VAM fungus *Glomus mosseae*. *New Phytol.* **80:** 135-141.

Mackkum, I.R., McNaughton, S.J., Raynal, D.J. and Leopold, D.J. 1993. Comparative foliage and twig chemistry of co-occurring *Myrica gale* and *Chamaedaphne calyculata*. *Can. J. Bot.* **71:** 129-135.

Madej, A. and Haggblom, P. 1985. Radioimmuno assay for determination of Indole-3 - acetic acid in fungi and plants. *Physiol. Plant.* **64:** 389-392.

Mahadevan, A., Raman, N. and Natarajan, K. 1988. *Mycorrhizae for Green Asia,* University of Madras, India, Madras. 351 pp.

Malajczuk, N., Molina, R. and Trappe, J.M. 1984. Endomycorrhiza formation in *Eucalyptus*. II. The ultrastructure of compatible and incompatible mycorrhizal fungi and associated roots. *New Phytol.* **96:** 43-53.

Marcar, N.E., P. Dart and C. Sweeney. 1991. Effect of root zone salinity on growth and chemical composition of *Acacia ampliceps,* A. cunn ex. Benth. and *A. mangium* Willd. at two nitrogen levels. *New Phytol.* **119:** 567-573.

Marschner, H. 1986. Mineral nutrition of higher plants. Academic press. New York.

Marschner, H. 1998. Role of root growth, arbuscular mycorrhizae and root exudates for the efficiency in nutrient acquisition. *Field Crops Res.* **56:** 203-207.

Marschner, H. and Dell, B. 1994. Nutrient uptake in mycorrhizal symbiosis. *Plant Soil* **159:** 89-102.

McVean, D.N. 1955. Ecology of *Alnus glutinosa* (L.) Gaertn. I. Fruit formation. *J. Ecol.* **43:** 46-60.

Megahed. M.M. and El-Lakamy, A.M.H. 1986. Biomass characteristics of young *Casuarina* plantations in Northwestern region of Egypt. *Alexandria J. Agric.* **31:** 411-422.

Meharg, A.A. 1994. Integrated tolerance mechanisms-constitutive and adaptive plant responses to elevated metal concentrations in the environment. *Plant Cell Environ.* **17:** 989-993.

Menge, J.A., Sterile, D., Bagyaraj, D.J., Johnson, E.L.V. and Leonard, R.T. 1978. Phosphorus concentration in plant responsible for inhibition of mycorrhizal infection. *New Phytol.* **80:** 575-578.

Meysselle, J.P., Gay, G. and Debaud, J.C. 1991. Intraspecific genetic variation of acid phosphatase activity in monokaryotic and dikaryotic population of the ectomycorrhizal fungus, *Hebeloma cylindrosporum*. *Can. J. Bot.* **69:** 808-813.

Midgley, S.J., Turnbull, J.W. and Johnson, R.D. 1983. *Casuarina ecology, management and utilization.* (Eds. Midgley, S.J., Turnbull, J.W. and Johnson, R.D.). Commonwealth Scientific and Industrial Research Organization (CSIRO), Melbourne, Australia. 1 pp.

Minchin, F.R., Summerfield, R.J., Halley, P., Roberts, E.H. and Rawsthrone, S. 1981. Carbon and nitrogen nutrition of nodulated roots of grain legumes. *Plant Cell Environ.* **4:** 5-26.

Moezel, P.G., Van der, P.G., Walton, C.S., Pearce-Pinto, G.V.N. and Bell, D.T. 1989. Screening for salinity and water logging tolerance in five *Casuarina* species. *Landscape and Urban Planning* **17:** 331-337.

Mortin, F. and Hilbert, J.L. 1991. Morphological, biochemical and molecular changes during ectomycorrhizal development. *Experimentia* **47:** 321-331.

Mosse, B. 1957. Growth and chemical composition of mycorrhizal and non-mycorrhizal apples. *Nature* **179:** 922-924.

Mosse, B. 1959. The regular germination of resting spores and some observations on the growth requirements of an *Endogone* sp causing VAM. *Trans. Br. Mycol. Soc.* **42:** 273.

Mosse, B. and Bowen, G.D. 1968. A key to recognition of some *Endogone* spore types. *Trans. Br. Myco. Soc.* **51:** 469.

Murry, M.A., Fontaine, M.S. and Tjepkema, J.D. 1984. Oxygen protection of nitrogenase in *Frankia* sp. HFPAr 13. *Arch. Microbiol.* **139:** 162-166.

Mylona, P., Pawloski, K. and Bisseling, T. 1995. Symbiotic nitrogen fixation. *Plant cell* **7:** 869-885.

Nagy, S., Nordby, H.E. and Nemec, S. 1980. Composition of lipids in roots of six citrus cultivars infected by the vesicular arbuscular mycorrhizzal fungus *Glomus mosseae. New Phytol.* **84:** 377-384.

Nap, J.P. and Bisseling, T. 1990. Developmental biology of a plant prokaryote symbiosis: The legume root nodule. *Science* **250:** 948-954.

Ndioye, P., Mailly, D., Pineau, M. and Margolis, H.A. 1993. Growth and yield of *Casuarina equisetifolia* plantation on the coastal sand dunes of Senegal as a function of microphotography. *For. Eco. and manag.* **56:** 13-28.

Nemec, S. and Guy, C. 1982. Carbohydrates status of mycorrhizal and non-mycorrhizal *Citrus* root stocks. *J. Am. Soc. Hortic. Sci.* **107:** 177-180.

Nemec, S. and Meredith, F.I. 1981. Amino acid content of leaves in mycorrhizal and non-mycorrhizal *Citrus* root stocks. *Ann. Bot.* **47:** 351-358.

Newcomb, W., Syono, K. and Torrey, J.J. 1977. Development of an ineffective pea root nodule. Morphogenesis, fine structure and cytokinin biosynthesis. *Can. J. Bot.* **55:** 1891-1907.

Nordby, H.E., Nemec, S. and Nagy, S. 1981. Fatty acids and sterols associated with *Citrus* root mycorrhizal. *J. Agric. Food. Chem.* **29:** 396-401.

Noridge, J.A. and Benson, D.A. 1986. Isolation and nitrogen fixing activity of *Frankia* sp. strain CPI 1 vesicles. *J. Bacteriol.* **166:** 301-305.

Normand, P., Orso, S., Cournoyer, B., Jeannin, P., Chapelon, C., Dawson, J., Evtushenko, L. and Misra, A.K. 1996. Molecular phylogeny of the genus *Frankia* and related genera and emendation of the family Frankiaceae. *Int. J. Syst. Bacteriol.* **46:** 1-9.

Ocampo, J.A. and Azcon, R. 1985. Relationship between the concentration of sugars in the roots and VA mycorrhizal infection. *Plant Soil* **86:** 95-100.

Ohwaki, Y. and Hirata, H. 1992. Difference in carboxylic acid exudation from P - starved leguminous crops in relation to carboxylic acid content in plant tissues and phospholipid level in roots. *Soil Sci Plant Nutr.* **38:** 235-243.

Olson, J.B., Steppe, T.F., Litaker, R.W. and Paerl, H.W. 1998. N_2 – fixing microbial consortia associated with the ice cover of Lake Bonney, Antartica. *Microb. Ecol.* **36:** 231-238.

Pacovsky, R.S. and Fuller, G. 1988. Mineral and lipid composition of *Glycine - Glomus - Bradyrhizobium* symbiosis. *Physiol. Plant.* **72:** 733-746.

Pais, M.S., Barroso, J. and Chavensneves, H. 1986. Production of indole - 3 - ethanol and indole - 3 - acetic acid by mycorrhizal fungus of *Oprys lutea* (Orchidaceae). *New Phytol.* **103:** 145-149.

Parsons, R., Silver, W.B., Harris, S., Gruijtens, W.T.M. and Bullivant, S. 1987. *Frankia* vesicles provide inducible and absolute protection for nitrogenase. *Plant Physiol.* **83:** 728-732.

Pate, J.S., Layzel, D.B. and McNeil, D.L. 1979. Modelling the transport and utilization of C and N in a nodulated legume. *Plant Physiol.* **63:** 730-738.

Pawlowski, K. and Bisseling, T. 1996. Rhizobial and actinorrhizal symbiosis: What are the shared features. *Plant cell* **8:** 1899-1913.

Pawlowski, T., Chaney, R.L., Chin, L. and Charvat, I. 2000. Effect of metal phytoextraction practices on the indigenous community of arbuscular mycorrhizal fungi at a metal contaminated landfill. *Appli. Environ. Microbiol.* **66:** 2526-2530.

Pawlowski. K., Twigg, P., Dobritsa, S., Guan, C. and Mullin, B.C. 1997. A nodule - specific gene family from *Alnus glutinosa* encodes glycine and histidine-rich proteins expressed in the early stage of actinorhizal nodule development. *Mol. Plant-Microbe Interact.* **10:** 656-664.

Pearson, V. and Read, D.J. 1975. The physiology of the mycorrhizal endophyte of *Callua vulgaris. Trans. Br. Mycol. Soc.* **64:** 1-7.

Periasamy, K. and Raman, N. 1995. Effects of Cu and Ni on acid phosphatase activities of an ectomycorrhizal fungus, *Laccaria laccata*. *Ind. J. Exp. Biol.* **33:** 537-538.

Perinet, P., Bronillett, J.G., Fortin, J.A. and Lalonde, M. 1985. Large scale inoculation of actinorhizal plants with *Frankia*. *Plant Soil* **87:** 175-183.

Perry, D.A., Molina, R. and Amaranthus, M.P. 1987. Mycorrhizae, mycorrhizosphere and reforestation: Current knowledge and Research Needs. *Can. J. For. Res.* **17:** 929-940.

Phillips, D.A. and Torrey, J.D. 1970. Cytokinin production by *Rhizobium japonicum*. *Physiol. Plant.* **23:** 1057-1063.

Phinney, B.O. 1984. Gibberellin A_1, dwarfism and the control of shoot elongation in higher plants. In: *The biosynthesis and metabolism of plant hormones.* (Eds. Crozier, A. and Hillman, J.R.). pp. 17-41. Cambridge University Press, Cambridge, UK.

Prin, Y., Maggia, L., Picard, B., Diem, H.G. and Goullet, P. 1991. Electrophoretic comparison of enzymes from 22 single spore cultures obtained from *Frankia* strains ORS 140102. *FEMS Microbiol. Lett.* **77:** 223-228.

Quispel, A. and Burggraaf, A. 1981. *Frankia* the diazotrophic endophyte from actinorhizas. In: *Current perspectives in nitrogen fixation.* (Eds. Gibson, A. and Newton, W.E.). pp. 229-236. Elsevier/North - Holland Biomedical Press, New York.

Quesada, A. and Valiente, E.F. 1996. Relationship between Abundance of N_2 fixing cyanobacteria and Environmental features of Spanish Rice Fields. *Microb. Ecol.* **32:** 59-71.

Quesada, A., Leganes, F. and Valiente, E.F. 1997. Environmental factors controlling N_2 fixation in Mediterraean Rice fields. *Microb. Ecol.* **34:** 39-48.

Quesada, A., Nieva, M., Leganes, F., Ucha, A., Martin, M., Prosperi, C. and Valiente, E.F. 1998. Acclimation of cyanobacterial communities in rice fields and response of nitrogenase activity to light regime. *Microb. Ecol.* **35:** 147-155.

Racette, S. and Torrey, J.G. 1989. The isolation, culture and infectivity of a *Frankia* strain from *Gymnostoma papuanum* (Casuarinaceae). *Plant Soil* **118:** 165-170.

Raman, N. and Elumalai, S. 1991. Studies on mycorrhizal and actinorhizal association in *Casuarina equisetifolia* in Coramandal coast region. *J. Trop. For.* **7:** 138-150.

Raman, N. and Elumalai, S. 1992. Influence of phosphorus on cluster root formation by *Casuarina equisetifolia* in water culture. *Ind. J. Expt. Biol.* **30:** 928-929.

Raman, N. and Elumalai, S. 1996. Presence of cytokinin in the root nodules of *Casuarina equisetifolia. Ind. J. Expt. Biol.* **34**: 577-579.

Raman, N. and Elumalai, S. 1998. Gibberellic acid in mycorrhizal and actinirhizal inoculated *Casuarina equisetifolia. Ind. J. Expt. Biol.* **36**: 703-708.

Raman, N. and Mahadevan, A. 1996. Mycorrhizae research – a priority in agriculture. In: *Concepts in mycorrhizal research.* (Ed. Mukerji, K.G.). pp. 41-75. Kluwer Academic Publishers, The Netherlands.

Raman, N. and Sambandan, K. 1998. Distribution of VAM fungi in tannery effluent polluted soils of Tamil Nadu, India. *Bull. Environ. Contamin. Toxicol.* **60:** 142-150.

Raman, N., Nagarajan, N. and Sambandan, K. 1992. Vesicular arbuscular mycorrhizae of tree species of Mamandur forest of Tamil Nadu. *J. Tree Sci.* **11:** 135-139.

Raman, N., Ravi, I. and Gnanaguru, M. 1994. Enhancement of indole-3-acetic acid in nodules of *Prosopis juliflora* inoculated with *Glomus mosseae* and *Rhizobium. Ind. J. Microbiol.* **34:** 33-35.

Raman, N., Ravi, M. and Srinivasan, V. 1998. Effects of copper and nickel on the axenic growth and phosphatase activity of an ectomycorrhizal fungus, *Amanita muscaria. Indian J. Environ Health* **40:** 197-202.

Raman, N., Srinivasan, V. and Ravi, M. 2002. Effect of chromium on the axenic growth and phosphatase activity of Ectomycorrhizal fungi, *Laccaria laccata* and *Suillus bovinus. Bull. Environ. Contam. Toxicol.* **68:** 569-575.

Raman, N., Nagarajan, N., Gopinathan, S. and Sambandan, K. 1993. Mycorrhizal status of plat species colonization a magnesite mine spoil in India. *Biol. Fert. Soils* **16:** 76-78.

Ramesh, C. 1997. Studies on the development of VAM fungi on *Pennisetum pedicellatum*. Ph.D., thesis, University of Madras. India.

Ramesh, C., Chellappan, P. and Mahadevan, A. 2000. Comparison of protein profiles and enzymes in non-mycorrhizal and mycorrhizal roots of *Pennisetum pedicellatum. Ind. J. Expt. Biol.* **38:** 483-487.

Rangarajan, M., Kandasamy, D., Oblisami, G. and Narayanan, R. 1988. Variation in responses to inoculation with *Frankia* and two ectomycorrhizal fungi on growth and nodulation in *Casuarina equisetifolia*. In: *Mycorrhiza for Green Asia, First Asian conference on Mycorrhizae, Jan. 29-31, Madras. India.* (Eds. Mahadevan, A., Raman, N. and Natarajan, K.). pp. 291-293. University of Madras, Madras.

Ratnayake, M., Leonard, R.T. and Menge, J.M. 1978. Root exudation in relation to supply of phosphorus and its possible relevance to mycorrhizal formation. *New Phytol.* **81:** 543-553.

Razak, A.A., Ramadan, S.E., Haron, B.M. and Lashile, I. 1985. Influence of osmotic stress on amino acid proline and protein contents in *Aspergillus tamari*. *J. Coll. Sci. King Sand Univ.* **16**: 41-48.

Reddell, P. 1986. In soil and plant factors affecting nodulation and nitrogen fixation in Casuarinaceae *Frankia* symbiosis. Ph.D., thesis, University of Western Australia, Perth.

Reddell, P. and Bowen, G.D. 1985. *Frankia* source affects growth, nodulation and nitrogen fixation in *Casuarina* sp. *New Phytol.* **100:** 115-122.

Reddell, P. and Bowen, G.D. 1986. Host - *Frankia* specificity within Casuarinaceae. *Plant Soil* **93:** 293-298.

Reddell, P., Bowen, G.D. and Robson, A.D. 1985. The effects of soil temperature on plant growth, nodulation and nitrogen fixation in *Casuarina cunninghamiana* Miq. *New Phytol.* **101:** 441-450.

Reddell, P., Bowen, G.D. and Robson, A.D. 1986. Nodulation of Casuarinaceae in relation to host species and soil properties. *Aust. J. Bot.* **34:** 435-444.

Ribeiro, A., Akkermans, A.D.L., Van Kammen, A., Bisseling, T. and Pawlowski, K. 1995. A nodule - specific gene encoding a subtilin - like protease is expressed in early stages of actinorhizal nodule development. *Plant Cell* **7:** 785-794.

Righetti, T.L., Chard, C.H. and Backhaus, R.A. 1986. Soil and environmental factors related to nodulation in *Casuarina* and *Purshia*. *Plant Soil* **91:** 147-160.

Robson, A.D. and Loneragan, J.F. 1978. Responses of pasture grasses to soil chemical factors other than nitrogen and phosphorus with particular emphasis on the legume symbiosis. In: *Plant relations in pasture.* (Ed. Wilson, J.R.). pp. 128-142. CSIRO – Melbourne.

Rodriguez - Barrucco, G. 1968. The occurrence root nodule endophytes of *Alnus glutinosa* and *Myrica gale* in soils. *J. Gen. Microbiol.* **52:** 189-194.

Rodriguez - Barrucco, G. 1972. Effecto del - N - amonico sobre la fijaconde N-atmosferico en nodules de *Casuarina*. *Anal. Edafol. Agrob.* **31:** 905-916.

Rosbrook, P.A. and Bowen, G.D. 1987. The abilities of three *Frankia* isolates to nodulate and fix nitrogen with four species of *Casuarina*. *Physiol. Plant.* **70:** 373-377.

Rose, S.L. 1980. VA Mycorrhizal association of some desert plants of Baja California. *Can. J. Bot.* **59:** 1056-1060.

Rose, S.L. and Youngberg, C.T. 1981. Tripartite association in Snowbrough (*Ceanothus relutinus*): effect of VAM on growth, nodulation and nitrogen fixation. *Can. J. Bot.* **59:** 34-39.

Ross, J.J., Murfet, I.C. and Reid, J.B. 1993. Distribution of gibberellins in *Lathyrus odoratus* L. and their role in leaf growth. *Plant Physiol.* **102:** 603-608.

Sambandan, K., Kannan, K. and Raman, N. 1991. Effect of VAM fungus, *Glomus geosporum* on effluent tolerance of *Casuarina equisetifolia*. *Poll. Res.* **10:** 55 - 67.

Sambandan, K., Kannan, K. and Raman, N. 1992. Distribution of vesicular arbuscular mycorrhizal fungi in heavy metal polluted soil of Tamil Nadu. *J. Environ. Biol.* **13:** 159-167.

Sambandan, K., Kannan, K. and Raman, N. 1994. Vesicular arbuscular mycorrhizae of *Casuarina equisetifolia* in four different soil types in Tamil Nadu. *Indian Forester* **120:** 510-516.

Same, B.L., Robson, A.D. and Abbott, L.K. 1983. Phosphorus, soluble carbohydrates and endomycorrhizal infection. *Soil Biol. Biochem.* **15:** 593-597.

Sanders, F.E. and Tinker, P.B. 1971. Mechanism of absorption of phosphate from soil by endogone mycorrhizas. *Nature.* **233:** 278-279.

Saxena, D., Amin, M. and Khanna, S. 1996. Modulation of protein profiles in *Rhizobium* sp. under salt stress. *Can. J. Microbiol.* **42**: 617-620.

Scheres, B., Van de Wiel, C., Zalensky, A., Horvath, B., Spaink, H.P., Van Eck, H., Zwartkruis, F., Wolters, A.M., Gloudemans, T., Van Kammen, A. and Bisseling, T. 1990 a. The ENOD12 gene product is involved in the infection process during pea - Rhizabium interaction. *Cell* **60:** 281-294.

Scheres, B., Van Engelen, F., Van der Knaap, E., Van de Wiel, C., Van Kammen, A. and Bisseling, T. 1990 b. Sequential induction of nodulin gene expression in the developing pea nodule. *Plant Cell* **8:** 687-700.

Schwintezer, G., Berry, A. and Disney, L. 1982. Seasonal pattern of root nodules growth endophyte morphology, nitrogenase activity and shoot development in *Myrica gale. Can. J. Bot.* **60:** 746-757.

Shaul-Keinan, O., Gadkar, V., Ginzberg, I., Grunzweig, J. M., Chet, I., Elad, Y., Wininger, S., Belausov, E., Eshed, Y., Atzmon, N., Ben-Tal, Y. and Kapulnik, Y. 2002. Hormone concentrations in tobacco roots change during arbuscular mycorrhizal colonization with *Glomus intraradices. New Phytol.* **154:** 501-507.

Shetty, K.G., Banks, M.K., Hetrick, B.A.D. and Schwab, A.P. 1995. Effects of mycorrhizae and fertilizer amendments on zinc tolerance of plants. *Environ. Pollut.* **88:** 307-314.

Shipton, W.A. and Burggraaf, A.J.P. 1983. *Frankia* growth and activity as influenced by water potential. *Plant Soil* **69:** 293-297.

Silver, W.S. 1971. Physiological chemistry of non - leguminous symbiosis. In: *The chemistry and biochemistry of nitrogen fixation.* (Eds. Gray, T.R.G. and Postgat, J.R.). pp. 246-282. Plenum press, London.

Silver, W.S., Bendana, F.S. and Powell, R.D. 1966. Root nodule symbiosis. II. The relation of auxin to root geotropism in roots and root nodules of non-legumes. *Physiol. Plant.* **19:** 207-218.

Silvester, W.B. 1977. Dinitrogen fixation by plant associations excluding legumes. In: *"A treatise and dinitrogen fixation, Section 4, Agronomy and Ecology"*. (Eds. Hardey, R.W.F. and Gibson, A.H.). pp. 141-190. John Wiley & Sons, New York.

Sloger, C. and Silver, W.S. 1965. Note on nitrogen fixation by exposed root nodules and nodular homogenates of *Myrica cerifera* L. In: *Non - heme iron proteins: role in energy conversion.* (Ed. San-Pietro, A.). pp 299-302. Antioch Press, Yellow Springs, Ohio.

Small, E. 1972. Ecological significance of four critical elements in plants of raised *Sphagnum*, peat bogs. *Ecology* **53:** 498-503.

Smith, S.E. 1966. Physiology and ecology of orchid mycorrhizal fungi with reference to seedling nutrition. *New Phytol.* **65:** 488-499.

Smith, S.E. 1967. Carbohydrates translocation in Orchid mycorrhizal fungi. *New Phytol.* **66:** 371-378.

Smith, S.N., Ince, E. and Armstrong, R.A. 1990. Effect of osmotic and matrix potential on *Saprolegnia diclina. Mycol. Rev.* **94**: 71-77.

Smolander, A., Van Dijk, C. and Sundman, V. 1988. Survival of *Frankia* strains introduced into soil. *Plant Soil* **106:** 65-72.

Snellgrove, R.C. Splittstoesser, W.E., Stribley, D.P. and Tinker, P.B. 1982. The distribution of carbon and demand of the fungal symbiont in leek plants with VA mycorrhiza. *New Phytol.* **92:** 75-87.

Snellgrove, R.C., Stribely, D.P., Tinker, P.B. and Lowlor, D.W. 1986. The effect of vesicular arbuscular mycorrhizal infection on photosynthesis and carbon distribution in leek plants. In: *Physiological and Genetical Aspects of Mycorrhizae.* (Eds. Gianinazzi-Pearson, V. and Gianinazzi, S.). pp. 421-424. INRA, Paris.

Sougoufara, B., Diem, H.G. and Dommergues, Y.R. 1989. Response of field grown *Casuarina equisetifolia* to inoculated with *Frankia* strain ORS 021001 entrapped in alginate beads. *Plant Soil* **118:** 133-137.

Spratt, E.R. 1912. The morphology of the roots tubercles of *Alnus* and *Elaegnus* and the polymorphism of the organism causing their formation. *Ann. Bot.* **26:** 119-128.

Spray, C.R., Kobayashi, M., Suzuki, Y., Phinney, B.O., Gaskin, P. and Mac Millan, J. 1996. The *dwarf-1* (*d1*) mutant of *Zea mays* blocks three steps in the gibberellin - biosynthetic pathway. *Proccedings of the National academy of Sciences, USA.* **93:** 10515-10518.

Srinivasan, V. 2000. Investigations on bioaccumulation of aluminium and chromium by ectomycorrhizal fungus, *Pisolithus tinctorius*. Ph.D. thesis, University of Madras, India. 158 p.

Stachurski, A. and Zimka, J.R. 1975. Methods of studying forest ecosystems; leaf area, leaf production and withdrawal of nutrients from leaves of trees. *Ekologia Polska* **23:** 637-648.

Stele, D.B., Ramirez, K. and Strowers, M.D. 1989. Host plants growth response to inoculation with *Frankia*. *Plant Soil* **118:** 139-143.

Straker, C.J. and Mitchell, D.T. 1986. The activity and characterization of acid phosphatases in endomycorrhizal fungi of the Ericaceae. *New Phytol.* **104:** 243-256.

Streeter, J.G. 1980. Carbohydrates in soybean nodules. *Plant Physiol.* **66:** 471-476.

Subramanian, K.S., Charest, C. Dwyer, L.M. and Hamilton, R.I. 1997. Effects of arbuscular mycorrhizae on leaf water potential, sugar content and P content during drought and recovery of maize. *Can. J. Bot.* **75:** 1582-1591.

Sward, R.J. 1978. Studies of VA mycorrhizae of some Australian heath land plants. Ph.D., thesis. Monash University, Victoria, Australia.

Sylvia, D.M. and Williams, S.E. 1992. Vesicular arbuscular mycorrhizae and environmental stress. In: *Mycorrhizae in sustainable agriculture.* ASA Special publication no. 54, ASA, CSSA, SSSA, Madison. WI. pp. 101-124.

Tarafdar, J.C. and Marschner, H. 1994. Phosphatase activity in the rhizosphere and hyposphere of VA mycorrhizal wheat supplied with inorganic and organic phosphorus. *Soil Biol. Biochem.* **26:** 387-395.

Theodorou, C. and Reddell, P. 1991. *In vitro* synthesis of ectomycorrhizas on Casuarinaceae with a range of mycorrrhizal fungi. *New Phytol.* **118:** 279-288.

Thoen, D., Sougoufara, B. and Dommergues, Y. R. 1990. *In vitro* mycorrhization of *Casuarina* and *Allocasuarina* species by *Pisolithus* isolates. *Can. J. Bot.* **68:** 2537-2542.

Tjepkema, J.D. 1979. Oxygen regulation in leguminous and actinorhizal nodules. In: *Symbiotic nitrogen fixation in the management of temperate forests.* (Eds. Gardon, J.C., Wheeler, C.T. and Perry, D.A.). pp. 175-176. Oregon state University, Corvallis.

Tjepkema, J.D. 1988. Oxygen regulation and energy usage in actinorhizal plants. In: *Biological nitrogen fixation: Recent developments.* (Ed. Subba Rao, N.S.). pp. 311-333. Oxford and IBA Publishing Co, Pvt Ltd., New Delhi, India.

Tjepkema, J.D., Ormerod, W. and Torrey, J.G. 1980. Vesicle formation and acetylene reduction activity in *Frankia* Cp I1 cultured in defined nutrient media. *Nature* **187:** 633-635.

Tjepkema, J.D., Ormerod, W. and Torrey, J.G. 1981. Factors affecting vesicle formation and acetylene reduction in *Frankia* sp. Cp I 1. *Can. J. Microbiol.* **27:** 815-823.

Torelli, A., Trotta, A., Acerbi, L., Arcidiacono, G., Berta, G. and Branca, C. 2000. IAA and ZR content in leek (*Allium porrum* L.), as influenced by P nutrition and arbuscular mycorrhizae, in relation to plant development. *Plant Soil* **225:** 29-35.

Torrey, J.G. 1978. Nitrogen fixation by actinomycetes nodulated angiosperms. *Bioscience* **28:** 586-592.

Torrey, J.G. 1982. *Casuarina*: actinorhizal N_2 - fixing tree of the tropics. In: *Biological nitrogen fixation technology for tropical agriculture.* (Eds. Graham, P.H. and Harris, S.C.). pp. 427-439. Centro International de Agricultura. Tropical (CIAT) Cali, Columbia.

Torrey, J.G. 1983. Root development and root nodulation in *Casuarina*. In: *Casuarina ecology, management and utilization.* (Eds. Midgley, S.J., Turnbull, J.W. and Johnson, R.D.). pp. 180-192. CSIRO, Melbourne.

Torrey, J.G. and Callaham, D. 1982. Structural features of the vesicles of *Frankia* Cp I1 in culture. *Can. J. Microbiol.* **28:** 749-757.

Torrey, J.G. and Racette, S. 1989. Specificity among the Casuarinaceae in root nodulation by *Frankia*. *Plant Soil* **118:** 157-164.

Torrey, J.G. and Tjepkema, J.D. 1979. Symbiotic nitrogen fixation in actinomycete nodulated plants. *Bot. Gaz.* **140:** (Suppl.). I-II.

Torzilli, A.P. 1997. Tolerance to high temperature and salt stress by a salt marsh isolate of *Aureobasidium pullulans*. *Mycologia* **89**: 786-792.

US National Research Council. 1984. Casuarinas. In: *Nitrogen – fixing trees for adverse sites.* National Academy Press, Washington, D.C. 94 p.

Van Straten, J., Akkermans, A.D.L. and Roelofsen, W. 1977. Nitrogenase activity of endophyte suspensions derived from root nodules of *Alnus, Shepherdia* and *Myrica* sp. *Nature* (London) **266:** 257-258.

Van Dijk, C. 1979. Endophyte distribution in the soil. In: *Symbiotic nitrogen fixation in the management of temperate forests.* (Eds. Gordon, J.C., Wheeler, C.T. and Perry, D.A.). pp. 84-94. Oregon State University Press, Corvallis.

Verghese, S.K. and Misra, A.K. 2000. PCR – RFLP based screening of *Frankia* genotypes in alder nodules having different levels of nitrogenase activity. *Symbiosis* **28:** 337-350.

Verghese, S.K. and A.K. Misra. 2002. *Frankia* - actinorhizal symbiosis with special reference to host - microsymbiont relationship. *Current Science* **83:** 404-408.

Vernand, L., Chaboud, A., Prin, Y. and Rougier, R. 1985. Preinfection events in the establishment of *Alnus – Frankia* symbioses: Development of a spot inoculation technique. *Plant Soil* **87:** 67-78.

Wall, L.G. 2000. The actinorhizal symbiosis. *J. Plant Growth Regul.* **19:** 167-182.

Weissenhorn, I. and Leyval, C. 1995. Root colonization of maize by a Cd - sensitive and a Cd - tolerant *Glomus mosseae* and cadmium uptake in sand culture. *Plant Soil* **175:** 233-238.

Wheeler, C.T., Henson, I.E. and Mclaughlin, M.E. 1979. Hormones in plants bearing actinomycete nodules. *Bot. Gaz.* (Chicago) suppl. **140:** S 52-S 57.

Williams, P.M. and Sicardi de Mallorca, M. 1982. Abscisic acid and gibberellin - like substances in roots and root nodules of *Glycine max. Plant Soil* **65:** 19-26.

Wong, K.K.Y. and Fortin, J.A. 1990. Root colonization and interspecific mycobiont variation in ectomycorrhiza. *Symbiosis* **6:** 191-231.

Young, J.L. and Trappe, J.M. 1972. Endomycorrhizal invasion and effect on free amino acid content of corn roots. *Agron. Abstr.* pp. 102.

Zeng, S., Tjepkema, J.D. and Berg, R.H. 1989. Gas diffusion pathway in nodules of *Casuarina cunninghamiana. Plant Soil* **118:** 119-123.

Frontiers in Plant Sciences (Pages 701-718)
Eds. K.G. Mukerji, K.V.B.R. Tilak, S.M. Reddy, L.V. Gangwane, P. Prakash, I.K. Kunwar

Mycorrhizae as Biocontrol Agent in Disease Control

Rajni Gupta & K.G. Mukerji

ABSTRACT

The use of Vesicular Arbuscular Mycorrhizal fungi in biological control of plant pathogens is a fascinating and challenging field of research. There lies tremendous potential in use of mycorrhiza in biological control of root diseases. The exact mechanism of disease control as a result of mycorrhizal symbiosis is difficult to ascertain. The mycorrhizal fungi provide disease resistance to host plant by early colonization of root system, increased nutrient uptake especially phosphorus and improved growth. An association significantly reduced the incidence and intensity of wilt caused by *Fusarium oxysporum* and *Verticillium albo-atrum*. VAM fungi also reduced the number of pathogen propagules in soil. Introduction of *Glomus* sp. in soil helps in increasing the plant dry mass and save loss due to infection. Phenols, Phytoalexins, HRGPs and PR proteins help in controlling the plant pathogens.

Key words: Biocontrol, VAM fungi - *Glomus* sp., Phytoalexins, Phenol, PR proteins, *Fusarium oxysporum* and *Verticillium albo-atrum.*

INTRODUCTION

Microorganisms are essential for the functioning and sustainability of all natural ecosystems but are frequently ignored due to their small size and consequent methodological difficulties in detecting cells and assessing their activity. The central role of microorganisms in ecosystem processes, including biogeochemical cycling of nutrients and biodegradation, makes it both unwise and dangerous to ignore any aspect of their ecology. Analysis of diversity is particularly important when considering the ability of

Author's address: Deptt. of Botany, University of Delhi, Delhi-110007

ecosystems to respond to changing environmental conditions, the need for conservation of the microbial gene pool and the links between diversity, ecosystem processes, functional and physiological diversity and sustainability (Prosser, 2002). Successful infection by microbes require surface attachment, degradation of host chemical and physical barriers, production of toxins and inactivation of plant defenses (Lamb et al., 1989). Regulation of this phenomenon involves a complex interplay of signals between host and microbe. Molecular signaling events must occur that lead to various physiological and anatomical changes in both symbionts. Recognition is defined as an interaction between factors that lead to a result, the plant and fungal molecules that participate in the recognition events are beneficial for mycorrhizal formation. The complexity of the life cylce of the fungus within the plant indicates that compatibility cannot be a result of a single recognition event or a unique gene (Koide and Scheiner, 1992). The molecular communication may be exchanged between the fungus and host plant in the rhizosphere, the rhizoplane, the root epidermis and cortex (Azcon and Ocampo, 1984). Spores are important propagules for many arbuscular mycorrhizal fungi, and their value as inoculum depends on their capacity to germinate and infect plants. Signals diffused from host plants affect spore germinated, germ tube extension and its direction in the rhizosphere.

Around two-thirds of known plants form arbuscular mycorrhizas, a symbiosis between plant roots and glomalean fungi that leads to the formation of intraradical organs of nutrient exchange and an extraradical network of fungal hyphae effectively extending the plant health by increasing acquisition of immobile mineral nutrients, particularly phosphate (Smith and Read, 1997), enhancing drought tolerance (Subramanian et al., 1995) and promoting resistance in plants to root pathogens (Cordier et al., 1998). On the other side, the fungal partner profits from carbon supplied by the plant. Depending of mycorrhizal fungal diversity, mycorrhiza formation influences the function of ecosystems as well as the biodiversity of plant communities (Helgason et al., 1998). Researchers estimate that from 85-90% of the approximately, 2,31000 species of angiosperms form this symbiosis despite there being only approximately 120 described species of AM fungi (Schenck and Perez, 1987).

VAM AS A BIOPROTECTION AGENT

When both the VAM fungi and pathogens are free living in the soil before penetration, pathogen often have competitive advantage because of their saprophytic ability. Many of them are less strictly biotrophic than the mycorrhizal fungi. But situation is different when mycorrhizal fungus is already established the mycorrhizal fungus does not have to compete by pathogens. In many cases the fungal diseases are completely controlled in VA mycorrhizal plants (Table 1).

Fusarium sp. are potential pathogens found in all types of soils of the world causing considerable damage to cultivated crops in the form of vascular wilts, root rots and damping off. Much of the early work on biocontrol using AMF was against *F. oxysporum* of sp. *lycopersici* the pathogen which causes vascular wilt in tomato. Successful biocontrol of the disease was, achieved by pre-inoculation with *G. mosseae* (Dehne and Schoenbeck, 1975, 1979a) which was further tested and proved by Schenck and Kellam (1978) using *G. fasciculatum* and *G. mosseae* and Al-Momany and Al-Raddad (1988) using seven isolates of *Glomus* sp.

Mycorrhizal root with *Glomus intraradices* exhibited reduction in root necrosis in tomato due to *F. oxysporum* of sp. *radicis lycopersici* independent of the sequence of inoculation of the two fungi. The effect was more pronounced when *Glomus intraradices* was inoculated four weeks prior to *Fusarium oxysporum*. VAM fungi also reduced number of propagules of pathogenic fungi. Inoculation by *G. intraradices* increased plant dry mass and saved loss due to infection of *F. oxysporum* even when VAM fungus was inoculated later (Caron *et al.*, 1986 a,b)

AM association significantly reduced the incidence and intensity of wilt of cucumber incited by *F. oxysporum* f. sp. *cucumerinum* (Dehne, 1977 and Dehne and Schoenbeck, 1979b). The incidence of wilt of chick pea incited by *F. oxysporum* f. sp. ciceri was significantly reduced by preinoculation with *G. fasciculatum* and *G. mosseae* (Jalali and Thareja, 1981). Caron et al., (1986) also obtained significant reduction in the incidence and development of crown and root rot of tomato incident by *F. oxysporum* f sp. *radices* lycopersici using AMF. The pathogen population and root necrosis were also significantly reduced. Reduction of population of *Fusarium* sp. and disease suppression in legumes (Jalali, 1983) *F. solani* in groundnut (Parvathi et al., 1985). *Fusarium* sp. in tomato (Caron, 1989) and *F. moniliformae* causing damping off of cardomom (Thomas et al., 1994) were also reported due to AM symbiosis.

The wilt caused by *Verticillium albo atrum* was found to be significantly reduced by *Gigaspora* sp. (Melo et al., 1985). Melo et al. (1985) further observed that interaction between different AMF species and the pathogen was highly specified and the ability of disease suppression by AMF varied greatly with species. Hwang et al., 1992 also showed that inoculation with *G. fasciculatum* and *G. mosseae* significantly reduced the population of *V. albo atrum* in alfalfa.

TABLE 1 *Influence of VAM on the control of soil borne pathogens (Source : Singh et al., 2000).*

Host	*Disease*	*Pathogens*	*VAM fungi*	*(Effect of VAM fungi on host)*
Allium cepa	Wilt	*Fusarium oxysporum* f.sp. cepi		Delayed disease development
	White rot	*Sclerotium cepivorum*	*Glomus* sp.	Delayed disease epidemic and increased the yield by 22%
Arachis hypogea	White rot	*Sclerotium rolfsii*	*Gl. fasciculatum*	Reduction in the severity of disease.
Brassica sp.	Damping off	*Rhizoctonia solani*	*Gl. fasciculatum*	Survival percentage increased and plant growth increased
Cajanas cajan	Pigeon pea blight	*Phytophthora drechsleri f. sp. cajani*	*Gigaspora calospora*	Inhibition effect on the development of disease and improve plant growth.
Citrus sp.	Wilt	*Phytophthora parasitica*	*Gigaspora margarita Glomus constrictum Gl. fasciculatum Gl. mosseae*	Significant increase in root dry weight and better growth of plant.
Euphorbia pulcherima	Wilt	*Pythium ultimum*	*Gl. fasciculatum*	Increased plant height and lower final pathogen population.
Glycine max	Root rot	*F. solani R. solani*	*Gl. mosseae*	Reduced disease symptom
	Root rot	*Phytophthora megsperma var. sojae*	*Gl. macrocarpum*	Increased severity of disease, internal stem discoloration.
Hordeum vulgare	Root rot	*Bipolaris sorokiniana*	*Glomus sp.*	Severity of disease reduced.
Linum usitatissimum	Wilt	*Fusarium oxysporum oidium lini*	*Glomus intraradices*	VAM plants showed increased tolerance against *F. oxysporum* but become susceptible against.
Lycopersicon esculentum	Root rot	*F. oxysporum f. sp. radices lycopersici*	*Glomus intraradices*	Reduction in disease incidence and the population of pathogen and improvement in plant growth.
Nicotiana tabacum	Root rot	*Pythium sp.*	*Glomus intraradices*	Reduction in disease and population of pathogen inoculum and improved plant growth
Nicotiana tabacum	Root rot	*Thielaviopsis basicala*	*Glomus monosporum*	Better tolerance to pathogen higher root and leaf dry weight.
Oryza sativa	Root rot	*Rhizoctonia solani*	*Glomus mosseae*	Increased root dry weight.

(Table 1 Contd.)

(Table 1 Contd.)

Host	Disease	Pathogens	VAM fungi	(Effect of VAM fungi on host)
Phaseolus aureus	Root rot	*Macrophomina phasceolina*	*Glomus mosseae*	Disease incidence reduced.
Pisum sativum	Root rot	*Aphanomyces euteicaes*	*Glomus fascicuculatum*	Infection suppressed
Solanum melongena	Wilt	*Verticillium albo atrum*	*Glomus fascicuculatum*	Increased plant growth
Triticum sp.	Root rot	*Fusarium sp.* *Pythium sp.*	*Glomus fascicuculatum*	Increased Plant growth
Triticum sp.	Take all disease	*Guemannomy-ces graminis*	*Glomus fascicuculatum*	Reduction of infection.

Phytophthora sp. are one of the most serious groups of pathogens causing severe economic losses to important tropical and temperature crops. The significance of AM associations in reducing the incidence of *Phytophthora* incited disease was initially recorded in papaya (Ramirez, 1974) and later in citrus in which root rot incidence and intensity caused by *P. parasitica* were significantly reduced (Davis and Menge, 1980). Bisht et al., (1985) proved the potential inhibitory effect of *Gigaspora calospora* on the development of pigeon pea blight (*P. drechsleri* f.sp. *cajani*). An AMF conferred 'P' nutritional advantage in sweet orange seedlings was recognised by Graham and Egel (1988) which increased the resistance/ tolerance of seedling to *Phytophthora* root rot.

Mycorrhizal tomato plants were more resistant to *Phytophthora nicotianae* var. *parasitica* than controls. Resistance was shown by a reduction in both root necrosis and plant growth inhibition caused by the pathogen (Trotta et al., 1996). Reduction in tomato root necrosis in mycorrhizal plants with *Glomus intraradices* may be due to induced resistance with biochemical modifications in the host e.g. production of defense related molecules or modifications in enzymatic activity of the pathogen (Caron et al., 1986). The extensive network of external mycelim could also (i) supply other nutrients than P or (ii) prevent water shortage, so compensating for the loss of active roots rotted by the pathogen (Keen and Yoshikawa, 1983).

Another economically important disease which has effectively been suppressed by mycorrhizal fungi is 'foot rot' of black pepper incited by *P. capsci*. Initial report of disease suppression of foot rot with *G. fasciculatum* was made by Anandraj et al., 1993. Subsequent reports further established the suppressive effect of mycorrhizal fungi on foot rot incidence and stimulation of root regeneration (Sarma et al., 1996). There was considerable reduction in foot rot incidence, yellowing and defoliation under natural conditions. When mycorrhization was with native isolates of AMF which are more adapted to local soil colonization.

HOST PHYSIOLOGY

Mycorrhizal association results in altered host physiology. Suppression or promotion of a plant disease as affected by mycorrhizal development may ultimately be determined by several biotic and abiotic interacting forces involved in the host-pathogen mycorrhiza system. Changes in soil environment, particularly in terms of its nutrients and water status, significantly affect the germination, growth, development and pathogenic behaviour of many soil organisms. Early establishment of mycorrhizal symbiont appears to be vital if the microbial system is to be employed effectively in the control of diseases (Kaushik and Mukerji, 1996). Mycorrhizal plants develop extensive root system as compared to non-mycorrhizal plants, which ensures the plant with increased availability of nutrients, thereby helping the plant for better growth and development (Young et al., 1984; Mukerji, 1999). Mycorrhizal plants usually have more vascular bundles, hence lignification in the xylem is greater (Dehne and Schonbeck 1979) which strengthens the cell walls. The growth of pathogen is restricted/inhibited because of this (Dehne and Schonbeck, 1979). A stronger vascular system will increase the flow of nutrients, impart greater mechanical strength and diminish derogatory effects from vascular pathogens (Hussey and Roncadori, 1982). The size of the nuclei increases in host cells and accumulation of ER, cisternae and ribosomes in the vicinity of the fungus is known. Mycorrhizal roots show increased respiration (Schonbeck, 1979)

DEFENSE MECHANISM

Development and functioning of arbuscular mycorrhiza is governed by morphogenetic characters of both partners and evidence for an interplay of signalling/recognition events can be found at different stages during plant—fungal interactions (Roussel et al., 2001). Signalling steps have been traced in made over recent years towards unravelling mechanisms governing arbuscular mycorrhiza symbiosis, particularly in relation to molecular host responses to the myco-symbionts and the plant genes involved (Harrison, 1998). Some 50 different plant genes have been analysed for expression in arbuscular mycorrhiza. Studies targetting known genes have focussed mostly on those encoding defense response molecules or nodulins in legume–rhizobium interactions. The localised and/or low activation of such gene suggests that some signalling pathways in arbuscular mycorrhiza and other plant—microbe interactions may be shared (Albrecht et al., 1998).

Hypotheses to explain attenuated plant defense responses to symbiotic fungi include elicitor degradation cell signalling through altered sugar fluxes and modified hormone balance (Blee and Anderson, 2000). Cytokinins or chitooligosaccharides, released from glomalean walls, have been proposed as components of the signal transduction pathway mediating interactions (Gianianzzi-Pearson and Denarie, 1997). Differential regulation has also been reported for plant genes encoding membrane proteins which may be involved in transport processes with in the symbiosis (Gianinazzi–Pearson et al., 2000) and a number of other plant genes with putative functions in the symbiosis have been identified by non-targeted molecular analysis (Lappopin et al., 1999). Porcel et al. (2003) studied the mechanism by which the mycorrhizal symbiosis protects soyabean plants against premature nodule senescence induced by drought stress. This was done by evaluating the activity of a set of antioxidant enzymes in relation to nodule senescence. Superoxide dismutase (SOD), catalase (CAT), ascorbate peroxidase (APX) and glutathione reductase (GR) activity was determined in well watered and drought stressed soyabean plants inoculated with *Bradyrhizobium japonicum* alone or in combination of *Glomus mosseae*. In roots GR activity was found to be higher in mycorrhizal than non-mycorrhizal plants.

Molecular research on arbuscular mycorrhizal symbiosis is based on the concept that the expression of sets of genes underlies the phenotype and behaviour of organisms and tht evolution involves the recruitment and shuffling of different combinations of genes for new purposes. AM symbiosis is highly beneficial for the survival of plants in the natural ecosystem (Selosse and LE Tacon, 1998) and therefore, that AM symbiosis is an underlying genetic state, fundamental to root biology (Singh et al., 2002). Molecular researchers have identify genes that are involved in mycorrhizal symbiosis by their changed transcript abundance when comparing different stages of colonisation or equivalently grown mycorrhizal and non-mycorrhizal plants.

Rosewarne (1998) found five tomato genes that were differentially expressed during establishmenkt of the symbiosis with *Glomus intraradices*. Four of these *Les* 1, 2, 4 and 5 were increased in transcript abundance as a consequence of mycorrhizal colonisation, while the fifth gene, *DB297*, was decreased in abundance.

Certain genes could also be specifically involved in arbuscule development and its degradation. Gene products of defense related genes involved in cell wall metabolism as a hydroxyproline rich glycoprotein were shown to be expressed in arbuscule containing cells and localized, in the matrix of symbiotic interface (Balestrini et al., 1997). The gene *Prp*-I of potato encoding a glutathione—S-transferase has been shown by *in situ* hybridization to be expressed only in certain arbuscule containing cells (Franken et al., 2000) as this enzyme is involved in the degreadation of arbuscules. A number of genes are distinctly expressed during the whole process of mycorrhiza development. The role of such genes could be the control of hyphal spread in the roots or a reaction to the stress induced by fungal colonization (Franken and Requend, 2001). Recently β-tubulin gene sequences have been isolated from several AM fungi. On the basis of these sequences a glomalean—specific primer pair was designed to amplify exclusively the β-tubulin gene from AM fungi, and not from other fungi, including zygomycetes, or from plants. Other targets for PCR cloning were genes involved in nutrient transport as well as in cell wall metabolism. A fragment of the gene encoding for the nitrate reductase apoprotein from the AM fungus *Glomus intraradices* was PCR amplified and subsequently cloned.

Several copies of Chitin synthase genes from *Gigaspora margarita* were found after PCR with degenerate primers. A recent finding suggests that the failure of AM fungi to complete their life cycle in the absence of the plant could be due to lack of storage lipid biosynthesis during the asymbiotic growth (Vanloon et al., 1999).

Delp et al. (2000) using differential display PCR with AMF colonised and control barley roots, identified three *G. intraradices* genes with putative regulatory functions *Ginmyc* 2 was expressed both in extraradicle hyphae and in colonised roots. Whilst expression of *Ginmycel* and *Giri* HB1 was only detectable in external hyphae. Other genes identified include chitin synthase, β-tubulin and metabolic enzymes (Harrier, 2001). A further fascination of AMF biology have been the identification and gene characterisation of bacteria like organisms (BLO's) present through out the life cylce of some AMF isolates of obligate endo-symbiotic bacteria, belonging to the genus *Burkholderia* (Biancioth *et al.*, 1996). The bacteria cause a complication in the analysis of fungal DNA from spores as well as making an unknown contribution to gene expression and the functional contribution of AMF to the symbiosis. The presence of nitrogen fixation genes have also been detected (Minerdi *et al.*, 2001).

There is considerable evidence for the role of arbuscular mycorrhizal fungi in the control of root pathogens (Caron, 1989, Starnaud et al., 1995). The possible mechanisms involved, include competition for colonization sites, direct antibiosis, nutritional aspects and plant defense reactions (Azcon-Aguilar and Barera, 1996).

1. The colonization of roots by AM fungi induces biochemical changes within host tissues.

2. Stimulation of the phenylpropanoid pathway (Harrison and Dixon, 1993), change in levels of aliphatic polyamines (El-Ghachtouli et al., 1995).
3. Synthesis of proteins of unknown function (Dumas-Gaudot et al., 1994).
4. Activation of defense related genes (Franken and Gnadinger, 1994) enhancement of certain hydrolase activities (Dumas-Gaudot et al., 1992).

Plant hydrolytic enzymes i.e. chitinase and $\beta_{1,3}$ glucanases, appear good candidates amongst molecules with a potential role in bioprotection. Chitinases glycanohydrolase catalyze the hydrolysis of chitin a linear homopolymer of $\beta_{1,4}$ linked N-acetylglucosamine resideues, $\beta_{1,3}$ glucanases. Since the cell walls of many fungi contain chitin on β-D glucans as major structural components, chitinases and β-1,3 glucanses were proposed early as potential antifungal compounds in plant disease control (Mauch et al., 1988; Faull, 2000). Chitinase activity was also found to increase prior to colonization of alfalfa roots inoculated by the mycorrhizal fungus *G. intraradix* and again this was interpreted as a host defense response which was subsequently suppressed (Volpin et al., 1984).

Mycorrhizal fungi and root pathogens compete for the carbon compounds reaching the root (Linderman 1994). When mycorrhizal fungus has primary access to photosynthates, the higher demand may inhibit pathogen growth (Azcon-Aguilar and Barera 1996). If VAM fungus colonizes the root then this would limit the colonization of the pathogenic fungi to only areas of the root which had not been colonized, thus affording protection (Goncalves et al., 1991). Cordier et al. (1996) showed that *Phytophthora* does not penetrate arbuscule containing cells and its development is also reduced in adjacent uncolonized regions. Mycorrhizal symbiosis increases the nutrient uptake and results in more vigorous plants, the plant itself may thus become more resistant to pathogen attack (Faull, 2000). Benefits achieved by mycorrhizal inoculation against pathogens could not be duplicated by adding P fertilizer (Cooper and Grandison, 1997). Mycorrhizal fungi compensate for the root disruption caused by the pathogen allowing the plant to continue functioning inspite of damaging levels of the fungi and the nematodes (Cordier et al., 1996; Pinochet et al., 1996). Colonization with VAM fungi also results in increased amino acid production especially arginine which inhibits pathogen spore formation (Bartuschat and Schonkbeck, 1975). Exudation pattern also changes both qualitatively as well as quantitatively (Graham et al., 1981). Changes in soil microorganisms community induced by VAM formation may lead to stimulation of those microorganisms which may be antagonistic to root pathogens.

PRODUCTION OF SECONDARY METABOLITES

The term secondary metabolite was formalised by Bu'Lock (1961) is generally accepted to describe compounds that are naturally produced by microbes, but not important in the organisms primary metabolism. It is important to consider the origins of these compounds. The pathways of secondary metabolite production arose initially by duplication of the genes for the primary metabolic pathways. Duplication was followed by changes to the genes of the secondary primary pathway which lead to new products accumulating in the cell. These events are summarised in Fig. 1

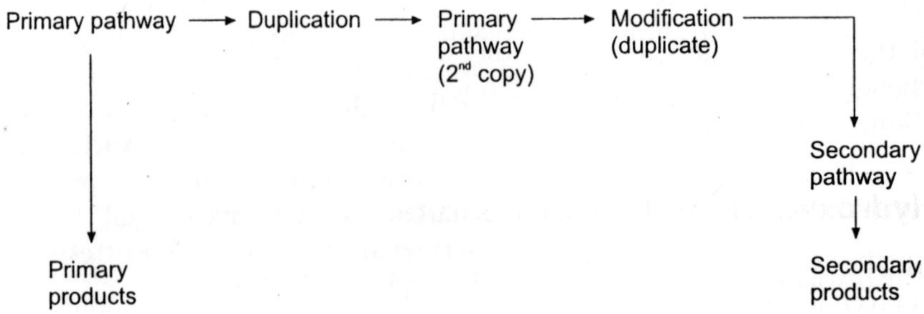

Fig. 1 Pathway of secondary metabolite production

Among the various secondary metabolites, phytoalexins, phenols and HRGPs are very common antifungal compounds.

Phytoalexins: These are low molecular weight isoflavoid compounds derived by phenyl propanoid pathway. These are the toxic compounds released at the site of infection. The isoflavonoids, glyceollin which is fungitoxic (Kaplan *et al.*, 1980) and coumestrol which is nematostatic, together with the coumestan isojagol, showed greater amounts of accumulation in mycorrhizal roots than in non-mycorrhizal roots (Morandi and Lequerre, 1991). Enhanced levels of glyceolin have only been detected in later stages of symbiosis and not during early stages of root colonization. The level of phytoalexin medicarpin, medicarpin-malonyl glucoside, daidzein, coumestrol, formononetin, formononetin-malonylglucoside and 4',7-dihydroxy inoculation (Harrison and Dixon, 1993) showing that phytoalexins and their precursors are activated by VAM inoculation.

Phytoalexins are not only antifungal (Burden et al., 1974) but also interfere with the integrity of bacterial membranes (Smith and Banks, 1986). There is considerable evidence to suggest that in addition to pathogens, beneficial organisms such as rhizobia and mycorrhizal fungi elicit phytoalexins (Wyss et al., 1991). In contrast phytoanticipins are polycyclic compounds which are chemically identical to phytoalexins but are stored in plant cells in anticipation of or prior to pathogen attack and released at different times and in different concentrations, and are solely produced from pre-existing constituents (Vanetten et al., 1995). function of an isoflavonoid molecule acting as a phytoalexin or phytoanticipin can be predicted based on the cellular location or the molecule (Stafford, 1992). Increase in the activities and transcripts of two enzyme : phenylalanine ammonia lyase (PAL) the first in the phenyl propanoid pathway and chalcone isomerase (CHI) specific to the isoflavanoid biosynthesis and followed by decrease in both the levels below those in *Medicago sativa* (Volpin et al., 1994, 1995).

Phenolics: These are aromatic compounds having various attacked substituent group such as hydroxyl, carboxyl and methoxyl groups. Many of these compounds appears to be simply by products of metabolism. Phenolics may have a role in protecting VAM roots against pathogenic fungi (Grandmaison et al., 1993).

Hydroxyproline Rich Glycoproteins (HRGP)

HRGPs are structural cell wall glycoproteins which accumulate in intercellular spaces or as wall appositions and papillae (Benhamou *et al.*, 1990). significant amounts of HRGP-encoding mRNA has also been found to accumulate in mycorrhizal parsley roots (Franken and Gnadinger, 1994). maize roots, corresponding glycoprotein has been immunolocated around arbuscule, hyphae both in the structured wall material at the point of cell penetration and in the interfacial contact zone (Bonfante et al., 1991).

Pathogenesis Related (PR) Proteins

PR proteins are induced in plants during resistance to pathogen infection (Carr and Klessig, 1989). These PR proteins have been grouped into eleven families designated PR-I to PR-II proteins are hydrolytic enzymes with β-1,3 glucanase or chitinase activities (Stintize et al., 1993).

CONCLUSION

Selection of effective mycorrhizal fungi. The selection of mycorrhizal fungi for biocontrol purposes must be based on their ability to survive in the rhizosphere and in soil, aggressively colonize host root and be highly effective on the host, selection of highly effective mycorrhizal fungi may enhance plant tolerance to a pathogen, or compete effectively against indigenous mycorrhizal species, even if it was not an effective biocontrol agent. Organic amendments such as forest humus and charcoal compost appear to enhance AMF density and augment mycorrhizal benefits including a significant reduction in disease severity.

Bibliography

Albrecht, C., Genrts, R., Lapeyrie, F. and Bisseling, T. 1998. Endomycorrhizae and rhizobial Nod factors both require sym8 to induce the expression of the early nodulin genes P_sENOD5 and P_sENOD12A. Plant. J-15, 605-614.

Al-Momany and Al-Raddad, A.M. 1988. Effect of vesicular arbuscular mycorrhizae on *Fusarium* wilt of tomato and pepper. J. Agric. Rec. 33: 249-261.

Anandaraj, M., Ramana, K.V. and Sarma, Y.R. 1993. Suppressive effects of VAM on root pathogens of black pepper, a component of Western Ghats forest ecosystem. In: IUFRO Symposium, Kerala Forest Research Institute Peechi, p. 64.

Azcon, R. and Ocampo, J.A. 1984. Effect of root exudation on VA mycorrhizal infection and mycorrhizal dependency of thirteen wheat cultivars. New Phytologist. 87: 677-685.

Azcon-Aguilar, C. and Barea, J.M. 1996. Arbuscular mycorrhizas and biological control of soil-borne plant pathogens – an overview of mechanisms involved. Mycorrhiza, 6: 457-464.

Balestrini, R., Jose Estanyol, M., Pingdomenech, P., Bonfanta, P. 1997. Hydroxyproline rich glycoprotein mRNA accumulation in maize root cells colonized by an arbuscular mycorrhizal fungus as revealed by in situ hybridization. Protoplasma. 198: 36-42.

Baltruschat, H. and Schonbeck, F. 1975. The influence of endotrophic mycorrhiza on the infestation of tobacco by *Thielaviopsis basicola Phytopathol*. Z-172-188.

Benhamou, N., Mazau, D., Esquerre, Tugaye, M.T. and Asselin, A. 1990. Immunogold localization of hydroxyproline rich glycoproteins in necrotic tissue of *Nicotiana tabacum*. L. Cu. *Xanthine* infected by tobacco mosaic virus. Physiol. Mol. Plant Pathol. 36: 129-145.

Biancioth, V., Bandi, C., Minerdi, D., Sironi, M., Tichy, H.V. and Bonfante, P. 1996. An obligately endosymbiotic mycorrhizal fungus itself harbors obligately intracellular bacteria. Appl. Environ. Microbiol. 62: 3005-3010.

Bisht, V.S., Krishna, K.R. and Nene, Y.L. 1985. Interaction between vesicular arbuscular mycorrhiza and *Phytophthora drechsleri* f. sp. cajani, Internat. Pigeon Pea News Lett. 4: 63-64.

Blee, K.A. and Anderson, A.J. 2000. Defense responses in plants to arbuscular mycorrhizal fungi. In: Current Advances in Mycorrhizae Research. Eds. G.K. Podila and D.D. Douds pp.27-43, APS Press, St. Paul, Minnesota.

Bonfante, P., Tanagrove, C., Peretto, R., Esquerre-Tugaye, M.T., Mazau, D., Mosinik, M. and Viam B. 1991. Immuno-chemical location of hydroxyproline rich glycoproteins at the interface between a mycorrhizal fungus and its host plant Protoplasma. 165: 127-138.

Burden, R.S., Rogers, P.M., and Wain, R.L. 1974. Investigations on fungicides, XVI. Natural resistance of plant roots to fungal pathogens. Annals Applied Biology, 78: 59-63.

Caron, M. 1989. Potential use of mycorrhizae in control of soil borne disease. Can. J. Plant Pathol. 11, 177-179.

Caron, M., Fortin, J.A. and Richard, C. 1986. Effect of inoculation sequence on the interaction between *Glomus intraradices* and *Fusarium oxysporum* f. sp. radicis – lycopersici in tomatoes. Can. J. Plant Pathol. 8: 12-16.

Caron, M., Fortin, J.A. and Richard, C. 1986a. Effect of *Glomus intraradices* on infection by *F. oxysporum* f. *radicis lycopersici* in tomato over a 12-week period. Can. J. Bot. 64: 552-556.

Caron, M., Fortin, J.A. and Richard C. 1986b. Effect of phosphorus concentration and *Glomus intraradices* on *Fusarium* crown and root rot of tomatoes, Phytopathology 78: 942-946.

Carr, J.P. and Klessig, D.F. 1989. The pathogenesis related proteins in plants. In : Genetic engineering: Principles and Methods, Plenum Press, New York, pp.65-109.

Copper, K.M. and Grandison, G.S. 1987. Effects of vesicular arbuscular mycorrhizal fungi on infection of tomarillo (*Cyphomandra betacea*) by *Meloidogyne incognita* in fumigated soil. Plants Dis. 71: 1101-1106.

Cordier, C., Gianinazzi, S. and Gianinazzi Pearson, V. 1996. Colonization patterns of root tissues by *Phytophthora nicotianae* var. *parasitica* related to reduced disease in mycorrhizal tomato. Plant Soil 185: 223-232.

Cordier, C., Pozo, M.J., Barea, J.M., Gianinazzi, S., Gianinazzi-Pearson, V. 1998. Cell defense responses associated with localized and systemic resistance to *Phytophthora parasitica* induced in tomato by an arbuscular mycorrhizal fungus. Molecular Plant Microbe Interactions 11: 1017-1028.

Dehne, H.W. and Schoenbeck, F. 1979b. The influence of endotrophic mycorrhiza on plant disease, 2, Phenol metabolism and lignification, Phytopath. Z-95: 210-216.

Davis, R.M. and Menge, J.A. 1980. Influence of *Glomus fasciculatum* and soil phosphorus on *Phytophthora* root rot of citrus. Phytopath. 68: 1614-1617.

Dehne, H.W. 1977. Untersuchungen uber den Einfluss der endotrophen mycorrhiza anf die *Fusarium* Welke an Tomato und Gurke, Dessertation, Univ. Bonn., W. Germany, 150 p.

Dehne, H.W. and Schoenbeck, F. 1975. The influence of endotrophic mycorrhiza on the *Fusarium* wilt of tomato. Z. Pflanzentir. Pflazenchutz. 82: 630.

Dehne, H.W. and Schoenbeck, F. 1979. The influence of endotrophic mycorrhiza on plant diseases, 1. Colonization of tomato plants by *F. oxysporum* f. sp. lycopersici. Phytopath. Z-95: 105-110.

Dehne, H.W. and Schonbeck, F. 1979a, Untersuchungen Zum Einflu - b-der-endotrophen Mykorrhiza auf Pflanzenkrankheiten, II Phenol stoffwechsel und Lignitizierung, Phytopathol. Z-5: 210-216.

Dehne, H.W. and Schonbeck, F. 1979b. The influence of endotrophic mycorrhizae on plant disease, I, colonization of tomato plants by *Fusarium oxysporum* f. sp. lycopersici, Phytopathol Z. 95-105-109.

Delp, G., Smith, S.E. and Baker, S.J. 2000. Isolation by differential display of three cDNAs coding for proteins from the VA mycorrhizal fungus *G. intraradices*. Mycol. Res. 104: 293-300.

Bu'Lock, J.D. 1961. Advances in Applied Microbiology. 3, 293-242.

Dumas Gaudot, E., Grenier, J., Furlan, V. and Asselin A. 1992. Changes I polypeptide pattern in tobacco roots colonised by two *Glomus* species. Mycorrhiza, 4, 215-221.

Dumas Gaudot, E., Guillaume, P., Tahiri-Alaoui, A., Gianinazzi, S. 1994. Chitinase-chitosanase and β-1,3 glucanase activities in Allium and *Pisum* roots colonized by *Glomus* species. Plant Science, 84, 17-24.

El-Ghachtouli, N., Paynot, M. Morandi, D., Martin, T.J. and Gianinazzi, S. 1995. The effect of polyamines on endomycorrhizal infection of wild type *Pisum sativum* C.V. Frisson and two mutants. Mycorrhiza, 5: 189-192.

Faull, J.L. 2000. Secondary metabolism in Fungi: A Determinant of Substrate Possession and Intermycelial Competition? In Glimpses in Plant Sciences (Eds. K.R. Aneja, and Charaya, A.). APH Publishing House, 200-207.

Franken, P. and Gnadinger, F. 1994. Analysis of parsley arbuscular mycorrhiza: infection development and mRNA levels of defence related genes. MPM I 7, 612-620.

Franken, P. and Requend, 2001. Analysis of gene expression in arbuscular mycorrhizas : new approaches and challenges. New Phytologist: 150 : 517-523.

Gianinazzi-Pearson, V. and Denarie, J. 1997. Red carpet programmes for root endosymbiosis. Trends Plant Sci. 2, 371-372.

Gianinazzi-Pearson, V., Arnould, C., Oufattole, M., Arango, M. and Gianinazzi, S. 2000. Differential activation of H^+ ATPase genes by an arbuscular mycorrhizal fungus in root cells of transgenic tobacco. Plants 211, 609-613.

Gonclaves, E.J., Muchovej, J.J. and Muchovej, R.M.C. 1991. Effect of kind and method of fungicidal treatment of bean seed on infection by the VA mycorrhizal fungus *glomus macrocarpum* and by the pathogenic fungus *Fusarium solani*, I, Fungal and Plant parameters. Plant Soil 132: 41-46.

Graham, J.H. and Egel, D.S. 1988. *Phytophthora* root rot development on mycorrhizal and phosphorus fertilized non-mycorrhizal sweet orange seedlings. Plant Dis. 72: 611-614.

Graham, J.H., Leonard, R.T. and Menge, J.A. 1981. Membrane mediated decrease in root exudation responsible for phosphorus inhibition of vesicular arbuscular mycorrhizae formation. Plant Physiol. 68: 548-552.

Grandmaison, J., Olah, G.M., Van Calsteren M.R. and Furlan, V. 1993. Characterization and localization of plant phenolics likely involved in the pathogen resistance expressed by endomycorrhizal roots. Mycorrhiza 3: 155-164.

Harrier, L.A. 2001. The arbuscular mycorrhizal symbiosis : a molecular review of the fungal dimension. J. Exp. Bot. 52: 469-478.

Harrison, M. J. 1998. Molecular and cellular aspects of the arbuscular mycorrhizal symbiosis. Ann. Rev. Plant Physiol. Plant Mol. Biol. 50, 361-389.

Harrison, M.J. and Dixon, R.A. 1993. Isoflavonoid accumulation and expression of defence gene transcripts during the establishment of vesicular arbuscular mycorrhizal associations in roots of *Medicago truncatula*, MPMI, 6, 643-654.

Helgason, T., Daniell, T.J., Husband, R., Fitter, A.H., Young, J.P.W. 1998. Ploughing up the wood-wide web ? Nature. 394-431.

Hussey, R.S. and Rancadori, R.W. 1982. Vesicular arbuscular mycorrhizae may limit nematode activity and improve plant growth. Plant Dis. 66: 9-13.

Hwang, S.F., Chang, K.F. and Chakravarty, P. 1992. Effect of vesicular arbuscular mycorrhizal fungi on the development of *Verticillium* and *Fusarium* wilt of alfalfa. Plant Dis. 76: 239-243.

Jalali, B.L. 1983. Biological suppression of root disease fungi by vesicular arbuscular mycorrhizal system. In: Proc. 4th Inter Congr. Plant Pathol. Melbourne, pp. 46.

Jalali, B.L. and Thareja, M.L. 1981. Suppression of fusarial wilt of Chicken pea in VA mycorrhiza inoculated soils. Intern. Chickpea Newsl. ICRISAT, 4: 21-23.

Kaplan, D.R. Keen, N.T. and Thomson, I.J. 1980. Study of the mode of action of glyceollin in soyabean, incompatibility to the root knot nematode *Meloidogyne incognita*. Physiol, Plant Pathol. 16: 319-325.

Kaushik, A. and Mukerji, K.G. 1996. Mycorrhiza in control of Disease in : Disease scenario in crop plant. Vol. II, cereals, pulses, oilseeds and cash crops. V.P. Agnihotri, O. Prakash, R. Kishun and A.K. Mishra, eds., International Books and Periodical Supply Service, Delhi, India, pp. 217-231.

Keen, N.T. and Yoshikawa, M. 1983. Physiology of disease and nature of resistance the *Phytophthora*. In: *Phythophthora* its biology, taxonomy, ecology and pathology, D.C. Erwin, S. Bartnicki-Garcia and P.H. Tsao eds. pp. 279-287, APS Press St. Paul, U.S.A.

Koide, R.T. and Scheiner, R.P. 1992. Regulation of the Vesicular Arbuscular Mycorrhizal Symbiosis. Annhual Review of Plant Physiology and Plant Molecular Biology, 43: L 557-581.

Krishna, K.R. and Bagyaraj, D.G. 1984. Phenols in mycorrhizal roots of *Arachis hypogea*. Experimentia 40: 85-86.

Lamb, C.J., Lawton, M.A., Dron, M. and Dixon, R.A., 1989. Signal and transduction mechanisms for activation of plant defenses against microbial attack. Cell. 56: 215-224.

Lapopin, L. Gianinazzo-Pearson, V. and Franken, P.1999. Comparative differential RNA Display analysis of arbuscular mycorrhiza in *Pisum sativum* wild type and a mutant defective in late stage development. Plant. Mol. Biol. 41, 769-867.

Linderman, R.G. 1994. Role of VAM fungi in biocontrol. In: Mycorrhiza and Plant Health. F.L. Pleger and R.G. Linderman. eds. APS. St. paul, pp. 1-26.

Mauch, F., Manch Mani, B. and Boller, T. 1988. Antifungal hydrolases in Pea tissue II. Inhibition of fungal growth by combinations of chitinase and β-1,3 glucanase. Plant Physiol. 88, 936-942.

Melo, I.S., Costa, C.P. and Silveria, A.P.D. 1985. Effect of vesicular arbuscular mycorrhizas on auber gine wilt caused by *Verticillium albo atrum* Reinke and Berth, Summa Phytopath, 11: 173-179.

Minerdi, D., Fani, R., Gallo, R., Boarino, A. and Bonfante, P. 2001. Nitrogen fixation genes in an endosymbiotic *Burkholderia* strain. Appl. Env. Microbiol, 67 : 725-732.

Morandi, D. and Le querre, J.L. 1991. Influence of nitrogen on accumulation of isosojagol and associated isoflavanoids in roots and nodules of mycorrhizal and non-mycorrhizal soyabean. New Phytol. 117: 75-79.

Mukerji, K.G. 1999. Mycorrhiza in control of plant Pathogens: Molecular Approach. In: Biotechnological Approaches in Biocontrol of Plant Pathogens. eds, K.G. Mukerji, B.P. Chamola and R.K. Upadhyay, Kluwer Academic/Plenum Publishers, New York, USA pp. 135-156.

Parvathi, K. Venkateswarlu, K. and Rao, A.S. 1985. Influence of root infecting fungi on development of *Glomus mosseae* in groundnut. Current Sci. 54: 1006-1007.

Peterson, R.L. Piche, Y. and Planchette, C. 1989. Mycorrhiza and their potential use in the agricultural and forestry industries. Biotechnical. Advan. 2: 101-120.

Pinochet, J., Calvet, C. Canprubi, A. and Fernandez, C. 1996. Interactions between migratory endoparasitic nematodes and arbuscular mycorrhizal fungi in perennial crops. A. Review, Plant Soil 185: 183-190.

Porcel, R., Barea, J.M. and Lozano, J.M.R. 2003. Antioxidant activities in mycorrhizal soyabean plants under drought stress and their possible relationship to the process of nodule senescence. New Phytologist. 157: 135-143.

Prosser, J.I., 2002. Molecular and functional diversity in soil microorganisms. Plant and Soil. 244: 9-17.

Ramirez, B.N. 1974. Influence of endomycorrhizae on the relationship of inoculum density of *Phytophthora palmivora* in soil to infection of papaya roots. M.Sc. Thesis Univ. Florida, Gainsville USA.

Rosewarne, G.M. 1998. Molecular characterisation of the vesicular arbuscular mycorrhizal symbiosis in *Lycopersicon esculentum* Mill. Ph.D. Thesis. The University of Adelaide, South Australia.

Roussel, H., Tuinen, D.V., Franken, P., Gianinazzi, S. & Gianinazzi Pearson, V. 2001. Signalling between AM fungi and plants : identification of a gene expressed during early interactions by differential RNA display analysis. Plant and Soil. 232: 13-19.

Sarma, Y.R., Anandraj, M. and Venugopal, M.V. 1996. Biological control of disease of spices, In: Biological control of spices, eds. M. Anandaraj and K.V. Peter, Indian Institute of Spices Research (ICAR) Calicut, Kerala, India, pp. 1-19.

Schenck, N.C. and Kellam, M.K. 1978. The influence of vesicular arbuscular mycorrizae on disease development. Fla Agric. Exp. Stn. Bull. p.799.

Schenck, N.C. and Perez, Y. 1987. Manual for the identification of VA fungi. Gainsville, Florida, INVAM.

Schonbeck, F. 1979. Endomycorrhiza in relation to plant diseases in soil borne plant pathogens, B. Schippens and W. Ganis eds., Academic Press, New York, USA, pp. 271-280.

Singh, R., Adholeya, A. and Mukerji, K.G. 2000. Mycorrhiza in control of soil borne pathogens. In; Mycorrhizal Biology (eds. K.G. Mukerji, B.P. Chamola and Jagjit Singh) Kluwer Academic/Plenum Publishers, pp. 173-196.

Smith, D.A. and Banks S.W. 1986. Biosynthesis, elicitation and biological activity of isoflavonoid phytoalexins. Phyto-chemistry, 25: 979-995.

Smith, S.E., Read, D.J. 1997. Mycorrhizal symbiosis. 2nd edn. San Diego CA, USA, Academic Press.

Stafford, H.A. 1992. Flavonoid Metabolism. CRC Press, Boca Raton, F.L. pp. 202-207.

Starnaud, M., Hamel, C., Caron, M. and Fortin, J.A. 1995. Inhibition of *Pythium ultimum* in roots and growth substrate of mycorrhizal *Tagetus patula* colonized with *Glomus intraradices*. Can. J. Plant Pathol. 16, 187-194.

Stintize, A., Heitz, T., Prasad, V., Wiedemann – Meidino Gulu, S., Kaufmann, S., Geoffroy, P., Legrand, M. and Fritig B. 1993. Plant pathogenesis – related proteins and their role in defence against pathogens, Biochimie, 75: 687-706.

Subramanian, K.S., Charest, C., Dwyer, L.M., Hamilton, R.I. 1995. Arbuscular mycorrhizas and water relations in maize under drought stress at tasselling. New Phytologist 129: 643-650.

Thomas, L., Mallesha, B.C. and Bagyaraj, D.J. 1994. Biological control of damping off of Cardamon by VA mycorrhizal fungus *Glomus fasciculatum*. Microbiol. Res. 149: 413-417.

Trotta, A., Varese, G.C., Gnavi, E. Fusconi, A., Sampo, S. and Berta G. 1996. Interactions between the soil borne root pathogen *Phytophthora nicotianae* vas. parasitica and the arbuscular mycorrhizal fungus *Glomus mosseae* in tomato plants, Plant and Soil, 185: 199-209.

Van Etten, H.D., Mansfield, J.W., Bailey, J.A. and Former E.E. 1995. Two classes of plant antibiotics; phytoalexins versus "phytoanticipins". Plant Cell, 6: 1191-1192.

Van Loon, L.C., Piperpoint, W.S., Boller, T. and Conejero, V. 1994. Recommendations for naming plant pathogenesis related proteins. Plant Mole. Biol. Rep. 12: 245-264.

Volpin, H., Elkind, J., Okon, Y. and Kapulnik, Y. 1994. A vesicular arbucular mycorrhizal fungus *Glomus intraradix* induces a defuse response in alfalfa roots. Plant Physiol. 104, 683-689.

Wyss, P., Boller, T. and Wiemken, A. 1991. Phytoalexin response is elicited by a pathogen (*Rhizoctonia solani*) but not by a mycorrhizal fungus (*Glomus mosseae*) in soybean roots. Experimentia, 47: 395-399.

Young, C.C., Juang, T.C. and Guo, H.Y. 1984. Effects of VA mycorrhizal inoculation of the soyabean growth, yield, N-fixation and absorption of mineral phosphate. J. Agric. Assoc. China 128: 29-40.

SECTION 7

LICHENS

Frontiers in Plant Sciences (Pages 719-730)
Eds. K.G. Mukerji, K.V.B.R. Tilak, S.M. Reddy, L.V. Gangwane, P. Prakash, I.K. Kunwar

Phytosociology of Macrolichens Occurring on Pine Tree of Nainital Kumaun Himalaya

S.C. Sati, B. Pathak and M. Belwal

ABSTRACT

The Himalayan region harbours lush green vegetation and support a good lichen flora. The present paper deals with an altitudinal distribution and phytosociology of epiphytic macrolichens occurring on the pine trees (*Pinus roxburghii*), a common gymnosperm of the locality. The study was carried out using a quadrate technique (25 × 25 cm in size) at six sites of different altitudes (1250 to 2300 m asl) of Nainital, Kumaun Himalaya. The data were analyzed for frequency, density, abundance, biomass and IVI of the species. Altogether, 13 macrolichens belonging to 6 genera were encountered. The number of species was found varied site to site. A maximum number of species (11species) were recorded at site III (1700-1800 m. asl), while a minimum species number (6 species) were found at site V (2000-2100 m asl). Four lichen species namely *Parmelia cristifera, P. isidoisa, P. letissima,* and *Physcia setosa* were found the most common and frequently occurring species. Foliose forms were found in greater number and were common in occurrence to all altitudinal range.

Key words: Lichens phytosociology, ecology, phorophytes, symbiosis

INTRODUCTION

Lichens are unique and highly diverse organisms. They make specific demands on the environment and respond critically to change in it. Now they are treated as nature's most remarkable alliance as they can even grow in such climatic conditions where normally no other plant can thrive.

Author's address: Department of Botany, Kumaun University, Nainital - 263002 India

The environmental response and effects on lichen have been studied by Kappen (1973), the substrate ecology and lichen in relation to atmospheric pollution and growth have been studied (Brodo, 1973; Gilbert, 1973 and Hale, 1973). Lichens are preferred in assessment of atmospheric pollution levels as are more effected by pollutants. At a point source of pollution e.g. an industrial area, sometimes resulted to form a 'lichen desert' (Gilbert, 1973). At higher levels of SO_2 in the atmosphere, even highly resistant lichen species may disappear.

Ecological studies on lichen communities have been mainly carried out by descriptive (Alvin, 1960; Brown and Brown, 1969; Sheard, 1968), numerical (Fletcher, 1973b; Yarranton, 1967) and phytosociological (Coker, 1968; Hawksworth, 1969, 1972 a, b; Rose and James, 1974) methods. Some recent reviews available concerned with the theory and practice of phytosociological approaches to the study of plant communities (Barkman, 1973; Hawksworth, 1974; Shimwel, 1971).

Air pollution surveys and bioindicational importance of lichen have been well documented by Ferry et al (1973) and Hawksworth and Rose (1976). Kappen (1976) studied lichens in environmental extremes while, extensive quantitative estimation of corticolous epiphytes including bryophytes and lichens were brought out by Culberson (1973). Modern analytical methods like ordinance technique and multifarious analysis have also been employed for lichen ecology (Beals, 1965; Jesberger and Sheard, 1973).

India has great lichen diversity and stands 5th in term of lichen genera in the world. At present the Indian flora represents about 10% of the world lichens, as it consist of about 1,850 species belonging to 234 genera and 80 families, of which 700 species of 71 genera belonging to macrolichens and 1150 species of 163 genera are of microlichens. Roughly, the Himalayan region represent about 20% of the Lichen flora of India (Upreti, 1997; Singh and Sinha, 1997). In the recent past some taxonomic and phytosociological studies on lichens of Kumaun Himalaya were carried out by Sati et al (1992), Sati et al (1992), Sati and Usman (1992) and Sah (1995) however, the work on lichen ecology in India is quite meager.

The present study describes an altitudinal distribution, phytosociology of some macrolichens occurring on the pine tree (*Pinus roxburghii)*, at six different altitudinal zones at Nainital, Kumaun Himalaya.

SITE DESCRIPTION

The study site Nainital is situated in the outer part of central Himalaya, commonly known as Kumaun Himalaya and lies in between 29°22′- 24′N latitude and 79°28′-29′ E longitude at an elevation of 1200 m to 2600 m height (asl). The area comprises of mostly hilly areas and some plains in the foothills. It comes under the temperate climate. The temperature rises maximum in the summer months (May-mid June) upto 30°C, while in winter (December-March) it slumps to 5° or 0°C. There is monsoon pattern of rain fall being heaviest in July to August and from October to mid December it is usually dry period, however occasional showers are also not uncommon in these months.

Six sites situated at different elevational range viz., Site I- Jeolikot (1250-1350 m. asl), Site II- Veerbhatti (1500-1600), Site III- Alookhet (1700-1800), Site IV –Kailakhan (1850-1950), Site V- Kilburry (2000-2100) and Lariakanta (2200-2300) were selected for the study. All the sites had well marked vegetation of pine, alongwith other common tree plants of this locality. Site I and II were dry and illuminated throughout the day while site III, IV, V and VI were humid and received sunlight either in the morning hours or at evening. At site I, II, III and V pine trees were found in pure patches whereas at sites IV and VI pine trees were dominated with oak, deodar and other trees.

METHODOLOGY

For the present study a thorough survey of all the selected sites was made. At each site 10 trees of Pine (*Pinus roxburghii*) having same girth size i.e., 70-100 cm were selected for lichen vegetation. The trees selected were of almost same age group. Sampling was done after rainy seasons in September and October months. For vegetational analysis quadrate technique was employed. Three quadrates of 25 × 25 cm size were laid randomly on each tree at the breast height. Thus 30 quadrates in all were studied at each site. Lichen thalli were carefully removed with the help of a sharp scalpel and stored in separate paper packets. The material was brought to the laboratory for identifications and quantitative estimation. Identification of lichen was done with the help of available literatures and concern authorities.

Frequency, density, abundance, A/F ratio, biomass and IVI (Importance Value Index) of each lichen species was determined following Mishra (1968).

Importance Value Index

The IVI shows a relative contribution of species to the entire community and was determined by summing up the values of relative frequency, relative density and relative biomass (Curtis and Mc Instosh, 1961). The various expression used are as follows:

$$\text{Relative Frequency (RF)} = \frac{\text{Frequency of the species}}{\text{Sum of the frequency of all species}} \times 100$$

$$\text{Relative density (RD)} = \frac{\text{Total no. of individuals of the species}}{\text{Total no. of individuals of all species}} \times 100$$

$$\text{Relative biomass (RB)} = \frac{\text{Biomass of individual species}}{\text{Biomass of all species}} \times 100$$

$$\mathbf{IVI = RF + RD + RB}$$

RESULT AND DISCUSSION

The phytosociology and altitudinal variation of epiphytic microlichens occurring on pine tree bark with emphasis on the different altitudinal zones at Nainital Kumaun Himalaya was carried out. Vegetational analysis of lichen community at six sites along a definite altitudinal gradient ranging from 1250-2300 (m. asl) was done by using quadrate technique. A total of 180 quadrates (30 at each site) were studied. The data were analyzed employing different ecological parameters viz. frequency, density, abundance, A/F ratio, IVI index and biomass for each species at each site. Analyzed data are tabulated in the Table 1.

Altogether 13 species of lichen were encountered across all sites studied. Obviously the number of species is less and it is because only macrolichens were taken into consideration. Out of these 13 species, 11 were of foliose lichens, while only two species i.e., *Ramalina fraxinia* and *Usnea florida* were of fruticose lichens (Table 2). Foliose forms were greater in number indicating slow succession, as foliose forms are supposed to be more primitive than fruticose. The paucity of fruticose lichens indicates that pine bark was not well suited for the growth of fruticose lichens.

Hoffman (1971) suggested that the floristic richness is a function of available species, microclimatic conditions, physical factors including incidents of dispersals and biotic interrelations. Under the influence of similar environmental factors generally a substratum tend to bear uniform and remarkable assemblage of lichen species (Hawksworth and Hill, 1974). Floristically there was high degree of resemblance between all the sites of study area. *Parmelia* species appeared one of the most important lichen at all the elevations studied. Out of total 13 species, *Parmelia* was represented by 7 species. Further on relying IVI, this species was found dominant in the community except at two sites i.e. Site IV and VI, where *Physcia* and *Ramalina* were found dominant.

TABLE 1 *Frequency, Biomass and IVI values of Lichen species at different altitudinal sites of Nainital*

S. No.	Species	Frequency						Biomass						Importance Value Index (IVI)					
		I	II	III	IV	V	VI	I	II	III	IV	V	VI	I	II	III	IV	V	VI
1.	*Anaptychia firmula* Nyl.	–	–	6.67	–	–	–	–	–	0.23		–	–	–	–	5.85	–	–	–
2.	*Lecidia subkochiana* Ach.	–	–	–	3.33	–	–	–	–	–	0.30	–	–	–	–	–	3.63	–	–
3.	*Parmelia caperata* (L) Ach	–	13.33	3.33	13.33	–	–	–	1.28	0.42	0.73	–	–	–	29.08	6.25	15.64	–	–
4.	*P. cetrarioides* Del.	36.67	10.0	10.0	–	–	23.33	2.13	0.41	0.26	–	–	2.33	75.29	15.40	8.66	–	–	37.17
5.	*P. conspersa* Ach.	3.33	6.67	3.33	6.67	–	16.67	0.50	0.33	0.75	0.78	–	1.04	3.99	9.62	9.48	13.52	–	22.42
6.	*P. cristifera* Tayl.	46.67	13.33	66.67	63.33	26.67	26.67	1.58	0.29	5.82	5.15	1.20	1.25	87.06	15.67	100.64	56.84	41.68	31.74
7.	*P. isidiosa* Mull.	26.67	50.0	16.67	26.67	23.33	6.67	1.57	6.85	3.82	2.64	4.14	0.73	56.65	131.60	41.58	48.61	87.18	12.0
8.	*P. letissima* Hue.	16.67	30.0	16.67	40.0	60.0	30.0	0.86	2.73	2.02	3.62	3.72	4.0	36.76	66.20	31.04	64.08	113.33	55.40
9.	*P. tinctorum* Despr.	–	–	–	–	33.33	–	–	–	–	–	1.34	–	–	–	–	–	49.54	–
10.	*Physcia hispida* Tuekern.	–	–	36.67	6.67	–	13.33			4.53	0.41	–	0.56	–	–	62.94	9.56	–	12.95
11.	*Physcia setosa* (Ach.) Nyl	10.0	10.0	20.0	60.0	3.33	6.67	0.16	1.32	0.69	2.60	0.04	0.37	12.14	24.69	18.75	65.87	3.16	6.98
12.	*Ramalina fraxinia* Ach.	10.0	3.33	6.67	10.0	–	73.33	1.07	0.47	0.26	1.34	–	7.58	28.06	7.67	6.25	17.46	–	118.10
13.	*Usnea florida* web.	–	–	10.0	–	3.33	3.33	–	–	0.38	–	0.22	0.10	–	–	8.58	–	5.08	2.76

Site I = Jeolikot (1250-1350 m asl) Site II = Veerbhatti (1500-1600 m asl) Site III = Alookhet (1700-1800 m asl)
Site IV=Kailakhan (1850-1950 m asl) Ste V = Kilburry (2000-2100 m asl) Site VI= Lariakanta (2200-2300 m asl)

TABLE 2 *Species composition of Lichen at different altitudinal sites*

S. No.	Species	Sites					
		I	II	III	IV	V	VI
1.	*Anaptychia firmula* Nyl.	–	–	+	–	–	–
2.	*Lecidia subkochiana* Ach.	–	–	–	+	–	–
3.	*Parmelia caperata* (L) Ach	–	+	+	+	–	–
4.	*P. cetrarioides* Del.	+	+	+	–	–	+
5.	*P. conspersa* Ach.	+	+	+	+	–	+
6.	*P. cristifera* Tayl.	+	+	+	+	+	+
7.	*P. isidiosa* Mull.	+	+	+	+	+	+
8.	*P. letissima* Hue.	+	+	+	+	+	+
9.	*P. tinctorum* Despr.	–	–	–	–	+	–
10.	*Physcia hispida* Tuekern.	–	–	+	+	–	+
11.	*Physcia setosa* (Ach.) Nyl	+	+	+	+	+	+
12.	*Ramalina fraxinia* Ach.	+	+	+	+	–	+
13.	*Usnea florida* web.	–	–	–	–	+	+
	Total Species	7	8	11	9	6	9

Site I = Jeolikot (1250-1350 masl) Site II= Veerbhatti (1500-1600 masl)
Site III = Alookhet (1700-1800 masl) Site IV= Kailakhan (1850-1950 masl)
Site V = Kilburry (2000-2100 masl) Site VI = Lariakanta (2200-2300 masl)

In term of total dry matter (g/m^2), the communities growing over the trees at higher elevations (Site III, IV and VI) had maximum biomass (Table 1). This may be due to the fact that all these sites had greater number of individuals of lichens. Comparing the number of individuals and biomass of lichens, site VI (2200-2300 m) had maximum number of individuals, while total dry matter was higher in site III. It may be due to the fact that site III (1700-1800 m) was dominated by *Parmelia cristifera* and *Physcia setosa* with maximum number of individuals and both the type had bulky thalli compared to the lighter thalli of *Ramalina fraxinia* which was found dominated in site VI (2200- 2300 m altitude).

The assemblage of lichen and their richness on phorophytes are largely under the control of various factors viz., pH of the bark surface, nutrient status of the bark, moisture retaining and absorbing capacity, degree of illumination and the humidity of surrounding atmosphere (James et al, 1971). Though the species composition of all sites was not much variable (Table 1), the species number varied from one altitudinal zone to another. The maximum species richness was found at an altitude between 1700-1800 m (site III), followed by sites IV and VI (Table 2). The variation in species may be due to prevailing climatic conditions and status of host tree species in the forest. As in sites IV and VI, there was moist and humid climate coupled with low temperature and less illumination. Here the pine trees were intermixed with broad leaved evergreen species like *Quercus, Rhododendron* and other shrub species. Slightly decreasing in number at site V (2000–2100 m) may be due to less humid condition as represented a negligible under vegetation (shrub and herbs) at this site. Another important observation to note that this site had sign of forest fire, which might have broken few months before the collection (summer fire).

Three species namely *Anaptychia firmula, Lecidia subkochiana* and *Parmelia tinctorum* were found restricted in distribution. The former two species were confined at altitudes 1700-1800 m (site III) and 1850-1950 m (site IV), while the last one was found only at site V (2000-2100 m).

Collection of lichens for commercial purpose and vehicle fumes of heavy traffic seemed to be responsible for low number of lichens species at sites I which is at lowest altitude (1250-1350 m) as this site is close to the main road with 24 hours traffic. Moreover the scrapping off of lichens from trees is a common practice in these forest because of easy transportation facilities. From climatic point of view Site I and V were much illuminated, exposed and dry.

It is remarkable to note that distribution pattern of lichen species at all altitudes was contagious. This indicate the poor distribution of lichens in the particular sites as the number of individual collected for most of the species was too low. In some cases e.g., *Usnea florida*, it was even a single individual per quadrate and frequency was also poor. The contagious distribution pattern is quite relevant for such conditions. This may be due to various climatic factors that prevail on chir pine forest which is characteristically dry. Moreover, as the collection was made during September – December months, the lichens may have started shrinking and decaying by the time due to onset of winter and autumn.

From the vegetational analysis it is evident that though the pine bark in all study sites situated at different altitudes, supported more or less similar type of lichen flora, yet the various climatic as well as biotic factors played a significant role in occurrence, growth and distribution of lichen species.

Out of total six sites *Parmelia cristifera* was found the most frequent and common species at 3 sites (site I, II and IV) having maximum frequency and density. At sites II and V, *P. issidosa* and *P. letissima* were found most frequent. While *Ramalina fraxinia* was found frequent only at site VI (2200-2300 m) with maximum frequency and density. Though with increasing altitude, upto site V (2000-2100 m) only the foliose forms (*Parmelia sp.*) occupied the status of most frequent species. However sudden appearance of *Ramalina fraxinia*, a fruticose form as dominant species at the higher elevation may be due to much favourable conditions for the lichen growth. The frequency of all species including *Ramalina* was found maximum at site VI (Table 1). The A/F ratio of all species at each altitudinal sites exhibit a contagious pattern of distribution. Mostly the foliose lichen forms had maximum IVI in all the sites except at site VI. The lower site were dominated by *Parmelia cristifera* (at sites I, III and IV), *P. isidoisa* (at site II) and *Parmelia letissima* (at site V) respectively.

Persistence of substrate is also a key factor for occurrence and distribution, short-lived habitats are less easily colonized by lichens (Hawksworth and Hill, 1984). The chir pine bark is not persistent as it shades off regularly in chips from the trunk. Thus pine host itself is not ideal for rich lichen diversity. Upreti (1997) also suggested that the common pine (*Pinus roxburghii*) bears a few species of *Parmelia* and *Physcia* in dry places in lower altitude (1500-2000 m. asl).

As evident from the available literature, the earlier workers have not made altitudinal distribution and phytosociological study of lichens. Therefore the present work provides new information on the occurrence and distribution of lichens.

Acknowledgement

The authors are thankful to Prof. S. P. Singh, Head, Department of Botany, Kumaun University, Nainital for providing lab facilities.

References

Alvin, K.L. 1960. Observations on the Lichen ecology of South Haven Peninsula, Studland Heath, Dorset. *J. Ecol.* 48: 331-339.

Barkman, J.J. 1973. Synusial approaches to classification. In: *Handbook of Vegetation Science* (R.H. Whittaker, Ed.), 5: 435-491p., W. Junk, The Hague.

Beals, E.W. 1965. *Ordination of some Corticolous Cryptogamic communities in south Central Wisconsin*, Oikos, 16: 1-8.

Brodo, I.M. 1973. Substrate Ecology. In: *The lichens* (V. Ahmadjain and M.E. Hale, eds.), pp. 401-439. Academic Press, New York and London.

Brown, D.H. and Brown, R.M. 1969. Lichen Communities at Balkeney Point, Norfolk. Trans. Norfolk Norwich Nat. Sci. 21: 235-250pp.

Coker, P.D. 1968. The epiphytic communities of the Ruislip district. *E. Ruislip District Nat. Hist. Soc.* 17: 26-37p.

Culberson, W.L. 1955. Qualitative and quantitative study on the distribution of Corticolous lichens and bryophytes in Wisconsin. *Lloydia,* 18:25-36.

Ferry, B.W. Baddeley, M.S. and Hawksworth, D.L. 1973. *Air pollution and Lichens.* Athlon press, London.

Fletcher, A. 1973b. The ecology of maritime (supralittoral) lichen on some rocky shores of Anglesey. *Lichenologists,* 5: 368-400p.

Gilbert, O.L. 1973. Lichens and air pollution. In: *The Lichens (V. Ahmadjian and M.E. Hale, eds.)*, pp. 443-472. Academic press, London and New York.

Hale, M.E. Jr. 1973. Growth *In: Lichens (Ahmadjian, V. and Hale, M.E. eds)*, pp. 473-492, Academic press New York.

Hawksworth, D.L. 1969. The lichen flora of Derbyshire. *Lichenologists*, 4, 105-193.

Hawksworth, D.L. 1972. The natural history of Slapton ley Nature Reserve. IV. *Lichens Fld. Stud.* 13, 535-578.

Hawksworth, D.L. and Hill, D.J. 1984. *The lichen forming Fungi*, Blackie, Glassow and London. 158 pp.

Hawksworth, D.L. and Rose, E. 1976. *Lichens as Pollution Monitors*. Arnold London.

Jesberger, J.A. and Sheard, J.W. 1973. A quantitative study and multivariate analysis of corticolous lichen communities in the Southern boreal forests of Saskatchewan. Can. J. Bot., 51: 185-201.

Kappen, L. 1973. Response to Extreme Environments. In: *The Lichens* (V. Ahmadjian and M.E. Hale, eds.), pp.311-380. Academic press, London and New York.

Mishra, 1968. *Ecology Workbook*, Oxford and IBH Publishing, New Delhi.

Rose, F. and James, P.W. 1974. Regional studies on the British lichen flora. The corticolous and Lignicolous species of the New Forest, Hampshire, *Lichenologists*, 6: 1-72.

Sati , S.C., Joshi, P. and Tiwari, P. 1992. Some corticolous lichen occurring in Nainital, Kumaun Himalaya. In: *Microbial Activity in Himalaya.* (Ed. R.D. Khulbe), Sree Almora Book Depot. Almora. 273-277.

Sati, S.C., Tiwari, P. and Joshi, P. 1992. Some saxicolous lichen occurring in Nainital, Kumaun Himalaya. In: *Microbial Activity in Himalaya*. (Ed. R.D. Khulbe), Sree Almora Book Depot. Almora. 264-272.

Sati, S.C. and Usman, S. 1992. Lichen from Sat-Tal (Nainital) of Kumaun Himalaya, India. In: *Microbial Activity in Himalaya*. (Ed. R. D. Khulbe), Sree Almora Book Depot. Almora. 278-284.

Sah, S. 1995. *Phytosociology, Growth Rate and Environmental Response of Epiphytic Macrolichen Vegetation in Ranikhet (Almora), Kumaun Himalaya*. Ph. D. Thesis, Kumaun University, Nainital, p.103.

Sheard, J.W. 1968. In zonation of Lichens on three rocky shores at Inishowen. *Proc. R. Acad*. B.66: 101-102.

Shimwell, D.W. 1971. *The Description and Classification of vegetation.* Sidgwick and Jackson, London.

Singh, K.P., and Sinha, G.P. 1997. Lichen diversity of the Eastern Himalaya and its conservation. In: *Himalayan Microbial Diversity*. XI (2): 349-359. Eds. S.C.Sati, J. Saxena and R.C. Dubey. Today and Tomorrow printers and Publishers, New Delhi.

Uprebi, D.K. 1997. Diversity of Himalayan lichens. In "*Himalayan Microbial Diversity*". XI (2): 339-347. Eds. S. C. Sati, J. Saxena and R.C. Dubey. Today and Tomorrow printers and Publishers, New Delhi.

Yarranton, G.A. 1967. A quantitative study of the bryophyte and macrolichen vegetation of the Dart-moor sranite. *Lichenologist*, 3: 392-408.

SECTION 8

GENETICS

Frontiers in Plant Sciences (Pages 731-746)
Eds. K.G. Mukerji, K.V.B.R. Tilak, S.M. Reddy, L.V. Gangwane, P. Prakash, I.K. Kunwar

Induced Genetic Variability in Green Gram. I. Correlation Studies

T.N. Mary[1] and H.R. Pulivarthi[2]

ABSTRACT

Two varieties of green gram viz., Pusa 105 and Neelalu were subjected to gamma irradiation at doses ranging from 10 kR to 50 kR. The nature and amount of variability induced in quantitative and biochemical characters were studied from M_1 to M_3 generations. Based on the results of enhancement of correlations, it was evident that yield can be improved if selection is made for number of pods/plant and 100 seed weight in Pusa 105 and number of pods/cluster and pods/plant in Neelalu.

Key words: Pusa 105, Neelalu, Phenotypic correlation, Genotypic correlation

INTRODUTION

Pulses play an important role in meeting the dietary requirement of proteins for vegetarians in India. They also form a unique part of our farming system, as they provide protein for human consumption, green nutritious fodder for cattle and enrichment of soil through biological nitrogen fixation. Andhra Pradesh is one of the important pulse growing states of India with a production of over 5 lakh tonnes annually. Though green gram occupies first place in area under cultivation (5.27 lakh ha), it

Author's address: [1]Cytogenetics Laboratory, Acharya Nagarjuna University, Nagarjuna Nagar-522 510, Guntur, A.P.

[2]Baylor College of Medicine, Hematology/Oncology Section, Department of Pediatrics, Houston, Tx 77030 2399, U.S.A.

is next only to black gram in production (170,000 tonnes annually). The yield and quality are important attributes of any crop. Success in breeding programme depends on the presence of sufficient variability for characters of economic importance. The natural variability is very meagre in a predominantly self-pollinated crop like green gram, hence induction of mutations has been accepted as a useful tool for creating variability and enhancing the scope for selection. there have been no concrete efforts at genetic improvement of yield in green gram and available information on the mutation studies in this crop is scanty (Dahiya, 1973; Rajput, 1974; Khan & Hashim, 1978; Shakoor *et al.*, 1978; and Khan 1982, 1983, 1984).

Correlation studies are an important asset to the breeder of any crop to determine the relationship between yield and various factors contributing to yield. Different characters of plants are often correlated with each other. This may be either due to pleiotropy or due to genetic linkage (Harland, 1939). Therefore, the present study deals with the correlation studies between yield and various morphological and biochemical parameters in green gram.

MATERIALS AND METHODS

The material used in the present study in Pusa 105 and Neelalu and green gram (*Vigna radiata* (L.) Wilczek) obtained from Regional Agricultural Research Station, Lam, A.P. Dry seeds of above green gram cultivars were gamma irradiated (Cobalt^{60} source) from 10 to 50 kR doses in the Botany Department of Andhra University, Waltair. The treated and control were grown in the field at Botanical Garden, Acharya Nagarjuna University, Nagarjuna Nagar. Each plot comprised six rows and the spacing was 20 cm between plants and 40 cm between lines, 8 morphological and few biochemical characters were chosen for the study.

Seeds from each M_1 treatment of both the varieties were harvested individually and used in a randomized blok design to study the quantitative variability in the M_2 generation. For studying the quantitative variability in M_3 generation, M_2 plants were harvested individually and grown with three replications. The total phenotype variance (σ^2p) was partitioned into environmental variance (σ^2e) and radiation induced genetic variance (σ^2g). The genetic variance was further used to determine the genotypic coefficient of variation, heritability and genetic advance.

RESULTS AND DISCUSSION

The present study discloses the enhancement of phenotypic correlations between number of pods/plant and yield/plant, 100-seed weight and yield/

plant and number of seeds/pod and pod length in Pusa 105 in the M_3 generation (Table 1) and between pod length and number of seeds/pod, number of pods/plant and yield/plant and number of pods/cluster and yield/plant in Neelalu (Table 2). In some of the character combinations, the association was non-significant in the control but was significant and negative in different treatments of Pusa 105 and Neelalu. Such character combinations are plant height, number of pods/plant, plant height and seed weight, number of pods/cluster and number of seeds/pod and number of pods/plant and yield/plant in Pusa 105, whereas in Neelalu they included plant height with number of pods/cluster and 100 seed weight, number of branches with yield/plant and pod length with yield/plant. The present results revealed the negative association between protein content and yield/plant in most of the treatments in Pusa 105 and Neelalu (Table 3). Significant and positive correlations were obtained between protein and sugar content in both the varieties.

In general, genotypic correlation coefficients were higher than the corresponding phenotypic correlations. The genetic correlation in Pusa 105 between yield and various morphological traits are presented (Table 4). Significant positive correlations were obtained for plant height with days to flowering in 10, 30 and 50 kR doses, number of branches with number of seeds/pod in 40 kR and 50 kR doses, number of branhces with 100 seed weight in all the treatments except 10 kR, days to flowering with number of pods/plant in all the treatments except in 10 and 30 kR doses, number of seeds/pod with pod length in all the treatments, number of seeds/pod with yield/plant in 50 kR dose and 100-seed weight with yield/plant in 30 kR and 40 kR treatments.

Significant, negative genetic correlations were recorded in Pusa 105 for plant height with yield/plant except in 30 kR; number of seeds/pod, 100-seed weight in 40 and 50 kR treatments and number of branches in all the treatments, 100-seed weight with yield/plant in 10 and 20 kR doses, number of pods/plant with number of seeds/pod, plant height with pod length, number of branches with pod length and days to flowering with 100-seed weight in all the treatments except in 40 and 50 kR treatments.

Significant positive genetic correlations were obtained in Neelalu for plant height with yield/plant (10 and 40 kR) days to flowering with yield/plant in all the treatments except in 10 and 30 kR treatments, number of pods/plant with yield/plant (30 and 50 kR), number of seeds/pod with yield/plant (20 and 30 kR) and number of seeds/pod with pod length in all treatments except 30 kR dose. A noticeable negative correlation was observed for plant height with days to flowering (30 and 50 kR), plant height with number of pods/plant, number of seeds/pod with 100-seed weight (except 20 and 30 kR), 100 seed weight with yield/plant (50 kR) and number of pods/plant with pod length except in 10 kR treatment (Table 5).

TABLE 1 *Phenotypic correlations between yield and various morphological characters in the M_3 generation of green gram cultivar Pusa 105*

Character	Dose	NB	DF	NP/C	NP/P	NS/P	PL	100-SW	Yield/Plant
Plant height (PH)	0 kR	0.3836**	−0.2220	0.1416	0.2388*	−0.0133	0.4159**	0.0382	0.2373
	10 kR	0.0765	−0.1085	−0.5216**	−0.2064	0.0673	−0.1032	−0.2851*	0.0502
	20 kR	0.1092	−0.2067	−0.0048	−0.1045	0.0515	−0.0143	−0.0609	−0.2129
	30 kR	−0.1761	0.2186	0.1455	−0.1444	0.0670	−0.1880	0.2860*	0.2082
	40 kR	0.0830	−0.1334	−0.1040	0.0906	−0.1663	−0.0787	−0.3634**	−0.3903**
	50 kR	0.0012	0.1694	0.1815	0.2007	−0.1493	−0.0334	−0.0680	0.0392
Number of Branches (NB)	0 kR		−0.2130	0.3439**	−0.1215	−0.3148**	0.3345**	−0.2141	0.0254
	10 kR		−0.0802	0.0983	0.0557	−0.0957	0.0363	−0.2060	0.2829*
	20 kR		−0.0268	0.1908	−0.0658	−0.3180**	−0.3027**	0.0387	−0.4275**
	30 kR		−0.2093	−0.1744	0.2535*	−0.0141	0.0562	0.2746*	−0.1583
	40 kR		0.0000	−0.0200	−0.1799	0.2721	0.0377	0.2478*	0.2470*
	50 kR		0.0559	−0.0281	−0.0197	0.0370	−0.0272	0.2606*	0.0460
Days to Flowering (DF)	0 kR			−0.1969	−0.0841	0.3499**	−0.4270**	0.4076**	0.1188
	10 kR			0.0277	0.0211	−0.0947	0.2519*	−0.1070	−0.0594
	20 kR			−0.2746*	0.1700	−0.0052	0.1390	0.2734*	0.2232
	30 kR			−0.1305	−0.0548	−0.0674	−0.0874	0.0793	−0.0579
	40 kR			−0.0079	0.1044	−0.0787	−0.0414	0.3126**	0.3536**
	50 kR			−0.0663	0.1073	−0.2135	−0.3746**	0.6296**	−0.1130
Number of Pods/ cluster (NP/C)	0 kR				0.1974	0.0157	−0.1162	0.0937	−0.0437
	10 kR				0.1801	0.0037	−0.0116	0.2080	−0.1135
	20 kR				0.1496	0.1267	0.1334	0.1502	−0.3360**
	30 kR				−0.2148	0.1162	−0.0532	−0.1723	0.1818
	40 kR				−0.1057	0.1083	−0.1970	−0.4593**	−0.0668
	50 kR				−0.2254	0.2036	−0.0727	0.1319	0.0557
Number of pods/plant (NP/P)	0 kR					0.1363	−0.0174	0.1850	−0.2102
	10 kR					−0.4935**	−0.1125	−0.1062	0.4020**
	20 kR					0.0921	0.2946*	0.0509	0.4364**
	30 kR					−0.1619	0.1293	−0.2597*	0.6302**
	40 kR					−0.1188	0.1871	0.1719	0.4303**
	50 kR					−0.4684**	0.0259	−0.1247	0.5265**

Number of seeds/pod (NP/P)	0 kR						0.1242	0.4133**	-0.0255
	10 kR						0.3001*	0.0917	-0.4166**
	20 kR						0.2660*	0.2830**	0.0406
	30 kR						0.4800**	-0.0466	-0.2900*
	40 kR						0.5842**	0.0282	0.0830
	50 kR						0.4954**	-0.2053	0.1219
Pod length (PL)	0 kR							-0.3134**	0.2162
	10 kR							-0.3646**	0.0716
	20 kR							-0.4270**	0.1243
	30 kR							0.4541**	-0.3154**
	40 kR							-0.0377	-0.0033
	50 kR							-0.4586**	0.0458
100-seed weight (100-SW)	0 kR								0.0667
	10 kR								0.3536**
	20 kR								0.2459*
	30 kR								0.3438**
	40 kR								0.3268**
	50 kR								0.4412**

*Significant at 5% level **Significant at 1% level

TABLE 2 *Phenotypic correlations between yield and various morphological characters in the M_3 generation of green gram cultivar Neelalu*

Character	Dose	DF	NP/C	NP/P	NS/P	PL	100-SW	Yield/ Plant
Plant height (PH)	0 kR	0.1185	−0.1328	0.2215	−0.0457	0.1145	0.0288	0.1221
	10 kR	0.0384	−0.3551**	0.0151	−0.0463	0.2645	−0.3051**	0.5106**
	20 kR	−0.0067	−0.0190	0.1067	−0.0827	−0.0849	−0.0894	−0.0727
	30 kR	−0.1620	−0.0568	0.0443	0.0477	−0.2577	−0.1930	−0.1610
	40 kR	0.2573*	0.0623	−0.4032**	0.1287	−0.0956	−0.0827	0.3630**
	50 kR	−0.1078	0.0405	−0.0155	−0.2109	0.1315	0.3540**	−0.1976
Days to Flowering (DF)	0 kR		−0.0365	0.0282	−0.1385	−0.0722	−0.0339	−0.1245
	10 kR		0.2026	−0.0602	0.0265	0.0506	−0.1263	−0.1095
	20 kR		−0.1199	−0.0100	−0.2931*	−0.1229	0.0086	0.1264
	30 kR		−0.1419	0.3368**	−0.0612	0.3461**	0.1820	−0.3734**
	40 kR		−0.1969	0.0817	−0.0447	0.1030	−0.0284	0.2999
	50 kR		−0.3859**	0.2076	0.1536	−0.0555	−0.3099**	0.3327**
Number of Pods/cluster (NP/C)	0 kR			0.1182	0.2340	−0.3575	−0.0169	0.0394
	10 kR			0.1330	−0.3895**	−0.4293**	0.0574	0.2840*
	20 kR			0.4456**	0.1683	0.2967*	0.1034	0.4437**
	30 kR			0.0758	0.2393*	−0.1181	0.0188	0.2390*
	40 kR			−0.1421	0.1273	−0.1115	−0.0263	0.1660
	50 kR			−0.1754	−0.0655	0.0510	0.1373	0.0491
Number of pods/plant (NP/P)	0 kR				0.0910	0.1689	−0.0914	0.2847*
	10 kR				−0.1564	0.2713*	0.1437	0.0067
	20 kR				−0.0182	0.2400	0.0430	0.0934
	30 kR				−0.0408	−0.2289	−0.1966	0.4160**
	40 kR				−0.1823	0.0647	0.0858	0.4494**
	50 kR				0.1128	0.0082	−0.1126	0.4790**
Number of seeds/pod (NP/P)	0 kR					0.2230	−0.1099	0.0252
	10 kR					0.4265**	−0.2679*	−0.1584
	20 kR					0.6826**	−0.0879	0.1800
	30 kR					0.1369**	−0.0584	0.1456
	40 kR					0.3035**	−0.2149	0.1256
	50 kR					0.3064**	−0.1930	0.0741

Pod length (PL)	0 kR						-0.0259	−0.1691
	10 kR						0.0781	0.0747
	20 kR						-0.2225	0.0774
	30 kR						0.0945	−0.4162**
	40 kR						-0.1944	−0.0698
	50 kR						0.1691	−0.0422
100-seed weight (100-SW)	0 kR							−0.0389
	10 kR							−0.0809
	20 kR							−0.1282
	30 kR							−0.0317
	40 kR							−0.0430
	50 kR							−0.1877

*Significant at 5% level **Significant at 1% level

TABLE 3 *Phenotypic and genotypic correlations between yield and various biochemical parameters in the M_3 generations of Pusa 105 and Neelalu*

r/rg	0 kR	10 kR	20 kR	30 kR	40 kR	50 kR
Pusa 105						
r PR-SW	−0.2326*	0.0780	0.1348	−0.2007	0.0184	−0.0358
r PR-YP	−0.1302	−0.273*	0.0431	0.3392**	−0.1446	−0.2741*
r PR-SU	−0.0986	0.7487**	0.5890**	0.3734**	0.5084**	−0.0586
r SU-SW	0.0472	0.0127	0.2082	−0.1728	0.0990	−0.1382
r SU-YP	−0.1302	0.2373	0.0431	0.3392**	−0.1446	−0.2741*
rg PR-SW	–	0.1668	0.4302**	−0.2384*	0.1714	0.0209
rg PR-YP	–	−0.4240**	0.1833	0.6752**	−0.1664	−0.3477**
rg PR-SU	–	1.1431**	0.9173**	1.5931**	1.8540**	−0.2280
rg SU-SW	–	−0.0107	0.4799**	−1.3419**	0.3277**	−3.6162**
rg SU-YP	–	−0.0026	0.5420**	2.7218**	0.1442	1.2187**
Neelalu						
r PR-SW	0.2791*	0.1710	−0.0597	−0.3865**	0.1485	−0.1629
r PR-YP	0.0021	−0.5283**	−0.1653	−0.2483*	−0.3147**	0.3684**
r PR-SU	0.2124	0.2684*	0.1392	−0.0148	−0.1229	0.3969**
r SU-SW	−0.0639	0.2499*	−0.0472	−0.0160	0.1774	0.0374
r SU-YP	0.0593	0.0817	0.0894	−0.4776**	−0.2232	0.2458*
rg PR-SW	–	0.0933	−0.5527	−0.8858**	−0.2580*	−0.6857**
rg PR-YP	–	−0.8101**	−0.3069**	−0.5199**	−1.2808**	−0.7060**
rg PR-SU	–	0.3932**	−0.1799	−0.4477**	−1.6952**	0.6456**
rg SU-SW	–	0.7954**	0.0079	0.0072	0.3808**	0.0229
rg SU-YP	–	0.2683**	0.7404**	−0.9363**	−0.4969	0.7814**

*Significant at 5% level **Significant at 1% level
r/rg = phenotypic and genotypic correlation coefficients respectively
PR, Protein Content; SW, Seed Weight; YP, Yield/Plant; SU, total sugar content

Based on the results of enhancement of correlations, the yield can be improved if selection is made for number of pods/plant and 100-seed weight in Pusa 105 and number of pods/cluster and pods/plant in Neelalu. This increase in the correlations among traits may be utilized to enhance the rate of selection response on a primary trait (Moll and Stubber, 1976). The usefulness of a correlated selection response in strengthening character association has been demonstrated by Krishnaswami and Rathinam (1982) in green gram. The present results revealed the negative association between protein content and yield/plant in Pusa 105 and Neelalu. Similar results were observed by Papa *et al.*, (1961) in Soybeans; Sandhu *et al.*, (1978) in black gram; Sandhu *et al.*, (1979) in green gram. Lelji *et al.*, (1972) postulated that many plants produced seed with high percentage of protein because they have a limited seed production or sink capacity. Therefore, it may be necessary to construct plant types with a larger sink capacity before high seed yield and high crude protein can be combined with a larger sink capacity.

TABLE 4 Genotypic correlations between yield and various morphological characters in the M_3 generation of green gram cultivar Pusa 105

Character	Dose	NB	DF	NP/C	NP/P	NS/P	PL	100-SW	Yield/ Plant
Plant height (PH)	0 kR	–	–	–	–	–	–	–	–
	10 kR	–0.4865**	0.2373*	–1.4691**	–1.2186**	0.2689*	–1.0931**	–0.8149**	–0.5781**
	20 kR	–0.2596*	–0.9431**	–0.1012	–0.8844**	0.1451	–0.4647**	–0.2249	–1.0030**
	30 kR	–0.6383**	0.6616**	0.1624	–1.5727**	0.1323	–0.6791**	0.6910**	0.1727
	40 kR	–0.1405	0.0203	–0.2246	0.0094	–0.3016*	–2.2588**	–0.87098*	–0.9944**
	50 kR	–0.3639**	0.4046**	0.2090	0.1637	–0.2593*	–0.5172**	–0.1209	–0.1097
Number of Branches (NB)	0 kR		–	–	–	–	–	–	–
	10 kR		0.1189	–0.0362	0.2558*	0.1736	–0.2463*	–0.2043	0.7504**
	20 kR		2.2348**	0.0730	0.1010	–0.3243**	–1.2495**	0.6118**	–1.5251**
	30 kR		–0.2083	–0.3745**	1.4683**	0.1689	–0.1156	0.8979**	–0.3485**
	40 kR		0.3770	–0.1602	–0.2097	0.7058**	-1.3923**	0.3207*	0.4280**
	50 kR		0.2405	–0.2710*	0.1003	0.3627**	–0.4725**	0.5361**	0.0643
Days to Flowering (DF)	0 kR			–	–	–	–	–	–
	10 kR			0.2427*	0.2157	–0.9961**	1.3217**	–0.7138**	–0.5535**
	20 kR			–2.5932**	1.8883**	–1.6532**	3.5721**	–1.8370**	2.1524**
	30 kR			–0.1109	0.0069	–0.3791**	0.1680	–0.4018**	–0.2733*
	40 kR			0.1940	0.4137**	–1.0000**	2.5473**	0.0678	0.9688**
	50 kR			–0.0139	0.2543*	–0.5402**	0.3638**	0.7071**	–0.2378*
Number of Pods/cluster (NP/C)	0 kR				–	–	–	–	–
	10 kR				0.1857	–0.0047	0.0473	0.2573*	–0.2196
	20 kR				0.1452	0.2520*	0.3181**	0.2441*	–0.7607**
	30 kR				–1.1547	0.1563	–0.0318	–0.4131**	0.3476**
	40 kR				–0.1953	0.1577	–1.0694**	–0.9006**	–0.0844
	50 kR				–0.5384**	0.3114**	–0.0472	0.1446	0.1073
Number of pods/plant (NP/P)	0 kR					–	–	–	–
	10 kR					1.4247**	–0.2208	–0.3547**	1.8475**
	20 kR					–0.0433	1.0990**	–0.4394**	0.1470
	30 kR					–1.0757**	0.6249**	–1.6304**	0.7195**
	40 kR					–0.2666*	1.4994**	0.1846	–0.2546*
	50 kR					–1.0792**	0.0835	–0.3167**	–0.8332**

Number of seeds/pod (NP/P)	0 kR						–	–	–
	10 kR						0.8326**	−0.1958	−1.3000**
	20 kR						0.7176**	−0.0220	0.2061
	30 kR						0.8415**	−0.5916**	−0.5609**
	40 kR						1.9917**	−0.4415**	0.1862
	50 kR						0.5325**	−0.4926**	0.2294*
Pod length (PL)	0 kR							–	–
	10 kR							−0.4073**	−0.1663
	20 kR							1.7456**	−0.0364
	30 kR							0.7293**	−0.8508**
	40 kR							1.2479**	−1.4538**
	50 kR							−0.6073**	−0.1181
100-seed weight (100-SW)	0 kR								–
	10 kR								−0.7923**
	20 kR								−0.7567**
	30 kR								1.1317**
	40 kR								0.8556**
	50 kR								0.2330

*Significant at 5% level **Significant at 1% level

TABLE 5 *Genotypic correlation between yield and various morphological characters in the M_3 generation of green gram cultivar Neelalu*

Character	*Dose*	*DF*	*NP/C*	*NP/P*	*NS/P*	*PL*	*100-SW*	*Yield/ Plant*
Plant height (PH)	0 kR	–	–	–	–	–	–	–
	10 kR	–0.1489	–1.0144	–0.8022**	–0.0516	0.4333**	–0.7551**	0.9382**
	20 kR	–0.1732	0.0905	–0.6221**	–0.1080	–0.1715	–0.1934	-0.1557
	30 kR	–0.8280**	–0.0019	–0.0694	0.1675	–0.4794**	–0.3120**	-0.3749**
	40 kR	0.8728**	0.6706**	–2.7271**	0.5362**	–0.3610**	–0.2044	0.7032**
	50 kR	–0.3456*	0.2212	–1.0476**	–0.3763**	0.1596	1.0192**	-0.6983**
Days to Flowering (DF)	0 kR		–	–	–	–	–	–
	10 kR		1.0566**	–0.5043**	0.2369	0.1984	–0.2904*	–0.1166
	20 kR		–0.3944**	–2.0089**	–0.6149**	–0.2255	0.1062	0.4642**
	30 kR		–0.3656**	0.9253**	0.1434	1.0303**	0.5083**	–0.9154**
	40 kR		–0.7641**	0.3051**	0.1843	0.3686**	–0.0524	1.2500**
	50 kR		–0.5956**	0.8289**	0.2993*	–0.0464	–0.6795**	0.7657**
Number of Pods/cluster (NP/C)	0 kR			–	–	–	–	–
	10 kR			0.2288	–1.4522**	–0.7005**	0.2174	0.1791
	20 kR			3.2622**	0.1030	0.9292**	0.3097**	–0.1238
	30 kR			0.0565	0.2701*	–0.0280	0.0304	0.3436**
	40 kR			–0.9045**	–0.0397	0.0610	–0.0436	–0.3150
	50 kR			–1.0662**	–0.2109	0.2781*	0.3404**	–0.1330
Number of pods/plant (NP/P)	0 kR				–	–	–	–
	10 kR				–0.7838**	0.6737**	0.8620**	–0.5434**
	20 kR				–2.9337**	–2.0756**	4.7269**	–3.6296**
	30 kR				–0.1496	–0.3638**	–0.2284**	0.4775
	40 kR				–0.7766**	–0.0114	0.2308	–1.7868**
	50 kR				0.2683*	–0.3615**	–0.2481*	0.3772**
Number of seeds/pod (NP/P)	0 kR					–	–	–
	10 kR					0.5227**	–0.3819**	0.2504*
	20 kR					0.9709**	0.0648	0.2745*
	30 kR					0.1004	0.1588	0.2646*
	40 kR					0.3852**	–0.3035**	0.2320
	50 kR					0.3452**	–0.3130**	0.1125

Pod length (PL)	0 kR							–	–
	10 kR							0.1444	0.1829
	20 kR							–0.3905**	0.1727
	30 kR							0.1217	–0.5162**
	40 kR							–0.2232	–0.0250
	50 kR							0.4108**	0.0627
100-seed weight (100-SW)	0 kR								–
	10 kR								–0.1103
	20 kR								–0.2062
	30 kR								–0.0306
	40 kR								–0.0513
	50 kR								–0.4646**

*Significant at 5% level **Significant at 1% level

The genetic correlations between various quantitative characters are of interest to practical plant breeding. In general, the genetic correlation coefficients were observed higher than the corresponding phenotypic correlations indicating the strong inherent association between the characters studied, the phenotype expression is reduced under the influence of the environment. Sakai and Suzuki (1964) discussed the importance of genetic correlations between different quantitative characters due to pleiotropy of polygenes in relation to plant breeding. The genetic correlations among traits measured in the present study did not deviate much in sign or magnitude. This suggests indirectly that irradiation caused variation of the point or micro-mutation type.

Acknowledgements

The authors are grateful to Head of the Department of Botany, Acharya Nagarjuna University for necessary facilities and one of us (H.R. Pulivarthi) is thankful to CSIR for financial assistance.

References

Dahiya, B.S. 1973. Improvement of mung bean through induced mutations. *Indian J. Genet.* 33: 460-468.

Falconer, D.S. 1985. *Introduction to quantitative genetics* Longman Inc., New York.

Harland, S.C. 1939. *The genetics of cotton.* Jonathan cape, London.

Khan, I.A. 1982. Variation in quantitative characters of mung bean (*Phaseolus aureus* Roxb.) after irradiation. *Bot. Bull. Acad. Sci.* 23: 105-118.

Khan, I.A. 1983. Mutation studies in mung bean (*Phaseolus aureus* Roxb.) V. Induced polygenic variability after seed irradiation. *Can. J. Gent. Cytol.* 25: 298-303.

Khan, I.A. 1985. Induction of genetic variability in mung bean (*Vigna radiata* (L.) Wilczek). *Indian Bot. Repotr.* 4(2): 148-150.

Khan, I.A. and Hashim, M. 1978. Radiation induced variability in traits of mung bean (*Phaseolus aureus* Roxb.) *J. Cytol. Genet.* 13: 12-15.

Krishnaswami, S. and Rathinam, M. 1982. Studies on mutagenic effect on genetic variability in green gram (*Vigna radiata* (L.) Wilczek). *J. Nuclear Agric. Biol.* 11: 25-27.

Leleji, O.I., Dickson, M.H., Crowder, L.V. and Bowlee, J.B. 1972. Inheritance of crude protein percentage and its correlation with seed yield in soyabean, *Phaseolus vulgaris* L. *Crop. Sci.* 12: 168-171.

Moll, R.H. and Stubber, C.W. 1976. Quantitative genetics. Empirical results relevant to plant breeding. *Adv. Agron.,* 26: 279-313.

Papa, K.E., Williams, J.H. and Hanway, D.G. 1961. Effectiveness of selection for quantitative characters in the third generation following irradiation of soybean seeds with x-ray and thermal neutrons. *Crop Sci.* 1: 87-90.

Rajput, M.A. 1974. Increased variability in the M_2 gamma irradiated mung bean (*Phaseolus aureus* Roxb.). *Rad. Bot.* 14: 85-89.

Sakai, K. and Suzuki, A. 1964. Induced mutation and pleiotropy of genes for quantitative characters in rice. *Rad. Bot.* 4: 141-151.

Sandhu, T.S., Bhullar, B.S., Cheema, H.S. and Brar, J.S. 1979. Grain protein, yield and its components in urd bean. *Indian J. Genet.* 38: 410-415.

Sandhu, T.S., Bhullar, B.S., Cheema, H.S. and Gill, A.S. 1979. Variability and interrelationships among grain protein yield and yield components in mung bean. *Indian J. Genet.* 39: 480-484.

Shakoor, A., Ahsanulhaq, M. and Sadiq, M. 1978. Induced variation in mung bean (*Vigna radiata* (L.) Wilczek). *Environ. Exp. Bot.* 18: 169-175.

Frontiers in Plant Sciences (Pages 747-756)
Eds. K.G. Mukerji, K.V.B.R. Tilak, S.M. Reddy, L.V. Gangwane, P. Prakash, I.K. Kunwar

Origin of B Chromosome–Evidence from Fluorescence *In–situ* Hybridisation

A.K. Koul

ABSTRACT

The paper reviews the different hypotheses forwarded, from time to time, to explain the origin of enigmatic B chromosomes. The one weakness all these hypotheses suffer is that none of them dilates on "---------the really rare event where a fragment is born which simultaneously seizes immortality". Although different Bs may have originated differently, their derivation from the additional chromosomes of trisomics has attained great deal of acceptance. The paper documents the various processes that led to the origin of B chromosome from chromosome no. 2 of a triplo – 2 plant of *Plantago lagopus*. Fluorescence in-situ hybridization studies have generated unflinching evidence illustrating stepwise changes involved in the origin of B chromosome in this species.

Key words: B chromosome, Trisomic, Centric fragment, 5s, 18s, 5.8s and 25s rDNA, Telomeric repeats, Fluorescence *in-situ* hybridization.

INTRODUCTION

The origin of enigmatic B chromosomes has remained a vexed problem since their discovery during early 20th century. It continues so even today (Jones, 1995; Laugdon *et al.*, 2000). About a decade back we reviewed the various hypotheses that have been advanced, from time to time, to explain the origin of B chromosomes (Koul, Dhar & Langer, 1995). Realizing that chromosomes cannot arise *de novo*, all these hypotheses suggest that Bs have

Author address: University of Jammu, Jammu – 180 006
E-mail: akkoul2003@yahoo.com

arisen from A chromosomes at some stage in evolution. For this reason they must have initially shared homology with them but lost it gradually over a long period of time through structural alterations and/or heterochromatinization. The latter must also have led to lack of any major genes, because of which presence or absence of Bs makes no difference to the phenotype and functioning of the carrier plants.

One weakness that is common to all the hypotheses forwarded hitherto is that, none of them has been able to dilate on "---------the really rare event where a fragment is born which simultaneously seizes immortality" (Jones, 1995). On account of this short coming all the known hypotheses involve some degree of conjecture. Recently, some evidence of this rare event has been generated by application of Fluorescence in-situ hybridization technique (Dhar, Friebe, Koul & Gill, 2002) which has helped to finally resolve the problem.

The origin of B chromosome has to be explained against the backdrop of such of their characters that distinguish them from A chromosomes. These characters are:

(i) Exceptions apart, B chromosomes are smaller in size and more densely heterochromatic than As.

(ii) They display non- Mendelian inheritance.

(iii) With rare exceptions they carry no nucleolus organizer.

(iv) They often display non – disjunction during mitosis which leads to inconstancy in their number in different organs of a plant.

(v) They also display non-disjunction during meiosis or pre- or post meiotic mitosis which gets reflected in their transmission.

(vi) Ordinarily, Bs do not affect the phenotype of carrier plants but depress fertility when they accumulate in high numbers.

The paper will first review the few major hypotheses advanced to explain the origin of B chromosomes, and then discuss implications of the results of Fluorescence *in-situ* hybridization studies.

1. Hypotheses about origin of B chromosomes.

1. (i) Inter and Intra – chromosome aberrations as source of B chromosomes.

a. **Interchanges:** The origin of B chromosomes in *Paspalum stolonifera* and *Haplopappus gracilis* (Avdulove & Titova, 1933 and Jackson, 1960) has been attributed to unequal reciprocal translocations which lead to formation of heterochromatic centric fragments. On account of small size and predominance of heterochromatin these centric fragments do not pair or pair very rarely with the chromosomes of which they have been parts. Their heterochromatic nature also accounts for their ineffectiveness to modify the phenotype and functioning of plants in which they accumulate.

b. **Paracentric inversions**: Meyer (1944) reported the origin of B chromosomes during meiosis of paracentric inversion heterozygotes of *Phlox*. Crossing over in the inversion loop formed at prophase of such heterozygotes results in the formation of the dicentric chromosome bridge and an acentric fragment at anaphase – I. Breakage of the dicentric chromosome bridge close to one of the poles results in the inclusion of a whole chromatid and a centric fragment in one of the two daughter nuclei. The centric fragment functions as a chromosome after losing homology with the particular A chromosome it has been derived from.

c. **Deletions**: The origin of Bs in *Oenothera*, *Tradescantia paludosa* and *Haplopappus gracilis* has been attributed to deletion of large chromatin segments from either side of the centromere of some A chromosome (Cleland, 1951). Since the leftover parts in the immediate neighbourhood of centromere are largely heterochromatic the centric fragment exhibits little tendency to pair with the chromosome of which it has been a part, and functions as B chromosome.

1 (ii) Disarticulation of satellites: Satellite is the small chromatin segment distil to the secondary constriction of the nucleolar chromosome. Several authors have proposed that B chromosomes represent such satellites as have separated from the main chromosome by a break at the secondary constriction which is weak and therefore vulnerable to fracture (Battaglia, 1964). The disarticulated chromatin segment is acentric. The hypotheses assumes that the satellites possess centromere which remains latent under the influence of the main centromere so long as they are part of the intact chromosome. On detachment of the satellite, 'latent centromere' is assumed to get activated and become functional. Thereafter, it functions as independent chromosome. Markarian and Schultz – Schaeffer (1958) and Schultz – Schaeffer (1966) have furnished evidence from *Zea mays, Alopecurus pratensis* and *Haplopappus spinulosus* in support of this hypothesis. In all these species the B chromosomes compare well with satellites in size which provides support to the hypothesis. However, the explanation offered regarding origin of centromere in the disarticulated satellite is the major weakness of this hypothesis.

Sato *et al.*, (1983) recovered plants of *Brodiara uniflora* with different combinations of disarticulated satellite and nucleolar chromosome with terminal secondary constriction. He employed the NOR specific silver staining test to label and locate the two. While this observation emphasizes the process of separation of satellite and the possibility of its survival and functioning as B chromosome it leaves the question of origin of centromere in the disarticulated satellite unanswered. For this reason, Sato *et al.*, (1983) desisted from giving the name B chromosome to the centric fragments considered to represent the fractured satellite.

1 (iii) Intergenomic elements functioning as B chromosomes: Battaglia (1964) proposed the origin of B chromosomes through inter and intraspecific hybridization for which Sapre and Deshpande (1987) and Camancho *et al.*, (2000) furnished supporting evidence. The former located spontaneous hybrids between two sympatric grasses *Coix gigantea* (2n = 20) and *C. aquatica* (2n = 10). Investigation of the interspecific hybrid and its progeny revealed that 1 or 2 *gigantea* chromosomes get incorporated into the *aquatica* genome where they display B chromosome like behaviour. These observations demonstrate how homologous and non-homologous chromosomes can flow between related species, particularly those not isolated reproductively, and settle down as B chromosomes in the recipient species. Plants with 2n = 11 (10 *aquatica* + 1 *gigantea*) and 12 (10 *aquatica* + 2 *gigantea*) chromosomes conveniently pass for *C. aquatica* with 1 or 2 B chromosomes.

Sapre and Deshpande (1987) opined that B chromosomes may have originated through this method in many other taxa also. This mode of origin of B chromosome is likely to be more common at the centers of origin where many related species coexist and grow side by side. Once incorporated, the alien chromosome becomes part of the recipient's genome. Such chromosomes are, however, dispensable; their loss causing no effect on growth, development and fertility of the carrier plants.

1 (iv) Origin of B from trisomic fragments: This hypothesis has gained greater acceptability than others as it explains the origin of Bs in many unrelated angiosperms. All primary trisomics have one chromosome of their somatic complement in triplicate. During meiosis, the additional chromosome either pairs and forms a trivalent or remains unpaired as univalent. The disjunction of trivalent and movement of the univalent at anaphase – I is highly erratic. In some meiocytes the univalent divides precociously into chromatids and in others it suffers breakage which leads to the formation of fragments of varying size. The centric fragments get incorporated in some gametes. Such gametes on fusion with haploid gametes of opposite sex result in offsprings carrying one or more centric fragments besides the standard complement. If the fragment is sufficiently long it pairs with the corresponding homologue during meiosis; the small fragments do not. They are often lost, but such of them as are retained diverge further through heterochromatinization, gene mutations or structural aberrations.

Actual evidence in support of the hypothesis is provided by Frost and Ising (1964) who worked on trisomic barley (*Hordeum vulgare*). The trisomic plant on selfing produced offsprings some of which carried 14 chromosomes and a small centric fragment. From these plants Frost and Ising (1964) raised individuals carrying 14 chromosomes and 2 fragments. The latter paired between themselves but not with any of the 14 chromosomes. For this reason the fragments were suspected to be fore runners of B chromosomes in this species. A few years later, Tsuchiya (1969) also isolated a barley line with 14 + 2 centric fragment chromosomes. Being small and carrying a paracentric inversion these fragments did not pair with any A chromosome. Wiebe *et al.*, (1974) raised a trisomic from a translocation heterozygote of the 16-chromosome barley. Repeated selfing of the trisomic-yielded offsprings carrying 14 A chromosomes and 2 small, structurally altered chromosomes. The latter paired with each other but never with any A chromosome (Fedek, 1976). Structural alterations are suspected to have contributed to the divergence of the centric fragments and formation of B chromosomes.

The origin of nucleolus organizing B chromosomes, which are very rare, is believed to have taken place the same way in *Calycadena* (Carr & Carr, 1982).

2. Fluorescence *in-situ* hybridization – an evidence favouring origin of B chromosomes from trisomics.

Sharma and Koul (1984) and Bhan (1990) raised trisomics for five of the six chromosomes comprising the haploid complement of *Plantago lagopus*. In triplo-2 plants the additional chromosome underwent rapid structural changes which culminated in the formation of an iso-chromosome (Dhar *et al.*, 1990; Dhar, 1991) that acquired all characters of B chromosome over a period of few generations.

The origin of B chromosomes from trisomic fragments involves a number of distinct processes which are enumerated below:

(a) The first is the formation of aneuploid by any one of the several known processes (Khush, 1973).

(b) Fragmentation of the additional chromosome of the aneuploid to make raw material – centromere and autonomously replicating DNA sequences (ARS), available for formation of B chromosomes.

(c) Amplification of DNA sequences to impart body mass to centric fragments so that they assume certain minimum size required for functional chromosome (Schubert & Oud, 1977).

(d) Acquisition of Telomeric DNA sequences to stabilize the ends of newly formed chromosome.

All these steps have been properly documented in *P. lagopus* (Dhar, 1991; Dhar et al., 2002). To begin with, a trisomic for chromosome 2, called triplo-2 was isolated which triggered the chain of events leading to formation of a stable B chromosome. Because chromosome 2 was present as a univalent in the aneuploid, it was more vulnerable to breakage. The chromosome suffered two breaks, one in each arm that resulted in the formation of one centric and two acentric fragments. Observation of a ring and double minutes in some progeny plants of triplo-2 is an evidence of the breakage suffered by the univalent chromosome of the trisomic (Dhar, 1991). The nascent minichromosome contained both the 5s and 18s, 5.8s and 25s rDNA arrays fused to the centromeric sequences.

This was followed by gene amplification. The unscheduled DNA replication probably through Onion – Scale replication followed (Schimke *et al.*, 1986), which is suspected to have played role in amplifying 5s rDNA arrays. Misdivision of this chromosome followed by opening out of the chromatids led to the formation of iso-chromosomes. Southern blot analysis of the disomic and 12 + i chromosome plants using 5s rDNA clone p Pov 1 revealed a distinct band at top of the lane at about 23 kb in 12 + i chromosome plants. Differentiation of the additional 5s rDNA band in the trisomic and its absence from the disomic plants suggest its derivation from the isochromosome (Dhar *et al.*, 2002).

Fluorescence *in-situ* hybridization, using 5s rDNA clone p Pov 1 as a probe, detected one FISH site close to centromere on one chromosome pair of the standard chromosome complement of *Plantago lagopus.* The isochromosome was painted over its entire length, which confirms that the bulk of it consist of 5s DNA as was also indicated by Southern blot analysis (Dhar *et al.*, 2002).

Sequential probing of mitotic metaphase chromosomes with 18s, 5.8s and 25s rDNA clone p Ta 71 detected two nucleolar organizer region (NOR) loci on two different chromosome pairs of the standard complement (Dhar *et al.*, 2002). The major NOR site is located at the terminal region of the short arm of one pair (Bhan, 1990). The minor NOR site mapped adjacent to the 5s rDNA site close to the centromere on another chromosome pair. Additional NOR loci are present at the termini of each arm of the isochromosome.

The structural organization of the isochromosome was further confirmed by performing rGISH, using total genomic DNA of *Plantago lagopus* as a probe and either the complete clone p Pov 1 or only the coding part of the repeat unit as a blocker. In both cases, all chromosomes of the standard complement were completely labeled, but the entire isochromosome had reduced labeling, suggesting that the latter mostly consists of tandomly repeated 5s rDNA units (including the coding and non transcribed spacer) (Dhar *et al.*, 2002).

In order to establish the derivation of the isochromosome from the nucleolar chromosome of the standard complement somatic metaphase plates of the disomic and 12 + i chromosome plants were subjected to silver staining. Two pairs (4) of Ag-NOR bands were detected in the disomic and 6 in the 12 + i plants. The additional Ag-NOR bands were present at each terminal end of the isochromosome. Silver staining has confirmed the derivation of the isochromosome from one of the two nucleolar chromosomes of the standard complement (Dhar *et al.*, 2002).

The 12 + i chromosome plants were compared with their disomic counterparts for the number of nucleoli organized per nucleus. Whereas nuclei of disomic plants had 4, those of the 12 + i chromosome plants had 6 nucleoli suggesting that the additional NORs located on the isochromosome are transcriptionally active in organizing nucleoli (Dhar *et al.*, 2002). Small size, totally heterochromatic nature, inconstancy in number, and lack of any effect on phenotype are features which the isochromosome of *P. lagopus* shares with B chromosomes. The one feature that still required testing was the presence of telomeres which impart functional and structural stability to chromosomes. Fluorescence in-situ hybridization analysis, using telomere specific clone pAt T4 as probe, revealed expected FISH sites at telomeres of all chromosomes of the standard complement. FISH signals also appear at terminal ends of both arms of isochromosome suggesting presence of telomeres. These observations suggest that the additional chromosome of aneuploid *P. lagopus* plant fulfill nearly all criteria used to identify B chromosomes.

The isochromosome is cytogenetically unique. It is the largest chromosome of the aneuploid more than twice the average of size of the A chromosomes. C-banding analysis has shown that unlike the 12 A chromosomes which are mostly euchromatic except for the centric bands, the isochromosome is completely heterochromatic. It has NOR loci at the ends and bulk of the body mass comprises 5s rDNA sequences. The ends of the chromosome carry telomeric sequences and are therefore stabilized. The chromosome has a functional centromere which accounts for its regular behaviour during mitosis and meiosis. It does not pair or form chiasmate association with any A chromosome during meiosis. Instead, its arms pair internally and form ring univalent. The isochromosome undergoes normal disjunction at meiosis I and II.

The structure and behaviour of the isochromosome is typical of naturally occurring B chromosome. The much studied B chromosomes of *Crepis capillaris* exist as isochromosome and also carry NOR and 5s rDNA loci (Jones, 1995). The only difference is that the isochromosome of *P. lagopus* has its entire body mass derived from 5s DNA sequences (Dhar *et al.*, 2002) which could be because very few Bs have been analyzed for the presence of

5s rDNA sequence. Nevertheless, Stitou *et al.* (2000) reported that in B chromosomes of *Rattus* ribosomal cistrons are dispersed throughout. For reasons spelt out above "the isochromosomes of *P. lagopus* is a bonafide B chromosome with fully documented mode of origin".

References

Avdulov, N. and Titova, N 1933 Additional chromosomes in *Paspalum stolonifera*. Bosco. Bull. Appl. Bot. Genet. & Plant Breeding. 11d series 2. 165–172.

Battaglia, E 1964 Cytogenetics of B chromosomes. Caryologia. **17:** 245–249.

Bhan, A 1990 Cytogenetic analysis of aneuploids of *Plantago lagopus*. Ph. D. thesis, Jammu University, Jammu, India.

Camancho, J. P. M., Sharbel, T. F., Baukeboom, L. W 2000 B chromosome evolution. Philos. Trans. R. Soc. London B. Biol. Sci. **355:** 163–178.

Carr, G. D and Carr, R. L 1982 Micro and nucleolar organizing B chromosomes in *Calycadenia* (Asteraceae). Cytologia. **47:** 79–87.

Cleland, R. E 1951 Extra dimunitive chromosomes in *Oenothera*. Evolution. **5:** 165–176.

Dhar, M. K 1991 Cytogenetic studies on some telo aneuploids of *Plantago lagopus* and their progeny. Ph. D. thesis, University of Jammu, Jammu, India.

Dhar, M.K., Koul, A.K and Langer, A 1990 Genetic diversity among Plantagos XVII A novel trisomic in *Plantago lagopus*. Theor. Appl. Genet. **79:** 216–218.

Dhar, M. K., Friebe, B, Koul, A. K and Gill, B. S 2002 Origin of an apparent B chromosome by mutation, chromosome fragmentation and specific DNA sequence amplification. Chromosoma. **111:** 332–340.

Fedak, G 1976 Cytogenetics of Wiebe's 16-chromosome barley. Can. J. Genet. Cytol. **18:** 763–768.

Frost, S and Ising, G 1964 Cytogenetics of fragment chromosomes in barley. Hereditas. **52:** 176–180.

Jackson, R. C 1960 Supernumerary chromosomes. *Haplopappus gracilis*. Evolution. **14:** 135.

Jones, R. N 1995 B chromosomes in Plants (Tansley Rev. No. 85) New Phytol. **131:** 411–434.

Khush, G. S 1973 Cytogenetics of aneuploids. Academic Press, London.

Koul, A.K., Dhar, M.K and Langer, A 1995 Origin of B chromosomes: Present Status. In Botany Towards 2000 A.D. (ed. M. D. Padhya, P. K. Mukherjee and A.S. Khalatkar. Dattsons), Nagpur.

Laugdon, T., Seago, C., Jones, R. N., Ougham, H., Thomas, H., Foster, J. W and Jenkins, G 2000 *De novo* evolution of satellite DNA on the rye B chromosome. Genetics. **154:** 869–884.

Markarian, D and Schulz-Schaeffer, J 1958 A possible origin of supernumerary fragment chromosomes. J. Hered. **49:** 3–7.

Meyer, J. R 1944 Chromosome studies in *Phlox*. Genetics. **29:** 199–216.

Sapre, A. B and Deshpande, D. S 1987 Origin of B chromosome in *Coix* through spontaneous interspecific hybridization. J. Hered. **78:** 191-196.

Sato, S., Sato, M and Kuroki, Y 1983 Unstable behaviour of the chromosome satellite combined with an elongated connected thread during mitosis in *Brodiaea uniflora* Engl. Protoplasma. **114:** 141-150.

Schubert, I and Oud, J. L 1997 There is an upper limit of chromosome size for normal development of an organism. Cell. **88:** 515-520.

Schulz – Schaeffer, J 1966 A possible origin of B chromosomes. Z. Pflzucht. **55:** 21-28.

Sharma, P. K and Koul, A. K 1984 Genetic diversity among Plantagos. III Primary trisomy in *Plantago lagopus*. Genetica. **64:** 135-138.

Stitou, S., Jimenez, R., Diaz de la Guardia, R and Burgos, M 2000 Inactive ribosomal cistrons are spread throughout the B chromosomes of *Rattus rattus* (Rodentia, Muridae) Implications for their origin and evolution. Chromosome Res. **8:** 305–311.

Tsuchiya, T 1969 Cytogenetics of a new type of barley with 16 chromosomes. Chromosoma. **26:** 130-139.

Wiebe, G. A., Ramage, R. T and Eslick, R. F 1974 Eight paired barley lines. Barley Genet. Newslett. **4:** 93-95.

SECTION 9

CYNOBACTERIA

Frontiers in Plant Sciences (Pages 757-768)
Eds. K.G. Mukerji, K.V.B.R. Tilak, S.M. Reddy, L.V. Gangwane, P. Prakash, I.K. Kunwar

Heterocyst Differentiation and Function in Cyanobacteria

D.N. Tiwari*, A.K. Mishra and A.P. Singh

ABSTRACT

The differentiation of heterocysts in filamentous cyanobacteria i.e. *Anabaena* sp. is a devise to protect nitrogen-fixation apparatus from oxygen which is generated in vegetative cells during oxygenic photosynthesis. The separation of two biological functions in two different cell types i.e. nitrogen fixation in heterocysts and carbon dioxide fixation in vegetative cells, is characteristically observed in certain cyanobacteria under aerobic nitrogen-fixing conditions. The development of regularly spaced heterocysts in N_2-grown filaments is possibly regulated by intercellular exchange of metabolites between heterocysts and vegetative cells through the cytoplasmic connection (pore). The HetR and PatA genes have been recently identified as components of the developmental genetic circuit, which are required for heterocyst differentiation and pattern formation in *Anabaena* 7120, respectively. Developmental mutants impaired in heterocyst differentiation have been found useful tools to identify and characterize these genes. Initiation of the pattern in *Anabaena* filaments is still a mystry, but the role of cyanophyin metabolism is speculated.

Key words: Cyanobacteria, developmental mutants, heterocyst differentiation, nitrogen-fixation, pattern formation.

INTRODUCTION

Cyanobacteria, capable of heterocyst differentiation, belong to the order Nostocals and Stigonematales of the Cyanophyta (Desikachary, 1958). These organisms are morphologically more advanced than bacteria and

Author's address: Algal Research Laboratory, Centre of Advanced Study in Botany, Banaras Hindu University, Varanasi-221 005
E-mail: dntiwari@bhu.ac.in

their further evolution along this morphological line seems to have stopped at heterotrichous habits formed by the members of Stigonematales. Because of their procaryotic characteristics like bacteria (Ris and Singh, 1961) and their photosynthetic features identical to those of chloroplasts, cyanobacteria occupy an unique position in the continuum of life. The newly discovered unicellular green procaryote *Prochloron* (Lewin, 1977) which contains two photosystems with chlorophyll a and chlorophyll b as major photosynthetic pigments and which carries out oxygenic photosynthesis like chloroplasts has given added impetus and credibility to the theory of endosymbiotic origin of chloroplasts, according to which cyanobacterial cells are supposed to have given rise to chloroplasts of algae and higher plants.

Since last one decade, heterocystous cyanobacteria have attracted attention of leading cyanobacteriologists for using them as model systems for understanding the biology of cellular differentiation at molecular level because some genera like *Anabaena* and *Nostoc* are simple filamentous forms with mainly three type of cells i.e., vegetative cells, heterocysts and akinetes. The last two differentiated cells are vastly distinct structures formed from vegetative cells under special growth conditions, and which have different features and roles. Heterocysts are essentially produced as nitrogen-fixing cells during the active growth phase in nitrogen-limiting conditions (Fogg, 1949; Wolk, 1982) while akinetes are formed as asexual reproductive structures during the post exponential growth phase possibly in phosphate and energy limiting growth conditions (Wolk, 1965). Heterocyst function is finally settled i.e. it serves as the anaerobic site for nitrogen fixation under external aerobic conditions while the function of akinetes is assigned to tide over the period of vegetative cell multiplication (Adams and Carr, 1981). Based on the evolutionary and biochemical relationship between these cell types, Wolk *et al.* (1994) has recently proposed that heterocysts may be considered as evolutionary derivatives of akinetes. Several excellent reviews on heterocyst differentiation have been published in the past that deal with environmental or genetical control of heterocyst differentiation and pattern formation (Wolk, 1973, 1982, 2000; Haselkorn, 1978; Buikema and Haselkorn, 1993; Koksharova, 2002). In this brief article, some of the recent and important findings reported on heterocyst differentiation especially in *Anabaena* sp. PCC 7120 have been discussed.

HETEROCYST DIFFERENTIATION AND NITROGEN FIXATION

In *Anabaena*, heterocysts are formed at regular intervals along the filament with a major function to provide an anaerobic environment necessary for the production and proper functioning of nitrogen fixing enzyme complex (nitrogenase) and other proteins related to nitrogen fixation. Nitrogen fixation, the conversion of dinitrogen (N_2) to ammonia, is catalyzed by the enzyme nitrogenase. This is energetically expensive process and known to be extremely sensitive to even low level of oxygen. These cyanobacteria overcome this complicated problem by differentiating a fraction of vegetative cells (5-10%) into heterocysts which provide an anaerobic site for nitrogenase activity. Photosystem II is dismantled during the differentiation of heterocysts which are also simultaneously invested with the specialized envelope that limits the influx of gases. Two layers within the envelope have been implicated to protect heterocyst-based nitrogenase from oxygen damage. The outer layer is composed of polysaccharide while the inner layer is composed of glycolipid (Murry and Wolk, 1989) and both layers are impermeable to gases (Fay, 1992). Vegetative cells lack these specialized layers. Much of the oxygen that overcomes these barriers enters the heterocyst through cytoplasmic connections (pores) and is consumed by the rich oxidase activity localised mainly at the honeycomb like membranous plug in the heterocysts. This separation of functions i.e. nitrogen fixation in heterocysts and carbon dioxide fixation in vegetative cells, is needed as such nitrogenase is sensitive to oxygen. The cytoplasmic connection between heterocysts and vegetative cells possibly regulate the intercellular communication or transport of metabolites between the two cells. Since the process of nitrogen fixation requires reductants (fixed carbon), the vegetative cells that serve as the site of oxygenic photosynthesis, export fixed carbon (possibly sucrose) to heterocysts (N_2-fixing cells) and in return, receive fixed nitrogen (possibly glutamine) from heterocysts (Wolk, 1968; Thomas *et al.*, 1977). The burden of providing the *Anabaena* filament carbon and nitrogen is thus divided between these two cell types, and under the conditions of azotrophic growth, the filament might be considered a simple multicellular organism, with each cell type dependent on the other for growth. Nitrogen fixed within heterocysts as ammonia is converted into glutamine by high glutamine synthetase (GS) activity associated with heterocysts. The intercellular exchange of metabolites by these cells is not only required for growth on dinitrogen but possibly also for regulating heterocysts spacing pattern in filaments.

A source of fixed nitrogen, such ns ammonia or nitrate, in the medium inhibits heterocyst formation in most *Anabaena* strains (Fogg, 1949) and a nitrogenous product of heterocyst metabolism has been suggested to inhibit vegetative cells adjacent to heterocysts from heterocyst differentiation (Fay *et al.*, 1968). Methionine sulfoximine (MSX), specific inhibitor of ammonia assimilating enzyme (GS), is able to provide relief to the inhibitory effect of fixed nitrogen both on heterocyst differentiation and nitrogen fixation, since these processes are tightly linked (Ownby, 1977).

Although the mechanism by which heterocyst formation is repressed during growth on a source of fixed nitrogen is not known, it has been suggested that glutamine or a product of glutamine metabolism, rather than ammonia, is a regulatory molecule (Wolk *et al.*, 1994). In this respect, the reported germination of heterocysts in *Gloeotrichia* formed during growth in ammonium medium appears quite interesting (Singh and Tiwari, 1970). In the absence of nitrogenase activity of such cultures and our knowledge with regard to regulation of nitrogen fixation by ammoniacal nitrogen, such germinating heterocysts could not be treated equivalent to heterocysts at least in functional sense. May be that basal cell in *Gloeotrichia* is so morphologically designed as to function like a nitrogen-fixing cell under aerobic nitrogen-fixing conditions and to function like akinetes when nitrogen source is no limiting factor for growth. However, this view is subject to confirmation as similar observations also exist for the lateral heterocysts of *Nostochopsis lobatus* (Tiwari, 1978).

HETEROCYST DIFFERENTIATION AND NIF GENE REARRANGEMENT

Adaptive response of heterocystous cyanobacteria to nitrogen starvation does not lead to changes only in the cell structure but also in reorganisation of nif genes occurring specifically within the developing heterocysts. The exciting discovery is that a developing heterocyst of *Anabaena* 7120 is accompanied by the reorganisation of nitrogen-fixing genes of an operon nif HDK (Golden *et al.*, 1985). vegetative cells contain two DNA elements which interrupt operons including the genes necessary for encoding the active nitrogenase components and assembly proteins (Haselkorn, 1992). These elements are excised from the main genome allowing the nif genes (Operon HDK) to fuse together and to synthesize active nitrogenase complex protein in the heterocyst. Since this process is restricted mainly to developing heterocysts it appears to be a developmentally controlled process, although nitrogen-limitation and anaerobic conditions are reported to be the minimum requirement for nif gene rearrangement which occurs within a 10-14 hour old developing heterocyst. The absence of this process in vegetative cells does not allow the nif operon (HDK) to synthesize active components of nitrogenase and thus, renders the vegetative cells incapable of nitrogen fixation. An 11 Kb DNA element is present within nifD gene, which is excised/removed by a process of site-specific recombination between two direct repeated ends of 11-base pair sequence involving the enzyme recombinase. The gene xisA which is present in the element itself and which encodes, the recombinational enzyme, is normally active in developing heterocysts but remains tightly repressed in vegetative cells. The excised 11 Kb element from the main genome remains in the heterocyst in the form of a circular and non-replicating molecule with unknown function. The simultaneous excision of another DNA element of 55 Kb also occurs during heterocyst differentiation involving a similar site specific recombination process. This element has been reported to interrupt a gene called fdxN of an operon that includes nifB, nifS and nifU, the product of which may be acting as electron donor to nitrogenase (Haselkorn, 1992). The removal of these DNA elements from the main genome in the hetcrocyst may be one of the important factors for the terminal differentiation of heterocysts. This discovery has opened for the possibility of such process also occurring during akinete differentiation although akinetes are not terminally differentiated cells.

HETEROCYST DIFFERENTIATION AND PATTERN FORMATION

Morphological polarity among filamentous blue green algae has been characteristically observed in heterocystous forms which are uniquely associated with nitrogen fixation. Whether this is merely a coincidence or interdependence of two features remains to be finally settled, although several workers have suggested that the formation of heterocysts is linked with the onset of nitrogenase activity (Haselkorn, 1978). Where apparent, polarity of algal trichomes has definite correlation with the position of heterocysts, and is interpreted as being due to a gradient of diffusible mtetabolites emanating from the latter (Fay *et al.*, 1968). Directionality of expression of polar morphology imparts significance to the corresponding modes of cellular growth and division as being equally important determining factors for the morphology ultimately assumed by the algal trichomes. Accordingly, algal trichomes, and often the heterocyst, can be categorized as (1) unidirectional and monopolar as in rivularian trichomes with a basal heterocyst and a gradual tapering end; (2) unidirectional but counterpolar as in *Cylindrospermum* and *Anabaenopsis* with filaments having two terminal heterocysts with centripetally faced opposite poles; (3) bipolar as in the rest of the nostocacean filaments possessing several intercalary heterocysts connected with vegetative cells at both the poles; and (4) complex, as in branched stigonematalean filaments with the heterocysts placed either at the base of the branch (tripolar) or in close proximity with it in the form of intercalary heterocysts (bipolar) and sometimes, lateral heterocysts (monopolar), either sessile or pedicellate, are also formed in *Nostochopsis*. While heterocysts of the *Cylindrospermum* type are monopolar like those of rivularian filaments, the potency to grow and divide is more uniformally distributed along their trichome length than the latter where actively dividing cells form a sort of meristematic zone in the sub-median part of the trichome. The four developmental patterns in these algae are produced under limited availability of combined nitrogen whose presence disturbs such patterns by inhibiting heterocyst formation along the filaments (Fogg, 1949).

Anabaena filaments grown on N_2 form intercalary heterocysts at more or less equidistant from the existing heterocysts and maintain a regularly spaced pattern throughout the active growth phase. Although such pattern is formed long before nitrogen fixation begins, the nitrogen fixing product of heterocysts has been implicated to control the pattern formation (Fay *et al.*, 1968; Tiwari, 1977). The spacing pattern can be changed by physical fragmentation of filaments (Wolk, 1967; Singh *et al.*, 1972b) or by treatment of filament with low concentration of rifampicin, an inhibitor of RNA polymerase (Wolk and Quine, 1975; Tiwari *et al.*, 1981). Apparently, most of the cells in the filament are genetically potential to develop into heterocysts but the mystery of pattern formation is still unresolved. Two general models to explain spacing pattern have been proposed on the basis of cell-cell interaction (Wolk, 1989). One is based on 'diffusible inhibitor' model in which a cell starved for a source of fixed nitrogen releases a metabolite, the amino acid, that inhibits adjacent cells from differentiation. The second model is the 'induced pump' in which cells of the filament that responds to nitrogen deprivation respond by inducing high affinity uptake system for amino acids (Haselkorn, 1978).

Although starvation for a source of fixed nitrogen is known to be primary signal for differentiation, the biochemical mechanisms by which nitrogen starvation mediates control over differentiation are not fully understood. One of the earliest events in nitrogen starved cells is degradation of protein reserves, the cyanophycin. Two enzymes are known to degrade proteins during nitrogen starvation: a) protease that acts early in the process (Wood and Haselkorn, 1980) and b) cyanophycinase that specifically degrades cyanophycin in the differentiating heterocyst (Gupta and Carr, 1981). The role of protein turnover may be attributed to change the C:N ratio of cells, a condition possibly required for the regulatory proteins to switch on the genes required for heterocyst differentiation. In this respect, the product of NtcA gene reported by Wei (1994) may act as a nitrogen senser and help in mechanism of the differential expression of glnA gene.

Developmental mutants of cyanobacteria have been useful tools in analysing various cellular processes especially related to control mechanism and pattern formation (Singh et. al., 1972a; Tiwari et. al., 1979; Mishra and Tiwari, 1986; Spence and Stewart, 1987). Recently, Buikema and Haselkorn (1991) have marked a gene, called HetR, in the mutant of *Anabaena* 7120 that acts early in the process of heterocyst differentiation. Mutation to HetR gene results in Het$^-$ phenotype. Cells carrying extra copies of HetR on plasmids are induced to form multiple heterocysts even in the presence of nitrate. Another gene (HetN) has been identified to control heterocyst differentiation in N10 mutants of *Anabaena* (Ernst *et al.*, 1992). This gene is reported to be involved in production of a secondary metabolite that inhibits heterocyst development (Black and Wolk, 1994). The most exciting report on the mutant of *Anabaena* with multiple phenotypes has been reportedly linked with NtcA gene whose product is required for growth on nitrate and heterocyst formation (Wei *et al.*, 1994; Frias, 1994). The NtcA gene requirement for heterocyst differentiation suggests that it may have a regulatory function than nitrogen control. Identification of DevA gene has been linked with the development of mature heterocysts (Maldener *et al.*, 1994). The existence of more than one genes affecting heterocyst differentiation indicates that the genetic circuit of the development program is highly complicated. The Het$^-$ class of mutants are important because most of them are incapable of nitrogen fixation even under anaerobic conditions. This shows for a direct linkage of nif gene transcription to heterocyst development as suggested by Buikema and Haselkorn (1993).

Mutants of *Anabaena* species altered in the pattern formation of heterocysts are of interest because they are likely to have defects in genes that control the development of intercalary heterocysts in the filament. A recent report on mutation to PatA gene in *Anabaena* 7120 indicates the involvement of PatA gene product only in the development of intercalary heterocysts (Liang *et al.*, 1992). Such mutants are able to produce only terminal heterocysts in filaments. The presence of HetR gene in terminal heterocysts of PatA mutants does indicate that HetR and PatA genes are the components of the same regulatory circuit of the developmental program. The product of PatA gene is possibly required to interpret the environment only in intercalary cells. Development of contiguous heterocysts, as suppressed by PatA gene, has been observed in some mutants and the entire filament of certain mutants is transformed into heterocysts. Such mutants might be defective in producing a morphogen which is released from heterocysts and inhibits the neighbouring vegetative cells from differentiation (Tiwari, 1977). Mutants that fragment and fail to differentiate have been isolated (Ernst *et al.*, 1992) suggesting that two phenotypes are probably the result of a single genetic lesion. The failure of such mutants to differentiate into heterocysts may be due to an intercellular communication gap. Studies on the isolation and characterization of such mutants are of interest and promise to reveal more about the mystery of pattern formation in the heterocystous cyanobacteria.

Acknowledgement

We are thankful to the Head and Program Coordinator of Botany Department for providing necessary research facilities.

References

Adams, D.G. and Carr, N.G. 1981. The developmental biology of heterocyst and akinete formation in cyanobacteria. *CRC Critical Reviews in Microbiology.* 9 : 45-100.

Black, T.A. and Wolk, C.P. 1994. Analysis of a Het⁻ mutation in *Anabaena* sp. strain PCC 7120 implicates a secondary metabolite in the regulation of heterocyst spacing. *Journal of Bacteriology.* 176 : 2282-2292.

Buikema, W.J. and Haselkorn, R. 1991. Characterization of a gene controlling heterocyst differentiation in the cyanobacterium *Anabaena* 7120. *Genes and Development.* 5 : 321-330.

Buikema, W.J. and Haselkorn, R. 1993. Molecular genetics of cyanobacterial development. *Annual Review of Plant Physiology and Plant Molecular Biology.* 44 : 33-52.

Desikachary, T.V. l959. Cyanophyta. Indian Council of Agriculture Research Publication, New Delhi.

Ernst, A., Black, T., Cai, Y., Panoff, J.M., Tiwari, D.N. and Wolk, C.P. 1992. Synthesis of nitrogenase in mutants of the cyanobacterium *Anabaena* affected in heterocyst development or metabolism. *Journal of Bacteriology.* 174 : 6025-6032.

Fay, P. 1992. Oxygen relations of nitrogen fixation in cyanobacteria. *Microbiology Reviews.* 56 : 340-373.

Fay, P., Stewart, W.D.P., Walsby, A.E. and Fogg, G.E. 1968. Is the heterocyst site of nitrogen fixation in blue green algae? *Nature.* 220 : 810-812.

Fogg, G.E. 1949. Growth and heterocyst production in *Anabaena* cylindrica Lemm. II. In relation to carbon and nitrogen metabolism. *Annals of Botany.* 13 : 241-259.

Frias, J.E., Flores, E. and Herrero, A. 1994. Requirement of the regulatory protein NtcA for the expression of nitrogen assimilation and heterocyst development gene in the cyanobacterium *Anabaena* 7120. *Molecular Microbiology.* 14 : 823-832.

Golden, J.W., Robinson, S.J. and Haselkorn, R. 1985. Rearrangement of nitrogen fixation genes during heterocyst differentiation in the cyanobacterium *Anabaena. Nature.* 314 : 419-423.

Gupta, M. and Carr, N.G. 1981. Enzyme activities related to cyanophycin metabolism in heterocysts and vegetative cells of *Anabaena*. *Journal of General Microbiology.* 125 : 17-23.

Haselkorn, R. 1978. Heterocysts. *Annual Review of Plant Physiology.* 29 : 319-344.

Haselkorn, R. 1992. Developmentally regulated gene rearrangements in procaryotes. *Annual Review of Genetics.* 26 : 111-128.

Koksharova, O.A. and Wolk, C.P. 2002. Genetic tools for cyanobacteria. *Journal of Applied Microbiology and Biotechnology.* 58 : 123-137.

Lewin, R.A. 1977. *Prochloron*, type genus of the Prochlorophyta. *Phycologia*. 16 : 217.

Liang, J., Scappino, L. and Haselkorn, R. 1992. The PatA gene product controls heterocyst pattern formation in the cyanobacterium *Anabaena* 7120. *Proceeding of the National Academy of Sciences (U.S.A .)*. 89 : 5655-5659.

Maldener, I., Fielder, G., Ernst, A., Fernandez-Pinas, F. and Wolk, C.P. 1994. Characterization of Deva, a gene required for maturation of proheterocysts in the cyanobacterium *Anabaena* 7120. *Journal of Bacteriology.* 176 : 7543-7549.

Mishra, A.K. and Tiwari, D.N. 1986. Mutagenesis and characterization of morphological mutants impaired in nitrogen-fixing capacity from a cyanobacterium *Gloeotrichia ghosei*. *New Phytologist.* 103 : 69-77.

Murry, M.A. and Wolk, C.P. 1989. Evidence that the barrier to the penetration of oxygen into heterocysts depends upon two layers of the cell envelope. *Archives of Microbiology.* 151 : 469-474.

Ownby, J.D. 1977. Effects of amino acids on methionine-sulfoximine induced heterocyst formation in Anabaena. *Planta.* 136 : 277-279.

Ris, H. and Singh, R.N. 1961. Electron microscopic studies on blue-green algae. *Journal of Biophysics Biochemistry and Cytology.* 9 : 63-80.

Singh, R.N., Singh, S.P. and Singh, P.K. 1972a. Genetic regulation of nitrogen fixation in blue green algae. In : *Taxonomy and Biology of Blue-green algae* (Ed. Desikachary, T.V.). pp. 264-268. University of Madras, India.

Singh, R.N., Tiwari, D.N. and Singh, V.P. 1972b. Genetic basis of cellular differentiation in blue-green algae. In : *Taxonomy and Biology of Blue-green algae* (Ed. Desikachary, T.V.). pp. 27-37. University of Madras, India.

Singh, R.N. and Tiwari, D.N. 1970. Frequent heterocyst germination in the blue-green alga *Gloeotrichia ghosei* Singh. *Journal of Phycology.* 6 : 172-176.

Spence, D.W. and Stewart, W.D.P. 1987. Heterocystless mutants of *Anabaena* 7120 with nitrogenase activity. *FEMS Microbiology Letters.* 40 : 119-122.

Tiwari, D.N. 1977. Pattern of heterocyst differentiation as a function of nitrogen-fixing ability in the blue-green alga *Nostoc linkia*. *Beitraege zur Biologie der Pflanzen.* 53 : 177-186.

Tiwari, D.N. 1978. The heterocysts of blue-green alga *Nostochopsis lobatus* : effects of cultural conditions. *New Phytologist.* 81 : 853-856.

Tiwari, D.N., Yadav, P.K. and Mishra, A.K. 1979. Developmental behaviour in mutants of *Gloeotrichia ghosei* defective in nitrogen fixation. *Microbios Letters.* 12 : 121-131.

Tiwari, D.N., Pandey, A.K. and Mishra, A.K. 1981. Action of 2,4-D and rifampicin on heterocyst differentiation in blue-green alga *Nostoc linkia*. *Journal of Bioscience.* 3: 33-39.

Thomas, J., Meeks, J.C., Wolk, C.P. Shaffer, P.W. Austin, S.M. and **Chien, W.S.** 1977. Formation of glutamine from (^{13}N) ammonia, dinitrogen, and (^{14}C) glutamate by heterocysts isolated from *Anabaena cylindrica*. *Journal of Bacteriology.* 129: 1545-1555.

Wei, T.F., Rama Subramanian, T.S. and Golden, J.W. 1994. NtcA gene required for growth in nitrate and heterocyst development in *Anabaena* 7120. *Journal of Bacteriology*. 176: 4473-4482.

Wolk, C.P. 1965. Control of sporulation in the blue green alga. *Developmental Biology.* 12: 15-35.

Wolk, C.P. 1967. Physiological basis of the pattern of vegetative growth of a blue green alga. *Proceedings of the National Academy of Sciences (USA).* 57 : 1246-1251.

Wolk, C.P. 1968. Movement of carbon from vegetative cells to heterocysts in *Anabaena cylindrica*. *Journal of Bacteriology.* 96 : 2138-2143.

Wolk, C.P. 1973. Physiology and cytological chemistry of blue-green algae. *Bacterial Reviews.* 37 : 32-101.

Wolk, C.P. 1982. Heterocysts. In: *The Biology of Cyanobacteria* (Eds. Carr, N.G. and Whitton, B.A.). pp. 359-386. *Blackwell Scientific Publishers*, Oxford.

Wolk, C.P. 1989. Alternative models for the development of the pattern of spaced heterocysts in *Anabaena* (cyanophyta). *Plant Systematics and Evolution.* 164: 27-31.

Wolk, C.P. 2000. Heterocyst formation in *Anabaena*. In : *Prokaryotic development* (Eds. Brun, Y.V. and Shimkets, L.J.). pp. 83-104. *American Society for Microbiology,* Washington. ISBN 1-55581158-2.

Wolk, C.P., Ernst, A. and Elhai, J. 1994. Heterocyst metabolism and development. In: *The Molecular Biology of Cyanobacteria* (Ed. Bryant, D.A.). pp. 769-823. Kluwar Academic Publishers, Netherland,

Wolk, C.P. and Quine, M.P. 1975. Formation of One-dimensional patterns by stochastic processes and by filamentous blue-green algae. *Developmental Biology.* 46: 370-382.

Wood, N.B. and Haselkorn, R. 1980. Control of phycobiliprotein proteolysis and heterocyst differentiation in *Anabaena*. *Journal of Bacteriology.* 141: 1375-1385.

Frontiers in Plant Sciences (Pages 769-780)
Eds. K.G. Mukerji, K.V.B.R. Tilak, S.M. Reddy, L.V. Gangwane, P. Prakash, I.K. Kunwar

Cyanobacteria in the Soil Crusts and Their UV-sunscreen Pigments

Johnnita Tirkey and Siba Prasad Adhikary

ABSTRACT

The bluish-green and blackish-brown biological crusts occurring on different kinds of soil beds in the Utkal University Campus, Orissa state in India was studied. A total number of 28 species attributed to 14 different genera of cyanobacteria have been isolated, described and identified. The microorganisms in these biological soil crusts were mostly composed of the cyanobacterial genera *Oscillatoria, Phormidium, Lyngbya, Microcoleus, Plectonema, Anabaena, Cylindrospermum, Nostoc, Calothrix, Scytonema, Tolypothrix, Westiellopsis, Fischerella* and *Stigonema*. The species belonging to *Nostoc* were the widely occurring forms followed by *Phormidium, Scytonema, Fischerella, Calothrix* and *Tolypothrix* species.

Key words: Soil crust, cyanobacteria, UV-sunscreen pigments, scytonemin, MAAs

INTRODUCTION

Biological crust on soil beds forms an intimate association between soil particles and cyanobacteria, algae, micro-fungi, lichens and bryophytes (in different proportions) that live within, or immediately on top of the uppermost millimeters of soil. Binding of these organisms with soil particles is because of their by-products (extracellular carbohydrate matrix) forming a surface layer is a common phenomenon in these crusts and such layers are called "films" or "biofilms" (Castenholz and Garcia-Pichel 2000).

Author's address: P.G. Department of Botany & Biotechnology,
Utkal University, Bhubaneswar, Orissa 751004
E-mail: adhikary2k@hotmail.com

Biological soil crusts occur in different substrata, e.g. soil surface, desert soils, bark of trees, inselbergs etc. Several authors have described the world wide distribution of soil crust communities *viz.* Belnap and Rosenstreter (2001) for North America, Büdel (2001 a,b,c) for South America, Europe, Mediterranean regions and for Asia, Hansen (2001) for Artic Greenland, Turk and Gartner (2001) for the Alps, Galun and Garty (2001) for the Middle East, Büdel (2001a) for Africa, Eldrige (2001) for Australia and Green and Broady (2001) for Antarctica, Reynaud and Lumpkin (1988) for the Lanzhou region in China, Dor and Danin (1996) for the Dead Sea Valley in Israel, Malam Issa (1998), Malam Issa *et al.* (1999) and Hahn and Kusserow (1998) for the Sahel region in Africa, Eldrige and Koen (1998) and De Winder (1990) for coastal dunes in Northen Europe and Paus (1997) for Northen Germany. We analysed for the first time the cyanobacterial composition of the crusts occurring in soil beds of certain locations in the eastern region of India. Attempt was taken to analyze the cyanobacterial composition at different locations on biological soil crusts. Since these organisms in the crusts inhabit exposed to intense solar radiation, the presence of UV sunscreen pigments in their cell was also determined.

MATERIALS AND METHODS

Biological soil crust samples were collected from six different locations of Bhubaneswar in the Orissa state. Blackish-brown soil crust/mat were collected from the surface of soil using forceps, scalpel and needle. These were kept in sterilized petriplates. They were assigned with a collection number followed by date of collection and stored at Utkal University. The first letter of the collection number denotes the state, second denotes site and third denotes the types of soil from which the crust had been collected. The cyanobacterial forms occurring as crust or mat on the soil were collected in dried state. Even on wetting these soil crusts did not reveal any distinct morphological feature of the organism present in it. In many cases only sheaths or a few cells were observed. Thus materials collected from the field could not be used for taxonomic analysis. To identify the organism up to species level, materials were cultured in a defined medium. For culture a pinch of the soil crust/mat collected from each site was transferred to agarised (1% w/v, agar-agar) BG-11 (± N) medium (Rippka *et al.* 1979) and incubated at 25 ± 1°C under continuous light from fluorescent tubes at an intensity of 7.5 W/m^2 up to 30 days. The organism appeared in the culture were isolated into unialgal state, their morphometric analysis was done and identification was made following Desikachary (1959), Komark & Anagnostidis (1989) and Anagnostidis & Komark (1990).

Camera Lucida diagram of each organism was drawn and microphotograph was taken in a Meiji ML-TH-05 trinacular research microscope using an F-108 Nikon Camera. The unialgal culture of the each species was given a specific isolate number (UU-Utkal University) and maintained on agar slants in test tubes in the laboratory.

Cyanobacterial cell suspension was centrifuged in a Remi 8 RC centrifuge at 3,000 rpm for 15-20 minutes. The supernatant was discarded and pellet was extracted using 90% methanol (BDH, AR) for 15 minutes in dark at room temperature followed by incubating at 60°C for 1 min. The methanol extract was separated by centrifugation at 3,000 rpm for 15 minutes. The process was repeated until all the lipid soluble pigments were extracted Absorbance of the methanol extract was measured at 665 nm in a Hitachi U-2000 UV visible spectrophotometer. The amount of chlorophyll-a was determined using the extinction coefficient of Mackinney (1941).

Known quantity of crusts containing cyanobacteria and cells from exponential growth phase of the corresponding organisms were taken for extraction of UV absorbing pigments. These were extracted with 20% (v/v) methanol at 45°C, 1 h for mycosporine amino acid like substances (MAAs) and with 90% (v/v) methanol at 45°C, 1h for scytonemin pigment. The contents were centrifuged at 3,000 rpm for 20 minutes and the supernatant was used for measurement of absorption spectra in a Hitachi U-2000 UV visible spectrophotometer in the wavelength range of 250-750 nm.

RESULT AND DISCUSSION

Dried soil crusts were collected from 6 different locations of Bhubaneswar. The substratum at their sites was mostly laterite soil with scanty vegetation. Soil crusts appeared blackish in colouration and were irregular in shape. Most of the samples were collected during the month of August – September 2003. Visible appearance of crusts on the soil surface was distinctly observed after a day of receiving monsoon rain. However, when dried, these crusts form flakes on the soil surface.

When a pinch of the wet crust was mounted on a slide and observed under microscope, few cells of algae and cyanobacteria with or without sheath layer were observed. In many cases several empty sheath layers were found embedded together showing occurrence of *Microcoleus* species. These sheath layers are known to retain moisture in desiccated state (Adhikary, 1998; Grilli-Caiola *et al.*, 1993, 1996). Such microscopical observation of the wetted crust did not show about the exact organism inhibiting the crusts. However, presence or absence of sheath, unisereate, false–branch or branching pattern showed that the organism in the crust belongs to cyanobacteria. Cyanobacteria have been observed as the most abundant organisms in diverse extreme biotopes, e.g. in high humidity, low temperature and low light intensity (Grilli-Caiola *et al.*, 1987; Ortega-Calvo *et al.*, 1993) in hot and cold deserts (Büdel and Wessels, 1991) and also under desiccated conditions (Potts, 1994).

All the crust samples were washed properly and pinch of it was transferred to BG-11 medium containing nitrogen or without, and were grown in defined condition in laboratory for 2 weeks .The organisms appeared in the culture were sub-cultured repeatedly in agarised media until uniculture of these organism was obtained. These organisms belong to 14 different genera, mostly filamentous from, majority of which were heterocystous, hence nitrogen fixing. All together 28 species of cynobacteria were encounted in the biological soil crusts of Bhubaneswar (Table-1). These belong to the genera *Oscillatoria, Phormidium, Lyngbya, Microcoleus, Plectonema, Anabaena, Cylindrospermum, Nostoc, Calothrix, Scytonema, Tolypothrix, Westiellopsis, Fischerella* and *Stigonema*. Analysis of these results showed that the major component of these soil crusts were species belonging to *Microcoleus, Lyngbya* and *Nostoc* species, all those which have a well defined sheath or slime layer around their trichome. Members under branched from of cyanobactria such as *Stigonema tomentosum, Fischerella ambigua* and *Westiellopsis prolifica* together with many species of *Nostoc* and *Calothrix* appeared in the culture. These result suggests that filamentous from of cynobacteria, mostly sheath forming, are the major component of the soil crusts. The sheath layer of these organisms absorbed moisture soon after monsoon, swell and adhere the cells together, leading to visible crusts formation on the soil surfaces.

Growth pattern of 15 heterocystous cyanobacterial species isolated in unialgal culture were used in further experiments. The organisms were grown for up to a month, a batch of each was harvested at every four days interval and their generation time (doubling time in hours) was calculated. All these organisms grew extreme slowly in culture (Table-2). Relatively fast growing species showing lesser doubling time was observed in *Calothrix scytonemicola* (117 h) whereas *Nostoc microscopicum* with minute cells covered by distinct slime matrix showed the slowest growth rate (300 h). Invariably most of the organisms investigated grew in the range of 200-300 h of generation time suggesting that all these cyanobacteria in the biological soil crusts are slow growing forms.

The colours of the organisms isolated from soil crusts were not typical bluish-green and varied from bluish to greenish, brownish or reddish shades. Hence in order to check the chlorophyll content of these organisms, equal amount of cell suspension of each organism was taken and chlorophyll-a was extracted using 90 % methanol (v/v). The cells were taken from the exponential growth phase and the chlorophyll-a content was

TABLE 1 *List of the 28 Cyanobacterial species identified in the biological soil crusts at six different sites of Bhubaneswar, Orissa.*

Sl.no.	Cyanobacterial species	S1	S2	S3	S4	S5	S6
1	*Nostoc piscinale*	+	+	+	+	+	–
2	*Nostoc punctiforme*	+	–	+	–	–	–
3	*Nostoc commune*	+	+	–	+	++	–
4	*Nostoc microscopicum*	–	–	–	+	–	–
5	*Nostoc curviceps.*		+	+			
6	*Nostoc ellipsosporum*	+				+	
7	*Anabeana torulosa*	+	+	+		+	
8	*Anabaena unisporae*	+	+			+	
9	*Cylindrospermum indicum.*					+	
10	*Lyngbya estuarii*		+	+	+		
11	*Oscillatoria ornate*		+		+		
12	*Oscillatoria ordinary*	+					
13	*Calothrix marchica*						+
14	*Calothrix brevissima*						+
15	*Calothrix javanica*			+			
16	*Calothrix gloecola*		+				
17	*Calothrix fusa var. crassa*						+
18	*Calothrix scytonemicola*			+	+		
19	*Scytonema rivularae*	+	+				
20	*Plectonema hangsgiri*	+	+				
21	*Microcoleus lacustris*	+	+	+			
22	*Microcoleus sp.*				+		
23	*Tolypothrix crassa*			+			+
24	*Tolypothrix sp.*					+	
25	*Westiellopsis prolifica*	+	+	+	+		+
26	*Westiellopsis sp.*	+	+				
27	*Fischerella sp.*	+	+				+
28	*Stigonema sp.*	+		+			

estimated (Table-3). The result showed that the chlorophyll-content of the organisms differed widely which varied from 21.4 to 128 mg uniform quantity of cell of each organism (0.2 absorbance of culture suspension at 750 nm). Wide variation also occurred within the species under same genera from 40.1 to 128 mg and also among three species of *Calothrix*, varing from 30.6 to 95.2 mgl. This shows that colouration of organisms varied because of quantity of the difference in the chlorophyll-a with that of other photosynthetic pigment in same organism.

The crust and the corresponding organisms are exposed to high solar radiation in their natural environment; hence, they tend to develop efficient

TABLE 2 *Growth pattern (doubling time in hour) of various cyanobacterial species isolated from the biological soil crusts (BG-11 medium, 25 ± 1°C, 7.5 W/m² light intensity).*

Isolate no.	Cyanobacterial species	Doubling time in (h)
1	*Anabaena unisporae*	218
2	*Anabeana torulosa*	218
3	*Nostoc punctiforme*	240
4	*Nostoc curviceps.*	240
5	*Nostoc piscinale*	266
6	*Nostoc ellipsosporum*	240
7	*Nostoc commune*	200
8	*Nostoc microscopicum*	300
9	*Calothrix marchica*	240
10	*Calothrix gloecola*	218
11	*Calothrix scytonemicola*	171
12	*Tolypothrix* sp.	184
13	*Westiellopsis* sp.	184
14	*Fischerella* sp.	218
15	*Fischerella* sp.	240

TABLE 3 *Chlorophyll content ((mg/uniform cell suspension of each organism) of various cyanobacterial species isolated from the biological soil crusts in culture (BG-11 medium, 25 ± 1°C, 7.5 W/m² light intensity). Cultures were grown up to 30 days under continuous light.*

Isolate no.	Cyanobacterial species	Chlorophyll content (mg/ml culture)
1	*Anabaena unisporae*	21.4
2	*Anabeana torulosa*	81.24
3	*Nostoc punctiforme*	40.1
4	*Nostoc curviceps*	63.16
5	*Nostoc piscinale*	111.5
6	*Nostoc ellipsosporum*	89.25
7	*Nostoc commune*	58.9
8	*Nostoc microscopicum*	128.9
9	*Calothrix marchica*	40.1
10	*Calothrix gloecola*	95.2
11	*Calothrix scytonemicola*	30.6
12	*Tolypothrix* sp.	47.4
13	*Westiellopsis* sp.	13.9
14	*Fischerella* sp.	122.2
15	*Fischerella* sp.	31.1

mechanism to prevent harmful effects mainly caused by the deleterious UV–radiation of the spectrum (Garcia-Pichel and Castenholz, 1993). Cyanobacteria synthesize a variety of UV absorbing pigments to prevent UV-photodamage (Scherer *et al.*, 1988; Proteau *et al.*, 1993; Bohm *et al.*, 1995; Büdel *et al.*, 1997; Ehling-Schulz and Scherer, 1999). Also terrestrial cyanobacteria possess external envelope layer, which contain non-photosynthetic pigment such as scytonemin and mycosporine amino acid-like substances (MAAs) (Garcia-Pichel *et al.*, 1992; Proteou *et al.*, 1993; Adhikary and Sahu, 1998). These extra-cellular pigments absorb maximally in UV region of the spectrum and provide protection to the cell under high solar radiation. Scytonemin and mycosporine amino acid like substances (MAAs) are two major UV absorbing pigments appear to act as sunscreen and enable them to inhabit environment that are exposed to intense solar radiation (Garcia-Pichel *et al.*, 1992; Ehling-Schulz *et al.*, 1997; Xiong *et al.*, 1997).

For detection of these pigments and their quantity in relation to chlorophyll-a, all 14 different cyanobacteria isolated from crusts were used for the pigment extraction. Absorption spectra of methanolic extracts was taken within the wavelength range of 250-750 nm in a UV-visible spectrophotometer. Analysis of these spectra showed that all the 15 cyanobacteia absorbed maximally at 668-671 nm due to chlorophyll-a, 507 and 480nm due to two different forms of carotenoids in the visible range. In addition, many of the cyanobacteria, 11 out of 15 showed a distinct

TABLE 4 *Absorption maxima of methanolic extract (90% v/v) of cyanobacterial species isolated from biological soil crusts at different wavelengths (nm)*

Isolate no	Cyanobacterial species	Wavelength (nm)						
		671:	507:	480:	384:	344:	300:	266
1	*Anabaena unisporae*	0.32	0.26	0.34	0.45	0.35	0.34	0.56
2	*Anabeana torulosa*	0.32	0.26	0.34	0.45	0.35	0.34	0.56
3	*Nostoc puctiforme*	0.09	0.22	0.31		0.31	0.4	0.55
4	*Nostoc curviceps.*	0.45	0.35	0.46	0.49	0.39	0.37	0.51
5	*Nostoc piscinale*	0.05	0.14	0.18	0.13	0.14	–	0.32
6	*Nostoc ellipsosporum*	0.13	0.41	0.63	–	0.23	0.029	0.414
7	*Nostoc commune*	0.4	0.33	0.43	0.46	–	0.987	1.132
8	*Nostoc microscopicum*	0.24	0.19	0:24	0.21	0.19	0.19	0.26
9	*Calothrix marchica*	0.61	0.73	0.93	1.19	0.84	0.78	0.129
10	*Calothrix gloecola*	0.21	0.95	1.23	0.47	0.47	0.46	0.67
11	*Calothrix scytonemicola*	0.53	0.56	0.77	0.63	0.57	0.46	0.66
12	*Tolypothrix sp.*	0.47	0.38	0.54	0.94	0.7	0.62	0.93
13	*Westiellopsis sp.*	0.24	0.17	0.24		0.3	–	1.2
14	*Fischerella sp.*	0.1	0.09	0.14	–	0.23	–	0.65
15	*Fischerella sp.*	0.36	0.17	0.27	0.38	0.36	0.71	–

absorption peak at 384nm due to presence of scytonemin pigment. Similarly almost all except *Nostoc commune* showed an absorption peak at 344,300 or 266 nm due to presence of different MAAs (Table-4). The ratio of absorption of 671:507:480:384:300:266 is given in Table-5. The objective was to analysis the changes in the carotenoid, scytonemin and various MAAs with respect to chlorophyll-a concentration in each of the organisms. The result showed that carotenoid content, in addition to UV sunscreen pigment, was relatively higher in *Calothrix gloecola* and *C. marchica* with respect to chlorophyll-a pigment. *Tolypothrix* sp. also showed higher quantity of scytonemin and MAAs. However, scytonemin absorbing at UV-A was not detected in *Nostoc ellipsosporum, Calothrix* sp., *Westiolopsis* sp. and *Fischerella* sp. All other species did contain scytonemin or MAAs or both, which possibly protects their cells in the crusts from high solar radiation.

TABLE 5 *Ratio of the absorbance of methanolic extract (90% v/v) of cyanobacterial species at different wavelengths (nm)*

Isolate no	Cyanobacterial species	Wavelength (nm)						
		671	507	480	384	344	300	266
1	*Anabaena unisporae*	1	0.95	1.02	1.14	1.03	1.02	1.24
2	*Anabeana torulosa*	1	0.99	1.17	1.19	1.2	1.12	1.3
3	*Nostoc punctiforme*	1	1.13	1.22	–	1.22	1.31	1.46
4	*Nostoc curviceps.*	1	0.91	1.0	1.0	0.95	0.93	1.0
5	*Nostoc piscinale*	1	1.1	1.13	1.08	1.1	–	1.28
6	*Nostoc ellipsosporum*	1	1.28	1.5	–	1.01	1.16	1.28
7	*Nostoc commune*	1	0.99	1.09	1.12	–	0.99	1.13
8	*Nostoc microscopicum*	1	0.96	1.0	0.98	0.96	0.96	1.0
9	*Calothrix marchica*	1	1.12	1.32	1.58	1.23	1.17	1.68
10	*Calothrix gloecola*	1	1.74	2.02	1.26	1.26	1.25	1.46
11	*Calothrix scytonemicola*	1	1.03	1.24	1.10	1.01	0.93	1.13
12	*Tolypothrix sp.*	1	0.9	1.07	1.46	1.23	1.14	1.45
13	*Westiellopsis sp.*	1	0.93	1.0	–	1.06	–	1.96
14	*Fischerella sp.*	1	0.98	1.03	–	1.13	–	1.54
15	*Fischerella sp.*	1	0.82	0.92	1.03	1.0	1.35	–

References

Adhikary, S.P. 1998. Polysaccharides from mucilaginous envelop layers of cyanobacteria and their ecological significance. *J. Sci. Ind. Res.* 57:454-466

Adhikary, S.P. & Sahu, J.K. 1998. UV-protecting pigments of the terrestrial cyanobacterium *Tolypothrix byssoidea* . *J. Plant. Physiol.* 153:770-773

Anagnostidis, S.K & Komarek, J. 1990. Modern approach to the classification of cyanophytes. 5-Stigonematales. *Arch.Hydrobiol. Algol. Studs.* 59:1-73

Belnap, J 2001 Comparative structure of physical and biological soil crusts. In: Lange, O.L and Belnap, J (eds.) *Biological soil crusts: Ecological studies*. Springer, Berlin Heidelberg New York, pp 177-191

Böhm, G.A.; Pfleiderer, W.; Böger, P. and Scherer, S. 1995. Structure of a novel oligosaccharide mycosporine amino acid ultraviolet A/B sunscreen pigment from the teresstrial cyanobacterium *Nostoc commune. J. biol.Chem.* 270: 8536-8539

Büdel, B. & Wessels D.C.J. 1991. Rock inhabiting blue-green algae/cyanobacteria from hot arid region. *Algol. Stud.* 64: 385-398

Büdel, B. 2001. Synopsis: comparative Biography of soil crust Biota *Ecological Studies,* vol. 150: 141-152

Büdel, B. 2001a. "Biological soil crust of South America" In: Lange, O.L; Belnap, j. (eds) Biological soil crusts: *Ecological studies*. Springer, Berlin, Heidelberg & New York, vol. 150: 51-55

Büdel, B. 2001b. Biological soil crust of European temperate and medditerian regions. In: Lange, O.L & Belnap, J. (eds.) Biological soil crusts: *Ecological studies*. Springer, Berlin, Heidelberg & New York, pp 75-86

Büdel, B. 2001c. "Biological soil crust of Asia including the Don and Volga region. In: Lange, O.L& Belnap, J. (eds.) Biological soil crusts: *Ecological studies*. Springer, Berlin, Heidelberg & New York, pp 87-94

Büdel. & Wessels D.C.J. 1991. Rock inhabiting blue-green algae/cyanobacteria from hot arid region. *Algol. Stud.* 64: 385-398

Budel, B.; Karsten, U. and Garcia-Pichel, F. 1997. Ultra-violet absorbing scytonemin and mycosporine like amino acid derivatives in exposed rock inhabiting cyanobacterial lichens. *Oecologia.* 112: 165-172

Castenholz, R.W. and Garcia-Pichel, F. 2000. Cyanobacterial response to UV-response. In: Whitton, B. A. and Potts, W. (eds.) *The ecology of cyanobacteria. Kluwer, Dordrecht,* pp 591-611

De Winder. 1990. Ecophysiological strategies of drought tolerant phototrophic microorganisms in dune soils. *Academisch Proeschrift,* University of Amsterdam. Amsterdam

Desikachary, T.V. 1959: Cyanophyta p.686. I.C.A.R Monograph on Algae, New Delhi.

Dor, I. and Danin, A. 1996 Cyanobacterial desert crusts in the Dead Sea Valley, Israel. *Algol. stud.* 83: 197-206

Ehling-Schulz, M. and Scherer, S. 1999. UV protection in cyanobacteria. *Eur. J. Phycol.* 34: 329-338

Ehling-Schulz, M.; Bilger, W. and Schere, S. 1997. UV-B induced synthesis of photoprotective pigments and extracellular polysaccharides in the teresstrial cyanobacterium Nostoc commune *J.Bacteriol*. 179: 1940-1945

Eldridge, D.J. and Koen, T.B. 1998 Cover and floristic of microphysics soil crusts in relation to indices of heath. *Plant Eco*. 137: 101-114.

Eldridge, D.J. 2001. Biological soil crusts of Australia. In: Lange, O. L. and Belnap, J. (eds.) Biological soil crusts.*Ecological studies*. Springer. Berlin, Heidelberg & New York, pp 119-131

Galun, M, Garty 2001 Biological soil crusts of the middle East In: Lang, O.L.; Benap, (eds.) Biological soil crusts. *Ecological studies.* Springer, Berlin, Heidelberg & New York, pp 95 –106.

Garcia-Pichel, F. and Castenholz, R.W. 1991 Characterisation and biological implications of scytonemin, a cyanobacterial sheath pigment *J. Phycol*. 27: 395-409

Garcia-Pichel, F.; Sherry, N.D. and Castenholz, R.W. 1992. Evidence for a UV sunscreen role of the extra-cellular pigment scytonemin in the terrestrial cyanobacterium *Chlorogleopsis* sp. *Photochem.Photobiol.* 56: 17-23

Green. T. G. A and Broady, P. 2001. Biological soil crusts of Antarctica. In: Lange, O. L. & Belnap, J. (eds.) Biological soil crusts. *Ecological studies.* Springer, Berlin, Heidelberg & New York pp 142-139.

Grilli-Caiola, M.; Billi, D. & Friedmann, C.I. 1996. Effect of desiccation on envelops of the cyanobacterium *Chroococcidiopsis* sp. (Chrooccales) *Eur. J. Phycol.* 31: 97-105

Grilli-Caiola, M.; Forni, C & Alberano, P. 1987. Characterization of the algal flora growing on ancient Roman frescoes. *Phycologia* 26: 387-390

Grilli-Caiola, M.; Ocampo- Friedmann, R. & Friedmann, E.I. 1993. Cytology of long term desiccation in the desert cyanobacterium *Chroococcidiopsis* sp. (Chrooccales) *Phycologia* 32: 315-322

Hahn, A. and Kusserow, H. 1998. Spatial and temporal distribution of algae in soil crusts in the Sahel of W. Africa: preliminary results. *Wildenowia* 28: 227-238

Hansen, E.S. 2001. Lichen-rich soil crusts of Artic Greenland. In: Lange, O.L. and Belnap, J. (eds.) Biological soil crusts. *Ecological studies*. Springer. Berlin, Heidelberg & New York, pp 57-65

Kindron, G. 1995. The *impact of microbial crust upon rainfall rundruff sediment yield relationships on longitudinal dune slopes. Nizzana, Western Nagev Desert, Israel.* Ph.D. Thesis. The Hebrew University of Jerusalem (English Summary).

Komrek, J. & Anagnostidis, K. 1989. Modern approach to the classification on system of cyanophytes.4.Nostocales. -*Arch. Hydrobiol.Algolog.stud.* 56: 247-345

Mackinney, G. 1941. Absorption of light by chlorophyll solutions. *J. biol. Chem.* 140: 315-322

Malam Issa, O 1998. Role of microbiotic soil crust in two Sahelian ecosystems (fallow lands & tiger bush) of Niger. *Micromorphology, phycal & biogeochemical properties*. Diss. CNRS-Universite d' Orleans

Malam Issa, O.; Trichet, J; Defargee, C.; Coute, A and Valentin C. 1999. Morphological & micro-structure of microbiotic soil crusts on a tiger bush sequence (Niger, Sahel) *Catena* 37: 175-196

Ortega-Calvo, J.J.; Sanchez-Castillo, P.M.; Hernandez Marine, M. & Saiz-Jimenenz, C. 1993. Isolation and characterization of epilithic chlorophytes and cyanobacteria from two Spanish Cathedrals (Salamanca and Toledo) *Nova Hedwigia*. 57: 239-253

Potts, M. 1994. Desiccation tolerance of prokaryotes. *Microbiol. Rev.* 58: 755-805

Reynaud, P.A. and Lumpkin, T.A. 1988. Mikroalgae of the Lanzhou (China) cryptogamic crust. *Arid Soil Res. Rehab.* 2: 145-155

Rippka, R.; Dervelles, J. Waterbury; J.B; Herdman, M.A and Stanier, R.Y. 1979. Genric assignments, strain histories and properties of pure cultures of cyanobacteria. *J. Gen.. Microbiol*. 111: 1-61

Scherer, S.; Chen, T.W. and Böger, P. 1988. A new UV-A/B protection pigment in the teresstrial cyanobacterium *Nostoc commune. Plant Physiol*. 88: 1055-1057

Turk, R and Gartner, G. 2001. Biological soil crust of the Sub alpine and naval areas in the Alps. In: Lange, O.L.; Belnap, J (eds.) Biological soil crust, *Ecological studies*. Springer, Berlin, Heidelberg & New York, pp 67-73

Xiong, F. Komenda, J.; Kopecky, J. and Nedbal, L. 1997. Strategies of ultraviolet-B protection in microscopic algae. *Physiol. Plant*. 100: 378-388

Frontiers in Plant Sciences (Pages 781-796)
Eds. K.G. Mukerji, K.V.B.R. Tilak, S.M. Reddy, L.V. Gangwane, P. Prakash, I.K. Kunwar

Industrially Important Enzymes from Cyanobacteria

Padmapriya, V. and N. Anand

ABSTRACT

Cyanobacteria are the least exploited when compared to bacteria and fungi in the field of biotechnology. Cyanobacteria have an advantage over other microorganisms in that they are autotrophic and this may play a decisive factor in the cost economics at the higher level of production. This mini review deals with some of the important enzymes of commercial importance and biochemical importance from cyanobacteria.

Key words: Cyanobacteria, enzymes, proteases, antioxidants, restriction endonucleases.

Cyanobacteria are gram-negative prokaryotes possessing a simple cell structure but combine the unique property of carrying out oxygenic photosynthesis like the members of the higher plant kingdom. Their ability to inhabit almost all sorts of habitats coupled with a diversity of varying morphological features has enabled them to survive the race on earth. Many strains have proved their ability to fix atmospheric elemental nitrogen and alter their phycobiliproteins composition to optimize their harnessing potential of solar energy. As cyanobacteria are exposed to almost all known forms of habitat ranging from adverse extremes to favourable conditions, they have evolved alternate metabolic pathways and suitable regulatory mechanisms to respond to adverse environmental conditions and nutrient availability. One of the major biochemical contributing factors to their survival is their ability to synthesize enzymes such as nitrogenase, hydrogenase, phophatases, and antioxidants like superoxide dismutases, catalases and peroxidases.

Author's address: Centre for Advanced Studies in Botany, University of Madras, Chennai 600 025.

Enzymes are protein molecules with catalytic abilities. To date about 1500-2000 enzymes have been catalogued. Bacteria such as *E.coli* possess about 3000 different proteins per cell while higher eukaryotes may synthesize about 50,000 proteins of which the majority is enzymes (Karlson *et al.,* 1979). Microbial sources are considered as alternate to plant and animal sources of commercial enzymes.

Many industrially important microbial enzymes are exo enzymes since these enzymes are frequently excreted to catalyze the initial stage in utilization of naturally occurring polymers as sources of nutrients Enzymes have several advantages over chemical catalysts with their ability to function under relatively mild conditions of temperature, pH and pressure. Microbial enzyme production on an industrial scale has come a long way since Takamine who in 1890 started the production of the enzyme production Takadiastase, a preparation of α-amylase from *Aspergillus oryzae.*

Cyanobacteria constitute a vast potential resource but only few genera, rather species, have been studied exclusively and taken up for commercial exploitation. Though they have been an object of study for other extracellular products they have been neglected in the field of enzymes since bacteria and fungi have dominated the scenario from the inception of the concept of microbial enzymes. This review is an attempt to look into the potentials of this unique class of photosynthetic prokaryotes.

ENZYMES OF COMMERCIAL IMPORTANCE

Study of cyanobacteria for industrially important or commercial enzymes have been carried out in the past. Some of the enzymes that have been considered to be of commercial value are: β- lactamase, lycopene cyclase, ribonuclease P, desaturase, peptide synthetase, amylase, phosphatase, protease, antioxidants, restriction endonucleases and others.

β-Lactamase (Penicillin β Lactam Aminohydrolase)

Probably, the first antibiotic conversion was ironically enough observed with penicillin almost parallel with the demonstration of the chemotherapeutic value of this antibiotic. It is now well known that β-lactamase action transforms penicillins into their corresponding penicilloic acids (Citri and Pollock, 1966; Richmond & Sykes, 1973). This enzyme is presently being commercially produced only from bacterial sources (Richmond M.H. 1975; Thatcher D.R..1975).

Though β lactamases are constitutive enzymes in gram-negative bacteria, there are reports that they are inducible in cyanobacteria (Prabaharan *et al.,* 1994). Production and localization of β lactamase have been studied in a unicellular species *Coccochloris elabeus* and in the filamentous form *Anabaena sp.* (Kushner & Breuil, 1977). Prabaharan *et al.*, (1994) report the enzyme to be functionally optimum at 25°C unlike the bacterial enzymes which show activity ranging from 35°C-50°C (Novick, R.P. 1964) and extracellularly produced when induced by ampicillin in the marine form *Phormidium valderianum* BDU30501.The enzyme is reported to be thermolabile since storage above 0°C renders a loss in the activity of the enzyme. Maharana *et al.*, (1986) also report the effect of other antibiotics-amoxycillin and cephaloxin on the growth and nitrogen fixation by the cyanobacterium *Calothrix marchica.* Earlier reports by various workers have given an insight into the workings of this enzyme. Reports by Kumar (1964) give the activity of *Anabaena variabilis* in the presence of streptomycin and penicillin and of *Anacystis nidulans* when grown in the presence of antibiotics mitomycin C and neomycin (1968), Srivastava (1969) reported the effect of UV induced mutations on the growth and penicillin resistance of a cyanobacterium, Vance B.D. (1966) reported the sensitivity of *Microcystis aeruginosa* along with other cyanobacteria and their associated bacteria to some selected antibiotics and Reddy (1977) gave reports of the effect on the growth and survival of *Nostoc carneum* and *Anabaena doliolum* when grown in the presence of the antibiotics - streptomycin and penicillin, in the presence of different sources of nitrogen, carbon and phosphate.

Applications of the enzyme are found in testing in the dairy industries and biological specimens such as blood serum, as inhibitors of starter cultures or to inactivate penicillin in milk so as to prevent allergic reactions in consumers and inhibit micro flora in the soil and receiving water bodies.

Lycopene Cyclase

Beta carotene is particularly important as the main source of pro vitamin A required by animals for the synthesis of the visual pigment rhodopsin and in the regulation of genes involved in limb and skin development. β-carotene is synthesized from lycopene by the enzyme lycopene cyclase (LYC). The cyanobacterial gene for this enzyme, *crt L*, has been cloned and sequenced for the better understanding of carotenoid synthesis in higher plants (Cunningham & Gantt, 1998).

The research group at the Instituto de Bioquimica Vegetal y Fotosintesis, has focused on this aspect of biotechnological potentiality of cyanobacteria by studying the function of the genes involved and their regulation and characterization of the enzymes involved in the beta carotene synthesis such as phytoene synthase, phytoene -carotene desaturase, carotene isomerase, β-carotene hydroxylase, Δ desaturase, and β-carotene ketolase (Breitenbach, *et al.*, 1998).

Ribonuclease P

tRNA biosynthesis requires processing of precursors by specific nucleases to generate the mature 5′ and 3′ ends of the tRNAs. Ribonuclease P (RNase P) is an endonuclease responsible for generating the 5′ end of pre-tRNAs as well as for the processing of other precursor RNAs. The enzyme is a ribonucleoprotein. In bacteria, the RNA subunit is the catalytic component.

Agustín Vioque Peña and his group have studied in detail the catalytic mechanism of the RNase P ribozyme from cyanobacteria, the identification of its substrates, the characterization of the chloroplast enzyme and its evolutionary relationship with RNase P from cyanobacteria. They have also studied the functional and structural properties of RNase P from a large number of cyanobacteria, a contribution to the understanding of the evolutionary dynamics of this catalytic RNA (Vioque, 1997). They have identified new substrates for the enzyme, the properties of the enzyme-substrate interaction and have put forth the design of strategies for plant biotechnology applications of RNase P (Gopalan *et al.*, 2002).

Desaturase

Fatty acid desaturases may be isolated from a broad range of organisms though animal sources prove to be impractical for large scale productions. Cyanobacteria produce a variety of desaturases, including a Δ6-desaturase that uses acyl-lipid substrates (Murata *et al.*, 1995). Δ6-desaturase is needed for the synthesis of a valuable fatty acid- Y linolenic acid that is presently got only from black currant, borage and evening primrose plants. Overexpressing the cyanobacterial Δ6-desaturase in *E. coli*, as has been done with other desaturases from cyanobacteria, should be possible and would serve as an excellent source of enzyme. Photoautotrophic culture of *Synechocystis sp.* PCC 6803 and isolation of membrane fractions containing delta-6-desaturase activity have yet to be ascertained but promise to hold immense potential in the coming years.

Peptide Synthetase

Several cyclic, branched, or linear bioactive peptides of bacteria and lower eukaryotes are produced non-ribosomally by multidomain peptide synthetases, employing so called thiotemplate mechanism. Different domains of peptide synthetases act as independent enzymes whose function is to join one amino acid to the growing polypeptide chain and make possible modifications. Peptide synthetase genes are common in cyanobacteria. Rouhiainen and his group in Helsinki have studied the hepatotoxic strain *Anabaena* 90 for these enzymes so that a better understanding of the toxicological aspects of these strains may be had and methods to harness them biotechnologically in future (Rouhiainen *et al.*, 2000).

Amylase

Amylases represent one of the largest groups of industrial enzymes in terms of production and their uses, but the primary source of these enzymes are bacteria and fungi. Amylase activity in cyanobacteria has been reported as early as 1972 (Chao *et al.*) but not much has followed since then. Amylase activity has been reported in unicellular form-*Synechococcus aeruginosa* (Reddy *et al.* 1987) and filamentous forms- *Anabaena sp* (Sarma *et al.* 1977) and *Nostoc muscorum* (Chao *et al.* 1972).

Phosphatases

Non-specific phosphatases have been reported in microorganisms wherein these enzymes supply phosphates from phosphate esters under phosphate-deprived conditions. Cyanobacteria show the presence of both the types of phosphatases- acid and alkaline (Tiwari *et al.*, 2002). Both cell bound and extracellular phosphatase activities of cyanobacteria have been reported in *Rivularia sp.* (Livingstone&Whitton, 1984), *Anabaena cylindrica* (Sarma & Kanta, 1982) and *Synechococcus aeruginosa* (Reddy&Lakshminarayana, 1988). Whitton *et al.*, (1991) have also conducted a study on the cell bound and extracellular phosphatase activites of cyanobacterial isolates. In a comparative study between the cyanobacterium *Microcystis aeruginosa* and a diatom, *Synedra capitana*, carried out by Giraudet and his group (1994) it was seen that the cyanobacterium showed a large capacity of mineralizing organic phosphorus per unit biomass due to its alkaline phosphatase activity.

Proteases

Proteases are one of the most abundantly used enzymes in the industries. In fact commercial applications of microbial enzymes reached a boom phase when proteolytic enzymes produced by the strains of *Bacillus* began to be extensively used in the detergent industries. Besides the detergent industries, proteases have a very wide application in other industries such as the food industries, the beverage and baking industries, the dairy industries, tanning industries, medical industries and in various other applications. Though protease and their uses are quite abundant they are almost exclusively produced from bacteria and fungi for commercial purposes.

Reports on the proteases of cyanobacteria are now gradually increasing. Proteases have been found to play an active role in the heterocystous cyanobacteria during the differentiation of these specialized structures when environmental conditions demand (Wood and Haselkorn, 1979). Anantani and Vaidya (1982) reported the changing levels of protease in the halotolerant cyanobacterium - *Phormidium ambiguum* Gomont in response to growing saline environment. Lockau *et. al.,* (1988) have carried out the purification and characterization of a calcium simulated protease from *A. variabilis*. Maldener *et. al.,* (1991) give a detailed DNA and amino acid sequence of the *prcA* gene of *Anabaena variablis* for a Ca^{2+} dependent protease which was initially thought to regulate the heterocyst development but was proved otherwise. Dong *et. al.,* (2000) report the identification of the active site of the protease HetR from the strain *Anabaena sp*. PCC 7120. This protease activity is needed for heterocyst differentiation and for repression of the enzyme overproduction under conditions of nitrogen deprivation. Besides playing a major role in heterocyst development and regulation, proteases in cyanobacteria are also involved in the protein turnover and in degradation of the phycobillin proteins (Foulds & Carr, 1977). Studies on the different types of stress experienced by the cyanobacteria (nitrogen and salt) show an active role played by proteases (Allen & Hutchinson, 1980; Zuther *et al.,* 1998). A study on the protease production by the various classes of the cyanophytes in our laboratory have revealed that unicells show sporadic protease activity in the absence of nitrate in the medium while the best activity is shown by the members of Stigonematales. Sidoruk *et al.,* 1996 have reported a new class of carboxy –terminal protease from the unicellular cyanobacterium *Synechocystis sp* PCC 6803 which may give an insight into the physiological functions.

Anti oxidants

Cyanobacteria were one of the first forms on earth to undergo photosynthetic autotrophism with a release of molecular oxygen. Since they were the first to release molecular oxygen they must have also been the first to sense it (Obinger *et al.,* 1998). Singlet dioxygen is highly toxic to all cellular components and thus a system of enzymic defences has been evolved to combat these radicals. Of them, superoxide dismutases are active in the first rung of the antioxidant enzyme ladder.

Reports as early as in 1975 (Asada *et al.,*) have stated the presence of these enzymes in cyanobacteria like *Plectonema boryanum*. Other strains showing SOD activity have also been reported- *Anacystis nidulans* (Abeliovich *et al.,* 1974), *Spirulina platensis* (Cohen, Z. 1986), *Anabaena cylindrica* (Grilli Caiola *et al.,* 1991), *Anabaena variabilis* (Bagchi *et al.,* 1991),

Synechococcus sp. PCC7942 (Herbert *et al.*, 1992), and *Anabaena sp* PCC 7120 (Li *et al.*, 2002). Work carried out in our laboratory on the representative members of the different isolates of cyanobacteria has shown that there is generally a very high content of the enzyme during the initial growth phase which gradually peters out in the exponential phase and starts to reappear in the late exponential phase. The branched filamentous heterocystous forms show the higher enzyme activity.

Since the publications of the initial reports in the early 1970's there has been extensive work done in this field to harness these enzymes and a considerable breakthrough has been achieved by Takeshima *et al.*, (1994) where they have succeeded in high level expression of the human superoxide dismutases in the cyanobacterium*Anacystis nidulans* 6301. SOD's are known to play an important role in the protection of the organism. against oxidative stress (Thomas *et al.*, 1998), chilling damages (Thomas *et al.*, 1999) protecting the proton donating systems of nitrogen fixation (Henry *et al.*, 1978) and help tide over the dessication periods (Shirkey *et. al.*, 2000).

Catalases and ascorbate peroxidases follow the superoxide dismutases in the cascade of antioxidant enzymes. These enzymes are a group of oxido-reductases which find applications in industries. Prokaryotic catalase – peroxidase system have been characterized in *Synechococcus* PCC7942 (Mutsuda *et al.*, 1996) and in *Synechococcus* PCC 6301 (Obinger *et al.*, 1997).

Ascorbate peroxidases have been reported from various cyanobacteria. *Tolypothrix tenuis, Aphanothece halophytica, Fremyella diplosiphon* (Miyake *et al.*, 1991), *Nostoc muscorum* and *Synechococcus* 6311 (Tel-Or *et al.*, 1985) show the presence of dehydroascorbate reductases while monohydro ascorbate reductases have been reported in forms which show ascorbate peroxidase activity. Glutathione reductase has been reported in *Synechococcus lividus* (Dupoy *et al.*, 1985), *N.muscorum* (Bhunia *et al.*, 1993) *A. variabilis* (Bagchi *et al.*, 1991), *Spirulina maxima* and *Synechococcus* 6301 (Tel-Or *et al.*, 1985)

The economic potential of these enzymes are still to be tapped as they are valuable in the field of pharmaceuticals and cosmetology where they are employed as anti aging, anti wrinkle drugs and in the treatment of diseases like gout, rheumatism etc.

RESTRICTION ENDONUCLEASES

Restriction endonucleases have proved to be of importance in the emerging highly technologically enhanced field of molecular research where they are deployed as molecular scissors. Since these enzymes facilitate cleavage of DNA into characteristic fragments these foreign DNA fragments can be easily carried into a vector DNA molecule (Puhler & Human 1981, Mayer,

1981) thus bringing about recombinant technology into reality. The more the specific recognition sites the more advantage a molecular biologist has. Thus the search for novel sequence recognizing restriction enzymes from various sources wherein the cyanobacteria have proved to be a vital source of specific endonucleases and periplasmic nucleases (Duyvesteyn *et al.*, 1981).

In cyanobacteria there are nearly 100 isoschizomers from about 50 strains (Houmard & Marsac, 1988) and a few of them have been cited below.

Endonuclease	*Organism*	*Recognition sequence*
Ava I	*Nostoc sp* PCC7118	CPyCGPuG
Ava II	*Nostoc sp* PCC7118	GGCC
Acy I	*Anabaena sp* PCC 7122	GPuGCPyC
Mla II	*Fischerella sp* PCC 7414	TTCGAA
Mst II, Mst I	*Microcoleus sp* UTEX2220	CCTNAGG
Asu I	*Anabaena sp* PCC 6309	GGNCC
Nsp CI	*Nostoc sp* PCC 7524	Pu CATGPy

ENZYMES OF BIOCHEMICAL IMPORTANCE

There have been certain enzymes that have been identified in the biochemical reactions. These are of good commercial value in analytical kits. These are phosphoglycerate kinase, polyphenol oxidases, laccases, aminoacid oxidases and hydrogenases.

Phosphoglycerate Kinase

The currently available phosphoglycerate kinases, which are used in the determination of ATP, are not specific to ATP alone but respond to GTP and ITP also (Bergmeyer, 1974). The phosphoglycerate kinase got from *Spirulina platensis* is highly specific to ATP and promises to be of good commercial value if exploited (Krietsch & Kuntz, 1978).

Polyphenol Oxidases and Laccases

Sashirekha *et al.*, (1997) have shown the activity of the strain *Phormidium valderianum* BDU 30501 to effectively remove and degrade the toxic component in effluents- phenol, due to the active role of the enzymes polyphenol oxidases and laccases.

Aminoacid Oxidase

Aminoacid oxidase activity has been studied in the cyanobacterium *Anacystis nidulans*. Both free and immobilized cells of the strain show the activity of the enzyme, which increases when the quality of the incident light is red and when grown along with the bacteria *Providenea* (Wikstrom *et al.*, 1982).

Hydrogenases

Hydrogenases can serve both for the uptake of hydrogen into the cell and the disposal of electrons via hydrogen production. Since electrons can be used for energy production in electron transport chains, some organisms evidently recover and reoxidize hydrogen, which has escaped during nitrogen fixation by the inherent hydrogen producing activity of nitrogenase (Tamagini *et a*l., 2002). It is of fundamental ecological and biotechnological interest to understand the functions of uptake vs. hydrogen producing hydrogenases, especially in cyanobacteria. Genetic engineering has become increasingly important with the establishment of molecular biological tools and techniques for cyanobacteria. A few unicellular strains, including *Synechococcus* strains PCC 6301 and PCC 7942 and *Synechocystis* strain PCC 6803, are naturally transformable. These methods have been used with success in filamentous genera such as *Anabaena* and *Nostoc*, which are interesting candidates for future photo biotechnological applications (Hansel and Lindblad, 1998; Miyake *et al.*, 2001). The introduction of the structural genes encoding a clostridial Fe hydrogenase into the unicellular cyanobacterium *Synechococcus elongatus* PCC 7942 and their subsequent transcription and translation into a functional hydrogen-evolving enzyme in a photosynthetic microorganism (Asada *et al.*, 2000) is interesting and deserves further studies.

Elastase

Pabbi & Singh (1998) give an account of the presence of an anti fertility enzyme, elastase, from a cyanobacterium, *Oscillatoria sp.*

ENZYMES INVOLVED IN TRANSGENIC CYANOBACTERIA

Novel Enzymes

Genes responsible for regulating various processes in cyanobacteria have been used in numerous laboratories for the production of transgenic cyanobacteria, which are used in various fields and amongst them even in enzymology. (Tyagi & Kaushik, 2003)

Some genes cloned in cyanobacteria for expression of some desired enzymes are:

Ezpression of the human carbonic anhydrase (Price & Badger, 1989)

lac Z –expression of acetyl transferase (Subramaniam & Kaushik, 2001)

luc – expression of beta-galactosidase (Subramaniam & Kaushik, 2001)

lux A, lux B- expression of luciferase. (Schmetterer *et al.*, 1986)

CONCLUSIONS

Cyanobacteria have so far not been exploited for their commercially important intracellular and extracellular enzymes even though they are very good sources of such enzymes. The only reason, perhaps, is their slow growth rates. Study of the dynamics and proper scaling up the biomass of the suitable strains will be required for using cyanobacteria for commercial enzyme production.

Acknowledgments

The Director, Centre for Advanced Studies in Botany, University of Madras, is duly acknowledged for the facilities provided. Thanks are also due to the Council for Scientific and Industrial Research, India, for the award of Senior Research Fellow to Padmapriya. V.

References

Allen, M.M. and Hutchinson, F. 1980. Nitrogen limitation and recovery in the cyanobacterium *Aphanocapsa* 6308. *Arch. Microbiol.* **128**: 1-7.

Anantani A.S and Vaidya, B.S. 1982. RNA, RNASE and Protease levels in a halotolerant cyanophyte. *Phykos.* **21**: 105-107

Asada K, Kyoko Yoshikawa, Masa-Aki Takashi, Yutaka Maeda, and Koe Enmanji. 1973. Superoxide Dismutases from a Blue-Green Alga, *Plectonema boryanum. J. Biol. Chem.* **8**: 2801-2807.

Asada, Y., Koike, Y., Schnackenberg, J., Miyake, M., Uemura, I., and Miyake, J. 2000. Heterologous expression of clostridial hydrogenase in the cyanobacterium *Synechococcus* PCC7942. *Biochim. Biophys. Acta.* **1490**: 269-278

Bagchi, S. N., Ernst, A., and Borger, P. 1991. The effect of activated oxygen species on nitrogenase of *Anabaena variabilis. Z. Naturforsch. Sect. C* **46**: 407-415

Bergmeyer, H.U. 1974. *Methoden der enzymatischan analyse.* Verlag Chemie, Weinhom Bergstrasse. 2142

Bhunia, A.K., Roy, D., and Banerjee, S.K. 1993. Carbaryl induced effects on glutathione content, glutathione reductase and superoxide dismutases activity of the cyanobacterium *Nostoc muscorum. Lett. Appl. Microbiol.* **18**: 10-13.

Breitenbach, J., Fernández-González, B., Vioque, A. and Sandmann, G. 1998. Higher plant type -carotene desaturase in the cyanobacterium *Synechocystis* PCC6803. *Plant Mol. Biol.,* **36**: 725-732.

Chao, I. and Bowen, C.C. 1972. Characterization of alpha-granules and ribosomes in the bluegreen algae. *In*: T.V. Desikachary (Ed.), *Taxonomy and Biology of blue-green algae,* University of Madras.

Citri, N. and Pollock, M.R. 1966. The biochemistry and function of b-lactamase (penicillinase). *In*: Nord, F. F. (ed) *Advances in Enzymology,* Vol **28**. New York, Inter Science Publishers. Pp 287-323.

Cohen, Z. 1986. Products from Microalgae. *In: CRC Handbook of Microalgal Mass Culture*. (Ed.). Richmond, A. CRC Press,Florida, USA. pp 421-456.

Cunningham F.X., and Gantt, E. 1998. Genes and enzymes of carotenoid biosynthesis in plants. *Annual Review of Plant Physiology and Plant Molecular biology.* **49**: 567-583.

Dong. Y, Xu Huang, Xiang –Yu Wu and Jindong Zhao 2000. Identification of the active site of Het R protease and its requirement for heterocyst differentiation in the cyanobacterium *Anabaena sp*. strain PCC 7120. *J. Bacteri*. **182**: 1575-1579.

Dupoy, D., Conter, A., Croute, F., Murat, M. and Planel, H. 1985. Sensitivity of *Synechococcus lividus* to hydrogen peroxide. *Environ. Exp. Bot*. **25**: 339-347.

Foulds I.J. and Carr, N.G. 1977. A proteolytic enzyme degrading phycocyanin. *FEMS Microbiol Lett* **2**: 117-119.

Giraudet, H. Berthon, J., and Buisson, B. 1997. A comparison of daily alkaline phosphatase activity of a cyanobacterium (*Microcystis aeruginosa*) and a diatom (*Synedra capitana*). *Comp.Ren.Acade.Scien.Serie. III Scien.Vie Life Sciences.* **320**: 451-458.

Gopalan, V., Vioque, A. and Altman, S. 2002. RNase P: Varieties and uses. *J. Biol. Chem*. 277, 6759-6762.

Grilli Caiola. M, Canini, A., Galiazzo, F., and Rotilio, G. 1991. Superoxide dismutases in vegetative cells, heterocysts and akinetes of *Anabaena cylindrica* Lemm. *FEMS Microbiology Letters* **80**: 161-166.

Hansel, A. and Lindblad, P. 1998. Towards optimization of cyanobacteria as biotechnologically relevant producers of molecular hydrogen, a clean and renewable energy source. *Appl. Microbiol. Biotechnol.* **50:** 153-160

Herbert S.K., Guy Samson, David Fork and David. E. Laudenbach 1992. Characterization of damage to photo systems I and II in a cyanobacterium lacking detectable iron superoxide dismutases activity. *Proc.Natl.Acad.Sci.USA* **89**: 8716-8720.

Houmard, J. and Tandeau de Marsac, N. *Meth.Enzymol.* **167**: 808-847,1988.

Iris, M., Wolfgang Lockau, Yuping Cai and Peter Wolk 1991. Calcium dependent protease of the cyanobacterium *Anabaena*: molecular cloning and expression of the gene in Escherichia coli, sequencing and site-directed mutagenesis. *Mol. Gen. Genet.* **225**: 113-120.

Karlson, P., Bielka, H., Horecker, B.L., Jacoby, W.B., Keil, B., Lliebecq, C., Lindberg B., and Webb, E.C. 1979. *Enzyme Nomenclature, Recommendations of the*

Nomenclature committee of the International Union of Biochemistry, Academic Press, New York, San Francisco, London.

Krietsch, W.B. and Kuntz, G.W. 1978. Specific enzymatic determination of adenosine triphosphate. *Anal. Biochemistry*. **90**: 829

Kumar, H.D. 1968. Inhibitory action of the antibiotics mitomycin C and neomycin on the blue-green alga *Anacystis nidulans. Flora Jena* **159**: 437-444.

Kushner, D.J. and Breuil, C. 1977. Penicillanase (β-Lactamase) formation by blue-green algae. *Arch. Microbiol.* **112**: 219-223.

Landis E.A. Henry, Ivan N. Gogotov and David O. Hall. 1978. Superoxide Dismutase and Catalase in the Protection of the Proton-Donating systems of Nitrogen Fixation in the Blue-Green Alga *Anabaena cylindrica. Biochem.J.* **174**: 373-377.

Li, T., Xu Huang, Zhou, R., Liu, Y., Li, B., Nomura, C. and Jindong Zhao 2002. Differential Expression and localization of Mn and Fe superoxide dismutases in the heterocystous cyanobacterium *Anabaena sp*. strain PCC 7120. *J. Bacteriol.* **184**: 5096-5103.

Livingstone, D. and Whitton, B.A. 1984 Water chemistry and phosphatase activity of the blue-green alga, *Rivularia* sp.in upper Teesdale streams; *J. Ecol.* **72**: 405-421.

Lockau. W, Massalsky, B., and Dirmeier, A. 1988. Purification and partial characterization of a calcium stimulated protease from the cyanobacterium, *Anabaena variabilis. Europ. J. Biochem.***172**: 433-438

Maharana, R.K., Pradhan, J.M. and Padhi S. 1986. Effect of amoxicillin and cephaloxin on growth and nitrogen fixation of *Calothrix marchica. IBC* **3**: 45-47.

Mayer H. 1981. Restriction Endonucleases. *Biochemica Hand Book of Microalgal Mass Culture*. (Ed. Richmond). CRC Press, Florida, USA. Pp 437.

Miyake C, Fumi Michihata and Kozi Asada.1991. Scavenging of Hydrogen Peroxide in Prokaryotic and Eukaryotic Algae: Acquisition of Ascorbate Peroxidase during the Evolution of Cyanobacteria. *Plant Cell Physiol.* **32**: 33-34.

Miyake, M., Schnackenberg, J., Nakamura, C., Asada, Y., and Miyake, J. 2001. Molecular handling of hydrogenase, p. 205-219. *In* J. Miyake, T. Matsunaga, and A. San Pietro (ed.), *Biohydrogen II*. Elsevier Science Ltd., Oxford, United Kingdom.

Murata, N., Higashi, S., Wada, H., Sakamoto, T., Macherel, M.-H., Macherel, D., Tasaka, Y., and Los, D. The cyanobacterial desaturases: aspects of their structure and regulation. In: *Plant Lipid Metabolism* (J.-C. Kadar and P. Mazliak, eds.), Kluwer Academic Publ., Dordrecht, 1995, pp. 3-8.

Mutsuda, M., Ishikawa, T., Takeda, T., and Shigeoka, S. 1996. The catalase-peroxidases of *Synechococcus* PCC 7942: Purification , nucleotide sequence analysis and expression in *E. coli*. *Biochem. J.* **316**: 251-257.

Novick, R.P. 1964. Analysis by transduction of mutation affecting penicillanase formation in *Staphylococci aureus. J. Gen. Microbiol* **33**: 121-136.

Obinger, C., Regelsberger, G., Pircher, A., Strasser, G., and Peschek, G.A. 1998. Scavenging of superoxide and hydrogen peroxide in blue-green algae (cyanobacteria). *Physiol. Plant.* **104**: 693-698.

Obinger, C., Rgelsberger, G., Strasser, G., Burner, U. and Peschek, G.A. 1997. Purification and characterization of a homodimeric catalase-peroxidase from the cyanobacterium *Anacyatis nidulans*. *Biochem. Biophys. Res. Commun.* **235**: 545-552.

Pabbi, S. and Singh, P. K. 1998. Conservation and exploitation of cyanobacteria. *In*: Trends in microbial exploitation (Eds) Bharat Rai, R.S. Upadhyay and N.K. Dubey. International Society for Conservation of natural Resources, Dept. of Botany, BHU, Varanasi, India. pp 27-37

Prabaharan, D., Sumathi, M. and Subramanian, G. 1994. Ability to Use Ampicillin as a Nitrogen Source by the Marine Cyanobacterium *Phormidium valderianum* BDU 30501. *Curr. Microbiol.* **28**: 315-320.

Price, G.D. and Badger, M.R. 1989. *Plant Physiol.* **91**: 505-513.

Puhler, A. and Heumann, W. 1981. Genetic Engineering in Biotechnology. Vol I (Eds. Reh, M.J. and Reed, G.). Verlag Chemie, Wenheim, Deerfield Beach.

Reddy, T. R. 1977. Toxicity of antibiotics on *Nostoc carneum* in presence of different nutrients. *Phykos.* **16**: 31-42.

Reddy, T.R.K. and Lakshminarayana, R. 1988. Properties of phosphatases from green alga *Scenedesmus incrassatulus* Bahlin and blue-green alga *Synechococcus aeruginosus* Nag. *Proc. Indian Acad. Sci. (Plant Sci.)* **98**: 475-482.

Reddy, T.R.K., Rao, K.S.N., and Jayadeva, V. 1987. Amylase from Green Alga *Scenedesmus incrassatulus* Bohlin and Blue Green Alga *Synechococcus aeruginosus* Nag. *Indian J. Exp. Biol.* **25**: 389-391.

Richmond, M.H. 1975. β-lactamase (*Staphylococcus aureus*). *Methods Enzymol.* **43**:664-672.

Richmond, M.M., and Sykes, R.B., 1973. The b lactamases of gram negative bacteria and their possible physiological role. *Advanc. Microbiol. Physiol.* **9:** 31-88.

Rouhiainen, L., Paulin, L., Buikema, W., Suomalainen, S., Hyytiäinen, H., Haselkorn, R. and Sivonen, K. 2000. Genes encoding synthetases of cyclic depsipeptides, anabaenopeptilides, in *Anabaena* strain 90. *Molecular Microbiology* **37(1)**: 156-167

Shashirekha, S., Uma, L. and Subramanian, G.1997. Phenol degradation by the marine cyanobacterium *Phormidium valderianum* BDU 30501. *J of Industr. Microbiol.* **19**: 130-133

Sarma, T., Swarnakanta, A., Ahuja, B.S. and Usha Kiran 1977. *Curr Sci,* 46: 609.

Sarma, T.A. and Kanta, S. 1982 Variation in enzyme levels during sporulation in blue-green algae. *Biochem. Physiol. Pflanz.* **177**: 278-282.

Schmetterer, G., Wolk, P. and Elhai, J. 1986. Expression of luciferases from *Vibrio harveyi* and *Vibrio fischeri* in filamentous cyanobacteria. *J. Bacteriol.* **167**: 411-414.

Shirkey, B., Kovarcik, D.P., Wright, D.J., Wilmoth, G., Prickett T.F., Helm R.F., Gregory, E.M. and Potts, M. 2000. Active Fe-containing superoxide dismutases and abundant *sodF* mRNA in *Nostoc commune* (cyanobacteria) after years of dessication. *J. Bacteriol.* **182**: 189-197.

Sidoruk, K., Shestopalov, V., Babykin, M. and Shestakov, S. 1996. Nucleotid sequence of a new member (Accession No. X96490) of the carboxyl-terminal processing protease family in the cyanobacterium *Synechocystis sp*. PCC 6803. *Plant Physiol.* **111**: 947.

Srivastava, B.S. 1969. Ultra-violet induced mutation to growth factor requirement and penicillin resistance in a blue-green alga. *Arch Microbiol* **66**: 234.

Subramanian, S. and Kaushik, B.D. 2001. Development of genetic markers in cyanobacteria and stability of genetically marked strains in soil. *World J. Microbiol. Biotech.* **17**: 535-544.

Takeshima, Y., Noboru Takatsugu, Masahiro Sugiura and Hideaki Hagiwara 1994. High-Level expression of human superoxide dismutases in the cyanobacterium *Anacystis nidulans* 6301. *Proc Natl Acad.Sci.USA* **91**: 9685-9689.

Tamagnini, P., Rikard Axelsson, Pia Lindberg, Fredrik Oxelfelt, Röbbe Wünschiers, and Peter Lindblad 2002. Hydrogenases and Hydrogen Metabolism of Cyanobacteria. *Microbiology and Molecular Biology Reviews.* **66**: 1-20.

Tel-Or, E., Huflejt, M.E., and Packer, L. 1985. The role of glutathione and ascorbate in hydroperoxide removal in cyanobacteria. *Biochem. Biophys. Res. Commun.* **132**: 533-539.

Thatcher, D.R. 1975 a. β-lactamase (Bacillus cereus). *In*: *Methods Enzymol.* **43**: 640-652.

Thatcher, D.R. 1975 b. β-lactamase (Bacillus licheniformis). *In*: *Methods Enzymol.* **43**: 653-663.

Thomas, D.J., Thomas, J.B., Prier, S.D., Nasso, N.E. and Herbert, S.K. 1999. Iron superoxide dismutases protects against chilling damage in the cyanobacterium *Synechococcus sp*. PCC 7942. *Plant Physiol.* **120**: 275-282.

Thomas, D.J., Avenson, T.J., Thomas, J.B. and Herbert, S.K. 1998. A cyanobacterium lacking iron superoxide dismutases is sensitized to oxidative stress induced with methyl viologen but is not sensitized to oxidative stress induced with norflurazon. *Plant Physiol.* **116**: 1593-1602.

Tiwari, D.N., Singh, S.K. and Pandey, M. 2002. Alkaline phosphatase activity in cyanobacteria. *In*: *Algological Research In India* (Ed. N. Anand). B.S. M. Pal Singh. Dehra Dun. pp 345-350

Tyagi, R. and Kaushik, B.D. 2003. Production of transgenic cyanobacteria and their applications. *In: Phycology: principles, processes and applications.* (Ed. A.S. Ahluwalia). Daya publishing house. Delhi. Pp 312-329.

Vance, B.D. 1966. Sensitivity of *Microcystis aeruginosa* and other blue-green algae and associated bacteria to selected antibiotics. *J. Phycol.* **2**: 125-128.

Vioque, A. 1997. The RNase P RNA from cyanobacteria. Short tandemly repeated repetitive (STRR) sequences are present within the RNase P RNA gene in heterocyst-forming cyanobacteria. *Nucleic Acids Res.* **25:** 3471-3477.

Whitton, B.A., Grainger, S.L.J., Hawley, G.R.W. and Simon, J.W. 1991. Cell bound and extracellular phosphatase activities of cyanobacterial isolates. *Microbial Ecology*: **21**: 85-98.

Wikstrom, P., Szwajcer, E., Brodelius, P., Nilsson, K.N. and Mosbach, K. 1982. Formation of alphaketo acids from amino acids using immobilized bacteria and algae. *Biotech. Lett.* **4**: 153.

Wood, N. B. and Haselkorn, R. 1979. Proteinase activity during heterocyst differentiation in nitrogen-fixing cyanobacteria. In: Cohen, G.N. and H. Holzer (Eds.). *Limited proteolysis in microorganisms.* US Dept. of Health, Education and Welfare Publication # NIH. 79-1591, US Printing Office, Washington DC, pp 159-166.

Zuther, E.H. Schubert and Hagemann, M. 1998. Mutation of a Gene encoding a putative glycoprotease leads to reduced salt tolerance, altered pigmentation, and cyanophycin accumulation in the cyanobacterium *Synechocystis sp.* strain PCC 6803. *J. Bact.* **180:** 1715-1722.

Frontiers in Plant Sciences (Pages 797-806)
Eds. K.G. Mukerji, K.V.B.R. Tilak, S.M. Reddy, L.V. Gangwane, P. Prakash, I.K. Kunwar

Impact of Electric Field on *Spirulina platensis* (Lonar Lake Isolate)

Pushpa Srivastava and Md. Akhtar Khan

ABSTRACT

Lonar lake isolate of *Spirulina platensis* when subjected to 1 and 5 kv electric field, growth and cellular constituents were enhanced. Highest yield was procured at 40 min exposure at 1 kv and 10 min at 5 kv. Protein, amino acids, chlorophyll$_a$, biliproteins were enhanced. But carbohydrates, lipids and marginal reduction in carotenoids was observed. Electric field is suggested as cheapest source for biomass production of commercial alga *spirulina*.

Key words: *Spirulina platensis*, electric field, biomass, cellular metabolites.

INTRODUCTION

The electric field is defined as the space property generated by a charge, by virtue of which it modifies the space around itself, so as to exert force on any charge or material to bring into the space. The work on the exciting emotions of electric field are of greatest importance on human beings, laboratory animals, organ culture, plants and certain individual cells. The use of such a field in the area of medicine and psychotherapy have been innumerable, but the latter half of the century attracted much more attention than ever before on other objects of biology. Biological structure, at all levels of organization of organism have electric multipoles and therefore can be affected by an electric field (Behari, 1994).

Author's address: Algal Biotechnological Laboratory, Department of Botany, University of Rajasthan, Jaipur – 302004
E-mail: pushpashyam@rediffmail.com

Initially, it was being used to set some disorder, but additional interest has been stimulated by the discovery of the fact that extremely low frequency of electric field affected RNA transcription and protein synthesis (Polk and Soong, 1990). Frazier *et al.* (1990) further enhanced our knowledge by their findings that DNA damage caused by Co^{60} in the isolated human peripheral lymphocytes can be repaired by a dose of 60 Hz electric field. However, Angell *et al.* (1990) have made significant contributions in this field and found that high voltage direct current transmission upto ±500 kv was found to be effective on beef cattle production.

Nevertheless, Nordersen *et al.* (1984, 1988) reported increased frequency of chromosomal aberrations in peripheral blood cells of workers, who worked under high voltage switch yard. However, Kowallcizuk and Sounders (1989) found that electric field of 50 Hz could not cause any serious lethal impact on male mice, although they were exposed continuously for a period of 2-weeks at 20 kv.

Besides, various plants and plant parts were also put under different types of electric field impacts. Ascensor *et al.* (1990) employed high spherical symmetry electric field at a frequency of 50 to 75 Hz to *Saccharomyces cerevisiae, Kloeckera apiculata* and bacterium *Citrobacter fruendii* to find out the change in growth pattern but the overall growth was found to be unaffected. Amongst the higher plant-groups, root tips of *Zea mays* was also subjected to a 60 Hz electric field by Brayman *et al.* (1990) and observed an inhibition in the growth and correlated it with the field strength rather than the current density in the medium. Souda *et al.* (1990) studied the relationship between growth and electric oscillation in bean-roots and found that root elongation was proportional to the magnification of electric oscillation frequency.

The impact of electric field on enzyme had been tried by Guido *et al.* (1990). They exposed acid phosphotase to 50 Hz current at different densities of 10-60 mA/cm^2, but could not find any change in structural properties of the enzyme. Significant contribution has been on record of Gosztola (1995), who pointed out the effect of electric field on photo generated ion pairs on nearby molecules, which resulted in the carotenoid bands shift during photosynthesis.

Electric field application were also tried on different algal forms by Walker (1960) on Chara and Nitella and Williams and Spanswick (1964) on Nitella, where electrodes were used to observe membrane resistance. Kumar et al. (1981) found flocculation in the cultures, when electrodes were directly inserted in the cultures of Chlorella. In 1992, Srivastava exposed two species of *Spirulina* i.e. *S. subsalsa* and *S. platensis* to electric field of 1 kv and 5 kv and registered an increment in growth and protein contents. Singh (1992) and Sharma (1995), also found a qualitative and quantitative increase in the protein contents of *A. fusiformis* and *S. subsalsa* respectively. Malhotra (1999), further recorded similar observation on *Phormidium*. These workers found that low voltage of electric field i.e. 1 kv was more advantageous for cellular protein contents rather than the higher voltage of 5 to 25 kv.

In the present study, cultures of *S. platensis* (Lonar lake isolate) were subjected to a range of electric field 1 kv and 5 kv for a period of 5-70 min and 5-60 min respectively, in order to register its impact on its growth, morphology and cellular constituents, particularly the protein percentage.

MATERIALS AND METHODS

Spirulina platensis was collected and isolated from Lonar lake of Maharashtra and raised into unialgal cultures in improvised CFTRI medium. Electric field was provided to the cultures by self designed apparatus, emitting an electric field within voltage range of 1-5 kv. It supplied a stable voltage continuously across the aluminium electrodes. One of the electrodes was earthed and other one was held parallel to it. The distance between the two electrodes could be adjusted with the help of a screw management. In order to avoid interaction between electric field and the ambient atmosphere, the electrode along with the culture sample was covered with a wooden cover.

5 ml of freshly growing culture was taken in a petriplate (17 × 50 mm) which formed a uniform depth of 2 mm was exposed to 5-70 min at 1 kv and 5-60 for 5 kv electric field. After the treatment with electric field for the stipulated time, the culture sample was transferred to a culture tube containing 10 ml of minimal medium. Duplicate sets of test tubes were prepared for each treated sample. These cultures were raised into 250 ml cultures in flasks of 500 ml capacity. The biochemical analysis of the treated cultures were carried out. Untreated culture were maintained as control. The controlled and treated cultures were subjected to a light period of 12 h/d at 30 ± 2°C.

Test tube culture was exclusively used for optical density record and the other one was employed for measuring trichome breadth and microscopic observations. Data was collected for a period of 5 weeks at weekly intervals. At the end of the experiment, flask cultures were used for the analysis of cellular metabolites i.e. proteins, carbohydrates, lipids, carotenoids, chl.-a along with phycobiliproteins.

RESULTS

Spirulina Exposed to 1 kv Electric Field

With 5 min exposure, the growth of the cultures had an edge over unexposed cultures, so also the breadth of the cultures. Further treatment of 10-40 min led to the enhancement in optical density, which increased four times the initial value (Fig. 1). Similarly trichome breadth showed an increment of 2.2 mm. Cultures were brilliant bluish green with healthy trichomes having intact cell contents. 50-60 in exposed cultures were on par with unexposed cultures with marginal increase in optical density. However, 70 min treatment retarded the growth and cultures turned dull blue green.

Spirulina Exposed to 5 kv Electric Field

The cultures exposed to 5 and 10 min showed higher rate of growth and breadth of the trichomes compared unexposed controlled cultures (Fig. 3 and 4). Apparently, culture samples were lush bluish green expressing healthy status of the cultures. With 20-60 min treatment, reduced growth through optical density was observed, when compared to 10 min exposed cultures. However, it was still comparable to unexposed cultures upto 40 min treatment of electric field. But, the cultures were unhealthy dull yellowish green in colour with reduced breadth of the trichomes. Most of the trichomes showed damage caused by electric field with scattered chromatophores (Plate 1).

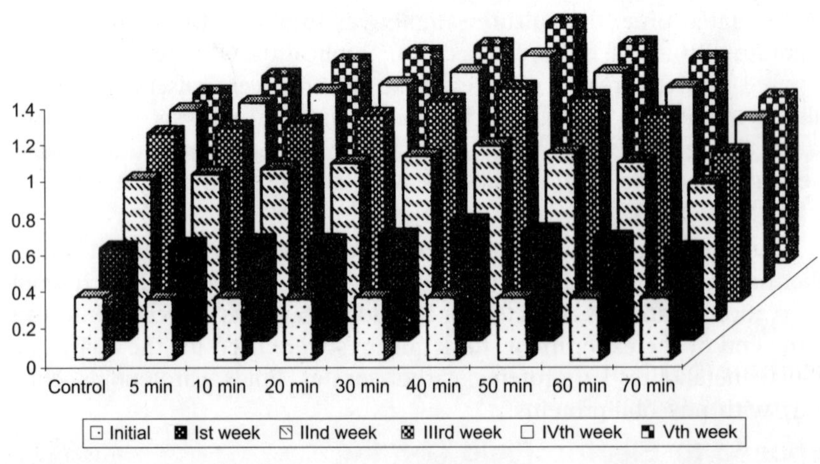

Fig. 1 *S. platensis* (Lonar lake isolate) Optical density records after 5-70 min exposures

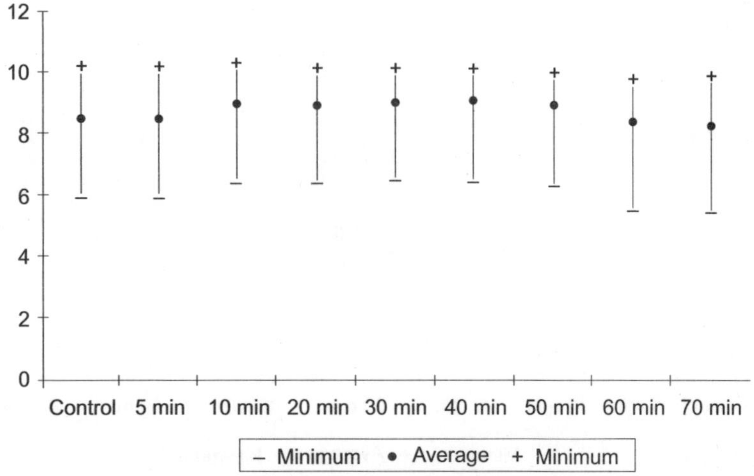

Fig. 2 *S. platensis* (Lonar lake isolate) Trichome breadth (μm) after 5-70 min exposure to 1 kv electric field

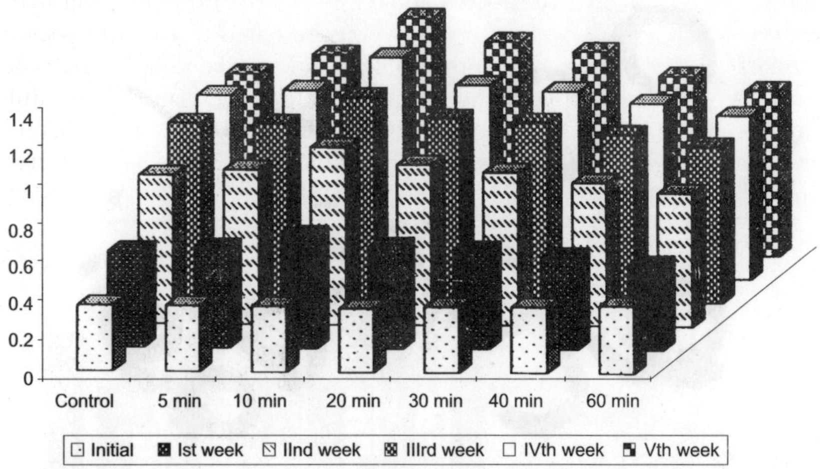

Fig. 3 *S. platensis* (Lonar lake isolate) Optical density records after 5-60 min exposures to 5 kv electric field

Biochemical Estimation of Cellular Metabolites of *S. platensis* Exposed to Electric Field (1-5 kv)

The growth was maximum in cultures subjected to 1 kv for 40 min and 5 kv for 10 min. The optical density records, trichome breadth and bright blue

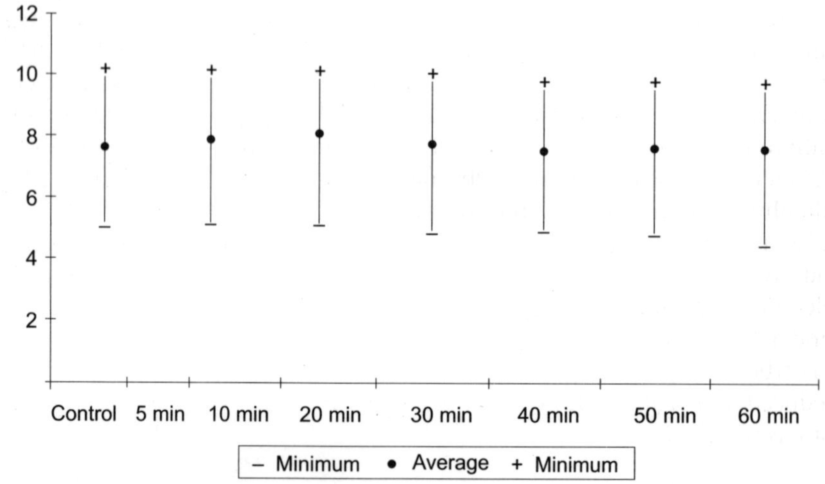

Fig. 4 The growth of *S. platensis* (Lonar lake isolate) through trichome breadth (μm) after 5-60 min exposure to 5 kv electric field

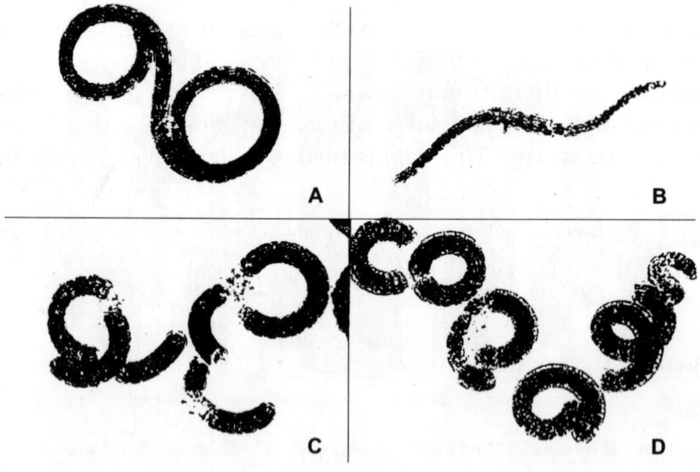

Plate 1 *Spirulina platensis* under 5 kv electric field x 800 X

A. 'S' shaped coiling of trichomes under 20 min exposure.

B. Straightening of trichome under 30 min exposure.

C&D. Fragmentation of trichomes with disintegrating cell contents under 40 & 60 min exposure.

green biomass were supportive of this fact. Therefore, these cultures were employed for biochemical analysis for estimation of cellular metabolites along with controlled cultures.

Table 1 clearly denoted that the 48% protein of untreated cultures was raised to 50 and 50.5% in cultures exposed to electric field for 40 and 10 min at 1 and 5 kv respectively. Amino acids also followed the same trend and enhanced from 6084 to 6190 and 6395 □g/g of *Spirulina* serially. Biliproteins showed an increment of 8.0 to 8.56% under two exposures from 7.43% of unexposed cultures. Phycocyanin was raised to 4.12 and 4.22%, allophycocyanin 2.80 to 3.46% and phycoerythrin 1.08 to 0.88% from 3.79, 2.78 and 0.86 sequentially. The universal algal pigment chl_a also on higher side was 1.21 and 1.26% compared to initial 1.09% under the 2 doses of electric field. However, surprisingly carotenoids were reduced in this alga under the impact of electric field. But significant reduction was observed in carbohydrates and lipids. Carbohydrates were initially 13.0% which were reduced drastically to 10.6 and 10.8% under the influence of two exposures of electric field.

DISCUSSION

The effect of electric field on growth, morphology and cellular constituents of *S. platensis* (Lonar lake isolate) was found to have enhanced as compared to unexposed cultures. The highest yield and healthy status of the alga has been procured under the influence of electric field of 1 kv exposure for 40 min and at 5 kv after 10 min exposure. Similar observations have been recorded by Singh (1992) on *Ankistrodesmus fusiformis* and Srivastava (1992) on two *Spirulina* species. This data is further comparable with the findings

TABLE 1 *Biochemical Analysis of cellular metabolites of S. platensis under electric field*

Cell contents	*Control %*	*Exposed for 40 min at 1 kv*	*Exposed for 10 min at 5 kv*
Chl_a	1.09	1.21	1.26
Carotenoids	0.170	0.154	0.159
Biliprotein	7.43	8.0	8.56
Phycocyanin	3.79	4.12	4.22
Allophycocyanin	2.78	2.80	3.46
Phycoerythrin	0.86	1.08	0.88
Lipid	7.1	6.81	6.62
Carbohydrate	13.0	10.6	10.88
Amino acid	6084 □g/g	6190 □g/g	6394 □g/g
Protein	48	50	50.5

of Behari (1994) who found the effect of electromagnetic field to depend upon the age, sex and pre-existing stress. It initiated a complex of integrated and interrelated series of physiological response in the cells. This response to the field seemed to be initiated possibly by a central mechanism in response to the information eyed by receptors located in the cell membrane.

Frasier *et al.* (1990) found that low voltage of electric field repaired DNA damage in mammalian cells and activated the enzymatic activity in them. Angell *et al.* (1990) also reported significant increase in body weight by 58 kg and calving capacity of cows by 0.93 kg per head, per day, while growth and survival were unaffected. Present alga also had enhanced growth and health of the trichomes under low voltage of 1 and 5 kv electric field.

Earlier workers i.e. Gajraj (1994), Bhatia (1996) on *S. platensis* and *S. labyrinthiformis* respectively suggested that higher voltage of electric field produced detrimental impacts on these algae. The fact was further supported by the findings in plants by Brayman *et al.* (1990), who reported the reduction in growth of root tips of *Zea mays* subjected to high voltage of electric field. On the other hand, the low voltage shock of electric field has been found to be beneficial for the physiology of living organisms (Allipov, 1994 and Brewer *et al.*, 1995). In the present experimental alga, the longer duration of exposures beyond 40 min at 1 kv and 10 min at 5 kv produced detrimental impacts on the growth and also on the morphology of the trichomes. Reduced density and less number of healthy forms have been on record along with an uneven and irregular morphology.

However, present findings are not in agreement with the findings of Nordenson *et al.* (1984 & 1988), who found an increase in frequency of chromosomal aberration in peripheral blood cells of workers, working under high voltage switch yard. But, Bauchinger *et al.* (1981) have refuted this observation as they found that 50 Hz alternating electric field at 380 kv switch yard could not increase or inflict any change in chromosomes of workers exposed for more than 20 years. According to Foster and Schwany (1986) direct effect of 50 Hz electric field on chromosomes would be unexpected as subcellular components are shielded by the membranes.

In view of these findings, the suppressed growth and aberrant forms may be attributed to the field strength, rather than to the current density. However, present study on *Spirulina platensis* is not in conformity with the result of Kowallczuk and Saunders (1989) on mice and Rommereim *et al.* (1989) on rats that there were no mutational or lethal impact of charge in body weight or reproductive capacity under the voltage of 65-150 kv.

The cellular constituents of alga, when biochemically analysed have further shown that 40 min exposure to 1 kv and 10 min exposure to 5 kv were suited for the growth and morphology of *Spirulina platensis*. Protein was increased by 2% at 1 kv and 2.5% at 5 kv, while free amino acids were more by 106 □g/g and 310 □g/g dry weight of the alga, under these voltages serially. The carbohydrate and lipid contents were reduced. Carbohydrates were down by 2.4% at 1 kv treatment for 40 min and 2.12% at 5 kv for 10 minutes. Lipids were low by 0.29% at 1 kv and by 0.48% at 5 kv. The electric field induced changes in lipid layers have been reported earlier by Sokirko *et al.* (1994).

The carotenoid contents of the alga was slightly reduced after both the exposures. Gasztola *et al.* (1995) also supported this view as they found, the electric field exposure have its impact on photogenerated ion pairs on nearby molecules, which resulted into carotenoid band shift in photosynthesis. The percentage of chlorophyll-a was enhanced by 0.12% in 40 min exposure at 1 kv and 0.17% after 10 min at 5 kv. Similarly total biliproteins were high by 0.57% at 1 kv and 1.13% at 5 kv.

However, it is clearly indicated that electric field having low voltage of 1 and 5 kv at 40 min and 10 min exposure respectively increased the rate of growth, protein contents and pigments along with reduced percentage of lipids, carbohydrate and carotenoids. It may be concluded that the lower doses of low magnitude of electric field have been found to be effective in increasing protein percentage of biomass. The data stands a witness to the fact that electric field is a novel and cheapest source for the enhancement of biomass production with enhanced cellular metabolites and can be a yet another profitable tool in commercial production of *Spirulina*.

Acknowledgements

The authors are grateful to Head, Department of Botany, for facilities and Department of Biotechnology, Govt. of India, Delhi for financial assistance.

References

Allipov, Y.D., Belyaev, I.Y. and Aizenberg, O.Z. 1994. Systematic reactions of Escherichia coli cells to weak electromagnetic field of extremely low frequency. *Bioelectrochem* and *Bioenergetics*, 24.

Angell, R.F., Scott, M.R., Raleigh, R.J. and Bracken, T.C. 1990. Effect of a high voltage Direct Current Transmission line on Beef Cattle Production. *Bioelectromagnetics*, 11 : 273-283.

Bauchinger, M., Hauf, R., Schmid, E. and Dresp, J. 1981. Analysis of exposure to electric and magnetic fields from 380 KV Systems. *Radiat. Environ. Biophys*. 19: 235-238.

Behari, J. 1994. Biohazards of Electromagnetic fields. *Everyman's Science Vol.* XXX No. 5, pp. 128.

Bhatia, R. 1996. *Biochemical and nutritional studies on Spirulina labyrinthiformis.* Thesis submitted to University of Rajasthan for award of Ph.D. degree.

Brayman, A. Andrew and Morton, W. Miller 1990. 60 Hz electric field exposure inhibits net apparent H^+ ion excretion from excised roots of Zea mays. *Radiation Research*, 123 : 22-31.

Brewer, J.E., Tuedt, M.A., Adams, T.P. and Krall, M.W. 1995. Low voltage shocks have a significantly higher tilt of internal electric field than do high voltage shocks. *Pl. Cell Physiol*. 18(1) : 214.

Foster, K.R. and Schwany, H.P. 1986. Dielectric permittivity and electric conductivity of biological materials. In: *Handbook of biological effects of electromagenetic fields* (Eds. Polk, C. and Posoro, E.) CRC Press, Boca Raton, Florida, pp. 27 – 98.

Frazier, M.E., Reese, J.A., Morris, J.E., Jostes, R.E. and Miller, D.L. 1990. Exposure of mammalian cells to 60 Hz magnetic and electric fields. Analysis of DNA repair of induced single strand breaks. *Bioelectromagnetics* 11 : 229-234.

Gajraj, R.S. 1994. Blue green algae as a protein source for animal feed. Thesis submitted to the University of Rajasthan, Jaipur for the award of Ph.D. degree.

Gosztola, D., Yamada, H. and Wasielew 1995. Electric field effect of photogenerated ion pairs on nearby molecules. A model for the carotenoids band shift in phatorythesn. *J. Ame. Soc.* 117(7) : 2041.

Kowallczuk, C.I. and Saunders, R.D. 1989. Dominant lethal studies in male mice after exposure to a 50 Hz electric field. Wiley-Liss Inc.

Nordenson, I., Hansson, M.K., Nordstrom, S., Sweins, A. and Birke, E. 1984. Clastogenic effect in human lymphocytes of power frequency electric fields; *in vivo* and *in vitro* studies. *Radiat. Environ. Biophys*. 23 : 191-201.

Nordenson, I., Hansson, M.I., Östman, U., and Lzungberg, H. 1988. Chromosomal effects in lymphocytes of 400 KV sub-station workers. *Radiat. Environ. Biophys*. 27 : 39-47.

Rommereim, D.N., Kaune, W.T., Anderson, L.E. and Silov, M.R. 1989. Rats reproduce and rear litters after chronic exposure to 150-KV/m, 60 Hz electric fields. *Bioelectromagnetics* 10 : 285-349.

Singh, G.P. 1992. Chlorococcealean alga for utilization as source of protein: *Ankistrodesmus fusiformis* (Corda). *Thesis submitted to Univ. of Raj.,* Jaipur.

Sokorko, A., Pastushenko, N., Svelina, S. and Zeks, B. 1994. Deformation of a lipid vesicle in a electric field. A theoretical study. *Bioelectrochem and Bioenergetics*, 34(2) : 101.

Srivastava, P. 1992. The impact of electric field on the protein contents of *Spirulinas.* In : *Spirulina* ETTA (ed : CV Seshadri and N. Jeeji Bai, Madras).

SECTION 10

AQUATIC ANGIOSPERMS

Frontiers in Plant Sciences (Pages 809-862)
Eds. K.G. Mukerji, K.V.B.R. Tilak, S.M. Reddy, L.V. Gangwane, P. Prakash, I.K. Kunwar

Morphology and Distribution of Podostemaceae in India

Sanavar, Prem L. Uniyal and Shantanu Suman

ABSTRACT

Podostemaceae are represented by 49 genera and 270 species mainly of tropical distribution. Twenty eight species have been recorded from India of which 26 are endemic. Podostemads grow attached to water worn rocks, boulders and wood pieces located under water falls and in rapidly flowing streams where only certain mosses and algae can survive. The gummy substance synthesized by the plant enables their adhesion firmly in these extreme conditions. A combination of characters that make Podostemads unusual among angiosperms are: lack of distinct root-shoot axis and plumule, perplexing polymorphism, structural reductionism, presence pseudo-embryo sac, absence of double fertilization, submerged vegetative and emerged anthesis. During the early seed germination, the post-developed plumule produces a few leaves and then a prostrate thallus-like body is developed from the hypocotyls region that produce linear leaves and flowers on the dorsal surface. There is a deep concern for the conservation of biodiversity all over the world. There is a need of protecting of sites and *ex situ* conservation of these rare and threatened plants.

Key words: Podostemaceae, aquatic angiosperms, seed germination, thalloid plant body, adaptation, morphology, endemic, endangered, *Polypleurum, Willisia, Indotristicha, Zeylainidium,*

INTRODUCTION

The pantropical family Podostemaceae (the riverweeds) is an unique family of aquatic angiosperms consisting of rheophytic herbaceous members, which inhabit an ecological niche on water washed rocks in water falls,

---

Author's address: Department of Botany, University of Delhi, Delhi 110 007, India
E-mail: uniyalpl@rediffmail.com

cascades, wooden logs and thrive under extreme conditions for angiosperms. The plant body is a highly reduced thallus with no clear demarcation into root and shoot, often resembling algae, lichens or bryophytes. They have rhizoids and hapteras for firm fixation to the substratum which probably function as roots. They lack aerenchyma. All combined features make them an unusual family among angiosperms. They exhibit wide range of phenotypic plasticity or perplexing polymorphism. They show unique embryological features such as lack of antipodals, triple fusion and endosperm, presence of single fertilization and pseudo-embryo sac. Taxonomists believe that Podostemads have affinities with Crassulaceae, Saxifragaceae and Hydrostachydaceae.

The Podostemaceae consists of 270 species spread over 49 genera (Mohan Ram and Sehgal, 2001). Out of these, 23 genera are monotypic and endemic to different parts of the tropics. They are distributed in South Africa, Australia, Myanmar, Central and South America, India, Java, Japan, Malaysia, Sri Lanka, Thailand and West Indies. It is noteworthy that out of 28 species in India, 26 are endemic. The Indian Podostemads are concentrated mainly in the Western Ghats in extremely limited areas of Maharashtra, Kerala, Karnataka, Tamil Nadu, Andhra Pradesh, Madhya Pradesh, Orissa and also in the hills of Assam and Meghalaya. (Table-3)

The plants are restricted to rushing mountain streams, rivers or cataracts in submerged conditions (Figs. 1, 2, 3) attached to rocks by means of rhizoids or/and haptera. The occupation of this strange habitat coincides with their unique nature, like thalloid body resembling an alga, lichen or bryophyte, without truly recognizable stems and roots. They produce fruits only when the water level begins to recede in the streams or rivers after monsoon. [Figs. 1, 4)

ORIGIN AND EVOLUTION

Bryophytes were the first plants to establish themselves to land habit. During the process of evolution, some of the angiosperm groups migrated to more moist environment in which large leaf surface would have been advantageous. Mauseth (1988) reported that in some plants, mutations affecting the stem were selected, which made the stem flat, expanded and leaf-like. It may not be surprising if the thalli of the podostemads also fit into Mauseth's (1988) hypothesis and turn out to be stem in nature.

Willis (1915, 1926) wrote that the ancestors of Podostemaceae must have been land plants on the bank of rivers that sent out creeping roots along the stones of waterfalls. These plants might have shifted to water at one time and undergone whole of the evolution in water. As running water provided most uniform conditions, it is difficult to account for the origin of diverse morphological forms. Willis (1926) explained the origin of Podostemaceae as a result of single massive mutation without natural selection. Goldschmidt (1933) and by Van Steenis (1981) put forth the ideas about the evolution of a separate line of these plants and absence of any close relative of Podostmaceae. According to Cusset and Cusset (1988a), the Tristichoidae members are neither dicotyledons nor monocotyledons but are an ancient group that has evolved from terrestrial Gondwanian plants. They viewed that *Indotristicha* might be quite primitive and suggested *Indotristicha ramosissima* deviates the least from a hypothetical ancesstor called *Prototristicha,* followed by the new world taxon *Tristicha trifaria.* Rutishauser and Huber (1991) have postulated the stepwise structural changes required to arrive at the present day podostemads from a putative ancestor of the podostemads that had a vegetative plant body of conventional root-shoot model, typical of most angiosperms. The seeds of this *Indotristicha*–progenitor might have developed copious mucilage for instant fixation to rocks in respect of moving water and the seedling axis lost its potential to develop completely. Roots arose from the hypocotyl. They were slightly dorsiventral and had root caps. The creeping roots adhered to the stones/rocks with the help of mucilage secreting rhizoid, hapters and holdfasts. Several shoots developed endogenously from the roots. The vegetative shoots developed scale leaves and ramuli for photosynthesis. Thus the plant body began to float in running water without the development of aerenchyma. However, in the absence of intermediates to corroborate these facts, it is still unknown that how the podostemads settled on the submerged rocks in running water of rivers.

From a study of Podostemaceae, Willis (1902a, 1914, 1915) arrived at certain views that these plants have been derived from some terrestrial group. Since the structure of the flower and fruit is typically adapted to land habit. Willis (1902b) suggested that a possible origin for the family is from plants already growing on the banks of mountainous streams with creeping adventitious roots (thallus) upon which secondary shoots (leaves and flowers) were regularly developed. These secondary shoots might provide the opportunity for an entrance into aquatic life. Most of the peculiarities of the group can be traced to the remarkable plasticity of the skeleton-less root and to the parallel dorsiventrality of the vegetative organs and flowers. The dorsiventrality is associated with their plagiotropic method of growth, forced upon by the fact that they live only upon an unyielding substratum.

Arber (1920) while agreeing with Willis (1915) on the concept of natural selection does not explain the origin of specific and distinctive entities, urged caution in accepting Willis' assertion that natural selection is unimportant in contributing to adaptation. The Podostemaceae as it seems, afford an example of evolution unhindered by the limiting effects of competition and natural selection. Arber (1920) surmised that the absence of adaptation may in part be attributed to the absence of competition and therefore of natural selection. A study of Podostemaceae yields important negative evidence of the general effectiveness of natural selection in the inception of adaptations, since in this group, the absence of competition and natural selection is correlated with the absence of special adaptations.

The Podostemaceae thus exhibits a great variety and marked specific differentiation, but the features in which the genera and the species differ from one another cannot be explained as adaptational. The natural selection can scarcely be a factor of any great importance in Podostemaceae and yet there is a quite extraordinary variety of specific forms, many of which are confined to extremely limited areas. Among the Podostemads, presumably all variations have had an almost equal chance of perpetuation provided they did not interfere with those general features which gave the group its special capacity for growth in rapidly running water, which is so inimical to the most forms of plant life.

CLASSIFICATION AND RELATIONSHIP OF PODOSTEMACEAE

Podostemaceae are represented by 49 genera and 270 species of which 28 species occur in India (Table-3). The family is divided into two sub-families

TABLE 1 *Relationships between Podostemoideae and Tristichoideae*

Characters	*Tristichoideae*	*Podostemoideae*
Flowers	Trimerous	Dimerous
Perianth	Present	Absent
Spathe	Absent	Present
Stamens	1-25	Many (1 or 2 in Indian taxa)
Staminodes	Absent	Present
Pollen	Monads	Dyads
Ovary	Tricarpellary	Bicarpellary
Ovules	Many, bitegmic	Few to many, bitegmic
Pseudo-embryo sac	Organized after fertilization	Arises before fertilization
Embryo	Plumule differentiated (?)	Plumule not differentiated
Similarities		
Embryo sac	*Apinagia* type *Podostemum* type	*Apinagia*, *Podostemum* and *Polypleurum* type
Fertilization	Single (only syngamy)	Single (only syngamy)
Endosperm	Absent	Absent
Embryogeny	Solanad type	Solanad type
Embryonal haustorium	Present	Present

TABLE 2 *Embryological relationship with related families*

Character	*Podostemaceae*	*Hydrostachydaceae*	*Saxifragaceae*	*Crassulaceae*
Anther wall	4 or 5 layered	–	4 layered	5 layered
Anther tapetum	Secretory	Secretory	Secretory or amoeboid	Secretory
Microsporogen-esis	Simultaneous or successive	Simultaneous	Simultaneous	Simultaneous
Pollen grains	Monads or dyads, shed at 2-celled stage	Compound, permanent in tetrads	Monads or tetrads	Monads shed at 2-celled stage
Gynoecium	2 or 3 carpellary, ovary syncarpous 2 or 3 locular	Bicarpellary, syncarpous ovary unilocular	2-5 carpellary syncarpous, 2-5 locular	Apocarpous
Ovules	Many, tenuinucellate bitegmic, on axile placenta	Many, tenuinucellate unitegmic, anatropous on parietal placenta	Many crassinucellate or tenuinucellate bitegmic/ unitegmic on axile placenta	Numerous on marginal placenta, anatropous crassinucellate, bitegmic
Megaspore tetrad	No linear tetrad formed	Linear	Linear/bisporic	Linear/T-shaped/isobilateral
Embryo sac development	Bisporic, *Apinagia*, *Podostemum* or *Polypleurum* type, mature embryo sac 4 or 5 nucleate	Polygonum type	Polygonum type and Allium type	Polygonum type
Antipodal cells	Absent (2 in *Dicraea?*)	Present	Present	Present
Pseudo-embryo sac	Present	Absent	Absent	Absent
Embryogeny	Solanad, suspensor haustoria present	Onagrad type-Capsella variation, suspensor haustorium absent	Caryophyllad type, Sedum variation	Caryophyllad type, suspensor haustorium absent.

TABLE 3 *Occurrence and status of Podostemaceae in India*

S.No.	*Taxa*	*Distribution*	*Status*
	Sub-family Podostemoideae		
1.	*Cladopus hookerianus*	Kerala, Karnataka	Endemic
2.	*Farmeria metzgerioides*	Kerala	Endemic, Rare, Threatened
3.	*Hydrobryopsis sessilis*	Western Ghats of Karnataka Maharashtra and Andhra Pradesh	Endemic, Rare, Threatened
4.	*Hydrobryum griffithii*	North-Eastern Himalaya in Arunachal Pradesh, Meghalaya, Assam, Manipur, Sikkim and Garhwal.	Endemic, Rare, Threatened
5.	*Maferria indica*	Tamil Nadu	Endemic, Rare, Threatened
6.	*Podostemum barberi*	Karnataka, Madhya Pradesh	–
7.	*Podostemum munnarense*	Western Ghats in Kerala	Endemic, Rare, Threatened
8.	*Podostemum subulatum*	Western Ghats of Kerala, Tamil Nadu and Maharashtra	Endemic
9.	*Polypleurum agharkarii*	Assam and Meghalaya	Endemic and Rare
10.	*Polypleurum dichotomum*	Maharashtra and Tamil Nadu	Endemic
11.	*Polypleurum disciforme*	Kerala	Endemic and Rare
12.	*Polypleurum elongatum*	Kerala and Tamil Nadu	Endemic and Rare
13.	*Polypleurum filifolium*	Kerala and Orissa	Endemic and Rare
14.	*Polypleurum minor*	Karnataka, Kerala, Tamil Nadu, Maharashtra, Rarely in Assam, Meghalaya.	Endemic and Rare
15.	*Polypleurum munnarensis*	Kerala	Rare
16.	*Polypleurum prostratum*	Kerala	Endemic and Rare
17.	*Polypleurum stylosum*	Maharashtra, Kerala, Karnataka, Tamil Nadu, rarely occur in Assam, Meghalaya.	Endemic

(Table 3 Contd.)

18.	*Polypleurum wallichi*	Meghalaya	Endemic and Rare
19.	*Willisia arekaliana*	Silent valley, Kerala	Endemic and Rare
20.	*Willisia selaginoides*	Western Ghats of Kerala and Tamil Nadu	Endemic, Rare, Threatened
21.	*Zeylanidium jhonsonii*	Karnataka, Kerala, Andhra Pradesh.	Endemic and Rare
22.	*Zeylanidium lichenoides*	North Eastern India, Western Ghats including Maharashtra, Kerala, Karnataka, Tamil Nadu, Madhya Pradesh, Orissa, Andhra Pradesh.	Widely distributed
23.	*Zeylanidium maheshwarii*	Kerala	Endemic and Rare
24.	*Zeylanidium olivaceum*	Western Ghats of Tamil Nadu and Kerala, North Eastern India in Arunachal Pradesh	Endemic and Rare
	Sub-family Tristichoideae		
25.	*Dalzellia gracilis*	Kerala	Endemic
26.	*Dalzellia zeylanica*	Western Ghats of Karnataka, Kerala and Tamil Nadu	Endemic
27.	*Indotristicha ramosissima*	Western Ghats of Karnataka, Kerala and Tamil Nadu	Endemic
28.	*Indotristicha tirunelveliana*	Tamil Nadu	Endemic, Rare, Threatened

namely Podostemoideae and Tristichoideae (Table 1). Tristichoideae are represented by four genera- *Indotristicha, Dalzellia, Malcotrisicha,* and *Tristicha* of which former two are endemic to India. The Podostemoideae includes 45 genera.

Members of the subfamily Tristichoideae are simple, less-specialised and more wide-spread (Willis, 1902b). They have greater tolerance to environmental fluctuations and thus occupy more trophic niches. In Tristichoideae one finds transition from regular erect branches to prostrate, dorsiventral and flat thalli. In contrast to Tristichoideae, Podostemoideae are highly specialized having dorsiventral thalloid plant body on which secondary shoots (leaves and flowers) arise. It consequently occupies narrow niches. The seemingly simple structure of Podostemaceae appears to be the case of reduction.

Relationship of Podostemaceae with Other Families

Taxonomists have held various views concerning the relationships of the Podostemaceae with other angiosperm families. For example, Podostemaceae have been related to Saxifragaceae (Warming, 1888; Engler, 1930); Crassulaceae (Hutchinson, 1959); Hydrostachydaceae (Subramanyam, 1962). The embryological features of families related to Podostemaceae are presented in Table 2.

On the basis of embryological studies Maheshwari (1945) observed that the Posostemaceae are much reduced apetalous derivatives of the Crassulaceae.

LIFE HISTORY

The Podostemaceae with rare exceptions are confined to tropics. They are essentially plants of hot regions as indicated by the forms living in the South India, inhabiting water which maintains a very constant temperature at 27°C. The members form an anomalous group characterized as regards to their morphology, by remarkable variety but as regards to their ecology agreeing in one feature–for inhabiting water which flows rapidly or even torrentially over a rocky substratum. (Figs. 1, 3, 4).

The majority of the peculiarities of the Podostemads are closely related to the nature of their habitat. A creeping thread-like organ is formed which though morphologically a root is dorsiventral in structure and gives rise to leafy shoots endogenously in acropetal succession. The thread-like root is not apparently competent to anchor the leafy shoots with the necessary firmness and additional organs called 'haptera' are formed. They are produced exogenously from the creeping root and by their positive

Fig. 1

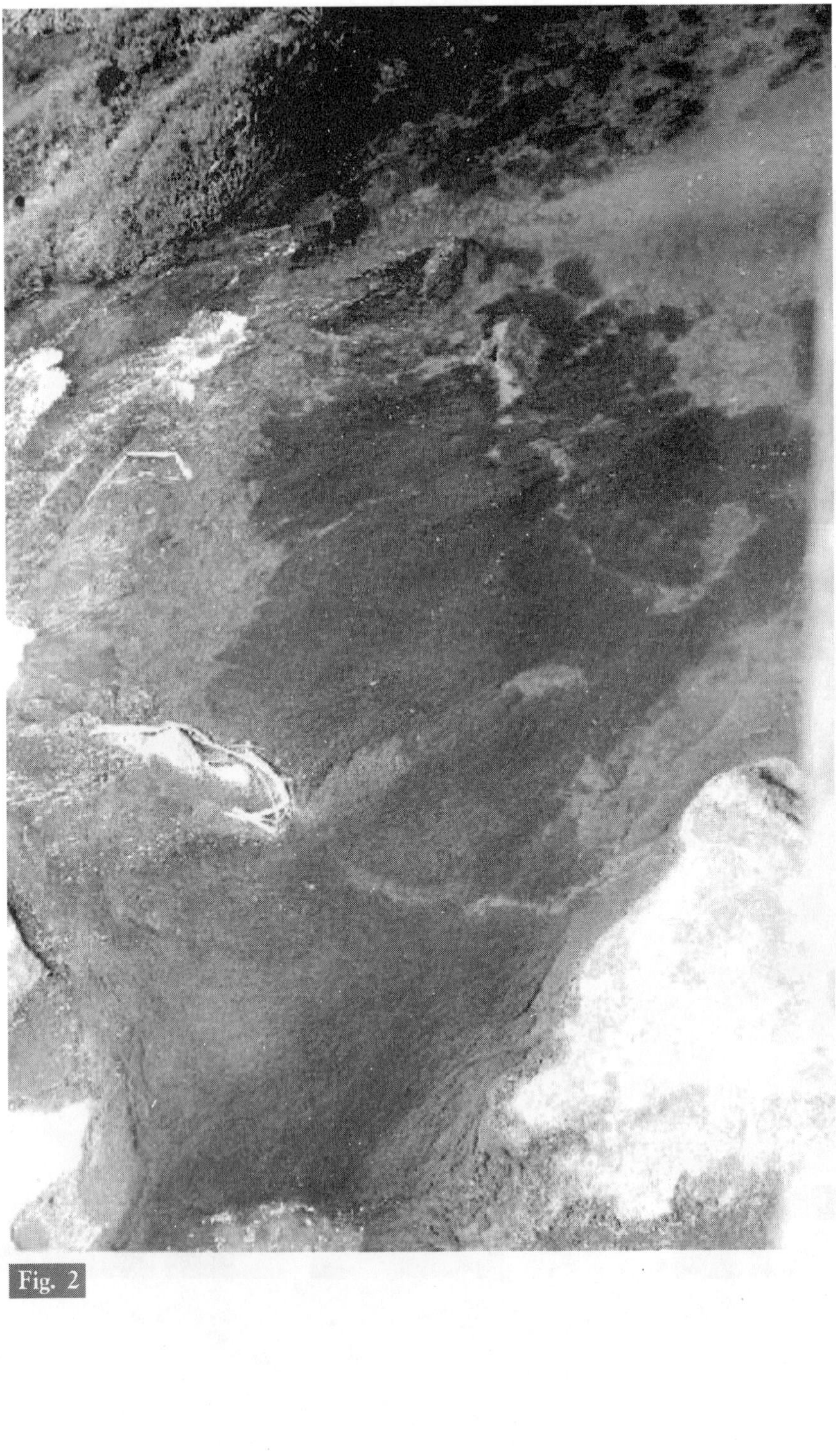

Fig. 2

Fig. 3

Fig. 4

geotropism and a power of flattening themselves against the substratum form firm attachment organs. They also secrete a kind of cement, which renders their adhesion to the rock very close and permanent. These haptera are found in many Podostemaceae. In many of the members, creeping root discards its root characteristics even more completely and it becomes converted into a thallus, which either follows out every irregularity in the substratum or remaining more or less free, develops into all sorts of curious shapes. It still produces secondary shoots becomes less so, until in such genera as *Polypleurum, Griffithella, Hydrobryum, Farmeria,* they are much reduced and assimilation is mainly performed by thallus. The thallus of the Podostemads is sometimes amazingly polymorphic its capacity for developing in exceptional forms depends apparently on the fact that it is not restricted by rigid skeletal system and nearly all cells possess the capacity for renewed meristematic activity. *Griffithella hookeriana,* for instance, has a thallus which may develop into various shapes recalling different algae that grow in moving water, one of its forms resemble the basal cup of *Hirnanthalia lorea* and also sometimes form a ribbon-like structure of a plant body.

Farmeria metzerioides recalls *Delesseria leprieurii* while *Podostemun subulatus* simulates such an alga as *Bostrychia moritziana,* which also grows in rapids. Willis (1902a) draws attention to these cases of simulation, alludes to the great difficulty of interpreting such resemblances between plants for distant relationship from one another, and adds, "It is impossible at present to do more than point out these very suggestive analogies to form which accompany analogy of the conditions of life and which seem to indicate that an experimental and comparative morphological studies of the forms of the Algae and Podostemaceae should be attended with interesting results". Tristichaceae do not possess polymorphism and show remarkable similarities in morphological features and in the arrangement and anatomy of leaves to many mosses or liverworts especially to those of wet situations (Willis, 1902b).

The genus *Dalzellia (Lawia)* differs from those hitherto mentioned in having a thallus which is not of root nature, which originates by the fusion of flattened, dorsiventral shoots. The internal structure of podostemads is similar to that of many other submerged plants in reduction of xylem, absence of stomates, and the presence of chlorophyll in the epidermis. On the other hand, a character in which these plants diverge from other hydrophytes is the presence of large quantities of silica in the cells (Goebel, 1891-93). It seems on the whole most probable that this silica is merely useless by-product of the plants metabolism. It has been suggested that it serves as a protection against the attacks of animals (Schnell and Cusset 1963) but there seems little evidence for this view. The most important anatomical peculiarity of the Podostemads is the extreme reduction of the intercellular spaces (Warming, 1881) in the respect that the members of these families contrast most remarkably with other water plants. This feature is probably associated with the thorough aeration of the torrential water which they frequent. Certain species, however, possess delicate outgrowth from the surface of the leaves which have been interpreted as "gill tufts". Possibly these structures to some extent compensate for the lacks of an internal aerating system. The vegetative parts of the Podostemads die very quickly. When out of their element and the flowering and seed-setting both of which take utmost rapidity.

Entomophily and hydrophily occurs among the Podostemads. The peduncles of the podostemads contain little water conducting tissue and possibly in correlation with this the seed development proves to be of a decidedly xerophilous type (Magnus 1913). By disappearance of nuclear tissue, a cavity is formed beneath the embryo-sac which at the time of fertilization, is filled with fluid. This cavity is bounded by strongly cuticularized inner wall of the inner integument and suberized cells of the chalaza. It is open only on the side towards the developing embryo and is described as an "ideal water reservoir". Mucilage is often present in the neighboring cells of the inner integument and this may perhaps form an additional protection against loss of water. The seeds of Podostemads are often small and numerous.

The adaptive modifications of the structure, such as the gradual reduction through a series of forms, of the shoots and leaves, the increased multiplication of the shoots by vegetative budding, the reduction in the number of flowers per shoot, and the change to anemophily, the increased dorsiventrality of characters were shown to be rather correlated with the rise and fall of the water than with the velocity of the stream. The extremely local distribution of many forms, their anomalous morphology and progressive dorsiventrality and great variety of types of structure which they present, offer every incentive to speculation.

DEVELOPMENTAL MORPHOLOGY AND VARIABILITY

Two general pattern of development–CRS (conventional Root Shoot) verses thalloid are evident in the sub-family Tristichoideae. *Indotristicha* represents CRS type and *Dalzellia zeylanica* the thalloid type. Detailed description of *I. ramosissima* are presented by Cusset and Cusset (1988); Rutishauser (1997); Rutishauser and Huber (1991) and Jager-Zurn (1992, 1997). Podostemaceae show the thalloid pattern of development (Uniyal and Mohan Ram 2001, Uniyal 2001).

The plant body of Podostemads resemble some of the bryophytes and phaeophyeae in general appearance. They are mostly annuals which grow attached to rocks and stones usually in rapid flowing water (Figs. 1, 2, 3). Flower buds appear in submerged condition but anthesis and fruiting occur only when the plants are exposed after rainy season.

Thallus (Prostrate Axis)

The vegetative body consists of a thallus which roughly corresponds to stems and leaves in superficial appearance. The thallus which adheres firmly to the substratum either by masses of hair-like structures which originate as extension of the epidermal cells or by more complex organs which superficially resemble the haptera. The air spaces which usually occur in aquatic plants, is totally absent in podostemads. Mechanical tissue when present is usually in the form of collenchyma which is especially well developed around the vascular elements. In the plant body of the large-sized species, the collenchyma becomes converted to sclerenchyma in the form of lignified, prosenchymatous elements with slit-like pits. The occurrence of silica bodies especially in the superficial tissues of the thallus is an important feature. In some species these are sufficiently numerous to form a kind of armour or carapace which prevents the plant from collapsing during the lowering of water level. In other species including the silica bodies are found to be much less numerous, so that the plant contact considerably before drying. Hairs have been recorded on the leaves of species of some Podostemads like *Apinagia, Polypleurum, Podostemum, Zeylanidium* and *Willisia.* Stomata are unknown, and the epidermis, which is well differentiated, as well as the subjacent cell contains chlorophyll. The outer wall of each cell of the mature epidermis is composed of thick cellulose. It appears that these cells serve for the rapid absorption of salts and gases dissolved in the water as found in *Podostemum atrichus* where the epidermis is provided with a circular or ellipsoidal disk of cutin which appears to have the form of a ring because it is particularly thick at the periphery. Secretory cells and cavities are known to occur in certain genera. Fungal hyphae have also been reported to occur in the secretory cavities of *Mourera aspera.*

The plant body of the Podostemaceae can most conveniently be regarded as a thallus which originates either from the middle, form the base of the hypocotyl as an outgrowth and sometimes from the cotyledons. Sculthorpe (1967) opinioned that the prostrate body is morphologically a flattened root with floriferous shoots. Imaichi et al (1999) stated that the ribbon-shaped, flattened structures in *Malaccotristicha* are roots, which are the main photosynthetic organs. The ramuli are laterally on either side of the roots. Rutishauser (1997) used various terms such as flattened roots, crustaceous roots and root-thallus as the structure produce endogenous shoots but no exogenous leaves. They observed an asymmetrical root cap on this thallus like organ in the members of Tristichoideae. Jager-Zurn (2000a, b) has also reported root with distinct root cap and root born shoots (leaves and flowers) in the members of Podostemoideae, *Zeylanidium olivaceum* and *Podostemum subulatum*. Engler (1930) accepts the view that the fundamental morphological nature of thallus is a horizontal assimilatory root which, in many species, resembles a rhizome. Uniyal (2001) used the term prostrate body for the thallus. Rutishauser (1997), Jager-Zurn (2000) preferred to call it a root or prostrate axis instead of thallus.

Root

The roots like-nature of the thallus in *Indotristicha ramosissima* is indicated by the possession of a somewhat abnormal root cap. In some species, the dorsiventrally compressed, the ventral surface being flat and the dorsal convex. A root of this kind contains a central Vascular System including two xylem groups, the latter contains few vessels. According to "Engler" the xylem is sometimes completely absent, in which case the vascular tissue consists wholly of phloem, but the vascular stand is supported by collenchyma especially on the dorsal side. The remainder of root consists of large, amyliferous cells. In some species like *Mniopsis saldanhana* and similar species the roots are attached to the substratum by hair like organs. The roots possess considerable powers of regeneration and small pieces of them can survive under stones and during dry periods. The type of root just described may be regarded as fundamental but there are several distinct modifications of it in different members of the family. In species of *Polypleurum* some of the roots are of the creeping kind just described but others are free floating and ribbon-like. The flattened roots exhibit pallisade tissue towards the upper surface. In *Zeylanidium* and *Hydrobryopsis* the root is a flattened thallus which adheres closely to the substratum (Figs. 10, 20). The roots contain the chloroplasts in some species. The roots bear organs which are sometimes described as shoots, leaves and flowers (often called as secondary shoots) but it is often difficult to distinguish between these various morphological categories especially as the so-called branches of the shoots do not arise from the axillary buds.

Shoot

All the shoots are produced from the creeping root-thallus. These flattened dorsiventral roots are firmly fastened to the rocks and produce adventitious shoots. The shoots arise laterally at the ventral sides of the roots, generally one on each side, breaking through the outer tissues.

The floating stems of some species like *Mourera aspera* exhibits a central cylinder containing a number of vascular bundles in which large air canals develop where the vessels collapse. The vessels are at first provided with spiral and annular thickenings. Sieve tubes were observed as a thin-walled tissue of narrow elements around the vessels as phloem in which the presence of callose is confirmed by staining with aniline blue. Silicious bodies are present in large numbers in many species, especially in the peripheral tissues and in the hairs of all parts of the plant. They are, however, absent from free floating leaves. They enable the plants to withstand mechanical damage and periods of desiccation. *Podostemum galvonis* is thickly coated with these bodies so that the dried and fresh plants look very much alike. *P. subulatum*, on the other hand, in which silica bodies do not occur shrinks very considerably when dried.

Leaf

Generally in Podostemaceae, the chloroplasts are restricted almost entirely to the epidermis, coating the inner cell walls, whereas the outer ones all coated with light green leucoplasts. Very few chloroplasts are found in the sub-epidermal layer of cells, whereas the inner parenchyma is destitute of green colour. The absence of stomata in this family is generally known. The hairs are present on the leaves. The hairs are easily thrown off. The hairs have small basal cell and a long apical one and probably are the principle absorbing organs.

ADAPTATIONS

With the exception of some of the parasitic orders, such as the Balanophoraceae, there are probably no families in flowering plants, which are so completely transformed from the average or mesophytic type of the phanerogams into types which are so completely unique and peculiar, as the Podostemaceae. There is a remarkable variety in morphological structure in this family in the manner:

(i) That the families contain the most astonishing variety of morphological structure (ii) that all different stages, so as to speak, in the attainment most bizarre of these forms, live together and that least bizarre are the most widespread and common, and in consequence (iii) that as there are no changed conditions to which to be adapted, there can be no adaptation to conditions other than to the general conditions which are common to all, and have been common to them since family began and

(iv) that therefore can be in all possibility, be no selection of infinitesimal variations, but progress must have been by something of the nature of mutations or fixed changes without natural selection. There are some factors which appear to have had an influence in the evolution of these remarkable families. The structure of the plant body as well as the conditions under which they live is being unique and highly adapted to such habitats. The flat thallus-like expansions of stem or root being looked upon as admirably suited to the rushing water in which they live.

Evolution is usually supposed to have produced the extraordinary forms now existing by adapting members of very different families to very different conditions of life. In this family the evolution can be studied with many of complications of factors that are removed or simplified. The conditions of their life are almost absolutely uniform. There is no competition as the family has its habitat to itself. They can grow naked on water worn rock in similar climate that grow in running water in sub-tropical or tropical zone, and are represented by seed in colder weather. Perhaps these plants took to water at different stages of evolution into a family upon land, they must obviously have gone through the whole of their evolution in water forms at an extremely early stage, during which the ancestors probably retained a certain power of surviving upon land. Now the family consists of 49 genera and over 250 species of the most varied morphological structure possible under perfectly uniform conditions which cannot have varied except for all alike, since the evolution of families began. The evolution consequently cannot have been any sort of response to a necessity of adaptation to different conditions, for there are and have been no different conditions to which to be adapted. Since the first members of this family began to live in running water.

Conditions of Life

The conditions of life of these plants must have been uniform as those of other water plants. They grew as a rule, only upon rocks or any immovable substratum without any specific chemical or physical composition of the substratum. The plants take nothing from it except, silica. They fix themselves on rocks by means of creeping roots, which give lofty secondary shoots. They are thus absolutely compelled from the start to a plagiotrophic mode of life to face the speed of water. The members grow in more uniform temperature and light condition than the other water plants. Their vegetative phase pass at 50-75 cm below the water level. The medium in which they live, the running water of tropical, is very uniform but the speed of water current varies in different localities.

Morphological Structure

If such simple forms suffice to occupy the positions, there can be no adaptational need for the complex forms of this family. Also there is increasing dorsiventrality in the thallus. Once placed on a rock, in which downward growth is impossible, then dorsiventral or plagiotropic direction of growth is ensured by creeping roots. So, the whole further evolution seems to be in the direction of increasing dorsiventrality both in the vegetative and the floral organs. The flowers are steadily more and more zygomorphic a condition usually regarded as an adaptation to insect visits and accompanied by horizontal position of the open flower, they at the same time stand stiffly erect and are more and more anemophilous and autogamous. The dorsiventrality of flowers may has been forced upon them by the steadily increasing dorsiventrality of the vegetative systems, due to the general effect of the total conditions of life acting on the hereditary peculiarities of the ancestral forms, whether directly or indirectly. By no stretch of imagination can the dorsiventrality of the flowers be regarded as adaptational as the least dorsiventral and the least modified species of least modified genus of Tristichoideae can and do live in nearly all places occupied by other members of the family. The whole differentiation of the family is merely without any adaptational significance. The stigma lobes are unequal and lie appressed and bent over the surface of the ovary at young stage lie below the stigma. In the subsequent stages the filaments elongate, curve inward and bring the anthers to the same level as that of the stigma lobes. At this stage, in a majority of the plants, the tips of shoots remain at the level of the flowing water. The andropodium elongates and the anthers extend beyond the stigmas causing removal of the spathe.

The seeds are numerous and minute. The cells of the seed coat swell and become mucilagenous when the seeds come in contact with water. The mucilage helps the seeds to become attached to the substratum. The embryo consists of two well-developed cotyledons without a distinct plumule. The remnants of the suspensor are also observed at the radicular end of the embryo. In some genera the number is much reduced especially in *Farmeria indica* where a dehiscent fruit has about four seeds and in *F. metzgeriodes*, an indehiscent fruit has only two seeds, which germinate *in situ*. This peculiar feature of indehiscent fruit is by no means an adaptation but may be merely another expression of the increasing dorsiventrality which runs throughout the Podostemaceae family. After consideration of all the features of this family, Willis (1902a) concluded that there are practically no characters in the orders that can possibly be regarded as adaptational as the members display following features:-

1. The regular development of secondary shoots upon creeping closely attached roots (thallus).
2. The development of haptera, adhesive hairs or climbing organs to attach the plants to rocks.
3. The absence of intercellular spaces.

On the whole, the evolution of these families was by indiscriminatory mutation or mutation in every direction, without natural selection (Willis 1902a).

SEED GERMINATION PATTERNS

Interpretations of basic form of plant are hindered by lack of knowledge of how seedlings develop into mature plants under extreme conditions. Early observations were based on study of field collected seedlings (Warming, 1881; Willis, 1902b; Hammond, 1937; Schnell and Cusset, 1963). However, botanists have been dissuaded from detailed field based study of seedling biology and development because seedlings are inconspicuous, difficult to identify to taxon and occur in often inaccessible habitats.

Nagendran(1975) succeeded in germinating seeds of nine Indian Podostemads in petri plates, but was unable to sustain their growth beyond 30 days. Grubert (1976) observed germination in Venezuelan podostemads. Philbrick and Novelo (1994) recorded a high percentage of germination in laboratory studies of four Mexican species and discussed issues such as dormancy, viability and the role of seed biology in population dynamics. In vitro studies of germination and seedling development in sterile culture have revealed considerable insight into the developmental morphology of Podostemads (Mohan Ram and Sehgal 1997; Uniyal and Mohan Ram 1996; 2001, Uniyal 2001)

In Podostemaceae the mature embryo does not show histological zonation. The polar ends, i.e. hypophysis and epiphysis, can be recognized only on topographic grounds (Mukkuda and Chopra 1973; Nagendran et al, 1981). The life cyle of Podostemads start in the beginning of monsoon. It is difficult to study the seed germination and establishment of plants in the nature as seeds are minute, and the germination occur under the flowing water. The events of seed germination in Indian Podostemaceae were studied using in vitro techniques (Vidyashankari and Mohan Ram 1987; Sehgal et.al 1993; Uniyal and Mohan Ram 1996, 2001; Sehgal and Mohan Ram 1997; Uniyal 1999-2001).

The embryo consists of two cotyledons only without distinct plumule and radicle. The radicular pole produces rhizoids for fixation to substratum after a week of germination. The plumular pole produces 3-4 leaves and then a lateral outgrowth arises at the hypocotyl region of the seedling. This lateral outgrowth creeps over the substratum, take a form of dorsiventral thallus (prostrate axis) and throw the branches in all directions. Leaves and flowers appear on the dorsal side of the thallus. In the members of Tristichoideae the plumule produces leaves and buds consisting of green scale leaves (ramulus) and this activity continue unlike that of podostemaceae. At the base of the hypocotyls, 3-4 root primorida originate and creep horizontally on the substratum and adhere to it by means of hairs and discoid hapters. These creeping roots grow apically and poduces shoot buds on their dorsal surface and root primordia on the ventral surface.

PROSPECTS

Podostemads with their fuzzy morphology, perplexing polymorphism, organ identity crisis, unique embryology and unexplored ecology offer fascinating plant systems for understanding the evolution and survival strategy. In the concern of enormous biodiversity displayed by plants and their ecological functions, there is a justifiable requirement to mount explorations in the biodiversity hot spots, especially in the Western Ghats and the North-eastern regions of India for the yet unknown species of the Podostemaceae. The mode of fertilization in the rare and endemic taxa needs to be studied thoroughly. It is also important to establish the morphological nature of the prostrate axis by means of careful ontogenic and histological studies and through the techniques of molecular biology to understand the organ specific homeotic gene expression in the stem, root and leaf. The cellular mechanisms involved in the adaptation of the plants to the stresses caused by rapid water flow, high pressure, low light intensity, high O_2 and low CO_2 content of water are to be understood.

The unique combination of characters presented by this family is unparalleled among the angiosperms, leading to recent resurgence of world wide interest. One major thrust is the recent attempt to establish phylogenetic relationships of the Podostemaceae by comparing DNA sequences of the chloroplast – encoded rbcL gene by Les et al. (1997).

The ecological role of podostemads and other plants associated with them in riverine system are largely unknown, especially when river rapids (Home and Goldman 1994) are highly productive. The microsite variation leading to distribution of various species in a river or cohabitation of different species at any given site is little understood phenomena. It will also be worthwhile to conduct population studies of different species to determine their stability and evolutionary trends in the family.

Podostemads are subjected to a wide range of anthropogenic disturbances. Rivers are severely affected by construction of dams and bridges that impound a large body of water, causing diminution of the rate of flow down below, which has taken a heavy toll of biota including fishes, prawns and plants. Toxic discharge of effluents from industries and agrochemical residues are serious threat to Podostemads that have unique ecological requirement. Cross Bell (1990) studied the effects of the effluents of a rubber factory in Tamil Nadu, India and observed the elimination of three species of Podostemaceae in the down stream due to acidic discharges. The disappearance of *Willisia selaginoides* from Kerala state and *Hydrobryopsis sessilis* from a particular site in Killoor (Uniyal, 2001) are glaring examples of the consequences of perturbations. Predictions regarding global warming may also adversely affect Podostemads as they occur in seasonally pulsating river (Grimm 1993, Philbrick 1997). There is an urgent need to evolve methods of *ex situ* and *in situ* conservation. Relocation of endangered species to alternate sites has often been suggested but such steps are never taken in developing countries even with displaced tribal and the rural poor. It is a pity that plants such as those of Podostemaceae with no proven economic value have few takers for conservation. Though the economic value of podostemaceae is not yet discovered but they are the part of bio-diversity and aquatic ecosystem, provide a platform to complete the life cycle of many aquatic animals and contribute gaseous balance. Many species specifically occur at the frequently disturbed sites.

SUBFAMILY: PODOSTEMOIDEAE

Cladopus

Total number of species: 07

Number of species in India: 01

Cladopus hookerianus **(Tul.) Cusset**

[= *Griffithella hookeriana* (Tul.) Warming]

DISTRIBUTION: Endemic to Western Ghats. Distributed in Kerala, Karnataka..

HABITAT: The plant appears to grow in moderately rapid water.

HABIT: Thalloid plant body in the shapes of ribbon, disc or cup-like.

Attachment to the rocks: Roots are thalloid, pinnately or irregularly branched or ribbon like, closely attached to rocks.

Thallus: The thallus is highly polymorphic, occur in various forms such as cup form, ribbon form and disc form, looks like a liverwort. In slow flowing shallow water the thallus is fucoid, slightly raised from susbstrate. It is profusely lobed and studded with flowering buds. The thallus along exposed boulders having periodic splashes of water is distinctly cup-shaped with strong holdfast and minute filamentous roots (Vidyashankari, 1988a).

Shoots: Vegetative shoots (group of leaves) are very short; consist of a few leaves arising on the thallus margins. The vegetative shoots change into floriferous shoots.

Leaves: Leaves 4-6, appear in acropetal manner on the margins of thallus that form vegetative shoots. The leaves have sheathing bases that become bracts to protect the young flower buds.

Flowers: Flowers arise in between the rosette of leaves (vegetative shoots) are terminal, pedicellate, closely crowded in a tubular spathe. Flower is protected by funnel shaped spathe. Stamens are two and raised by a divided androphore, anthers are bilobed and pollen occurs in dyads. Style is reduced, and stigma bifid. Capsule smooth, two valved with dehiscence ribs.

Farmeria

Total number of species: 01

Number of species in India: 01

Farmeria metzgerioides (Figs. 9)

DISTRIBUTION: Endemic to Kerala.

HABITAT: It grows on smooth rocks in rapids in more shady places than other species.

HABIT: Narrow ribbon-shaped thalloid body

Attachment to rocks: The thalli are attached to the substratum by the adhesive hairs that arise from the ventral surface of the thallus.

Thallus: There is narrow ribbon-like thallus (root), upcurved at edges. The secondary shoots arise behind the point of branching. Thallus regularly branched.

Leaves: Leaves appear on the dorsal surface in the groups of 4-6 per shoot leaves are subulate and having sheathing bases, bear hair upon upper surfaces.

Flower: Flowering shoots arise in between the groups of flowers terminal, solitary, sessile. Tepals are long, stamen is one, pollen in dyads. Ovary is broad, elliptical, stigma bilobed (Arekal and Nagendran, 1974; Rutishauser 1997).

Hydrobryopsis

Total number of species: 01

Number of species in India: 01

***Hydrobryopsis sessilis* (Willis) Engl.** (Figs. 10, 11)

[= *Hydrobryum sessile* Willis]

Fig. 5-8

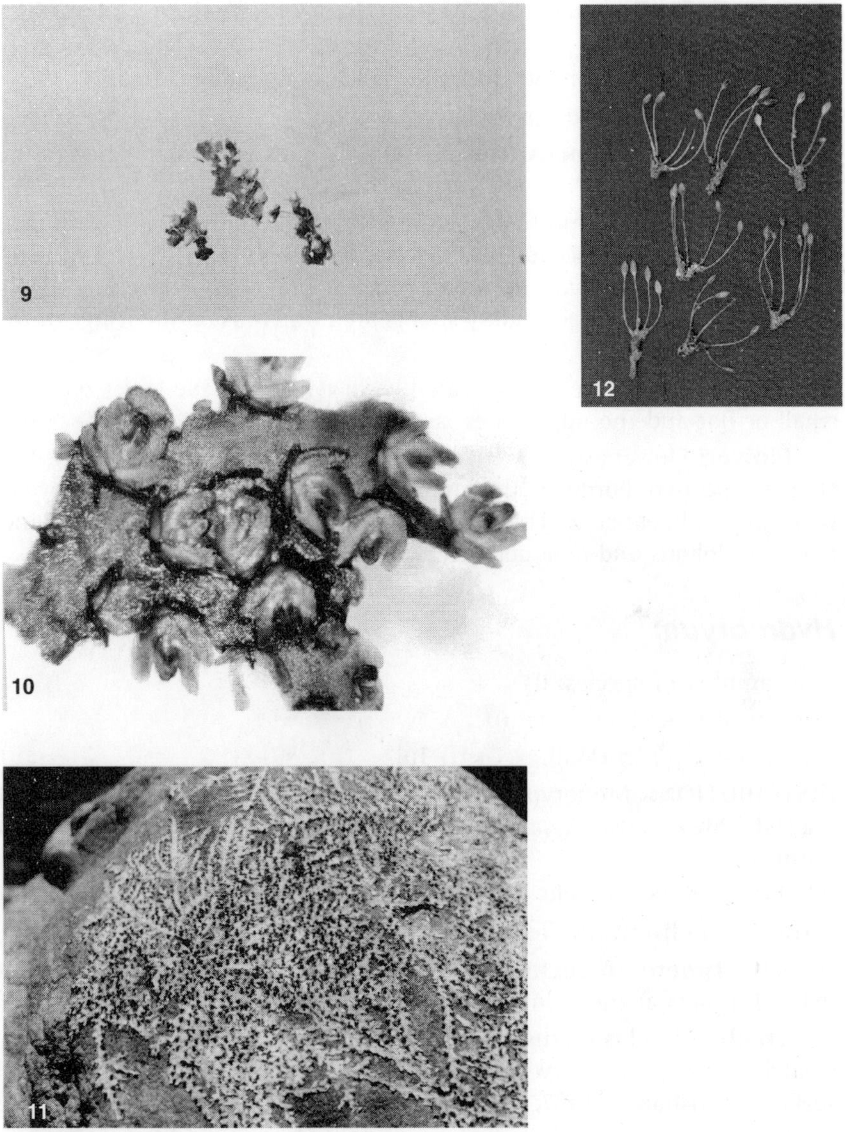

Fig. 9-12

DISTRIBUTION: Endemic to the Western Ghats of Karnataka, Maharashtra and Andhra Pradesh.

HABITAT: Found on smooth rocks in rapids and shallow streams.

HABIT: narrow, thalloid plant body

Attachment to rocks: Thallus tightly appressed and attached to the rocks by rhizoids.

Thallus: Thallus profusely branched and sometimes more or less confluent into a crustaceous disc. Secondary shoots consists of 5-6 linear leaves that are marginal and along sides of branches. Flowering shoots arising in between the secondary shoots are numerous on the fringe of the thallus (Uniyal 2001).

Leaves: Leaves are linear, scale-like in two rows, the lower ones are small or flat and the upper ones larger.

Flower: Flowers solitary covered by a boat-shaped spathella (Fig 40), stamens are two, borne on the andropod of same length. Ovary is ovoid, with smooth linear style. The fruit is a smooth, ball shaped sessile capsule. It is anisolobous and obliquely dehiscing.

Hydrobryum

Total number of species: 01

Number of species in India: 01

Hydrobryum griffithii (Wall ex Griff) Tul.

DISTRIBUTION: Monotypic, North Eastern Himalaya in Arunachal Pradesh, Meghalaya, Assam, Manipur, Sikkim and also reported from Garhwal.

HABITAT: Grows on rocks in fast running shallow waters.

HABIT: small liverwort-like thalloid plant body.

Attachment to rocks: Hard filamentous structures called Hapters that form pads at the points of contacts on the rocks.

Thallus: Thallus is crustaceous, more or less circular, twisted variously in all directions. Secondary shoots many, irregularly scattered only on dorsal surface (Rutishauser 1997; Ota et al 2001).

Leaves: Vegetative leaves linear, scattered in groups of 2-6 on the upper surface of the thallus. Leaves on flowering shoots are scale-like, distichous and imbricate.

Flowers: Solitary, sub-sessile, remaining with the leaves, spathella usually splitting down one side. Tepals two, linear. Stamens two, borne on andropodium. Capsule 12 or 14-ribbed, opening by two equals, stigma two, entire or lobed. Seeds elliptical, surface granular.

Maferria

Total number of species: 01

Number of species in India: 01

Maferria indica (Willis) C. Cusset

[= *Farmeria indica* Willis emend Arekal et Nagendran]

DISTRIBUTION: Endemic to Tamil Nadu.

HABITAT: It grows on smooth rocks in rapids.

HABIT: narrow ribbon-shaped plant body

Attachment to rocks: The thallus creeps in looping motion superficially on rocks, and fasten to the substratum by prominent discoid hapters which are cortical outgrowths of the thallus. A well developed root cap is always found. The thallus (root) borne shoots are initiated near the root tip and arise endogenously.

Thallus: Thallus is narrow, dorsiventrally flattened closely attached to rocks at intervals by haptera. Thallus is ribbon-like, and regularly branched. The secondary shoots consist of a few leaves from which single, bisexual flower arise. The plants occur sympatrically along with *Dalzellia zeylanica.*

Shoots: The secondary shoots arise on the thallus and terminate in a solitary, terminal flower.

Leaves: Leaves occur in the groups of 4-6 in distichous manner. These are subulate and have sheathing bases that become bract on floriferous shoots. The phyllotaxy patterns change in the course of foliation.

Flowers: Flowers are terminal and occur singly and lack sepals and petals or perianth. The flower is enclosed in a spathe at young stages. Androecium consists of only one stamen. Ovary is ellipsoidal with two stigma. Fruits are oblique, 10-12 ribbed capsules with 8 small seeds.

Podostemum

Total number of species: 10

Number of species in India: 03

***Podostemum barberi* Willis** (Fig 12)

DISTRIBUTION: Karnataka, Madhya Pradesh.

HABITAT: It occurs in smoothly flowing rapids

HABIT: narrow creeping thallus and erect bunch of long leaves.

Thallus: The thallus is flat, thick in middle and tapering off to the sides. It is frequently branched, in the horizontal plane, which are exogenous in origin. The thallus is attached to the rocks by the holdfast that arises along the middle of the lower side of thallus. The adhesive hairs arise from ventral surface of the thallus closely matted and mixed with a gummy excretion.

Shoot axis: The secondary shoots arise on the upper margins of the thallus, in the angles of branching. The lower leaves remain persistent.

Leaves: The leaves are smaller than that of *P. subulatum* but are similar in structure and arrangement on stem. The branching of the stem takes place from the lower axils of the dithecous leaves.

Flower: Flowers are cleistogamous, usually arising in axils of the dithecous leaves. The young flower is covered by long and narrow spathella until fertilisation. There is only one stamen, which stands closely up against the stigmas. Ovary is with six broad wings. The Ovary soon expands into the fruit after fertilization. The pedicle elongate and the spathe splits and falls away (Satheesh 1996).

podostemum munnarense

DISTRIBUTION: Endemic to Western Ghats in Kerala.

HABITAT: Mountain streams in Munnar (Kerala).

HABIT: narrow creeping thallus and erect bunch of long leaves

Attachment to the rocks: *P. munnarense* is fixed to substratum at the base through the roots. The roots are broadened and ribbon-like with distinct margins at both sides, separating an upper from the lower side. Thus established dorsiventrality of roots is additionally emphasised by so-called hapters (holdfasts) developed from the lower sides of roots (which face the substratum) as well as by the generation of numerous root-borne shoots arising near the flanks on the upper sides. Though roots (thalli) are mainly floating, they are also creeping upon stones, being here and then attached to the substratum. They are slightly groove-shaped towards the upper side, and their margins slant to and fro in accordance with the alternate shoot position.

Thallus: The thallus (root) attached at the base and other points near the base with free floating upper parts of the thallus.

Shoots: Root-borne shoots arise alternately from enlarged plate-like outgrowths of roots which develop in connection with shoot initiation at the root margins, surrounding the shoot bases like a collar. Shoots are short and bear six leaves in distichous foliation. Most shoots terminate in a single flower. Shoots are dorsiventral in structure (Jager-Zurn 2000).

Leaves: Leaves are linear, laterally flattened, enclosed in a prominent sheath and grass-like blade, which show plications. In many leaves, the two sides of plicate blade are repeatedly twisted up to the top. Hairs do not occur. Leaves are laterally flattened.

Flower: Flower is borne singly. Young flower protected by spathella, and by prominent leaf sheaths. Spathella is long, membranous and splitting at tip and is boat-shaped. Stigma is trifid on the wide ovary. Stamens are two at the top of the androphore. Andropodium is shorter than gynoecium. Capsule is long.

Podostemum subulatum **Gardn. (Fig. 13)**

DISTRIBUTION: Western Ghats of Kerala, Tamil Nadu and Maharashtra

HABITAT: It occurs in great masses in the more quiet bays in the rocks, where the eddy keeps the water in constant slow movement waving the long red or green leaves to and fro.

HABIT: narrow creeping thallus and erect bunch of long leaves

Attachment to the rocks: Holdfasts also called haptera are other means to fix the plants to the solid ground. Holdfast are often disc-like lateral outgrowths of the roots (more rarely shoots), with epidermal cells that secrete a gum to make the firm fixation.

Thallus: The thallus is seen as thin thread on the rocks between the stems. The secondary shoots are 1-3 cm high and fallen leaves make the scars. Primary axis creeps over the rocks, attached by adhesive hair or haptera, branched alternately with no root cap. Secondary axis stem is sub-erect and covered by leaves shrivel and drop off at once due to exposure to the air (Jager-Zurn 2000).

Shoot axis: Vegetative shoots are 2-8 mm long, arise along the margin of the thallus in the axils. The arm of the branch usually grows much more rapidly than the other and further branching appear. The thallus becomes more or less zig-zag in its construction, the two branches diverge at every node.

Leaves: Leaves are 6-15 cm long, subulate, distichous double sheathed.

Flowers: Flowers arise in axils of distichous leaves, are terminal erect. Hard spathella 3-6 mm long form protective covering of young flowers, is membranous and splitting irregulary at the tip. Two tepals and two stamens are present. Pollen are in dyads. Ovary is long, ovoid, with stigma lobes. The capsule is long with six ribs and dehisce by unequal valves.

Polypleurum

Total number of species: 09

Number of species in India: 09

It comprises of nine species distributed in India. The thalloid plant body is polymorphic and is believed to be of more taxonomic value than the flower characters. The characters of stigma distinguish the species. (Figs. 14-23).

***Polypleurum agharkarii* (Nandi) Nagendran** (Fig. 14)

[= *Diacrea agharkarii* Nandi]

DISTRIBUTION : Assam, Meghalaya.

***Polypleurum dichotomum* (Gardn.) Hall** (Fig. 15)

[= *Diacrea dichotoma* Tul.]

DISTRIBUTION: Maharashtra and Tamil Nadu

HABITAT: grows attached to rocks in fast running water.

HABIT: Thalloid body is floating. Thallus is ribbon-shaped zig-zag in appearance (10-30 cm long). Stigma sub-conical as compared to lobed found in *P. stylosum.*

Polypleurum disciforme (Fig. 16)

DISTRIBUTION: Kerala

HABITAT: Grows in moderately flowing shallow water.

HABIT: small disc shaped thalli

Attachment to rocks: The thalloid plant body is attached to the substratum by means of hapters present at the points of contact to the substratum.

Thallus: Thick, disc-shaped plant body, rarely ribbon-shaped. Vegetative and floral shoots are in marginal or sub-marginal position. 2-4 symmetric or asymmetric boat-shaped bracts in which the lower ones are smaller are present at the base of the single pedicellate terminal flower.

Flowers: The spathella covering the flower bud is funnel shaped, 4-5 mm long, symmetric or asymmetric with tendency to be boat-shaped, splitting into two small lobes when open. The two tepals are shorter than or equal to the length of the ovary. The androphore and 2 stamens are usually equal to the length of the ovary (1.5 – 3 mm). The elliptical ovary is 1.5 – 3 mm long, isolobous with numerous ovules on an axile placenta. The bilobed stigmas are 0.75 – 1 mm long and subulate. The fruit is 8-ribbed dehiscent capsule, long, 2-valved.

Polypleurum elongatum (Fig17)

[= *Diacrea elongata* Tul.]

DISTRIBUTION: Kerala, Tamil Nadu.

HABITAT: Grows in moderately flowing shallow water

HABIT: Thalloid plant body is cylindrical, floating, 50-60 cms long, stigma filiform.

Polypleurum filifolium (Fig. 18)

DISTRIBUTION: Kerala and Orissa.

HABITAT: grow attached to rocks in fast flowing water.

HABIT: Thalloid body short and filamentous attached to rocks wholly and not floating as found in other species of *Polypleurum*. Upper pair of leaves of flowering shoots with 2 long apical filaments (tapering tips) (Nagendran and Arekal 1981).

***Polypleurum minor* (Wedd.) Nagendran**

[=*Diacrea minor* Wedd]

DISTRIBUTION: Karnataka, Kerala, Tamil Nadu, Maharashtra, rarely in Assam, Meghalaya.

HABIT: Similar to *P. stylosum*. The thallus is very small, less branched and have short pedicel (2-5 mm) (Nagendran and Arekal 1981).

***Polypleurum munnarensis* Nagendran and Arekal** (Fig. 19)

DISTRIBUTION : Munnar, Kerala

HABITAT: Grow in fast flowing montane streams in Kerala

HABIT: long, flattened, large floating thallus

Thallus: The ribbon-like thalloid plant body attached to the rocks by means of haptera at the proximal part of the thallus. Long hair-like leaves (2 cm long) arise in groups on the margins of the thallus. Flowers arise in between these leaves at the time when the water level start lowering. Rest of the features show close resemblance to *P. dichotomum*.

Polypleurum prostratum (Fig. 20)

DISTRIBUTION: Kerala

HABITAT: grow nearly adpressed to the rocks in slow moving water.

HABIT: narrow elongated thalli.

Thallus: Plant body is ribbon-shaped, brown to green coloured, 2-7 cm long and 2-11 mm broad prostrate plant body, attached to the substratum at the base. The plant body is interpreted as a root, from which numerous vegetative and floral secondary shoots arise.

Leaves: Vegetative shoots occur in marginal or submarginal positions, remain small and developed 2-4 subulate structures. Floriferous shoots are sheathed by 2-4 stout bracts, the lower ones smaller and terminate in a single pedicellate flower.

Flower: The spathella is funnel-shaped, 3-5.5 mm long, symmetrical, 2-5 lobed. The two tepals are 1.25 – 2.5 mm long, shorter than the ovary. The androphore is usually as long as the ovary, bearing 2 stamens with short filaments, overtopping the ovary. The bi-carpellate ovary is elliptical and 1.5 – 3.5 mm long, terminates in two subconical stigmas. The fruit stalk is 5 mm – 2.7 cm long. The fruit is 2-4 mm long dehiscent capsule, the two valves of which possess three ribs each.

***Polypleurum stylosum* (Wight) Hall** (Figs. 21-23)

[= *Diacrea stylosa* Wight]

DISTRIBUTION: Maharashtra, Kerala, Karnataka, Tamil Nadu, rarely Occur in Assam, Meghalaya.

HABITAT: Occurs exclusively in gushing waters of rivers streams and cataracts in the Western Ghats.

HABIT: large frequently branched ribbon-shaped thalli.

Attachment to the rocks: The thalloid plant body is attached to the rocks by means of hold-fast (hapteron) at the base and rest of the body float with the water.

Thallus: The dorsiventrally compressed thalloid plant body is attached to the substratum by means of rhizoids and haptera. The thalloid bodies are ribbon-like, dichotomously branched and bear flower buds along margins. Thallus bears group of leaves along the margins (called Vegetative shoots). (Khosla 1996; Khosla et al 2000).

Shoots: Vegetative shoots consists of a few leaves, are very short, marginal, sub-marginal on the dorsal surface. Many of them develop into floriferous shoots by giving rise to a floral bud and subsequently long pedicellate fruit.

Leaves: Four to six leaves appear in alternate manner on the dorsal side of the thallus are distichous, delicate, subulate with persistent sheathing bases.

Flower: Flowers arise along the thallus margin in between the leaves, secondary shoot or vegetative shoots. Flower is enclosed by a hard protective covering of spathe. The sepal like staminodes arise at base of filament of stamen. In young ovary, the short style bears two lobes of stigma which are bent over each other. Of these one is bifid and covers other entire stigma at its base. Stamen are two and are raised by a common filament called andropodium Pollen occur as dyads. Fruit pedicallate, eight ribbed, isolobous capsule. Seeds are brownish orange in colour, non-endospermous. After pollination the flower is raised considerably due to the rapid elongation of pedicel.

***Polypleurum wallichi* (R. Br. Ex. Griff.) Warming.**

DISTRIBUTION: Meghalaya.

HABITAT: long filiform thalli

HABIT: Thalloid body short and ribbon-like. Thallus attached to substratum by hapters at the base as well as at points of contacts only. Upper pair of leaves of flowering shoots with one apical filament. Thallus body of *P. wallichi* is foliose, stigma subulate (Rao and Hajra 1975, Raveendran and Mathew, 1994).

Willisia

Total number of species: 02

Number of species in India: 02

Willisia arekaliana Shivamurthy and Sadanand (Fig. 24)

DISTRIBUTION: Silent Valley, Kerala.

HABITAT: Submerged rocks in fast-flowing stream.

HABIT: narrow, prostrate, elongated vegetative thalli on which short erect shoots appear

Attachment to rocks: Thallus closely attached to the rocks by rhizoids (adhesive hairs).

Thallus: Submerged herb. Thallus crustose, appressed closely to rocks, extensively branched. Secondary shoots arising mostly near the margin of the thallus, singly or rarely 2-3 together, 1-1.8 cm long, simple or rarely branched (Shivamurthy and Sadanand 1997).

Leaves: Leaves on secondary shoots, in 6 rows. Apical leaves triangular, apically unequally tridentate with a filiform extension arising between the three apical teeth.

Flowers: Flowers terminal on secondary shoots, solitary. Spathella with a cap-like tip unequally bilobed at apex which becomes detached at anthesis. Androecium consists two lateral staminodes and two stamens on a divided androphore. Ovary sub-spherical, stigmas apical. Fruit a sub-spherical capsule, dehiscing into two valves one of which persists. Seeds numerous.

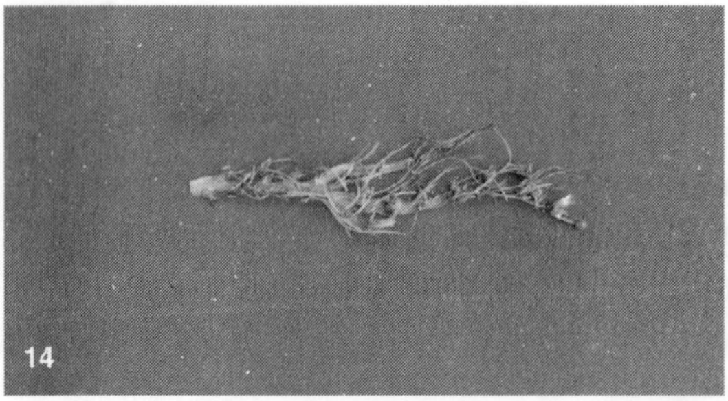
14

15

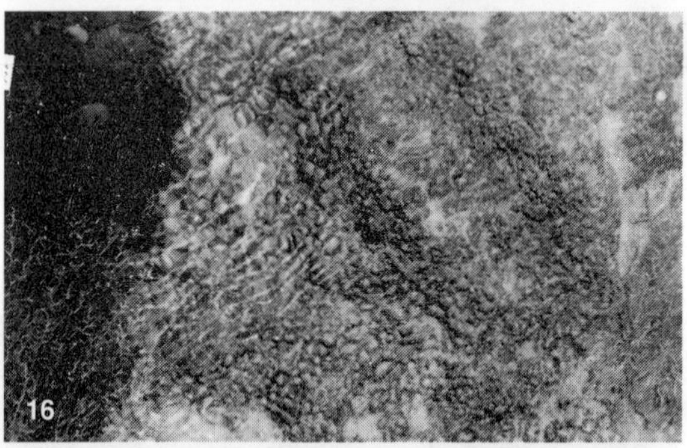
16

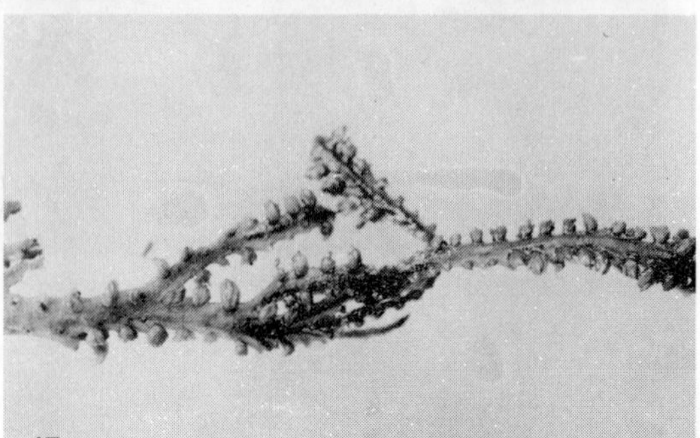
17

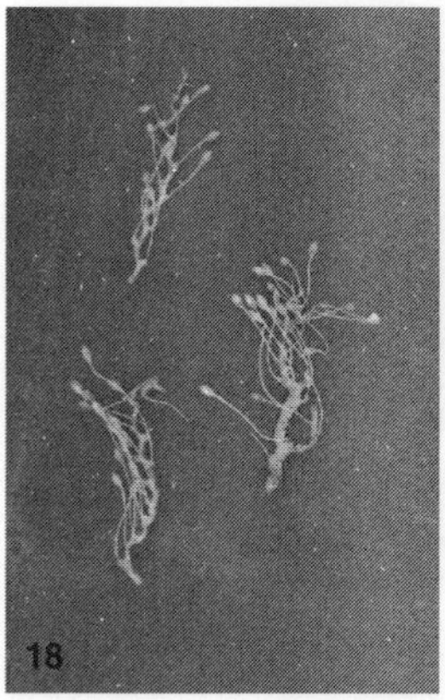
18

19

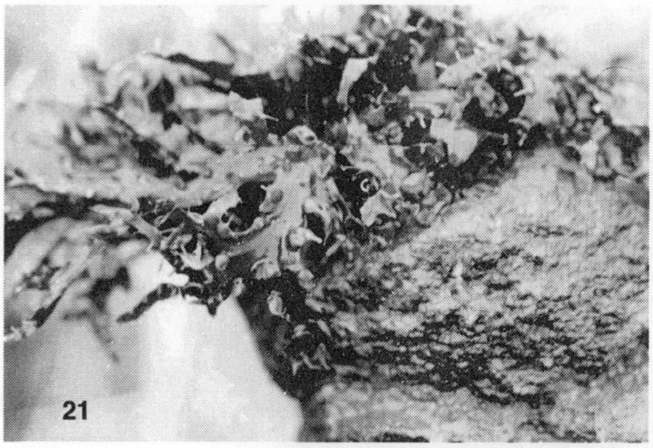
21

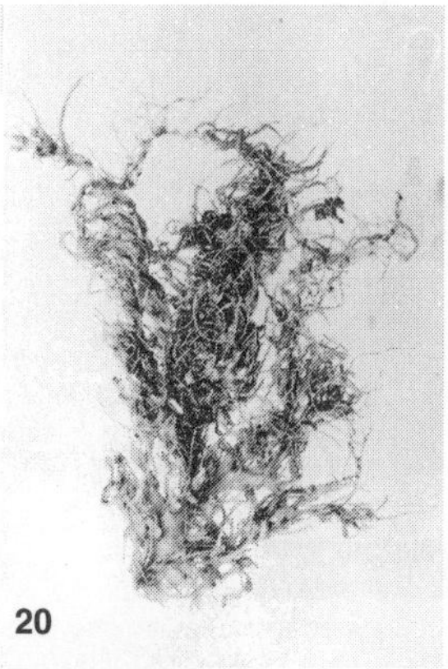
20

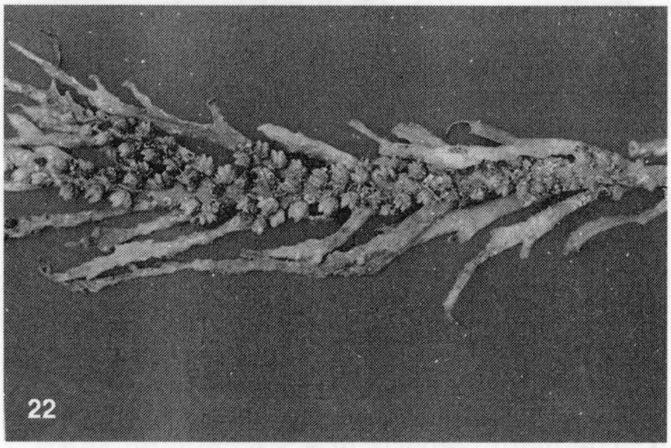
22

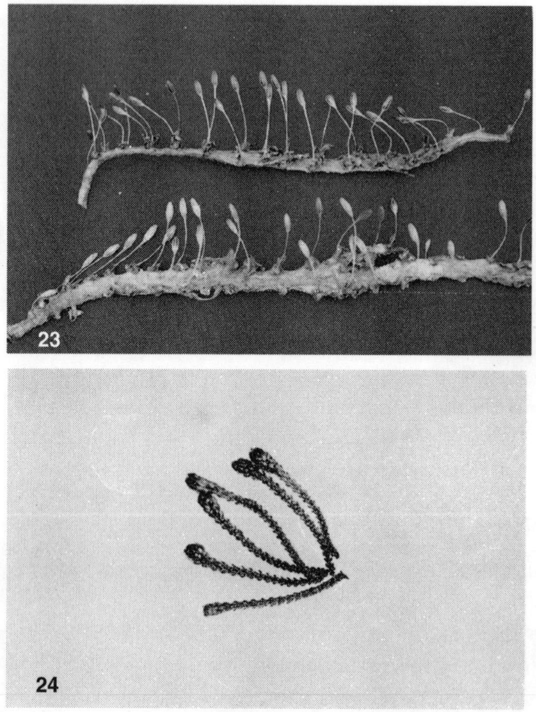

Fig. 13-24

***Willisia selaginoides* (Bedd.) Warm. ex Willis** (Figs. 25, 26, 44)
[= *Podostemon selaginoides* (Bedd) Benth]
DISTRIBUTION: Endemic to Western Ghats of Kerala.
HABITAT: Grows on rocks and woods in places with a rapid flow of water current.
HABIT: large, prostrate, elongated vegetative thalli on which long erect shoots appear

Attachment to rocks: Thin-walled, unicellular rhizoids are restricted to the ventral surface of the thalloid body. The tips of rhizoids secret a gum that fills spaces between the rhizoids. The gum helps in the firm attachment of the thalloid body to rocks.

Thallus: The plant consists of a 5-6 cm long and 3-4 cm wide thalloid body, attached to rocks by means of rhizoids. Leaves (5-6 cm long) and the tufts of shoots appear on the dorsal surface of the thalloid body. The thalloid body originates as a lateral appendage from the base of the seedling.

Shoots: Flowering shoots arise from the dorsal surface of the thalloid plant body. The shoots attain a maximum length of 10 cm, are branched or unbranched and bear leaves all along their length. On the development of shoots, the thallus disappears. The shoot axis is cylindrical (2.0 – 2.5 mm diameter) and terminates in a flower bud. Certain elongated, multicellular structures termed haptera arise from the base of individual shoots and creep down on the rocks and become firmly attached to them providing further anchorage (Uniyal and Mohan Ram 2001).

Leaves: Vegetative thalli bear long linear leaves and the flowering shoots arise from the point of origin of vegetative leaves. The flowering shoot is covered by 4-ranked, opposite-decussate, overlapping, triquetrous leaves which are closely appressed to the stem. The leaf has a hairy surface and an elongated tip. The abaxial surface of the leaves is hard and brittle due to the presence of silica deposition. The elongated tips of leaves gradually drop off an exposure to air.

Flowers: A solitary flower arises at the tip of the flowering shoot and is enclosed in a spathe. The spathe is enclosed up to half of its length by leaves, and this part is fleshy. The exposed apical part of the spathe is hard and terminates in two projections, encrusted with silica. The elongating stamens push the spathe upwards, which splits longitudinally at the base and drops off. The flowers lack sepals and petals or perianth. The androecium is Y-shaped with the andropodium terminally carrying two stamens. A pair of hook-like staminodes arises laterally. The anthers are introrse, have four locules each and dehisce longitudinally. The pistil consists of an egg-shaped ovary and a bipartite stigma. However, a distinct style is lacking. The stigma lobes are unequal and lie appressed and bent over the surface of the ovary at young stage and lie below the stigma. In the subsequent stages the filaments elongate, curve inward and bring the anthers to the same level as that of the stigma lobes. In the older flowers the staminodes and anthers are shed and the shrivelled andropodium persists at the base of the ovary. The gynoecium is bicarpillary and the ovary is bilocular. The average total number of ovules per ovary is 270 ± 20, bringing the pollen (dyad): ovule ratio to approximately 27: 1.

Fruits: The fruit is oval-elliptic (4.0 × 3.0 mm) with a sharp tip. It has four longitudinal ribs along the wall. Two of these that run along the septum are thin and the other two running along the middle of the locules are prominent. At maturity the fruit splits longitudinally in line with the septum (septicidal capsule) and the seeds are dispersed singly. On average (n = 25), a fruit contains 200 ± 20 seeds, arranged on an axile placenta. The ovule: seed ratio is 9: 7. The brown, orbid seeds (0.4 × 0.2 mm) are non-endospermic.

Zeylanidium

Total number of species: 04

Number of species in India: 04

***Zeylanidium. johnsonii* (Wight) Engl.** (Fig. 27)

[= *Podostemon johnsonii* (Wt.)Wedd, = *Hydrobryum johnsonii* (Wt.) Willis**]**

DISTRIBUTION: Karnataka, Kerala and Andhra Pradesh.

HABITAT: thalli closely adpressed to rocks in rapids.

HABIT: Similar to *Z. lichenoides* and characterized by the vegetative shoots scattered over the upper surface of the thallus. Shoots are of one kind: fertile and prostrate with 4-6 leaves. Stamens 4-8 mm long, gynophore is absent and stigma bilobed.

***Zeylanidium lichenoides* (Kurz) Engl. (**Fig. 28)

[= *Hydrobryum lichenoides* Kurz.]

DISTRIBUTION: Endemic to India, distributed in North Eastern India, Western ghats including Maharashtra, Kerala, Karnataka, Tamil Nadu, Madhya Pradesh, Orissa, Andhra Pradesh.

HABITAT: It occurs very commonly in shallow streams, and in rapids along the ghats during monsoon.

HABIT: Thalli closely adpressed to rocks in rapids

Attachment to rocks: There is a continuum of exogenous adhesive hairs (rhizoids) on the ventral surface of the thalloid body.

Thallus: The plant body is highly reduced as there are no distinct stem and roots. It consists of a light green, crustaceous, branched, dorsiventral thallus found attached to rocks. The margins of the thalloid body bear linear leaves.

Shoots: In *Z. lichenoides* two kinds of shoots are present. The upright vegetative shoots in branch axils with narrow leaves are the primary axis. The secondary axis fertile shoots in along the sides of thallus with distichous leafy shoots which terminate in a flower (Satheesh and Mathew 1997).

Leaves: Alternate, bract-like, distichous.

Shoot branching: The shoot axis elongates and becomes prostrate on the thallus (or may be erect), pointing in all cases towards the tip. The bracts from on each shoot are 2-8 in number and are developed by enlargement of the sheathing bases of leaves.

Flower: Flowers terminal, solitary, pedicellate. Each flower enclosed in a boat-shaped protective covering called spathella. Stamens are two with a fleshy andropodium. Ovary is elliptical to ovoid, syncarpous, anisolobous, stigmas two in number. Capsule is long, anisolobous, distinctly and broadly

8-ribbed. Seeds are numerous, pale orange in colour and show elliptical shapes.

Zeylanidium maheshwarii **Mathew et Satheesh** [Figs. 29-31)

DISTRIBUTION: Endemic to State of Kerala.

HABITAT: Occur on the bank of streams in the splash zone and slow moving water.

HABIT: small disc-shaped thalli

Thallus: The young thallus is thin, smooth, prostrate, even and light olive green in colour whereas older parts are dull green in colour. The thallus grows in patches and tightly attached to rocks by the adhesive hair (rhizoids) on lower surface (Mathew and Satheesh 1997).

Shoots: Vegetative shoots consists of a few linear leaves. Most of the vegetative shoots change into floriferous shoots and bear sub-sessile terminal flower.

Leaves: Four to six leaves appear in alternate distichous manner as shoots. Leaves are subulate and deciduous with persistent sheathing bases. The leaf bases form bracts where the flower appear on the shoots.

Flowers: Young flowers are protected by boat-shaped spathella. Flowers are solitary, pedicellate and terminal. Tepals are two in number present one on either side of the andropodium which are attached to the pedicel. The two stamens are raised on andropodium which is shorter than gynoecium. Pollen occurs as dyads. Ovary is ovoid, stigma is multilobed.Fruit is an obovoid capsule and is broadly 3-ribbed.

Zeylanidium olivaceum **(Gardn.) Engl.** (Figs. 32, 33)

[= *Podostemum olivaceum* Gardn.,= *Hydrobryum olivaceum* Tul.]

DISTRIBUTION: Endemic to Western Ghats of Tamil Nadu and Kerala. Also present in North Eastern India in Arunachal Pradesh.

HABITAT: *Z. olivaceum* is often found on the edges of water-falls and shorter places of very strong current. This species is able to inhabit much smaller streams and shallower water than other species.

Habit: Cup-shaped thalli on the margins of which small shoots arise

Attachment to the rocks: The ability to develop endogenous shoots is the characteristics of roots.

Thallus: Thallus is crust-like closely attached to the rocks. The thallus form cup shape structure has an irregular lobed margin. This is the only crustose-rooted species of the genus that still develops primary shoots from the seedlings in addition to secondary (root borne) shoots forming plant body. The endogenous development of floriferous shoots as well as vegetative shoots is an indication of its root nature of the thalloid body which has root cap at tip.

Shoots: The primary shoots are articulated into an up to 8.5 cm long and 4 mm thick stalk and is sympodially branched. The short stems within leafy head do not separate, but are fused to a dense aggregate (coenosome). Branches are mainly vegetative with rosette of 20 long subulate leaves. Floriferous shoots are secondary shoots developed from the crust. They are extremely short, terminating in a single flower. This shoot bears 6-9 leaves, which are alternately arranged (Jager-Zurn 2000).

Leaves: The vegetative shoots have narrowly elongate-subulate leaves with hairs. The floriferous shoots have broad leaves. Leaves below the flower bud of primary shoot are either boat shaped or broadened at the base only, but distally elongated into hairy subulate blade.

Flower: Flower is terminal, zygomorphic, lack sepals and petals and covered by spathella. Stamens two develop from common androphore. Ovary is elliptical, stigma bifid. Capsule is nearly elliptical. Each valve is broadly 3-ribbed.

SUBFAMILY: TRISTICHOIDEAE

Dalzellia

Total number of species: 02

Number of species in India: 02

Dalzellia gracilis Mathew, Jager-Zurn, Nileena, sp. nov. (Fig. 34)

DISTRIBUTION: Kerala

HABITAT: *D. gracilis* is a rheophytic plant with a creeping thalloid plant body that is closely attached to rocks.

HABIT: small narrow ribbon shaped thalli

Attachment to rocks: *D. gracilis* characterized by a narrow short thallus (roots) from which numerous shoots develop. Thick hapters also develop from shoots. Thallus develops adhesive hairs on their lower side.

Thallus: The shoots develop alternately or in pairs along the flanks of the (thallus) root. The shoot axis is first visible as a globular bulge arising from the thallus. While elongating the young shoot axis broadens as soon as the first leaves arise.

Leaves: Each shoot consists of an apical tufts of leaves, whereas the lower part of the crust is mainly leafless. The floriferous shoots show a distinct row of two to three stout and considerably elongated leaves on either side. These leaves are upto 1.6 cm long and have a prominent midrib.

Shoot axis: Branches are flattened and develop separately from their mother shoot axis, i.e. they are not fused to the mother shoots as in *D.*

zeylanica. At frequent ramification the single, slender, nearly ribbon – like crustose root–borne shoots finally attain an irregular shape. As a result of elongation growth of shoot axis, alternately positioned leaves become separated. Leaves of floriferous shoots are more closely positioned to each other than leaves of vegetative region.

Flowers: The floriferous shoot is only basely foliated. The pedicel attains a length of 7-15 mm, only after that, the single flower opens. The perianth is three lobed, tepals are united at the base. The three stamens alternate with the perianth lobes. Stamens are 1.75-2 mm long. Before anthesis, they are equal to the gynoecium in length and filaments are short and less than 0.25-0.75 mm but elongate much at anthesis. The ovary is elliptical, 0.75-1.25 mm long, tricarpellate, syncarpous, trilocular with numerous ovules, ends in three hairy long stigmas. The fruit is an elliptical, 1.25-3 mm long, three valved capsule. Each valve is 3 ribbed.

***Dalzellia zeylanica* (Gardn.) Wight** (Figs. 35, 36)

[= *Lawia zeylanica* Tul., *Terniola zeylanica* Tul.]

DISTRIBUTION: Western Ghats of Karnataka, Kerala and Tamil Nadu.

HABITAT: *Dalzellia zeylanica* grows on smooth rocks in places where the flow of water is very swift. Most often the plants cover the rock to the exclusion of other species.

HABIT: large disc shaped thallus.

Attachment to rocks: Thallus attached to rocks by sticky hairs or rhizoids which cover the lower surface of the prostrate crusts. Neither roots non holdfasts are found in *D. zeylanica*.

Thallus: The plant body consists of green to reddish prostrate photosynthetic crusts. The crust is composed of shoots only and show branching in dichasial order; the consecutive branches remain combined and form sympodial coenosom. They carry exogenously borne leaves on the upper surface. The "rosettes" at the upper surface of the crust are short shoot systems.

Leaves: Leaves are dimorphic.They are either in rosettes on a short stem or in longitudinal rows on dorsal side and margin of crust. Each rosette leaf is long and linear with narrow and elongated cells in the middle are shorter and lanceolate in the margins of the crust. The upper surface is partially covered with green scales. The scales are only 1-2 mm long, about one cell layer thick. The scales along the thallus margins are provided with the midrib and those on upper surface lack midrib. The endogenous scales fall down by flowing water during maturation of thalli. But the thalli are able to produce another set of photosynthetic scales from endogenous rosettes which rupture the cortical cells of the upper surface. Towards the end of rainy season flowering shoots appear from closely packed each dwarf (Jager-Zurn, 1995).

Shoot axis, branching: The shoot axis (thallus or crust) is characterised by lateral symmetry (dorsiventrality) and longitudinal symmetry. The "rosettes" at the upper surface of crusts are short shoot system, composed of three axes (main axis and 2 branches) of a construction in diachasial and sympodial order. They develop as dormant buds and derived from the axillary meristem of the lateral leaves of older "rosettes". They grow out only after some distance from the top of the crust and can be called catleptic bud.

Flower: Single flower form at the tip of short shoots (pedicels) which bear scales. The young flower is surrounded by a cup-like cover "cupule" that consists of many ligulate scales that are fused at the base (Jager-Zurn 1995). The spathella is absent. Flowers have perianth whorl with 3 or 5 (-6) overlapping segments that protects young flower bud until anthesis. There are three introrse stamens that are considerably flattened. Pollen are dispersed as single grains (monads). The ovary is syncarpous provided with three stigma lobes. The seeds are yellowish brown 0.1-0.2 mm long ovate, and become sticky when moistened.

Rejuvenescence: The thallus has a remarkable capacity for rejuvenescence from any detached part which is alive on the substratum - during growing season.

25

26

27

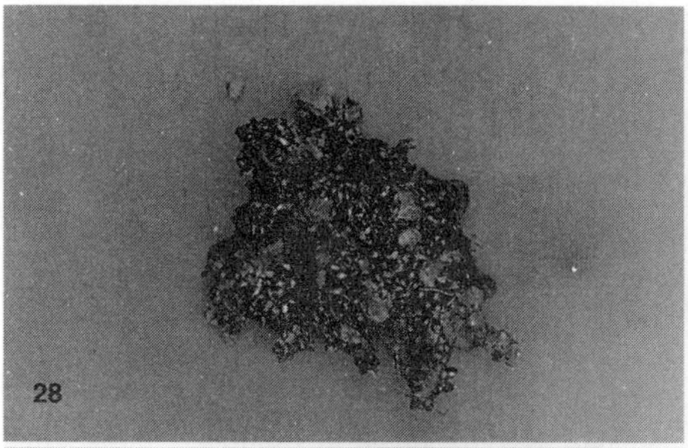
28

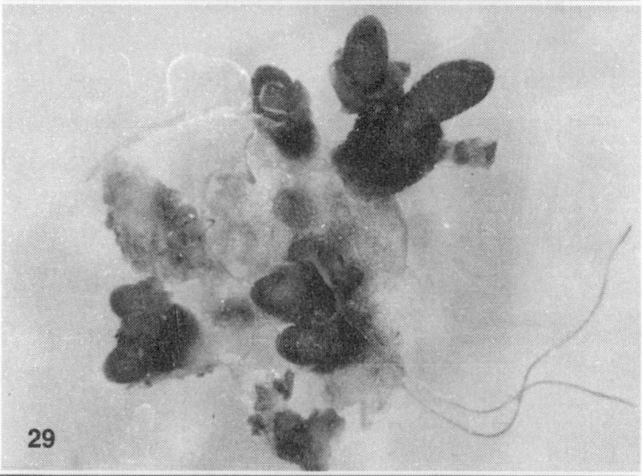
29

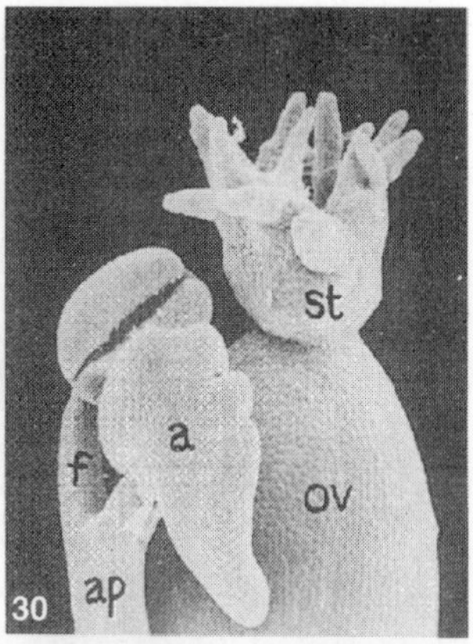
st
a
f
ov
ap
30

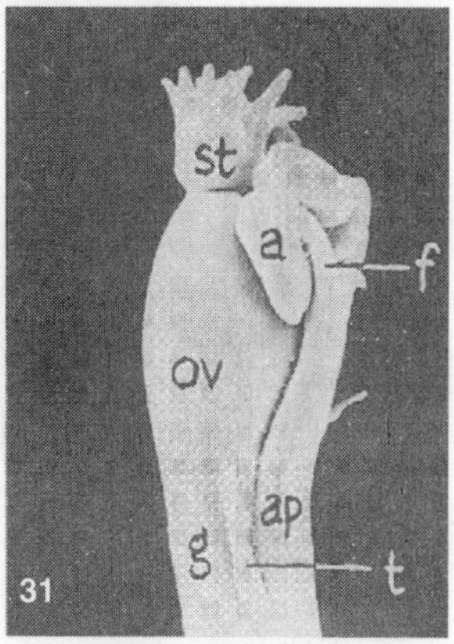
st
a
f
ov
ap
g
t
31

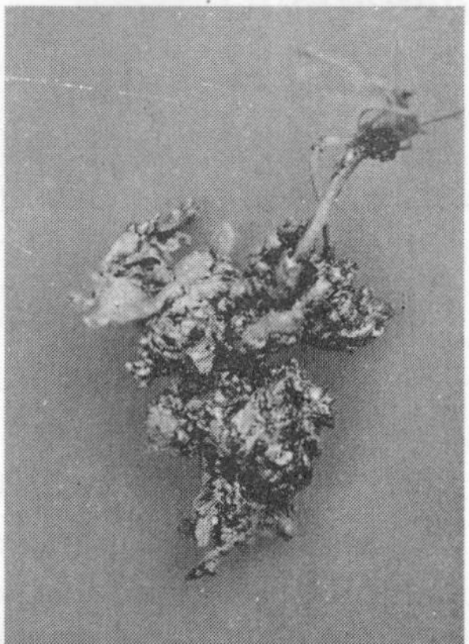
32

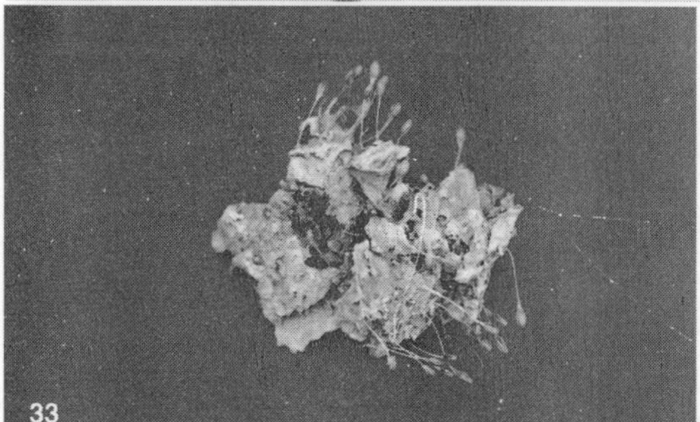
33

34

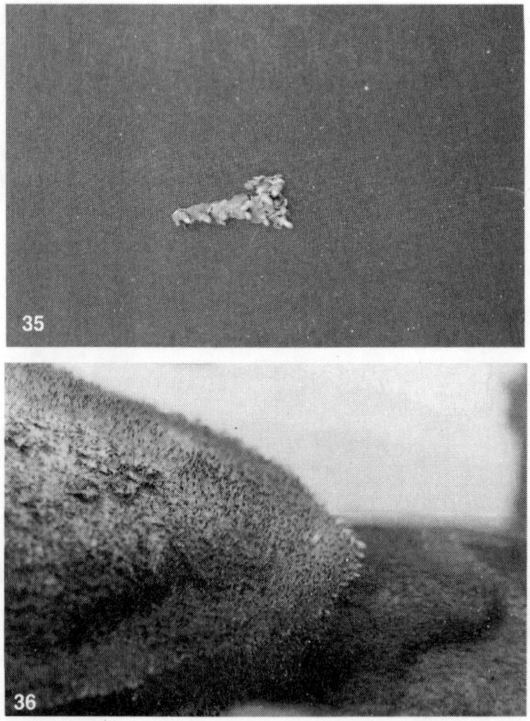

Fig. 35-36

Indotristicha

Total number of species: 02

Number of species in India: 02

Indotristicha ramosissima **(Wight) Van Royen (**Figs. 37, 38)

(= *Dalzellia ramosissima* Wight.,= *Terniola ramosissima* (Wight) Wedd. = *Tristicha ramosissima* (Wight) Willis)

DISTRIBUTION: Western ghats of Karnataka, Kerala, Tamil Nadu.

HABITAT: The plant lives in slow-moving as well as rapid water, red or green forming tangled masses.

HABIT: Frequently branched stem, attached to rocks from the base and rest lond branches float on the water

Mature structure: Thread-like thallus attached to rocks by disc-like feet with many free floating frequently branched secondary shoots which are upto 60°cm long.

Dry season appearance: The rocks above water are covered with a muddy tangle of dead branches. The frequently branched rhizome-like thalli creeps on the rocks in all directions. The branches are usually at right angles to the main thallus. At frequent intervals along the thalli, are thin disc-like organs springing from the sides of the thalli and spreading upon the rock, by which they fastened to the rocks very firmly. There discs are mostly in pairs, one on either side.

Roots: The creeping organ 'roots' are nearly cylindrical to slightly flattened, with prominent, asymmetrical multicellular cap. It consists of epidermis paranchymatous cortical cells and the central vascular bundle with inconspicuous xylem elements. Rhizoids grow out on the root side which is attached to the rock. The lateral roots originate as endogenous buds (Rutishauser and Huber, 1991).

Holdfast: Holdfasts also called haptera are the other means of firm attachment to the substratum. They are disc-like or finger-like lateral outgrowths of the root with epidermal cells that secrete the gum once they make contact with the substratum. The endogenous buds rupture the root cortex and then develop into disc-like or finger-like holdfasts. The shoot bud arises from the upper proximal area of the hold fast. Additional holdfasts grow out from the stem base in order to serve as a stable anchorage .

Shoots and ramuli: The complex photosynthetic appendages are called ramuli. The shoots of the second order grow very rapidly and cover the main apex with their leaves. These are shoots of limited growth and known as "ramuli". The ramuli show determinate growth and are normally unbranched appendages of 2-4 cm in length. They consist of a thin vasularised axis and transversally inserted unistratose photosynthetic scales. The photosynthetic "scales" are irregularly or spirally arranged along the ramuli axis. These first formed ramuli show a distichous to spiral arrangement along the main stem. Additional ramuli which are formed in a later stage, arise in the axils of stem scales (Rutishauser and Huber 1991, Rutishauser 1997).

The vegetative shoots are up to 60 cm long. They are submerged in the flowing water and fixed to the substratum (rock or pebbles) by a holdfast. The shoots are monopodial systems. Ramuli are positioned homologous to subtending leaves in typical flowering plants.

Leaves: The scales which are appendages of shoot axis and ramuli are called leaves.

Flower: Single flowers are formed at the tip of the short shoots which bear scales as well as ramuli below the flower. Additional floral shoots may arise as endogenous buds along the vegetative shoot axes (Rutishauser and Huber 1991). The flowers lack spathella and covered by a sheathing perianth with three overlapping segments. An additional cover cupule encloses a young flower. Cupule consists of 5-7 ligulate scales and 1-3 ramuli that are basally united to form a cup like structure. Tepals are free or fused at the base. The pre-anthesis anthers are introrse and considerably flattened. They are provided with a connective tip and a sagittate to cordate base. The pollen is monad, spheroid and panporate with about 12 or less pore.The syncarpous ovary is provided with three linear stigmas, depending on number of carpel primordial. The capsules are 3-locular, 9-ribbed, open by 3 septicidal slits i.e. with three equal persistent valves. There are +100 seeds per capsule which is 2.0-2.5 mm long.

Fig. 37-40

Indotristicha tirunelveliana **Sharma, Karthik Shetty** (Fig. 39)

DISTRIBUTION: Endemic to Tamil Nadu.

HABITAT: The plants are found attached to rapids of rocks at their bases whereas the rest of the body floats downwards.

HABIT: A creeping sub-erect, long rhizome on which small, dense leafy shoots arise

Attachment to rocks: Haptera emerge endogenously from the basal portion of the rhizome. They may branch and spread over substratum.

Plant Body: The plant body consists of 7-25 cm long, sub-erect rhizome, attached to rocks with the help of haptera. The rhizome bears densely crowded shoots with four rows of spiral leaves (Uniyal 1999).

Shoots: The shoots arise from the dorsal surface of the rhizome, are long (upto 3 cm), unbranched and densly covered with leaves. The shoot axis is cylindrical.

Leaves: The shoot axis is closely covered with 4-ranked spirally arranged, ovate-lanceolate leaves (1-2 mm long).

Floral shoots: arise as buds on the dorsal surface of rhizome. Flower bearing leafy shoots are often branched at base. The upper leaves become densly crowded to form a cup-like structure which covers the young flower bud.

Flowers: Plants bearing flower buds appear when the plants are 10-20 cm below the water level and they become partially exposed as the level of flowing water drops. Each floral shoot bears a solitary flower. Floral organs are completely covered by three pale green, membranous tepals. During flower maturation the pedical elongates considerably (upto 2.5 cm), the tepals open and flowers emerge above the water surface. The androecium consists of 3 fertile stamens (upto 8 mm long) alternating the tepals. Anthers are pale-red, introrse, four loculed and dehisce longitudinally. The pistil consists of an elongated-ovoid ovary (2-3 mm long) and bears three deep red, papillose stigma lobes. A distinct style is lacking. The ovary is three carpellate with thin septa connecting the large, fleshy and axile placenta.

Fruits: Fruiting occurs from January onwards and depends on depth of water in stream. Fruit is a 9-ribbed capsule (2-4 mm long) and it opens by 2-3 septicidal slits. On an average, the fruit contains 100 ± 20 seeds arranged on axile placenta. Seeds are brownish orange in colour, non-endospermic.

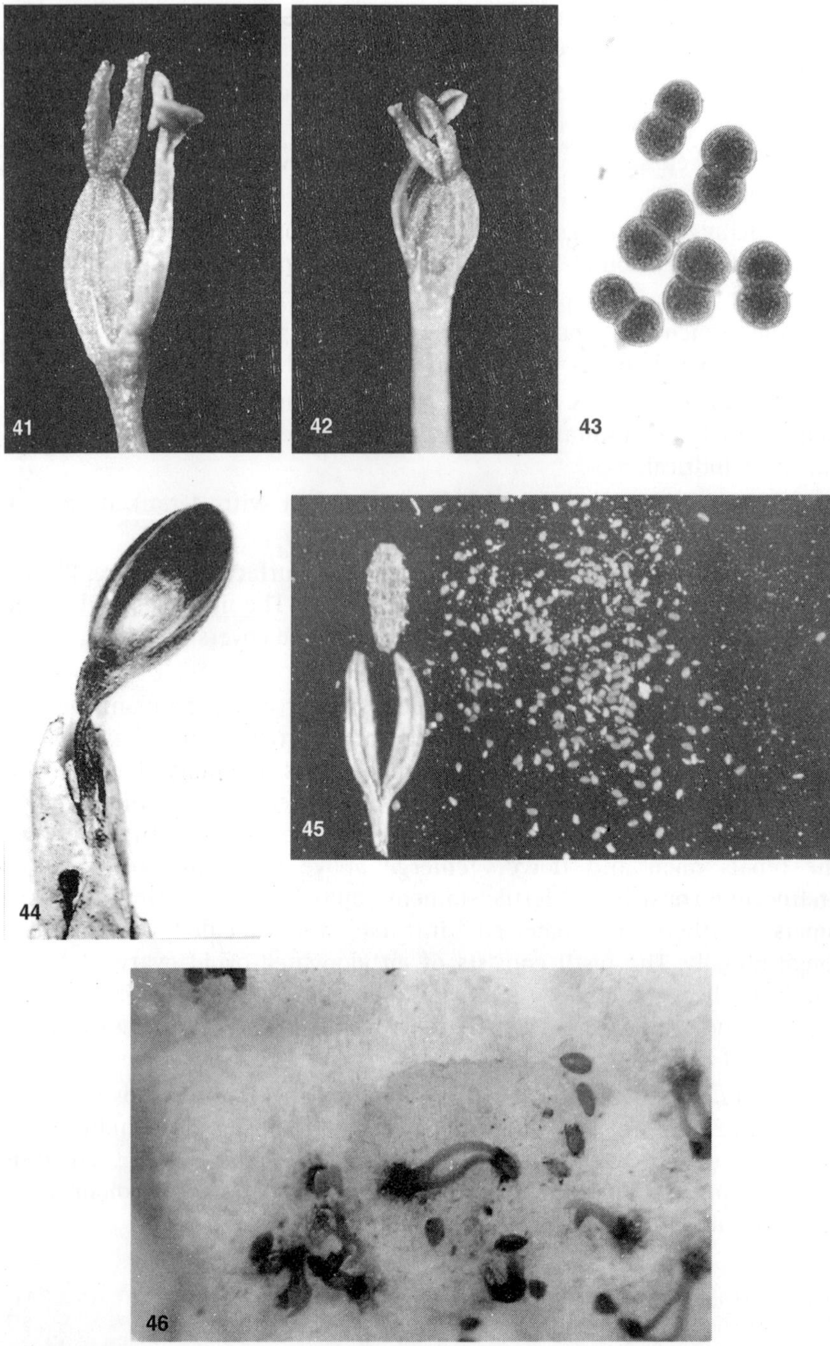

Fig. 41-46

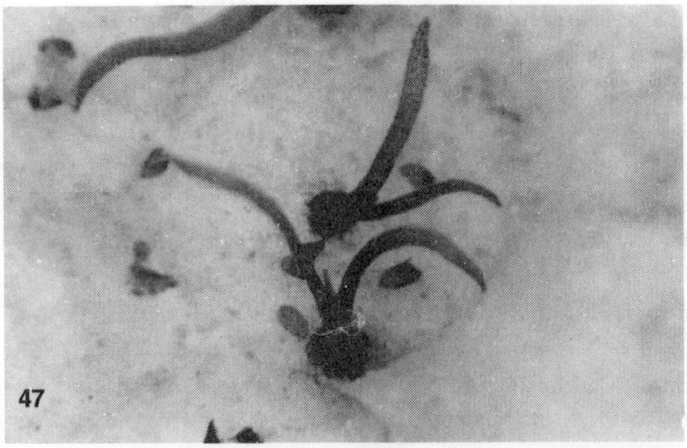

47

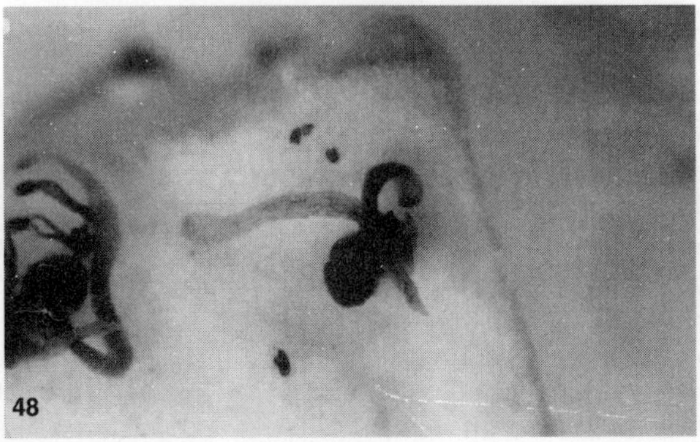

48

Fig. 47-48

References

Arber A 1920 *Water Plants: A Study of Aquatic Angiosperms* (Cambridge University Press: Cambridge, UK).

Arekal GD and Nagendran CR 1974 Additional notes on *Farneria* Willis (Podostemaceae), *Proc. Indian Acad. Sci.* **B80** 2266-228.

Arekal GD and Nagendran CR 1975a Is there a Podostemum type of embryo sac in the genus *Farmeria? Caryologia* **28** 229-235.

Arekal GD and Nagendran CR 1975b Embryo sac of *Hydrobryopsis sessilis* (Podostemaceae) – origin, organization and significance, *Bot. Natiser* **128** 332-338.

Arekal GD and Nagendran CR 1976 A new type of embryo sac organization in angiosperms, *Curr. Sci.* **45** 717-719.

Arekal GD and Nagendran CR 1977a The female gametophyte in two Indian genera of Tristichoideae (Podostemaceae) - a reinvestigation, *Proc. Indian Acad. Sci.* **6D** 287-291.

Arekal GD and Nagendran CR 1977b Female gametophyte of *Zeylanidium* (Podostemaceae) – a clarification, *Phytomorphology* **27** 123-129.

Battaglia E 1971 The embryo sac of Podostemaceae – an interpretation, *Caryologia* **B24** 403-420.

Bell A 1991 *Plant form: An Illustrated Guide to Flowering Plant Morphology* (Oxford University Press: Oxford, UK).

Cook CDK 1990 *Aquatic Plant Book* (SPB Academic Publishing: The Hague, the Netherlands).

Cook CDK 1996 *Aquatic and Wetland Plants of India* (Oxford University Press, Oxford, UK).

Chopra RN and Mukkada AJ 1966 Gametogenesis and pseudo-embryo sac in *Indotristicha ramosissima* (Wight) van Royen, *Phytomorphology* **16** 182-188.

Cross Bell D 1990 Biomonitoring the effect of rubber factory effluent on hill stream in Kanya Kumari district, *Geobios* **17** 220-222.

Cusset C 1972 Les Podocemaceae de Madagascar, *Adansonia* **12** 557-568.

Cusset C 1987 Podostemaceae/Tristichaceae, In *Flore du Cameroun*, eds B Sabatie and Ph Morat **30** 51-105.

Cusset C 1992 Contribution a Petude des Podostemaceae. 12. Les genres asiatiques, *Bull. Mus. Natn. Hist. Nat., Paris, B. Adansonia* **14** 13-54.

Cusset C and Cusset G 1988a Etude sur les Podostemales. 9. Delimitations taxinomiques dans les tristichaceae, *Bull Mus. Natn. Hist. Nat., Paris b Adansonia* **10** 179-218.

Cusset C and Cusset G 1988b Etude sur les Podostemales 10. Structurese florales et. Vegetative des Tristichaceae, *Bull. Mus. Natn. Hist. Nat., Paris B Adansonia* **B10** 179-218.

Cusset C and Cusset G 1989 Etude sur les Podostemales 12. Biogeographic evolutive de *Tristicha trifaria* (Bory. Ex Willd.) Sprengel, *Bull. Mus. Natn. Hist. Nat., Paris* **11** 39-70.

Esau K 1965 *Plant Anatomy*, 2nd edn (John Wiley: New York, USA).

Goldschmidt R 1933 Some aspects of evolution, *Science* **78** 539-547.

Grant V and Grant KA 1965 *Flower Pollination in the Phlox Family* (Columbia University Press; New York, USA).

Grimm NB 1993 Implications of climate change on stream communities, In *Biotic Interactions and Global Change*, pp. 293-314 eds PM Karieva, JQ Kingslover and RB Huey (Sinaeur, Sunderland: MA).

Hammond BL 1937 Development of *Podostemum ceratophyllum*, *Bull. Torrey Bot. Club* **64** 17-36.

Horne AJ and Goldman CR 1994 *Limnology*, 2nd edn (McGraw Hill: New York, USA).

Imaichi R, Ichiba T and Kato M 1999 Developmental morphology and anatomy of the vegetative organs in *Malaccotristicha malayana* (Podostemaceae), *Int. J. Plant Sci.* **160** 253-259.

Jäger-Zurn 1992 Morphologie der Podostemaceae II. *Indotristicha ramosissima* (Wight) van Royen (Tristichoideae), *Trop. Subtrop. Pflanzenwelt* **80** 7-48.

Jäger-Zurn I 1995 Morphologie der Podostemaceae III. *Dalzellia ceylanica* (Gard.) wight (Tristichoideae), *Trop. u. Subtrop. Pflanzenwelt* **92** 7-77.

Jäger-Zurn I 1997 Comparative morphology of the vegetative structures of *Tristicha trifaria, Indostristicha ramosissima and Dalzellia ceylanica* (Podostemaceae, Tristichoideae): a review, *Aquat. Bot.* **57** 71-96.

Jäger-Zurn I 2000 Developmental morphology of the roots and root-borne shoots of *Podostemum subulatum* as compared with *Zeylanidium olivaceum* (Podostemacee-Podostemoideae) part VII of the series 'Morphology of Podostemaceae', *Plant Syst. Evol.* **220** 55-67.

Jäger-Zurn I and Gubert M 2000 Podostemaceae depend on sticky biofilms with respect to attachment to rocks in waterfalls, *Int. J. Plant Sci.* **161** 599-607.

Johri BM, Ambegaokar KB and Srivastava PS 1992 *Comparative Embryology of Angiosperms,* Vols 1 and 2 (Springer-Verlag: Berlin, Germany).

Kapil RN 1970 Podostemaceae, *Bull. Indian nat. Sci. Acad.* **41** 104-109.

Khosla C 1996 Reproductive biology of three aquatic angiosperms–*Polypleurum, Utricularia and Ceratophyllum,* Ph.D. Thesis, University of Delhi, Delhi, India.

Khosla C, Shivanna KR and Mohan Ram HY 2000 Reproductive biology of *Polypleurum stylosum* (Podostemaceae), *Aquat. Bot.* **67** 143-154.

Khosla C, Shivanna KR and Mohan Ram HY 2001 Cleistogamy in *Griffithella hookeriana* (Podostemaceae), *S. Afr. J. Bot,* **67** 324-328.

Les DH, Philbrick CT and A Novelo R 1997 The phylogenetic position of river-weeds (Podostemaceae): Insights from *rbc* L sequence data, *Aquat. Bot.* **57** 5-27.

Magnus W 1913 Die atypische embryonal entwicklung der Podostmaceae, *Flora* **105** 275-336.

Maheshwari SC 1995 The occurrence of bisporic embryo sacs in angiosperms–a critical review, *Phytomorphology* **5** 67-99.

Maheshwari Devi H, Johri BM, Rau MA, Singh D, Dathan ASR and Bhanwara RK 1995 Embryology of Angiosperms, In *Botany in India : History and Progress,* pp. 58-146 ed BM Johri (Oxford and IBH Publishing Co.: New Delhi, India).

Mathew CJ and Satheesh VK 1997 Taxonomy and distribution of the Podostemaceae in Kerala, India, *Aquat. Bot.* **57** 243-274.

Mathew CJ, Jager-Zurn I and Nileena CB 2001 *Dalzellia gracilis:* a new species of Podostemaceae (Tristichoideae) from Kerala, India, *Int. J. Plant. Sci.* **162** 899-909.

Mauseth JD 1988 *Plant anatomy* (The Bengamin/cummings Pub. Co., Inc: Menlopark, California, USA).

Metcalfe and Chalk 1957 Anatomy of the Dicotyledons Vol **2** *Oxford* (1101-1104).

Mohan Ram HY and Sehgal A 1992 Podostemaceae–the strange family of aquatic angiosperms, *Palaeobotanist* **41** 192-197.

Mohan Ram HY and Sehgal A 1997 In vitro studies on developmental morphology of Indian Podostemaceae, *Aquat. Bot.* **57** 97-132.

Mohan Ram HY and Sehgal A 2001 Biology of Indian Podostemaceae, *Phytomorphology* Golden Jubilee Issue 365-391.

Mukkada AJ 1962. *Morphological and Embryological Studies on Some Indian Podostemaceae,* Ph. D. Thesis, University of Delhi, Delhi, India.

Mukkada AK 1969 Some aspects of the morphology, embryology and biology of *Terniola zeylania* (Gardner) Tulasne, *New Phytol.* **68** 1145-1157.

Mukkada AJ and Chopra RN 1973 Post-fertilization development in *Indotristicha ramosissima* (Wight) van Royen, *New Phytol.* **72** 639-646.

Murashige J and Skoog F 1962. A revised medium for rapid growth and biassays with tobacco tissue cultures, *Physiologia Pl.* **15** 473-497.

Nagendran CR 1975 *Studies on Podostemaceae.* Ph.D. Thesis, University of Mysore, India.

Nagendran CR and Arekal GD 1976 Embryo sac of *Griffithella hookeriana* - a reinvestigation, *Phytomorphology* **26** 359-363.

Nagendran CR and Arekal GD 1981 Studies on Indian Podostemaceae, *Buill. Bot. Surv. India* **23** 228-234.

Nagendran CR, Anand VV and Arekal GD 1980 The embryo sac of *Podostemum subulatus* (Podostemaceae)–a reinvestigation, *Pl. syst. Evol.* **134** 121-125.

Nagendran CR, Subranmanyam K and Arekal GD 1981 A morphogentic approach to the embryogeny of *Indotristicha* (Podostemaceae), *Ann. Bot.* **47** 799-804.

Nair PKK 1965 Pollen morphology of Indian Podostemaceae, *Curr. Sci.* **34** 381-382.

Nayar MP and Sastry ARK 1987 *Red Data Book of Indian Plants,* Vol. I (Botanical Survey of India: Calcutta, India).

Ota MR, Imaichi R and Kato M 2001 Developmental morphology of the thalloid *Hydrobryum:* implications for colonization of glaciated North America, *Aquat. Bot.* **43** 311-325.

Pannier F 1960 Physiological responses of Podostemaceae in their natural habitat, *Int. Revue. Ges. Hydroboil.,* **45** 347-354.

Philbrick CT 1984 Aspects of floral biology, breeding system, and seed and seedling biology in *Podostemum ceratophyllum* (Podostemaceae), *Syst. Bot.* **9** 166-174.

Philbrick CT and Novelo AR 1995 New World Podostemaceae: Ecological and evolutionary enigmas, *Brittonia* **47** 210-222.

Philbrick CT 1997 Introduction, *Aquat. Bot.* (special issue on podostemaceae) **57** 1-4.

Rao AS and Hajra PK 1970 *Zeylanidium olivaceum* (Gardn.) Engler–Podostemaceae: first report of its presence in Kameng district, Arunachal Pradesh, *Bull, Bot; Surv. India* **12** 271-272.

Rao AS and Hajra PK 1975 *Polypleurum wallichii* (R. Br. Ex Griff.) Warming and *Zeylanidium lichenoides* (Kulz.) Engl.: two interesting Podostemaceae from Meghalaya, *Bull. Bot. Surv. India* **17** 207-209.

Rao AS and Hajra PK 1979 *Polypleurum stylosum* (Wt.) J.B. Hall (Podostemaceae) In Kamrup District, Assam with taxonomic notes and a new combination, *Bull. Bot. Surv. India* **27** 102-195.

Raveendran TP and Mathew P 1994 *Polypleurum wallichii* (R. Br. Ex Griff.) Wrm. (Podostemaceae) A new plant record for South India, *Rheedea* **2** 106-108.

Razi BA 1949 Embryological studies of two members of Podostemaceae, *Bot. Gaz.* **111** 211-218.

Razi BA 1955. Some aspects of the embryology of *Zeylanidium olivaceum* (Tul.) Engl. and *Lawia zeylanica* Tul., *Bull. Bot. Soc. Beng.* **9** 36-41.

Richards AJ 1986 *Plant Breeding Systems* (George Allen and Unwin: London).

Rutishauser R 1997 Structural and developmental diversity in Podostemaceae (river – weeds), *Aquat. Bot.* **57** 29-70.

Rutishauser R 1999 Polymerous leaf whorls in vascular plans: developmental morphology and fuzziness of organ identities, *Int. J. Plant. Sci.* **160** 81-103.

Rutishauser R and Grubert M 1994 The architecture of *Mourera fluviatilis* (Podostemaceae): mature structure and leaf development, *Bot. Helv.* **104** 179-194.

Rutishauser R and Huber KA 1991 The developmental morphology of *Indotristicha ramosissima* (Podostemaceae-Tristichoideae), *Plant Syst. Evol.* **178** 195-223.

Satheesh VK 1996 *A Study of the Vegetative and Reproductive Features of the Podostemaceae of Kerala Rivers,* Ph.D. Thesis, Mahatma Gandhi University, Kottayam, Kerala, India.

Schnell R and Cusset G 1963 Remarques sur la structure des plantules des Podostemacees, *Adansonia* **3** 358-369.

Sculthorpe CD 1967 *The Biology of Aquatic Vascular Plants.* (Edward Arnold: London, UK).

Sehgal A, Mohan Ram HY and Bhatt JR 1993 In vitro germination, growth, morphogenesis and flowering of an aquatic angiosperm *Polypleurum stylosum* (Podostemaceae), *Aquat. Bot.* **45** 269-283.

Sharma BD, Karthikeyan S and Shetty BV 1974 *Indotristicha tirunelveliana* Sharma, Karthik and Shetty – a new species of Podostemaceae from South India, *Bull. Bot. Surv. India* **16** 157-161.

Shivamurthy GR and Sadanand KB 1997 A new species of *Willisia* Warm. (Podostemaceae) from the Silent Valley, Kerala, India, *Kew Bulletin* **52** 243-245.

Tulasne LR 1852 Monographia Podostemacearum, *Arch. Mus. Hist. Nat.* **6** 1-208.

Tur NM 1997 Taxonomy of Podostemaceae in Argentina, *Aquat. Bot.* **57** 225-253.

Uniyal PL 1999 Studies on *Indotristicha tirunelveliana* Sharma Karthik and Shetty (Podostemaceae): an endemic, rare and enigmatic taxon, *Flora* **194** 169-178.

Uniyal PL and Mohan Ram HY 2001 Studies on the morphology and in vitro germination in *Willisia selaginoides* (Bedd.) Warm and Willis (Podostemaceae), *Flora* **196** 370-380.

Uniyal PL 2001 Some aspects of biology of *Hydrobryopsis sessilis* (Willis) Engl.

Van Steenis CGGJ 1981 *Rheophytes of the World* (Sijthoff and Noordhoff: Alphen aan den Rijin, The Netherlands).

Vartak VD 1974 Contribution to the palynological study on the family Podostemaceae from western India, *1st Indian Palynological Congress,* Chandigarh, India.

Vartak VD and Bhadbhade M 1973 Family Podostemaceae from Mahrashtra and Goa, *J. Univ. Poona* **44** 181-197.

Vartak VD and Kumbhojkar MS 1984 Palynological study of the family Podostemaceae from Western India, *Biovigyanam* **10** 89-92.

Vidyashankari B 1988a *Developmental Biology of Griffithella hookeriana,* Ph.D. Thesis, University of Delhi, Delhi, India.

Vidyashankari B 1988b Seed germination and seedling morphology in *Indostristicha ramosissima* (Podostemaceae), *Aquat. Bot.* **28** 161-169.

Vidyashankari B and Mohan Ram HY 1987 In vitro germination and origin of thallus in *Griffithella hookeriana* (Podostemaceae), *Aquat. Bot.* **28** 161-169.

Wardlow CW 1965 *Organisation and Evolution in Plants,* Longsman Green and Co. Ltd.

Warming E 1881 Familien Podostemaceae I, Kgl Danske Vid Selsk Skr 2 1-34.

Warming E 1882 Familien Podostemaceae II, Kgl Danske Vid Selsk Nat Math Afd **2** 77-130.

Warming E 1888 Familien Podostemaceae III, Kgl Danske Vid Selsk Skr **4** 443-514.

Warming E 1891 Familien Podostemaceae IV, Kgl Danske Vid Selsk Skr **7** 136-144.

Warming E 1899 Familien Podostemaceae V, Kgl Danske Vid Selsk Skr **9** 107-154.

Warming E 1901 Familien Podostemaceae VI, Kgl Danske Vid Selsk Skr **11** 2-67.

Went FAFC 1908 The development of ovule, embryo sac and egg in Podostemaceae, *Rec. Trav. Bot. Neeri.* **5** 1-16.

Went FAFC 1926 Morphological and histological peculiarities of the Podostemaceae, *Proceeding of the International Congress of Plant Sciences* **1** 351-358.

Willis JC 1902a A revision of the Podostemaceae of India and Ceylon, *Ann. R. Bot. Gdns Peradeniya* **1** 181-250.

Willis JC 1902b Studies in the morphology and ecology of the Podostemaceae of Ceylon and India, *Ann. R. Bot. Gdns. Peradeniya* **1** 267-465.

Willis JC 1902 On the dorsiventrality of podostemaceae with reference to current views on evolution, *Ann. Bot.* **16** 593-594

Willis JC 1914 On the lack of adaptation in the Tristichaceae and Podostemaceae. *Proc. R. Soc.* **B57** 532-550.

Willis JC 1915 The origin of Tristichaceae and Podostemaceae, *Ann. Bot.* **29** 299-306.

Willis JC 1926 The evolution of Tristichaceae and Podostemaceae, *Ann. Bot.* **40** 349-367.

Willis JC 1949 *The Birth and Spread of Plants* (Chronica Botanica Co: Watham, Mass, USA).

Index

The Editors

Prof. K.G. Mukerji, retired Senior Professor, University of Delhi has published over 600 research papers. He has co-authored 'Taxonomy of Indian Myxomycetes' (J. Cramer, Germany), 'Index to Plant Diseases of ,India' (Tata-McGraw Hill, India), 'The Haustorium' (Research Studies Press Limited, England and John Wiley and Sons Inc., USA) and edited 'Biocontrol of Plant Diseases' Vols I and II (CRC Press, USA), 'Handbook of Applied Mycology', Vols I-IV (Marcel and Dekkar, USA), Concepts in Mycorrhizal Research', 'Techniques in Mycorrhizal Studies", 'Fruit and Vegetable Diseases' (Kluwer Academic Publishers, The Netherland), 'Recent Developments in Biocontrol of plant Diseases', 'Current Concepts in Seed Biology', 'Advances in Medical Myucology', 'IPM Systems in Agriculture Vols I to VII' (from Aditya Book Pvt. Ltd., India) 'Progress in Microbial Ecology', 'Frontiers in Applied Microbiology—Vols I to IV' (Print House India Ltd., Lucknow, India), 'Mycorrhiza Round Table' (IDRC, Canada), Advances in Botany', 'The Mushrooms of India—Boletaceae', 'Red Algae', 'Brown Algae', 'Integrated Pest and Disease Management', 'Glimpses in Botany' (APH Publishing Corporation, New Delhi), 'Biology of lichens' (Aravali Books International, New Delhi), 'Biotechnological Approaches in Biocontrol of Plant Pathogens', 'Mycorrhizal Biology', 'Biocontrol Potential and Its Exploitation in Sustainable Agriculture—Vols I and II' (Kluwer Academic/Plenum Publishers, USA). He is on the editorial board of several National and International Journals dealing with Botany, Mycology, Plant Pathology, Microbiology and Biotechnology. He is a Honorary Member of The Research Board of Advisors of the American Biographical Institute, North Carolina, USA. More than 60 students have got Ph.D. in Botany under the supervision of Professor Mukerji. He is a member of numerous societies and associations concerning mycology, plant pathology and microbial ecology. Professor Mukerji is a distinguished mycologist and microbial ecologist and he is well reputed for his research contributions all over the world.

Prof. K.V.B.R. Tilak, has a very distinguished background in the field of research of soil microbiology, biological nitrogen fixation, biofertilisers and plant microbe interactions. He was a coordinator of Indo-US Science & Technology Initiative (Senior Scientific Panel) program during 1984-91 and visited various universities n USA in a bilateral program on biological nitrogen fixation and biofertilisers including mycorrhizae. Dr. Tilak held various positions at Indian Agricultural Research Institute, New Delhi as Professor of Microbiology, National Coordinator (ICAR) of Biological Nitrogen Fixation program, Project Director, National Facility for Blue-Green Algal Collections and as Head of the Division of Microbiology besides serving a number of organisations in the country on programs dealing with biofertilisers and soil microbiology. Presently he is working as an Emeritus Scientist (CSIR) in the Department of Botany, Osmania University, Hyderabad, A.P. For his outstanding contributions towards research in agricultural microbiology and biofertilisers, he was awarded the prestigious Prof. S.R. Vyas Memorial Award by the Association of Microbiologists of India. He was the recipient of the Best Teacher award of the Indian Agricultural Research Institute in the Year 2000. He is Fellow of the National Academy of Sciences, India, Fellow of National Academy of Agricultural Sciences, India, Fellow of A.P. Academy of Sciences, India, Fellow of Indian Botanical Society and Fellow of Association of Microbiologists of India. He was the President of Association of Microbiologists of India for the year 1997. He has guided 25 students for Ph.D degree in Microbiology, besides teaching at IARI, New Delhi. He has published over 200 research papers in journals of repute.

Dr. I. V. Gangawane is Professor and Head, Department of Botany, and Soil Microbiology at Dr. Babasaheb Ambedkar Marathwada University, Aurangabad, Maharashtra. During his 35 years of teaching and research experience in Soil Microbiology and Plant Pathology he has guided 35 PhD. Students and has published more than 215 research papers in national and International journals in the field of Plant Pathology, pesticides, biofertilizers, microbial physiology and rhizosphere pathology. He has authored seven books in the above-mentioned fields. The book 'Biofertilizer Technology Transfer' is well known in India and abroad. He has participated in more than 90 national and international conferences in India and abroad and acted as Chairman many a times. Dr. Gangawane has visited several countries and completed international post-graduates courses in fungicide resistance in Plant Pathologens under the guidance of Prof. J. Dekkar at Agricultural University, Wageningen, Netherlands. He is associated with many professional bodies. He is a fellow of Indian Phypathological Society (FIPS) and Plant Protection Association of India, Councillor and President West Zone (FPPAI). He is also a recipient of Prof. M.S. Pavgi Award. IPS, New Delhi, and P.R. Verma Award, SMPP, Udaipur for his excellent work in management of fungicide resistance and application of biopesticides. Recently he was awarded best teacher award by government of Maharashtra.

Dr. P. Prakash did his postdoctoral research work on Biological control on Soil Borne Microbes/Pathogens at Humboldt University, Berlin Germany. He was recipient of Young Scientist award of Indian Botanical Society. Published many scientific papers in Peer reviewed Journals and chaired many National/International Symposia/Conferneces/Workshops apart from participation in Radio and Television Programmes. Dr. Prakash began his humble career teaching Post graduate classes at Osmania University, Hyderabad before joining UGC as an Education Administrator. Contributed immensely on tertiary education–Policies, Planning, framing subjects curricula, particularly in the context of maintenance of quality education

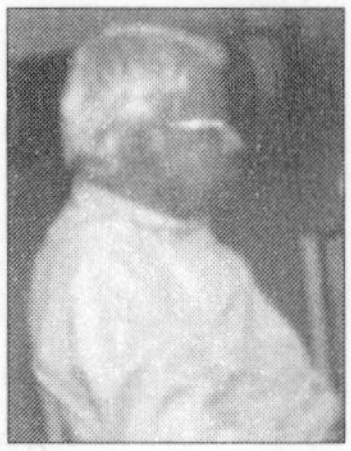

Dr. S.M. Reddy, Professor Emeritus (UGC), Department of Botany, Kakatiya University, is a recipient of Best Teacher Award by the Govt. of A.P. He authored/edited bout books of university education. His research work spans over wide areas of microbiology right from microbial biotechnology. He guided students for their Ph.D. degrees and published many original research papers in national and international journals and successfully completed many research projects sponsored by different funding agencies. He is a fellow of many academic societies and an active member several academic bodies.

Dr. (Mrs.) I.K. Kunwar D. Phil. from Allahabad University, Allahabad and worked as Lecturer in Jumaon University, Nainital. She was Visiting Research Associate in the University of Illinois, Urbana Champaign, U.S.A. Presently she is working as Research Scientist in the Botany Department of Osmania University, Hyderabad. She has published more than 60 research papers in National and International journals and contributed 9 chapters in various books. Her Areas of interest are AMF interactions, storage fungi, seed pathology and histopathological studies. She has discovered five new genera and 12 new species and is presently actively engaged in studies on biodiversity and taxonomy of fungi.